Evolution or Creation?

Evolution or Creation?

A Comparison of the Arguments

Third Edition, 2014

Albert DeBenedictis

To order additional copies of this book, contact:
Xlibris
844-714-8691
www.Xlibris.com
Orders@Xlibris.com
552822

Table of Contents

Dedication

I would like to thank my wife, Linda, for her numerous comments and suggestions. Linda has been a great help, tirelessly proofreading the manuscript and working with me through the review process. I would also like to thank her for working on the cover design and the graphic images that appear throughout the text. Linda has always been an inspiration and an encouragement to me.

~~~

Preface

As we look around the great outdoors, we see there is an abundance of life all around us. On a clear night there are millions of stars in the sky. There are multiple types of plants, trees, shrubs, flowers and bushes, as well as a tremendous amount of animal and insect life. Where did it all come from? How did it all get here? Some believe that everything we see around us is the result of nature, while others believe that some form of intelligence is responsible for the universe and life. Who is right and who is wrong? Is there any way one can know for sure?

This book was developed to help you answer these questions by presenting the arguments equally and fairly, to enable you to form your own conclusions. Two people can look at the same evidence and come up with completely different conclusions. It depends on one's perception and perspective.

In a court of law a judge will ask the jurors to consider all the evidence before reaching a decision. The judge will instruct the jurors to keep an open mind and disregard any personal prejudices, biases and preconceived ideas before rendering a verdict. Similarly, I am asking you to set aside any personal biases you may have and consider all the various views, arguments and evidence included in this book before arriving at any conclusions regarding origins. I sincerely hope this book will be a valuable aid in your search for answers.

~~~

Introduction

The subject of origins often causes a great deal of controversy. Why are there different views? What do those of opposing views base their beliefs on? Is there any reconciling the different views? The differences, I believe, are in part due to one's perceptions regarding science (evolution) and theology (creation).

 Versus

My purpose in writing this book is to help you determine whether the universe and life originated as a result of supernatural causes or whether everything originated by natural means. Some believe it doesn't matter what one believes about origins, while others believe that one's perception of origins greatly influences one's mores, ethics and life-style. Some believe we are just another species of animals while others believe we are special created beings somehow answerable to a deity. Therefore what you believe about origins will affect your views of life.

Whatever your views are regarding the origins of the universe and life, this book is about considering various sides of the issue of evolution. The question of whether a Creator exists or whether natural processes caused the universe and life to appear was the question I asked myself many times while conducting my own research before writing this book. Is there a Creator?

Could the universe and life have originated by natural causes? Could the abundance of life we see around us (plus those creatures that became extinct) have evolved from simpler life forms? The ultimate question is: What does the evidence indicate? This will be for you to determine.

The Author

Opening Comments

Evolutionist Comments

Theodosius Dobzhansky once said: "Nothing in biology makes sense except in the light of evolution." This is so true. Evolution provides a framework that explains how organisms adapt to challenges and changes in the environment. The fossil record and diversity of organisms we see today, along with modern techniques of molecular biology, taxonomy and geology, give evidence for current evolutionary theory. By these methods of research, scientists have firmly established evolution as an important natural process.[OC-0]

The theory of evolution is more than simply stating that evolution took place or that evolution occurs. A theory is an extensively documented set of principles that explain how and why something may have happened or happens. A theory is considered to be valid when it can be both tested and verified. Observations about the real world need to be conducted that either support or disprove it. Good theories are formed from predictions about natural occurrences in the real world. A theory can be scientific even if phenomena are not directly observable. Evolutionary theory is built in the same way it is in any field that uses indirect testing, and some aspects of evolutionary theory can be directly tested. Any theory must be verified continually over time. A scientific theory becomes a fact when repeated confirmations are made and no discrepancies are found.[OC-1]

The modern theory of evolution teaches that life on Earth evolved gradually beginning with one primitive species, perhaps a self-replicating molecule (Chapters 5 & 6), that lived more than 3.5 billion years ago; it then branched out over time, producing many new and diverse species (Chapters

3, 7, 8); and the mechanism for most (but not all) of evolutionary changes is natural selection (Chapter 9).[OC-2]

Evolution is just a theory, in the same way that the atomic theory of matter is just a theory, the Copernican theory of the solar system is just a theory, or the germ theory of disease is just a theory. (Nicolaus Copernicus developed the view that the solar system centered on the Sun, with Earth and other planets moving around it (as opposed to the belief that the Sun and other planets revolve around the Earth). The germ theory, developed by Louis Pasteur, states that certain diseases are caused by the invasion of the body by microorganisms, organisms too small to be seen except through a microscope.) But theories are not hunches, they are not unproven speculation. Theories are systems of explanations which are strongly supported by factual observations and which explain whole sets of facts and experimental results. The difference between a theory and a fact is that a fact is a repeatable, verifiable observation or a result. A theory takes the facts and develops them into an explanation.[OC-3]

A theory can be verified by a mass of facts, but it becomes a proven theory, not a fact. For example, parts of the Copernican world model, such as the contention that the earth rotates around the sun, and not the other way around, have not been verified by direct observations. Yet scientists accept the model as an accurate representation of reality because it makes sense of a multitude of facts which are otherwise meaningless or extravagant.[OC-4]

The proof of evolution is based on multiple lines of evidence, including (1) the fossil record, (2) the geologic record, (3) comparative anatomy, (4) comparative embryology, (5) systematic (classification) work, and (6) molecular phylogenies.[OC-5]

Many objections against evolution are based on a misunderstanding of some aspects of evolutionary theory. Some of the common objections include: 1) Evolution cannot be proven. The question, instead, is: What is the best position to take given the available evidence? 2) Evolution cannot be falsified. Actually, it can. Because evolution can explain a lot does not mean it can explain any conceivable data. If ape fossils were found in Jurassic rocks (Chapter 3), evolution would be falsified. 3) Evolution is only a theory. Theories are not conjectures, but well-substantiated explanations. Evolution is a good theory. 4) Evolution is not a fact. It is not a fact in the same sense of something being obvious. It is a fact in that it is the best explanation for a huge body of data. 5) Natural selection is needlessly repetitious—it says that the fittest are those that survive and those that survive are the fittest (Chapter 9).

Fitness (or lack of it) can be defined independently of survival. For example, animals that are not able to avoid predators are unfit. The consequence is extinction. **6)** Many scientists doubt the theory of evolution. Astronomers and engineers do not count. The overwhelming consensus among palaeontologists and neonatologists is that evolutionary theory is very well established. **7)** If evolution is true, why have some species remained unchanged for supposedly millions of years? There is nothing in evolutionary theory that requires species to evolve. For example, earthworms have not evolved legs because they do not need them.[OC-6]

The proof of evolution is based on multiple lines of evidence, including the fossil record (Chapters 3, 7, 8, 16), the geologic record (Chapter 3), comparative anatomy, comparative embryology (Chapter 11), systematic (classification) work, and molecular phylogenies. There are seven main lines of evidence for evolution. They are: **(1)** Similarities—All organisms show similarities at many levels that would not be expected if they had independent origins. **(2)** Direct Observation—Changes in gene composition between generations can be seen in rapidly reproducing organisms. **(3)** Transitional Fossils—Fossils of forms intermediate between one type of organism and another type of organism living today have been found. (Chapters 3, 7, 8, 16) **(4)** Logical Inference—Evolution is the inevitable consequence of natural selection acting upon the effects of mutations in DNA. (Natural selection and mutations are primarily discussed in Chapter 9.) **(5)** Hierarchical Classification—Organisms can be classified into groups within groups, as expected if they are all related by descent. **(6)** Biogeography—Location provides a better index of biological similarity than does similarity of climate because geography reflects descent from common ancestors. **(7)** Vestigial Organs And Functionless Genes—Features that become useless are slowly lost, becoming vestiges of ancestral forms. (Chapter 12 discusses vestiges and Chapter 13 discusses functionless or pseudo genes.)[OC-7]

The theory of evolution must always be subject to new evidence as it is obtained. A scientific theory is always subject to revision or abandonment if new facts contradict the theory. The theory of evolution is still a theory, just as the theory of gravity is a theory, however, it is a theory based on factual testing. Scientists believe a theory is not a speculation or a guess, but a logical explanation of a collection of experimental data. The theory of evolution is a theory that has been thoroughly tested and has been accepted by the majority of scientists. Creation is based on a religious concept and is not scientific nor can it be tested.[OC-8]

Creationist Comments

Science is both a great body of knowledge and a great method of investigation. Most people just assume evolution can be studied scientifically but creation cannot. Many people believe that evolution is science and creation is religion. The question is whether or not it is possible to talk honestly and fairly about scientific evidence of creation. To recognize creation as plausible, researchers utilize the same scientific methods that are used to study anything else. Logic and observation is used to determine whether something was created or whether it occurred naturally. A rock that resembles weathering and erosion is considered to be the result of nature. An arrowhead, for example, shows marks of withering and chipping. The object appears to have been shaped and molded according to a plan by a designer that gives the rocky material a special purpose (Chapter 14 discusses intelligent design). If one would agree that the creation (the model, the process, and the products) could be studied scientifically, it does not mean that one would need to believe in creation. It only means that the models of origin, creation and evolution, can be compared on the basis of scientific merit.[OC-9]

Science involves observation, theory, design, testing, and predictable outcome. The theory of evolution would then need to be observed, tested and proven true. Darwin imagined that there should be many transitional species that could be observed presently. Scientists make different assumptions before evaluating the scientific evidence. It depends on what one assumes is true. Therefore, even science is based on faith. Evolutionists believe that evolution is true just as creationists believe everyone was created. Evolutionists do not test the theory of evolution because they assume it is true. They only try to find evidence that proves the theory is correct, just as creationists only try to find evidence that disproves the theory.[OC-10]

The debate regarding origins is not one sided in favor of creation by a Creator or evolution by natural means. When it comes to origins, it is not possible to utilize direct observation because no one was around when it all began. We also cannot run experiments on what occurred in the past (Chapter 6). Therefore, we only have circumstantial evidence, which is subject to more than one interpretation.[OC-11]

Creationists are not against science, nor are they afraid of scientific data. While no generalization can characterize all individuals who believe in creation any more than one statement can describe all who believe in evolution, all knowledgeable, scientifically minded creationists fully welcome new scientific data. Creationists and evolutionists have the same data. Reality is the same for both. Perception of that reality and interpretation of that data can, however, be remarkably different for both, depending on the individual's perspective, or assumptions, world-view or even bias.[OC-12] The ultimate question is, Who has the better case? The only way to answer that is to look honestly at the evidence. The task of science, after all, is to search for truth wherever it leads. [OC-12a]

Evolutionists claim that Intelligent Design as well as Creation Science are not science because they are not testable by the methods of science. From a creationist perspective, Intelligent Design is indeed falsifiable. Intelligent Design is probably more falsifiable than Darwinism. To falsify an Intelligent Design claim, all that would be needed would be for an evolutionist to produce evidence that falsifies the creationist's claim, that is, by providing evidence as to how an organism originated by natural means. To falsify a Darwinism claim, it would be more difficult because one would have to prove that the system being discussed could not have originated by any number of possible natural processes.[OC-13]

Science is defined as "the observation, identification, description, experimental investigation, and theoretical explanation of phenomena." Nothing requires science, in and of itself, to be naturalistic. Naturalism, like creationism, requires a series of presuppositions that are not generated by experiments. They are not extrapolated from data or derived from test results. These philosophical presuppositions are accepted before any data is ever taken. Because both naturalism and creationism are strongly influenced by presuppositions that are neither provable nor testable, and enter into the discussion well before the facts do, it is fair to say that creationism is at least as scientific as naturalism. Creationism, like naturalism, can be "scientific," in that it is compatible with the scientific method of discovery. These two concepts are not, however, sciences in and of themselves, because both views include aspects that are not considered "scientific" in the normal sense. Neither creationism nor naturalism is falsifiable; that is, there is no experiment that could conclusively prove or disprove the existence of a Creator, just as there are no experiments that could prove or disprove that everything in existence is a result of natural occurrences. Neither one is predictive; they do not generate or enhance the ability to predict an outcome. Solely on the basis of these two points, there is no logical reason to consider one more scientifically valid than the other.

Naturalists claim that miracles are impossible because miracles violate the laws of nature, which (the laws of nature) have been clearly and historically observed. Such a view is ironic on several counts. As a single example, consider abiogenesis, the theory of life springing from non-living matter. Abiogenesis is one of the most thoroughly refuted concepts of science (abiogenesis is discussed in chapters 5 & 6). A truly naturalistic viewpoint presumes that life on earth's self-replicating, self-sustaining, complex organic life "arose by chance from non-living matter." Yet, no one has ever seen life develop from non-life. The beneficial evolutionary changes needed to cause a creature to become a more complex form (macro-evolution) have also never been observed. To label creationism as unscientific on account of miracles demands a similar label for naturalism.[OC-14]

Evolutionists claim that a creation concept for the origin of the universe and life is not science because creation cannot be tested or falsified. This claim is not necessarily true. The idea that creation as being an explanation for the origin of life that is based on Biblical passages that describe God's creative actions (as indicated in the Bible – Genesis 1 & 2 and other passages) is just as valid as naturalistic explanations. The methods to confirm creation also utilizes scientific research to demonstrate and support the accuracy of the Biblical accounts. These views can indeed be tested and falsified (if the evidence leads to this conclusion). Applying scientific research to Biblical accounts allows for any creation claim to be tested. Creation then becomes testable and falls within the domain of science. A creation view that states that (1) life appears early in Earth's history while our planet was still in its primordial state, (2) life originated and persisted through the hostile conditions of early Earth, (3) life originated abruptly on Earth, and (4) Earth's first life displays complexity, can certainly be tested (Chapter 5, section "Life Originated Too Early in Earth's History To Have Originated by Natural Causes"). Rather than just relying on a single biochemical feature such as irreducible complexity (Chapter 14) to argue for a Creator's role in life's origin, the case for biochemical intelligent design is built on multiple arguments of evidence (Chapter 5). Throughout this book a great deal of scientific evidence and research will be discussed that substantiates the claim that the universe and life were created and was not the result of natural selection working on random genetic changes (see "Old Universe Progressive Creationist View"). [OC-15]

~~~

Chapter 1

The Origin of the Universe

In the beginning God created the heaven and the earth . . .

The earth was without form, and void; and darkness was on the face of the deep. And the Spirit of God was hovering over the face of the waters . . . Then God said, "Let there be a firmament in the midst of the waters, and let it divide the waters from the waters." Thus God made the firmament, and divided the waters which were under the firmament from the waters which were above the firmament; and it was so . . . Then God said, "Let the waters under the heavens be gathered together into one place, and let the dry land appear"; and it was so. And God called the dry land Earth, and the gathering together of the waters He called Seas. [1-1]

The Big Bang was the beginning of the universe as we know it . . . [1-2]

Evolution had no room for the supernatural. The earth and its inhabitants were not created, they evolved. [1-3]

Where were you when I [God] laid the foundations of the earth? Tell Me, if you have understanding. Who determined its measurements? Surely you know! Or who stretched the line upon it? To what were its foundations fastened? Or who laid its cornerstone? [1-4]

Introduction

This chapter (1) deals with the origin of the universe. Some believe the universe always existed, while others say it originated by natural causes and still others believe in a creation account.

The various views presented in this chapter regarding the origin of the universe are not exclusively of strict evolutionists and strict creationists. Some views are of those who believe in variations of the two extremes. See "Various View Descriptions" section in the back of this book for descriptions of the views presented in this chapter.

Bosons, Fermions, Quarks, and Leptons

Before beginning our discussion on the origin of the universe, a few terms need to be defined, such as bosons, fermions, quarks, and leptons. In particle physics, a **boson** is a subatomic particle whose spin quantum number has an integer value (0,1,2 ...). Bosons control the interaction of physical forces, such as electromagnetism. Bosons form one of the two fundamental classes of subatomic particle, the other being **fermions**, which have odd half-integer spin (1/2, 3/2, 5/2 ...). Every observed subatomic particle is either a boson or a fermion. Fermions include all quarks and leptons and all composite particles made of an odd number of these, such as all baryons and many atoms and nuclei. A **quark** is a type of elementary particle and a fundamental constituent of matter. Quarks combine to form composite particles called **hadrons**, the most stable of which are **protons** and **neutrons**, the components of atomic nuclei. All commonly observable matter is composed of up quarks, down quarks and **electrons**. In particle physics, a **lepton** is an elementary particle of half-integer spin (spin 1/2) that does not undergo strong interactions. Leptons are elementary particles with half integral spin(½) and it does not go under any strong interaction and it is divided into two branches that are: **(1)** Electrically charged lepton (also called electron like lepton) and **(2)** Neutrally charged lepton (also called neutrinos). The charged leptons can combine with other particles to form composite particles such as **atoms** (See Figure 1.1 in the "Origin of Matter" section.) and positronium (it is the reverse of electron it has the same mass but with a positive charge) while the neutrino rarely reacts with any other particle and they are rarely taken into observation and the best known lepton is electrons (electrons also have least mass of all charged leptons). (See Figure 1.1a in the "Origin of Matter" section for illustration of quarks and leptons.) In particle physics, a **baryon** is a type of composite subatomic particle which contains an odd number of valence quarks (at least 3). Baryons belong to the hadron family of particles; **hadrons** are composed of quarks. Baryons are also classified as fermions because they have half-integer spin.

Why is There Something Rather Than Nothing?

Evolutionist View

> . . . in answer to the question 'Why is there something instead of nothing?,' it is okay to say "I don't know" and keep searching. There is no need to turn to supernatural answers just to fulfill an emotional need for certainty and comfort. Science's uncertainty is its greatest strength. We should embrace it.[1-5]

> Because there is a law such as gravity, the universe can and will create itself from nothing. Spontaneous creation is the reason there is something rather than nothing, why the universe exists, why we exist. It is not necessary to invoke God to light the blue touch paper and set the universe going.[1-5a]

Why is there something rather than nothing? This question has been posed many times by Christian creationists who claim that "God did it" whenever they cannot explain something scientifically. We should not rely on supernatural explanations for anything we cannot fully explain. The Christian creationist who tries this line of attack is merely revealing an utter lack of knowledge about science itself.[1-6]

Many conceptual problems are associated with the question of why there is something rather than nothing. How do we define nothing? What are its properties? If it has properties, does not that make it something? The theist claims that God is the answer. But, then, why is there God rather than nothing? Assuming we can define nothing, why should nothing be a more natural state of affairs than something? In fact, we can give a plausible scientific reason based on our best current knowledge of physics that something is more natural than nothing. Of course, that requires providing a physical definition of nothing. Can one imagine a physical system that has no properties? Yes, as long as one does not call the lack of properties a property.

Suppose we remove all the particles and any possible non-particulate energy from some unbounded region of space. Then we have no mass, no energy, or any other physical property. This includes space and time, if you accept that these are relational properties that depend on the presence of matter to be meaningful. While we can never produce this physical nothing in practice, we have the theoretical tools to describe a system with no particles. In the current universe, bosons outnumber fermions by a factor of a billion. This has led people to conclude that the vacuum energy of the universe, identified with the zero point energy remaining after all matter is removed, is very large, although this estimate is wrong. Since a non-particulate vacuum's energy density is proportional to Einstein's cosmological constant, this is called the cosmological constant problem.

Instead of using numbers from the current universe, we can visualize a vacuum with equal numbers of bosons and fermions. Such a vacuum might have existed at the very beginning of the big bang (discussed later). Indeed this is exactly what is to be expected if the vacuum out of which the universe emerged was super-symmetric, made no distinction between bosons and fermions. This suggests a more precise definition of nothing. Nothing is a state that is the simplest of all conceivable states. It has no mass, no energy, no space, no time, no spin, no bosons, no fermions, nothing.

Then why is there something rather than nothing? Because something is the more natural state of affairs and is thus more likely than nothing; more than twice as likely according to one calculation. We can infer this from the processes of nature where simple systems tend to be unstable and often spontaneously transform into more complex ones. Theoretical models such as the inflationary model of the early universe bear this out. Since "nothing" is as simple as it gets, we would not expect it to be completely stable. In some models of the origin of the universe, the vacuum undergoes a spontaneous phase transition to something more complicated, like a universe containing matter. The transition nothing-to-something is a natural one, not requiring any external agent (God). The answer to the ancient question 'Why is there something rather than nothing?' would then be that 'nothing' is unstable.[1-7]

Creationist and Intelligent Design View

The existence of the universe has raised many questions. What set it in motion to begin with? What is the source of energy that keeps it going?[1-8] We know something exists, but why does something exist? Why is there something rather than nothing at all? Cosmologists believe that the physical universe came into being 'ex nihilo' (out of nothing) a finite time ago. Matter, space, and time all had their beginning at an absolute point of origin, before which there was no physical reality. While the scientific evidence does point to an absolute origin of physical reality, it does not preclude the possibility of a pre-existent, immaterial reality from which the physical universe emerged and thus does not require that physical existence emerge from absolute nonexistence. That question is left open, as it is beyond the realm of scientific inquiry. Why and how did something emerge from nothing? The most basic ontological principle is that out of nothing, nothing comes; and yet in the case of the universe, out of nothing something came. There must be a sufficient cause for the universe to come into being, and that requires that something exists external to the universe.[1-9]

Where did particles of matter and matter come from? Many claim that all matter is eternal, or that everything appeared by chance out of nothing. The idea that matter always existed is difficult to believe. If one believes that all matter appeared by chance out of nothing, an obvious question arises: What caused the first event that originally set the evolutionary process in motion? Those who believe that everything originated from nothing must believe that particles of matter and matter literally came from nothing. This concept is also difficult to comprehend.[1-10]

What was the first cause that caused everything else? Where did particles of matter and matter come from? Where did energy come from? What holds everything together and what keeps everything going? There are a few different explanations as to how the universe might have originated. Some would say that something came from nothing. The question is: How do you get something from nothing?[1-11] Some cosmologists believe that the universe could have developed from nothing. Cosmologists argue that quantum mechanics predicts that a vacuum can, under some circumstances, give rise to matter. But the problem with this line of reasoning is that a vacuum is not nothing; it is something, it is a vacuum that can be made to appear or disappear, as in the case of the Torricellian vacuum, which is found at the sealed end of a mercury barometer.[1-12]

The Big Bang (discussed later) is said to have set off a chain reaction and the universe was formed.[1-13] Some believe that somehow particles of matter and matter came into existence and then exploded (commonly known as the Big Bang) which became what is now known as the universe.[1-14] (See "The Higgs Particle and the Origin of Matter" section for further discussion of the origin of matter from particles.) Others, however, believe that matter has always existed. According to this view, there was no Big Bang because the universe always existed. For some, there seems to be no rational explanation for the universe, and since it is beyond human's comprehension, some believe a god must have created it.[1-15] If one believes in the last possibility, then the following question might be raised: Where did God come from? Many theists answer this question by claiming that God has no beginning and no end; He is beyond time and human comprehension.[1-16]

What is the meaning of "Nothing"?

Creationist View

Could the universe spontaneously come into existence from nothing? Many have asked: Why is there something rather than nothing? And what is this "nothing"? Is it actually no thing? Do those who believe that something could develop from nothing eliminate the theists' claim that the Big Bang beginning of the universe requires a transcendent cause that is not bound by space and time? At one time scientists would have considered a vacuum in deep space to be a perfect definition of nothing. However, it is known that the space-time fabric of our universe is made up of a quantum vacuum that is quite different from what was previously conceived. The elementary particles and forces in the universe are described by a relativistic quantum field theory (QFT).[1-16a]

Quantum Mechanics describes how the universe works at very small scales, about the size of atoms or smaller. When describing the universe at small scales and at very high energies, a particular subset of Quantum Mechanics called a Quantum Field Theory (QFT) is needed. The Standard Model (discussed at the end of this chapter) describes the most fundamental particles in the universe and three of the four fundamental forces in the universe (electromagnetism, the weak, and strong force) is a QFT. QFT's are important when describing the "beginning" of the universe because the visible universe was very small and very hot (at an extremely high energy) so that it should be able to be described by some kind of QFT. [1-16b] Included in the QFT theory, the quantum vacuum can be thought of as a bubbling sea of many different kinds of fields. These fields have the potential of creating particles out of nothing. (Virtual particles do not actually come into existence from nothing, but come into existence from the underlying fields in the space-time fabric of the universe.) For instance, an excitation, or perturbation, of an electron field can produce an electron. Within QFT a particle and its antiparticle (e.g. an electron and a positron) can come into existence from "nothing" where the "nothing" is the underlying quantum vacuum. Except for near a strong gravitational field like a black hole, these particles from nothing will quickly annihilate each other and cease to exist, and are thus called "virtual" particles (described below in section "The Origin of Matter" sub-section "Virtual Particles"). Everything described to this point is well known and well tested science. However, at this point this more recent view deviates from known science and begins to speculate about how the universe functions based on our lack of understanding of any quantum theory of gravity. This view (quantum vacuum) proposes that, just as our previous definition of nothing (a vacuum in deep space) turned out to have the potential for creating particles from the "nothingness" of the vacuum, maybe there is an analogous underlying reality in which virtual space-time universes like ours can be created from nothing.

Just as a space-time quantum vacuum can create particles, maybe another type of unknown "nothing" has the potential to create space-time universes. Thus, the idea of "nothing" is not the underlying space-time structure of our universe, but a broader definition of nothing that has the potential of spawning universes. This more recent proposal raises a number of questions that need to be answered. **First**, is this new definition of nothing really nothing? **Second**, is this more recent theory scientific and does it really propose anything new? **Finally**, if this more recent theory were found to be true, what affect would it have on arguments for theism, particularly regarding the Christian idea of a Creator God?

What exactly does "nothing" mean? Just about every scientist and philosopher has argued that this more recent definition of nothing is not nothing but something similar to the underlying structure of our universe. Although it is definitely not the same as our known space-time quantum vacuum universe, an environment that can spawn virtual universes in the same way our universe can spawn virtual particles can hardly be called "nothing." Those who accept the more recent view of nothing claims that even the laws of physics that we know would be initiated with our universe from nothing, but this idea still does not seem to recognize the fact that these ideas still require some set of physical laws to pre-exist in order to bring our universe into existence. Although this more recent explanation of "nothing" would expand our understanding of what underlies the reality of our universe, just as a quantum vacuum expanded our understanding of what really comprises the vacuum of space, it would not qualify as no-thing as most scientists, philosophers, or other humans would use the term.

Like many ideas about the origin of our universe, this more recent idea is not based on any known science but rather only on what is not known. Because there is no theory of quantum gravity, we do not know what laws of physics governed the first 10-35 seconds or so of our universe or what laws, if any, existed beforehand. Those who proclaim such views are attempting to propose a solution that they believe removes the need for any act of a Creator God to begin our universe. Those who want to remove God from the equation must appeal to what is not known, rather than to what is known. This may be called an "atheism of the gaps". Everything that is known about the origin and design of the universe looks a lot like there is a creator God, so to remove God, any proposal must appeal to what is not known.

It is ironic that atheists have for years claimed that Creationists appeal to a "god of the gaps" to explain things that are not known, but many of the current arguments from atheists against God can only appeal to gaps in our present level of knowledge. There have been many proposals that claim our universe started from some kind of quantum fluctuation. This most recent claim that attempts to explain the origin of the universe just proposes a particular kind of fluctuation and one that currently has no basis in any known science. This is not science, since science is based on observations and measurements. It is simply speculation based on philosophical naturalism. Since this latest definition of "nothing" (regarding what existed prior to the formation of the universe and what caused the universe to come into existence) is clearly not no-thing, those who advocate such views are actually claiming that this universe came into existence through an external mechanism that can spawn universes from nothing. Creationists believe that this universe began from nothing by a mechanism that came from God's creative character. It is unknown what mechanism God used to create this universe, and whether it was through some "natural" or "supernatural" means. However, it is known that God created this universe "in the beginning" (Genesis 1:1) which would be in complete agreement with this most recent explanation. This most recent claim made by scientists adds little to previous speculation about the origin of the universe, is not based on any confirmed scientific facts, and does nothing to remove the need for an ultimate cause.[1-16a]

Critical Creationist Response

Many creationists argue that the universe and life are so complex they require a Creator. Of course, this claim raises the question: "Where did God come from?" Since the existence of a Creator cannot be proven, it is very difficult to convince an atheist that a Creator exists or that creation science is valid. It is not fair to evolutionists that creationists argue that evolutionists need to provide a comprehensive explanation of how the universe and life could have originated, while creationists simply claim a Creator existed from eternity and has no beginning, without providing any proof. The "God who has no beginning and no end" argument does not prove anything scientifically. Claims need to be supported by convincing evidence.[1-17] If creationists are going to prove there is a Creator, they need to do more than simply make claims that God created everything. They need to provide convincing scientific evidence that supports the claim that a Creator exists and that this Creator created everything (or at least whatever they are claiming this God created).

The Origin of Matter

Evolutionist View

Where did matter come from? Why is there something rather than nothing? To explain this, a few words need to be said about antimatter. Antimatter behaves almost exactly the same as ordinary matter. It has the same mass, same amount of spin, etc., but it has the opposite charge and other quantum numbers. A positron, for instance, behaves just like an electron, but has a positive charge rather than a negative one. An anti-proton has a negative charge, and so on. If there is enough energy in a particle accelerator, or in the very early universe, one can create a particle-antiparticle pair out of thin air (and energy). This sort of thing happens in the vacuum of space all the time. It is claimed that particles and antiparticles get created and destroyed in perfect harmony with one another. This, however, is not totally accurate. If matter and antimatter are always created and destroyed in equal quantity, then there should not be any of either around today. As near as can be determined, matter came from a symmetry violation in the universe from very near the beginning. It was a very small effect. For every billion antiparticles that were created, there were a billion and one particles. Eventually, all of the antiparticles annihilated with almost all of the particles, leaving the one part in a billion to make all of the "stuff" that we now see. To put it another way, everything we see is a round-off error from around 10^{-35} seconds after the big bang.[1-18] (The Big Bang Theory is discussed later in this chapter.)

Physically, matter is composed of particles, which are made up of electrons, protons, and neutrons, which are composed of particles of matter. Atoms are composed of matter, and matter is composed of particles. According to modern physics, matter consists of various types of particles, each with mass and size. The most familiar examples of material particles are the electron, the proton and the neutron. Combinations of these particles form atoms. There are more than 100 different kinds of atoms, each kind constituting a unique chemical element. A combination of atoms forms a molecule. Atoms and/or molecules can join together to form a compound. Matter can exist in three forms: solid, liquid or a gas.

Molecules are the smallest bits of compounds. Particles are not matter, but the building blocks of it. For example, copper is an element. A copper atom is the smallest piece of copper that exists. Hydrogen is also an element; two hydrogen atoms and one oxygen atom combine to form a molecule of water, which is a compound. Atoms are made of the sub-atomic building blocks of matter—protons and electrons, revolving around a nucleus. The "atomic number" of an element, as seen on a periodic chart, refers to the number of protons contained in one atom of that element. Quarks are the subatomic particles that compose the protons and neutrons that create the nucleus of an atom. The electrons that rotate around the nucleus of the atom belong to a group of fundamental subatomic particles known as leptons (discussed shortly). [1-18-1]

In February 2020, scientists measured a property of the neutron, a fundamental particle in the universe, more precisely than ever before. Their research is part of an investigation into why there is matter left over in the universe, that is, why all the antimatter created in the Big Bang didn't just cancel out the matter. The researchers wanted to know why the Universe contains so much more matter than antimatter, and, why it now contains any matter at all. They wanted to know why the antimatter didn't cancel out all the matter and why there is any matter left at all. The researchers have determined that the answer relates to a structural symmetry that should appear in fundamental particles like neutrons. This is what scientists have been looking for. They found that the "electric dipole moment" (EDM) is smaller than previously believed. From this experiment scientists were able to rule out theories about why there is matter left over.

Physicists were searching for asymmetry in the neutron (that is extremely tiny) which would show that it is positive at one end and negative at the other. They were looking into whether or not the neutron acts like an "electric compass". Neutrons are believed to be slightly asymmetrical in shape, being slightly positive at one end and slightly negative at the other, something like the electrical equivalent of a bar magnet. This is the so-called "electric dipole moment" (EDM), and is what the researchers were looking for. This is an important piece of the puzzle in the mystery of why matter remains in the Universe, because scientific theories about why there is matter left over also predict that neutrons have the "electric compass" property, to a greater or lesser extent. Measuring the neutrons helps scientists to get closer to the truth about why matter remains. The team of physicists found that the neutron has a significantly smaller EDM than predicted by various theories about why matter remains in the universe; this makes these theories less likely to be correct, so they have to be altered, or new theories found. These EDM measurements, considered as a set, have probably disproved more theories than any other experiment in the history of physics. [1-18-2]

Virtual Particles

Creationist View

One of the most unusual features of our universe may be entities that scientists call "virtual particles," fundamental particles that apparently come into existence from nothing for a brief period of time, then disappear. (As will be discussed later, virtual particles do not actually come into existence from nothing, but come into existence from the underlying fields in the space-time fabric of the universe.) Since virtual particles are so small and exist for only short periods of time (usually about a trillionth of a second or less), it is impossible to visually see any fundamental particle studied by particle physicists including electrons, protons, and the constituents of protons, quarks (defined shortly). Instead, what is seen is an effect in some kind of macroscopic detector. All of the existing information about the fundamental structure of matter comes from the interaction of particles too small to be seen with our eyes because they interact with some kind of macroscopic detector. Since no one can see these subatomic particles, but can only detect their effect, it is believed they exist because mathematical theories have been developed that predict certain outcomes for our experiments. When the experimental outcome is accurately described by the theory it is usually claimed that the theory is valid and postulate that the entities implied by the theory really exist.

No one has ever seen a virtual particle since they are infinitesimal small fundamental particles and only exist for very brief periods of time, usually about a trillionth of a second or less. (See section "The Higgs Particle and the Origin of Matter" below for update.) However, mathematical calculations can be obtained to determine what the outcome of experiments would be if virtual particles exist and what the outcome would be if they did not exist. All experiments performed only match the theoretical predictions when virtual particles are included in the math. Thus, scientists conclude that these virtual particles are real and that they do actually exist for these extremely brief fractions of a second. In essence, virtual particles are the usual particles of nature that are created for a brief moment, but must quickly be annihilated or conservation of energy would be permanently violated. Nature allows a quick violation of conservation of energy, but not a permanent violation. As stated, currently existing mathematical calculations only agree with experimental results when these virtual particles in the calculation are included. In addition, our current theory states that certain processes, like natural radioactive decay, only occur through virtual particles. It seems these particles must actually exist in nature and perform valuable functions like radioactive decay.

It has been observed that the mass of the proton seems to be finely tuned. If it were to change slightly there would be many life-destroying consequences. Stars like our sun that have a stable burn for a long time would not be possible. The ratio of the mass of the proton to the electron is fine tuned to allow appropriate neutron decay. Neutron decay and other factors would radically affect chemistry and biology if the proton's mass were to change slightly. At a very simplistic level every proton is made of 3 quarks. But actually the proton is very complicated and is made of not just 3 quarks, but also gluons and even virtual particles. (*As will be discussed in more detail later, a **quark** is a set of six hypothetical elementary particles together with their antiparticles thought to be fundamental units of all baryons and mesons but unable to exist in isolation. A **baryon** is a class of elementary particles, including the proton and neutron that take part in strong interactions. Baryons are a class of elementary particles, including the proton and neutron that take part in strong interactions. Baryons are composed of a triplet of quarks. A **Meson** is any hadron, or strong interacting particle, other than a baryon. Mesons are bosons, having spins of 0, 1, 2, …, and, unlike baryons, do not obey a conservation law.*)

On rare occasions a top quark and a top-antiquark will momentarily be created and destroyed in the proton as virtual particles. This occurs even though the top quark is the heaviest known fundamental particle and about 175 times heavier than the proton itself. So ultimately, the mass of the proton is partially determined by the mass of the virtual top quarks that momentarily and very rarely are created then destroyed inside the proton. Consequently, if the mass of the top quark were to change by a few percent, virtual top quarks in the proton would change the mass of the proton as well, ultimately leading to a universe in which life could not exist. So the exact mass of the top quark and its virtual existence inside the proton is required for our existence in this universe. Virtual particles do not come into existence from nothing. Virtual particles actually come into existence from the underlying fields in the space-time fabric of the universe. Since no one has actually seen a virtual particle, the only way one can accept their existence is from experimental results that only agree with theoretical calculations when virtual particles are included in the theory. Virtual particles are responsible for phenomena like radioactive decay, and their existence in the proton slightly affects the proton's mass which puts the proton mass into the appropriate range to allow life-necessary features of our universe.[1-18a]

Atoms and Particles

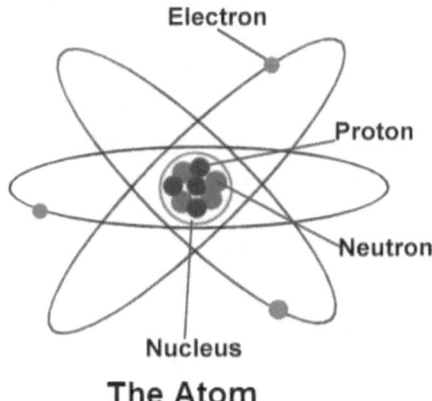

Figure 1.1—The Atom

Atoms are composed of various subatomic particles. Subatomic particles are the particles smaller than an atom (although some subatomic particles have mass greater than some atoms). There are two types of subatomic particles: **(1)** elementary particles, which according to current theories are not made of other particles; and **(2)** composite particles. An elementary particle or fundamental particle is a particle whose substructure is unknown, thus it is not known to be composed of other particles. Known elementary particles include the fundamental fermions (quarks, leptons, antiquarks, and antileptons), which generally are "matter particles" and "antimatter particles." A particle containing two or more elementary particles is a composite particle. An atom is composed of particles. These particles consist of electrons, protons and neutrons (Figure 1.1). (The hydrogen-1 atom, however, does not have any neutrons in the nucleus.) (Quarks and leptons are discussed in more detail in section "Quarks and Leptons" below.) The electron has a negative electrical charge. Protons have a positive charge. Neutrons have no electrical charge.[1-19] Atoms exist in the air and in living things. The Earth also consists of atoms. The protons and neutrons make up the central core of the atom, while the electrons circle the core in defined orbitals (Figure 1.1). Very few atoms have the quantity of electrons they need, so to get their full complement of electrons, they will bond with other atoms to form molecules.[1-20]

No one really knows for sure where atoms come from since they have been around as long as anything has existed. One theory states that atoms started as a cloud of Hydrogen and Helium, which condensed into a star. This star then exploded and atoms were created.[1-21] Atoms are made from protons, neutrons, and electrons (Figure 1.1). A proton will connect to a neutron and create a nucleus. Once the nucleus is formed the electron then circles it. That is how an atom is formed. Electrons group themselves in pairs in their energy levels. Hydrogen was the first basic atom. Gravity pulled the hydrogen atoms to form clouds and then condensed into balls of hydrogen gas as the hydrogen gas condensed.[1-22]

Elementary Particles

Elementary particles are the most fundamental elements of the universe. They are not, as far as we know, made up of other particles. The proton, for example, is not an elementary particle, because it is made up of three quarks, whereas the electron is an elementary particle, because it seems to have no internal structure. There are 31 elementary particles, however, most of what we think of as matter consists of just 3 of them: **(1)** the up quark, **(2)** the down quark and **(3)** the electron (protons and neutrons are both made of up and down quarks; atoms are made of protons, neutrons and electrons; most of what we think of as matter is made of atoms). (Quarks are defined above in section Virtual Particles.) Much of the mass of atoms is due to the gluons that bind the quarks. The photons bind the electrons to the quarks. Two of the elementary particles are speculative. No one has ever observed a Higgs boson or a graviton; physicists imagined them in an attempt to explain mass and the gravitational force. (For an update, see section "The Higgs Particle and the Origin of Matter" for further discussion of this subject.) Where do these 31 elementary particles come from? Nobody knows.[1-23]

Quarks and Leptons

Creationist View

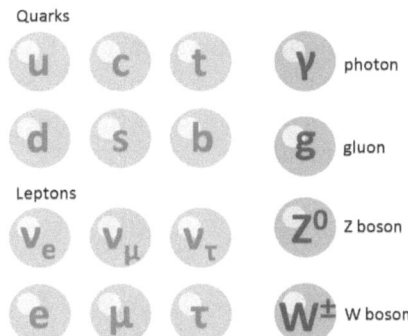

Figure 1.1a Quarks & Leptons

All the known matter in the universe is composed of two classes of particles: quarks and leptons. There are six types of quarks and six types of leptons. The figure above (Figure 1.1a) shows these fundamental particles. Three of the leptons, the electron (e), the muon (μ), and the tau lepton (τ) have an electrical charge that is a factor of –1 that of a proton, and three of the leptons, called the electron neutrino (ve), the muon neutrino (vμ), and the tau neutrino (vτ) have zero electrical charge. Quarks (defined previously) are named (in order of increasing mass) up (u), down (d), strange (s), charm (c), bottom (b), and top (t). The up, charm, and top quark have a charge that is +2/3 that of a proton, and the down, strange, and bottom quarks have a charge that is –1/3 that of a proton. Therefore, in an atom composed of a nucleus surrounded by electrons, the electrons are fundamental particles, which means they are not composed of anything smaller, as far as is known. But the nucleus is composed of neutrons and protons, which are themselves composed of quarks. At a very basic level a proton is made up two up quarks and a down quark with electric charge +2/3 + 2/3 – 1/3 = 1 while a neutron is made up of one up quark and two down quarks with an electric charge of +2/3 – 1/3 – 1/3 = 0. The two quarks and two leptons in the **first column** in the figure above (Figure 1.1a) are called the first generation of particles, the second column is the **second generation** of particles, and the **third column** is the third generation of particles. Most of all the matter we know of is made of the first generation of particles since atoms are made of neutrons and protons and electrons with the neutrons and protons made of up and down quarks.

Quarks are unusual. **First**, as already indicated, they have a charge that is a fraction of the proton's charge rather than an integer value of the proton charge. **Second**, as far as is known, they have no size since our theoretical idea of fundamental particles is that they do not have any size. **Third**, they cannot be isolated to exist by themselves. Though leptons, like an electron, can be found by themselves quarks are always bound together in groups of three called a baryon (protons and neutrons are examples of baryons), or bound together in a quark-antiquark pair called a meson (baryons and mesons were defined previously). Every particle in the first three columns of the figure above (Figure 1.1a) has an antiparticle partner with the same mass but opposite charge. The antiparticle of the electron is a positron with the same mass as an electron but an electric charge of +1. The other antiparticles are just named with an anti as an anti-up anti-quark with the same mass as an up quark but an electric charge of $-2/3$ rather than $+2/3$.

The top quark is the heaviest fundamental particle so far ever discovered. It has a mass equal to about 184 times that of a single proton. This is an unbelievably large mass for a fundamental particle. It is approximately the mass of a single tungsten nucleus which has an isotope composed of 552 up and down quarks. In other words, one top quark has the same mass as about 550 up and down quarks. When a top quark is created in the laboratory it only exists for about 5×10^{-25} seconds before it decays to other particles. This is such a short length of time that it does not have time to actually bind together with other quarks. Thus, it is the only known quark that exists apart from other quarks, (although it does not exist very long). The large mass of the top quark along with the large mass of the Higgs Boson (see section at the end of this chapter "The Higgs Particle and the Origin of Matter") plays a role in allowing our universe to have a stable or meta-stable vacuum. That is, the masses of these two very heavy particles are correctly tuned to allow our universe to exist over a long period of time. As discussed in the section "Virtual Particles," top quarks are produced virtually inside protons that actually fine-tune the mass of the proton to give its life-permitting value. If the mass of the proton were to change slightly there would be many life-destroying consequences. Stars like our sun that burn stably for a long time would not be possible. If the proton's mass were to change slightly neutron decay would be altered and similar changes would radically affect chemistry and biology. Very rarely inside the proton, a top quark and a top-antiquark will momentarily be created and destroyed as virtual particles. It's like two pickup trucks being created and destroyed inside your bicycle while you're riding it.[1-23a]

The **fourth column** consists of other elementary particles. The photon is a type of elementary particle. It is the quantum of the electromagnetic field including electromagnetic radiation such as light and radio waves, and the force carrier for the electromagnetic force (even when static via virtual particles). The invariant mass of the photon is zero; it always moves at the speed of light in a vacuum. A **gluon** is an elementary particle that acts as the exchange particle (or gauge boson) for the strong force between quarks. It is analogous to the exchange of photons in the electromagnetic force between two charged particles. The two charged particles "glue" quarks together, forming hadrons such as protons and neutrons. The **W** and **Z bosons** are together known as the weak or, more generally, as the intermediate vector bosons. These elementary particles mediate the weak interaction; the respective symbols are W+, W–, and Z0 . The W± bosons have either a positive or negative electric charge of 1 elementary charge and are each other's antiparticles. The Z0 boson is electrically neutral and is its own antiparticle. The three particles have a spin of 1. The W± bosons have a magnetic moment, but the Z0 has none. All three of these particles are very short-lived, with a half-life of about 3×10^{-25} s. The W bosons are named after the weak force, and the Z bosons were named for having zero electric charge. [1-23b]

Atoms and Particles - Continued

Creationist View

Everything in the universe is made of atoms, from the stars in the farthest heavens to the cells in the human body. (Atoms are elements that make up minerals, and minerals make up rocks.) Within the atom, the neutron is just slightly more massive than the proton, which means that free neutrons (those not trapped within an atom) can decay and turn into protons. Within the atom, the neutron is just slightly more massive than the proton, which means that free neutrons (those not trapped within an atom) can decay and turn into protons. If the make-up were reversed (if the proton was larger and had a tendency to decay) the very structure of the universe would be impossible. This is because a free proton is simply a hydrogen atom, and if free protons had a tendency to decay, then everything made of hydrogen would decay. The sun, which is made of hydrogen, would melt away. Water, a liquid oxide of hydrogen (H_2O) would be impossible. In fact, the universe itself would decay, since about 74 percent of the observed universe consists of hydrogen.

No one knows why the neutron is larger than the proton. There is no physical cause to explain why the neutron is larger. The difference in size between neutrons and protons allows the universe to exist and to support life. Not only do atomic particles (electron, proton, and neutron) have a size, but they also have an electrical charge. Within the atom, electrons have a negative charge, and protons have a positive charge. Most objects that are commonly observed have no electrical charge. The charges in electrons and protons are not very noticeable because the charge of the proton exactly balances that of the electron. If the electron carried more charge than the proton, all atoms would be negatively charged. If this were the case, since identical charges repel, all the atoms composing all the objects in the universe would fly apart in a catastrophic explosion. On the other hand, if the proton carried more charge than the electrons, all atoms would be positively charged, which would result in the same disastrous consequences that would result if the electron carried more charge than the proton.

There is no known physical reason, that is, there is no natural explanation, for the precise balance in the electrical charges of the proton and the electron. This is particularly interesting in considering when you consider that the two particles differ from one another in all other respects: in size, weight, magnetic properties, and so on. There is no natural explanation for this make-up. It is also interesting that if the values of the fundamental forces of physics (gravity, electromagnetism, that is, the strong and weak nuclear forces) were different, it would have been impossible for the universe to have sustained life. The physics in our universe are just right for what is needed for our universe to be capable of supporting life. (See section "The Fine Tuning of the Universe" for further discussion of the properties necessary to sustain life.) Could this have been the result of coincidence?[1-24]

Attempts to Determine the Origin of the Universe

Evolutionist View

Albert Einstein was one of the first to attempt to explain the origin of the physical universe. He believed at first that the universe had a static, uniform, isotropic distribution of matter. In 1915, Albert Einstein developed his general theory of relativity and started applying it to the universe as a whole. Since he believed that the universe had always been in existence and remained unchanged, he was perplexed when his calculations did not allow for a static universe. According to his equations, the universe should either be exploding or imploding. In order to make the universe static, he had to alter his equations by putting in a factor that would hold the universe steady.[1-25] Einstein's own calculations resulted in the exact opposite of what he had thought. Since the expansion of the universe had not yet been discovered Einstein arbitrarily introduced a special term "the cosmological constant" into his equations to make the universe static.[1-26] "The cosmological constant" developed the concept of a spherical, four-dimensional closed universe.[1-27] Then, in 1917, Albert Einstein revised his own set of equations from his general theory of relativity that predicted the nature of the universe. He came to the conclusion that the universe was in reality unstable, that it was constantly expanding or contracting. Later, during the 1920's, further solutions to Einstein's equations were worked out. These solutions indicated that the center of the universe was hot and dense, and was indeed expanding.[1-28]

A Look at the Big Bang Theory

Evolutionist View

The Big Bang theory makes an attempt to explain the origin of the universe. The idea of a "Big Bang" started in 1929 when astronomer Edwin Powell Hubble proved that the universe is constantly expanding, verifying Albert Einstein's theory. Prior to this time, scientists believed that the universe was always there, that is, static, with no change. The discovery by Hubble caused scientists to come to the conclusion that the universe had a beginning at a single point in space and time. Just as Galileo Galilei and Nicolaus Copernicus in the 1500s and 1600s met with a great deal of opposition when they declared the Earth was not the center of the universe, the Big Bang theory also experienced much opposition. After a great deal of observation and study, cosmologists came to the conclusion that the universe was not steady. Every prediction that quantum physics and the theories of relativity have made regarding the origin and the state of the universe have either been observed and confirmed or proven to be false and discarded. This is basically why this view is now so widely accepted. Scientists are confident that if time could be rewound, one could look back fifteen billion years and see the birth of our universe.[1-29]

Scientists now believe that the universe had a beginning of time and that approximately twelve to fifteen billion years ago there was nothing; but out of nothing matter and radiation developed. How matter came from nothing is unknown. All matter was at first concentrated into a tiny region of space smaller than the nucleus of a single atom. Within a millionth of a second after this, a great deal of energy converted into protons (the nuclei of hydrogen atoms). Then, within the next millisecond electrons formed, and these electrons collided with the protons which became neutrons. (Neutrons will only survive a few minutes as independent particles. A neutron has a mean lifetime of approximately 1.0×10^3 seconds as a free particle.)[1-30]

Scientists believe that perhaps within the first fifteen minutes after the "Big Bang" occurred the protons reacted with the neutrons that were decaying, thus developing into the nuclei of helium atoms. Then, within a relatively short period of time, the universe cooled and expanded. During this cooling and expansion, the universe converted a quarter of its matter to form hydrogen into helium. The hydrogen that was left became stars.[1-31] Astrophysicists used the Big Bang theory to calculate that the conditions of the universe about three minutes after the Big Bang were very similar to the interior of a star and just right for fusion reactions. The temperature and density of the hydrogen gas allowed it to fuse into helium and into trace amounts of deuterium and lithium. The calculations of the Big Bang model indicated that about 24 percent of the gas would have been helium.[1-32]

The Origin of the Universe—The Big Bang

Evolutionist View

It is commonly understood that the universe started off with a "Big Bang." It was called the Big Bang because the universe has been expanding like the sound waves of an explosion ever since. No one knows what caused the Big Bang. Before the Big Bang occurred, nothing that can be measured existed—not time, space, energy, or matter. Suddenly, the universe existed. It has been estimated that the universe originated about 15 billion years ago. (See Chapter 2 for further discussion of the age of the universe.)[1-33] When the universe originally formed, it was made up entirely of hydrogen and helium that was too hot for matter to exist, so pure energy filled the universe. As the universe kept expanding, its energy spread thinner and thinner. Its temperature began to cool. When it cooled down to several millions of degrees, matter appeared. It first appeared as the particles that make up atoms and then, with further cooling, as atoms themselves. The universe is still cooling.

Hydrogen and helium are the simplest elements in the universe. These gases are lighter than air. In the beginning, the universe was made up entirely of these two gases (hydrogen and helium). Stars are mainly great combinations of hydrogen and helium that have been condensed and compacted by the force of gravity. Gravity is a little-understood force of attraction that acts between every atom in the universe. It increases with mass (the amount of matter) and decreases with distance, but regardless of masses or distance, it is always present.

In the early universe the effects of gravity attracted hydrogen and helium atoms to mass together and condense into clouds. These clouds attracted other clouds. The force of gravity produced clouds with a condensed core called proto-stars. As the space between atoms in a proto-star decreased, the atoms began to hit each other more often. Eventually, they were bouncing off each other. Each bounce created a little bit of heat energy. The denser the gas became, the hotter it got. Then, these lighter-than-air gases were compressed until they became denser than any metal. The atoms went wild and the temperature soared.

As mentioned previously, elements are made up of atoms. An atom is the smallest particle of an element that still behaves like the element of which it is a part. Atoms, in turn, are made up of subatomic particles. The three important ones are protons, neutrons, and electrons. The number of protons in an atom determines which element it is. Nature uses only 6 elements to make up 99 percent of all living matter. These 6 elements are (1) hydrogen, (2) oxygen, (3) carbon, (4) nitrogen, (5) sulfur, and (6) phosphorus. Molecules are combinations of atoms. Chemicals are combinations of elements or the elements themselves. Molecules do not form randomly but in ordered, patterned ways. In water, for example, each molecule is lopsided because the two hydrogen atoms always line up on one side of the oxygen atom at a 105-degree angle from one another. Without the natural attraction that certain atoms have to one another and the precise patterns in which they align themselves, life itself would not have been possible. Molecules are like a puzzle with interlocking pieces. Shortly after Earth's crust solidified and its surface cooled to just below water's boiling point, rainwater began filling in the low spots. As it fell, rain picked up carbon dioxide, nitrogen, hydrogen, and sulfur from the atmosphere. As this chemical solution splashed on the rocks, it dissolved more elements. The highlands eroded. Their sediments slowly settled in the lowlands. Sands and clays formed from matter that settled out of floodwaters. With complex proteins being formed and held in position on clay, life itself was soon to develop.

Life will develop wherever it is given a chance to do so. Amino acids and other organic chemicals found in human cells have also been found in meteorites and comets exposed for eons to the hazards of space. Some say that meteors brought life to this barren planet (discussed in Chapter 5). It is more likely, however, that the early Earth had just as many amino acids as any meteor did. All the matter in this solar system came together at the same time and under roughly the same conditions. The Earth was just wet enough and warm enough for complex chemicals to become living chemicals.[1-34] (The origin of life is discussed in Chapters 5 & 6.)

Creation Science and Intelligent Design View

The concept of a Big Bang would require that an equal amount of matter and antimatter be originally made at the time of the initial explosion; yet the universe only has matter, and lacks the antimatter. (Anti-matter is discussed in section "The Origin of Matter" also in section "The Quantum Theory.") This is strong evidence against the Big Bang and any other theory of initial self-creation of matter in the universe. There is no physical mechanism in the near vacuum of outer space to compress gas into a ball. A cloud of hydrogen gas must be compressed to a small enough size so that gravity can dominate it. For example, our own sun is a stable sphere of gas. But what force could initially press it into a ball? Scientists have no answer.

Experiments indicate that it would be next to impossible for floating gas molecules out in space to clump together. There is nothing to compress it. How could the stars evolve from floating gasses? Gravity is not a sufficient mechanism to do this. In outer space, the gas is millions of times more expansive than the critical compressed size needed for gravity to hold it as a stable star. Because of this, outward gas pressures cause these clouds to keep spreading outward. They do not pull together, but instead gradually move outward. In spite of all the theories of the evolutionists, the fact remains that gas in outer space always has a density so rarified that it is far less than the emptiest atmospheric vacuum bottle in any laboratory in the world. How could such rarified hydrogen develop into planets and stars? If hydrogen gas blew outward after an explosion in outer space, there would be no way to slow it. An explosion of matter would cause an outward spray of gas and energy. It would continue to move outward in space forever. Space is frictionless. There would be no way to slow the gas, that is, there would be no way to stop it. In addition, there would also be no way for a gas to clump into a solid.

On earth, gas never clumps into a solid. Out in space, where everything is a near vacuum, it would be totally impossible for this to occur. Throughout the voids of space between the stars are various gases, the primary one is hydrogen. These gaseous compounds never move away from an area of vacuum into an area of congestion or density. The hydrogen gas observed by astronomers through telescopes is gradually expanding. None of it is packing together. In reality, it would have been physically impossible for gas from a Big Bang to have stopped and clumped. Gas floating in a vacuum of outer space cannot form itself into stars. Once a star is formed, it can hold itself together by gravity, but there is no way that gas in outer space can get the mechanisms started. Once a star exists, it will absorb gas into it by gravitational attraction. But before the star exists, gas will not push itself together and form a star—or a planet, or anything else. It would remain just loose, floating gas.

It is physically impossible for a random explosion such as a Big Bang to produce intricate orbits. There are extremely complicated factors involved just in maintaining the proper rotations and revolutions of galaxies, stars and planets. The careful balancing of gravity versus centrifugal force that occurs throughout the universe in the orbits of the spheres is a marvel. Random explosions never produce orbits. No type of explosion can produce the intricate, balanced orbits of the stars, planets, and moons.

There is no way that stars could spin (rotate) or orbit (revolve) if, before star formation, there was only outward exploding gas or even randomly floating gas. Either way, there would be no means by which the turning movements could start.[1-35] At some point after the explosion (the Big Bang), as temperatures cooled, physicists claim that somehow hydrogen was formed. Then, at some point immediately or sometime afterward (opinions differ), some of the hydrogen changed to helium. Both hydrogen and helium are gases. Physicists claim that the gases spread outward throughout the universe for about ten billion years, and then the hydrogen and helium gas gradually pushed itself into chunks. Then more and more clumps formed, until soon gigantic pieces of it had formed. These became stars and galaxies with their orbits.

Physicists believe that the initial "Bang" explosion produced only hydrogen and perhaps helium, but after the stars had pushed themselves together, they began exploding like strings of firecrackers. Then, upon reforming, large numbers exploded a second time. Then, somehow, all 90 elements in existence were produced by the second wave of explosions. For the Big Bang to have happened, it would have required a process to develop atoms after the initial explosion. This initial atom-building process would need successive neutron-capture reactions to achieve elements of increasing atomic weights in a stepwise manner, starting with (according to the Big Bang theory) a 100 percent neutron content of the primordial (super dense core) ylem. (*A ylem is defined as the initial substance of the universe from which all matter is said to be derived. It is the original matter from which the basic elements are said to have been formed following the explosion postulated in the big bang theory of cosmology.*)

According to the Big Bang theory, at the end of the first 30 minutes slightly more than half of the (super dense core) ylem was converted into hydrogen, with slightly less than half into helium. This may sound feasible; however, it is quite another thing to actually happen. How could the other elements (besides hydrogen and helium) have developed? Physicists know that, among nuclides that can actually be formed, a gap exists at mass 5 and 8. The first gap is caused by the fact that neither a proton nor a neutron can be attached to a helium nucleus of mass 4. Because of this gap, the only element that hydrogen can normally change into is helium. It is true that some scientists believe that a hydrogen bomb explosion can produce elements beyond helium, but there is also evidence which would indicate that this is not so. It would have taken a tremendous amount of energy for the Big Bang to have actually occurred. Without energy there can be no heat, no explosion. The Big Bang theory is supposed to explain the origin of matter, however, it would also need to explain the origin of energy, as the two are variant forms of one another. An explosion could not occur without energy, and without matter there would be nothing to explode outward. Without pre-existing matter and energy, there could be no explosion. Some who support the Big Bang theory claim that perhaps there may originally have been an immense concentration of energy, but they have no idea where it came from or how it got there. Others believe that an explosion without energy initially created energy. Without energy no explosion could have occurred.[1-36]

For the Big Bang to have been successful in developing the universe, the Big Bang had to have exploded with just the right degree of vigor for our present universe to have formed. If it had occurred with too little velocity, the universe would have collapsed back in on itself shortly after the big bang because of gravitational forces; if it had occurred with too much velocity, the matter would have streaked away so fast that it would have been impossible for galaxies and solar systems to subsequently form. In other words, the force of gravity must be fine-tuned to allow the universe to expand at precisely the right rate (accurate to within 1 part in 10^{60}). The fact that the force of gravity just happens to be the right number is one of the great mysteries of cosmology.[1-37] Normally, an explosion causes disorder rather than organization. The most basic principles of thermodynamics, physics, and biology suggest that chance simply cannot be the determinative force that has brought about the order and interdependence we see in our universe, much less the diversity of life we find on our own planet.[1-38]

Evolutionist View

The Big Bang Theory has caused many misconceptions and misunderstandings. The concept of a Big Bang causes many to think of a giant explosion. Scientists do not believe there was an explosion. Rather, they believe there was an expansion, which continues even today (See section "The Expansion of the Universe"). Another misconception is that the Big Bang began with a small fireball traveling through space. Many scientists, however, believe that space didn't exist before the Big Bang. It has been theorized that time and space had a finite beginning that began around the same time as the origin of matter and energy.[1-39] The Big Bang is properly envisioned as the expansion of space, with matter carried along, rather than an explosion hurtling matter into empty space.[1-40]

Other Possible Theories

The Inflationary Universe Theory states that the universe did not begin from a single big bang, but rather, matter appeared as fluctuations in a vacuum. The Oscillating Model theory attempts to eliminate the need for an absolute beginning of the universe by suggesting that at first the universe expanded, then collapsed, then expanded again and continues in this cycle indefinitely.

Oscillating Model

Evolutionist View

The Oscillating Universe Theory is a cosmological model that combines both the Big Bang and the Big Crunch as part of a cyclical event. That is, if this theory holds true, then the Universe in which we live in exists between a Big Bang and a Big Crunch. In other words, our universe can be the first of a possible series of universes or it can be one in a number of universes. As previously discussed, according to the Big Bang Theory, the Universe is believed to be expanding from a very hot, very dense, and very small entity. If we extrapolate back to the moment of the Big Bang, we are able to reach a point of singularity characterized by infinitely high energy and density, as well as zero volume. This description would mean that all the laws of physics would need to be discarded. This is understandably unacceptable to physicists. To make matters worse, some cosmologists even believe that the Universe will eventually reach a maximum point of expansion and that once this happens, it will then collapse into itself. This will essentially lead to the same conditions as when we extrapolate back to the moment of the Big Bang. To remedy this dilemma, some scientists are proposing that perhaps the Universe will not reach the point of singularity after all. Instead, because of repulsive forces brought about by quantum effects of gravity, the Universe will bounce back to an expanding one. An expansion (Big Bang) following a collapse (Big Crunch) such as this is aptly called a Big Bounce.[1-41] This idea that Big Bangs follow Big Crunches in a never-ending cycle is known as an oscillating universe.[1-42] The bounce marks the end of the previous universe and the beginning of the next. The probability of a Big Bounce, or even a Big Crunch for that matter, is however becoming negligible. The most recent studies indicate that the Universe will continue on expanding and will most likely end in what is known as a Big Freeze or Heat Death. It is therefore highly unlikely that future findings will deviate much from what has been discovered regarding the Universe's expansion.[1-41]

Creationist and Intelligent Design View

There are several problems with the Oscillating Model. First, this theory contradicts the known laws of physics.[1-43] The Oscillating Model theory ignores the second law of thermodynamics, which requires usable energy to continually decrease and for the universe to become more random and disorganized.[1-42] There is no known laws of physics that could reverse a contracting universe and cause it to return to a singular state.[1-43] It has been determined that as long as the universe is governed by general relativity, the existence of an initial singularity (or beginning) is inevitable, and that it is impossible to pass through a singularity to a subsequent state. The whole theory was simply a theoretical abstraction that physics never supported.

In order for the universe to oscillate, it has to contract at some point. For this to happen, the universe would have to be dense enough to generate sufficient gravity that would eventually slow its expansion to a halt and then, with increasing rapidity, contract it into a big crunch. But estimates have consistently indicated that the universe is far below the density needed to contract, even when you include not only its luminous matter, but also all of the invisible dark matter as well. Tests calculated a ninety-five-percent certainty that the universe will not contract, but that it will expand forever.[1-44] This means that the universe is not closed and therefore will expand forever.[1-42] These tests indicated that the expansion is not decelerating, but is actually accelerating, thus disproving the Oscillating Model.[1-44] Another reason why the Oscillating Model is no longer considered is that this theory really does not provide an explanation of the initial origin of the universe, but instead only pushes it back further in time.[1-42]

The Quantum Theory

Evolutionist View

Earlier versions of the Big Bang theory stated that the universe originated from a point of zero volume and infinite density, where the laws of physics have no meaning. This has been replaced by the idea that the universe originated from literally nothing at all. According to the quantum theory, matter and anti-matter particles are created in pairs all the time out of nothing (i.e. vacuum) and cancel each other out with no effect on the universe. They are therefore called virtual particles. At the Big Bang, however, massive amounts of matter and antimatter were created, and although much of it was similarly canceled out with a huge release of energy, matter succeeded and produced the universe as we know it.[1-45] (Anti-matter is discussed in section "The Origin of Matter" and section "The Higgs Particle and the Origin of Matter.")

Creationist View

The theory of anti-matter developed in the 1920's, from the idea that if the theory of relativity and quantum mechanics were true, then for every particle of matter a corresponding particle of antimatter also existed. Science has been experimenting with anti-matter ever since. In order for the Big Bang to have occurred, the question remains, what produced the particle of matter and the anti-particle? The quantum theory basically states that our universe is a part of a bigger mother universe that is made up of a quantum vacuum where fluctuations occur that turn into new universes. While spin-offs of the mother universe expand, the bigger mother universe is infinite and eternal. It appears that there is no scientific evidence to support this theory. This theory does not explain how the mother or main universe developed. In contrast to the Big Bang Theory, which is said to have a beginning, the quantum theory does not have a beginning point. In the Big Bang Theory, you could determine the beginning, whereas in the quantum theory, if you attempted to go back to the beginning, you would only end up going forward again. It would be like traveling around the world. From any starting point you would eventually end up back where you started. In the quantum theory, there is no beginning and no end. Thus the universe always existed. The quantum theory does, however, stipulate that the universe had an origin out of nothing, in the sense that there is nothing that comes before it.[1-46]

Many-Universe Generator Machine

Creation Science and Intelligent Design View

The Many-Universe Generator Theory stipulates that if there are an infinite number of universes, with enough variation in atmospheric conditions, the odds are that at least one would become a viable planet. If there are many universes, each governed by different atmospheric conditions; there would be at least one that would have the right conditions to sustain life, such as the Earth. Many physicists believe in some variation of the Multiple Universe or Multiverse Theory. The most popular theory, inflationary cosmology, is the most credible. The Self-reproducing Inflationary Universe model is based on advanced principles of quantum physics. The existence of many universes is consistent with all that is known about physics and cosmology.[1-47] The Many-Universe Generator theory can only produce life-sustaining universes if it has the right components and mechanisms. Right now the Many-Universe Generator theory is just a theory whose main qualities are that it is mathematically possible and that it holds the promise of unifying quantum mechanics and general relativity, two branches of physics that physicists have struggled to reconcile for over fifty years. **First**, a mechanism to supply the energy would be needed for the multiple universes. That would be the inflation field that has been hypothesized, which effectively acts like a reservoir of unlimited energy. **Second**, a mechanism would be needed to form the universes. This would be Einstein's equation of General Relativity. Because of its peculiar form, this would supposedly cause the multiple universes to form and the space of the universe generator to keep expanding. **Third**, a mechanism would be needed to convert the energy of the inflation field to the normal mass/energy that we find in our universe. **Fourth**, a mechanism would be needed to allow enough variation in the constants of physics among the various universes. In other words, a way to vary the constants of physics would be needed so that by random chance some universes would be created like ours that would have the right fine-tuning to sustain life. The problem with the Many-Universes Generator theory is that it would need all four of the factors if it ever hoped to produce a functioning universe.

If the universe functioned according to Newton's theory of gravity instead of Einstein's theory of relativity, it would not work. Without the so-called principle of quantization that explains the behavior of matter and its interactions with energy on the scale of atoms and subatomic particles, all of the electrons in an atom would be sucked into the atomic nuclei. That would make atoms impossible. (*Quantization means to restrict a variable quantity to discrete values rather than to a continuous set of values. It is to change the description of a physical system from classical to quantum-mechanical, usually resulting in discrete values for observable quantities, as energy or angular momentum.*) Without a universally attractive force between all masses, such as gravity; stars and planets could not form. If just one of these components was missing or different, it is highly improbable that any viable universes could be produced. Also, to make unlimited amounts of universes in order to increase the odds that the cosmological constant would come out right at least once is nearly impossible, because the universe is finely tuned to an incomprehensible degree. And that is just one parameter. The problem is that it is highly unlikely that such a universe-generating system would have all the right components and ingredients in place by random chance.[1-48]

The Fine Tuning of the Universe

Creationist and Intelligent Design View

In the late 1950s, astronomers described the precise process by which carbon and oxygen are produced in a certain ratio inside stars. If the resonant states of carbon are changed even slightly, the materials for building life will not be obtained. Recent studies show that just one-percent change in the strong nuclear force would have a thirty to a thousand-fold impact on the production of oxygen and carbon in stars. Since stars provide the carbon and oxygen needed for life on planets, if you throw that balance off, conditions in the universe would be much less optimal for the existence of life. The atmosphere of the Earth consists of precisely the right gases necessary for life to flourish. The Earth's atmosphere is composed of 78 percent nitrogen, 21 percent oxygen, a small amount of other gases, and water. These gases have precisely the correct ratio to permit the complex biological processes that are required for plants, humans and animals to exist. For the universe to exist as it does, requires that hydrogen be converted to helium in a precise manner, specifically, in a way that converts seven one-thousandths of its mass of energy. Lower that value very slightly from 0.007 percent to 0.006 percent, for example, and no transformation could take place. The universe would consist of hydrogen and nothing else. If the value is raised very slightly to 0.008 percent, for example, and bonding would be so wildly abundant that the hydrogen would long since have been exhausted. In either case, with the slightest tweaking of the conditions, the universe as we know it would not exist.[1-49]

If the percentages are changed even slightly, it is likely that animals anywhere near the size of human beings would be crushed. Any planet with a gravitational pull of a thousand times that of the Earth would have a diameter of only forty feet, which wouldn't be enough to sustain an ecosystem. Also, stars with lifetimes of more than a billion years couldn't exist if the gravity is increased by just three thousand times. Our sun is estimated to sustain life for ten billion years. When the various force strengths in nature are compared, gravity has a very narrow range for life to exist as we know it. Our planet has been so precisely fine-tuned, almost too precise to comprehend.[1-50]

The mass and size of the Earth are just right. If the Earth's diameter were 7,200 miles instead of 8,000, almost the whole Earth, due to a lessening of its atmospheric mantle, would be reduced to a snow and ice waste. If there were a variation of only 10 percent, either in the increase or decrease of the size of our world, no life as we know it on Earth would be possible. If the Earth were smaller, it would not have the gravitational pull necessary to retain the water and atmosphere necessary for life. A smaller Earth would have a thinner atmosphere that would diminish our protection from the multiple meteors that bombard our planet daily. A thinner atmosphere would also diminish our protection from the Sun and would result in a warmer Earth. If the Earth were much larger, it would have retained a large percentage of gases unfavorable to life. If it were much smaller, its gravitational forces would have been insufficient to retain virtually any atmosphere at all. The smaller planets with smaller gravitational fields lose a large proportion of their lighter elements during the cooling process. The larger planets retain most of their original atmosphere. If the average temperature of the Earth were raised by two or three degrees, water from glaciers would flood the Earth. This would also inundate hundreds of thousands of square miles of our most fertile lands.

The Earth's axis, which now points toward the North Star, is tilted just right—at an angle of 23 degrees from the perpendicular, that is, in relation to the plane of its orbit (Figure 1.3). Because of this tilt the sun appears to go north in the summer and south in the winter, giving us four seasons in the temperate zone. For the same reason, there is twice as much of the land area of the Earth that can be cultivated and inhabited as there would be if the sun were always over the equator, with no change of seasons.

The amazing accuracy with which the universe revolves can be seen in the perfection that characterizes the journey of our Earth around the sun. It takes Earth 365 days, 5 hours, 48 minutes and 48 seconds to make the journey around the sun. And in this circuit the Earth has varied in only the slightest degree.[1-50a]

Of all the planets in our Solar System, Earth has one of the most circular orbits around the sun, although it is not perfectly circular. No planet's orbit is. Each planet revolves around the Sun in an elliptical orbit. Venus has the most circular orbit of any of the planets in the Solar System. The Earth has the next most circular orbit. Seasons occur on the Earth because Earth's axis is tilted at an angle of about 23.4 degrees and different parts of Earth receive more solar energy than others (Figure 1.3). The Earth orbits the Sun on a slant which means different areas of Earth point toward or away from the Sun at different times of the year.[1-50b] The Earth is 93 million miles from the sun. This is the exact distance necessary for life, as we know it, to exist here.[1-51]

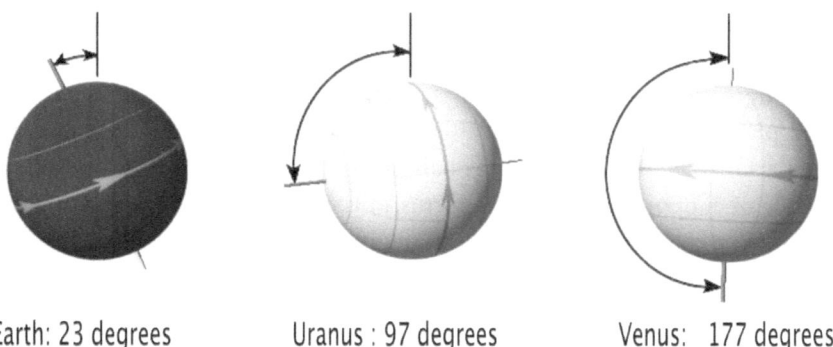

Earth: 23 degrees   Uranus : 97 degrees   Venus: 177 degrees

Figure 1.3 Tilt of the Axis of Planets Earth Uranus and Venus

We live miraculously on this planet, protected from eight killer rays from the sun, by a thin layer of ozone high up in our atmosphere. If that little belt of ozone, approximately forty miles up and only one eighth of an inch thick (if compressed), should suddenly drift into space, all life on Earth would perish. The first miracle, in the light of what the rest of the universe is like, is that there is an ocean here. In the universe as a whole, liquid water of any kind, sweet or salt, is an exotic rarity. Contrary to common belief, the liquid state is exceptional in nature; most matter in the universe seems to consist of flaming gases, as in the stars, or frozen solids drifting in the abyss of space.[1-50a] If our planet was just a few hundred thousand miles closer to the sun, or just a few hundred thousand miles farther away from it, all life on earth would quickly perish from a variety of causes, only one of which would be temperature.[1-52]

Water has a host of unique properties absolutely indispensable for life. For example, it is the only known substance whose solid phase (ice) is less dense than its liquid phase. This is why ice forms on the top of oceans and lakes instead of on the bottom, allowing fish and other marine life to survive the winter. On the microscopic level, water molecules exhibit something called the hydrophobic effect, which gives water the unique ability to shape proteins and nucleic acids in DNA.[1-53]

To explain our existence on the planet Earth, evolutionists simply claim that there are many planets so one must have the conditions necessary to support higher life forms. This belief is both naive and unscientific for there is enough information about the requirements necessary for a planet to support higher life forms that are known to able to do a rough estimate of the probability of finding even a single planet like the earth. Creationist researchers have developed a rough estimation of the probability of finding a single earth-like planet by chance based on 322 parameters known to be necessary if a planet is to support higher life forms. They took correlations and longevity factors into account as well as the fact that there are at least 1022 planets in the visible universe. These researchers' order-of-magnitude calculation comes up with a probability of 10^{-282} for finding one planet capable of supporting higher life forms in the entire visible universe. Those who believe that because there are so many planets in the universe, that one must be just right for life, are wrong. Even with a lot of planets one should not expect to find one suitable for our existence purely by chance. [1-53a] The degree of fine-tuning favorable for life indicates that this planet must have been created by a divine Creator. Even if the universe contains as many as ten billion trillion planets (10^{22}), we would not expect even one, by natural processes alone, to end up with the surface gravity, surface temperature, atmospheric composition, atmospheric pressure, crustal iron abundance, tectonics, volcanism, rotation rate, rate of decline in rotation rate, and stable rotation axis tilt necessary for the support of life.[1-53a-1]

Evolutionist View

The claim made by creationists that the universe is fine-tuned by a creator is simply not true. After the universe developed, anyone could make just about any type of comment they wish about how miraculously it was made. After the universe developed, the odds that it is constructed as it is, is 100%. The problem with the creationists' fine-tuning reasoning is that the odds were determined after the universe came to its present state. The fundamental physical constants are the charge of the electron and the proton, the mass of the electron, the mass of the proton, the gravitational constant, and others. The charge of the electron is exactly equal to that of the proton but opposite in sign. The proton carries a positive charge, and the electron carries a negative charge. The mass of the earth and the acceleration due to gravity are not fundamental constants because, in principle, they can be expressed in terms of one or more of the fundamental constants. The values of the fundamental constants themselves cannot be expressed in terms of any other constants, as is currently known. This is the reason why they are referred to as fundamental. There is no reason to assume that the universe got its fundamental constants by some miraculous event. The universe may have gotten their constants during the big bang by a process that is yet unknown or not currently understood. There is no reason to assume that the fundamental constants are independent of each other. If they are not independent of each other, then simple probability theory does not apply. Possibly the universe assembled itself in the way it did because of a series of causal relationships.

It has been suggested that it is not accurate to examine the possibilities of a universe developing by natural means to support life by altering only one constant at a time. It would be more reasonable to speculate about how the universe turned out by making changes in all the fundamental constants, by varying them randomly and then observe what universes develop. If a substantial fraction of the universes developed from the experiments have the properties necessary to support life, then our universe cannot be considered fine-tuned by some Creator.

Heavy elements are formed in the interiors of stars. Iron and heavier elements are formed in supernovas. For our discussion, let's assume that stars are necessary for life and that only long-lived stars can form the heaviest elements. By varying the values of the constants randomly, for example, it can be demonstrated that various universes could be developed with lifetimes in excess of 10 billion years, many in excess of 1 billion years. Our sun is estimated to be about 4.5 billion years old, therefore it can be assumed that 5-10 billion years is long enough for intelligent life to develop. Because over half of the universes that developed from experiments could have developed with lifetimes in excess of 10 billion years, the logical conclusion is that there is nothing special about our universe and its fundamental constants. The fine-tuning argument may be faulty because its proponents have not examined enough alternatives.

What would creationists have claimed if the universe (or some other possible universes) had a great deal of life on other planets in addition to life on earth? What if creationists thought that the conditions favorable for life on Earth were also common on other planets? They would have concluded that the universe was designed for life. Since life appears to be uncommon in the universe, creationists conclude that the earth was designed for life. This goes back to the days before the 1500 and 1600's, before Copernicus and Galileo, when it was believed that all the stars and planets revolved around the earth, that the earth was the center of the universe. Creationists are again attempting to claim that the earth is something special but it is not.[1-54]

The M-Theory (a version of String Theory)

Creationist View

Regarding the laws of physics that seem to be finely-tuned to allow life to exist, evolutionists often appeal to M-theory, the most recent and encompassing string theory. (According to Edward Witten, the originator of the M-theory, M could stand for "magic", "mystery", or "membrane" according to one's preference. He felt that the true meaning of the letter M should not be decided until a more fundamental formulation of the theory is known.) (*The string theory is a theoretical framework in which the point-like particles of particle physics are replaced by one-dimensional objects called strings. It describes how these strings propagate through space and interact with each other.*) String theory proposes that the fundamental entities that make up our universe are "vibrating strings of energy."[1-53a] There are several versions of string theory: type I, type IIA, type IIB, and two flavors of heterotic string theory (SO(32) and $E_8 \times E_8$). The different theories allow different types of strings, and the particles that arise at low energies exhibit different symmetries. For example, the type I theory includes both open strings (which are segments with endpoints) and closed strings (which form closed loops), while types IIA and IIB include only closed strings. Each of these five string theories arises as a special limiting case of M-theory. The M-theory, like its string theory predecessors, is an example of a quantum theory of gravity. It describes a force just like the familiar gravitational force subject to the rules of quantum mechanics. In string theory, spacetime is *ten-dimensional* (nine spatial dimensions, and one time dimension), while in M-theory spacetime is *eleven-dimensional* (ten spatial dimensions, and one time dimension).[1-53b]

Evolutionists believe that M-theory holds a lot of promise as a scientific theory, including the development of a consistent quantum theory of gravity, which has been an elusive goal for about 100 years. M-theory requires that there are 11 dimensions of space-time. M-theory has about 10,500 possible configurations, and allows for the possibility that there are many universes. If ours is just one of many universes (a multiverse), with different laws and parameters of physics in every different universe, then just by chance one of the universes would have the laws and parameter falling in the necessary range to be able to support life. Evolutionists believe we are here because we happen to be in the right universe.

There are many problems with M-theory. The **first** problem is that, as with the no-boundary condition, there is no scientific evidence that M-theory is true, so a belief in M-theory is not based on science at all. **Second**, there are few, if any, definitive predictions of M-theory. For instance, it is unknown if any of the "other" universes would actually be created or just have the potential of being created. Some evolutionists believe in an almost infinite number of versions of M-theory, which "predict" an almost infinite number of possible universes. M-theory is just another unproved hyphothesis.[1-53a]

The Expansion of the Universe

Creationist and Intelligent Design View

It has been known for quite some time that our universe is expanding. This led astronomers to believe that the universe had a spectacular beginning at some point in time, which evolutionists claim occurred at least 10 billion years ago. At this point, it is believed that all matter and energy in the universe would have been concentrated in one spot. Then a type of king-sized atomic explosion occurred, and everything blew apart with enormous violence and at extremely high temperatures. As the expansion continued, things gradually cooled down, until eventually galaxies of stars were able to form. Recent mathematical studies have shown that this is almost certainly what happened. These calculations are based on data obtained in atomic energy laboratories as well as astronomical data. If some sort of 'big bang' is to produce a habitable universe, then the explosive power has to be exactly matched to the power of gravity. If the Big Bang that formed our universe had not been quite big enough, the expanding matter would soon have collapsed back on itself. In that case the universe would have ceased to exist, long before any planets suitable for supporting life could have formed. On the other hand, if the Big Bang had been too big, it would have caused the universe to expand so much that it would have contained nothing but gas. Stars and planets would never have formed. If the moment of expansion had been reduced by only one part in a thousand billion, then the universe would have collapsed after a few million years. On the other hand, if the rate had been even slightly greater, then the expansion would have reached such magnitudes that no gravitationally bound system (galaxies and stars) could have formed.[1-55]

There are fundamental forces in our universe that are essential for the existence of the universe and the development of life within it. These forces are such that if they were different in any way, then life could not exist. The force of gravity is set so precisely that the minutest change would have prevented the expansion of the Big Bang. A famous professor stated: *"If the rate of expansion one second after the big bang had been smaller by even one part in a hundred thousand million million it would have re-collapsed before it reached its present size."*[1-56]

Regarding the nuclear forces within atoms and for the strength of electromagnetism, if these forces were different by a small percent, then no other atoms besides hydrogen would have been formed in the universe, including the carbon atom that is essential for life. This same professor also stated: *"The odds against a universe like ours emerging out of something like the Big Bang are enormous. I think there are clearly religious implications."*[1-56]

Astronomers suggested that the galaxies are moving away from each other. They also stated that the farther away the galaxies are from each other, the faster they recede. Scientists use the term "redshift" to describe the experimental finding that suggests increase in speed. As a galaxy moves away from earth, its color becomes redder, and the degree of color change is supposedly directly proportional to the speed of the galaxy moving away from earth. This effect is called "redshift." Galaxies that are far away from earth display this redshift. This is allegedly due to the fact that stars that move away from an observer emit light of a slightly longer wavelength, the faster they move, the greater the change in the wavelength. Young-earth creationists have some doubts with the idea that the universe is expanding. For example, this expansion is not directly observed but is rather an inference from the observation that light from distant galaxies has a longer wavelength (it is red). In recent years, some have questioned whether the redshift proves the galaxies are in fact moving away from each other. They suggest that perhaps the redshifts are due to something far simpler, such as light losing some of its energy as it crosses billions of miles of space.[1-57] If the rate of the expansion of our universe were any different, life on Earth would not be possible. If it expanded too rapidly, matter would disperse so fast that it could not clump together to form galaxies (with stars and planets). If the universe expanded too slowly, matter would end up clumping together and collapse into a dense lump before individual stars could form.[1-58]

Evolutionist View

Creationists today use the expanding universe and the Big Bang theory and the view that the universe had a beginning to support their view that a creator developed it. The expansion of the universe, however, implies that the universe has a finite age and will someday die out as the nuclear fuel in the stars is used up. It has been observed that when the Big Bang theory was being developed, many theologians and some philosophers argued in favor of a steady-state universe, partly because they did not like the concept that the universe has a finite age. Then, when the evidence in favor of the Big Bang theory and against the steady-state universe and continuous creation became overwhelming, theological arguments favoring the steady-state theory ceased. Theology and philosophy are extremely flexible and can be made to adapt with either theory, just as the special-earth hypothesis can be made to adapt with a universe that is either teeming with life or in which life is exceedingly uncommon.[1-59]

The Higgs Particle and the Origin of Matter

Intelligent Design and Creationist View

In the early 1960's a theory of the elementary particles and forces in the universe, what eventually became known as "The Standard Model of Particles and Fields," was being developed by theoretical physicists (The Standard Model is defined at the end of this chapter). One of the problems was that within the theoretical framework the truly fundamental particles did not have any intrinsic mass, though it was known that they must have mass. In 1964 three independent groups proposed a model which would give a mass to the fundamental particles, but required that a new particle should exist: the Higgs boson. Over the last 50 years, all of the predictions of the Standard Model have been confirmed. Like putting a jigsaw puzzle together, the pieces of the model that were discovered fit perfectly. But one piece, the Higgs boson, was still missing. (*A boson is a subatomic particle with an intrinsic quantum mechanical property of spin equal to an integer value times the Planck constant.*) That one important piece which would finish the puzzle could not be found. The European particle physics laboratory, CERN, built a particle collider called the Large Hadron Collider, with the primary purpose of discovering the last piece of this beautiful puzzle. (*A hadron is a composite particle made of quarks held together by the strong force in a similar way as molecules are held together by the electromagnetic force.*[1-59b]) The discovery of the Higgs boson (or Higgs particle) in 2012 affirmed the validity of the Standard Model and confirmed our understanding of the mechanism which gives mass to fundamental particles.

There has been much controversy about whether or not the Higgs particle provides any insight regarding the existence of God. In 1964 six physicists did mathematical calculations about the universe. Their mathematics predicted that an undiscovered particle should exist. Nearly 50 years later (in 2012), the particle was discovered. It is interesting that mathematics describes our universe and that we can trust the results of this theoretical mathematics enough to confidently believe that a physical particle actually exists. Although it may not be properly called "The God Particle," the mathematical description and complexity of our universe, along with its actual existence, gives a clear indication of a true deity, a Creator God, who has designed and created what we now have the privilege to observe and study.[1-59a]

In 2012, "The God particle" (also known as "Higgs boson" or "Higgs particle") made national headlines. Peter Higgs is the scientist who predicted its existence over 40 years ago in 1972. These experiments were conducted at the CERN research center (European Organization for Nuclear Research) in Geneva, Switzerland. These experiments were designed to locate a particle that is believed to have been responsible for developing all matter. The apparatus is located inside a tunnel, called the Large Hadron Collider (LHC). The LHC (Large Hadron Collider) was designed to create a tiny, compact wad of energy from which a Higgs particle might spark into existence long enough and vivaciously enough to be recognized. By smashing pieces of matter together, creating energies and temperatures not seen since the universe's earliest moments, the LHC could reveal the particles and forces that wrote the rules for everything that followed. Inside the LHC tunnel there are more than 1,600 magnets, most half the length of a basketball court and weighing more than 30 tons. None of those magnets will accelerate particles. The acceleration will come from electrical waves in a separate apparatus that boosts particles around the ring. The magnets were to nudge the beams of particles to bend ever so slightly around the ring. A tremendous amount of particles moving at nearly the speed of light have only one desire in life, and that is to keep moving straight ahead. So the bend needs to be gradual—thus the 17-mile circumference of the ring.

The "Higgs boson" is presumed to be massive compared with most subatomic particles. It might have 100 to 200 times the mass of a proton. That's why a huge collider to produce a Higgs particle is needed—the more energy in the collision, the more massive the particles in the debris. Jumbo particles like the Higgs particle would be, like all oversize particles, unstable. It is not the kind of particle that sticks around in a manner that we can detect—in a fraction of a fraction of a fraction of a second it will decay into other particles. When the particles collide, they'll produce showers of debris as their energy gets transformed into mass. The physicists won't see the Higgs particle itself in that shower, but two of the four major experiments that will be performed at the LHC (Large Hadron Collider) are capable of recording the detritus of the disintegrating Higgs particle—the telltale signal that a Higgs particle is decaying. And the assumption is that only the rare collision—one among many trillions—will produce a Higgs particle.

The discovery of the "God particle" (Higgs boson or Higgs particle) helps to explain how all matter, including creatures, in the universe came to have mass. The "God particle" also provides evidence for the Standard Model of physics (see below), a theory that explains the Big Bang. The discovery of the "God particle" helps to explain how matter could have developed, but it does not explain how protons, electrons, and neutrons developed. It does not even explain how atoms could have developed in a natural environment where nothing exists but elements. Protons, electrons, and neutrons are not nothing. They are something. So for one to claim that everything originated by natural means, one must explain how these elements came into existence. Before anything existed, there would not have been a magnetic field for the particles to bend to form atoms, that is, matter. And without this magnetic field, the particles would have continued flying through space without forming anything. The particles were not accelerated by the magnets. A different apparatus was used to accomplish this. Without the apparatus to move the particles, they would have drifted aimlessly through space. Even if one explains how protons, electrons, and neutrons, came into existence, there is still no valid explanation as to how atoms and the universe could have developed in a natural setting, outside of the experiments, that is, without sophisticated equipment.

The "Higgs boson" is important because without the Higgs boson (or Higgs particle), particles would just have remained traveling through the universe at the speed of light. The question is: How could the "Higgs particle" have developed in nature? What force would have been available to bend the particles before the universe was developed? How could an atom have developed by natural forces prior to the existence of the universe?[1-60] Even those who discovered the "Higgs particle" are hesitant about its discovery. CERN (European Organization for Nuclear Research) called this discovery a "historic milestone" but cautioned that much work lies ahead as physicists attempt to confirm the newfound particle's identity and further probe its properties. For example, though the teams are certain the new particle has the proper mass for the predicted Higgs boson, they still need to determine whether it behaves as the God particle is thought to behave, and therefore what its role in the creation and maintenance of the universe is.

Higgs's idea was that the universe is bathed in an invisible field similar to a magnetic field. Every particle feels this field, now known as the Higgs field, but to varying degrees. If a particle can move through this field with little or no interaction, there will be no drag, and that particle will have little or no mass. Alternatively, if a particle interacts significantly with the Higgs field, it will have a higher mass. The idea of the Higgs field requires the acceptance of a related particle: the Higgs boson. According to the standard model (discussed below), if the Higgs field did not exist, the universe would be a very different place. It would be very difficult to form atoms. So our orderly world, where matter is made of atoms, and electrons form chemical bonds, we wouldn't have that if we did not have the Higgs field. In other words: no galaxies, no stars, no planets, no life on Earth. There are also lingering questions that will require years of follow-up work, such as what the God particle's "decay channels" are, that is, what particles the Higgs transforms into as it sheds energy. The answer to that question will allow physicists to determine whether the particle they have discovered is the one predicted from theory or something more exotic.[1-61]

The Standard Model—The Standard Model explains the relationship between three of the four fundamental forces, namely **(1)** electromagnetism, **(2)** strong interaction (the strong nuclear force that holds atomic nuclei together) and **(3)** weak interaction (the weak nuclear force that has to do with radioactivity.) The Standard Model, however, leaves out that fourth force, **(4)** gravity, a non-trifling omission that needs to be included.[1-62]

In the next chapter (2) we will discuss whether or not the universe (including the Earth) is billions of years old or whether the universe and the Earth are less than 10 thousand years old. If the universe is less than 10,000 years old (as some Creationists contend) then there would not be ample time for slow, gradual evolution to have occurred.

~~~

Chapter 2

The Ages of the Universe and Earth

Introduction

In Chapter 1 we discussed various theories about how the universe may have originated. In this chapter (2) we will discuss whether the evidence supports an old or a young univese (and in particular the Earth).

The various views presented in this chapter regarding the age of the universe are not exclusively of strict evolutionists and strict creationists. Some views are of those who believe in variations of the two extremes. See "Various View Descriptions" section in the back of this book for descriptions of the views presented in this chapter.

How Old is the Universe?

Evolutionist and Old-age Creationist View

The exact age of the universe is not known, but astronomers believe that it is somewhere around 13 billion years old, give or take a few billion years. Astronomers estimate the age of the universe in two ways: **(a)** by looking for the oldest stars; and **(b)** by measuring the rate of expansion of the universe and extrapolating back to the Big Bang.[2-1] (The Big Bang is discussed in Chapter 1.)

Young-age Creationist View

According to evolutionists the age of the universe is 15 billion years.[2-2] The universe is not as old as evolutionists claim. Studies have been performed to determine the amount of energy given off by our sun. Some stars are so large and bright that they radiate energy anywhere from 100,000 to 1 million times as fast as our own sun. These stars could not have contained enough hydrogen at such rates for millions or billions of years because their initial mass would have been absolutely implausible. Therefore, the universe cannot be billions or even millions of years old, but rather only thousands of years old.[2-3]

Methods Used to Determine the Age of the Universe

Evolutionist and Old-age Creationist View

Globular Clusters

Even though the Universe's exact age cannot be determined by knowing the ages of the oldest stars in the universe, this information does provide a minimum age for the Universe. Globular clusters are very dense collections of a large number of stars, usually about a million stars. All of the stars in a globular cluster were formed at about the same time. The older a globular cluster is, the less mass its hydrogen-burning stars will have. The masses of the oldest globular clusters show that they are about 12 billion years old.[2-4]

White Dwarfs

White dwarf stars are very small, dense stars, about a million times as dense as water. They are about as massive as the Sun but with a radius only the size of that of the Earth. A white dwarf is the last stage in the evolution of a star that is not extremely massive. White dwarfs come from red giant stars. A red giant is a large, glowing giant star in the later stages of evolution. When a red giant reaches the end of its life, it sheds its outer layers, ejecting an envelope of gas and plasma known as a planetary nebula. The star's hot core remains. This core becomes a white dwarf. A white dwarf's glow comes from the heat left behind by the red giant. As time goes by, the white dwarf cools down. The older a white dwarf is, the cooler and fainter it will be. By studying faint white dwarfs, astrophysicists have been able to estimate the age of the oldest white dwarfs. These stars are at least 10 billion years old.[2-4]

The Hubble Constant

The age of the Universe can also be calculated by using the Hubble Constant, which is a measure of the rate at which the Universe is expanding, and then extrapolating backwards to the Big Bang (discussed in Chapter 1). There have been different estimates of the Hubble constant. Depending on which estimate is used, the age of the Universe has been calculated to be between 10 billion and 20 billion years. However, since some of the oldest stars in the Universe are more than 10 billion or 12 billion years old, there is a problem with some of the lower estimates.[2-4]

The Hubble Constant refers to how fast the velocities of the galaxies increase with their distance from the Earth. There is a dispute regarding the value of this constant. It ranges from 50 Km/sec per Mpc (Mpc is a Megaparsec, about 3 million light years) to 100 Km/sec per Mpc. This explains the discrepancy in the ± 5 billion year estimates for the age of the universe. The other constant that defines the deceleration of the expansion of the universe is known as q. Depending on the critical density of the universe that this q constant is based upon, the universe is either infinitely expanding as in the flat and open models, or is an oscillating closed universe (See Chapter 1). The Hubble Constant and the q constant remain the two most important unanswered problems in modern cosmology.[2-5]

By studying the temperature of the Cosmic Microwave Background Radiation (CMBR), scientists have been able to come up with a more accurate estimate of the rate at which the Universe is expanding. Right after the Big Bang, the Universe was made up of hydrogen plasma that was filled with radiation. This plasma was so hot that atoms were unable to form. Eventually, the Universe cooled down enough for atoms to develop. However, the atoms could not absorb the radiation and it lingered in the Universe. In 1964, Arno Penzias and Robert Wilson detected this Cosmic Microwave Background Radiation (CMBR) through a radio antenna.[2-6]

Young-age Creationist View

Globular Clusters

Globular clusters are some of the most spectacular astronomical objects observed through the telescope. Like a glittering jewel in the optical field of view, a globular cluster contains a myriad of stars. Because the stars of a globular cluster are closely associated visually, astronomers believe they are closely associated in space and the same distance away from Earth. Evolutionists further assume that all the stars formed from the same collapsing cloud of gas at about the same time and have held together subsequently by mutual gravitational attraction. Thus, the stars all began with a similar chemical composition and share a common evolutionary history. Their main differences are believed to be due to their different masses. Because the stars have so many similarities, evolutionists believe that globular clusters provide a straightforward way of testing their stellar evolutionary theory, which seeks to explain the behavior of stars over billions of years.

According to evolutionary theory, the energy emitted by each star is derived almost entirely from thermonuclear fusion. The section at the bottom of the vertical sequence is called the '**main sequence**' and each point represents a star that has supposedly been 'burning' hydrogen steadily for millions of years, not changing much over all that time. According to theoretical calculations, the more massive the star, the faster it 'burns' its fuel. So the points near the bottom of the vertical sequence are for the smallest and least luminous stars that burn their hydrogen the slowest. The points near the center of the vertical sequence represent the largest and brightest stars that are converting their hydrogen into helium most quickly. Calculations indicate that once a significant portion of the star's hydrogen has been converted to helium, the temperature and luminosity of the star changes drastically and it becomes larger and redder, and no longer plots on the main sequence. The points from the very top of the vertical sequence are interpreted as such stars. The brightest and reddest stars, called red giants, are located near the vertical center of the sequence. The vertical center of the sequence is called the '**red giant branch**.' All the stars are assumed to have formed at about the same time. Thus, if stars on the red giant branch are much more evolved, they then must have been much more massive than the stars still on the main sequence.

A third sequence of stars, called the '**horizontal branch**,' extends from the point along the top arranged in a horizontal formation, run across in relation to the vertical sequence. These stars are interpreted to be the most evolved, having passed through the red giant phase. They must have been the most massive of all the stars that originally formed in the globular cluster. In this way the color-magnitude plot is interpreted to represent progressive stages in stellar evolution, starting with the least evolved stars at the bottom of the vertical sequence and moving through the center of the vertical sequence and along the top area in a horizontal formation, run across in relation to the vertical sequence to the most evolved stars located along the left side of the horizontal sequence. Of particular interest is the area located near the center of the vertical sequence, the turn-off point, which represents the most massive star still on the main sequence. According to evolutionary theory, this area would gradually move downwards over millions of years as stars of successively lower mass burn up their hydrogen and evolve away. In picturesque language, this point is often described as 'burning down like a candle'.

Significantly, the main sequence turn-off can be used to estimate the age of the globular cluster from stellar evolution theory. Once the magnitude or luminosity of the star at the turn-off point has been determined, the mass of the star can be estimated from a mass-luminosity relationship using our Sun as a reference. Similarly, from the mass of the star, an 'age' can be calculated from nuclear fusion models for main-sequence stars. In this way, evolutionists have determined the 'age' of globular clusters surrounding our Galaxy is 13–15 billion years old.

This view sounds convincing until we realize that there is no way of testing it. By adding secondary explanations, the story can accommodate any astronomical observation. Indeed, the situation is not as simple as the simple theory claims. For example, most globular clusters are revolving around the Galaxy in highly eccentric orbits with a period of some 100 million years. In 15 billion years, each cluster would have orbited the Galaxy over one hundred times, passing though the Galaxy disk twice each time.

This raises the question of how the star clusters could have remained together and compact for all that time. Furthermore, stars toward the end of the red giant branch and in the horizontal branch have masses that are much lower than those of stars once on the main sequence from which they have supposedly evolved. Again, explanations are devised for how stars lose mass as they evolve from the main sequence. Also, the age of the globular clusters, calculated from the main sequence turn-off, has been the subject of ongoing revision. The problem is that the age of the stars in the globular clusters must be less than the age of the universe as calculated from the Hubble constant. The compact size of globular clusters reflects their abrupt, rapid formation process, and their youthfulness.[2-7]

Star clusters serve to indicate a young age for the universe. A star cluster contains hundreds or thousands of stars moving, as it has been described, 'like a swarm of bees.' These star clusters are held together by gravity, but in some star clusters, the stars are moving so fast that they could not have held together for millions or billions of years. Therefore, the presence of star clusters in the universe indicates that the age of the universe is numbered in the thousands of years, rather than millions or billions.[2-8]

White Dwarfs

Astronomers also attempt to determine the age of the universe by using another method called 'white dwarfs.' This method is not reliable for the following reasons. (White dwarfs were defined in the "Evolutionist and Old-age Creationist View" above.) The observed rapid rate of change in stars contradicts the vast ages assigned to stellar evolution. For example, studies in 1994 indicated that this star was most likely a white dwarf in the center of a planetary nebula; by 1997 it had grown to a bright yellow giant, about 80 times wider than the sun.[2-9] In 1998, it had expanded even further, to a red super-giant 150 times wider than the sun. But then it shrank just as quickly; by 2002 the star itself was invisible even to the most powerful optical telescopes, although it is detectable in the infrared, which shines through the dust.[2-10]

The Hubble Constant Hypothesis

The Expanding Universe Theory should not be used to determine the age of the universe, as research has not only disproved the speed theory of redshift, but also another theory based on it. Edwin Hubble's "Hubble hypothesis", which suggests that objects outside our galaxy are receding from the earth at speeds proportional to their distance from us. That is the basis of the "expanding universe" theory. So, if Hubble's theory is incorrect, we would then have a smaller, non-expanding universe. The "expanding universe" theory is based on the currently accepted, but very doubtful, interpretation of the redshift. That speed theory of the redshift requires that **(1)** all the stars in the universe are expanding outward, and **(2)** there are great distances between stars and galaxies. If, instead of the speed redshift theory, one or more of the three alternate explanations of the redshift were adopted, then the universe would be smaller, and it need not take so many light-years to travel across it. These three alternate explanations are: **(1)** gravitational redshift (gravity bends light, a fact predicted by Einstein and now known to be true), **(2)** second-order Doppler redshift (the entire universe is rotating, and this affects the shift), and **(3)** energy-loss or tired light redshift (light loses energy as it travels). Researchers have found evidence that photons slow down in transit from stars to us. (A photon is a single "piece" of moving starlight.) The evolutionary theory assumes photons never slow down and that they are never shifted in their spectra toward the red by gravity. Evidence has shown this assumption to be untrue. The conclusion is that the universe is not as old as most astronomers claim.[2-11]

Questions for Creationists

Evolutionist View

Young-earth creationists claim that the creation occurred suddenly, that God created the Earth with an appearance of old age. The question then is: Why would God create the geologic world with an appearance of old age? Why would rock formations appear to be billions of years older than creationists claim? [2-12]

The Age of the Earth

The Earth

Evolutionist View

The age of the Earth has been determined to be 4.54 billion years (give or take 0.05 billion years) (4.54×10^9 years ± 1%). (That is 4,500,000.000 years before present.) This age is based on evidence from radiometric age dating of meteorite material and is consistent with the ages of the oldest-known terrestrial and lunar samples.[2-13] According to reliable radiometric dating techniques used by geologists, the Earth is at least 4.5 billion years old (that is 4,500,000,000 years before present). This is based on measurements on outcrops of very ancient rocks (not fossils as some believe). But these rocks themselves were probably derived from even older rocks that formed the first-formed continental crust.[2-14]

According to the theory of evolution, the Earth was formed approximately 4.6 billion years ago. Its atmosphere probably contained little free oxygen, but a lot of water vapor and other gases, such as carbon dioxide and nitrogen, and was extremely hot. Approximately 3.9 billion years ago the Earth cooled enough for water vapor to condense, allowing millions of years of rain that formed Earth's oceans. Most scientists believe the age for the Earth and the rest of the solar system is about 4.55 billion years (plus or minus about 1%). This value is derived from several different lines of evidence. Unfortunately, the age cannot be computed directly from material that is solely from the Earth.

There is evidence that energy from the Earth's accumulation caused the surface to be molten. Further, the processes of erosion and crustal recycling have apparently destroyed all of the earliest surfaces. The oldest rocks that have been found so far on the Earth date to about 3.8 to 3.9 billion years ago according to several radiometric-dating methods. (See section "Radiometric Methods (Absolute Dating) Used to Determine the Age of the Earth" for further discussion of dating methods.) Some of these rocks are sedimentary, and include minerals that are themselves as old as 4.1 to 4.2 billion years. Rocks of this age are relatively rare, however rocks that are at least 3.5 billion years in age have been found on North America, Greenland, Australia, Africa, and Asia. While these values do not compute an age for the Earth, they do establish a lower limit as the Earth must be at least as old as any formation on it.[2-15]

Young-earth Creationist View

Old-age geologists believe that all geological features and formations, once attributed to geologic cataclysms, can now be satisfactorily explained by ordinary processes functioning over long periods of time. Geologists believe that changes that occur are due to steady, consistent methods. No consideration is given to the possibilities that climatic conditions or other unusual considerations could either speed up or slow down a process. Although the uniformitarian assumption appears to be reasonable, it must be remembered that it is merely an assumption and not a fact.[2-16]

The belief in an ancient earth has become so entrenched in our modern culture that advocates of a relatively young earth are considered as antiquated as if they had proposed that the earth was flat. In fact, it may even seem ridiculous or ludicrous to question the validity of various methods used to determine the age of the earth, such as radioactive dating. Most people believe that scientists have proven that the earth is billions of years old by various methods that are used. This however, is not necessarily true. The study of origins is beyond the ability of any scientific method to determine. Therefore dates obtained from these various techniques are merely circumstantial and are necessarily based on numerous assumptions, which may or may not be true. Thus, it is impossible to determine or prove that the earth is billions of years old.[2-17]

Methods Used to Determine the Age of the Earth

Young-earth Creationist View

There are several scientific methods to verify that the Earth is less than 10,000 years old. Below are some methods used to support this view:

Decaying magnetic field of the earth: We know that the earth's magnetic field has been decaying since the time it was first measured in 1835. Given the most plausible model of magnetism being generated by circulating electric currents that are decaying within the earth, and projecting the numbers backwards, 10,000 years ago the earth would have a field as strong as a magnetic star that utilizes thermonuclear processes to maintain a field of that strength. Critics of this theory insist on the existence of an electric generator ("dynamo") inside the earth, without theoretical or empirical evidence to justify this. (Paleomagnetic anomalies are presented as evidence, but are inferior to the global statistically averaged data used to justify the young-earth model. These paleomagnetic artifacts are dated using old-earth metrics and assumptions.)[2-18]

The strength of the Earth's magnetic field is decaying at a rate of a half-life of 1,400 years. This means that 1,400 years ago the magnetic field of the Earth was twice as strong as it is now. If we calculate back 10,000 years, the Earth would have been a magnetic field as strong as the magnetic field of a magnetic star. This is most unlikely. Therefore, based on the present decay rate of the Earth's magnetic field, 10,000 years appears to be the maximum age of the Earth.[2-19]

Helium rising into the atmosphere: One of the decay products of Uranium and Thorium is Helium-4. Given the estimated concentrations of Uranium and Thorium in the earth's surface, current decay rates and the estimated helium content of the atmosphere, the implication would be that this could not have been going on for millions of years. Based on the numbers used, the calculations that have been seen range from thousands of years to millions of years.[2-20]

The small amount of helium in our present-day atmosphere is also used to indicate the Earth is not millions of years old. Old age scientists claim that the radioactive decay processes of uranium and thorium that produce helium have been occurring in the Earth's crust for billions of years. But if this decay has been going on for billions of years, the Earth's atmosphere should contain much more than the present 1 part in 200,000 of helium. The common explanation offered for the absence of the required helium is that it has been escaping out through the exosphere. But there is no evidence to support this assumption and, in fact, recent data indicate that helium cannot escape into space the way hydrogen does. Realistic calculations based on available figures disclose that the amount of time required for natural alpha decay processes to have produced the observed helium is approximately 10,000 years.[2-21]

Helium is a lightweight, fast-moving, and slippery atom, not sticking to other atoms. Because of these qualities, scientists believe that helium works its way through solids rather rapidly. The leakage rates are so large that scientists had expected that most of the helium produced through radioactive decay had escaped out of the Earth's atmosphere (crust), into the upper atmosphere, and out into space. Research, however, shows that helium does not escape into space to any great degree as previously thought. [2-22]

Evolutionist and Old-age Creationist View

Decay of the Earth's magnetic field. The young-Earth creationists claim that the dipole component of the magnetic field has decreased slightly over the time that it has been measured. (A dipole is a pair of equal and oppositely charged or magnetized poles separated by a distance.) They believe that the generally accepted "dynamo theory" for the existence of the Earth's magnetic field is wrong, the mechanism might instead be an initially created field which has been losing strength ever since the creation event. While there is no complete model to the geo-dynamo, certain key properties of the core are unknown, there are reasonable theories that can explain this. There are no good reasons for rejecting these theories without any evidence. If it is possible for energy to be added to the field, then the extrapolation is useless. There is overwhelming evidence that the magnetic field has reversed itself, rendering any unidirectional extrapolation on total energy useless.

Accumulation of helium in the atmosphere. The young-Earth creationists believe that helium-4 is created by radioactive decay (alpha particles are helium nuclei) and is constantly added to the atmosphere. They believe that helium is not light enough to escape the Earth's gravity (unlike hydrogen), and it will therefore accumulate over time. Young-Earth creationists believe that the current level of helium in the atmosphere would accumulate in less than two hundred thousand years; therefore the Earth is young. Helium can and does escape from the atmosphere, at rates calculated to be nearly identical to rates of production. In order to obtain a young age from their calculations, young-Earth creationists attempt to dismiss mechanisms by which helium can escape.[2-23]

Radiometric Methods (Absolute Dating)
Used to Determine the Age of the Earth

Young-earth Creationist View

The following are the major radioisotope dating methods and their associated problems.

Carbon-14: Cosmic rays hit Nitrogen-14 in the earth's atmosphere, producing radioactive Carbon-14. Plants absorb the Carbon-14. Animals eat the plants. Animals eat animals. Eventually all living things are supposed to have the same amount of Carbon-14 in them. When the animal or plant dies, it quits eating and so takes in no more Carbon-14. The Carbon-14 decays back to Nitrogen-14 over time. Measuring the amount of Carbon-14 left in the animal remains, is supposed to tell you how long it has been since the animal or plant died. It is universally accepted, even among evolutionists, that Carbon-14 is only useful for dating the organic remains of living tissue and that it only works up to about 20, 30, maybe 60,000 years. So Carbon-14 dating is irrelevant to the discussion of the time frame of macroevolution, which is supposed to have occurred over hundreds of millions of years. It is assumed that the level of atmospheric Carbon-14 has been constant for tens of thousands of years, when it has only been measured since the early part of this century. This is a ratio of 1/1000 over the span of the proposed measurement period. (Tree ring dating and other methods of historical dating have provided some corroborating data for some samples, however.)

Things like the strength of the earth's magnetic field affect how much cosmic radiation gets through to the atmosphere (which affects how much Carbon-14 is produced.) The strength of the earth's magnetic field has declined since it was first measured in 1835. It is assumed that the rate of radioactive decay of Carbon-14 has never changed. However, in the laboratory, it has been demonstrated that the rate of decay of Carbon-14 can be significantly changed by application of an electric potential (specifically, 9 standard deviations for a potential difference of 180 volts in one particular experiment.) It is assumed that no exchange of Carbon-14 between the animal remains and the environment has occurred since the animal died. Successive Carbon-14 measurements of individual specimens have been shown to produce conflicting results, the differences amounting to about a 1:2 ratio. And dating of specimens of known age has produced erroneous results.

Potassium-Argon: Potassium-40 decays into Argon-40. When molten lava solidifies, it has some Potassium-40 in it. Potassium-40 trapped in the rock decays into Argon-40. The amount of Argon-40 that has formed in a rock since it solidified is supposed to tell how long it has been since the rock was formed. Potassium-40 also decays into Calcium-40. The rate of decay into Argon-40 vs. Calcium-40 is not accurately known. Uranium dating methods are used to "calibrate" the Potassium-Argon method. So to begin with, Potassium—Argon dating cannot be more accurate than Uranium isotope dating. It is assumed that no Argon was originally trapped in rock when it solidified. It is assumed that there was no exchange of either Potassium or Argon between the specimen and its environment since it solidified. It is assumed that the rate of decay of Potassium-40 has not changed since the formation of the rock. The strength of neutrino flux from cosmic radiation, which is affected by things like supernovas and the strength of the earth's magnetic field, which is known to change, are known to affect decay rates. (Although in this case this does not necessarily explain sufficient measurement error, it does demonstrate again that the rates are not necessarily constant.)

Successive measurements of individual specimens have produced different results, representing inconsistencies on the order of hundreds of millions or billions of years. The difference can be on the order of a ratio of 1:10. Measurements using Potassium-Argon have produced results inconsistent with those obtained using other radioisotope methods. Measurements of rocks of known age obtained from recent volcanoes using the Potassium-Argon method have produced erroneous results. Rocks known to be less than a couple hundred years old have been dated at billions of years old.[2-24]

Uranium-235: Similar principles and problems as shown above. Uranium-235 decays into Lead-207, and the amount of Lead-207 is supposed to tell how old the rock is. The original content of Uranium-235 vs. Lead-207 is not known. (It is simply assumed that there was no Lead-207 to begin with.) It is assumed that no Uranium-235 or Lead-207 is exchanged with the environment over the life of the rock. Laboratory experiments have leached Uranium out of some specimens with a weak acid. It is assumed that the decay rates have always been constant. Successive measurements of the same sample often produce different results. Measurements by this method often disagree with measurements using other methods.

Uranium-238: Similar principles and problems as shown above. Uranium-238 decays into Lead-206.[2-25] The uranium-to-lead dating method is used primarily on igneous rocks (rocks produced under conditions involving intense heat) and is used to date objects thought to be quite old. Uranium-238 has a half-life of about 4.5 billion years. This means that if a sample of U-238 is observed for 4.5 billion years, half of it would be gone, having decayed to lead (Pb)-206. Uranium-lead (U-Pb) dating is based upon three assumptions: 1.) A constant decay rate—this is a reasonable assumption based upon observed physical properties. 2.) No loss or gain of uranium or lead during the "life" of the rock. To avoid this problem, paleontologists choose specimens that appear to have no erosion forces acting on them. This is difficult to objectively guarantee, but it is nonetheless a reasonable assumption. 3.) No lead was in the specimen when it was formed. This assumption is questionable, especially since it is actually the entire basis for U-Pb dating. Why would there not be lead in the specimen when it was formed? Why would uranium be present but no lead? How is it known for sure that there was no lead in the specimen when it was formed?[2-26]

Thorium-232: Similar principles and problems as shown above. Thorium-232 decays into Lead-208.

Lead-Lead: Similar principles and problems as shown above. Lead-207 decays into Lead-206.

Rubidium-Strontium: Similar principles and problems as shown above. Rubidium-87 decays into Strontium-87. (It should be noted that the "Isochron" nature of this method eliminates only some of the unsubstantiated assumptions.) The magnitude of the problem can be easily seen. Many assumptions are made about decay rates, initial conditions, environmental influences, and etc. The results obtained are inconsistent with successive measurements made using the same and different dating methods. Measurements made of specimens of known age produce erroneous results.

The dating procedures are not testable under controlled, laboratory conditions over the period of time they are supposed to measure. It should be noted that dating of fossils is almost never done by measuring the fossil itself, but by measuring rocks in the vicinity of the fossil. So it is assumed that a rock in the vicinity of a fossil is the same age as the fossil. It can be concluded that radioisotope-dating methods lack the theoretical and experimental foundation that are needed to be considered reliable indicators of the age of the specimens being dated.[2-27]

Evolutionist and Old-age Creationist View

Uranium is an element that is usually used in nuclear reactors. It is radioactive, and decays into lead at a very slow rate. Two rates, that is, two different isotopes of uranium decay into two different isotopes of lead, one with a half-life of 700 million years, and the other with a half-life of 4.5 billion years. Potassium decays to argon with a half-life of 1.3 billion years, and rubidium decays to strontium with a half-life of 50 billion years, which is far older than any existing fossil. Dating methods are used on the rocks where fossils are found. Fossils exist in very specific and discrete layers of rock strata, and so all a geologist has to do is date the strata layers using one of the radiometric methods (also known as "Absolute Dating") (carbon-14, uranium, potassium, etc), and then any fossils found within that layer are placed roughly within that time frame (known as "Relative Dating"). The problem, however, is that none of these methods (carbon-14, uranium, potassium) are totally reliable, and of course no measurement is. But when scientists make measurements, they use the power of statistics in determining the age of a rock or fossil. If a measurement accurately reflects a particular object being studied, then multiple, independent measurements of that same object being studied should obtain similar results, with minimal variation. Many measurements are made when dating a particular stratum, or organic sample, and only if those measurements show a clear consensus, is the date accepted as reliable.[2-28]

Is Radioisotope Dating Reliable?

Young-earth Creationist View

For years, old-earth scientists have claimed that the earth is billions of years old. The question is, are their experiments reliable? Young-earth creationists have disputed the claims of old-earth scientific claims based on radioisotope methods because of the unjustified dependence upon these assumptions. It appears that much larger quantities of nuclear decay have occurred in most nuclear processes than would be expected for a few thousand years of radioactivity at the currently observed rates. A group of young-earth creationists hypothesized that sometime in the past much higher rates of radioisotope decay may have occurred, leading to the production of large quantities of daughter products in a short period of time. Young-earth creationists believe that these increased decay rates may have been part of the rock-forming process on the early earth and/or one of the results of God's judgment upon man following the Creation (Genesis 1-2), the Curse after Adam and Eve sinned (Genesis 3), or during the Flood (Genesis 6-8). (Young-earth Creationists believe in a global flood.) The amount of decay products which should exist, given the conventional old-earth model, differs from that expected by the young-earth creationist model. The young-earth researchers believe this large difference may make it possible to validate a true model of earth history. For example, the expected rate of escape of helium (^4He) from minerals in rocks which have undergone a large degree of nuclear decay only a few thousand years ago would seem to indicate that much of the decay-produced helium should still be trapped in the rocks. The observation of minimal helium in the rocks would support the old-earth model, but helium still in the rocks would validate the young-earth model. This difference should be capable of confirmation in the laboratory. Initial observations conducted by young-earth creationists indicate that a vast amount of helium has been produced by radioactive decay but most has been retained in the rocks. It is possible that geophysical and geochemical evidences could have been part of the first day of Creation (Genesis 1).

The distribution of parent and daughter elements in the stratigraphic record must occur in a manner which would validate the theory, and any deviations should be examined by geological/geochemical processes. For example, if accelerated decay occurred only during the Flood, then strata which were laid down before the Flood should show different ratios of radioisotopes and daughter products than strata laid down during or following the Flood. Radioisotope dating assumes that the decay of nuclei has been predictable and uniform over a vast timescale. The half-lives of radioactive parent atoms are taken as constants. Particular half-lives indeed show little variation today since the nucleus is well-shielded from the external atomic environment. However, the young-earth creationist model allows a possible large scale change in radioactive decay during a limited period of time in the past. Accelerated decay may have been a supernaturally-directed part of Creation (Genesis 1), the Curse after Adam and Eve sinned (Genesis 3), or possibly a part of the Genesis Flood catastrophe (Genesis 6-7). At whatever point in history this accelerated decay may have occurred, it would invalidate all traditional radioisotope ages from that time forward.

Three major weaknesses of radioisotope dating by old-earth scientists are apparent from research conducted by young-earth researchers. Each involves a major assumption for the isotopes. The **first** assumption is that of a constant nuclear decay rate of half-life in the past. From all evidence, half-lives are indeed quite constant today and largely independent of changes in the chemical environment. However, it has been proposed that decay rates were temporarily and dramatically accelerated at some time in the past. This acceleration may or may not have a natural explanation. The hypothesized temporary change gave rise to a large, rapid accumulation of daughter products, but not to a large timescale. **Second**, it is assumed that the isotope composition of rock samples has not been changed by fractionation over time. (Fractionation refers to a separation of isotopes based on their weights, due to various chemical interactions.) The conclusion is that current isotope ratios may have been quite different in the past, thus invalidating previously accepted rock ages. **Third**, it is often assumed that rock samples have been closed systems over eons of time. This claim is questionable. Modern science has stated that no part of nature is completely isolated. Over time, parent or daughter atoms may move into or out of rocks. Therefore, radioisotope data often may be describing atom migration rather than sample age. Radioisotope dating has not been proven with absolute certainty that the earth is old. In summary, nuclear decay acceleration would cause a general trend in the data. That is, the age of the earth calculations derived from radioisotope experiments would generally reflect the order and timing of events during the Genesis Flood. The timescale would shrink from millions of years down to months.[2-28a]

Old-earth Creationist View

In an attempt to reconcile young-earth beliefs with geologic data, a group of young-earth creationists, who believe the earth is less than 6,000 years old, did extensive research to disprove the radiometric dating methods used by most scientists in determining the age of the earth. After conducting an extensive research program, these young-earth researchers acknowledged that much larger quantities of nuclear decay have occurred in most geological processes than could be explained by an earth that is only a few thousand years old. As a result of their study, the young-earth creationists accepted what mainstream science had known since the early 1900s that nuclear decay was the best and perhaps only possible explanation for the isotopic patterns observed in rocks and minerals today. The young-earth scientists who performed the study eventually admitted that there were billions of years' worth of nuclear decay, which created a major dilemma for people believing in a 6,000-year-old earth. The only possible solution, apart from abandoning the young-earth position altogether, was to claim that nuclear decay rates were accelerated in the recent past.

One experiment that the young-earth scientists conducted was a helium diffusion experiment using zircon mineral samples from deep geothermal wells in Fenton Hill, New Mexico. When the young-earth researchers compared the nuclear decay clock with their helium diffusion clock, they found a large discrepancy. Apparently, the nuclear decay clock recorded an elapsed time of over a billion years, whereas their helium diffusion clock recorded an elapsed time of only a few thousand years. Using the helium diffusion time as the more reliable measurement, the young-earth researchers claimed that they had found convincing evidence for accelerated nuclear decay. However, this apparent result is not as simple as merely reading time from a stopwatch. The helium diffusion clock used by the young-earth team was actually a complex mathematical model describing the process of helium diffusion from zircon crystals. It is questionable as to how they read their diffusion clock. Recent (geologically speaking) volcanic activity has raised the ground temperature at the site (Fenton Hill, New Mexico) to over twice the typical value across the continent. These elevated temperatures have been sustained for a relatively short period of time on a geologic timescale. Therefore, the use of a constant temperature by the young-earth team demonstrates a misunderstanding of the thermal history of the site.

During another laboratory experiment, the young-earth researchers from the same team used data from a laboratory experiment in which gas released from a zircon sample was measured at different temperatures. They extracted the parameters for a simple kinetic model. The problem with this model is that it treated all helium atoms the same, regardless of whether they were in the bulk crystal or near a defect. Most helium atoms will lie in portions of the undisturbed crystal, whereas only a small fraction will lie in the vicinity of a defect. At low temperatures, the small fraction of atoms near a defect will be mobile, whereas the vast majority of atoms will only begin to move at higher temperatures. Essentially, there are distinct populations of helium atoms in the solid, each with different diffusion properties. This type of model ignores the possibility that helium atoms behave differently depending upon their location in the crystal, with atoms in the vicinity of defects moving more readily than those that are in the bulk crystal. In contrast, a multi-domain diffusion model used by several leading scientists, takes this effect into account. The revised model conducted by an old-earth researcher agrees well with the measured helium retention data. The old-earth model matches the revised measurements better than the young-earth model. The young-earth team claimed that essentially no helium would be left in these zircons if they were more than a few thousand years old. However, by direct computation, the results demonstrated otherwise. The helium content and the 1.5 billion year radiometric age of these zircons are in agreement. Since no anomaly exists, there is no scientific need to assume the existence of outlandish physics, like accelerated nuclear decay, to explain the phenomenon. (Young-earth creationists believe in a global flood, whereas Old-earth Creationists believe in a local flood.) This result disproves the accelerated nuclear decay hypothesis proposed by the young-earth researchers.[2-28b]

Other Methods Used to Determine the Age of the Earth

Evolutionist and Old-age Creationist View

Geologists have determined over 250 years ago that the Earth is much older than 6,000 years. Geologists have discovered that there are multiple layers of materials everywhere on the Earth. The major materials are rock strata. Layers also appear with (1) sediments found in water, (2) deposits that were excreted by animals, and (3) seasonal accumulations of snow. These materials are dependable to use to determine the age of the Earth because these activities are occurring today. Therefore the rate of deposits made for each layer can be measured and then the age of the Earth can be calculated by counting the number of layers. Measuring the thickness of all the layers determines the time it took to form the layers.

We will now discuss three examples of how layers can be measured to determine the age of the Earth. **The first method** to determine the age of the Earth involves seasonal runoff in freshwater lakes. The bottoms of lakes accumulate layers each year. Dark colored band sediments are deposited during the fall and winter and light colored band sediments are deposited in the spring and summer. **The second method** involves excretions from tropical saltwater organisms. Coral reefs are made up of limestone that is deposited by the coral animals. These deposits have annual layers of banding. The rate of accumulation has been determined to be between 5 and 8 millimeters per year, and the thickness of a reef can determine its approximate age. **The third method** involves snowfall in a polar region. Polar ice sheets feature tens of thousands of annual snow layers. Because there is no sunlight in this region during the winter, no daily evaporation occurs on the surface of the snow. This produces heavy and fine-grained snow crystals that appear as a dark band. When summer comes, the snow is warmed during the day and is cooled at night. This results in a light-colored layer with low density, coarse-grained crystals. The Greenland ice sheet is over two miles thick in some areas and has annual bands that have been estimated to be 100,000 years old. These three different methods of determining the age of the Earth have verified that the Earth is very old.[2-29]

In addition to the three methods mentioned above, there are many other methods used to determine the age of the earth. Some of them are a little complicated and some are not. Basically, all methods indicate that the earth is much greater than 10,000 years old. We will discuss two additional methods now that are easily understood. **One method** that indicates the earth is much greater than 10,000 years old is the presence of fossil fuels, such as coal and oil, below the earth's surface. It has been estimated that it took millions of years to deposit enough organic material necessary for all the fossil fuels found under the surface of the earth. If all known deposits of fossil fuel had been deposited in the year or so after the flood (discussed in Chapter 4), the earth would have needed to have a tremendous amount of organic matter to produce the amount of fossil fuels that are below the earth's surface, as the earth today does not produce enough organic matter to account for even a small amount of the known fossil fuel reserves.[2-30]

A second method that indicates the earth is much older than 10,000 years is the shale and claystone rocks that were formed from clay deposited on the bottom of some ancient lake or sea. In some places, these rocks are made up of many thin layers, which sometimes alternate between lighter and darker color. Layers look somewhat like the grain in a piece of wood that has been sawn at right angles to the rings. These layers, like tree-rings, are annual growth bands. When geologists first studied the clay settling to the beds of lakes, they found that often it was one color in summer and another color in the winter. Recent studies have found that fossil pollen is found in the summer layers but not in the winter layers. This means that by counting the number of layers or growth bands, the age of the rock can be determined. It usually takes two layers to make one annual growth band, or layer: one in the summer and one in the winter. Also, there is no way that pollen could have found its way into the darker bands, and only the darker bands, unless they were produced at yearly intervals over a vast period of time. The conclusion is that there is no possibility that a great flood (discussed in Chapter 4), or any other catastrophe, could have produced the millions of paper thin bands of alternating light and dark color in an extensive deposits of shale, which have been discovered in the Green River shale deposits in Wyoming, Utah, and Colorado.[2-31]

Different strata of the Earth reveal different types of plant and animal fossils. Near the bottom are simple life forms. As the layers progress toward the top more complex life forms are found. (See Chapter 3 for more details of the strata of the Earth.) The ages of the strata decrease from the bottom, which are the oldest to the top, which are the most recent. In other words, everything on the Earth was not created at one time, as claimed by young Earth creationists who interpret Genesis 1 literally (a day being a 24 hour period).[2-32]

Young-earth Creationist View

Tree Rings

Under normal circumstances, woody trees add one ring per year. A ring typically consists of a light-colored growth portion and a dark-colored portion produced in a stabilization season. However, some trees do not produce annual rings at all, especially those in temperate or tropical regions.

Overlapping and correlating rings have been used to produce "chronologies" of past years. Linear sequences of rings are obtained by cross-matching tree ring patterns from living trees as well as from older dead wood. Tree rings are more than a record of years. Year-to-year variation in the width of rings records information about the growth conditions in the particular year. Insect infestation, disease, and fire damage could all cause disruptions in the normal living conditions of trees. Each of these interrupts the normal growth cycle. (1) Day length, (2) amount of sunshine, (3) water potential, (4) nutrients, (5) age of tree, (6) temperature, (7) rainfall, (8) height above ground, and (9) proximity to a branch all impact tree growth and tree ring production. By assuming the outer ring records the most recent year and that each ring signals one year, a researcher can determine the "date" of a particular ring simply by counting rings.

But how valid is the assumption of one ring per year in a climate where tree-growing conditions are variable? That very assumption is regularly put to the test by research foresters. It has been determined that all trees, even slow-growing ones, respond dynamically to tiny environmental changes, even hourly changes in growing conditions. Scientists have observed that numerous "normal" conditions can produce an extra ring or no ring at all. Weather was fingered as the most "guilty" culprit. Unusual storms with abundant rainfall interspersed with dry periods can produce multiple rings, essentially one per major storm. Thus, the basic assumption of tree ring dating is demonstrably in error. Can we trust the overlapping calibration curves?

The centuries immediately following the Flood (Chapter 4) the Earth experienced the Ice Age. All trees growing on the continents were recently sprouted, actively growing trees. The still-warm oceans rapidly evaporated seawater, thus providing the raw material for major monsoon-type storms. Earth was ravaged by frequent and wide-ranging atmospheric disturbances, dumping excessive snowfall in northern regions and rainfall to the south. If ever there was a time when multiple rings could develop in trees, this was it. Those centuries probably produced tree ring growth that was anything but annual. Tree ring studies provide supportive and instructive information that supports the young earth view.[2-33]

Zircon Crystals

Young-earth Creationist View

The various age-dating methods are subject to interpretation. Consider the research from a creationist group concerning the age of zircon crystals in granite. Using one set of assumptions, these crystals could be interpreted to be around 1.5 billion years old, based on the amount of lead produced from the decay of uranium (which also produces helium). However, if one questions these assumptions, one should be motivated to test them. Measurements of the rate at which helium is able to "leak out" of these crystals indicate that if they were much older than about 6,000 years, they would have nowhere near the amount of helium still left in them. Hence, the originally applied assumption of a constant decay rate is flawed; one must assume, instead, that there has been acceleration of the decay rate in the past. Using this revised assumption, the same uranium-lead data can now be interpreted to also give an age that is less than 6,000 years.[2-34]

Analysis of zircon crystals, from five levels of hot rock in a 15,000-foot hole, revealed that almost no increase of lead escape had occurred at even the lowest level. This is powerful evidence in favor of a young earth and is consistent with a 6000-year age. Analysis of helium content in those small zircon crystals revealed amazingly high retention in 197° C. (386.6° F.) zircon crystals. This provides a double proof for a very young age for the earth. If the earth were millions of years old, that helium would have totally escaped from the zircon crystals.[2-35]

There is evidence that radioactive decay supports a young earth. One study involved the amount of helium found in granite rocks. Granite contains tiny zircon crystals, which contain radioactive uranium (^{238}U), which decays into lead (^{206}Pb). During this process, for each atom of ^{238}U decaying into ^{206}Pb, eight helium atoms are formed and migrate out of the zircons and granite rapidly. Within the zircon crystals, any helium atoms generated by nuclear decay in the distant past should have long ago migrated outward and escaped from these crystals. One would expect the helium gas to eventually diffuse upward out of the ground and then disappear into the atmosphere. This has not been the case, as large amounts of helium have been found trapped inside zircons. The decay of ^{238}U into lead is a slow process (half-life of 4.5 billion years). Since helium migrates out of rocks rapidly, there should be very little to no helium remaining in the zircon crystals.

Why is so much helium still in the granite? One likely explanation is that sometime in the past the radioactive decay rate was greatly accelerated. The decay rate was accelerated so much that helium was being produced faster than it could have escaped, causing an abundant amount of helium to remain in the granite. There is evidence that at some time in history nuclear decay was greatly accelerated. These experiments have clearly confirmed the numerical predictions of our Creation model. The data and our analysis show that over a billion years' worth of nuclear decay has occurred very recently, between 4000 and 8000 years ago. Confirmation of this accelerated nuclear decay having occurred is provided by adjacent uranium and polonium radiohalos that formed at the same time in the same biotite flakes in granites.

Radiohalos result from the physical damage caused by radioactive decay of uranium and intermediate daughter atoms of polonium, so they are observable evidence that a lot of radioactive decay has occurred during the earth's history. However, because the daughter polonium atoms are only short-lived (for example, polonium-218 decays within 3 minutes, compared to 4.47 million years for uranium-238), the polonium radiohalos had to form within hours to a few days. But in order to supply the needed polonium atoms to produce these polonium radiohalos within that timeframe, the nearby uranium atoms had to decay at an accelerated rate. Thus hundreds of millions of years' worth of uranium decay (compared to today's slow decay rate) had to have occurred within hours to a few days to produce these adjacent uranium and polonium radiohalos in granites. Creationist researchers have suggested that this accelerated decay took place during the Creation Week or during the Flood. Accelerated decay of this magnitude would result in immense amounts of heat being generated in rocks.[2-36]

Evolutionist View

Creationists are not able to explain how sediment and rock laid down in a mere year by a worldwide global flood can yield such fantastic, orderly differences in radiometric ages. (The possibilities of a global flood causing the Earth's geology are discussed in Chapter 4.) This poses a fatal problem whether one believes in the accuracy of radiometric dating or not. (Radiometric Dating is discussed in section "Radiometric Methods (Absolute Dating) Used to Determine the Age of the Earth.") One would think that the flood sediments (gathered from the four corners of the old antediluvian world) and their associated igneous rock (formed during the flood) would all register very little radiometric age. At the very least we would expect random fluctuations if the radiometric methods were totally at sea.

The percentage of lead to uranium in zircon crystals (the key to ordinary uranium-lead, radiometric dating) does not depend on which geologic period they are found in. If most of the geologic column were created during Noah's flood, it would really not matter whether a zircon crystal were found in Cambrian strata or Cretaceous strata, in Jurassic strata or Tertiary strata. (The geologic column and strata layers are discussed in Chapter 3.) Noah's flood might just as easily deposit the same crystal in one place as another. Pressure has nothing to do with zircon crystal formation, as zircon crystals all have about the same density, as their total lead content is small.

There is nothing in a Cambrian stratum that is also not in a Cretaceous stratum. The rock type in the Jurassic strata is no different than the rock type in the Tertiary strata. If rock type mattered then we would expect a zircon crystal's lead content to vary dramatically within the Cambrian or Cretaceous strata according to their local rock types. (The strata layers are discussed in Chapter 3.) However, this is not what we observe. The same thing goes for neutrinos and cosmic rays. *(See Chapter 1, sections "Bosons, Fermions, Quarks, and Leptons" and "Quarks and Leptons" for definition and discussion of neutrinos. **Cosmic rays** are high-energy protons and atomic nuclei that move through space at nearly the speed of light. They originate from the Sun, from outside of the Solar System in our own galaxy, and from distant galaxies. Upon impact with Earth's atmosphere, cosmic rays produce showers of secondary particles, some of which reach the surface, although the bulk is deflected off into space by the magnetosphere or the heliosphere.)* Neutrinos penetrate the earth so easily that they would affect all strata more or less equally, to the extent that they affect anything at all. Cosmic rays, on the other hand, do not penetrate that far into the earth to begin with, so they can be ruled out. The depth of burial, itself, has little to do with it. In some parts of the world the Cretaceous strata is found deeper than is the Cambrian strata in other parts of the world. The depth at which either is found can vary dramatically. In the Grand Canyon area the Cambrian strata lies beneath a huge column of strata; in California's Mojave Desert portions of the Cambrian strata are exposed at the surface. For the young-earth creationist, this is an unsolvable mystery, a mystery with parallels in each of the radiometric clocks used by geologists. The potassium-argon, rubidium-strontium, samarium-neodymium, luteium-hafnium, rhenium-osmium, thorium-lead, and the two uranium-lead dating methods all point to the same fact.

The ratio between tiny amounts of radioactive elements and their decay products can determine which strata a rock will appear in. There is nothing mysterious that affects rocks and zircon crystals. For those who believe that each of the geologic periods was laid down in days or weeks by Noah's flood, they cannot explain this. However, for those who do not believe that each of the geologic periods was laid down in days or weeks by Noah's flood, it is understood that time is what caused the geologic periods. The Cambrian strata has simply been around a lot longer than the Cretaceous strata, and the radioactive uranium in its zircon crystals has had more time to decay into lead. The same radioactive elements in different geologic periods will have decayed by different amounts. (Noah's Flood is discussed in Chapter 4.)

Even creationists realize that time is the only answer, but they provide an unrealistic explanation. They claim that the radioactive elements decayed much faster (that is, accelerated) in the past (see the Young-earth Creationist View above). However, such claims are false. There are multiple problems with this line of thinking. For instance, there are many boundaries (unconformities) in the geologic strata that exhibit a sharp change in radiometric age. Thus, zircons that are formed at about the same time in Noah's flood (from intruded magma close to each side of an unconformity, if such quick formation were even possible) would exhibit impossible differences in the decay of their uranium.

A few calculations will rule out a fast radioactive decay rate before Noah's flood occurred. Based on the present decay rate of U-238, the Cambrian period began about 570 million years ago. Since then the amount of uranium-238 has been reduced a bit (to 91.544% of itself) by radioactive decay. Had the decay rates remained high after the flood or in its later stages, the zircon crystals in the more recent strata (the last strata laid down by Noah's flood) would have "aged" considerably, which is not the case. Furthermore, the zircon crystals had to be created during Noah's flood in order to be "aged" according to the strata in which they were associated. (Noah's Flood is discussed in Chapter 4.)

It is too much to assume that each of the zircon crystals just happened to be deposited in the right strata (as mentioned above). Therefore, at the time of Noah's flood the decay rate had to be at least fast enough to reduce the amount of uranium-238 to 91.544% of itself in one year. If we generously take that minimum decay rate, with no thought of increasing it further as we look back into the past, we can calculate how much uranium-238 had to be present 1656 years before Noah's flood (when the earth was created, according to some young-earth creationists). It turns out that the amount of uranium-238 needed is 3.47 x 10^{63} times the amount of uranium-238 around at the start of Noah's flood. In other words, if our entire solar system were made of uranium-238 the quantity would not even begin to suffice.

These calculations clearly disprove the claims made by young-age creationists. It can be safely ruled out that the radioactive decay rates for uranium-238 (and, by quantum mechanical implication, all others) dwindled to their present values from high rates at creation time. An initial U-238 decay rate high enough to do young-age creationists any good also leads to an absurd conclusion. Young-age Creationists must now assume that the decay rates were low before Noah's flood, that they became phenomenally high during the start of Noah's flood, and that they dropped to normal after Noah's flood. Such tailor-made assumptions are not very scientific. (See section "Radiometric Methods (Absolute Dating) Used to Determine the Age of the Earth" for further discussion of dating of geologic strata.)

Some of the material that has been radio metrically dated, whose dates fully conform to the accepted ages of their place in the geologic column, comes from large masses of once-molten rock. Those samples could not possibly have cooled down in the course of a mere year no matter what. (Possibly a million years.) Thus, any "aging" done on their interior zircons had to occur, by creationist thinking, after Noah's flood. Only then did the inner rock cool enough so that those crystals finally formed. By young-age creationist reckoning, those crystals really formed after the flood and should reflect the normal decay rates. That is, the uranium-238 in zircon crystals formed after Noah's flood should show almost no decay at all. The young-age creationist concept, however, is not true. In reality, the radiometric age of zircons is in good agreement with the dating of the strata in which they were formed.[2-37]

In the next chapter (3) we will look at some of the fossil evidence to help determine whether or not the fossil evidence indicates that species gradually evolved into more complex and more diversified creatures. Additional fossil evidence will be discussed in chapters 7, 8 and 16.

~~~

Chapter 3

The Fossil Evidence in the Strata Layers of the Earth

Evolution is as well established a fact as gravitation. That living things evolve is as certain as a scientific fact can be.[3-1]

Why, if species have descended from other species by fine gradations, do we not everywhere see innumerable transitional forms? Why is not all nature in confusion, instead of the species being, as we see them, well defined?

• *Charles Darwin, Origin of Species,*
Chapter 6—Difficulties of the Theory

But, as by this theory innumerable transitional forms must have existed, why do we not find them embedded in countless numbers in the crust of the earth? . . . I believe the answer mainly lies in the record being incomparably less perfect than is generally supposed. The crust of the earth is a vast museum; but the natural collections have been imperfectly made, and only at long intervals of time.

• Charles Darwin, Origin of Species,
Chapter 6—Difficulties of the Theory

Why then is not every geological formation and every stratum full of such intermediate links? Geology assuredly does not reveal any such finely graduated organic chain; and this, perhaps, is the most obvious and serious objection which can be urged against my theory. The explanation lies, as I believe, in the extreme imperfection of the geological record.

• Charles Darwin, Origin of Species,
Chapter 10—On the Imperfection
Of The Geological Record
Sixth edition (1901)

The geological record is extremely imperfect and this fact will to a large extent explain why we do not find intermediate varieties, connecting together all the extinct and existing forms of life by the finest graduated steps. He who rejects these views on the nature of the geological record, will rightly reject my whole theory [3-2]

• Charles Darwin, Origin of Species,
Sixth edition (1901)

Introduction

In this chapter (3) we will look at the different strata layers within the earth and how the fossil evidence is interpreted. The various views presented in this chapter regarding the fossil evidence are not exclusively of strict evolutionists and strict creationists. Some views are of those who believe in variations of the two extremes. See "Various View Descriptions" section in the back of this book for descriptions of the views presented in this chapter.

The Strata Layers of the Earth

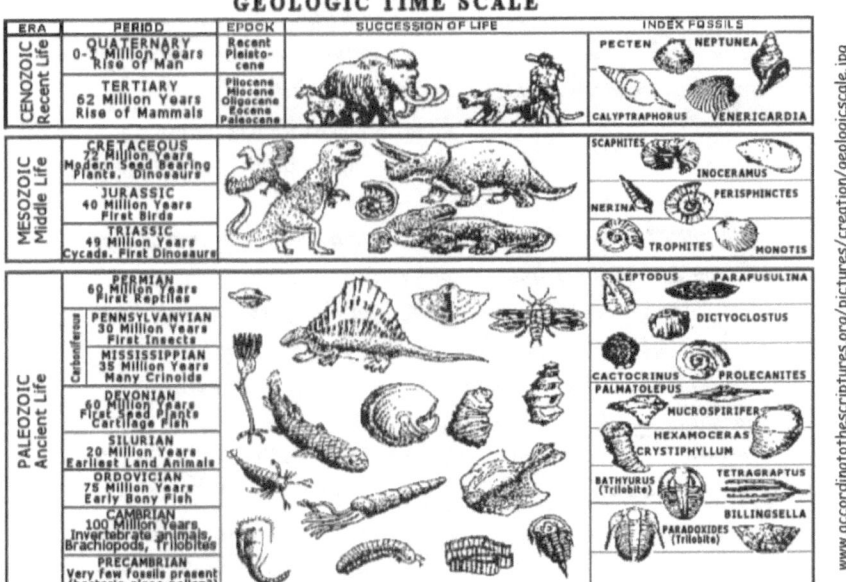

Figure 3.1 The chart above graphically illustrates the geological column under the Earth's surface. The chart begins at the top with the most recent strata layer and then continues down to describe the lower, older strata layers. To the right is a column of index fossils, discussed below in section "Relative Dating Method—Dating Strata and Fossils by Use of Index Fossils."

Evolutionist View

The descriptions of the strata layers of the Earth below will begin with the deepest, lowest, and oldest stratum and work up toward the surface of the Earth. Methods used to determine the age of a rock or fossil will be discussed later.

(bya = Billions of Years Ago   mya = Millions of Years Ago)

Archean Eon
(Beginning or Origin of Life) (3.8 bya-2.5 bya)

Fossils of cyanobacterial mats (stromatolites, which were instrumental in creating the free oxygen in the atmosphere) are found throughout the Archean Eon, becoming especially common late in the eon. A few possible small bacterial fossils have been found in the sedimentary rock. It appears from the fossil record that life may have been limited to simple non-nucleated single-celled organisms, called Prokaryota (See Chapter 5, Figure 5.1). No eukaryotic fossils (cells with a nucleus—See Chapter 5, Figure 5.2) appear to have existed during most of this eon, although they might have evolved during the later part. First organisms, simple photosynthetic bacteria, only about a billion years after the planet was formed (4.5 bya). These single cells were all that occupied the Earth for the next 2 billion years. Then the first simple eukaryotes that were organisms having true cells with nuclei and chromosomes appear (Chapter 5, Figure 5.2).

Proterozoic Eon
(Earlier Life) (2500 mya-570 mya)

During the Proterozoic Eon (2500 mya-570 mya), there was a great deal of oxygen buildup occurring in the Earth's atmosphere. Although oxygen was undoubtedly released by photosynthesis well back in Archean Eon. The first advanced single-celled, eukaryotes and multi-cellular life, roughly coincides with the start of the accumulation of free oxygen. This may have been due to an increase in the oxidized nitrates that eukaryotes use, as opposed to cynobacteria (a phylum of bacteria that obtain their energy through photosynthesis). It was also during the Proterozoic Eon that the first symbiotic relationships between mitochondria (for nearly all eukaryotes) and chloroplasts (for plants and some protists a diverse group of eukaryotic microorganisms only) and their hosts evolved.[3-3]

Paleozoic Era
(Ancient Life) (590-245 mya)

First signs of multi-cellular animals. Soft-shelled animals. First jawless fish. Animals develop hard shells and skeletons. First appearance of reptiles.

Precambrian/Cambrian Transitional Period—(590-580 mya) Imprints of algae have been found on all continents, but otherwise fossilized plant and animal remains are rare. During this time, life consisted entirely of soft-bodied forms that are rarely fossilized. First fossils of multi-cellular animals, mostly small, soft-bodied metazoans that are simple sponges, sea anemone-like organisms and jellyfish, appear.

Phanerozoic Eon
(Visible Life) (542 mya-0 mya)

The Phanerozoic Eon begins with the time when diverse hard-shelled animals first appeared. The time span of the Phanerozoic Eon includes the rapid emergence of a number of animal phyla. It also includes the evolution of these phyla into diverse forms, the emergence of terrestrial plants, the development of complex plants, the evolution of fish, the emergence of terrestrial animals, and the development of modern faunas. During this eon, continents drifted about, eventually collecting into a single landmass known as Pangaea and then splitting up into the current continental landmasses.

Cambrian Period—(580-505 mya) First jawless fishes. Animals develop hard shells and skeletons that remain as fossils. All invertebrates (creatures without a backbone.) phyla (organisms) appear. Land areas covered by sea during much of the period. Fossils of invertebrate sea organisms, such as trilobites, sponges, and echinoderms. 560 mya—Simple mollusks, worms, and echinoderms (organisms similar to starfish and sea cucumbers). 545 mya—First vertebrates that are small worm-like and primitive fish-like organisms, without bones, jaws, or fins (excepting a single dorsal fin).

Ordovician Period—(505-438 mya) More varieties of jawless fishes, corals, sponges, shellfish, and first vertebrates (fish) appear.

Silurian Period—(438-408 mya) First ray-finned & lobe-finned fish. Jawed fish, similar to current fish, appear.

Devonian Period—(408-360 mya) Fish are dominant. First amphibians appear. Ninety percent of the Earth's sediments, up until this period do not contain any land animals. The first evidence of insects and spiders appears.

Carboniferous Period—(360–3 mya) The first synapsids, such as pelycosaurs, evolve from stem reptiles. (Synapsids possess a skull design in the temple providing for better jaw muscle attachment, but otherwise possess few striking similarities with mammals. Synapsids are reptiles with one temporal opening on each side of the skull.) They evolved in the late Permian Period and were characterized by carrying their limbs under their body and developing front teeth that were different from their back teeth. One group of synapsids, the therapsids, gave rise to the mammals. (Therapsids are defined in the Late Permian Period.)

Mississippian Epoch—(360-320 mya) Big terrestrial amphibians, fishes.

Pennsylvanian Epoch—(320-286 mya) Amphibians dominant. First reptiles appear.

Permian Period—(286-248 mya) Amphibians dominant. First mammal-like reptiles appear. Mammal-like reptiles and conifer-like plants. Many species become extinct because of environmental changes after vast upheavals of the Earth's crust. There is a single great land mass called Pangaea or Pangea (all lands).

Early Permian Period—(286-276 mya) A synapsid line evolves into therapsids, which become dominant at this point. Beginning of jaw modification, an indication of a relatively lighter and more active body (also known as mammal-like reptiles).

Middle Permian Period—(275-256 mya) Major therapsid diversification, as well as large body size.

Late Permian Period—(255-248 mya) Therapsids known as cynodonts evolve. Therapsids are extinct mammal-like reptiles that are an advanced type of synapsid reptile that evolved in the Permian Period and existed during the Permian and Triassic Periods. Therapsids further differentiated their dentition into nipping, biting, and crushing teeth, and (unlike diapsids) had forelimbs that were more greatly developed than hind-limbs. Therapsids include the so-called mammal-like reptiles as well as mammals. Cynodonts are a division of the Therapsids from the Triassic period comprising small carnivorous tetrapod reptiles often with mammal-like teeth. Cynodonts possess secondary palates, greatly enlarged dentary and larger temporal openings implying complicated musculature.

Mesozoic Era
(Middle Life) (250-65 mya)

Life is dominated by large reptiles, i.e., dinosaurs. The predominant plants during this period are gymnosperms, like the cycads.

Triassic Period—(248-213 mya) Mammalian reptiles dominant. First dinosaurs become dominant. The land mass Pangaea or Pangea (Continental Drift— separation of major land area) begins to split into two great land masses, Leuvasele and Gondwana.

Late Triassic Period—(220-213 mya) Therapsid trend towards small size due to some unknown selective pressure. They remain diminutive until after the Cretaceous Period (144-65 mya).

Jurassic Period—(213-144 mya) Dinosaurs are dominant. First mammals that are small and appear half-reptile/half-rodent, then first birds, appear. First known flowering plants appear. Much of Europe and Asia submerged. North America drifts away from Africa and Europe. Africa and South America begin to separate.

Cretaceous Period—(144-65 mya) Dinosaurs continue to dominate. Small mammals and birds appear. Dinosaurs and many other species become extinct. Rocky Mountains form. South Atlantic Ocean opens as South America completes separation from Africa.

Cenozoic (Tertiary) Era
(Recent Life) (65-0 mya)

Mammals, birds and teleost fish dominate. Mammals and birds, similar to what appear on the Earth today, are present.

Paleogene Period—(66-23 mya) The Paleogene Period is most notable as being the time in which mammals evolved from relatively small, simple forms into a large group of diverse animals. Birds also evolved considerably during this period, changing into roughly modern forms. Some continental motion took place. Climates cooled somewhat over the duration of the Paleogene Period and inland seas retreated from North America early in the period.

Paleocene Epoch—(67-54 mya) Beginning of Age of mammals as first placental mammals. Flowering plants are dominant.

Eocene Epoch—(54-34 mya) Early horses, rhinoceroses, and camels appear.

Oligocene Epoch—(34-24 mya) Monkeys and apes appear.

Neogene Period—(23-2.6 mya) During the Neogene Period mammals and birds evolved considerably. Some continental motion took place, the most significant event being the connection of North and South America in the late Pliocene epoch (5.3-2.5 mya). Climates cooled somewhat over the duration of the Neogene Period.

Miocene Epoch—(24-5.3 mya) Tremendous volcanic activity. Spread of grasses and grazing animals. Birds appear.

Pliocene Epoch—(5.3-2.5 mya) Separation of Australia from Antarctica and of North America from northern Europe is complete. India collides with Asia, forming the Himalayas.

Quarternary Period—(2.5–0 mya) The Quaternary Period represents the time during which recognizable humans existed. Humans evolve to their modern form. At the start of the Quaternary Period, the continents were just about where they are today. Throughout this period, the planet drifted in and out of its orbit around the sun, causing ice ages. The last ice age ended about 10,000 years ago. Sea levels rose rapidly, and the continents achieved their present-day outline.

Pleistocene Epoch—(2.5-0.01 mya)—Evidence of plants and animals that resemble what is presently found on the Earth. The Ice Ages occur. Glaciers advance and retreat many times in North America and Europe. Early man appears.

Rocent or Holocane Epoch—(0.01-0 mya) Warm climate melts glaciers. Civilization spreads and flourishes. [3-4]

The Geologic Column and the Strata Layers of the Earth

Young-earth Creationist View

The strata layers, briefly described above, collectively are called 'the geologic column.' The Geologic Column, indicated in its ideal, continuous sequence, does not exist anywhere in the world, as portrayed by evolutionists. The entire geologic column was founded and built on the assumption that organic evolution was a fact. The only basis for placing rock formations in chronological order is their fossil content, especially index fossils (Figures 3.1 & 3.2). (For further discussion of index fossils see section "Relative Dating Method—Dating Strata and Fossils by Use of Index Fossils.") The only justification for assigning fossils to specific time periods in that chronology is the assumed evolutionary progression of life. Nowhere in the world does the geologic column actually occur. It exists only in the minds of the evolutionary geologists. It is simply an idea, an ideal series of geologic systems, and not an actual column of rocks that can be observed at a particular locality. Real rock formations are characterized by gaps and reversals of this ideal, imaginary sequence.[3-5]

Approximately 1% of all sites excavated contain 10 or 11 identifiable layers as described by the geologic column. The rest of the sites researched contain only a few of the 10 or 11 layers. This makes the existence of a reliable column, as is described, very questionable.[3-6] Although there are a number of different locations on Earth with the 'complete' (that is, not every strata layer included) column, it is rare that most strata are present. Regardless of whether there are 10 or 20 or even 50 locations on Earth where all ten Phanerozoic geologic systems are observed, there is no escaping the fact that this still totals less than 1% of the Earth's surface. Even this 1% does not include ocean basins. When the ocean basins are included (none of which have more than a few of the ten geologic systems in place), the global figure falls to less than 0.4%. If we include the various groups of animals as a basis for separating and correlating the physical characteristics of rocks or stratigraphic units (lithologies) into 'geologic periods', the percentage is even less. Only a small fraction of index fossils (Figure 3.2) are found at the same location on Earth. Therefore, scientific creationists are justified in concluding that the standard evolutionary-uniformitarian geologic column is essentially non-existent.[3-7]

Nowhere in the world are all the strata in the theoretical "geologic column" to be found in one place that is complete. Most of the time only two to eight of the 21 theoretical strata can be found. Even that classic example of rock strata, Grand Canyon, only has about half of them. In the Southwest United States, in order to find Precambrian or Paleozoic strata, we would need to go to the Grand Canyon.

To find Mesozoic stratum, one would need to go to eastern Arizona. To find Tertiary stratum, one would need to go to New Mexico. Nowhere in the world is the entire geologic column of the evolutionists to be found, for it is an imaginary column.[3-8]

In the Grand Canyon, for example, no Pennsylvania fossils are to be found, the Permian fossils resting upon the Mississippian, and yet there is no evidence of erosion during that assumed hiatus of perhaps some 30,000,000 years. Even in our brief time since measurements have been recorded, coastlines are rising and sinking, in Scandinavia, for instance. It is then inconceivable that the crust of the earth would remain so stable and at just the right elevation that it would be unaffected by either erosion a sedimentation of millions of years. The Grand Canyon is one of the best places in the world to study stratigraphy. Below the Mississippian stratum in most places the Devonian stratum is not present, and nowhere is the Silurian stratum or the Ordovician stratum present. This means that the Redwall formation, which is Lower Mississippian, actually rests upon the Cambrian Muav limestone, a time gap of over 50,000,000 years. Surely in this immerse space of time one would expect to find effects of very extensive erosion, perhaps warping and folding with angular discordance, but what is actually found is the appearance of perfectly conformable series of beds, laid down in fairly quick succession. Angular discordance or angular unconformity is a discontinuity in rock sequence indicating interruption of sedimentation, commonly accompanied by erosion of rocks below the break.[3-9]

Old-earth Creationist View

Young-earth creationists argue that there is nowhere in the Earth's crust where a complete geological column can be seen. It has to be built up from sections of it. Despite this objection, there is nothing illogical about its concept.

In one location, parts of the geological column appear, and in other locations of the world, other parts appear. There are so many of these fragmentary columns, and so much overlap between them, that it is quite easy to build up the complete geologic column. The reasons why a complete geologic column has never been found is because more often than not, a layer of rock that was deposited during one period eroded away during a later period. The rocks that are still in existence today are the ones that did not erode away. Also, sediments are mostly laid down under water, therefore, at any given location one would not expect to find sediments from those geological periods when that area was far above sea-level.

Young-earth creationists claim that geologists are guilty of circular reasoning in building up the geologic column. Paleontologists use fossils as time markers (index fossils—Figure 3.2) in determining the age of stratified rocks. Near the start of the nineteenth century independent workers in England and in France discovered that units of sedimentary rocks can be traced over wide areas by means of distinctive fossils (index fossils—Figure 3.2) in each unit. Some of the fossil species were found only in single beds or a few successive beds; and those species that were thicker were replaced in higher beds by different species. Such changes seem commonplace under present concepts of evolution; but the discoverers accepted the facts merely as a rule of thumb to aid in classifying and mapping the thick sections of sedimentary rocks. It is admitted that this method could be abused, and young-earth creationists often quote that evolutionists have used the geological column in an illogical way, by basing circular arguments upon it. These are rare cases and should not be attributed to the major study of geology. The geologic column idea itself was developed in a thoroughly logical way, long before the theory of evolution was developed. And many of those who contributed to the development of the geologic column idea, were creationists. There is nothing illogical or circular about the way that mainstream geology uses the geologic column today.[3-10]

Evolutionist View

Contrary to what young-earth creationists' claim, the entire geologic column does exist in certain parts of the world. The Bonaparte Basin of Australia and the Williston Basin of North Dakota are two examples of areas where the entire geologic column does exist. The entire column does not have to exist all in one place in order for it to be understood. There is enough overlap in many parts of the world to figure out the column anyway.[3-11] The entire geologic column is found in more than 26 basins around the world, piled up in proper order.[3-12]

Young-earth Creationist Response

The fact is that the geologic column is not found complete any place on Earth, it is a concept developed by evolutionists. While the geologic column consists of 10 basic layers, all 10 layers are found in very few places making up less than 1% of Earth's surface. The theory says it should be 100 miles thick. On average worldwide, the sediment layers are only 1 mile thick. The entire geologic column was patched together from various locations. Now there are places that are claimed to have the entire geological column. What they mean is that they have found layers that they can assign to all 10 geologic ages.

The Ghadames Basin in Libya does not contain the entire geologic column, as some claim. The only reference to fossils was a general reference to microfossils so this site seems to have little or no bearing on fossil order. Contrary to those who claim that this site represents the entire geological column, it is missing the Permian period. It is difficult to comprehend how one can claim the entire geological column is present when it is missing an entire period. The geological column labels seem to have been assigned based on the correlation with rocks from all across Libya. This means there was sufficient room for subjective analyzes, and that the labeling process assumed the geological column and thus cannot be used as evidence for its validity. The cross-section of the Ghadames Basin appears to indicate rather long ages brought these layers together as they are. The cross-section shows some 25 fracture lines. The pattern suggests that rather than being fault lines they are the result of compression stress. The Cretaceous, Jurassic, Triassic and Cretaceous strata seem to have been forced over the others. This seems to have occurred after they and the rock beneath them had buckled down and before both had hardened completely. Such an event could have occurred during the Flood while the layers where not yet completely solidified. (The Flood is discussed in Chapter 4.)

The Bonaparte Basin in Australia is fairly complete in that all "geologic period" layers are present; however, in no way does it qualify as a complete column, since there are numerous gaps when the strata are spread out on the geologic column chart. There seem to be cases of interbedding that go between periods. (Interbedding is rocks or minerals that appear between beds or strata.) They are at Devonian / Carboniferous and Triassic / Jurassic. There may be others as well. There are also cases of interbedding spanning large portions of periods that should still be separated by tens of millions of years. They occur in the Devonian, Carboniferous, Permian, Triassic, Jurassic, Cretaceous, and Tertiary. This means that the evidence suggests that this column formed a great deal more rapidly than the geologic column indicates. No references as to how any of the Rock layers were assigned to their respective ages nor have any references been found regarding what fossils were contained there. This means that this site shows nothing about fossil order.

The Williston Basin in North Dakota, also known as the North Dakota Column, is claimed to contain the entire geologic column. As stated earlier the, total theoretical column depth is 100 miles, but the depth of the Williston Basin is only 3.4 miles. This means that much of the column is missing. Such large amounts of sedimentation are possible during a yearlong global Flood, because it was laid down sideways, making it quite possible to lay down such large amounts of sediment very quickly. The main factors are available sediment and the rate of current flow.

What is needed to describe the strata layers is a statistical study of fossil locations that does not involve the geologic column classification system, but is based only on the actual three-dimensional locations of fossils, latitude, longitude, and depth relative to sea level. Surface altitude above sea level should be included as well; this would give a true picture of both local and global fossil distribution, not one based on a theoretical classification system.[3-13]

The Arrangement of the Strata Layers

Evolutionist View

The odds of arranging the Precambrian era, the seven geologic periods of the Paleozoic (Cambrian, Ordovician, Silurian, Devonian, Mississippian, Pennsylvanian, Permian), the three periods of the Mesozoic (Triassic, Jurassic, Cretaceous), and the two periods of the Cenozoic (Paleogene, Neogene or Tertiary, Quaternary), stratum during a global flood in the proper order by pure chance, is phenomenal. (The possibilities of a global flood causing the strata layers, is discussed in Chapter 4.) The chances are 6.2 billion to one of getting the right order for all thirteen. And, when one considers that each period (strata) can also be divided into "upper, middle, and lower," the odds of arranging them in the correct order by pure chance becomes astronomical.[3-14]

Radiometric dating methods (Chapter 2) have correctly placed the Cambrian Period stratum between the Precambrian and the Ordovician stratum, the Ordovician between the Cambrian and the Silurian stratum, the Silurian between the Ordovician and the Devonian stratum, and so forth.[3-15] From the evidence, it is quite clear that the young-earth creationist view does not support the scientific evidence.[3-16] (See Chapter 4, section "The Order of the Fossils" for discussion of the arrangement of the fossils.)

Methods Used to Determine the Age of Fossils

Evolutionist View

Scientists use basically two different methods for determining the age of a fossil: **(1)** Relative Dating (how old a fossil is in relation to other fossils or rock units) and **(2)** Absolute Dating (approximately how many years old a fossil is). One principle of relative dating is called superposition. The relative dating method is based on the concept that in any one place, the lower rock layers (and fossils in them) are older than higher ones, unless there is evidence that the layers have been overturned. This is logical, since the lower sediments would have been deposited first.

Relative Dating is called correlation. Rock layers (strata) are compared with others in another location on the basis of their mineral composition, fossil content, and other features. When a rock unit or series closely matches those of another area, they are said to be correlated and considered to be of similar age. Correlations are especially reliable when they include index fossils (Figure 3.1, right column & Figure 3.2), which are fossils with a limited age range. Good index fossils are easily recognized and widely distributed around the world, so that they aid both local and international correlations.

Absolute Dating works with relative dating by providing a specific (not necessarily precise) chronological age for a given specimen, such as "50 million years before present." In recent years reliable forms of absolute dating became available through the development of radiometric dating methods. These two methods (Relative Dating and Absolute Dating) are based on the known, regular (steady, continuous and uniform processes— called uniformitarianism) decay of certain radioactive elements (isotopes) into other isotopes or "child products." By measuring the amount of "parent" and "child" products in a rock sample, its approximate age may be calculated.[3-17] These dating methods will be discussed in this chapter.

Relative Dating Method—Dating Strata and Fossils by the Use of Index Fossils

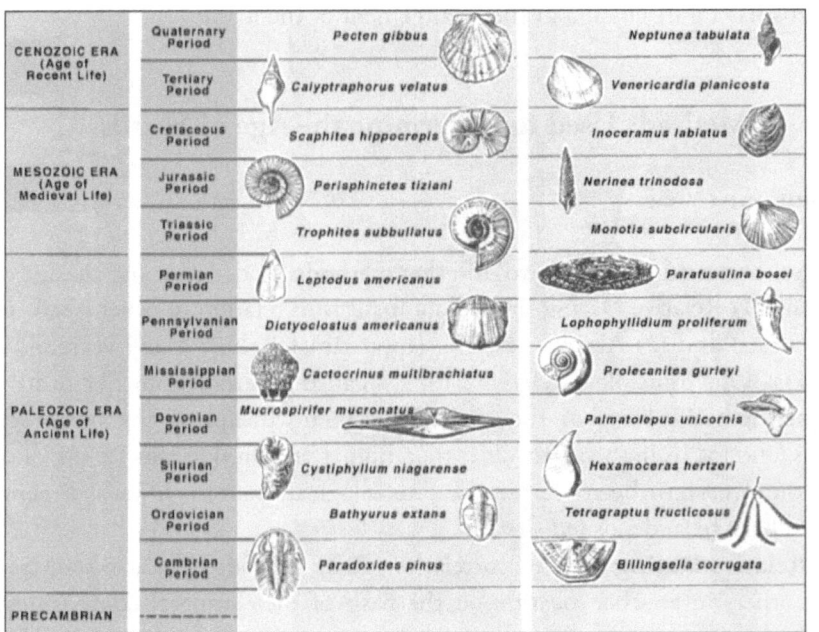

Figure 3.2 Index Fossils as they appear in major strata layers in the Earth

The above chart illustrates some of the index fossils found within various strata layers of the Earth, beginning with the most recent period at the top and progressing down to the deeper, older strata layers. (The eras and periods listed in the two left columns are described in section "The Strata Layers of the Earth.")

Evolutionist View

Relative dates of strata (whether layers are older or younger than others) are determined mainly by which strata are above others. Some strata are dated absolutely via radiometric dating, otherwise known as absolute dating. These methods are sufficient to determine a great deal of stratigraphy. Some fossils are seen to occur only in certain strata. When these fossils exist, they can be used to determine the age of the strata, because the fossils show that the strata correspond to strata that have already been dated by other means.[3-18] Such fossils can be used as index fossils (Figure 3.2).

Young-earth Creationist View

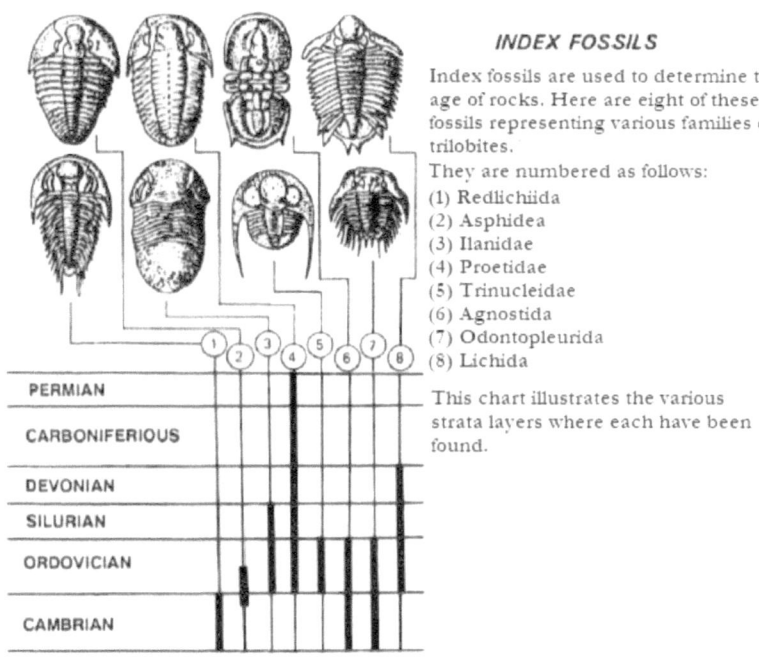

INDEX FOSSILS

Index fossils are used to determine the age of rocks. Here are eight of these fossils representing various families of trilobites.
They are numbered as follows:
(1) Redlichiida
(2) Asphidea
(3) Ilanidae
(4) Proetidae
(5) Trinucleidae
(6) Agnostida
(7) Odontopleurida
(8) Lichida

This chart illustrates the various strata layers where each have been found.

PERMIAN

CARBONIFEROUS

DEVONIAN

SILURIAN

ORDOVICIAN

CAMBRIAN

Figure 3.3 Samples of Various Trilobite Index Fossils[3-19]

The strata are dated by what the evolutionists call "index fossils" (Figure 3.2). In each stratum there are a few fossils that are not observed quite as often in the other strata. These are the fossils that are used to "date" that stratum and all the other fossils within it. It may sound strange, but that is the way it is done. These special index fossils are generally small marine invertebrates (Figure 3.2). Their presence in a sedimentary stratum is supposed to provide proof that stratum is just so many millions of years "younger" or millions of years "older" than other strata.

By just examining a particular index fossil there is no way to tell how old it is. For example, any rock that contains fossils of one type of trilobite (Paradoxides) is called a "Cambrian" rock, thus dating all the creatures in that rock to a time period beginning 580 million years ago and having existed some 75 million years. Rocks containing another type of trilobite (Bathyurus) are arbitrarily classified as "Ordovician," which is classified as beginning 505 million years ago and to have spanned a time period of some 67 million years (Figure 3.2). If index fossils are used to determine the age of rocks, it would result in confusion because there would be mixed-up or missing strata, each with fossils from a wide variety of ancient plants and animals that may also be found in still other rock strata.[3-20]

Absolute Dating Method—
Fossil Dating Obtained by Chemical Testing

Evolutionist View

The radioactive decay of uranium can be used to measure the geologic age of rocks. Radioactive decay is a process where an isotope of an element spontaneously loses particles from its nucleus to create a child isotope. This decay occurs at a specific rate. The half-life of a radioactive isotope is the amount of time it takes for one-half of the radioactive isotope to decay. By comparing the ratio of parent isotopes to the child isotopes, scientists are able to determine the age of the fossil and the rock.

Isotopes that have long half-lives are important to the dating of fossils because many fossils are millions of years old. Uranium-235 is used quite often in the absolute dating of fossils. Uranium-235 has a half-life of 704 million years. Therefore, Uranium-235 can be used to accurately date rocks that are hundreds of million years old. Contrary to what some people believe, carbon-14 is not used for dating of fossils. Carbon-14 has a half-life of 5,730 years, which is too short to be useful in dating fossils. Carbon-14 is quite useful, however, in dating artifacts.[3-21] (See Chapter 2 section "Radiometric Methods (Absolute Dating) Used to Determine the Age of the Earth" for further discussion of absolute dating methods.)

Evolutionist & Young-earth Creationist Views Compared

Critical Creationist Perspective

No one alive today was around to observe either creation or evolution happen. Prior belief in either evolution or creation determines how one interprets the data, whether it is eons of evolutionary history preserved in gradual deposition or catastrophic burial from a worldwide flood. The evolutionist and the creationist come up with entirely different stories from this picture, depending on the prior acceptance of either evolution or creation. The evolutionist pictures a gradual build-up of each stratum, or layer, over hundreds of millions of years of the accumulation of sediment, gradually fossilizing dead animals in the process. The oldest evolved life forms that supposedly arose out of the sea are logically to be found in the lowest layers. The most recently evolved life forms are to be found in the highest layers.

The issue of whether or not "transitional forms" exist is not a productive topic to debate in the creation/evolution controversy. Some evolutionists use similarities between three particular animals to argue that animal 'A' evolved into animal 'B' based on the fact that animal 'X' exists. Some creationists use the dissimilarities between these same animals to argue that animal 'A' did not evolve into animal 'X' and animal 'X' did not evolve into animal 'B.' Evolutionists keep seeking to justify their "transitional forms" on account of the similarities and despite the differences. Creationists keep seeking to rule out "transitional forms" on account of the differences and despite the similarities.

Anything is good enough for the evolutionist, and nothing is good enough for the creationist. Neither will ever satisfy the other or a discerning observer. A scientific theory is validated through experimental observation and/or theoretical evaluation. Neither party actually observed the origin of any specific species, so neither party is qualified to argue scientifically from an experimental perspective whether or not animal 'X' is a "transitional form."[3-22]

The Fossil Evidence

In primitive times, some of the Greeks thought that fossils were parts of animals formed in the lower parts of the earth by a process of spontaneous generation, which had died before they could make their way to the surface . . . In medieval times . . . [they thought] that the things called fossils were never the shells or bones of [actual] animals living in bygone times, but that they only simulate such things, but that they have been created as such together with the layers of rock from which they may have been taken.[3-23]

Evolutionist View

. . . the chance of discovering species with transitional grades of structure in a fossil condition will always be less, from their having existed in lesser numbers, than in the case of species with fully developed structures.

> • Charles Darwin, in Origin of Species,
> Chapter 6—Difficulties of the Theory

Fossils are the remains of plants or animals in which the tissues have been replaced by minerals. As time progresses, fossils are covered by sediment. The older fossils are buried deeper than more recent fossils. This means that the fossil record is layered, or stratified, because some fossils were deposited many thousands or millions of years before others. Radiocarbon dating and other forms of radiometric dating (discussed in Chapter 2) confirm that all fossils were not deposited during a single, short period of time, such a Noah Flood (discussed in Chapter 4).[3-24]

Looking at the strata layers, it may appear that no life existed before the Cambrian Period. However, there was no sudden explosion of life at the start of the Cambrian Period, as creationists claim. Fossils of Precambrian Period creatures have been found. Evidence of multi-cellular life dating back 590 million years has been found in China. One reason why Precambrian Period (4,600-542 mya) fossils are rare is because they were soft bodied. They had no hard body parts that could easily fossilize. Another reason why Precambrian period fossils are so rare is that in the Precambrian period rocks are so old many have been destroyed due to weathering, heat, pressure, and erosion. Also, when the Earth' crustal plates collided (Continental Drift and Plate

Tectonics), where rocks are pushed back into the mantle and melted, this caused the strata to be disturbed and much of it was destroyed. Many of the animals that lived at the beginning of the Cambrian period (590-505 mya) no longer exist. Most modern organisms, such as birds, mammals, reptiles, and amphibians, did not appear until after the Cambrian period (590-505 mya). The fish that existed then are totally different than those that exist in the world today.

Some creationists believe that the Cambrian Explosion happened all at once, however, this is not true. (The Cambrian Explosion is discussed in the next section.) The fossil evidence indicates that during the Pre-Cambrian Period, simple organisms appear. Examples of fossils dated prior to the Cambrian Explosion are Cnidarian dated around 565 mya and a mollusk, dated approximately 555 mya. (Mollusks are invertebrates having a calcareous shell that encloses the soft, unsegmented body. Examples of mollusks are chitons, snails, bivalves, squids, octopuses—see section "A Closer Look at the Cambrian Explosion" below.)[3-25]

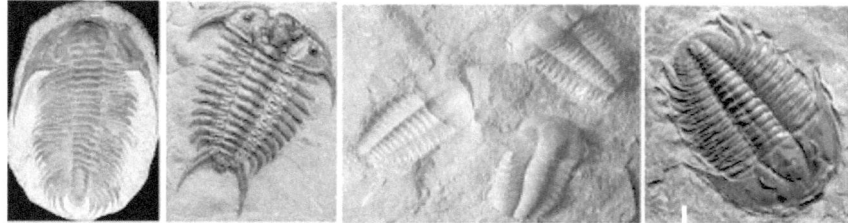

Trilobite Fossils

The strata layers indicate that in the lower, older strata, (in the Cambrian strata), trilobites appear. Trilobite fossils continue to be found in rock layers covering a period of approximately 300 million years. Then, beginning in the next layer above the Cambrian, the Permian strata, no trilobites are found. In strata further up, in the Triassic period, mammals appear. The fossil appearances and disappearances are called "geological succession" or "fossil succession." The fossil record clearly shows a trend from simple to complex creatures and plants. The more recent the strata layer, the more complex the plants and animals. This pattern is exactly what the theory of evolution would expect, a gradual progression of plants and animals.[3-26]

The apparent "sudden appearance" of body types is due to the imperfection in the fossil record. The Cambrian Explosion does not show that new forms of life arose suddenly. Instead, the fossil record shows that the fossil record has been poorly sampled and that the fossils themselves were not consistently preserved.[3-27]

Young-earth Creationist View

If the transitional changes occur [as] slowly [as evolutionists claim], then there should be vast numbers of transitional species living today, as well as etched into the fossil record. But they are not to be found. They do not exist; they have never existed.[3-28]

The creation account in Genesis and the theory of evolution could not be reconciled. One must be right and the other wrong. The story of the fossils agreed with the account of Genesis. In the oldest rocks we did not find a series of fossils covering the gradual changes from the most primitive creatures to developed forms, but rather in the oldest rocks, developed species suddenly appeared. Between every species there was a complete absence of intermediate fossils.[3-29]

The fossil record of life does not support evolution. The fossils that are found in what are usually considered the lowest deposits are alleged to belong to the Cambrian era of approximately 800 million years ago. In these rocks are found the fossils of various shellfish and crustaceans, sponges, worms, jellyfish, and various other complex invertebrate life forms. If you explored the bottom of the ocean today, and then explored a hypothetical ocean full of the life forms that are now represented by Cambrian fossils, you would probably not be able to tell the difference, except that many species have now become extinct (e.g. trilobites). In all, you would find fewer life forms today than you would in this "fossil ocean." This in itself would suggest the opposite of evolution.[3-30]

Evolutionists claim that the fossil record goes from the simple to the complex. The fact is the simple creatures found in the Cambrian strata were also complex. There are actually few examples in the fossil record of anything like "from simple to complex" progression. This is partly due to the fact that the fossils suddenly appear in great numbers and variety, too much so for much simple-to-complex progression to be sorted out.[3-31]

The Cambrian stratum is the lowest fossil strata level. It is extremely rich in fossilized life forms. There are at least 1500 different invertebrate species in the Cambrian strata. All of them are small very slow-moving sea creatures (see "Examples of Mollusks" below). Ranked by gross quantity, 60 percent of the fossils in the Cambrian strata are the tiny, complex trilobites (each one of which has multiple eyesight tubes), while 30 percent are various types of brachiopods. (Brachiopods are bivalves: shelled sea creatures with a top and a bottom shell.) Below the Cambrian strata is the Precambrian stratum, where there are few or no fossils at all. In the fossil record, life suddenly appears in widely varied profusion. Below the Cambrian stratum there is virtually nothing.

There are many complex trilobites in the Cambrian strata, yet below the Cambrian strata there is hardly anything that resembles a fossil. These little creatures had marvelously complicated eyes. (Trilobite eyes are discussed in Chapter 14, section "The Eye.") But they also had other very advanced features such as: **(1)** Jointed legs and appendages, which indicate that they had a complex system of muscles. **(2)** Chitinous exoskeleton (horny substance as their outer covering), which indicates that they grew by periodic ecdysis, a very complicated process of molting. **(3)** Compound eyes and antennae, which indicate a complex nervous system. **(4)** Special respiratory organs, which indicate a blood circulation system. **(5)** Complex mouth parts, which indicate specialized food requirements.

Many other types of creatures, found in great numbers in the Cambrian strata, are segmented marine worms. As with trilobites, they also had a complex musculature, specialized food habits and requirements, blood circulatory system, and advanced nervous system. The Cambrian rocks contain literally billions of the little trilobites, plus many, many other complex species. Yet below the Cambrian, called the "Precambrian," there is almost nothing in the way of life-forms. [3-32]

Complex creatures are found all through the Cambrian level. In the Cambrian stratum there are sponges, corals, jellyfish, mollusks, trilobites, crustaceans, and, in fact, every one of the major invertebrate forms of life. Included in the Cambrian stratum are complex organs, such as intestines, stomachs, bristles and spines. Eyes and feelers show the presence of nervous systems. For example, consider the specialized sting-cells (nematocysts) in the bodies of jellyfish, with their coiled, thread-like harpoons which are explosively triggered. How could this evolve?

Beginning with the very lowest of the fossil strata, the Cambrian, there are a great deal of fossil types. But each type (each species) of fossil in the Cambrian is different from the others. This diversity of life requires evolving (blending across species) to produce evolution, but this never occurs today, and it never occurred earlier. Looking at the fossils in the ancient world there were only distinct species. Looking at the modern world there are only distinct species. There are vast numbers (billions) of fossils of thousands of different species of complex creatures in the Cambrian, and below it is next to nothing. The vast host of transitional species, leading up to the complex Cambrian species, is totally missing.

Every major animal phylum group is included in the Cambrian rocks. That is, in the Cambrian fossil strata there is at least one species from every phyla of backbone-less animal (invertebrates) as well as animals with backbones (vertebrates). (Phyla are a group of species that have a genetic relationship.) Fully developed fish (heterostracan vertebrate fish fossils) were discovered in the Upper Cambrian strata of Wyoming. Not only complex animal life, but also complex plant life is represented in the Cambrian strata. Flowering plants are generally considered to be one of the most advanced forms of life in the plant kingdom. Spores from flowering plants have also been found in Cambrian strata.[3-33]

A Closer Look at the Cambrian Explosion

Evolutionist View

Mollusks, Lobopods, Arthropods, Arachnids

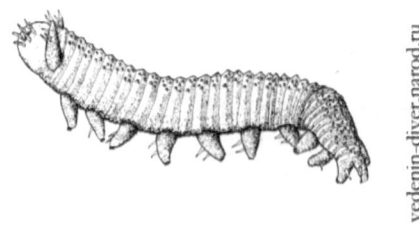

vedenin-diver.narod.ru

Lobopod

Within the Cambrian explosion there are transitional fossils. Lobopods (worms with legs) are intermediate between arthropods and worms.[3-34]

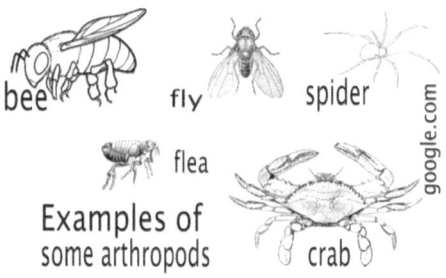

Examples of Arthropods

Arthropods are invertebrates having a segmented body, joined limbs, and a chitinous shell. Arthropods having a chitinous shell are hard-shelled aquatic creatures such as lobsters, shrimps, crabs, etc. Arthropods include the insects, crustaceans (lobsters and crabs), arachnids, myriapods, and extinct trilobites. Myriapods include creatures such as the centipede or millipede, having segmented bodies, one pair of antennae, and at least nine pairs of legs. Arthropods also include arachnids, including wingless (carnivorous) creatures, having a body divided into two parts, the cephalothorax (the anterior section of arachnids and many crustaceans, consisting of the fused head and thorax) and the abdomen, and having eight appendages (walking legs), no wings and no antennae.

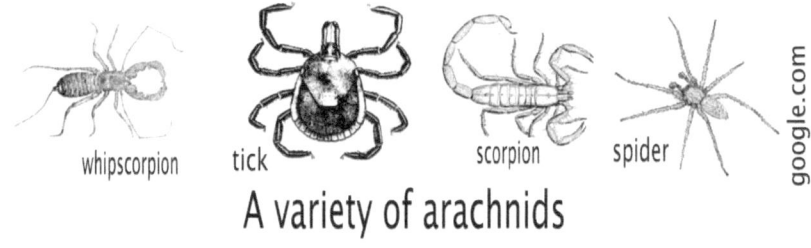

Examples of Arachnids

Arachnids include creatures such as spiders, scorpions, mites, ticks, daddy longlegs, tarantula, etc. By the end of the Cambrian Period, more complex organisms are present, such as Sirius Passet, which is dated around 535 mya.[3-35] Trace fossils indicate there existed organisms 60 mya prior to the Cambrian Period (590 mya). A sixty million year time period is plenty of time for evolution to have occurred.[3-36]

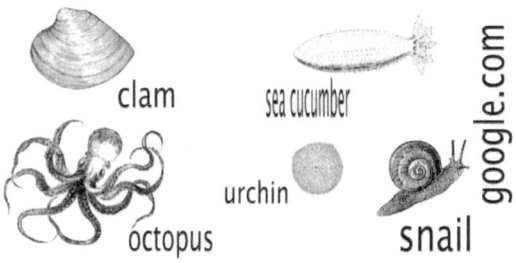

Examples of Mollusks

Soft parts of a creature do not preserve as well as hard, bony parts. This means that an organ in an offspring might have developed differently than the same organ in the parent, but it may not have been preserved in the fossil record. Therefore, just by examining hard part fossils, it is difficult to see many changes that could have developed in a creature. This would explain why a different species might appear as having a 'sudden appearance.'

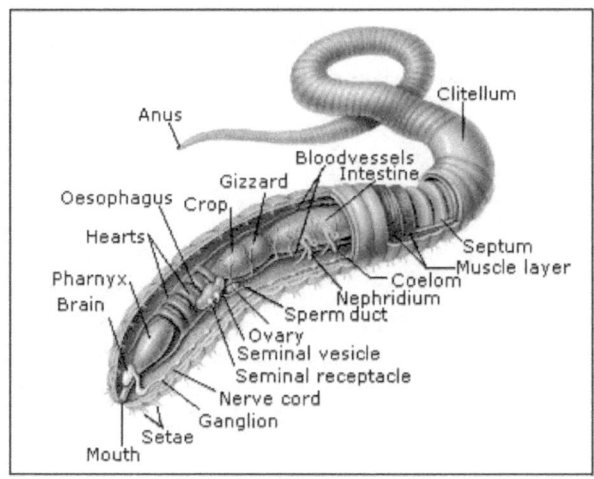

Figure 3.4 The Anatomy of an Earthworm
An Example of Annelids (ring worms)

The dates by which scientists estimate when the divergences between fossils occur, such as the annelids (ring worms—including earthworms, lugworms, leeches, marine worms—Figure 3.4) and the mollusks (see examples above), is done by molecular biology that looks at the configurations of genes on chromosomes (Chapter 5, Figure 5.5). By lining up the genes, the sequences of the genes are matched up with each other, and the closest matches and the more derived similarities, tell us which groups are most related to which (Genes and chromosomes are discussed in chapters 5 & 15). These are the unusual features of evolution. The progression from worm toward fish-like creatures begins with simple worms, which first appeared over 600 mya. (Worms and fish-like creatures are discussed in the next section, "Evolution From Worm-like Chordates to Fish-like Creatures.") The complex creatures that creationists refer to originally evolved through minor changes and continued to evolve into more complex animals throughout the Cambrian Period.[3-37]

Evolution From Worm-like Chordates to Fish-like Creatures

Evolutionist View

Lancelets   Pikaia   Haikouella   Haikouichthys   Anaspids
Worm-like creatures from the Cambrian Period

Figure 3.5 Worm-like Creatures from the Cambrian Period

Lancelets appear in the fossil record around 520 mya. They are known as amphioxus and are more developed than simple worms. They are somewhat fish-like but without any paired fins or other limbs (Figure 3.5). They have a primitive tail fin, but do not have a true skeleton, as they have mostly cartilage-like material. Lancelets are classified as chordates (animals having a notochord, or dorsal stiffening rod).

Pikaia appear in the fossil record during the Cambrian Period (545-490 mya), around 500 mya. It is classified as a worm with a head distinct from its tail. It had two forward-facing eyes and a notochord rather than a backbone (Figure 3.5).

Haikouella appear sometime during the Cambrian Period (545-490 mya). It is more developed than Pikaia. Haikouella does not have bones or a movable jaw (Figure 3.5). It does have a head, gills, brain, notochord, well-developed musculature, heart and circulatory system. It has what appears to be a primitive tail fin and may also have had a pair of lateral eyes.

Haikouichthys appear in the fossil record at about 530 mya, during the Cambrian explosion. It is believed by many to be one of the earliest fish because it has a backbone, a distinct head, and a well-defined skull (Figure 3.5).

Anaspid appear in the fossil record in the early Silurian Period (443-416 mya). They become more plentiful until the Late Devonian Period (408-360 mya). These creatures are small marine organisms that are more complex than Haikouichthys (Figure 3.5). Anaspid did not have scales or paired fins, however Anaspid had rows of holes that appeared gill-like.[3-38]

Peripatus

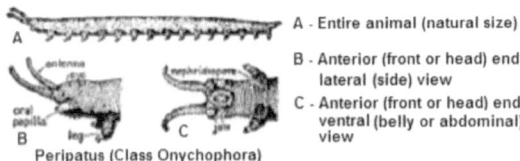

A - Entire animal (natural size)

B - Anterior (front or head) end lateral (side) view

C - Anterior (front or head) end ventral (belly or abdominal) view

Peripatus (Class Onychophora)

Figure 3.6 Peripatus

Evolutionist View

Peripatus (or the velvet worm) is a primitive tropical wormlike invertebrate with multiple pairs of legs (Figure 3.6). Peripatus is of the phylum (class) onychophora that gives birth to live young rather than laying eggs. It is said to be a living fossil because it has been unchanged for approximately 570 million years. Onychophora resemble slugs with legs and are sometimes described as the missing link between arthropods (See "Examples of Arthropods" above) and annelids (described above—see Figure 3.4).

Peripatus has been called the missing link between Annelid worms (segmented worms) (Figure 3.4) and the Arthropods (insects, crustaceans, etc.). Even though Peripatus does not provide all the answers to all the questions regarding origins, it is still useful in reconstructing the common ancestor between Annelid worms and Arthropods, due to its blend of annelid (ring worm) and arthropod characters. Its annelid characters include layers of circular and longitudinal muscles in the body wall, a flexible outer membrane and a simple head. Its arthropod characters include (1) a blood cavity, (2) dorsal blood vessel with slits and (3) respiration through trachease.[3-39]

Young-earth Creationist View

Peripatus looks like a caterpillar (Figure 3.6), however, upon close examination it looks like an annelid (earth) worm (Figure 3.4). The Peripatus is also called a 'velvet worm.' The head is continuous with the body. The outer cuticle is thin and flexible, with no external segmentation (divisions). The internal anatomy of the Peripatus is similar to the arthropods including, for example, lobsters, insects and spiders. The appendages (legs) are hollow, unjointed, cone-shaped structures with a retractable foot and hooked claw. Peripatus lives only in moist habitats, such as in tropical forest or rotting logs, as they need a lot of moisture in their environment. They eat small live insects. Peripatus has a very peculiar method for catching its prey. They have two slime glands, at the side of the mouth, which ejects a milky fluid up to 30 centimeters (11.8 inches). This milky fluid congeals on contact with air, entangling the prey. Peripatus is part of the group known as Onychophora. Peripatus range in size from 1.5 to 15 centimeters (5/8 of an inch to 6 inches).

Due to the Peripatus having multiple characteristics, many biologists believe that onychophorans are transitional between annelids and other arthropods. The Aysheaia is an organism in the fossil record that is extraordinarily similar. The Aysheaia was found in the Burgess Shale (dated at 530 million years old). The difference between Aysheaia and Peripatus is that Aysheaia was apparently marine, while Peripatus and all living onychophorans are terrestrial. According to evolutionary interpretations, organisms closely resembling Peripatus have existed for an extraordinary time interval. It seems most unlikely that an organism represented by such a restricted ecology should have survived so long.

Within Peripatus and other onychophorans there is almost no variation, however, there is a difference in reproductive methods. This diversity of reproductive styles (within different onychophorans) includes development like that of (1) monotremes (such as a platypus, which lay eggs), (2) marsupials (such as kangaroos, which protect tiny live-born young in a special pouch and nourish them on mother's milk) and (3) placentals (the young are nourished in the uterus by means of a placenta and are then born alive). It seems very unlikely that their reproduction process could become so varied while the rest of the organism stayed static and primitive for close to 600 million years. It is more likely that Peripatus demonstrates the variety of creation.[3-40]

A Closer Look at the Cambrian Explosion—continued

Evolutionist View

The reason why the "higher" taxonomic groups appear at the Cambrian Explosion is because the Cambrian Explosion organisms are often the first to show features that allow us to relate them to living groups. The Cambrian Explosion, for example, is the first time we are able to distinguish a chordate from an arthropod. This does not mean that the chordate or arthropod lineages evolved then, only that they then became recognizable as such. For a simple example, consider the turtle. A turtle is known by its shell. It would not be so obvious to recognize the ancestors of a living turtle before they developed a shell. This makes it more complicated to determine the ancestors. Because its ancestors would have lacked the diagnostic feature of a shell, ancestral turtles may be hard to recognize. In order to locate the remote ancestors of turtles, other, more subtle, features must be found.

Similarly, before the Cambrian Explosion, there were lots of "worms," now preserved as trace fossils (i.e., there is evidence of burrowing in the sediments). However, we cannot distinguish the chordate "worms" from the mollusk "worms" from the arthropod "worms" from the worm "worms." (Figure 3.4) Evolution predicts that the ancestor of all these groups was worm-like, but which worm evolved the notochord, and which the jointed appendages? If the animal does not have the typical diagnostic features of a known phylum, then we would be unable to place it and (by the rules of taxonomy) we would probably have to erect a new phylum for it. When paleontologists talk about the "sudden" origin of major animal "body plans," what is "sudden" is not the appearance of animals with a particular body plan, but the appearance of animals that we can recognize as having a particular body plan. Overall, however, the fossil record fits the pattern of evolution: we see evidence for worm-like bodies first, followed by variations on the worm theme.[3-41]

Creationist View

The theory of evolution claims that small, gradual changes over millions of years have transformed previously existing species into new ones. If evolution were true then numerous, intermediary species should have lived and left fossils behind. In fact, the number of fossils of intermediary species should be greater than that of the remains of present species of animals. For instance, many half-fish/half-reptile or half-ape/half-human fossils should have been found (See chapters 7, 8 & 16 for further discussion). Yet, more than 140 years of searching has not even revealed one transitional species. In contrast to evolutionists' claims, life has always appeared suddenly and fully formed in the fossil record.

Octopus   Starfish   Sea Urchins   Trilobite

Examples of Sea Creatures

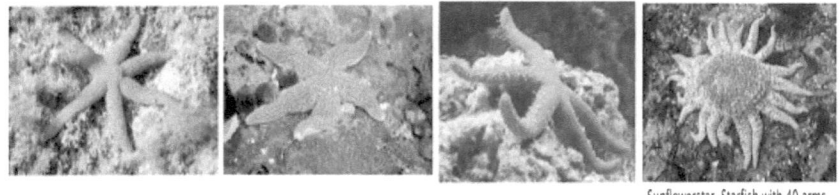

Sunflowerstar Starfish with 40 arms.

Examples of Starfish

The fossil record shows in the most primitive fossil strata, the Cambrian layer, a host of perfectly created sea creatures including octopi, starfish, sea urchins and trilobites. Below this stratum the fossil record is virtually blank, with only a few traces of plant algae in evidence. The primitive ancestors of these numerous and complex sea creatures do not appear in the fossil record. As the higher more recent fossil layers are examined, the same pattern is repeated. The remains go from invertebrates directly into fish, then to amphibians, later to reptiles and finally to mammals (Chapter 7). All these creatures, great and small, are perfectly formed without a fossil record of one species somehow evolving into another. The record also shows that many species, including insects, are the same throughout the entire geologic column (See section "The Geologic Column and the Strata Layers of the Earth").

Evolutionists claim that the fossil record indicates that all life evolved from a single cell and the more fossils that are unearthed, the more evidence is obtained to support molecules-to-man evolution. This claim is simply not true. The fossil record does not support the molecules-to-man evolution claim. Evolutionists basically believe that according to the fossil record, about 4 billion years ago, the first simple celled organisms (prokaryotes) appeared. The first simple organisms (prokaryotes) dominated until about 2 billion years ago (Archean period) when complex multi-cellular organisms appeared (eukaryotes), these include simple plants, fungi, and sponges. (Prokaryote and eukaryote cells are discussed in Chapter 5.) Then, about 530 million years ago, life exploded (the Cambrian explosion). In about a 5 million year period, most of the life, as we know it, suddenly appeared with no evolutionary ancestors. This means that entirely new and highly complex body plans appear in the fossil record with no ancestors. The fossil record literally goes from fungi and simple worms to the trilobite with an articulated body, complicated nervous system and compound eyes, fully formed and novel, in a relatively short period. (See Chapter 14 for further discussion of the trilobite eye.)[3-42]

Evolutionists expected to find fossils that showed stages through which one kind of animal or plant changed into a different kind. According to evolution, the boundaries between kinds should blur as we look further and further back into their fossil history. It should get more and more difficult, for example, to tell cats from dogs and then mammals from reptiles, land animals from water animals, and finally life from non-life.[3-43]

No transitional fossils, that is, no in-between form or common ancestor fossils have ever been found. There is no evidence that any creatures evolved from any other life forms. (See Chapters 7 & 8 for more details of the fossil evidence.) The sudden appearance of a multitude of complex and varied life forms at the very bottom of the fossil-rich portion of a geologic column is now routinely called the "Cambrian explosion." A great variety of basic body plans among the Cambrian fossils has been found along the Australian Great Barrier Reef. The Cambrian fossils are simply the descendants of the created (and corrupted) kinds first buried in the catastrophe of Noah's Global Flood. (See Chapter 4, section "The Strata Layers and a Global Flood".)[3-44]

The lowest strata level is called the Cambrian. Below this lowest of the fossil-bearing strata lies the Precambrian. The Cambrian Period stratum has invertebrate (non-backbone) animals, such as trilobites and brachiopods (marine animals having hard half shells that are hinged at the back end). These are both very complex little animals. In addition, many of our modern animals and plants are in that lowest level, just above the Precambrian. If these complex, multi-celled creatures found in the bottom of the Cambria Strata evolved from more ancient, simpler life forms, what creatures did they evolve from? Could they all have evolved from soft-shelled creatures that do not fossilize? Even though soft body parts do not normally fossilize, some do. Small bacterial fossils were found in the Archean and the Phanerozoic Eon strata. If small bacterial fossils can fossilize, soft-bodied creatures should also. If the complex, multi-celled creatures evolved from lower life forms, why is there no fossil evidence that shows a progression from primitive to more complex forms? In the very lowest fossil stratum, we find an abundance of complex plants and animals, with nothing to indicate that they evolved from anything lower or more ancient.[3-45]

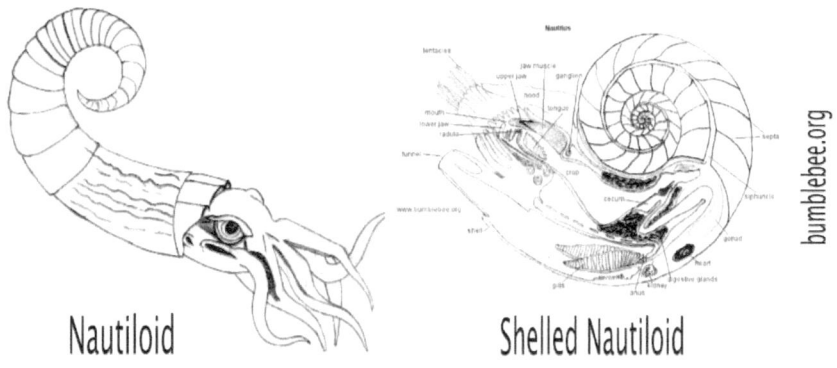

Nautiloid   Shelled Nautiloid   bumblebee.org

Examples of Nautiloids

At the Grand Canyon, as well as other sites around the world, the oldest or deepest layer to contain an abundance of fossil remains is called the "Cambrian geologic system." Rather than a few simple life forms that are difficult to classify, all types of complex life forms, such as clams, snails, lampshells, echinoderms (see examples below), and the most complex invertebrates, the nautiloids ("shelled squids"), with an eye that sees as humans do, and the trilobites, with their geometrically marvelous compound eyes, are found. (See Chapter 14 for further discussion of the trilobite eye.)[3-46]

Many evolutionists believe that life started slowly and gradually on Earth, but that the evidence rotted away since the early forms lacked the hard parts that make the best fossils. The Cambrian explosion, then, is simply an explosion of hard parts occurring simultaneously in many different animal groups. The hard-part hypothesis, however, contradicts the fossil evidence. Although it is rare, soft parts do preserve and have been found in Precambrian Period fossils. Evolutionists used to claim that geologists would have found the ancestors of Cambrian period life there if only the evidence hadn't rotted or been destroyed by heat in the rocks. That excuse is no longer valid. Although most Precambrian period rock is the igneous and metamorphic type unsuitable for fossil preservation, geologists have recently discovered great stretches of Precambrian period sedimentary rocks that could and should have preserved soft parts and the common ancestors of the diverse and complex Cambrian period life, that is, if any such evolutionary ancestors existed.[3-47]

Most people believe that the segmented earthworm is a simple creature, however, this is not true. The earthworm has five "hearts," a two-hemisphere brain, and a multi-organ digestive system (Figure 3.4). Fossils of these earthworms go back as far as the Cambrian explosion.[3-48]

Evolutionists believe that it would have taken 100 million years for a fish to evolve from an invertebrate. (See section "Evolution From Worm-like Chordates to Fish-like Creatures.") But there is no fossil evidence showing that this occurred. The evolutionists claim that it took perhaps 50 million years for a fish to evolve into an amphibian, but there are no transitional forms (Chapter 7). For example, not a single fossil with part fins and part feet has ever been found. And this is true between every major plant and animal kind. All higher categories of living things, such as complex invertebrates, fishes, amphibians, reptiles, flying reptiles, birds, bats, primates and man, appear abruptly (Chapters 7, 8 & 16).[3-49]

If evolution could possibly be true, then there should be millions upon millions of transitional creatures, all of which would be, at various steps in the imaginary evolutionary "trees," only part something, or part something else. There is no such thing as a part feather, a part scale or a part gill or a part lung (Chapter 7).[3-50]

There are very few fossils that could be considered intermediate. If intermediate creatures existed, their existence would have been short lived if they did not prove to be well fitted for survival, and therefore there would be very little fossil evidence. This would be especially true of a creature that was in the process of changing from fins to legs, legs to fins, or legs to wings. Natural selection would have eliminated the intermediate species, as they would have been impaired and therefore would not have been able to survive. (Natural selection is discussed in Chapter 9.)

Microscopic, single-celled, soft-bodied bacteria and algae, are in the fossil record. Since this is the case, it appears that transitional forms should also exist within the fossil record that would show a progression from those organisms to complex invertebrates. Evolutionists claim that all phyla share a common ancestor. (Phyla refer to the grouping together of classes of organisms that have the same body plan.)

A number of small bilateran fossils have been discovered in China that existed prior to the Cambrian period. Bilateran refers to an organism that has an axis of symmetry that goes right down the middle, including arthropoda (which includes insects) and chordata (which includes fish).

Examples of Echinoderms

The problem is that arthropoda and chordata, which have a two-fold symmetry, could not be ancestors of echinodermata because echinodermata (which includes starfish—see "Examples of Echinoderms" above), found near the start of the Cambrian period, have a five-fold symmetry. Evolutionists have no idea what the arthropod head could have evolved from nor do evolutionists have any idea how chordates (animals having a notochord, or dorsal stiffening rod) could have evolved.

The Cambrian Explosion occurred over a period of about 5-10 million years, about 530 million years ago. It has become known as Biology's Big Bang. During this period almost all living animals known today came into existence. Prior to the Cambrian Explosion there is evidence of jellyfish, sponges, and worms. Then suddenly, like a big bang, all of the major body plans show up in the fossil record. There is however, no evidence of transitional-like forms (also see Chapters 7 & 8).

If evolution were true, there would have to be far more of those halfway creatures than the "more advanced" ones. There would be thousands of creatures that would be part of one species and part of another. They would be even more bizarre creatures that would exist today if the "intermediate" species (which are missing in the fossil record) are missing because they were not so well "equipped to survive." Then it would naturally follow there would be far more of such creatures in the fossil record than the "equipped" or "fully developed" ones. The reason simply is because if they were not equipped to survive, they all died. And if they all died, there would be billions and billions of them, because there had to be a great deal more intermediate stages than the "final" or "well-developed" ones. Therefore, the fossil record would be reversed. Instead of perfectly formed fossils, looking, in most cases, exactly like life on Earth today, and no intermediate species, the fossils would abound with "intermediate" species; half this and half that, and there would be very few of the "developed species."[3-51] One of the most damaging evidences against evolution is the total absence of intermediate species, either living or dead.[3-52] The lack of evidence for evolution is clear from the fact that no one has ever seen it happen. If it were a real process, evolution should still be occurring, and there should be many "transitional" forms that we could observe. What we see instead is an array of distinct "kinds" of plants and animals with many varieties within each kind, but with very clear and— apparently—unbridgeable gaps between the kinds. Evolutionists claim that evolution goes too slowly for us to see it happening today. They used to claim that the real evidence for evolution was in the fossil record of the past, but the fact is that the billions of known fossils do not include a single unequivocal transitional form with transitional structures in the process of evolving. [3-52a]

Old Universe Progressive Creationist View

According to the fossil record, new life-forms increased through millions of years (approximately 3.5 and 3.86 billion years ago) before modern humans appeared (Flowering plants first appeared about 290 million years ago.). Prior to humans, frequent extinctions occurred, the introduction rate for new species matched or outstripped the extinction rate. (Dinosaurs appeared between 250 million to 65 million years ago.) Bipedal primate species appeared around five hundred thousand years to four million years ago. After humans began to appear (between 7,000 – 60,000 years ago), the appearance of new animal species immediately dropped to near zero. Botanists argue that speciation continues today (Chapter 9). Researchers have documented some distinguishable differentiation among plants, however, whether all of these new plants deserve distinct "species" labels remains debatable. Many scientists see most of the new plants merely as new breeds, or strains, of the old, rather than as truly new species. Research indicates that natural evolutionary processes, the observable microevolution, occurs at roughly the same rate today as it did before humans. Science offers no explanation for the sudden change in the speciation rate.[3-53]

In the next chapter (4) we will discuss whether the fossil layers developed quickly as a result of a global flood or whether the fossil layers developed gradually over millions of years by a slow, continuous process. This is important to our study because if the strata layers developed quickly as a result of a catastrophe, rather than by a slow gradual process, it could be an indication that the Earth is not as old as some believe. If the Earth is not billions of years old, there would not have been sufficient time for evolution to have occurred.

~~~

Chapter 4

How Did All the Strata Layers Form?

The Great Flood

And God said to Noah . . . Make yourself an ark of gopherwood; make rooms in the ark, and cover it inside and outside with pitch . . . The length of the ark shall be three hundred cubits[450 feet], its width fifty cubits[75 feet], and its height thirty cubits[45 feet] . . . You shall make it with lower, second, and third decks. Then the LORD said to Noah, "Come into the ark, you and all your household . . . You shall take with you seven each of every clean animal, a male and his female; two each of animals that are unclean, a male and his female; also seven each of birds of the air, male and female, to keep the species alive on the face of all the earth. And it came to pass . . . that the waters of the flood were on the earth . . . all the fountains of the great deep were broken up, and the windows of heaven were opened. And the rain was on the earth forty days and forty nights. The waters prevailed . . . and the mountains were covered. And all flesh died that moved on the earth: birds and cattle and beasts and every creeping thing that creeps on the earth, and every man . . . And the waters prevailed on the earth one hundred and fifty days.[4-1]

Introduction

Having discussed the fossil record in the different layers of strata of the earth in the previous chapter (3), in this chapter (4) we will discuss whether or not a global flood could have been the cause of the fossil record that exists. Was a global catastrophe ("The Genesis Noah Flood") responsible for the strata layers that exist in the Earth (catastrophism) or were the strata layers laid down gradually as the Earth aged over millions of years (uniformitarianism)? We will address these questions in this chapter.

The various views presented in this chapter are not exclusively of strict evolutionists and strict creationists. Some views are of those who believe in variations of the two extremes. See "Various View Descriptions" section in the back of this book for descriptions of the views presented in this chapter.

Were the Strata Layers the Result of a Global Flood?

Evolutionist View

The formation of certain kinds of rocks is particularly difficult to explain in short time scales (according to the global flood view). For example, sandstone requires four sequential processes, each taking a long time. **First**, hot magma cools to form quartz-rich rocks like granite, **second** the granite erodes over time to make quartz sand (a long slow process since granite is a hard rock that does not erode easily), **third** the sand is transported by wind or water to a place where it settles, and **fourth** the sand is compacted and chemically cemented into sandstone.

Some river valleys have carved through granite or other types of rock that are hard and difficult to erode. While flooding can explain the erosion of a valley if the rock is soft (like sandstone or limestone), granite and other hard rock types do not erode quickly unless they previously contained many fractures. A single year-long global Noah flood would have made only a small dent in unfractured hard rock; it would not have carved an entire river valley.

Some rocks indicate multiple floods or at least multiple wet periods. Samples of stratified conglomerate rock have been found in which smooth, rounded pebbles were embedded in the layers of fine-grained sediment. The pebbles themselves must have formed in an earlier wet period of sedimentation, dried and hardened into rock, and the rock broken apart into pebbles. Then in a later flood or streambed, water eroded the pebble to a smooth surface and it settled with other sediment to make the conglomerate rock. Rocks like this could not have formed during a single global flood.[4-2] Some stratified rock contains so many layers that it is hundreds of feet, even miles, deep. The sedimentary rocks in the central Appalachians of Pennsylvania are at least 40,000 feet thick. A single year-long global flood would not have eroded enough material to deposit layers that thick.[4-3]

Volcanic cones were discovered under grasslands in south central France. Since no human record or legend tells of volcanoes in that area, the last volcanic eruption must have been before human history. Upon close inspection, geologists were able to map multiple layers of lava flows, showing that the volcanoes in that area had erupted repeatedly, hardening after each eruption and forming additional structures. Evidence also shows significant water erosion taking place between the various volcanic eruptions. This area tells of a longer and more dynamic history than could be fit into a few thousand years, even with a flood.[4-4] The pattern that was discovered indicates that simpler life forms appear at the bottom of the layers of strata of the Earth, while more advanced life forms appear in most recent layers of strata.[4-5]

Old-earth Creationist View

Young-earth creationists realize that there are a great many fossils in the earth's crust. They even acknowledge that the Karroo Beds in Africa contain the remains of perhaps 800 billion vertebrates, yet they believe that all these animals died in the flood. They apparently have never thought about the enormous amount of animals that would have been in existence at one time. It has been calculated that if 800 billion animals were spread out over the entire land surface of the earth, there would have been an average of 21 animals per acre. The Karroo fossils range from lizards to animals the size of cows, with the average size about that of a fox. And this is only one fossil area. Other fossil areas have not even been included in this calculation. If all the other fossils of the world were included in this calculation, the number of animals would be much greater. Even if a very conservative figure of 2100 animals per acre were used, this would allow each animal a plot that would be the size of a hearthrug. This is certainly not enough space for an animal to survive. There would not have been enough grass for the grass-eating animals to survive. There would certainly not have been enough room for the animals to graze and walk around.[4-6]

Young-earth Creationist View

Evolutionists view the fossil evidence as proof that simpler life forms evolved into more complex forms because the simpler creatures are found in the lower layers while the more complex creatures appear higher. The first creatures were buried in greatest abundance in the rising Flood waters would be the heavy-shelled, bottom-dwelling sea creatures, and these would be followed successively by near shore forms and swimmers, then lowland plants and animals, and finally upland forms, with sea creatures found in all the systems of the geologic column as the waters finally covered everything. When the mountains rose and the valleys sank down (Psalm 104:8) at the end of the Flood, the continents were covered with layers of fossils formed as stages in the burial of eco-sedimentary zones during the catastrophe of Noah's Flood. Sediment does build up slowly at the mouths of rivers, such as the Mississippi delta, but slow sediment build-up could not possibly produce such widespread deposits, such broadly consistent sedimentary as is seen in the Morrison Formation, the St. Peter's formation, and also in the Karroo Beds in Africa.[4-7]

Evolutionists claim that fossil-bearing rocks were largely laid down in local floods and/or by rivers dumping sediments into lakes or seas. Those processes do build up sediment layers; the Mississippi River, for example, is continuing to build up its delta. When the Mississippi River is flowing rapidly, gravel is carried relatively far. When the flowing slows, sand is dumped where gravel was. These slow and gradual processes produce "lumpy" sediment layers that thicken and thin over short distances and contain virtually no fossils.

The Precambrian sedimentary rocks in the inner gorge probably do represent sediment laid down somewhat slowly and gradually. Like the Mississippi Delta deposits, these units thicken and thin, disappear and appear, over short distances, and they contain very few fossils. They do not have the layer-cake appearance, that is, they do not have deep and wide horizontal bands of fossil-rich rocks, characteristics of broad and rapid flood deposits. Instead, they have the swirl-cake appearance, with lumps of fossil-poor rock, like the sediment layers that are currently being produced at the mouth of the Mississippi River.[4-8]

Many people believe that the biblical account of the flood states that the waters covered the entire earth almost instantly, stirred everything up, and then suddenly dumped it all, however, this is not true. According to the biblical record (Genesis 7-9), Noah was in the ark for over a year. It was about five months (150 days) before "all the high mountains under the whole heaven" were covered (Genesis 7:19), and it took several more months (150 days) for the water to subside "The water receded steadily from the earth" (Genesis 8:3) at the end of the flood. As the Flood waters "slowly" rose over the earth, plants and animals were buried in a sort of ecologic series: sea-bottom creatures, near-shore forms, lowland plants and animals, then upland (with sea creatures deposited from bottom to top), as the sea eventually covered everything. The fossil-bearing rocks were laid down within minutes or months between the layers, not millions of years, as some suppose.[4-9]

One of the most startling facts about the sedimentary strata around the world is the vast quantities of fossils they contain. Without a worldwide Flood, it would be impossible for such huge amounts of plants and animals to have been rapidly buried. And without rapid burial they could not have fossilized. The Cumberland Bone Cave in Maryland was found to contain remains of dozens of mammal species together with reptiles and birds. The types represented include creatures native to Arctic, temperature, and tropical zones, and both dry and moist habitats; yet the fossils are all mixed together in one cave.[4-10]

When a fish dies its body normally floats on the surface or sinks to the bottom and is devoured rather quickly, actually in a matter of hours, by other fish. However, the fossil fish found in sedimentary rocks is very often preserved with all its bones intact. Entire shoals of fish over large areas, numbering billions of specimens, are found in a state of agony, but with no mark of a scavenger's attack.[4-11] (See section "How Fossils Develop" for further discussion of the formation of fossils.)

Examples That Disprove Uniformitarianism

Young-earth Creationist View

Uniformitarianism, as mentioned previously, is a theory developed by James Hutton and Charles Lyell in 1830. This theory is in opposition to the theory of Catastrophism, which states that a worldwide catastrophe (such as a global flood) caused the strata layers. The theory of Uniformitarianism stipulates that all geologic phenomena may be explained as the result of existing forces having operated uniformly from the origin of the earth to the present time.[4-12] According to Uniformitarianism, the geologic appearances are due to steady, continuous and uniform processes, rather than a result of a singular catastrophic event (such as a global flood). Catastrophism versus uniformitarianism describes alternative process which could be primarily responsible for the formation the geological strata and embedded fossils. Catastrophism was accepted as the only possible explanation until the about the 18th century (1700s). Catastrophism specifies that the geologic rock strata were primarily a result of catastrophes like the worldwide flood of Noah.

The theory of uniformitarianism stipulates that the same slow process that we see today is the process that is responsible for the formation of all the geological rock strata. Since deposition with the uniformitarianism theory was so slow, long eons of time were required. This meant that the current biblical beliefs at that time that the earth was young and that there was a worldwide flood of Noah that was responsible for the geologic conditions of the earth, were discredited. The theory of uniformitarianism, however, is not true. The Mount St. Helens eruption and subsequent erosion illustrates that rapid deposition and rapid canyon erosion is a fact. It does not take years to form. It is a scientific fact that life forms cannot be fossilized unless buried rapidly.[4-13]

Mount St. Helens' Eruption 5-18-1980

The uniformitarian assumption that millions of years of geological work (extrapolating from present, slow, natural processes) be required to explain structures such as the American Grand Canyon for instance, is called into serious question by the explosion of Mount St. Helens in the state of Washington on May 18, 1980. Massive energy equivalent to 20 million tons of TNT (trinitrotoluene—dynamite) destroyed 400 square kilometers (154.44 square miles) of forest in six minutes, changing the face of the mountain and digging out depths of earth and rock, leaving formations not unlike parts of the larger Grand Canyon. Recent studies of the Mount St. Helens phenomenon indicate that if attempts were made to date these structures (which were formed in 1980) on the basis of uniformitarian theory, millions of years of formation time would be necessarily claimed.[4-14] At the Mount St. Helens volcano site, geologists documented that up to 400 feet of new strata have been formed at the volcano since the first eruption in 1980. These deposits originated from air fall, pyroclastic flows, landslides, and stream water. This illustrates that laminated deposits (thin layers) can be produced quickly. It was previously thought that it took many years to form laminated strata, possibly one layer laid down each year. This view has been proven wrong, as one deposit at Mount St. Helens resulted in the creation of a 25-foot-thick finely laminated unit in a matter of hours.[4-14a]

If one looks at the Grand Canyon, the Muav layer supposedly represents evolution stage 1, while the Redwall layer represents the Mississippian or lower Carboniferous strata period, which is evolution stage 5 (The strata layers are described in Chapter 3, section "The Strata Layers of the Earth"). If the Grand Canyon represents the stages in evolution laid out, evolutionary stages 2 (Ordovician strata period), 3 (Silurian strata period), and 4 (Devonian strata period) are missing.

Evolutionists agree that there are 150 million years of strata missing from the Grand Canyon, however, they claim that stages 2, 3, and 4 were present at one time but these stages were uplifted and eroded away; then stage 5 rock (Mississippian strata period—Redwall layer) was laid down on top of stage 1 (Cambrian strata period—Muav layer). It is possible that erosion may have destroyed some of the strata layers, however, it has been determined that there was not enough evidence of erosion, at least on any sufficient scale.

When a rock layer is eroded slowly and gradually by streams and rivers, an irregular surface is normally produced. When sediment later accumulates on this surface and hardens, a wavy contact line is produced; and often old streambeds may be identified along its surface. This is not what is found at the Redwall layer (Mississippian strata period)/Muav layer (Cambrian strata period) contact. Over hundreds of miles of exposure in and out of various side canyons, the two rock layers are in smooth, horizontal contact. There are occasional small erosion dips, but the regional view looks like one rock layer that was deposited directly on top of the other with very little time break. The evidence does not appear to reflect evolutionary time periods. The evidence that is currently available indicates that the horizontal rock layers at the Grand Canyon were formed rapidly, not by a lot of time, but by a lot of water.[4-15]

The Channelled Scablands of eastern Washington State, an area of 15,000 square miles, have also been studied. It looks as if a giant, braided stream cut channels up to 900 feet deep in hard basaltic lava (much harder to cut than most of the Grand Canyon layers). Research has determined that a tongue of glacial ice blocked off what is now called the Columbia River near Spokane, Washington, damming up a huge body of water called the glacial Lake Missoula. Then the ice dam broke. The drainage from that lake cut the essential features of those channels 900 feet deep over 15,000 square miles has been estimated to have occurred in a day or two, not two million years, as evolutionists once supposed.[4-16]

How Fossils Develop

Young-earth Creationist View

The fossil record indicates that there was a global, catastrophic flood. The fossil records indicate that a large-scale fossilization is not occurring anywhere in the world today. When a fish dies, it does not sink to the bottom and become a fossil. Instead, it either decomposes or is destroyed by scavengers. In contrast to the virtual lack of fossilization transpiring today, there is an incredible amount of fossilization that occurred sometime in the past. The billions and billions of fossils we find preserved in the fossil record simply could not have been formed by processes observable in the world today. Such preservation is very abnormal, the exception, and not the rule. Our global fossil record, therefore, verifies a non-typical worldwide, cataclysmic, hydraulic event that is called the Noah Flood.[4-17]

Some believe that it takes millions of years and tremendous heat and pressure to turn sediments (such as sand, lime, or clay) into rock (such as sandstone, or shale). Time, heat, and pressure can and do alter the properties of rock, but the initial formation of most rocks is quite rapid. Many believe that fossilization takes hundreds of years to occur. In reality, this is not true. If a plant or animal just dies and falls to the ground or into the water, it is quickly broken up and decomposed by scavengers, wind and water currents, even sunlight. Most fossils are formed when a plant or animal is quickly and deeply buried, out of reach of scavengers and currents, usually in mud, lime, or sand sediment rich in cementing minerals that harden and preserve at least parts of the dead creatures. The ideal conditions for forming most fossils and fossil-bearing rock layers are flood conditions. Evolutionists believe there were many little floods while creationists believe there was just one big flood (Noah's Flood).[4-18]

Fossils usually refer to the petrified remains of animals that died a long time ago. It is often claimed that animals that have died fall to the ground and are slowly buried by the accumulation of sediment and fossilized in the process. This is not a reasonable assumption, nor is it supported by experimental observation. When an animal or plant dies, its remains are quickly eaten by scavengers and decomposed by bacteria, etc. Any remains are also affected by weather. Fish in the sea that have died usually float to the surface and are soon eaten (as opposed to settling down on the sea floor, waiting to be slowly buried by sediment and fossilized.) How then, should we expect a fossil to be formed? The most reasonable explanation involves a catastrophe.

To get such a fossil, you would have to suddenly and quickly bury the animal under tons of sediment, so that it would be isolated from scavengers and excluded from the effects of weather. Only then should you expect the petrification process to work. Also, these fossils in and of themselves do not give any indication of the age of the animals that they represent, for they are just impressions of once-living organisms that have died. Scientists who are not set on ignoring the biblical record generally agree that most fossils are most likely the result of the worldwide flood that is described in the Genesis record, with its cataclysmic geological implications.

Burial order does not imply ancestry. In many places in the world, you can find stratified layers of rock in which are embedded various fossils. The fossils found in each layer make up an approximately ordered sequence, from the fish in the lowest layers to the land-dwelling mammals in the highest.[4-19]

Polystrates

Young-earth Creationist View

Polystrate tree fossils (fossils of single organisms, such as a tree trunk, that extend through more than one geological stratum) have been found that pass vertically or diagonally through many layers of the Earth, which evolutionists claim represent different time periods. Polystrate fossils were found randomly tossed around, sometimes upside down, penetrating through many layers of the Earth. Many polystrates are found in Germany, France, the British Isles, Nova Scotia, California, and some eastern states. It is difficult for evolutionists to explain how huge falling trees could stay suspended for billions of years while the earth takes eons of time to creep up and form another layer around them.[4-20] Evolutionists believe that petrified wood proves that the earth has been here for millions of years. But there is clear evidence that petrification of wood can take place in a relatively short period of time. If petrification did not occur fairly rapidly, the tree would have rotted away before it could complete the petrification process, whereby silicates replace the wood cells.[4-21] Creation geologists have determined that the floating log mat and sunken logs at Spirit Lake in Washington State that resulted during Mount St. Helen's eruption in 1980, represents the type of devastation that occurred during the Great Flood. The sunken upright trees are used to explain the numerous polystrate trees that are often found extending through coal beds. If this log mat had been rapidly buried by subsequent sediments, it is likely that the trees that ended up at the bottom of the lake would have turned into a coal bed. Coal beds do not need vast amounts of time to form. They just need the right conditions.[4-21a]

The problem with evolutionary thinking is that fossils of various "evolutionary periods" are not consistently found in the proper strata. In many places, fossils representing "more recent" life forms are found in strata far below their supposed ancestors. The existence of polystratic fossils (fossil life forms that are found buried vertically through several layers of strata, such as trees and long cone-shaped mollusks) also disproves the evolution story, since this would require that the organic remains of such life forms remain intact and unfossilized for millions of years in place above the ground, awaiting the deposition of successive layers of strata.

For the evolutionist, the mere existence of polystrates and fossils of "recent" life forms below the fossils of their "ancestors" disproves their hypothesis. Evolutionists cannot explain polystrates at all, and they resort to theories of "overthrusting" to explain how older strata ends up over newer strata, even though such a phenomena has never been observed, and even though they cannot explain where the geologic forces should originate. Overthrust theories also demonstrate circular reasoning as evolutionists try to use the geologic column to support their theory, then they use their theory to explain away inconsistencies in the geologic column. However, the creationist acknowledges that the ordering would be approximate, based on the chaotic nature of the flood, and that different strata models would be found in different parts of the world, based upon the local ecosystem and what animals dwelt in it. And fossils buried through several layers of strata would obviously not be a problem.[4-22]

As just mentioned, polystrate fossils are fossilized trees that extend stratigraphically upward through several layers of rock. The roots are usually found in a coal seam, and the overlying deposits included bedded shale and thin carbon-rich layers. Evolutionists who believe in any form of uniformitarianism would believe that it took many, many years to deposit this sequence of layers (much longer than it takes for a tree to grow and eventually die and decay), yet "polystrate trees" are a direct contradiction to the evolutionary claims. According to uniformitarianism, many years are required for a thick layer of peat to accumulate in a swampy environment. This type of location is quite different from the marine environment in which tiny shale-sized particles are deposited. Over "millions and millions of years" of heat and pressure generated by the subsequently deposited overlying marine sediments, the peat is thought to have metamorphosed into coal. These "polystrate trees" were mature, yet they could not have grown in the location where the surrounding shale was deposited, since trees do not live long under the sea. Furthermore, the time required for shaley sediments to accumulate must be added to the tree's lifespan, as must the time to deeply bury the coal precursor and create the pressure to generate enough heat to alter the peat into coal.

No scenario possible today could account for this sequence of events if evolution's interpretation of earth history is true (that is, a slow, gradual process called uniformitarianism). An interesting event happened at Mount St. Helens in 1980, when an eruption toppled a standing forest. The tree trunks were deposited in Spirit Lake. After a few years of waterlogging, the trunks sunk roots down, in life's position (vertically) but not life's location (in a forest) but rather ended up in a lake. Today there are tens of thousands of upright trees standing on the bottom of the lake. They are being engulfed by fine particles of volcanic ash and clay, and if the underlying organic layer of bark were heated by a future eruption, it would likely metamorphose rapidly into coal and become "polystrate fossils." The eruption at Mount St. Helens demonstrates the effects of dynamic processes, such as a volcano eruption or a flood. The Mount St. Helens eruption provided a model for deciphering unseen past geologic cataclysms, and produced effects that before were unknown. Our understanding of possible events during the great Flood of Noah's day was substantially expanded, including that rapid deposition of sediments and burial of fossils could be expected during such a deluge.[4-23]

Old-earth Creationist View

Young-earth creationists point out that sometimes a fossil tree trunk is found projecting through two or more coal seams, and that sometimes a coal seam a stratum layer will fork into two seams separated vertically by a layer of rock. Although such 'polystrate fossils' are not rare, they affect only a relatively small number, perhaps one per cent, of the world's many coal seams. They show that a small minority of coal seams must have been formed in some exceptional fashion that is not yet understood, although research into the problem continues. Large local floods may have been involved, though this is not yet proved. The other ninety-nine percent of the coal seams present no great problem to the orthodox geologist. These coal seams fit quite well in the conventional explanation, of vegetation growing in a tropical swamp and then becoming deeply buried and, eventually, metamorphosed into coal.

But while conventional geologists cannot yet explain a few exceptional coal seams, young-earth creationists have a far greater problem. They are not able to explain the vast amount of coal in the earth. The first problem is, as with animal fossils, the sheer quantity existing. Well over a million million tons of coal have already been located. Nobody knows how much remains to be discovered, but one recent estimate calculates that the total world reserves of coal is 15.3 million million metric tons (approximately 16.9 million million American tons (15.3 x 1.1023). The most pessimistic person would at least agree that there are at least 5 million million tons, if you include seams too narrow to be worth mining. That would equate to be 65 pounds of coal for every square yard of the earth's land surface. The question would then be, where did all the vegetation come from to produce all that coal?

Considering from the fossil evidence, most coal is produced from the remains of large fern-like plants. Also, most likely the earth was not covered with just one type of vegetation in Noah's day. Fossil plants and animals show that there always were many different kinds of habitat in the past, just as there are now. Most wood floats and only becomes water-logged when it has been lying in water for years. The Flood could hardly be expected to bury more than a fraction of the vegetation of Noah's day. Much of it would have ended up on the surface of the water and decomposed. When these factors are considered, it is evident that the Flood could not possibly have produced as much coal as there is. Another thought that needs to be considered is that coal seams (strata layers) often occur in groups, one above another, with layers of rock between. How could the Flood have produced these series of coal seams?[4-24]

How Do Coal Seams Develop?

Evolutionist View

"In general, a high quality black coal seam would take millions of years, if not hundreds of millions of years to form" [4-24a]

The process of coal formation is still taking place today. The precursor to coal is called peat, and that is just uncompressed plant matter. Peat accumulates in wet swampy environments known as mires, and that process is taking place today in areas such as Indonesia and even the Antiplano in the Andes. Mires are swamps with trees growing in them, swamps with reeds, stagnant water into which pollen and plant matter fall, and coastal lagoons. Peat can even form in the highlands in rain-fed or glacier-fed lakes in mountain ranges. However, peat accumulates very slowly at about one millimeter (0.00328084 feet) a year on average, although it can happen faster, up to 2 to 3 millimeters (0.07874 - 0.11811 inches) per year in the tropics. At that rate, it would take about 12,000-60,000 years to accumulate enough peat to form a three-meter (9.8425 feet) coal seam.

The transformation from peat to coal takes even longer. It generally starts with burial of the peat by other sediments as a result of a volcanic eruption, migration of a river or a change in sea level. The pressure of overlying sediment squeezes the water out and causes the peat to compress. The thickness of the peat will be decreased by about ten to one during this process. The transformation from a plant substance to a metamorphic rock really starts once the peat is buried beneath 3—4 kilometers (9,842 – 13,123 feet) of sediment. At this depth, with an average rate of temperature increase of 30° Celsius (86° Fahrenheit) per kilometer, the temperature rises to over 100° Celsius (212° Fahrenheit) and sets off chemical reactions that transform the material into coal.

The chemical reactions release volatiles, which helps to compress the peat even more and it changes from being a plant substance, like lignin or cellulose, to a geopolymer that contains concentrated carbon. It's very different from peat or plant matter. The amount of transformation from peat to coal is described by a coal's rank. Brown coal and lignite are the lowest rank, then bituminous or black coal. As the temperature and pressure rises even more it changes to anthracite. And eventually some of the earliest coals that would have formed have been metamorphosed into graphite.

The formation of coal seams really kicked off with the diversification of land-based plants around 350 million years ago (mya). That was pretty much from the end of the Devonian period (408 – 360 mya) into the Carboniferous period (300 – 360 mya). Algae was around long before then in shallow seas, so there are coals made completely of algae that date back earlier than the Carboniferous. The Carboniferous period (300-360 mya) saw the evolution of tall lycopod trees that accelerated the rate at which peat could be formed in tropical equatorial mires. High sea levels and a warmer climate also encouraged coal formation, by extending the area of coastal mires and other wetlands.

Coal formation underwent a drastic change at the end of the Carboniferous period (around 300 mya). There was a global ice age at the end of the Carboniferous period, and the continents were drifting to new locations, so coal accumulation occurred in cold temperate places, closer to the poles. The cold favored a new type of plant called Glossopteris, dominated by gnarly little trees which lost their leaves in winter.

Plants grow more slowly in the cold, so this could have slowed peat accumulation, but frozen plant matter is less easily decayed and better preserved. It would be hard to distinguish any change in peat accumulation rate due to white rot fungi from the effects that climate change were having on peat. Peat continued to accumulate strongly throughout the Permian period (245-300 mya), when coalfields in the Hunter, Newcastle and Illawarra were forming. However it paused for a period at the Permian period (286 – 248 mya)—Triassic period (248 – 213 mya) boundary.

Coal formation stopped for about 15 million years at the end of the Permian period (286 – 248 mya), but this was due to a global extinction which wiped out most land plants. About 90 per cent of all species on Earth were wiped out at this time. Once the plants recovered, coal formation began again. This started with the recovery of spore-generated ferns, and a global "fern-spike". Land plants were unusually dominated by ferns until other plants regenerated. While the coal-forming process is still happening today, we interrupt that process when we mine coal, particularly of lower rank. If lower rank brown coal were left for a few more million years it would turn into black coal. Coal takes longer to form than any other rock type. Ironically, warming of the Earth's climate may increase the number of swampy coastal environments that are perfect for coal formation. But these coal seams won't be ready for a few million years.[4-24b]

Creationist View

Evolutionists and old-universe creationists believe that the material in coal beds accumulated over millions of years in swamp environments. Some geologists have claimed that even if all the vegetation on earth was suddenly converted to coal this would make a coal deposit only 1-3% of the known coal reserves on earth. Therefore at least 33 Noah's Floods would be needed, staggered in time, to generate our known coal beds. Therefore, as evolutionary (and old-universe creationist) geologists insist, a single Noah's Flood cannot be the cause of coal formation. This argument is based on valid estimates of the volume of vegetation currently on today's land surfaces. But it assumes that at least 12 meters (39.37 feet) of vegetation are needed to produce one meter (3.28 feet) of coal. Modern research shows that less than two meters (6.56 feet) of vegetation are needed to make one meter (3.28

feet) of coal. Some observations made by coal geologists working in mines (e.g. the compaction of coal around clay 'balls' included in some coal beds) suggest that the compaction ratio is probably much less than 2:1 and more likely very close to 1:1. These observations destroy this objection to coal bed formation during Noah's Flood, since instead of today's vegetation volume only compacting down to 1-3% of known coal reserves, today's vegetation volume would compact down to at least 30% of the known coal reserves. But where did the other 70% of the coal reserves come from?

It is possible that 60% of the coal reserves came from lands that no longer supports plant life. Evolutionists (as well as the old-universe creationists) base their calculations on the volume of vegetation on today's land surface and they ignore the fact that 60% of today's land surface is covered by deserts or only sparse vegetation. Also, there are the vast icy wastes of Antarctica beneath which are rock layers containing thick coal beds. So if all of today's land surface was covered with the lush vegetation suggested by Antarctica's coal beds, under the influence of a global sub-tropical greenhouse effect before Noah's Flood, then the volume of such vegetation on today's land surface would be sufficient to produce at least another 50% of the known coal reserves. The remaining 10% of the current coal reserves may have come from areas that were once covered by water.

Evolutionists (and old-universe Creationists) assume that the area of land surface available for vegetation growth has always been the same. It appears that when the earth was first created the earth was completely covered by water (Genesis 1:2, 6). Then, according to Genesis 1:9-10, God gathered the waters into one place so that dry land would appear. So, initially, there may have been one sea surrounded by one large land mass (Pangaea). It is therefore likely that there was at least twice as much land area available for vegetation growth in the pre-Flood world compared with today's world (i.e. at least 60% land versus 40% sea in the pre-Flood world compared with today's roughly 30% land verses 70% oceans). So, if this vast land area was under lush vegetation, then we can account for 100% of the known coal reserves.[4-24c]

No-one alive today has ever observed the process of coal formation. Evolutionary scientists develop possible explanations based on what they think may have happened. The presence of such great quantities of buried vegetation located beneath the earth's surface is easily explained by examining the effects of a global flood. By analyzing the contents beneath the earth's surface, the effects would be consistent with the devastation of Noah's Flood, which would have uprooted the entire pre-Flood biosphere and buried it with huge quantities of sand and mud. However, evolutionist (and old-universe creationist) geologists, who do not believe that a Global Flood (catastrophism view) could be the cause of coal seams beneath the earth's surface, try to explain everything by slow and gradual processes over millions of years (uniformitarianism view). For the brown coal deposits, evolutionist (and old-universe creationist) geologists believe that the vegetation accumulated as peat in a swamp during ideal climatic and geologic conditions. Evolutionist (and old-universe creationist) geologists believe the swamps formed on floodplains near the coast, which were slowly sinking and eventually inundated by the ocean.

However, the evidence indicates that these brown coal deposits did not accumulate in a peat bog or a swamp. **First**, there is no sign of soil under the coal, as there would be if the vegetation grew and accumulated in a swamp. Instead, the coal rests on a thick layer of clay and there is a 'knife edge' contact between the clay and the coal. This kaolin clay (*a type of clay primarily made up of kaolinite, which is a mineral that is found all over the earth. It's also sometimes called white clay or China*) is so pure that it could be used for high-class pottery. **Second**, there are no roots penetrating the clay. **Third**, there are a number of distinct ash layers that run horizontally through the coal. If the vegetation had grown in a swamp, these distinct ash layers would not be there. After each volcanic eruption, the volcanic texture of the ash would have been obliterated when the swamp plants recolonized the ash, turning it into soil. Not only is there no soil, but the vegetation found in the coal is not the kind that grows in swamps today. Instead, it is mostly the kind that is found in mountain rainforests. The kinds of plants that make up the coal did not grow in a swamp on a floodplain. **Fourth**, large broken tree trunks are found randomly distributed through the coal in many different orientations. Evolutionist and old-universe geologists are not able to explain how such large trees could have obtained an adequate root-hold in the 'very soft, organic medium', and how the roots could have breathed under water. These large trunks are not consistent with slow accumulation over thousands and thousands of years in a swamp, but indicate fierce and rapid transportation by water.

Obviously an environment conducive to prolific growth is needed for the formation of coal, but growth alone is not enough. Uniformitarian geologists believe that a mechanism is needed to conserve the vegetation for tens (or even hundreds) of thousands of years, until enough material has accumulated to slowly be converted to coal. Oxygen must be kept out to prevent decomposition, hence the need for stagnant water—that is, a swamp. These are the only places where vegetation accumulates today. In all other environments vegetation decomposes as quickly as it is produced. But how would such great thicknesses of peat accumulate in a swamp? Very precise geologic conditions would have been called for; namely that the swamp must have subsided slowly, at exactly the same rate as the vegetation was accumulating. If the vegetation had sunk too fast, the water would have drowned the plants, and growth would have been stopped. If the vegetation had sunk too slowly, the organic debris would have emerged above the water and decomposed. And these precise geologic conditions would be needed for tens, or hundreds of thousands of years. Geologically, the idea that thick seams of brown coal accumulated in a swamp is doubtful.

Not only does the swamp model have problems explaining the seam thickness, but it is also difficult to envisage how vegetation could have accumulated over such a large geographical area. We do not see peat swamps covering such extensive geographic areas today. Rather, peat only accumulates in relatively small, isolated swamps. Contrary to what some people believe, it does not take millions of years to produce coal and oil. Once we understand the conditions needed, it is clear that the 4,300 years since Noah's Flood is ample time for all the buried vegetation to have transformed into brown coal.

The presence of coal points to a global catastrophe, because huge quantities of vegetation have been uprooted, transported, and buried by water under great volumes of sediment all over the world. The Global Flood is, most likely, what caused coal to form. Most of the types of plants in the Latrobe Coal Measures still grow today. Although those who believe in the slow-and-gradual theory (uniformitarian view) that plants were fossilized in a swamp environment, the overwhelming majority of plants are not swamp-tolerant.[4-24d]

Two different theories have been developed to explain the formation of coal seams, 1) the autochthonous (growth in place) theory and 2) the allochthonous (water-transported) theory. The prevailing view among evolutionists supports the autochthonous (growth in place) theory, by a slow process. Young-universe Creationists believe that the global Noah Flood caused the formation of the coal seams. One may question whether plant remains, even if water-laid in the manner supposed in the allochthonous (water-transported) theory, could have been metamorphosed into coal in the relatively brief period time since the global Noah Flood. Somehow, the prevailing view is that enormous ages of time are necessary for coal to form, even after the materials had been deposited. This opinion is speculative, however, since the details of the carbonization process are not yet perfectly understood.

The various sources of energy necessary for the metamorphic processes that convert plant residues into high-rank coals leads to the conclusion that neither bacteria, hydrostatic head, nor localized high temperatures were the geologically active agencies. Thus, although bacterial activity, pressure and temperature have been generally assumed as the agents for converting peat-bog residues into coals, recent studies have demonstrated their inability to turn vegetation into coal. Apparently the most likely agent is the application of shearing forces, and these would have been quite high during the post-flood period of tectonic re-adjustment. Nor would they require long ages to produce coal.[4-24e]

Coal is the end product of the metamorphism of tremendous qualities of plant remains under the action of temperature, pressure and time. Coal has been found throughout the geologic column (discussed in Chapter 3) and in all parts of the world, even in Antarctica. Many coal fields contain great numbers of coal-bearing strata, interbedded with strata of other materials, each coal seam having a thickness which may vary from a few inches to several feet. And each foot of coal must represent many feet (just how many, no one knows) of plant remains, so that the coal measures testify of the former existence of almost unimaginably massive accumulations of buried plants.

As mentioned earlier, coal geologists are divided into two groups, **(1)** those favoring the autochthonous (growth-in-place) theory of coal origin and **(2)** those favoring the allochthonous (water transportation and deposition) theory. Those to hold to a consistent uniformitarianism view, of course, tend to favor the autochthonous (growth-in-place) theory and attempts to picture the coal-forming processes in terms of modern peat deposits forming under swamplands. The great thickness of the coal beds is contributed to the autochthonous (growth-in-place) theory by assuming a continuous subsidence of land more or less keeping up with the slow accumulation of plant remains. The interbedded strata of non-carbonaceous deposits are explained by alternating marine transgressions and resulting periods of sediment deposition. A wide variety of types of these intervening sediments have been noted and attempts made to explain them in terms of "cyclothems" (discussed shortly) or recurring cycles of deposition of different kinds of materials corresponding to the different stages of marine transgression and regression. The exact cycle, however, found at any one locality is always different from the cycle at any other locality.

If the autochthonous (growth-in-place) theory of coal bed origin is correct, it must have been accomplished by an amazing sequence of circumstances. One or two or three coal seams formed by alternate stages of **(1)** swamp growth, **(2)** peat accumulation, **(3)** marine transgression and emergence, etc., might be believable, but the assertion that this cycle was repeated scores of times on the same spot, over a period of perhaps millions of years, is questionable. And yet there are many sites where 75 or more such coal seams are found. Some seams, too, are up to 30 or 40 feet in thickness, representing perhaps an accumulation of 300 or 400 feet of plant remains for the one seam. This theory, which is purportedly uniformitarian (slow and steady) in essence, is doubtful, as there is no modern parallel for any of its major features. The peat-bog theory constitutes a very weak attempt to identify a modern parallel, but it will hardly suffice.

Except for uniformist (slow and steady growth-in-place process) preconceptions, it would seem that the actual physical evidence of the coal beds strongly favors the theory that the plant accumulations had been washed into place (by a violent global flood). The coal seams are almost universally found in stratified deposits. The non-carbonaceous sediments intervening between the coal seams are always said to have been water-deposited, and it would seem that consistency alone would warrant the conclusion that the coal seams were likewise water-borne and deposited. The great thickness of some seams and the great numbers of seams in a given locality also constitute convincing evidence of rapid and cyclic currents carrying and depositing heavy burdens of organic material.

The most important reason given for believing the coal seams to have been deposited in their original place with their root in the soil below the coal swamp (rather than after aqueous transport) is the evidence of the so-called stigmaria (discussed shortly). These are root-like fossils that project out under the coal seam into the "underclay" (discussed shortly) and have been interpreted as the roots of the trees which formerly grew in the peat-bog. This is thought to prove that the vegetation actually grew in the place where its remains now rest. However, other explanations are possible. It is conceivable that they were rhizomes (modified stems running underground horizontally) rather than true roots and were thus able to develop under water, independently of the plants to which they are attached. Or they may have simply been transported along with the plants and deposited together with them. [4-24f]

Stigmaria

The most important fossil relating to the controversy over the formation of coal is *Stigmaria,* a fossil root or rhizome. *Stigmaria* is frequently found in strata below coal seams and is commonly associated with upright trees. *Stigmaria* studied nearly 140 years ago by Charles Lyell and J.W. Dawson in the Carboniferous coal sequence of Nova Scotia was considered to provide unambiguous proof of the autochthonous (growth-in-place) theory. Many modern geologists still insist that Stigmaria represents a root that is situated in its original position or place in the soil below the coal swamp.[4-24g]

Related to the nature of the Stigmaria has been the significance of the "underclays" (discussed shortly), which are supposed to be the fossil soils in which the coal-swamp vegetation grew. However, recent careful studies on the chemical and physiological nature of the underclays show this to be highly improbable. The relationships between underclays and coals indicate that the underclays formed before the coals were deposited. Furthermore, the lack of a soil profile similar to modern soils and similarity of the mineralogy of all rock types below the coals indicate that underclay materials were essentially as they were transported into the basin. The underclays were probably deposited in a loose, hydrous, flocculated state, and slickensides developed during compaction.[4-24h]

Underclay

One of the most interesting portions of the *cyclothem* (discussed shortly) is the underclay. The non-bedded, plastic layer of clay often underlies the coal stratum and is considered by many geologists to be a fossil soil on which the swamp existed. The presence of underclay, especially when it possesses *Stigmaria,* is often claimed to be the presumed explanation of the evidence for the autochthonous (growth in place) origin of coal-forming plants. Modern research, however, has cast some doubt on the fossil soil interpretation of underclays. No soil profile similar to modern soils is evident in underclays. Some of the minerals found in the underclay are not the type which would be expected in a soil. Instead underclays commonly show graded bedding (coarser grained material at the base) and evidence of clay flocculation (*The process by which small particles of fine soils and sediments aggregate into larger lumps.*). These are simple sedimentary features which would form in any water accumulated layer. Many coal seams do not rest on underclays and little evidence of soil exists. In some cases coal strata rest on granite, schist, limestone, conglomerate or other rock unsuitable for soil. Underclay without a coal bed above is common as well as underclay resting on top of coal. The absence of recognizable soils below beds of coal shows the improbability of any type of luxuriant vegetation growing in place and argues for transportation of the coal-forming plants.

Cyclothems

Coal commonly occurs in a sequence of sedimentary strata called a cyclothem. An idealized Pennsylvanian *cyclothem* may have strata deposited in the following ascending order: **(1)** sandstone, **(2)** shale, **(3)** limestone, **(4)** underclay, **(5)** coal, **(6)** shale, **(7)** limestone, **(8)** shale. A *typical cyclothem* will normally be missing one or more of the component strata. In any one locality cyclothems commonly repeat tens of times with each cycle of deposition accumulated on a previous one. There are fifty successive cycles in Illinois and over a hundred in West Virginia.

If the autochthonous (growth in place) model for coal formation is correct, a very unusual set of circumstances must have prevailed. An entire region, often encompassing many tens of thousands of square miles, would have to be raised simultaneously relative to sea level to permit swamp accumulation, and then lowered to permit the ocean to flood the area. If the coal forest was raised too far above sea level, the swamp and its antiseptic water necessary for the accumulation of peat would have been drained. If during the peat accumulation time the sea invaded the swamp, the marine conditions would have killed the plants, and other sediment instead of peat would have been deposited. According to the popular autochthonous (growth in place) model, the formation of a thick bed of coal, would indicate the maintenance of an incredible balance over many thousands of years between the rate of peat accumulation and the rise of sea level. Such a situation seems very improbable, especially when the *cyclothem* is known to recur a hundred times or more in a vertical section. Could such cycles be better explained by accumulation during successive advances and retreats of flood waters? [4-24g]

It is not uncommon to find marine fossils such as fish, mollusks, and brachiopods in coal. Coal balls, which are rounded masses of matted and exceptionally well preserved plant and animal fossils (including marine creatures) are found within coal strata and associated with coal strata. The small marine tubeworm *Spirorbis* is commonly attached to plants in Carboniferous coals of Europe and North America. Since there is little anatomical evidence suggesting that coal plants were adapted to marine swamps, the occurrence of marine animals with nonmarine plants suggests mixing during transport, thus favoring the allochthonous (water transported) model.

Some of the most interesting types of fossils associated with coal seams are upright tree trunks which often penetrate tens of feet perpendicular to stratification (polystrates). These upright trees are frequently encountered in strata associated with coal, and on rare occasions are found in the coal. In each case the sediments must have amassed in a short time to cover the tree before it could rot and fall down.

Coalification

The nature of the process of metamorphosis of peat to form coal has been disputed for many years. **One theory** suggests that *time* is the major factor in coalification. The theory, however, has become questionable because it has been recognized that there is no systematic increase in the metamorphic rank of coal with increasing age. **A second theory** supposes *pressure* to be the major factor in coal metamorphosis. The theory is refuted by numerous geological examples where metamorphic rank does not increase in highly deformed and folded strata. Furthermore, laboratory experiments demonstrate that increase of pressure can actually retard the chemical alteration of peat to coal. **A third theory** (the most popular) suggests the temperature is the important factor in coal metamorphosis. Geological examples (igneous intrusions into coal seams and underground mine fires) demonstrate that elevated temperature can cause coalification. Therefore, the metamorphosis of coal does not require millions of years of applied pressure and heat, but can be produced by quick heating.[4-24g]

Regardless of the exact manner in which coal was formed, it is quite certain that there is nothing corresponding to it taking place in the world today. This is one of the most important of all types of geologic formations and one on which much of our supposed geologic history has been based. Nevertheless, the fundamental claim made by evolutionists and old-universe creationists of uniformity completely fails to account for the phenomena of how coal was formed. The evidence indicates that, most likely, the great global Noahic Flood caused the basis for coal seams to form. [4-24J]

Polystrates - Continued

Evolutionist View

Some creationists claim that "polystrate fossils" are proof that fossils formed rapidly, hence, proving Noah's Flood. The term "polystrate fossils" is used for fossils that intersect several beds (layers), usually in sedimentary rocks. It is not necessary for a tree to remain upright for slow accumulation to occur because individual beds can be deposited rapidly, such as by sands and mud during a levee breach, and then little deposition can occur for a long time (e.g., a soil horizon), as is observed in modern river floodplain environments where trees commonly occur. Polystrate fossil trees are probably one of the weakest pieces of evidence "young Earth global flood" creationists can offer for their interpretation of how fossils were formed. They were not formed by a global flood.[4-25]

The Strata Layers and a Global Flood

Evolutionist View

The strata layers took millions of years to develop (See Chapter 2, sections "Age of the Earth" and Chapter 3, section "Methods Used to Determine the Age of Fossils."). The fossil layers show simpler creatures becoming more and more complex in each subsequent stratum moving up towards the surface of the Earth. A global flood cannot explain this. Young-earth creationists who believe in a global flood will often claim that smaller creatures, like humans, floated to the top and larger creatures, like dinosaurs, sank to the bottom. Young-earth Creationists believe that the sequence of fossils indicates that helpless invertebrates succumbed first and are in the bottom layer; reptiles succumbed next and are in the middle layers. They believe that birds, being able to fly, and mammals and humans, with their intelligence, succumbed last and are found in the top layers. The average size of dinosaurs was that of a station wagon. Many were very small, no larger than modern birds. Why don't we find any of them at the top layer with humans? And why didn't any of the flying reptiles fly to higher elevations to avoid the rising water? And what about the creatures that lived before the dinosaurs, many of them were smaller than the large dinosaur species. Why did the big dinosaurs float higher than the smaller creatures that sank and became part of the lower strata? There has never been any fossil find that contradicted evolution.[4-26]

Old-earth Creationist View

Conglomerate is a rock that looks rather like a natural concrete. It is a matrix of sandstone or other fine-grated rock, but embedded in this are many rounded pebbles of various sizes, and even boulders. Rivers, under extreme flood conditions, carried these types of sediments down stream. Young-earth creationists claim that Noah's Flood was responsible for this, and for all the other great concentrations of conglomerates throughout the world. Young-earth creationists, however, do not admit to the problems associated with such thinking. One major difficulty is that many large deposits of conglomerate lie on top of great thicknesses—often several miles—of fine-grained sedimentary rock. The great conglomerate sea cliffs near Marseilles, for instance, are hundreds of feet high and contain boulders more than a foot in diameter.

What natural process would have enabled the Flood to deposit a thickness of several miles of fine-grained sediments first, and then place the boulder-laden conglomerate on top? If the Flood accomplished this process then the boulder-laden conglomerate should have ended up at the bottom and the fine-grained sediments should have ended up on top of the boulder-laden conglomerate. Another problem for young-earth creationists is the clean, sharp line often found at the boundary between a conglomerate and underlying sandstone. Clearly, the lower layer must always have hardened into rock when conglomerate was dumped on top, otherwise the stones would have sunk into it. If one Flood deposited both layers in quick succession, how could this be? Also, there is the fact that the boulders in conglomerates often contain fossils. How did they get there if fossils are the remains of creatures that died in the Flood? These boulders are nearly always rounded, as if they had been rolled around on a river or seabed for long periods before being dumped in their last resting place.

Another problem that young-earth creationists have not explained is the sheer volume of sedimentary rock in the earth's crust. The "Phanerozoic" stratum sedimentary rock, that is, the stratum rock containing fossils, is supposed by young earth creationists to have been deposited by the Flood within a single year's time. (The Cambrian stratum is included in this stratum. See strata descriptions in Chapter 3.) The problem with this line of thinking is that the ratio between the sedimentary rock and the amount of water combining would greatly affect the ocean water consistency and marine life. The result would not be just dirty water, but a rich, creamy mud, in which no fish life could possibly survive. Did Noah's ark float on water or on an earthly soup composed of mud? Young-earth creationists obviously do not consider looking at the difficulties involved in what they are claiming. They simply say that it happened.

Let's now consider what is viewed in the sedimentary rocks. Consider a place where the sedimentary layer is only 20,000 feet thick. (There are some places where it is twice that thickness.) Combining these figures gives a total of 80,000 strata in a typical column of sedimentary rock. The Flood is supposed to have lasted one year (Genesis 7:10-8:14), but during the first part of it the floodwaters were building up, so only a portion of the year would have been available for the deposition of sediments. If we allow 9 months (which is over-generous) for this to occur, 80,000 strata in 9 months works out to be one stratum being laid down every 5 minutes.

In each 5 minute intervals, then, the Flood had to **(1)** bring in a particular kind of sediment, **(2)** distribute it fairly uniformly over a wide area—often over many tens of square miles—and **(3)** deposit it on top of the previous layer. The two layers might sometimes be similar in composition, but would often be quite different. The Flood would have had to deposit the upper layer so gently (because it would not have hardened in such a short period of time) that the layers deposited in the previous 5 minutes was not disturbed, so that no mixing of the two layers could occur. And it would have had to be so firmly in place at the end of the 5 minutes (that is, hardened,) that the next layer could then safely be laid down, and so on, for over 5 minutes for 9 months (the time period between the beginning of the flood until after the waters supposedly subsided). Then there is the observation of geologists that the upper surfaces of strata often have fossil limpets or barnacles on them. (Limpets are marine gastropod mollusks having a conical shell and adhering to rocks of tidal areas.) This shows that those layers had time to harden into rock and attract rock-clinging shellfish before the next stratum was laid down; this is not likely to happen in 5 minutes.[4-27]

Young-earth Creationist View

Paleontologists call the sudden appearance of complex, multi-celled creatures, with no appearance that they evolved from anything lower, "the Cambrian Explosion," because vast numbers of complex creatures suddenly appear in the fossil strata, with no evidence that they evolved from any less complicated creatures. (The Cambrian Explosion is discussed in Chapter 3.) The Genesis Noah Flood caused the massive appearance of life-forms. The Genesis Flood (the one described in the Bible in Genesis 6 to 9) rapidly covered the Earth with water. When it did, sediments of pebbles, gravel, clay, and sand were laid down in successive strata, covering animal and plant life. Under great pressure, these sediments turned into what we today call "sedimentary rock." (Clay became shale; sand turned into sandstone; mixtures of gravel, clay and sand formed conglomerate rock.) All that mass of water-laid material successively covered millions of living creatures. The result is fossils, which today are only found in the sedimentary rock strata.[4-28]

The Evidence in the Strata Layers Indicates Species Evolved Over Great Periods of Time

Evolutionist View

The evidence found in the Earth's strata layers indicates that the first fish were jawless and they appeared about 500 million years ago (mya) [Precambrian/ Cambrian transitional period (580 mya); Silurian (438-408 mya); Ordovician Period (430 mya)]. About 100 million years later [Silurian period (~410 mya)], fish with jaws and teeth are found in the fossil record. Sharks have teeth that are shed throughout their lifetime. Young-earth creationists claim that shark teeth should be scattered across the original ocean floor, at the very lowest levels of the marine fossil record. But a great deal of these teeth only appears after 400 mya. Dinosaurs like the Tyrannosaurus rex continuously replaced teeth during their lifetimes. However, this reptile and its multiple shed teeth are confined to a narrow region that is relatively high in the geological record, between 85 and 65 mya (Mesozoic Period). If the universe was only created 6,000 years ago the Earths' fossil record should have the remains of T-Rex along with every other plant and animal ever made, including humans, at the base of the fossil record (i.e., in the creation basic layer.) But this view does not match up with the scientific facts.[4-29] Also, during the Silurian Period (around 420 mya) and Carboniferous Period (around 350 mya), plants appear. During the Mesozoic Period (250 to 65 mya), the predominant plants were unusual gymnosperms, like the cycads. Many of these land plants release millions of microscopic pollen grains during their reproductive cycle. Young-earth creationists believe that all plants were made on the third day of creation (Genesis 1:11-12). Therefore, one should expect to find every type of pollen in every layer of the Earth's crust. But this is not the case. Fruit-bearing (flowering) plants should also appear throughout the fossil record. However, they are only found near the top of geological strata, after 130 mya, from the Early and Late Cretaceous Epochs (146-66 mya).[4-30]

The Evidence in the Strata Layers Indicates a Recent Earth

Young-earth Creationist View

The strata layers occurred during and immediately following the Noah Global Flood, contrary to the claims made by evolutionists who believe the strata layers developed over millions and millions of years. Those who believe that the geologic column and fossil record developed over many hundreds of millions of years have difficulty explaining some anomalies that appear in the strata layers. The problem is that the fossil record appears to be too neatly sorted. Creatures suddenly appear without any prior record and then suddenly disappear without any additional appearance. Many layers have a limited number of preserved creatures, an extremely low number of different fossils to support a viable ecosystem. For example, in the fossil layers where meat-eating dinosaurs appear, there are no fossils of creatures that the dinosaurs would have eaten. There is no known reason why fossils of other creatures should not have appeared in these same strata layers.[4-31]

These smaller animals should be present if the strata had developed over millions of years. It is doubtful that all dinosaurs were plant eaters. If the fossil record is to be interpreted as having occurred slowly over millions of years, it is questionable as to why no smaller animal fossils are present in the same strata as the dinosaurs. The question would arise as to why dinosaurs were preserved while smaller creatures were not. It is not a matter that smaller animals do not preserve, as the fossil records include multiple smaller animal fossils. When bodies of large dinosaurs and whales are uncovered, the depth of the strata layer is not much thicker than the size of the fossil. It would seem highly unlikely that this could have occurred if the sedimentation slowly buried the dinosaurs over millions of years. Their bodies would have decomposed before they had a chance to fossilize. Therefore, it is more probable that their bodies fossilized fairly quickly.[4-32]

The fossil record indicates that simple organisms are buried in the lower levels and the more complex organisms are buried in higher levels. For example, single celled organisms first appear in the lowest layers followed by multi-celled ocean bottom-dwelling creatures like sponges and worms etc. In higher, more recent layers of strata, creatures such as bony fishes appear, followed by land plants and animals. Birds and larger land animals appear in higher, more recent strata layers. This would appear to support the view that simple organisms evolved into more and more complex organisms over time, and more complex organisms buried and fossilized above the earlier and simpler life forms. This appears to be a logical assumption, however the interpretation may be more complex than what it initially appears to be. This is a very generalized pattern and does not explain why certain creatures that lived on the bottoms of oceans, like trilobites, first appear in the Cambrian period (505-540 mya), while other creatures that live on ocean bottoms, like crabs and lobsters, do not appear until the beginning of the Cretaceous period (65 mya-145 mya).

It is questionable as to why creatures that would seem to share the same general environment while alive could be so widely separated in the fossil record if they did indeed live at the same time and in pretty much the same location. If the geologic column truly represents a series of closely spaced catastrophic burial events, such as a Global Flood, instead of occurring over long ages of time, it would explain these anomalies. It is possible that bioturbation may have mixed up the sediment and caused the strata layers to now appear as they do. For example, geologists noticed that hurricane Carla laid down a layer of sediment off the coast of central Texas in 1961. After twenty years, geologists re-examined the sediment and found that living creatures had burrowed into it and disturbed it and destroyed the layer. Where the layer could still be found, it was almost unrecognizable. The conclusion here is that it is very difficult to imagine how such layering of sediment found throughout the geologic column could have been kept in such excellent condition and been preserved for millions of years. There is also a problem explaining why many land animals, except for birds and mammals, do not generally have their footprints located in the same layer where their bodies are found. The footprints are located in lower layers than where the bodies are. If the assumption that lower layer creatures evolved into creatures found in higher layers, then the question would arise as to why animal bodies are found in layers higher than their footprints. This is also true for dinosaur fossils. The actual dinosaur fossils are located in layers above their footprints.[4-33]

The time frame in which dinosaurs are claimed to exist is questionable. In 2002, when a T. rex thigh bone was divided, flexible soft tissue was found inside. Microscopic examination revealed fine delicate blood vessels with what appear to be intact red blood cells and other type of cells like osteocytes, are bone forming cells. These vessels were still soft, translucent, and flexible. Subsequent examination of other previously excavated T. rex bones from this and other areas have also shown non-fossilized soft tissue preservation in most instances. This find calls into question not only the nature of the fossilization process, but also the age of these fossils. How such soft tissue preservation and detail could be realized after 68 million years is unbelievable. If the bone were 68 million years old, it would seem nearly impossible for soft tissue to have been preserved for such a long period of time. The discovery of soft tissue inside a dinosaur bone must raise the question of exactly how old these bones actually are.[4-34]

Old-earth Creationist View

In 2005 a team of paleontologists discovered 70 million year old fossilized dinosaur eggs that contained soft tissue. This controversial discovery has led to mixed reactions from the scientific community, since it has long been thought that no such tissue could survive such long periods of time. There are skeptics who think the discoveries reflect contamination, and therefore not tissue that was original to the fossilized organisms. There are those who accept the findings on their merits and are seeking to make sense of how it could happen. Another group are Young Earth Creationists (YECs) who accept the findings and see them as evidence that radiometric dating methods are unreliable when they date the fossils as older than a few thousand years. It is possible that soft tissue could be preserved over long time periods. Young Earth Creationists have used the discovery of soft tissue remnants (these are remnants, not large chunks of meat) when arguing their case for a young earth. Using radiometric dating techniques in an attempt to determine the age of the soft tissue is not totally reliable and can be misleading.[4-34a] It must be made clear that "red blood cells" or hemoglobin have not been found in dinosaur bone. [4-34b]

Scientists have demonstrated that some rock-hard fossils tens of millions of years old may have remnants of soft tissues hidden away in their interiors. When the specimen was first examined, it appeared that it contained red blood cells. However, upon further examination, it was determined that what appeared to be red blood cells were actually heme. (Heme is a part of hemoglobin, the protein that carries oxygen in the blood and gives red blood cells their color.) What researchers actually found was evidence of heme in the bones—additional support for the idea that they were red blood cells. [4-34c]

The Order of the Fossils

Evolutionist View

The strata layers contained in the Earth were laid down over millions of years rather than as young-earth creationists believe. They claim that most of the layers in the Earth's crust were made in one year during Noah's flood, which they believe was global. However the geological column indicates that this view is false. A global flood would have produced strata that contain a mixture of all living organisms. But scientific evidence indicates an orderly progression in the appearance of life forms indicative of evolution (See Chapter 3, section "The Geologic Column and the Strata Layers of the Earth").

For plants, the pattern of fossils indicates: single cells—> marine plants—> land plants—> seed bearing plants—> flowering plants. For animals, the progression appears as follows: single cells—> soft-bodied marine animals—> marine animals with skeletons—> jawless fish—> amphibians—> reptiles—> mammals—> primates—> pre-humans—> humans. (Chapter 3 discusses the evolution of worm-like creatures to fish. Chapter 7 discusses the evolution of fish to amphibians; amphibians to reptiles; and reptiles to mammals. Chapter 8 goes into more detail about the evolution of reptiles to mammals. Chapter 16 discusses the fossil record of mammalian primates leading up to humans.) To date, science has yet to find one fossil plant or animal in the Earth's crust that is outside this pattern in the geological column.[4-35]

It is a scientific fact that most of the layers (strata) and fossils in the Earth's crust were laid down in water. The parallel strata in the Earth's crust indicate this is the case. But a closer examination of the fossil pattern prediction based on a worldwide flood reveals an insurmountable problem. For there to have been a global flood, it would have had to produce a "global flood layer" with strata featuring the mixing of bones and teeth of every animal ever created. Creationists believe the Cambrian Stratum was laid down during a global flood. Depending on the turbulence of the flood, gravity should have placed heavier bones (e.g., dinosaurs, elephants) near the bottom of the flood layer and lighter ones at the top. But no such fossil pattern exists in the crust of the Earth. Instead, the geological record presents a very consistent pattern (consistent with the concept of uniformitarianism developed by Charles Lyell), with the sequential appearance of fish first, amphibians next, then reptiles, then mammals, then humans.[4-36] (For further discussion of the Cambrian Stratum, see Chapter 3, and also, in this chapter, see section "The Strata Layers and a Global Flood.")

The fossil record supports the view that fish evolved into amphibians, that amphibians evolved into reptiles and that reptiles evolved into mammals (Chapter 7). If fish, amphibians, reptiles, and mammals had been separately created, one would not expect them to appear in the fossil record in the exact order of their apparent evolution. Fish, frogs, lizards, and rats appear in the fossil record in evolutionary order. No fossil has ever appeared in the wrong order. For example, no rabbit fossil has ever been found in the Precambrian strata. The reason is that the rabbit, which is a fully formed mammal, must have evolved through reptilian, amphibiana, and piscine (fish-like) stages and should not therefore appear in the fossil record 100 million years or so before its fossil ancestors.[4-37] (See Chapter 3, section, "The Arrangement of the Strata Layers.")

Old-earth Creationist View

Young-earth creationists claim that most mobile creatures fled to the high ground and were the last to be drowned, while the least mobile creatures stayed where they were, making them the first to perish. They claim this is why shellfish are found in the lowest layers. The sluggish dinosaurs were only able to make it to the foothills, and so are concentrated in the middle layers. Human beings, being able to maneuver, were able to flee to the mountaintops to escape the oncoming flood. This sounds logical until further thought out. Why is there not a single human fossil below the topmost layer? Were there no inhabitants of the coastal plains overwhelmed in their sleep? Were there no cripples or sick folk unable to flee the oncoming waters? Why are the pterodactyl fossils all in the middle layers? You would think that at least one or two of them would have flown to the hilltops. (Pterodactyl is a flying reptile from the Jurassic and Cretaceous strata—See Chapter 7.) The logical conclusion is that a Noah Flood could not have been the cause of the strata record.[4-38]

Young-earth Creationist View

Ninety-five percent of the fossils found are marine invertebrates such as trilobites and clams. Of the rest of the fossils, ninety-five percent of those are plants and in the remaining five percent of those are amphibians, mammals, birds, and humans. This record shows a pattern: the immobile water-dwelling animals are at the bottom, and the smarter and more mobile land-dwelling beings are at the top. This order can be explained by the different natural habitats of each organism at the time of the flood. When the flood covered the Earth, the first to be covered were small, slow-moving animals, the next to be covered were somewhat larger, somewhat faster-moving animals. These rock strata indicate that the lowest stratum tends to have the slowest-moving creatures; above them are faster ones. When the "fountains of the deep" (underground springs—Genesis 7:11) broke open, the immobile water-dwelling animals would be the first to be covered with sediments and fossilized. (It also rained for forty days and forty nights—Genesis 7:12.) As the waters rose, humans and the intelligent free-moving animals would have tried to get away to higher ground or, in the animals' case, to the nearest cave. In order to survive, however, some humans would have built rafts or found other ways to float for a few weeks in order to survive. Unfortunately, these people would have either drowned or starved to death, their bodies floating on the top of the water; prey for predators both in the air and in the sea (and the bodies would therefore not have become fossilized). The young, the weak and the lame would have most likely drowned rather quickly, as they would not have been able to withstand the rapid waters.

During the Genesis Flood, plants would tend to have washed into higher strata, but their pollen could easily have been carried into the earliest alluvial layers: the Cambrian and even the Precambrian. It is true that, in a very few disputed instances, there may be a few items in the Precambrian, which some suggest to be life-forms. But a majority of scientists recognize that these are only algae. Blue-green algae, although small plants, are biochemically quite complex, utilizing an elaborate solar-to-chemical energy transformation, or photosynthesis. Such organisms could have been growing on the ground when the Floodwater first covered it.[4-39]

During the global flood, the fish are naturally to be found at the bottom because they dwelt in the lowest elevations, in ponds, lakes, and rivers. They were the first to be buried, and the least able to escape the deluge. The mammals are to be found at the top because they lived in the highest elevations in the region, and also were the best equipped to escape the deluge, resulting in them being the last and the fewest to be buried.[4-40]

There are many indications that a worldwide flood caused the sedimentary rock strata and the billions of fossils found in the strata. Animals living at the lowest levels would tend to be buried in the lowest strata. Creatures buried together would tend to be buried with other animals that lived in the same region or ecological community. Hydraulic forces (the suck and drag of rapidly moving water) would tend to sort out creatures of similar forms. Because of lower hydraulic drag, those with the simplest shapes would tend to be buried first. Sea creatures without backbones (marine invertebrates) would normally be found in the bottom strata, since they live on the sea bottom. Fish would be found in higher strata since they can swim up close to the surface. Amphibians and reptiles would be buried higher than the fishes, but as a rule, below the land animals, because the land animals live in higher elevations than fish. Few land plants or animals would be in the lower strata. The first land plants would be found where the amphibians were found. Mammals and birds would generally be found in higher levels than reptiles and amphibians.

Because many animals tend to go in herds in time of danger, herd animals are found buried together. In addition, the larger, stronger animals would tend to sort out into levels apart from the slower ones. For example, tigers would not be found in the same strata with hippopotamuses. Relatively few birds would be found in the strata, since they could fly to the highest points. Few humans would be found in the strata. They would be at top, trying to stay afloat until they died; following which they would sink to the surface of the sediments and decompose. (Humans, being more intelligent than other animals and having more of an ability to find a means to survive, would have tended to remain alive longer than animals. The weak and infirm humans would also be found in the higher strata since they dwell in higher elevations than fish and amphibians.) The above points reflect what is found in the sequence of fossils in the geologic column, which are consistent with the view that the strata layers were laid down during the Noah flood. (Compare with Chapter 3, section "The Strata Layers of the Earth.")[4-41]

Why are trilobite and dinosaur fossils not found together? Evolutionists say the Cambrian trilobites died out millions of years before the dinosaurs evolved. There is, however, another explanation that seems even more natural. After all, even if trilobites and dinosaurs were alive today, they still would not be found together because they live in different ecological zones. Dinosaurs are land animals, but trilobites are bottom-dwelling sea creatures. According to creationists, the strata layers of trilobites and dinosaurs represent two completely different ecological zones, the buried remains of plants and animals that once lived together in the same environment.[4-42]

Examples That Support a Catastrophic Global Flood

Young-earth Creationist View

There are numerous examples of fossilization that support the Genesis account of cataclysmic destruction (Genesis 7) by a worldwide flood. (1) Animals are commonly found buried with an appearance of terror with heads arched back, mouths open. (2) There are caves, fissures, and mass burial sites throughout the world that are literally packed with masses of fossils. (3) Often times the fossils of these various animals come from widely separated and differing climatic zones, only to be thrown together in disorderly masses. (4) There is also evidence that a great and sudden cataclysm by a worldwide flood that once struck the Earth is found in the millions of mammoths and other large animals that were killed instantly in the north Polar Regions (northern Siberia and Alaska). Many of these have been found preserved whole and undamaged (except for being dead, of course) with flesh and hair intact, and in some cases, either kneeling or standing upright with food on their tongues. The eyes and red blood cells were extremely well preserved, and the separating of water in the cells was only partial, which indicates extremely sudden and sustained freezing conditions.[4-43]

It has been observed that in many places the "Alaskan muck blanket" is packed with massive amounts of animal bones and debris. Masses of dead animals and trees have been discovered. They were frozen solid. Twisted parts of animals and trees intermingled with layers of ice and layers of peat and mosses were found. It appears to be the result of some catastrophic event where all living creatures were suddenly frozen. Tendons, ligaments, fragments of skin and hair, hooves—all are preserved in the muck. In some cases, portions of animal flesh have been preserved. Bones of mammoths, mastodons, bison, horses, wolves, bears and lions were found. Logs, twisted trees, branches and stumps were interlaced with the mammal menagerie.

The signs of sudden death were evident. For example, in this Alaskan muck, masses of frozen mammals have been discovered. Upon examination, their stomachs were found to have contained the leaves and grasses that the animals had just eaten before death struck.

Mammoth   Elephant

Mammoth & Elephant

The imperial mammoths, largest known members of the elephant family, roamed North America. In New England, the mastodon, another elephant cousin, roamed the countryside. Further north, another tusked relative, the wooly mammoth lived. Besides elephants, the woolly rhinoceros, giant ground sloths, giant armadillos, bear-sized beavers, saber-toothed tigers, camels, antelopes, giant jaguars all lived in the same geographic area. Then all of a sudden these creatures perished. In varying degrees, this is true on every continent all over the world. Across the vast stretches of Siberia, on the other side of the Arctic Ocean, the same type of monstrous mammal catastrophe is quite evident. Europe and Asia were also struck by this apparent tremendous catastrophe.

Evidence of a catastrophic worldwide flood includes fossils of African hippos and elephants found in England; cold-climate animals found mixed together with tropical warm-climate animals; crocodiles of the Nile found in the middle of Germany; and American moose-deer found in Ireland. Marine life has been found on the tops of mountains, such as a skeleton of a whale being 3,000 feet high on the top of the Sanhorn Mountain. Dinosaur fossils have been found in positions that suggest sudden violent death.[4-44]

As mentioned previously, millions of wooly mammoth bones, tusks, and a few carcasses have been found frozen in the surface sediments of Siberia, Alaska, and the Yukon Territory of Canada. It is possible that the wooly mammoths were part of a Northern Hemisphere community of animals that lived and died during the post-Flood Ice Age (rather than dying during the Flood). The wooly mammoths could have died after the Flood because there are thousands of carcasses scattered across Alaska and Siberia resting above Flood deposits. If the wooly mammoths died after the Flood (rather than during the Flood), there must have been sufficient time for the mammals to have repopulated these regions after the Flood before they died. The post-Flood

Ice Age provides another explanation as to why the wooly mammoths died so suddenly. The mammoths spread into these northern areas during early and middle Ice Age time because summers were cooler and winters were warmer. The mammoths died by the millions and were buried by dust, which later froze during the Ice Age, preserving the mammoths.[4-44a]

Apatosaurus   Triceratops   Stegosaurus   Tyrannosaurus Rex
(Brontosaurus)

PUBLIC DOMAIN GOOGLE.COM

Various Dinosaurs

The great dinosaurs vanished completely, leaving only a few small-scattered dinosaur-like creatures in existence today. In a relatively short period of time, the dinosaurs were exterminated.[4-45] (A dinosaur is a large reptile. Crocodiles and alligators are examples of large reptiles.)[4-46]

Many theories have been proposed to explain why the dinosaurs disappeared completely. These theories include: **(1)** climatic change, **(2)** food shortages, **(3)** great disease epidemics all over the Earth, **(4)** dinosaurs were poorly constructed, **(5)** other animals ate the dinosaur eggs, **(6)** multiple local catastrophes occurred throughout the Earth such as a volcano, **(7)** an asteroid altered the Earth's atmosphere, etc.[4-47] It has been claimed that the Chicxulub asteroid, over seven miles in diameter, collided with the Earth on the Mexican coast approximately 65 million years ago (the end of the Cretaceous Period) and killed the dinosaurs. The impact made an initial crater nineteen miles deep and between 110 and 170 miles wide, which also caused many earthquakes. This theory suggests that the impact penetrated the Earth's crust, causing dust and debris to scatter into the atmosphere, which in turn caused many fires, tsunamis, severe storms and high winds, as well as highly acidic rain. The impact caused the formation of concentrations of sulfuric acid, nitric acid, and fluoride. Heat caused by the impact of the asteroid burned up all plant and animal life that was anywhere near it. The plants and animals that could not adapt to the temperature change died out. The debris could have clouded the Earth so that no sunlight could reach the plants. Since plants need sunlight to survive, many plants and trees died. The result would be that the Earth's oxygen level greatly decreased, causing many plants and animals to die that were unable to adapt. Also, with less vegetation for animals to eat, many animals would have perished. As a result, the meat-eating animals, not having had sufficient food to eat, also died out.[4-48]

These are some of the theories that have been proposed to explain the sudden extinction of dinosaurs throughout the world. Each theory explains the death of some dinosaurs in some places, but attempts to apply any of them, or combinations of them, to worldwide extinction have failed.[4-49] No one really knows why the dinosaurs became extinct. At the time of the Flood (Genesis 6-8), many of the sea creatures died, but some survived. In addition, all of the land creatures outside the Ark died, but the representatives of all the kinds that survived on the Ark lived in the new world after the Flood. Those land animals (including dinosaurs) found the new world to be much different than the one before the Flood. Due to (1) competition for food that was no longer in abundance, (2) other catastrophes, (3) man killing for food (and perhaps for fun), and (4) the destruction of habitats, etc., many species of animals eventually died out. The group of animals we now call dinosaurs just happened to die out too. Quite a number of animals become extinct each year. Extinction seems to be the rule in Earth's history (not the formation of new types of animals as one would expect from evolution).[4-50]

The Development of the Grand Canyon

Evolutionist and Old-Age Creationist View

Young-earth creation scientists claim that most of the layers of stratification in the crust of the Earth, as seen in the sides of canyons and mountains, were deposited in one year, during and after Noah's flood. The Grand Canyon is over a mile deep. Young-earth creationists believe this all occurred during a universal flood. Young-earth creationists claim that the sequence of fossils happened because helpless invertebrates succumbed first and are in the bottom layer. They claim that reptiles were the next to succumb and are in the middle layers. Birds, being able to fly, and mammals and humans, with their intelligence, succumbed last and are found in the top layers.[4-51] In reality, the Colorado River eroded the canyon to the point where the strata that had developed over millions of years became visible.

Young-earth creationists believe that the Grand Canyon was carved out during a Noah Flood. The development of the Grand Canyon is easily explainable without the need for a catastrophic flood. Hundreds of millions of years ago the site of the Grand Canyon was under water. Later, the sea eventually receded and the region became a coastal zone, while later still the land rose higher and the Rockies were formed. Young-earth creationists are not able to explain how the flood managed to transport all those marine creatures more than four hundred miles, and spread them neatly over a great area of Arizona. There is also no explanation of how the layers of inter-tidal life were transported, along with life from the coastal plains, some of which were three to four hundred miles and were placed in successive layers. How the layers were topped off with living things from the uplands, all without scrambling the various layers is also unexplained by young-earth creationists.[4-52]

Young Earth Creationist View

Evolutionists claim that the Grand Canyon was caused by the Colorado River. This claim is false. The Colorado River starts out in the Rocky Mountains of western Colorado at about 12,000 feet above sea level. Then it progresses south toward the Grand Canyon. Just before the Colorado River reaches the Grand Canyon, the height is only 5,000 feet. This is a problem because the Grand Canyon begins at 8,000 feet high. For the Colorado River to have eroded the Grand Canyon area, it would have first had to flow uphill over a half mile.[4-53] The Grand Canyon appears to provide an excellent contrast between rocks that were laid down slowly and gradually on a local scale and those laid down rapidly and catastrophically on a colossal scale. The Precambrian rocks at the Grand Canyon (the pre-flood rocks that were laid down many centuries before the Flood) look like rocks that were formed by processes that are occurring today (slowly and gradual).[4-54]

Migration of Species and a Global Flood

Evolutionist and Old-Age Creationist View

An example that there was no global flood comes from biogeography, which presents a set of facts that young-earth creationists have particular difficulty explaining. There are about 2,000 known species of fruit flies and of these 700 species live only in the Hawaiian Islands. When the ark released the mass of animals on Mt. Ararat, how did those 700 species make it to the Hawaiian Islands and nowhere else? Many oceanic islands (islands that were never part of a continent) have many species of fruit flies that fit the evolutionary model of speciation precisely. Even though young-earth creationists try, they are not able to explain the distribution of species. (Speciation is discussed in Chapter 9.)

Kangaroo

Why did kangaroos make it only to Australia, humming birds only to the Americas, and hippos only to Africa?[4-55] Why are there creatures that are mammals without placentas that hop on two legs and carry their young in pouches and are only found in Australia? Why are all the mammals in Australia marsupials (including kangaroos) and why do they often look like mammals that live in the rest of the world despite the fact that they all lack placentas and also carry their young in pouches. (Marsupials are animals whose young are very undeveloped when born and continue developing outside their mother's body in pouches.) It is believed that approximately 250 million years ago, Australia was joined with the rest of the continents in a supercontinent called Pangaea. Then around 180 million years ago, the continents began to break apart due to continental drift. The animals in each continent evolved in isolation from one another. The evidence from fossils indicates that at the time of the separation of the continents, placental mammals had not reached Australia in any significant amount. This is the reason why kangaroos are only found in Australia. They did not migrate from Mount Ararat, where Noah's Ark was presumed to have landed.[4-56]

There are thousands of distinct life forms that exist in Australia and New Zealand alone and they live a long, long way from where the Ark came to rest. The three-toed sloth is an animal that only travels at, top speed, 0.068 M.P.H. and the fossil record says they have always been indigenous to South America only. Current evidence indicates that there is anywhere of up to one billion species that existed in the past but are now extinct. Assuming that there were a relatively small number of species in Noah's Ark, a very quick macroevolution would have to have occurred.[4-57]

If the story of Noah's Ark in the Bible were true (Genesis 6-8), how did kangaroos and giant earthworms make their way from Mount Ararat in Turkey (where the ark is presumed to have landed) across the oceans to their present home in Australia? Would not lions have quickly made a meal of the antelopes? Many questions arise regarding the distribution of species if a global flood actually occurred. If Noah's ark had landed at Ararat, how did Native Americans get to the New World? Also, if there were a global flood, it would be assumed that all insects would have been destroyed. What would the anteaters have eaten immediately after coming off the ark? [4-58]

Young-earth Creationist View

Evolutionists often claim that kangaroos could not have hopped to Australia, because there are no fossils of kangaroos along the way from Turkey to Australia. But the expectation of such fossils is unrealistic. Such an expectation is based on the assumption that fossils form gradually and inevitably from animal populations. In fact, fossilization is by no means inevitable. It usually requires sudden, rapid burial. Otherwise the bones would decompose before they were able to mineralize. One should ask why it is that, despite the fact that millions of bison used to roam the prairies of North America, hardly any bison fossils are found there. Similarly, lion fossils are not found in Israel even though we know that lions once lived there.

Many different theories have been developed to explain how animals could have migrated to other parts of the world after the flood. (1) Non-flying animals may have traveled to the outer parts of the world after the Flood. Many of them could have floated on vast floating logs, left-over from the massive pre-flood forests that were ripped up during the Flood and likely remained afloat for many decades on the world's oceans, transported by world current. (2) People could have later taken other animals to other parts of the world. (3) A third explanation of possible later migration is that animals could have crossed land bridges. This is, after all, how it is supposed by evolutionists that many animals and people migrated from Asia to the Americas, over a land bridge at the Bering Straits. For such land bridges to have existed, we may need to assume that sea levels were lower in the post-flood period.[4-59] How the kangaroos reached Australia from Mount Ararat can easily be explained. A great land bridge apparently connected Asia and Australia in the early post-Flood period. During this most intense phase of the 'ice age,' such vast quantities of water were locked in the polar regions that ocean levels were hundreds of feet lower than they are now. The National Geographic Atlas of the World clearly shows the shallow continental shelf that extends even now from Indochina almost to Australia.[4-61] Due to continental glaciers in the water, it is believed that the oceans have been as low as almost 150 meters (492 feet) below the present levels. Most likely there were four major lowerings of the sea level.[4-62] The water over most of the continental shelves of both Southeast Asia and Australia are much shallower than 400 feet (1,219 decimeters). These areas must have been land when the oceans were low and the islands were then mountains.[4-63]

Skeptic of Both Creation and Evolution

It may be possible that only representatives of different kinds of animals were included in the ark. It is possible that some species within kinds arose later by microevolution. (Microevolution is discussed in Chapter 9.) For example, it is possible that anteaters were not included in the ark but may have appeared later as microevolution developed additional creatures.[4-60]

In the next chapter (5) we will discuss the possibilities of how life could have originated by natural causes, from a biological perspective.

~~~

Chapter 5

The Origin of Life on Earth

A Warm Little Pond

It is often said that all the conditions for the first production of a living organism are present, which could ever have been present. But if (and Oh! what a big if!) we could conceive in some warm little pond, with all sorts of ammonia and phosphoric salts, light, heat, electricity, etc., present, that a protein compound was chemically formed ready to undergo still more complex changes, at the present day such matter would be instantly devoured or absorbed, which would not have been the case before living creatures were formed. It is mere rubbish thinking at present of the origin of life; one might as well think of the origin of matter.

> • Letter to botanist Joseph Hooker (1871),
> regarding the possibility of a chemical origin
> for life by Charles Darwin

Then God said, "Let the earth bring forth grass, the herb that yields seed, and the fruit tree that yields fruit according to its kind, whose seed is in itself, on the earth"; and it was so. And the earth brought forth grass, the herb that yields seed according to its kind, and the tree that yields fruit, whose seed is in itself according to its kind . . . And God said, Let the waters bring forth abundantly the moving creature [fish, sea creatures] that hath life, and fowl [birds] that may fly above the earth . . . Let the earth bring forth the living creature after his kind, cattle [livestock, tame domestic animals], and creeping thing [creatures that move along the ground, reptiles], and beast of the earth [wild animals—animals of the earth/land] after his kind.[5-1]

Yes, we are all animals, descendants of a vast lineage of replicators sprung from primordial pond scum.[5-2]

Introduction

In this chapter (5) we will take a look at the possibilities, from a biological perspective, of how life (the cell) could have evolved by natural means. The various views presented in this chapter regarding the origins of life are not exclusively strict evolutionists and strict creationists. Some views are of those who believe in variations of the two extremes. See "Various View Descriptions" section in the back of this book for descriptions of the views presented in this chapter.

What is life? For something to be considered living, it must demonstrate that it possesses certain characteristics. For something to be considered living, it must **(1)** be made of cells **(2)** obtain and use energy **(3)** grow and develop **(4)** reproduce **(5)** respond to its environment, and **(6)** adapt to its environment. To be considered alive, an object must exhibit all of the above characteristics of living things.[5-2a] Living things are further defined as those that display the following characteristics: **(1)** an organized structure, being made up of a cell or cells **(2)** requires energy to survive or sustain existence **(3)** ability to reproduce **(4)** ability to grow **(5)** ability to metabolize (Metabolism is the set of life-sustaining chemical transformations within the cells of living organisms. The three main purposes of metabolism are (a) the conversion of food/fuel to energy to run cellular processes, (b) the conversion of food/fuel to building blocks for proteins, lipids, nucleic acids, and some carbohydrates, and (c) the elimination of nitrogenous wastes.) **(6)** ability to respond to stimuli **(7)** ability to adapt to the environment **(8)** ability to move [animals, insects, etc., not plants, scrubs or trees] **(9)** ability to breathe.[5-2b]

In biology, an organism is a contiguous living system, such as an animal, plant, fungus, or bacterium. All known types of organisms are capable of some degree of response to stimuli, reproduction, growth and development and homeostasis (a system in which variables are regulated so that internal conditions remain stable and relatively constant). An organism consists of one or more cells; when it has one cell it is known as a unicellular organism; and when it has more than one cell it is known as a multicellular organism. An organism may be either a prokaryote (unicellular, single-celled organism) or a eukaryote (multicellular, an organism whose cells contain a nucleus and other organelles enclosed within membranes.).[5-2c] (Cells will be discussed in more detail later in this chapter.)

To the biologist, "life" encompasses a spectrum ranging from complicated organisms to bacteria and viruses. The simplest forms of life, the viruses and their relatives known as phages, are little more than macromolecules. The phage called OX174, for example, consists only of a length of DNA (deoxyribonucleic acid, the genetic material) 5, 375 units (nucleotides) long, that contains the information for producing only nine kinds of proteins. Such things are considered to be living because of two essential characteristics: **(1)** the ability to obtain energy from the environment, and **(2)** the ability to replicate themselves and produce new copies that are sometimes slightly altered. They have heredity, the essence of life. [5-2d]

How Life Could have Originated on Earth

Darwinist View

The origin of life on Earth is one of the most important, and elusive, problems in science. Efforts to understand the origin of life have been frustrated by the lack of evidence. In the face of this fundamental difficulty, the search for a scientific explanation for the origin of life has relied upon **(1)** speculative hypotheses, **(2)** observations from present conditions on Earth and elsewhere in our solar system, and **(3)** laboratory experiments that seek to simulate the conditions of the earliest period of Earth's history.

The best theory we currently have regarding the origin of life on Earth is that it first originated as the accumulation of organic compounds in a warm body of water. This hypothetical "warm little pond" has supporting evidence from a number of experiments (See Chapter 6). However, the origin of life is clouded in uncertainty, and the precise mechanisms by which basic chemicals came together to form complex organisms is not known. The lack of evidence from ancient Earth means we may never know precisely how life began. Nevertheless, of all the speculative theories, the "warm little pond" theory remains the most promising.

As just mentioned, a major problem with determining how life began on Earth is the lack of evidence. The fossil record is limited by the fact that almost all rocks over three billion years old have been deformed or destroyed by geological processes (Chapter 3). In addition to debating the issue of when life emerged, scientists also debate the conditions of ancient Earth. Some theories contend that early conditions on our planet were extremely cold, while other theories suggest that it was warm and temperate, and even boiling hot.

Life on Earth may have had a number of false starts. Early Earth was subjected to massive geological upheavals, as well as numerous impacts from space. Some impacts could have boiled the ancient oceans, or vaporized them completely, and huge dust clouds could have blocked out sunlight. Life may have begun several times, only to be wiped out by terrestrial or extra-terrestrial catastrophes.[5-3]

The process of how the chemical evolution of life from self-catalytic chemical reactions could have occurred and gradually developed into living organisms is still basically unknown. There is not very much known about the earliest developments of life. It is known, however, that all existing organisms share certain traits, including cellular structure and a genetic code. There is still no agreement among scientists as to how life could have begun. Some biologists have performed research on the functions of various macromolecules, especially RNA (ribonucleic acid), as well as on other systems in an effort to determine how life may have originated on Earth.

The development of all major types of organisms of modern animals began around 3 billion years ago during the development of oxygenic photosynthesis on the Earth. This period is called the Cambrian explosion (540 to 500 mya), described in Chapter 3. This period appears to have been activated by the development of the Hox genes. Hox genes are similar in many organisms and small changes in these genes could initiate evolutionary change. (See Chapter 9, section "Hox Genes" for further discussion of Hox genes.) Scientists believe that approximately 500 million years ago, plants and fungi colonized the land, and were soon followed by arthropods that were invertebrates such as insects, crustaceans, arachnids, myriapods, now extinct trilobites, as well as other animals, which lead to the development of land ecosystems (organisms that were able to adapt to their environment) that now appear on Earth (See Chapter 3 for further discussion of arthropods and arachnids).[5-4]

It has been theorized that complex biological compounds somehow assembled by some chance out of an organic broth on the early Earth's surface. Once assembled, they were able to make copies of themselves.[5-5] It is almost certain that direct fossil evidence will never be found to substantiate the claim that living molecular structures evolved from nonliving precursors. Such molecules surely could not have been preserved without degradation. But a combination of geochemical evidence and laboratory experiment shows that such evolution is not only plausible but almost undeniable.

While the primeval atmosphere of the earth lacked oxygen, it must have had methane, ammonia, water vapor, and other elementary gases such as amino acids that are the building blocks of proteins, and the nucleotide bases that are the building blocks of DNA. Moreover, these amino acids spontaneously assemble themselves into short proteins, which aggregate into spherical polymers that almost look like cells, and split into smaller spheres when they grow too large.

All living things require genetic information in the form of DNA (deoxyribonucleic acid) or RNA (ribonucleic acid). They also require proteins known as enzymes, which are encoded in the genetic information and in turn assemble free nucleotides into new copies of DNA or RNA. So far such a complete nucleic acid-plus-enzyme system has not been synthesized in the laboratory, but researchers have come very close. They have found an enzyme that replicates the RNA genetic information of a phage called Qx can string nucleotides together into short strands of RNA even in the absence of preexisting RNA. Many short sequences of RNA are formed in this way, some of which replicate faster than others, and replace the slower-replicating sequences: that is, they evolve by natural selection (Natural selection is discussed in Chapter 9). Moreover, different sequences, or "species," of RNA replicate fastest in different chemical environments. Thus, evolving genes can arise in the absence of preexisting life.

Researchers have found that even without a replicating enzyme short sequences of RNA can replicate themselves. The next step in the synthesis of life will be to develop RNA sequences that can produce their own replicating enzymes. This hasn't happened yet (as of 1992). (See section "Experiments Attempting to Synthesize Cells in the Laboratory" sub-section "Protocells created in a Lab" below for update.) It is important to realize that although human intelligence is guiding such experiments, chemists are not making RNA molecules by carefully stringing together nucleotides with sophisticated chemical techniques. They are simply providing in the laboratory the chemical and environmental conditions that are believed to have existed naturally billions of years ago. It is hoped that such experiments will one day produce RNA.

The genetic code of DNA and RNA is identical in all species from viruses to mammals. Thus all living things share fundamental biochemical characteristics which indicate that they have all evolved from a single form of life. It can be understood, from laboratory experiments, how the earliest form of life, perhaps a self-replicating nucleic acid, could have arisen spontaneously from simple chemical compounds.

Geologists have determined that life evolved more than 3 billion years ago and have diversified ever since. Many later forms of life are much more complex than the early single-celled forms, but life as a whole shows no consistent direction trends. Because evolution can occur rapidly at times (punctuated equilibrium), and because the fossil record is incomplete, there are many gaps between ancestors and their presumed descendants; but even so, the rocks reveal numerous instances of gradual evolutionary change and show that groups of animals that are quite distinct in the modern world become more and more indistinct as we pass back in time (Fossils are discussed in chapters 3, 7, 8, 16). Thus, together with the evidence of anatomy, embryology, biochemistry, and the geographic distribution of species, the fossils reveal a history of descent with modification.[5-5a] (Embryology is discussed in Chapter 11.)

Life May Have Originated in a Warm Pond

Darwinist View

Charles Darwin developed the modern theory of how life originated on Earth. While earlier natural philosophers considered the problem of the origin of life, it was Charles Darwin who first posed an explanation for life's origin that is consistent with the larger picture of evolution of life on Earth. Darwin suggested that simple chemicals in small or shallow bodies of water might spontaneously form organic compounds in the presence of energy from heat, light, or electricity from lightning strikes. These organic compounds could then have replicated and evolved to create more complex forms. Darwin's "little warm pond" remains one of the most suggestive explanations for the origin of life.

In a letter to fellow scientist, Joseph Hooker (quoted in the beginning of this chapter), Darwin speculated that life originated when chemicals, stimulated by heat, light, or electricity, began to react with each other, thereby generating organic compounds. Over time these compounds became more complex, eventually becoming life. Darwin imagined that this process might occur in shallow seas, tidal pools, or even a "little warm pond." While Darwin and his contemporaries saw life as a sudden spontaneous creation from a chemical soup, modern theories tend to regard the process as occurring in a series of small steps. Life either started in a "warm little pond" or it began in a protected environment away from the surface and the devastating elements that might prevent life from forming or continuing, such as under the sea, away from harmful radiation from the sun and other elements that bombarded the Earth.[5-6]

Scientists are conducting research to develop a feasible theory as to how life could have originated. This theory is sometimes called "chemical evolution" or "abiogenesis." It is a study of how the first living organism might have self-organized and assembled. Scientists are conducting research in an effort to determine if it could have been possible, given the amount of time and the types of conditions that existed then, for the existing chemicals to have organized themselves into simple living cells. The question is, "What are the chances that a simple living cell could have self-assembled on the Earth?" (A cell is the smallest organism that is capable of functioning on its own. Bacteria are one-celled organisms.)

One could imagine a warm pond of water with various simple organic molecules dissolved in it. One would then need to calculate the probability that millions of the right molecules could randomly collide together spontaneously to form a living cell. The chances of whether or not a simple living cell might self-assemble on the Earth, is debatable. It is admitted that the probability of that happening is extremely low. Some other theories are discussed below.[5-7]

In the early 1950s, Scientists believed the early Earth's atmosphere consisted of (1) hydrogen, (2) methane, and (3) ammonia. However, by the mid-1950s, some scientists came to believe that the early Earth's atmosphere consisted of different chemicals, such as an oxygen-rich mix of (1) carbon dioxide, (2) nitrogen, and (3) water vapor (Chapter 6). They began to look elsewhere to try to find alternative scenarios to explain the origin of life. Observations of the moon and other planets in the solar system have suggested that during the first billion years or so of Earth's history, the surface of our planet was extremely volatile and even hostile, bombarded by meteorites and intense solar radiation. These conditions make the formation of stable, warm, shallow ponds unlikely, but they suggest other interesting possibilities to explain the origin of life on Earth.[5-8]

Old Universe Progressive Creationist View

Geochemists have developed two significant methods for measuring the quantity of prebiotics on ancient Earth. One uses carbon and the other uses nitrogen. Both methods lead to the same conclusion.

1) Carbon ratio. Carbonaceous substances (the decay products of once-living organisms) manifest a distinctly lower ratio of carbon-13 to carbon-12 than do the same carbonaceous substances that chemically developed from inorganic compounds. Therefore, careful measurements of the carbon-13 to carbon-12 ratio in ancient deposits yield the quantity of prebiotics present on ancient Earth. The surprising result of carbon-13 to carbon-12 ratio measurements of carbonaceous deposits is that such deposits formed from the remains of once-living organisms. None of the deposits formed from prebiotic material. The researchers who made some of the most extensive carbon-13 to carbon-12 ratio measurements concluded that no known abiotic process can explain why this is so. With this accumulation of data, the primordial-soup hypothesis collapses. Given the measurable abundance of life on Earth 3.8 billion years ago, a primordial soup, if it existed, would be geochemically obvious. Naturalist researchers had concluded that the isotope enhancement of carbon-12 is the very old kerogen in the Isua rocks in Greenland is that there never was a primordial soup and that, nevertheless, living matter must have existed abundantly on Earth before 3.8 billion years ago. (The Isua Greenstone Belt is an Archean greenstone belt in southwestern Greenland. The belt is aged between 3.7 and 3.8 billion years. In 2016 melting snow revealed 3.7-billion-year-old stromatolite fossils, the oldest by several hundred million years thus far discovered on Earth. [5-8a])

2) Nitrogen ratio. A second geochemical tool, nitrogen isotope ratios, now provides independent confirmation that no primordial soup ever existed on (or in) Earth. The first of these confirmations comes from nitrogen-15 to nitrogen-14 ratio analysis. The same ancient carbonaceous filamentous microstructures in which researchers read the carbon-13 to carbon-12 ratio signature for post biotic decay (as opposed to prebiotic origin) also reveal a nitrogen-15 to nitrogen-14 ratio indicative of a biogenic origin.

Another confirmation arises from calculations of ammonia abundances. For any kind of primordial soup or mineral substrate to yield a chemical pathway for the development of complex life molecules, significant quantities of ammonia must be present. Laboratory simulations experiments, lacking significant quantities of ammonia, consistently fail to produce any amino acids. Several studies of the nitrogen-15 to nitrogen-14 ratio in ancient kerogens (carbonaceous deposits) show that while there may have been some ammonia in Earth's atmosphere at the time of life's origins 3.8 to 3.5 million years ago, the quantities would have been inadequate to sustain the prebiotic chemical pathways necessary for life's spontaneous origin. The conclusion is that there was no prebiotic soup billions of years ago when life began.[5-8b]

Other Theories About How Life Could Have Originated on Earth

Darwinist View

Because of the difficulties with the "warm little pond" theory and its variants (discussed in this chapter as well as Chapter 6), a number of other theories developed. Many of these theories are interesting, intriguing, and even possible. However, they all have unanswered questions, making them even more problematic than the idea of the "little warm pond."[5-9]

There are several other theories about the origin of small molecules that could lead to life in an early Earth. **(1)** One is that they came from meteorites (Murchison meteorites, which contain common amino acids such as glycine, alanine and glutamic acid). (See section below "Amino Acids Found in Meteorite") **(2)** Another theory is that small molecules were created at deep-sea vents (discussed below). **(3)** A third possibility is that small molecules were synthesized by lightning in a reducing atmosphere. But some other entity with the potential to self-replicate could have preceded RNA (ribonucleic acid), like clay or peptide nucleic acid.

Cells emerged at least 4.0-4.3 billion years ago. (Cells are discussed later in this chapter.) The current belief is that these cells were heterotrophs. (Heterotrophs are organisms that cannot synthesize their own food and are dependent on complex organic substances for nutrition.) An important characteristic of cells is the cell membrane, composed of a bi-layer of lipids. The early cell membranes were probably more simple and permeable than modern ones, with only a single fatty acid chain per lipid. Lipids are known to spontaneously form bi-layered vesicles in water, and could have preceded RNA. But the first cell membranes could also have been produced by catalytic RNA or by somehow acquiring structural proteins in order to form.[5-10] (See section "The Origin of the Cell Membrane" below that discusses the origin of the cell membrane.)

Geographical features such as ponds that repeatedly evaporate and then refill could have concentrated organic molecules. Mineral clays could have helped form long chain molecules and also could have held the long chain molecules in place long enough to assemble into larger structures. Deep underground fissures, regions near volcanoes, or deep ocean hydro-thermal vents might have provided more likely environments for life to form. These are different theories about how life could have originated, however none have yet been either verified or falsified.[5-11]

Life May Have Originated from Concentrated Clays

Darwinist View

Evaporation during droughts caused the development of concentrated clays and the other rainwater chemicals in mud-like puddles near the sea. The result was a gooey sludge at the water's edge. Sunlight and lightning bolts zapped the sludge. The simple elements within banded together to form a variety of more complex chemicals based on carbon, such as amino acids and simple sugars. In time, numbers of amino acids and sugars linked together to form long, complex chain molecules called proteins. Today, proteins make up a large part of the substances in each living cell. Proteins also act as enzymes to hasten chemical reactions.

The origin of life from simple elements could not have taken place without clay. Clay was the base on which the chains of amino acids slowly assembled into proteins. Clay is formed of microscopic crystals, usually of irregular and complex shape. These shapes continue growing in layers as more clay crystallizes on their surface. Each new layer "inherits" the shape of the one below it. As changes occur, the crystals "evolve." The early amino acids that existed within the framework of clay crystals also inherited the patterns of previous layers. When they finally formed certain patterns, some of these amino-acid chains became proteins. (Amino acids and proteins will be discussed later in this chapter and in Chapter 6.) With complex proteins being formed and held in position on clay, life was able to develop. The Earth was just wet enough and warm enough for complex chemicals to become living chemicals.[5-12] (See Chapter 1, section, "The Fine Tuning of the Universe" for further discussion of this topic.) When a clay-based concentration of amino acids was finally able to (1) absorb nutrients and grow, (2) reproduce on its own, (3) respond to stimuli, and (4) evolve, all by itself, then, by definition, it became alive.[5-13] A living thing is considered a living organism if it can (1) process energy, (2) construct other components (3) duplicate information (4) store information (5) duplicate itself.[5-13a]

Amino Acids Found in Meteorite

On September 28, 1969, a meteorite fell over Murchison, Australia. While only 100 kilograms were recovered, analysis of the meteorite has shown that it was rich with amino acids. Researchers so far have identified over 90 amino acids. Nineteen of these amino acids are found on Earth. The early Earth is believed to be similar to many of the asteroids and comets still roaming the galaxy. If amino acids are able to survive in outer space under extreme conditions, then this might suggest that amino acids were present when the Earth was formed. More importantly, the Murchison meteorite has demonstrated that the Earth may have acquired some of its amino acids and other organic compounds from other planets.[5-14] (Amino acids are discussed in more detail in Chapter 6.) Amino Acids and other organic chemicals found in human cells have also been found on meteorites and comets exposed for eons to the hazards of space. Some say that meteors brought life to this barren planet. It is more likely, however, that an early Earth had just as many amino acids as any meteor did. All the matter in this solar system came together at the same time and under roughly the same conditions. The Earth was just wet enough and warm enough for complex chemicals to become living chemicals.[5-15]

Old Universe Progressive Creationist View

The Murchison meteorite contained the nucleotide bases guanine, adenine, and uracil, but only in low abundances. Cytosine has not yet been detected in the Murchison meteorite or anywhere on Earth. Meteorites can absorb airborne contaminates in as little as one week upon entering the Earth's atmosphere. The Murchison meteorite, one with the highest amino acid abundance, contained less than fifteen parts per million of the amino acids found in proteins. Nevertheless, naturalists continue to consider meteors as a significant source of ingredients for the prebiotic soup.[5-15a] It is unknown to what extent, if any, the Murchison fragments suffered contamination or chemical change during and after their journey through Earth's atmosphere and onto its surface. This appears to be a possibility because the exterior samples had higher concentrations of amino acids than the interior samples. The exterior samples also had higher excesses of left-handed amino acids than the interior samples.[5-15b] (All known life uses amino acids that are exclusively of the "left-handed" form. – See Chapter 6, section "Left-handed and Right-handed Amino Acids" for further discussion.)

Life May Have Originated Deep within the Ocean

Darwinist View

Scientists discovered organisms that live in very hot conditions. These hot-living bacteria have been found in spring waters with temperatures of 144 degrees Fahrenheit (62.2 degrees Celsius), and some species near undersea volcanic vents at the boiling point of water. There is even some evidence of underground microbes at even higher temperatures of 366 degrees Fahrenheit (185.5 degrees Celsius). The discovery of such hardy organisms has led some to speculate that life originated not in a warm pond, but in a very hot one. A variant of this theory is the undersea volcanic vents as the birthplace of life, with the chemical ingredients literally cooked into life. There are even those who advocate a deeper, hotter, underground origin for life. Underwater and underground origins have some advantages over other theories. Such depth might make early life safe from the heavy bombardment of material from space the planet received, depending on the size of the object striking Earth and the depth of the water. They would also be safe from other surface dangers, such as intense ultraviolet radiation. There are some who question how hot organisms could have moved into cooler areas. Some theorists argue that it is easier to go from cool to hot, not the other way around. Also, environments such as undersea volcanic vents are notoriously unstable, and have fluctuations that can cause local temperature variation that would destroy rather than create complex organic compounds.

Some scientists have gone to the other extreme of the temperature scale and envision life beginning on a cold freezing ancient Earth. Just as hot microbes have been discovered, so have organisms capable of surviving the Antarctic cold have been discovered. Some suggest these as our common ancestors. Again, there are some advantages to such a theory. Compounds are more stable at colder temperatures, and so would survive longer once formed. However, the cold would inhibit the synthesis of compounds, and the mobility of any early life. Also, the premise that ancient Earth was a cold place is not widely accepted.[5-16]

Scientists have determined that approximately 4 billion years ago, somehow a set of molecular compounds combined to become life. Biologists believe that there is a possibility that life may have begun in an undersea microbe, called Methanosarcina acetivorans." This microbe eats carbon monoxide and expels methane and acetate (which is related to vinegar). It has been discovered that this primitive organism is able to obtain energy from a reaction between acetate and the mineral iron sulfide. Compared to other processes that require multiple proteins to acquire energy, this acetate-based primitive organism operates with just two very simple proteins. Biologists have stated that this simple geochemical cycle was what could have been the first organism's engine used to power its growth. The objection to this view is that something had to form the two proteins. Others believe that the idea of life beginning from a simple geo-chemical cycle appears to be a step in the right direction in determining how life actually began on the Earth.[5-17] (How a protein is developed is discussed in sections "Protein Synthesis (How a Protein is Developed Inside a Cell)," "The Process of Creating a Protein From DNA Instructions," and "How a Protein is Made – Step by Step.")

One current theory (briefly mentioned above) is that life originated deep beneath the surface of the ocean at deep sea hydro-thermal vents. These vents release hot gaseous substances from the center of the earth at temperatures in excess of 572 degrees Fahrenheit (300 degrees Celsius). Previously, scientists were sure that life could not exist deep beneath the surface of the ocean. After the discovery of hydrothermal vents, they found ecosystems thriving in the depths of the ocean. These ecosystems contained various types of fish, worms, crabs, bacteria and other organisms that had found a way to survive in a cold, hostile environment without energy input from sunlight. Because life had been found to exist where it previously was thought unable to, many scientists began to ask questions as to whether or not this was where life may have originated on the earth.

On the molecular level, the chances of life originating at deep-sea thermal vents are not likely. It is known that organic molecules are unstable at high temperatures, and are destroyed as quickly as they are produced. It has been estimated that life could not have arisen in the ocean unless the temperature was less than 25 degrees Celsius (77 degrees Fahrenheit). Supporters of this theory claim that the organic molecules at the thermal vents are not formed in 300 degrees Celsius (572 degrees Fahrenheit) temperatures, but rather in a slope formed between the hydro-thermal vent water, and the extremely cold water, 4 degrees Celsius (39.2 degrees Fahrenheit), which surrounds the vent at the bottom of the ocean. The temperatures at this slope would be suitable for organic chemistry to occur. Debates still remain, however, as to the slope's effectiveness in producing organic compounds.[5-18]

Life May Have Originated by Electromagnetism

Evolutionist View

The quantum field theory stems from the collective properties of microscopic components for the origin of life. This theory indicates that its electromagnetic field begins to behave as a particle of very small mass and rather peculiar form, namely, as very thin filaments. All these particles build a fine filamentous network within cells, which is highly stable since it survives even some time after destruction of the underlying order of dipoles (defined shortly). To achieve this filamentous structure, the correlations among dipoles are allowed to be neither too strong nor too weak. Life is thus seen as existing on the border between order and disorder or between differently ordered domains. (*In electromagnetism, there are two kinds of dipoles: (1) An electric dipole is a separation of positive and negative charges. The simplest example of this is a pair of electric charges of equal magnitude but opposite sign, separated by some (usually small) distance. A permanent electric dipole is called an electret. (2) A magnetic dipole is a closed circulation of electric current. A simple example of this is a single loop of wire with some constant current through it.*[5-18a])

If the frequency of an electromagnetic filament resonantly matches a neighboring molecule, the latter is attracted to its outer surface and is oriented at the same time. In the filamentous field, chemical interactions are therefore ordered and interconnected through the resonance induction. The filamentous field is also important from the thermodynamic standpoint:

the output energy of a chemical reaction is not dispersed since it continues traveling as a polarization wave.

From the molecular reductionistic standpoint of established, molecularly centered biology, the crux of the problem of the origin of life lies in finding possible or plausible chemical pathways for the synthesis of the relevant molecules and structures. Gradual evolution of biological order should come from a chemical self-organization of matter, based on the surplus of the free energy. The essential chemical process is that of autocatalysis. However, there is strong and convincing evidence that contemporary life functions on the basis of long-range correlations established through the endogenous coherent electromagnetic field, encompassing everything from the level of proteins and nucleic acids to the intercellular level.

Even from the strictly molecular standpoint, if life originated on Earth, then its molecular constituents had to be sufficiently concentrated. At the same time as the molecular precursors of life (being mostly electrically polarized) achieved a sufficient concentration to initiate chemical evolution leading to the beginning of life, a sufficient density of electrical dipoles should also have been achieved. However, there is an important difference between chemical and electromagnetic conditions for prebiotic evolution. In the **first case** (chemical conditions) there is a very narrow range of molecular processes and their constituent molecules suitable to progress to the origin of life (or the establishment of the genetic code); in the **second case** (electromagnetic conditions) it is enough just to have an adequate concentration of polar molecules with no special reference to their exact chemical constitution (a sufficient supply of free energy being a condition in both cases).

From all that is known about the prebiotic chemistry of the primordial Earth, it is highly probable that there are countless molecules belonging to various organic chemical species (amino acids, sugars, phosphates, lipids, etc.), each of them in a very small concentration. If such conditions existed during the primordial Earth, it is very possible that the electromagnetic origin of the long-range order within the primordial "soup" (not yet within organisms) occurred before any well organized molecular process took place; it is less demanding and therefore more probable. Consequently, there are two fundamental issues: **(1)** contemporary life seems to be based on the electromagnetic long-range (coherent) order; and **(2)** the chaotic conditions on the prebiotic Earth were much more inclined to the beginning of a coherent electromagnetic field than to the more exacting specific chemical pathways leading to the known genetic code.

The most suitable environment for the beginning of the coherent regime was not in the primordial sea or a lake, as there the solution of polar organic molecules was most probably highly diluted. There were other possibilities, however, like coacervates or a sort of microspheres or the surface of clays. (Coacervates is a colloid-rich viscous liquid phase which may separate from a colloidal solution on addition of a third component.) Also, there is the possibility that organic chemical systems, perhaps even if bounded within coacervate-like forms, came to the Earth from some other planet or solar system, or that life originated from outside the Earth and was transported to it later via meteorites or some other way.

These cellular structures are similar to coacervates or microspheres protocells. It may be assumed that within a protocell there was a high concentration of organic polar molecules of very different kinds. Some of them represented a constant supply of free energy, as they trapped it from the Sun, from abundant electric discharges, or from volcanic activity. Thus, within the protocells the conditions for the emergence of coherent oscillations were fulfilled, there was an emergent coherent electromagnetic field that established a long-range order.

Protocells represented by coacervates as well as microspheres divide and form buds if there is enough "food" in the vicinity. Thus they are able to multiply, which is essential for any natural selection process. In other words, the field (through its own self-organization, stability, and close interaction with the molecular stuff of protocells) has a hereditary role, quite different from contemporary genetic mechanisms but basically functioning in a parallel way. Without some sort of heredity mechanism then no movement in the direction of the origin of life can be stable. As soon as there is a constant supply of free energy and ensured continuity of organization through multiplication and a selection process, an evolutionary process could lead to more and more adapted forms. Where exactly life begins is a matter of definition. It is believed that as soon as an open evolutionary process exists in which no information (like error catastrophe) or other barriers are encountered, life will form.

In this origin of life scenario, life emerged in a specific evolutionary process involving primitive individualized cellular structures, which was allowed through a very close mutual interaction between the endogenous coherent electromagnetic field and the polar molecular stuff of the protocells. This may be called a field-molecular complex (or FM-complex). At the beginning of the selection process the complex had a somewhat randomly organized field and a chaotic mixture of molecules. Gradually, via the selection of more stable and better organized protocells, the field became more refined, the more the molecular population selected, and the chemical processes became better organized. Thus, both components of the FM-complex evolved in parallel. At a certain point of this electromagnetic-molecular self-organization or self-organization of matter-field, the genetic code was formed.[5-18b]

Life May Have Originated Elsewhere in the Universe

Darwinist View

One theory (briefly mentioned above) states that life did not originate on the Earth, but originated elsewhere in the universe. It is believed that cellular life reached the Earth hiding inside a meteor that hit the Earth long ago. Newly uncovered evidence suggests that this might be possible, since an organism inside a meteor would be safe from the high levels of radiation in space, and would be kept at a relatively low temperature. The odds of an organism surviving inside a meteor for thousands of years, however, are not high. It is even less likely that organisms would be able to withstand the high-energy impacts of bolides into the Earth or other planetary objects. (A bolide is a large, brilliant meteor, especially one that explodes. A bolide is also called a fireball.) Most scientists today do not look at this hypothesis as a very likely origin of life on the earth. However, it is considered possible, at least for now, and so is still a candidate for life's origin on earth.[5-19] Those scientists who considered the possibly that life originated in outer space and traveled to the Earth believe this is possible. The early solar system was swarming with meteors and comets, many of which plummeted to Earth. Surprisingly there are many organic compounds in space. One theory suggests that the compounds needed to form the primordial soup may have arrived from space, either from collisions, or just from near misses from comet clouds. Even today a constant rain of microscopic dust containing organic compounds still falls from the heavens. The question remains: Could the contents of the little warm pond have come from space?

There are others who have suggested that life may have arrived from space already formed. Living cells could possibly make the journey from other worlds, perhaps covered by a thin layer of protective ice. The recent uncovering of a meteorite that originated on Mars has provided support for this theory. There is some suggestion that the meteorite contains fossilized microorganisms, but most scientists doubt this claim. However, the collision of comets and meteors is far more likely to have hindered the development of life than help create it. Objects that would have been large enough to supply a good amount of organic material would have been very destructive when they hit. It seems more likely that life began on Earth, rather in space somewhere. Also, the idea that life may have traveled to Earth does not help explain its origin; it merely transposes the problem to some distant "little warm pond" on another world.[5-20]

All these theories may seem improbable, however, they are all being considered. It has been concluded that further research needs to be conducted to determine whether abiogenesis (origin of life) is possible or not. This seemingly impossibility for life to have formed on its own has led many creationists to believe that even the simplest life forms, such as a living cell, are too complex to have self-organized.[5-21] (Cells are discussed in this chapter (5) and Chapter 6.) While scientists may disagree on some aspects that define what life is, they agree on the general characteristics that define life from non-life. Early life must have had the ability to self-replicate, in order to propagate itself and survive. Self-replication is a tricky process, implying (1) a genetic memory, (2) energy management and internal stability within the organism, and (3) molecular cooperation. Just how the ingredients of the "primordial ooze" managed to go from simple chemical process to complex self-replication is not understood. Moreover, the process of replication could not have been exact, in order for natural selection to occur. Occasional "mistakes" in the replication process must have given rise to organisms with new characteristics.[5-22] (Natural selection is discussed in Chapter 9.)

The Conditions of Earth When Life First Began

Darwinist View

When life first began, the sea was full of sugars, amino acids, and other biological chemicals that had not yet become living things. These organic molecules, called the "primordial soup," may have originally made up 10 percent of the volume of the entire sea. The very first bacteria simply absorbed these molecules as they were needed. (Bacteria are examples of a prokaryotic cell—See Figure 5.1 Prokaryotic Cell.) As these primitive bacteria began eating, they began multiplying. It was not long before the sea was thick with primitive bacteria and they had eaten the last of their "soup." The fossil remains of bacteria date back 3.5 million years.

Some bacteria used the heat from hot springs as their energy source. Others, called photosynthetic bacteria, used the Sun. Sunlight would have killed the first bacteria. But photosynthetic bacteria required sunlight. Ultraviolet or UV light is part of sunlight that is particularly harmful to living things. It breaks down fragile chemical bonds. Some forms of photosynthetic bacteria survived by developing chemicals called pigments that absorbed the harmful rays. Others lived in the shade.

Floating mats of photosynthetic bacteria known as blue-green algae created shade. In some localities, these mat colonies became embedded with dissolved minerals. After decades, layer upon layer of these minerals shaped rocky mounds, known today as stromatolites. Stromatolites are the most ancient fossils the naked eye can see. Today, stromatolites form only in lagoons where it is too salty for bacteria-eating organisms to live.

Between 2.2 and 1.8 billion years ago, photosynthetic bacteria grew throughout the world. The waste product of photosynthesis is the gas oxygen. As stromatolites spread, oxygen levels in the atmosphere rose from one part in a million to today's level of one part in five. Most of Earth's fermenting bacteria were poisoned by oxygen. This caused a great deal of bacteria to become extinct. But just as certain bacteria had grown tolerant of sunlight, certain bacteria also became tolerant of oxygen. Others developed a cell chemistry that actually required oxygen. Respiring, or oxygen burning, releases energy from food many times more efficiently than does fermentation. Certain bacteria combined photosynthesis and respiration. By both making oxygen and using it, they became well adapted to this gas. With the introduction of oxygen into the atmosphere, a type of oxygen called ozone also appeared. In the upper atmosphere it formed a layer that acted to filter out much of the harmful UV light coming in. As those damaging rays lessened, life was finally able to come out of the shade.[5-23]

Could Life Have Begun By Natural Causes?

When it comes to the origin of life, we have only two possibilities as to how life arose. One is spontaneous generation arising to evolution; the other is a supernatural creative act of God. There is no third possibility – Spontaneous generation was scientifically [sic] [scientifically] disproved one hundred years ago by

Louis Pasteur, Spellanzani, Reddy and others. That leads us scientifically to only one possible conclusion – that life arose as a supernatural creative act of God… I will not accept that philosophically because I do not believe in God. Therefore, I choose to believe in that which I know is scientifically impossible, spontaneous generation arising to evolution.[5-23a]

Creationist View

The most common evolutionary concept of spontaneous biogenesis involves living matter coming about from non-living material by chance. For example, let's suppose that in a hypothetical primordial atmosphere, ammonia, water, methane and energy can combine to form amino acids. That this first step can happen is indisputable and has been verified through laboratory experiments such as in the famous Miller/Urey experiment of 1953. (See "Life May Have Originated in a Warm Pond" section Old Universe Progressive Creationist View above for further discussion.) However, to proceed beyond this point to living proteins by chance would involve a major miracle of such great proportion that one would think it easier to just accept the obvious, that it did not happen by chance. (Stanley Miller's 1953 Experiment is discussed in Chapter 6.)

Amino acids are molecules that have a three-dimensional geometry. Any particular molecule can exist in either of two mirror-image structures that we call left-handed and right-handed (See Chapter 6, Figure 6.2). Living matter consists only of left-handed amino acids. Right-handed amino acids are not useful to living organisms, and are in fact often lethal. The random formation of amino acids produces an equal proportion of left-handed and right-handed molecules. This has been confirmed by laboratory experiment and is essentially what Miller produced in his famous test-tube experiment (putting methane, ammonia, and water together and zapping them with electrical discharges.) Life as we know it cannot consist of a mixture of left-handed and right-handed amino acids. So it would take a great deal of trial and error for nature to develop only left-handed amino acids that would create a protein to be considered living matter. (Left and right-handed amino acids are discussed in Chapter 6.)

Proteins consist of amino acids linked together with only peptide bonds. Amino acids can also combine with non-peptide bonds just as easily. In fact, experiments conducted by naturalistic origin-of-life scientists in the laboratory yield only about 50% peptide bonds. So, it would take another enormous sequence of different combinations of amino acids to come up with a protein that could constitute living matter. Any particular protein contains amino acids that are linked together in a particular sequence geometrically. This sequence must be correct for any given protein at all the active sites which comprise about half of the amino acids in the protein. Proteins contain anywhere from 50 to as many as 1750 amino acids, depending on the particular protein. There are about 20 common amino acids that comprise the basic building blocks of life. Any particular protein must have all the correct left-handed amino acids joined with only peptide bonds with the correct amino acids at all the active sites.

Let's consider the sequence of chemical reactions necessary for us to produce one particular protein contained in living matter: One amino acid can combine with another amino acid in a condensation reaction to produce a peptide (two amino acids linked with a peptide bond) and water. One peptide can combine with another peptide in a condensation reaction to produce a polypeptide and water. (A polypeptide is a peptide, such as a small protein, containing many molecules of amino acids, typically between 10 and 100.) And so goes the sequence of chemical reactions that supposedly can produce one protein essential to living organisms that can reproduce.

Each condensation reaction described so far is reversible. That is, it can occur in either the forward or the reverse direction. That means that "randomness" would be consistent with things breaking down as they are being put together. But the popular scenario involves things happening in a primordial sea, implying an excess of water. Since a condensation reaction produces water, and there is already excess water in the presence of the chemical reaction, there is much more opportunity for any complex molecule to break down into the more simple ones. Thus, a polypeptide should combine with excess water to produce monopeptides, and a monopeptide should combine with excess water to produce amino acids. The initial reagents of the supposed equations that are given as a pathway to life are favored, in the presence of excess water. Amino acids can react and form bonds with other chemical compounds, and not just other amino acids. Assuming that there is more in our "primordial sea" than just amino acids and water, we will encounter scenarios where these other reactions will take place instead of the ones we want to produce a protein. An oxygen-rich atmosphere, such as we have today, is one example of what would ruin the chemical reactions proposed for the origin of life. It is believed that the atmosphere must have originally been reducing, rather than oxidizing, containing very little free oxygen and an abundance of hydrogen and gases like methane and ammonia.

The above description only considers the formation of a single protein, not to mention that there are many different kinds of proteins necessary to form the simplest single-cell organisms. (See section "Description of Proteins" subsection "Functions and Types of Proteins.") And we haven't even begun to address the formation of the various nucleic acids and other chemical constituents of life, which must be simultaneously present (by chance). (The process of how a protein is formed will be discussed later in this chapter.) Finally, all these factors must occur in a specific arrangement to form a complex structure that would make for a reproducing organism (by chance). Many evolutionists are now proposing that not proteins, but DNA or RNA occurred first. This is very doubtful since the same amount of information must be coded into the nucleic acid in order to synthesize a protein as is represented by design and structure of the protein itself. This makes such scenarios to be at least as unlikely. The spontaneous organization of nucleic acids into DNA or RNA suffers in concept from the same problems that the spontaneous organization of amino acids suffers from. All nucleic acids must **(1)** be right-handed, **(2)** form particular bonds, in a particular arrangement, in chemical reactions that proceed in a particular direction and **(3)** are not spoiled by other chemical reactions.

Some evolutionists are proposing that life originated not in a primordial sea but on some clay template. Again, this is doubtful, since the clay template must by necessity be as complex as what is formed on the template. This makes such scenarios to be at least as unlikely. Furthermore, the evolution of informational "defects" in the crystalline structures of clays has never been observed or demonstrated in theory. Shifting the medium for evolution from biological molecules to poly-aluminum silicates solves nothing. The classic examples given for the formation of some of the basic building blocks of life by chance therefore lacks substance on a theoretical basis both according to the principles of chemistry, the principles of probability and statistics, and the principles of basic information theory. Without proper theoretical or experimental basis, a scientific hypothesis cannot be supported. The formation of living matter from non-living matter by chance remains within the realm of speculation without foundation.[5-24] (See section "Other Theories About How Life Could Have Originated on Earth". Also see Chapter 6, Section "The Improbability of Life Forming Naturally by Chemicals" for further discussion of the impossibility of life forming naturally by chemicals.)

Old Universe Progressive Creationist View

Since it has been determined that life originated during an early Earth, many believe life developed shortly after the formation of the Earth. When the Earth was first formed, the conditions were very hostile to life. If life originated shortly after the origin of the Earth, life must have begun during the early Earth's hostile environment. Naturalistic origin-of-life researchers have therefore been studying more extreme environments in an attempt to determine how life originated on Earth.

Most extremophiles (extreme-loving) are single-celled microbes called archaea (a classification of bacteria). (Archaea (also known as Arechaeabacteria) is found in extreme environments such as in deep sea, hot springs, alkaline or acid water. The early planet had different environment composition from the environment of today. This oldest living organism (Archaea) had tolerance for that harsh environment. [5-24a]) Most bacteria are mesophiles (organisms requiring moderate conditions to survive). A few eubacteria (another classification of bacteria) life have been known to live at temperatures between 140° and 180° Fahrenheit (60° and 80° Celsius). These eubacteria are considered extremophiles. Extremophiles inhabit environments that simulate some of the extreme conditions of early Earth.

Life appeared early in Earth's history. Geochemical evidence indicates that life existed at or slightly before 3.8 billion years ago. The oldest rocks date 3.9 billion years ago. Prior to this time Earth existed (mostly) in a molten state unsuitable for living organisms. Life also appeared suddenly. Between 4.5 and 4.9 billion years ago, Earth experienced numerous impact events that sterilized its surface and subsurface. These impacts melted rock and volatilized oceans. Then, as soon as Earth's conditions were remotely able to support life, life appeared. The discovery of archaea (a classification of bacteria) and a few eubacteria (a classification of bacteria) in hostile environments suggests to some naturalistic origin-of-life researchers that life could have arisen under extreme conditions of early Earth prior to 3.9 billion years ago. They believe that the extremophiles originated first and provided a means for mesophiles to develop. The circumstantial evidence favoring any extremophilic origin-of-life scenarios may appear impressive, however, upon further investigation, this evidence appears less likely to be the origins of life. The discovery of extremophiles does not automatically explain how life originated on Earth. Just because life exists in extreme environments does not mean it originated there.

If extremophiles were the first life, at some point within the evolutionary paradigm they must have given rise to mesophiles (organisms that grow best in moderate temperatures, neither too hot nor too cold, typically between 20° and 45° Celsius (68° and 113° Fahrenheit)[5-24b]). This may not be as easy as it first appears. The fossil and geochemical record indicates that surface mesophiles appeared as soon as the Earth could possibly support life. This timing leaves only a few million years for subsurface or surface extremophiles, employing chemoautotrophic pathways (pathways that extract chemical energy from inorganic materials in the environment) to evolve into mesophilic photoautotrophs (organisms that rely of photosynthesis). Biochemical changes to support this transition would have had to occur all at once in practically every protein, transfer RNA, and ribosome RNA molecule. But biochemical alterations stabilizing extremophilic biomolecules typically make them unsuitable for mesophilic environments. Microbiology also causes doubt about an extremophile-to-mesophile transformation. Although extremophiles thrive under conditions inhospitable for most life, they are still particular. Extremophiles simply do not tolerate mesophilic conditions very well. While they can survive nonextreme conditions in a dormant state, they do not grow under mesophilic conditions. For example, Pyrolobus fumarii cannot grow below 94° Fahrenheit (34.4° Celsius). The need for wholesale biochemical changes, coupled with extremophile's inability to grow and flourish (and therefore to produce population numbers necessary to evolve) upon migration into mesophilic environments, raises questions as to the likelihood of mesophiles' rapid emergence from extremophiles.

Both temperature extremes, hot and cold, pose problems. At high temperatures the complex structure of proteins and RNA unfolds. This occurs because high temperatures disrupt the interactions that stabilize the biomolecules three-dimensional architecture. Ironically, temperatures approaching 32° Fahrenheit (0° Celsius, water's freezing point) also threaten protein and RNA structure. At cold temperatures, as water begins to develop a more orderly ice-like structure below 39° Fahrenheit (or 4° Celsius), membrane aggregation and the folding of protein, RNA, and DNA molecules become less likely because water already possesses a high degree of order. The hydrophobic effect ceases to operate at cold temperatures. Aggregation and segregation cannot produce disordered water because the reduced temperature has already forced water into an ordered state. This phenomenon has important implications for a cold origin of life. The three-dimensional structures of proteins, RNA, and DNA, critical to their function, cannot be maintained (without all the protective features of living cells already in place) at temperatures near or below 39° Fahrenheit (4° Celsius). Here again, the naturalistic emergence of life appears impossible. This was confirmed when biochemists recently demonstrated the cold denaturation of an RNA enzyme (the hammerhead ribozyme). This finding means that cold denaturation prevents any low temperature origin-of-life scenario that proceeds through any organism using RNA.[5-24c]

Life Originated Too Early in Earth's History to Have Originated by Natural Causes

Old Universe Progressive Creationist View

The fossil and geochemical data recovered from some of the world's oldest geological formations consistently indicate that life was present early in Earth's history. Prokaryotic microorganisms (discussed in more detail later) were definitely on Earth at 3.7 billion years ago. The record for ancient life may be as early as 3.8+ billion years ago. Naturalistic origin-of-life researchers acknowledge life's existence on Earth prior to 3.5 billion years ago (See "The Conditions of Earth When Life First Began" "Darwinist View" above). Several lines of fossil (stromatolites and microfossils) and geochemical (carbon-12, nitrogen-14, and sulfur-32 isotope enrichment and banded iron formations (BIFs)) research all independently indicate that life began when the Earth was young. (Note: in 2015 it was concluded that structures that were once thought to be Earth's oldest microfossils are actually multiple episodes of subsurface fluid flow over a long period of time that were formed by the redistribution of carbon around mineral grains during hydrothermal events.[5-24d]) Additionally, much of the evidence works together to substantiate life's ancient existence. For example, stromatolites, microfossils that resemble various types of cyanobacteria, BIFs (banded iron formations), and carbon-12-enriched kerogen not only indicate microbial activity but also reveal the type of validation scientists expect in the geological record if indeed photosynthetic bacteria were present early in Earth's history. For naturalistic origin-of-life researchers, the conclusion is inescapable: early Earth had an abundant variety of microbial life forms.

The fossil and geochemical records not only indicate the timing of life's appearance on Earth but also provide a means to assess the biochemical and metabolic properties of that life. Researchers now realize that Earth's first life, while morphologically simple, was biochemically complex. The cyanobacteria likely present on Earth 3.5 billion years ago appear to have been identical to the cyanobacterial forms on Earth today. Cyanobacteria are some of the most biochemically complex microbes known today. The fossil chronology indicates that by 3.5 billion years ago complex microbial ecosystems were already in place. The presence of stromatolite fossils imply that bacteria capable of both oxygenic and anoxygenic photosynthesis must have existed along with aerobic and anaerobic heterotrophs (creatures that feed on organic materials produced by other organisms). Naturalistic origin-of-life biologists believe that the existence of cyanobacteria on early Earth means that nearly all bacteria groups must have evolved by 3.5 billion years ago and that prokaryote (single-celled organisms) diversity was fully established by that time. This is because the placement of cyanobacteria (microbes) are considered the most recently evolved and advanced bacterial groups.

Origin-of-life scenarios must account not only for life's early appearance but also for life's diversification in a time frame of less than 400 million years. This is because naturalistic origin-of-life scenarios depend on the existence of anoxygenic photosynthesis well before 3.5 billion years ago since the fossil and geochemical evidence places cyanobacteria and oxygenic photosynthesis on Earth 3.5 billion years ago (See Chapter 3, Section "The Strata Layers of the Earth" sub-section "Archean Eon" for further discussion of when life originated on Earth). [5-24e]

The Cell—the Most Basic Form of Life

Darwinist View

Cells are considered the basic building units of life. A cell is the smallest structural unit of an organism that is capable of independent functioning, consisting of one or more nuclei, cytoplasm, and various organelles, all surrounded by a semipermeable cell membrane (See section "The Origin of the Cell Membrane" below for discussion of the origin of the cell membrane).[5-25] As mentioned previously, only a living thing such as a cell can: **(1)** absorb nutrients and grow, **(2)** reproduce on its own, **(3)** respond to stimuli, and **(4)** evolve.[5-26] A living thing is considered a living organism if it can **(1)** process energy, (2) construct other components (3) duplicate information (4) store information (5) replicate (duplicate itself). [5-26a]

The most important compounds for cell development are amino acids (also discussed in Chapter 6). Research has led to the theory that organic compounds could have been created in the early atmosphere. Further studies showed that some amino acids would have combined with hydrogen cyanide (HCN), which is a by-product of volcanic activity. This combination would form purines and pyrinidines, which are used to make nucleic acids, which in turn create DNA (deoxyribonucleic acid). After these compounds had been created on the early Earth, the Earth eventually began to cool. Water vapor then condensed, which formed vast oceans, seas and lakes, in which simple organic molecules began to accumulate for millions of years, producing an "organic soup" of sorts. The amino acids would have then polymerized (which means they formed chains, such as proteins). The most likely theory as to how the amino acids combined is that they were washed up into clay/rock depressions on land, where the water evaporated, leaving behind concentrated organic compounds in high heat.[5-27]

All known organisms need DNA (deoxyribonucleic acid) to replicate and proteins to run cellular functions. There are thousands of intertwining atoms that must function properly. The problem is that, most likely, there was no DNA in existence in the beginning for life to form. Scientists believe that the first life forms were small molecules that were self-contained chemical organisms that grew, reproduced and evolved without needing the complicated molecules that currently exist today. Possibly, molecules were much simpler in the beginning. Biologists have developed various ideas as to how this could have taken place. They have theorized that life could have begun in (1) tidal pools, (2) near underwater volcanic vents, (3) on the surface of clay sediments, or even (4) in outer space. Scientists do not know whether the first complex molecules were proteins or DNA or possibly something else. The problem is that proteins are needed to replicate DNA, but DNA is necessary to instruct the building of proteins. (DNA, RNA, and various types of amino acids are needed to instruct a cell as to how to make a specific protein. See sections "Protein Synthesis (How a Protein is Developed Inside a Cell)" and "The Process of Creating a Protein From DNA Instructions" below for further discussion.)

Many biologists believe that RNA (ribonucleic acid), similar to DNA, may have been the first complex molecule from which life is based upon. RNA carries genetic information like DNA, however, it can also direct chemical reactions as proteins do.[5-28] It was discovered that under some circumstances RNA can replicate on its own. Not only that, but it can store genetic information. RNA, in other words, can do it all. Scientists now believe that early life passed through a stage in which only RNA was present. They believe that all that is needed for life, besides those crucial amino acids, are the ingredients for RNA: ribose, a sugar; phosphate, a salt; and the four bases— adenine, cytosine, guanine, and uracil (thymine replaces the uracil in DNA). DNA and RNA have some differences. DNA contains the sugar deoxyribose, while RNA contains the sugar ribose. DNA is a double stranded molecule while RNA is a single stranded molecule. And as just mentioned, DNA uses the bases adenine, thymine, cytosine, and guanine, whereas RNA uses adenine, uracil, cytosine, and guanine. The question is, can these molecules be produced in an atmosphere where significant oxygen is present? Some believe that this may be accomplished by the use of methane.

Scientists have discovered that ribose is simply five molecules of formaldehyde strung together and formaldehyde is easy to make where there is carbon dioxide and light. Phosphate occurs routinely with the weathering of rocks. And all four bases, A, C, G, and U, (adenine, cytosine, guanine, uracil) can be synthesized from hydrogen cyanide, for which a sprinkling of methane is needed. The problem facing scientists now is to figure out if there was a good source for methane in the early atmosphere.[5-29]

Creationists and Intelligent Design advocates believe that even RNA (ribonucleic acid) is too complex to be the origin of life. RNA molecules that carry information are sequences of molecular bits. The primordial soup would be full of chemicals that would terminate these sequences before they grew long enough to be useful. The first life would not have the genetic material necessary to copy itself. It is believed that instead of complex molecules first forming to produce life, life started with small molecules interacting through a closed cycle of chemical interactions. These interactions would then produce compounds that would return to the cycle and create more complex molecular systems.[5-30] It is not exactly known for sure how RNA and DNA molecules evolved, however, laboratory experiments have been able to determine how some of the building parts were produced spontaneously under conditions that simulate the primitive earth. (Experiments that attempted to demonstrate how life could have originated on a primitive Earth are discussed in Chapter 6.) In current simulations of a primitive earth's atmosphere and under certain circumstances, nitrogenous bases were formed, which are found in RNA. It is rather easy to simulate experiments to produce adenine and cytosine, which are two of the bases for RNA. (The other two bases of RNA are guanine and uracil.)[5-31]

In 1961, researchers found that amino acids could be made from hydrogen cyanide (HCN) and ammonia (a compound made of Nitrogen and Hydrogen (NH_3)) in an aqueous solution. The key elements of an amino acid are carbon (C), hydrogen (H), oxygen (O), and nitrogen (N), although other elements are found in the side chains of certain amino acids. One experiment produced an amazing amount of the nucleotide base, adenine. Adenine is of tremendous biological significance as an organic compound because it is one of the four bases in RNA and DNA. (The other three DNA bases are guanine, cytosine, and thymine. For RNA, uracil is present instead of thymine.) Adenine is also a component of adenosine triphosphate (ATP), which is a major energy-releasing molecule in cells. Experiments conducted later showed that the other RNA and DNA bases could be obtained through simulated prebiotic chemistry with a reducing atmosphere. Scientists became very optimistic that the questions about the origin of life would be solved within a few decades. This has not been the case, however. Instead, the investigation into life's origins seems only to have just begun.[5-32]

Intelligent Design View

Living systems have the capacity for self-duplication. The construction of any sort of self-duplicating automation would necessitate the following functions: **(1)** storing information; **(2)** duplicating information; and **(3)** designing an automatic factory which could be programmed from the information stored to construct all the other components of the machine as well as duplicating itself. These three functions are found in all living things.[5-33]

Creationist and Intelligent Design View

Before the 1900s, it was thought that a cell was composed of a nucleus and a few other parts in a 'sea' of cytoplasm, with large spaces in the cell unoccupied. Now it is known that a cell is very complex and is full of important, functioning units necessary to the life of the cell and the body containing it. The theory of evolution assumes life developed from a simple cell, but science today demonstrates that there is no such thing as a simple cell.[5-34] It has been observed that self-replicating systems capable of Darwinian evolution appear too complex to have arisen suddenly from a prebiotic soup. This conclusion applies both to nucleic acid systems and to hypothetical protein-based genetic systems.[5-35] (Irreducible complexity is discussed in Chapter 14.)

The most difficult problem for evolutionists to explain is the origin of the apparatus of reproduction of early life, that is, a simple cell. Reproduction in present living cells is based on 2 groups of functions of macromolecules. Since no macromolecule seems to exist which can perform both functions at the same time, both are allotted separately to 2 substances, proteins and nucleic acids, in contemporary organisms. As a consequence, the synthesis of both polymers must be coupled mutually, proteins being formed due to information from nucleic acids (translation), and nucleic acids are replicated by catalysis due to proteins (replication and transcription). (Polymers are the simplest unit of a chemical compound that can exist, consisting of two or more atoms held together by chemical bonds.) The number of components making up the reproduction apparatus of present life is large, at least about 80 special proteins and 100 genes coding for them as well as for tRNA's and rRNA's are necessary. (Translation, polymers, proteins, nucleic acids, replication, transcription, tRNA's and rRNA's will be discussed later in this chapter.)[5-36]

Nucleic acid in DNA and protein need each other, but how could one originate without the other? Both depend on one other. Nucleotide sequences can be translated into protein synthesis only when mediated with enzymes, themselves proteins. So which came first? DNA or protein? Although Stanley Miller's experiments provided a way for demonstrating synthesis of amino acids, there has been no synthesis of nucleotides. (Stanley Miller's experiment is discussed in Chapter 6.) The prebiotic synthesis of nucleotides has not yet been accomplished, and this remains a problem for evolutionists.[5-37] Even if the spontaneous formation of nucleic acid could be explained, life today depends on both nucleic acids and proteins. Nucleic acids require proteins for their formation and proteins require nucleic acids. In addition, we can be certain that they did not both appear simultaneously, and thus a classic predicament occurs for which the evolutionist has no answer.[5-38]

There are many single-cell forms of life, but there are no known forms of animal life with 2, 3, 4, . ., or even 20 cells. If organic evolution happened, one would expect to find these forms of life in great abundance as transitional forms between one-celled and many-celled organisms.[5-39] In present-day organisms, DNA (deoxyribonucleic acid), RNA (ribonucleic acid), and proteins are mutually interdependent, with DNA storing the genetic information and copying it to RNA. The RNA in turn directs the synthesis of proteins. The proteins carry on the essential chemical work of the cell. They all work together within the cell. Evolutionary theory states that this complex system evolved from a much simpler system that most likely began with one of these three components (DNA, RNA, proteins). The question is which came first, the nucleic acids (DNA and RNA) or the proteins? Also, how could the first living molecule function and evolve without the other components? Some evolutionists believe that RNA may have developed first because it not only functions as the carrier of genetic information but also because it is capable of catalyzing some chemical reactions in the manner of proteins. The problem with the evolutionary theory is that there is a big difference between something being probable and something being experimentally verifiable.

Naturalistic origin-of-life chemists realize there are many problems with the idea that a self-replicating RNA molecule could have evolved from organic compounds on the early earth. It is understood that RNA would have had to evolve from some simpler genetic system that is no longer in existence. It has been proposed that clay crystals have qualities that might make possible their combination into a form of pre-organic mineral life. Then, the idea is that natural selection (discussed in Chapter 9) would then favor the more efficient clay replicators, preparing the way for an eventual genetic takeover by organic molecules that had evolved because of their increasing usefulness in the pre-organic process. Although this scenario regarding the origin of life is interesting, it is not scientifically proven. To date (1991), no verifiable explanation has been proposed how life could have originated on earth.[5-40]

As mentioned previously, evolutionists believe that the first information-containing molecules needed to create the first cell genome were composed of RNA, rather than DNA. This is highly improbable because of the need to have a molecule that could not only contain informational instructions or code, but could replicate itself and even perform other catalytic functions. In the typical cell scenario, genetic information in the form of a code is carried in the DNA molecule. When the cells want to make a protein, a copy of a gene (part of the DNA code) is made using another type of nucleic acid, using another type of nucleic acid, called a transcript or messenger RNA. The genetic code is never executed directly from DNA. Many evolutionists believe that RNA originated before DNA because RNA is simpler then DNA and RNA is also able to perform a variety of functions. It is true that RNA is utilized directly to create a protein (see section "How a Protein is Made—Step by Step " for more details), however, this is done generally in conjunction with one or more proteins and occurs as part of an intricate process involving a variety of other biomolecules. Currently (as of 2012), RNA has never been observed to replicate itself or any other molecule.

Another reason why RNA could not have originated prior to DNA is because RNA typically occurs as a single-stranded nucleic acid. The RNAs that causes some sort of catalytic activity do so by looping back on themselves and base-pairing with complimentary sequences. Thus, they assume unique shapes. The RNAs that perform this sort of task without being bound in some protein are very short, and for good reason. Because RNAs are single-stranded, they are very unstable, especially when they are not present in their highly engineered and protected cellular environment. The DNA molecule occurs in a double-stranded form and is packaged with a variety of protective proteins, making it considerably more stable. Thus, DNA has been chosen as the material for use in lipid packaging experiments that supposedly show how a first cell may have formed. However, since DNA is more complex than RNA, it is doubtful that DNA could have originated before RNA.[5-40a]

Description of a Cell

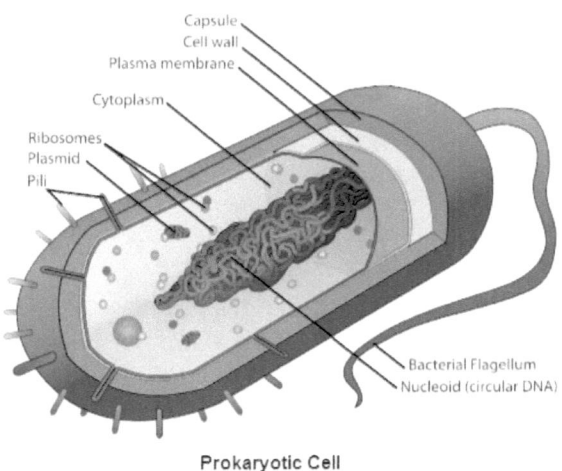

Figure 5.1 Prokaryotic Cell (structure of a bacterium)

- - - - - - - - - - - - - - - - - - - -

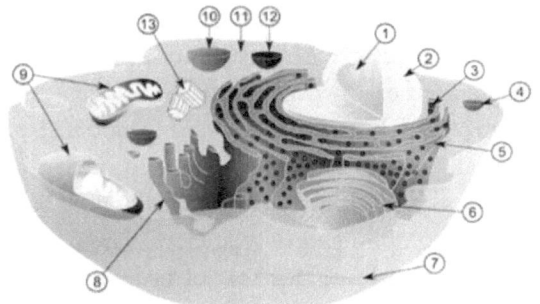

1. Nucleolus (See Figure 5.3 & 5.4)	8. Smooth Endoplasmic Reticulum
2. Nucleus (See Figure5.4)	9. Mitochondria (See Figure 5.3)
3. Ribosomes (See Figure5.3)	10. Vacuole
4. Vesicle	11. Cytosol
5. Rough Endoplasmic Reticulum	12. Lysosome (See Figure5.3)
6. Golgi apparatus (See Figure5.3)	13. Centrioles within Centrosome
7. Cytoskeleton	

Eukaryotic Cell

Figure 5.2 Eukaryotic Cell (plant and multi-celled animal)

- - - - - - - - - - - - - - - - - - - -

Cell Structure

Cilia

Lysosome

Centrioles

Microtubules

Golgi apparatus

Smooth endoplasmic reticulum

Mitochondrion

Rough endoplasmic reticulum

Cell membrane

Cytoplasm

Nucleolus

Chromatin

Ribosomes

Nuclear membrane

Eukaryotic Cell

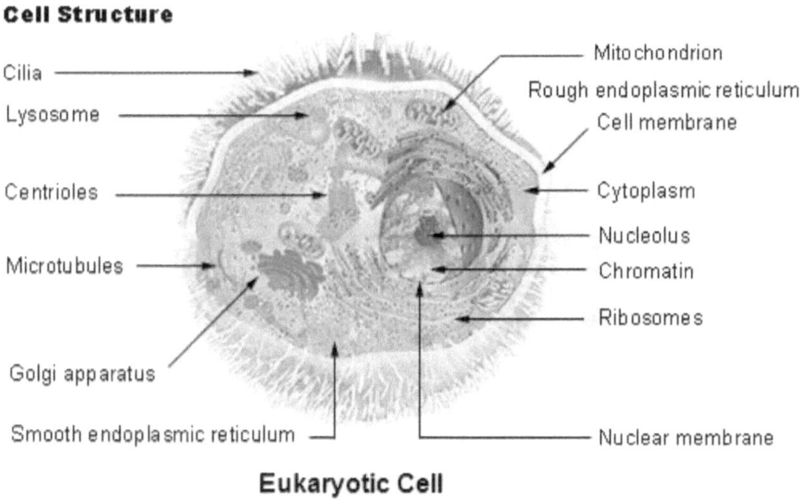

Figure 5.3 Eukaryotic Cell (plant and multi-celled animal)

There are two types of cells: (1) eukaryotic and (2) prokaryotic. Eukaryotic cells (Figures 5.2 & 5.3) are usually found in multicellular organisms (such as plants and animals), whereas prokaryotic cells (Figure 5.1) are usually found by themselves. Bacteria are examples of the prokaryotic cell type. In general, prokaryotic cells are those that do not have a membrane-bound nucleus. The eukaryotic cell is larger and more complex than a prokaryote cell. The prokaryote cell does not have a nucleus and most of the parts (organelles) included in a eukaryotic cell (Figure 5.4).[5-41]

The cytoplasm (made up of water, salts, and organic molecules) (Figure 5.3) forms the cell's internal area. Organelles are large structures embedded with the cytoplasm that are located within the eukaryotic cells. Organelles perform various functions within the cell (See figures 5.2 & 5.3 for locations in the cytoplasm of the organelles being discussed). The endoplasmic reticulum functions like an assembly line with a conveyor belt that prepares and processes proteins for export. The golgi apparatus makes biochemical modifications to the contents of vesicles that originate from the endoplasmic reticulum and then distributes these vesicles to different locations throughout the cell. Lysosomes break down food, unneeded molecules, and damaged cell components. Mitochondria, acting like the cell's power plant, carry out the biochemical reactions that extract energy from organic materials. This energy is stored in the form of high-energy chemical compounds used by the cell to power its processes.[5-41a]

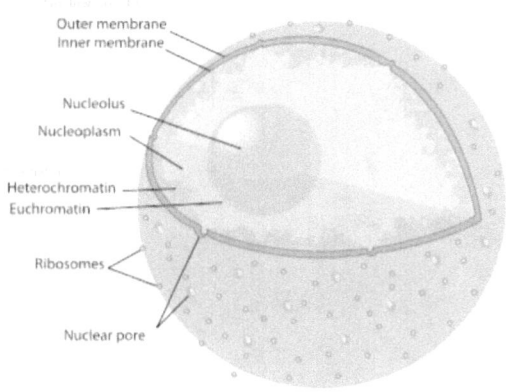

Cell nucleus of Eukaryotic Cell

Figure 5.4 Cell nucleus of Eukaryotic Cell (plant and animal)

The nucleolus is a non-membrane bound structure composed of proteins and nucleic acids found within the nucleus of a eukaryotic cell. Ribosomal RNA (rRNA) is transcribed and assembled within the nucleolus. Transcription is the process of creating a complementary RNA copy of a sequence of DNA (See figures 5.6 & 5.9, and sections "Protein Synthesis (How a Protein is Developed Inside a Cell)", and "How a Protein is Made—Step by Step" for further discussion on transcription.).[5-42] The nuclear pores (Figure 5.4) control the passage of materials in and out of the nucleus. Within the nucleolus (Figure 5.4), ribosomes (figures 5.2, 5.3, 5.4) are assembled, then they are exported from the nucleus to the cytoplasm and endoplasmic reticulum (figures 5.2 & 5.3).[5-42a]

The Origin of the Cell Membrane

Old Universe Progressive Creationist View

Biological membranes are very important to a cell. The cell membrane (also known as the plasma membrane or cytoplasmic membrane) is a biological membrane that separates the interior of all cells from the outside environment (Figure 5.3). The cell membrane consists of the phospholipid bilayer with embedded proteins (Figures 5.4a & 5.4b). Cell membranes are involved in a variety of cellular processes such as cell adhesion, ion conductivity and cell signaling and serve as the attachment surface for several extracellular structures, including the cell wall, glycocalyx, and intracellular cytoskeleton.[5-42b]

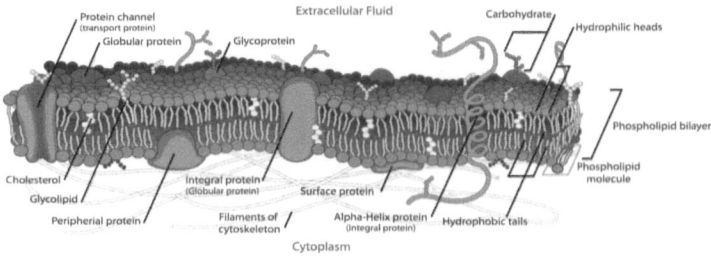

Figure 5.4a Cell Membrane

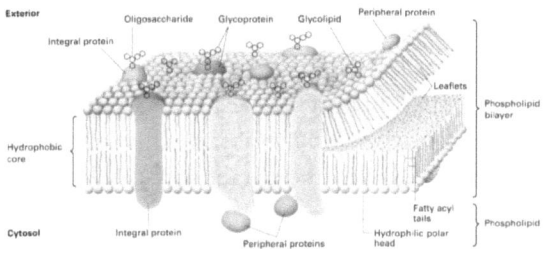

Figure 5.4b Cell Membrane Structure

Cell membranes (figures 5.3, 5.4a, 5.4b) keep harmful materials from entering the cell and isolates and protects the beneficial compounds inside it. Proteins embedded in the cell's membrane regulate the traffic of materials going into and out of the cell. These transport proteins ensure that the cell has the necessary nutrients and can efficiently expel waste products. Because the cell membrane defines life, determining whether it could have occurred through natural processes is important. If cell membranes occurred through natural processes, naturalists must explain how cell membranes originated in the early stages of the cell's origin. The beginning of cell membranes represents one of the first steps in the emergence of protocells. In spite of the cell membranes being so tiny (the membranes that form the cell's boundary are only 3.5 to 4 nanometers thick), cell membranes are very complex.

Two general classes of biomolecules, lipids and proteins, interact to form cell membranes. Lipids are a group of structurally dissimilar compounds that share water insolubility as a common and defining property. These compounds readily dissolve in organic solvents. Cholesterol, triglycerides, saturated and unsaturated fats, oils, and lecithin are some of the more widely recognized examples of lipids. Phospholipids (figures 5.4a & 5.4b) are the cell membrane's major lipid component. The phosopholipids are composed of basically two parts, the rounded portion and two tail-like portions. The rounded portion is water soluble (hydrophilic – water-loving) while the two tail-like portions are water insoluble (hydrophobic – water-hating). Phospholipids schizoid solubility properties cause them to organize into cell membranes. When added to water, phospholipids spontaneously organize into bilayers, sheets two molecules thick. In a bilayer, phospholipid molecules align into two monolayers with the phospholipid head groups (the rounded portion) adjacent to one another and the phospholipid tails packed close together. The monolayers, in turn, collect so that the phospholipid tails of one monolayer interface with the phospholipid tails of the bilayer's other monolayer. This tail-to-tail arrangement ensures that the water-soluble head groups contact water and the water-insoluble tails stay away from water.

Proteins are also very important to the cell membrane. These molecules (proteins) associate with the cell membrane in multiple ways. Peripheral proteins bind to the inner or outer membrane surfaces. Integral proteins embed into the cell membrane. Some integral proteins insert only partially into the membrane interior and still others span the entire membrane. (Proteins are further discussed in sections "Description of Proteins," "Protein Synthesis (How a Protein is Developed Inside a Cell)," "The Process of Creating a Protein From DNA Instructions" and "The Production of Proteins Within a Cell (The Protein Factory).")

Membrane proteins function as receptors, binding compounds that allow the cell to communicate with its external environment. They catalyze chemical reactions at the cell's interior and exterior surfaces. They shuttle molecules across the cell membranes and form pores and channels through the membrane. Some of these proteins provide structural integrity to the cell membrane. The inner and outer monolayers of cell membranes differ in composition, structure, and function. These differences make them asymmetric, that is, they are not similar in shape, size, or arrangement. The phospholipid classes on the inner and outer membrane surfaces are unique; the membrane proteins, likewise, are specific to either the inner or outer surfaces. Proteins that span the cell membrane possess a specific orientation. Because of protein dissimilarity, the functional characteristics of the inner and outer surfaces vary.

Most proteins are confined to certain areas with the membrane. Other proteins diffuse throughout the membrane, but instead of moving randomly, these proteins move in a directed fashion. Phospholipids, too, organize into domains (areas) with certain phospholipid classes laterally segregating in the bilayer. Bilayer fluidity also varies from region to region in the membrane. Cell membranes are not inert barriers. These intricate biosystems display a structure critical to life. Naturalistic origin-of-life researchers have difficulty in explaining the cell membrane's origin.

Because the cell membrane is so vital for a living cell to exist, evolutionists must explain how cell membranes composed of phospholipids developed by natural causes. Even if the first phospholipids on Earth were the ones that readily formed bilayers, they still would not have led to the spontaneous assembly of cell membrane systems. Bilayer-forming phospholipids have complex properties. Phospholipid bilayers spontaneously stack into sheets (multilamellar bilayers) or spherical structures that consist of multiple bilayer sheets. These groups only superficially resemble the cell membrane's single bilayer structure. Bilayer-forming phospholipids can form structures composed of a single bilayer. These particular groups arrange into a hollow spherical structure called liposomes or unilamellar vesicles. Liposomes do not form spontaneously but rather result only with laboratory manipulation. They exist for a limited lifetime and are considered a metastable phase. Liposomes fuse to revert to multilamellar sheets or vesicles. The question is, how is it that bilayer-forming lipids form multiple-bilayer sheets or relatively unstable vesicles (liposomes) when the cell membrane is made up of a stable single bilayer phase?

Single bilayer phases, similar to those that constitute cell membranes, are stable but form only under unique conditions. Formation of single bilayer vesicles occurs only at a specific temperature. Pure phospholipids spontaneously transform from either multiple bilayer sheets or unstable liposomes into stable single bilayers only at the critical temperature. This temperature depends on the specific phospholipid or on the bilayer's phospholipid composition. Research has determined that phospholipids extracted from rat and squid nervous-system tissue assemble into single bilayer structures at critical temperatures that correspond to the physiological temperatures of these organisms (rat and squid nervous-system tissue). It has been observed that for the cold-blooded sea urchin L. pictus, the cell membrane composition of the earliest cells in the embryo varies in response to the environment's temperature to maintain a single bilayer phase with a critical temperature matching the environmental conditions. Also, the bacterium E. coli also adjusted its cell membrane phospholipid composition to maintain a single bilayer phase.

Cell membranes are highly fine-tuned molecular structures dependent on a specific set of physical and chemical conditions. It is highly unlikely that chemical and physical processes operating on early Earth could have produced the precise phospholipid composition to form the stable single bilayer phase that universally defines cell membranes. Even if chance events arrived at this just-right phosopholopid composition, any fluctuations in temperature would have destroyed the single bilayer structure. With the loss of this structure, the first protocells would have fallen apart.[5-42c]

Description of a Cell – Continued

As discussed previously, a cell is the smallest structural unit of an organism that is capable of independent functioning, consisting of one or more nuclei, cytoplasm, and various organelles, all surrounded by a semipermeable cell membrane (See section "The Origin of the Cell Membrane" that discusses the origin of the cell membrane.). Cells are considered the basic building units of life. Every cell is made up of many smaller parts. There is the nucleus present in most cells, which contains DNA (deoxyribonucleic acid), which stores the blueprint of the cell (Figures 5.4, 5.5, 5.6, 5.9).

There are mitochondria, which break down compounds and produce energy. There are many other parts in a cell, each of which has a specific function (some of which are discussed above in section "Description of a Cell"). These parts of the cell are made up of proteins, which are made up of long strands of amino acids, which are made up of different combinations of the base elements of carbon (C), hydrogen (H), nitrogen (N), and oxygen (O), although other elements are found in the side chains of certain amino acids. Of all the possible types of combinations forming amino acids, only 20 amino acids are used in proteins. However, these 20 amino acids can form almost infinite numbers of combinations to create an almost infinite number of proteins. (Proteins are discussed in more detail in sections "Description of Proteins" and "Functions and Types of Protein".)

For cells to reproduce, they must divide. Cell division involves a single cell (called a mother cell) that divides into two daughter cells. (See section below "Types of Cell Division—Mitosis and Meiosis".) This leads to growth in multicellular organisms (the growth of tissue) and to procreation (vegetative reproduction) in unicellular organisms. The walls of a cell are made up of lipids in which are embedded certain protein molecules that determine what is allowed to leave and enter the cell.[5-43]

Types of Cell Division—Mitosis and Meiosis

As mentioned previously, for cells to reproduce, there must be cell division. Mitosis and meiosis refer to types of nuclear division. **Mitosis** is the process of cell division in which the nucleus of a cell normally divides into two identical nuclei. It is the process by which the nucleus divides in eukaryotic organisms to form new cells. **Meiosis** is the process of cell division in sexually reproducing organisms to form new reproductive cells. Meiosis only occurs in reproductive cells. This is also cell division, but in this process, the number of chromosomes in each sex cell is halved.[5-44] Chromosomes are microscopic rod-shaped structures that appear in a cell nucleus during cell division. Chromosomes are composed of DNA and a protein (Figure 5.5) that form in the nucleus when the cell begins to divide and that carry the genes which determine an individual's hereditary traits. A chromosome is a structure in all living cells that consists of a single molecule of DNA bonded to various proteins and that carries the genes determining heredity. In all eukaryotic cells, the chromosomes occur as threadlike strands in the nucleus. During cell reproduction, these strands coil up and condense into much thicker structures. Chromosomes occur in pairs in all of the eukaryotic cells (figures 5.2 & 5.3) except the reproductive cells, which have one of each chromosome, and some red blood cells (such as those in mammals) that expel their nuclei. In bacterial cells and other prokaryotic cells (Figure 5.1), which have no nucleus, the chromosome is a circular strand of DNA located in the cytoplasm.[5-44a]

Cell Division in Prokaryotes

Prokaryotes such as bacteria use a relatively simple form of cell division called binary fission. Typically, bacterial chromosomes consist of a single loop of DNA, often called circular DNA. (Eukaryotes have a linear DNA molecule.) During the duplication of the DNA in prokaryote cells, the bacterial chromosome replicates, leading to two identical chromosomes attached to separate points of attachment. This may seem like a simple process but it is not. As the cell begins to divide, each cell will have an identical chromosome. The result is two identical daughter cells.[5-45]

Description of a Cell – Continued

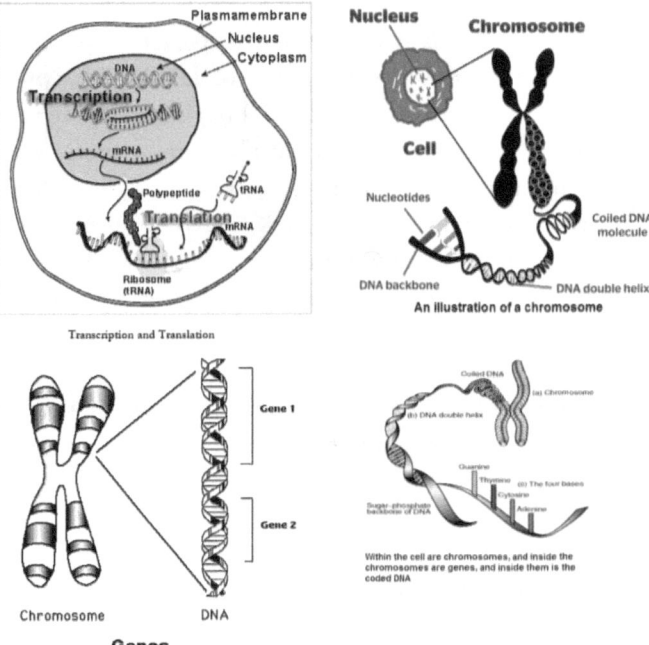

Figure 5.5 Chromosomes DNA & Genes

In the nucleus of each cell (Figures 5.5, upper left & 5.6), the DNA molecule is packaged into thread-like structures called chromosomes (Figure 5.5 lower left and upper right). Each chromosome is made up of DNA (deoxyribonucleic acid) tightly coiled many times around proteins called histones that support its structure. Chromosomes are not visible in the cell's nucleus, not even under a microscope, when the cell is not dividing.[5-46g] Genes are small sections of DNA (Figure 5.5 lower left) within the genome that code for proteins. They contain the instructions for our individual characteristics, such as eye and hair colour.[5-46h] Most DNA is located in the cell nucleus (where it is called nuclear DNA) (Figures 5.5, upper left & 5.6), but a small amount of DNA can also be found in the mitochondria (where it is called mitochondrial DNA or mtDNA) (Figure 5.2). The information in DNA is stored as a code made up of four chemical bases adenine (A), guanine (G), cytosine (C), and thymine (T) (Figures 5.5 lower right & 5.19).[5-46k] Genes are made up of DNA. Each chromosome contains many genes. [5-46j] A chromosome is a section of DNA that houses genes (Figure 5.5 lower left).[5-46]

A chromosome is a packaged and organized structure containing most of the DNA of a living organism. It is not usually found on its own, but rather is structured by being wrapped around protein complexes called nucleosomes, which consist of proteins called histones.[5-46a] A chromosome is a strand of DNA that is encoded with genes (Figure 5.5 bottom left).[5-46b] A chromosome is a part of a cell in an animal or plant. It contains genes which determine what characteristics the animal or plant will have.[5-46c] DNA in the cell's nucleus interacts with proteins to form chromosomes.[5-46d]

Old Universe Progressive Creationist View

The cell forms polynucleotide chains by linking together four different subunit molecules called nucleotides. A nucleotide is an organic compound made up of three subunits: a nitrogenous base, a five-carbon sugar, and a phosphate group. The sugar component may either be ribose or deoxyribose. The ribose sugar is the sugar component of the nucleotides that make up RNA (ribonucleic acid). The deoxyribose sugar is the sugar component of DNA. The four nucleotides used to build DNA chains are adenosine (A), guanosine (G), cytidine (C), and thymidine (T) (See Figure 5.19). DNA stores the information necessary to make all the polypeptides (proteins are made up of one or more polypeptides) used in the cell. The sequence of nucleotides in DNA strands specifies the sequence of amino acids in polypeptide chains. The nucleotide sequence that codes the amino acid sequence of a particular polypeptide (or other functional products) is known as a gene. Through the use of genes, DNA stores the information functionally expressed in the amino acids sequences of polypeptide chains. Proteins interact with DNA to make chromosomes. These structures only become visible in the cell nucleus when the cell divides. Each chromosome consists of a single DNA molecule that wraps around a series of globular protein complexes. The globular proteins are called histones. These structures repeat to form supramolecular structures known as nucleosomes. The nucleosomes coil to form a structure called a solenoid. The solenoid further condenses to form higher order structures that comprise the main part of the chromosome. Between cell division events, the chromosome exists in an extended diffuse form that is not detectable.

Prior to and during cell division, the chromosome condenses to form its readily recognizing compact structures. DNA does not leave the nucleus to direct the synthesis of polypeptide chains. Rather, the cellular machinery copies the gene's sequence by assembling another polynucleotide, messenger RNA (mRNA). The process of copying mRNA from DNA is called transcription (Figures 5.6 & 5.9). A single-stranded molecule, mRNA is similar, but not identical to the composition to DNA. One important difference between DNA and mRNA is the use of uridine (U) in place of thymidine (T) to form the mRNA chain. Once transcribed from the DNA, mRNA migrates from the nucleus of the cell into the cytoplasm (Figure 5.9). At the ribosome, mRNA directs the synthesis of polypeptide chains (Figure 5.10). The information content of the polynucleotide sequence is then translated into polypeptide amino acid sequences. The polypeptide chain then folds to form a fully functional protein.[5-46e]

There are four types of RNA, each encoded by its own type of gene: **1)** mRNA - Messenger RNA: Encodes amino acid sequence of a polypeptide. **2)** tRNA - Transfer RNA: Brings amino acids to ribosomes during translation. **3)** rRNA - Ribosomal RNA: With ribosomal proteins, makes up the ribosomes, the organelles that translate the mRNA. **4)** snRNA - Small nuclear RNA: With proteins, forms complexes that are used in RNA processing in eukaryotic cells. No snRNA is found in prokaryotic cells.[5-46l]

The sequences of nucleotides in the DNA strands specifies the sequence of amino acids in polypeptide chains. These coded instructions are called genes (like words in a book). Through the use of genes, DNA stores the messages functionally expressed (turned on) in the amino acid sequences of polypeptide chains (Figure 5.5). Nucleotides function as characters that build letters and the genes function like words. DNA is like the reference section of a library (the nucleus of a eukaryotic cell – Figure 5.6). The books (the DNA) can be read in the reference section but cannot be removed. The books in the reference section of the library must be copied, or translated, before it can be taken out of the library (that is, out of the nucleus of a eukaryotic cell). This is exactly what the cell does. DNA does not leave the nucleus to direct the synthesis of polypeptide chains. Instead, a process within the cell copies the gene's contents by assembling another polynucleotide, messenger RNA (mRNA).

Transcription occurs as the details in the DNA are copied or transcribed into mRNA (figures 5.6 & 5.9). Once assembled, mRNA migrates from the nucleus of the cell into the cytoplasm. At the ribosome, mRNA directs the synthesis of polypeptide chains. The information content of the polynucleotide sequence is translated into the polypeptide amino acid sequence. This process is like translating one human language into another (See section "The Process of Creating a Protein From DNA Instructions," and subsection "How a Protein is Made—Step by Step" and also see Figures 5.9 & 5.10). In other words, the nucleotide language of DNA and RNA is translated at the ribosome into the amino acid language of proteins (Figure 5.10). The analogical language used by molecular biologists to describe the flow of information in the biochemical systems is not arbitrary. The analogy between human language and the molecular-genetic language is quite similar. Biological information can adequately be illustrated by examples from human language. Biochemical systems are information systems that only comes from an intelligent mind.[5-46f]

The DNA spirals around the outside edge of the double-helix and the base pairs are horizontal that links the two DNA sides, like rungs on a ladder (see Figures 5.5 & 5.19). The base pairs adenine and thymine are on the same rung. Adenine is on one side and thymine is on the other side of the rung. Base pairs guanine and cytosine are on alternating rungs. Guanine is on one side and cytosine is on the other side of the rung. In a prokaryotic cell, DNA is located in the upper corner of the nucleoid, and then flows in the cytoplasm (Figure 5.1). As previously mentioned, prokaryote cells do not possess nuclei, therefore their DNA is organized into a structure called the nucleoid. The structure of a prokaryotic cell's DNA is actually circular DNA floating around the cell. This is because prokaryote cells do not have a nucleus to hold it in. The DNA in prokaryotic cells floats in the cytoplasm as a single circular chromosome.

In a eukaryotic cell (Figures 5.2 & 5.3), the DNA is located in the nucleus (Figure 5.6). DNA only exists as loose chromatin while in the nucleus and only forms chromosomes during meiosis or mitosis after the nuclear envelope has disappeared. (*Chromatin is a complex of DNA and protein found in eukaryotic cells. Its primary function is packaging long DNA molecules into a more compact, denser structures. This prevents the strands from becoming tangled and also plays important roles in reinforcing the DNA during cell division, preventing DNA damage, and regulating gene expression and DNA replication.*) A chromosome is an organized structure of DNA, protein, and RNA found in cells. DNA is actually located in two particular locations within a eukaryotic cell. It is located in the nucleus and also in the mitochondria (and chloroplast). The DNA found in the mitochondria (and chloroplast) are similar in structure to those found in prokaryotic cells.[5-47]

Darwinist View

The first cells on Earth did not have a nucleus or a membrane around their DNA. Bacteria carry out other cell functions and reproduce other bacteria without having a nucleus in their DNA. Cells with a nucleus (eukaryotic cells) (Figures 5.2 & 5.3) evolved around 1.5 billion years ago.[5-48]

Genes are a linear sequence of nucleotides along a segment of DNA (deoxyribonucleic acid) that provides the coded instructions for production of RNA (ribonucleic acid). (*A nucleotide is a group of molecules that, when linked together, form the building blocks of DNA or RNA. A group of molecules consists of the bases that are composed of a phosphate group. In DNA, these bases consist of adenine, cytosine, guanine, and thymine, (A, C, G, T) and a pentose sugar. In RNA the thymine base is replaced by uracil.*) The normal function of a gene is to make a protein. Genes are housed in molecular compartments known as chromosomes. If genes are thought of as being blueprints, chromosomes would be the binders that hold the pages together and to keep them organized. Chromosomes are housed in cells (Figure 5.5).[5-49] In DNA, there are long lines of adenine (A), cytosine (C), guanine (G), and thymine (T) that are precisely arranged to create proteins.

Description of Proteins

Proteins are very complex chemicals (molecules) constructed of amino acids (Figure 5.10 illustrates how a protein is made, based on the types of amino acids and the mRNA that carries the instructions to the ribosome for protein assembly.).[5-50] Proteins are very important molecules in our cells. They are involved in virtually all cell functions. Each protein within the body has a specific function. Some proteins are involved in structural support, while others are involved in bodily movement, or in defense against germs. Proteins vary in structure as well as function. They are constructed from a set of 20 amino acids and have distinct three-dimensional shapes.[5-51]

The chain of amino acids is called a "polypeptide," (Figure 5.10) and when it is very long, it is called a "protein." As the long chain forms, the sequence of amino acids determines the shape of the protein (discussed later in this chapter). Most polypeptide chains have hundreds or thousands of amino acids. (Amino acids are discussed in more detail below as well as in Chapter 6.) The shape that the chain forms is very important, because the shape of a protein determines how it functions. If a different sequence of amino acids is used, the protein chain would have a different shape, and therefore a different function. The protein can now be used to form structures, or be used as an enzyme to speed up cell reactions.[5-52]

A protein is a long linear array of amino acids structuring the proteins together in a line. Due to the forces between the amino acids, the proteins fold into very particular three-dimensional shapes. These three-dimensional shapes are highly irregular, something like the teeth in a key, and they have a lock-key fit with other molecules in the cell. Often, the proteins will catalyze reactions, or they'll form structural molecules, or linkers, or parts of the molecular machines. These molecular machines control every cellular process. This specific three-dimensional shape, which allows proteins to perform a function, is derived directly from the one-dimensional sequencing of amino acids. The sequence of amino acids is critical to getting the long chain to fold properly to form an actual functional protein. If the wrong sequence is made, the result would be no folding and the sequence of amino acids would be unable to serve its function.[5-53]

Proteins are organic compounds made of amino acids that are arranged in a linear chain and joined together by peptide bonds between the carboxyl and amino groups of adjacent amino acid residues. The sequence of amino acids in a protein is defined by the sequence of a gene, which is encoded in the genetic code. In general, the genetic code specifies 20 standard amino acids. Most proteins are linear polymers built from series of up to 20 different L-α-amino acids. Proteins are assembled from amino acids using information encoded in genes. Many proteins are the enzymes that catalyze the chemical reactions in metabolism. (Enzymes are proteins with special slots for selecting and holding other molecules for speedy reactions. Enzymes and metabolism are further defined below.) Other proteins have structural or mechanical functions, such as the proteins that form the cytoskeleton, a system of scaffolding that maintains the cell shape. Proteins are also important in (1) cell signaling, (2) immune responses, (3) cell adhesion, (4) active transport across membranes, and the (5) cell cycle. Amino acids also contribute to cellular energy metabolism by providing a carbon source for entry into the tricarboxylic acid cycle, especially when a primary source of energy, such as glucose, is scarce or when cells undergo metabolic stress.[5-54] Enzymes are the biological substances (proteins) that act as catalysts and help complex reactions occur everywhere in life.[5-55] Metabolism is the set of life-sustaining chemical transformations within the cells of living organisms. These enzyme-catalyzed reactions allow organisms to **(1)** grow and reproduce, **(2)** maintain their structures, and **(3)** respond to their environments.[5-56] Each protein has its own unique amino acid sequence that is specified by the nucleotide sequence of the gene encoding this protein. The genetic code is a set of three-nucleotide sets called codons and each three-nucleotide combination designates an amino acid (Figure 5.10a).[5-57]

Amino Acids

Amino acids are organic compounds containing amine and carboxyl functional groups, along with a side chain specific to each amino acid. The key elements of an amino acid are carbon, hydrogen, oxygen, and nitrogen, although other elements are found in the side chains of certain amino acids.[5-57a] Out of the 20 amino acids utilized by the body, 11 can be manufactured within the body. However, that means 9 amino acids cannot be manufactured within the body. A body cannot function without these 9 amino acids. It is essential that a living organism obtain these 9 amino acids from outside food sources.[5-57b]

Functions and Types of Proteins

Eukaryotic cells synthesize proteins for thousands of different functions. Some examples are: **(1)** to build the components of the cytosol (e.g. microtubules, glycolytic enzymes); **(2)** to build the receptors and other molecules exposed at the surface of the cell embedded in the plasma membrane; **(3)** to supply some of the components of the mitochondria and (in plant cells) chloroplasts; **(4)** proteins secreted from the cell to supply the needs of other cells and tissues (e.g. collagens to support cells, hormones to signal them).[5-58]

There are various types of proteins. They include: **(1)** Antibodies are specialized proteins involved in defending the body from antigens (foreign invaders). One way antibodies destroy antigens is by immobilizing them so that they can be destroyed by white blood cells. **(2)** Contractile proteins are responsible for movement. These proteins are involved in muscle contraction and movement. **(3)** Enzymes are proteins that facilitate biochemical reactions. They are often referred to as catalysts because they speed up chemical reactions. **(4)** Hormonal proteins are messenger proteins that help to coordinate certain bodily activities. **(5)** Structural proteins are fibrous and stringy and provide support. **(6)** Storage proteins store amino acids. **(7)** Transport proteins are carrier proteins that move molecules from one place to another around the body.[5-59]

Creationist and Intelligent Design View

Proteins are the key functional molecule in the cell; life could not exist without them. The functional attributes of proteins are derived from information stored in the DNA molecule. DNA is more like a library than a blueprint for how to build proteins as there are also additional sources of information in the cell and in organisms. As important as DNA is, it does not build everything. All that DNA builds are the protein molecules, but they are only sub-units of larger structures that are informatively arranged. DNA is like a library in that the organism accesses the information that it needs from DNA so it can build some of its critical components. And the library analogy is better than the blueprint analogy because of its alphabetic nature. In DNA, there are long lines of A, C, G, and T's (adenine, cytosine, guanine, and thymine) that are precisely arranged in order to create protein structure and folding. To build one protein, approximately 1,200 to 2,000 letters or bases are needed, which is a lot of information. The question is: Could this information have originated by random processes? If so, how?[5-60] (See section "The Origin of Biological Information Stored in DNA and RNA" for further discussion.)

Protein Synthesis
(How a Protein is Developed Inside a Cell)

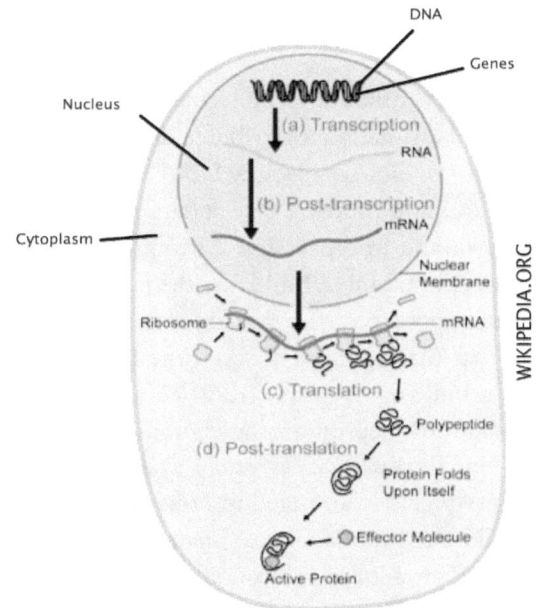

Protein Synthesis

Figure 5.6 How a eukaryotic cell produces a protein

Protein synthesis takes place within the nuclear membrane of a eukaryotic cell (Figure 5.2 & 5.6). Within the nuclear membrane of a cell, genes (within DNA) are transcribed into RNA (ribonucleic acid). (RNA is shorter than DNA because DNA contains the code for making multiple different proteins and mRNA includes only the information to make one single polypeptide chain.) This RNA then goes through post-transcriptional changes that result in a mature mRNA that is then transported out of the nucleus (Figure 5.4) and into the cytoplasm. The mature mRNA then goes through translation into a protein. The mRNA is translated by ribosomes (Figure 5.10) that match the three-base codons of the mRNA to the three-base anti-codons of the appropriate tRNA (transfer RNA) (Figure 5.10a). tRNA is responsible for carrying amino acids to the mRNA and then holding them in a way that enables them to join together. (See figures 5.9 & 5.10) (The role of ribosomes in developing a protein is discussed in more detail in section "How a Protein is Made—Step by Step.") Newly synthesized proteins (depicted in Figure 5.6 as black coils) may also go through additional changes, such as by binding to an effector molecule, before becoming fully active.

Protein Synthesis (Figures 5.6, 5.9 & 5.10) is the process of making proteins, using the information that is found in DNA (stored on chromosomes). Genes are stored in DNA (deoxyribonucleic acid) and DNA is stored in chromosomes (Figure 5.5 lower left). Proteins are very important molecules for a cell and are used to build cell structures and are used as enzymes. Proteins are long chains of small molecules, called amino acids. Different proteins are made by using different sequences of amino acids (Figures 5.6, 5.9 & 5.10). The pieces of information in DNA are called genes, which describe how to make proteins by putting the correct amino acids into a long chain in the correct order. The information of the DNA then needs to be converted to protein's amino sequence.

Protein synthesis begins with the stored genetic information of a DNA molecule. The DNA of a specific gene (enzyme called Helicase) will "unzip" (unwind) like the DNA does during replication (Figure 5.20). Only one side of the DNA will be used at this time. A single-strand of RNA (ribonucleic acid) forms, one subunit at a time, and transcribes (copies) the genetic information from the DNA. The new strand is an RNA molecule, which contains a uracil (U) instead of a thymine (T). The RNA now has copied the subunit sequence of the gene. The DNA is no longer needed in the process of protein synthesis and is closed and remains in the nucleus. The RNA molecule is called messenger RNA (mRNA) and will leave the nucleus to travel to a ribosome to build a protein molecule (Figures 5.6, 5.9, 5.10). The RNA inside the nucleus of the cell (Figures 5.4 & 5.6) was formed in the nucleus to get gene information from the chromosomes. The chromosomes never leave the nucleus (Figures 5.2, 5.3, 5.4, 5.5, 5.6, 5.9), however, the molecule of RNA will leave the nucleus and travel to a ribosome (Figures 5.6, 5.9), which are the cell parts that make proteins (Figures 5.6 & 5.10). Once the mRNA is at the ribosome, the genetic information will be translated by the ribosome to make a protein (Figures 5.6, 5.9 & 5.10).[5-61]

The genetic information in the mRNA is then interpreted by the ribosome, which is then used to assemble the protein (Figure 5.10). (This process is discussed in more detail in section "How a Protein is Made—Step by Step.") The mRNA is a sequence of subunits (like a chain) that tells how to build a protein. A protein is a sequence of subunits, that is, a chain of amino acids. The messenger RNA (mRNA) contains information in sets of three subunits called codons (See Figure 5.10a). Each set of three is the code for a particular amino acid. The information of the messenger RNA (mRNA) describes which amino acids should be in the protein chain.[5-62]

The Process of Creating a Protein
From DNA Instructions

Old Universe Progressive Creationist View

Proteins Make Other Proteins

When the organisms within the cell copies the genetic information stored in the DNA molecule, it sets the stage for protein synthesis. A single-stranded polynucleotide messenger RNA (mRNA) molecule is assembled using DNA as a template (see transcription – figures 5.6 & 5.9). After processing, mRNA migrates from the nucleus of the cell into the cytoplasm (figures 5.1 & 5.3). At the ribosome, mRNA directs the synthesis of protein molecules (translation). The translation process is somewhat different in the prokaryotic cells than it is in eukaryotic cells, both processes depend on proteins. In prokaryotic cells (bacteria), mRNA production requires RNA polymerase, a complex protein made from six polypeptide subunits. RNA polymerase consists of two subunits. It takes five subunits to form the core protein. The RNA polymerase core is capable of synthesizing mRNA on its own but cannot recognize the location along the DNA strand where the gene begins. This task is performed by a subunit. In eukaryotic cells, three different types of RNA polymerase transcribe genes. Like the prokaryotic cells (bacteria), eukaryotic RNA polymerases are composed of numerous subunits. However, in eukaryotic cells, RNA polymerases need transcription factors for this process of the cell to make a protein. Once this is established, proteins then recognize genes and begin the gene-copying process. Once mRNA is produced in eukaryotic cells, it undergoes extensive modification before going to the ribosome. The modification reactions all involve proteins.

The first set of reactions "caps" one end of mRNA. This capping process begins when an enzyme attaches a chemically modified guanine to the first nucleotide in the mRNA strand. After the guanine adds methyl groups to the nucleotides in the second and third positions of the mRNA chain. The next set of reactions modifies the opposite end of the mRNA strand. The poly (A) polymerase protein adds about two hundred adenine nucleotides to the last position of the mRNA molecule to form the poly tail. This tail provides stability to the mRNA molecule. The final modification to mRNA, the splicing reactions, also requires proteins. In eukaryotic cells, the sequences that make up genes consist of stretches of nucleotides that code for the amino acid sequence of polypeptide chains (exons). These exons (see definition in next paragraph) are interrupted by nucleotide sequences that do not code for anything (introns). After the gene is transcribed, the intron sequences are removed from the mRNA and the exons are spliced together. This process is facilitated by an RNA-protein complex called a spliceosome. Once synthesized and processed, mRNA migrates to the ribosome where it directs protein synthesis (see Figure 5.10 and sub-section "Ribosomes" below). In prokaryotes, the large subunit contains two rRNA (ribosome RNA) molecules and about thirty different protein molecules. The small subunit consists of a single rRNA molecule and about twenty proteins. The large subunit in eukaryotic cells is formed by three rRNA molecules that combine with about fifty different proteins. Their small subunit consists of a single rRNA molecule and over thirty different proteins. [5-62a]

(An exon is a part of a gene that will become a part of the final mature RNA produced by that gene after introns have been removed by RNA splicing. An intron is a nucleotide sequence within a gene that is removed by RNA splicing during maturation of the final RNA product. (One function of introns is to separate the exons so that one gene can code for different proteins.) A nucleotide consists of a base (one of four chemicals: adenine, thymine, guanine, and cytosine) plus a molecule of sugar and one of phosphoric acid. A nucleotide is regarded as the basic building block of nucleic acid polymers (e.g. DNA and RNA). That is, a nucleotide is one of the structural components, or building blocks, of DNA and RNA. It is an organic compound made up of three subunits: a nitrogenous base, a five-carbon sugar, and a phosphate group.[5-62b])

Proteins are critical in the initiation of translation. In both prokaryotic cells and eukaryotic cells, proteins called "initiation factors" help the ribosome during protein synthesis by binding to the small subunit. The small subunit and the initiation factors form the pre-initiation complex, which binds mRNA. Once the mRNA is bound, the initiation factors separate from the small subunit. The initiation complex then binds the large subunit to form the ribosome. Protein synthesis is then ready to begin. For protein production to proceed, RNA molecules called "transfer RNAs" (tRNAs) must bind amino acids, then transport them to the ribosome. Each of the twenty amino acids used by the cell to form proteins has at least one corresponding tRNA molecule. An activating enzyme links each amino acid to its specific tRNA carrier. Each tRNA and amino acid group has a corresponding activating enzyme specific to that pair. Another set of proteins help move the amino acid-tRNA pairs to the ribosome and properly position them for protein synthesis.

Protein Folding

Many coiled coil-type proteins are involved in important biological functions such as the regulation of gene expression, e.g. transcription factors.[5-62c] It is important that proteins take on certain basic shapes in their folded states because each protein's shape plays a large role in determining its function. Many different amino-acid sequences fold into the same or similar structures, which suggests that the structure may be of more fundamental importance than the amino-acid sequences. A protein's backbone needs to be compact in order to squeeze out water molecules from its central region. The backbone also needs to have enough room for its atomic components to fit along its winding path with a little extra space for movement. When the researchers calculated the shape that would result in just the right compromise between absolute compactness and maximum wiggle room, the result was a helix.[5-62d]

Many proteins need the assistance of other proteins to fold into the proper three-dimensional shape after they have been produced at the ribosome. The physico-chemical properties of amino acid sequences determine the way that the polypeptide chain folds into its complex three-dimensional shape. In a few cases, polypeptide chains will fold into the proper three-dimensional structure on their own. However, most proteins cannot, or if they can, the process is slow and inefficient. In the cell's environment, improperly folded proteins or proteins that fold slowly and inefficiently represent a potential catastrophe. In the crowded cell, improperly folded proteins tend to aggregate and form massive clumps that gunk up the cell's operations. To sidestep this potential disaster, virtually every cell throughout the biological realm, both prokaryotic cells (bacteria) and eukaryotic cells, rely on a group of proteins called "chaperones" to encourage efficient and accurate protein folding. Two types of chaperones exist in most organisms: molecular chaperones and chaperonins. Each category consists of numerous proteins that work together to assist folding. Once released from the ribosome, some proteins take on their native three-dimensional structure. Other proteins need additional help. Several different chaperones will bind to these polypeptides. They help stabilize the partially folded protein, preventing it from combining with other proteins in the cell. When these chaperones separate from the polypeptide chain, it folds into its intended three-dimensional shape. Other proteins need more help to fold than chaperones can provide. Once the chaperones disassociate from the partially folded polypeptide chain, these proteins are directed to chaperonins. These large complexes consist of several polypeptide subunits. Many proteins cannot fold without proteins. Even chaperones and chaperonins require ribosomes, chaperones, and chaperonins to fold.

The processes, DNA replication (figures 5.20 & 5.21), protein synthesis (figures 5.6, 5.9, 5.10, 5.11-5.18), and protein folding (Figure 5.6), raise questions about how life's chemistry came into existence. Molecules that comprise these operations cannot exist apart from one another, unless an Intelligent Designer created them at the same time. Like all irreducibly complex systems (see Chapter 14 for discussion on irreducibly complex systems), it remains difficult (if not impossible) for naturalistic origin-of-life researchers to provide a naturalistic evolutionary explanation as to how these processes could have developed by natural causes.[5-62a]

Intelligent Design View

Unlike proteins, nucleic acids do not fold up into complex 3-D conformations but remain as relatively simple long chain-like objects. Some DNA molecules may consist of several million subunits and when fully extended stretch for several centimeters. The linear sequence of subunits in the DNA molecule contains a series of encoded messages, genes, each of which is decoded by the cell and translated into the linear sequence of amino acids of a protein. Although the sequence of nucleotides in the DNA of the gene is the ultimate store of information necessary for the specification of the amino acid sequence of a protein, the nucleotide sequence of the DNA itself is not read directly into the amino acid sequence of a protein. Rather, the nucleotide sequence of the DNA is first copied into the nucleotide sequence of a particular type of RNA known as messenger RNA (mRNA). The process of copying the nucleotide sequence of the gene is known as transcription (Figures 5.6 & 5.9).

During transcription, one of the two strands of the DNA double helix is copied into RNA. **First**, the helix is unwound (that is, unzipped, by an enzyme called Helicase—Figure 5.20) and, **second**, one of the strands directs the synthesis of an RNA polymer of complementary nucleotide sequence

(Figures 5.20 & 5.21.). The transcription of mRNA is carried out by a complex of proteins known as RNA polymerase.

Most genes are about one thousand nucleotides long (this being the length of DNA necessary to specify for the average protein), each mRNA molecule, being merely a copy of a gene, consists of a long RNA chain about one thousand nucleotides in length.[5-66]

The process just described for the synthesis of the mRNA molecule applies mainly to the process as it occurs in bacterial cells (prokaryotic cells—Figure 5.1). The situation is somewhat more complicated in higher organisms (eukaryotic cells) because the coding sequences are separated by intervening sequences, or introns. After the transcription of the DNA, the initial RNA transcript is subjected to processing during which the neutrons are removed and the remaining coding sequences are spliced together to form the mature mRNA molecule.[5-67] In a prokaryotic cell, transcription and translation are coupled; that is, translation begins while the mRNA is still being synthesized. Because there is no nucleus in a prokaryotic cell to separate the processes of transcription and translation (Figure 5.1), when bacterial genes are transcribed, their transcripts can immediately be translated. In a eukaryotic cell, transcription occurs in the nucleus (Figure 5.4), and translation occurs in the cytoplasm (figures 5.3, 5.6, 5.9). In Eukaryotic Cells, transcription and translation occur separately (Figure 5.9); that is, transcription occurs in the nucleus to produce a pre-mRNA molecule. The pre-mRNA is typically processed to produce the mature mRNA, which exits the nucleus and is translated in the cytoplasm.[5-67a]

Protein Synthesis (How a Protein is Developed Inside a Cell)—continued

Creationist and Intelligent Design View

DNA stores information, the detailed instructions for assembling proteins, in the form of a four-character digital code. The parts (bases) of DNA are chemicals called (1) adenine, (2) guanine, (3) cytosine, and (4) thymine. These chemicals are represented by the letters A, G, C, and T. This is appropriate because these chemicals function as alphabetic characters in the genetic text. Properly arranging those four 'bases,' as they're called, will instruct the cell to build different sequences of amino acids, which are the building blocks of proteins. Different arrangements of characters yield different sequences of amino acids. RNA is made up of adenine (A), cytosine (C) and guanine (G), and includes uracil (U) instead of thymine (T), which is found in DNA.[5-68]

DNA does not possess the genetic code for making a protein, but only the genetic alphabet (A, C, G, T). The "alphabet letters" of DNA (the four bases, abbreviated GCAT or AGCT) are used in groups of three (triplet codons) as code names for the 20 different amino acids of proteins. But bases are equally spaced along DNA; there's nothing in the structure or chemistry that even hints why or which bases should be grouped as triplet codons. Three letter groupings (start codon: AUG, or GUG and stop codons: UGA, UAA, or UAG) are not natural in base sequences; they are forced on the base series by huge cellular particles called ribosomes (See Figure 5.6 & 5.9 & 5.10). (Ribosomes will be discussed shortly in sub-section "Ribosomes.") Ribosomes do not act directly on DNA, but on expendable "base pair copies" of DNA called "messenger RNA", or mRNA (Figure 5.10). The production of mRNA is complex, but it's a simple result of interlocking base shapes and ordinary chemical attraction (mediated by enzymes). The ribosomes only work on the mRNA and not the DNA to establish the genetic coding system for a specific protein. (The DNA helix remains in the nucleus while the mRNA travels outside the nucleus to meet up with the ribosome to be translated to produce a protein [figures 5.6 & 5.9].)[5-69]

It is not a simple matter for life forms to develop from DNA and protein, even though it is common for reactions to occur between acids and bases. One might think that if given enough time, acid-base reactions will get DNA and protein working together, and life will appear. But in reality, the opposite is true. The problem is that the properties of bases and acids produce the wrong relationship for living systems. Acid-base reactions would mess up DNA and protein units in all sorts of "deadly" combinations. These reactions would prevent, not promote, the use of DNA to code protein production. Since the use of DNA to code protein production is the basis of all life on Earth, these acid-base reactions would prevent, rather than promote, the evolution of life by chemical processes based on the inherent properties of matter.[5-70]

A cell needs molecules to make just one protein according to the instructions of just one DNA molecule. A cell needs over 75 "helper molecules," all working together in harmony, to make one protein (R-group series) as instructed by one DNA base series. A few of these molecules are RNA [messenger (mRNA), transfer (tRNA), and ribosomal RNA (rRNA) - See Figure 5.10]; most are highly specific proteins.[5-71]

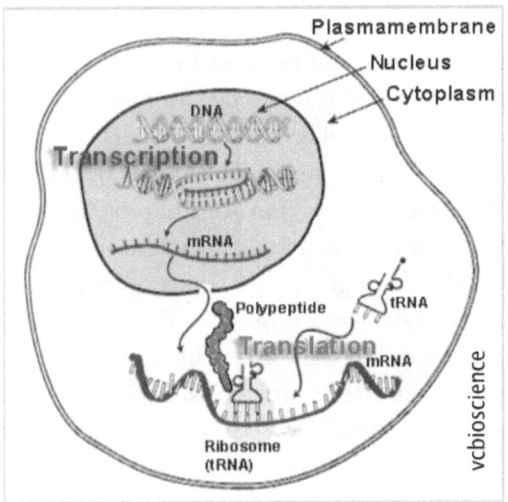

Transcription and Translation

Figure 5.9 Transcription and Translation within a eukaryotic cell

Figure 5.9 Illustrates how mRNA (developed from DNA within the nucleus of a eukaryotic cell) moves out of the nucleus of a cell to meet up with the ribosome and tRNA to develop a polypeptide chain.

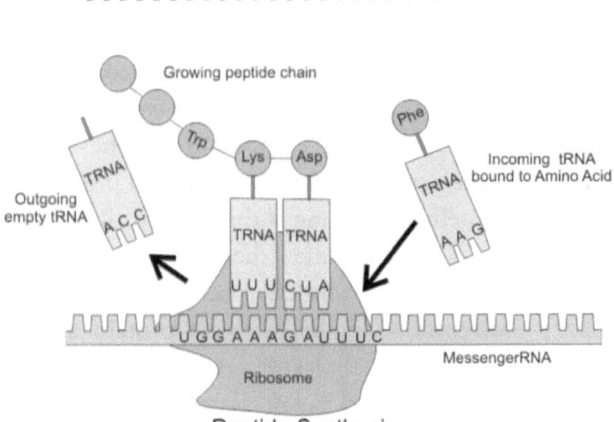

Peptide Synthesis

Figure 5.10 Illustrates in more detail how mRNA, ribosome and tRNA develop a peptide chain (Trp-Lys-Asp and soon to be added Phe) inside a cell. A chain of amino acids is called a "polypeptide," and when it is very long, it is called a "protein." (This process will be discussed in more detail in section "How a Protein is Made—Step by Step" below.)

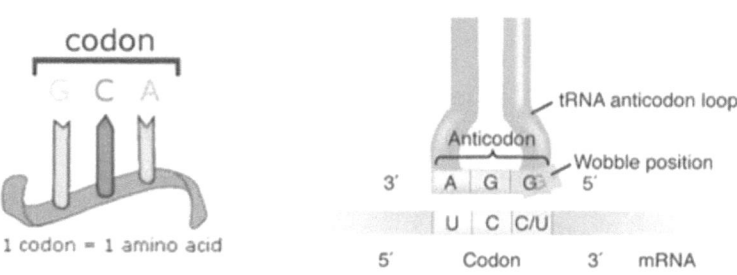

Figure 5.10a Codon

(Left) Illustration of a codon along a DNA strand. The codon GCA represents the amino acid Alanine. (Right) The anticodon AGG (Arginine) on the tRNA strand is linking up with the UCC (Serine) codon on the mRNA strand[5-71-1].

A codon is a triplet of adjacent nucleotides in the messenger RNA chain that codes for a specific amino acid in the synthesis of a protein molecule. A nucleotide is a of a group of molecules that, when linked together, form the building blocks of DNA or RNA: composed of a phosphate group, the bases adenine, cytosine, guanine, and thymine, and a pentose sugar, in RNA the thymine base being replaced by uracil.[5-71b]

The ribosome, mRNA, and the tRNA molecules work together to produce proteins (Figure 5.10). Using an assembly-line process, the protein manufacturing procedure forms the polypeptide chains (that consist of proteins) one amino acid at a time (Figure 5.10a). This protein synthetic apparatus joins together three to five amino acids per second. Ribosomes, in conjunction with mRNA and tRNAs, assemble the cell's smallest proteins, about one hundred to two hundred amino acids in length, in less than one minute.[5-71a]

Ribosomes

The ribosome is the molecular machine inside the cell that makes proteins from amino acids in the process called translation (See figures 5.6 & 5.9). It binds to a messenger ribonucleic acid (mRNA) and reads the information contained in the nucleotide sequence of the mRNA. Transfer RNAs (tRNAs) containing amino acids enter the ribosome in a special pocket, or binding site, called the acceptor site (Figure 5.10 - right side of the ribosome where the tRNA enters the ribosome). Once correctly bound, the ribosome can add the amino acid on the tRNA to the growing protein chain (Figure 5.10).[5-72]

Ribosomes are the parts of cells that make proteins from amino acids (Figures 5.6, 5.9, 5.10). Ribosomes are "molecular machines" each consisting of about 50 specific proteins and three large RNA molecules. Its overall 3-D shape gives a ribosome two adjacent slots each precisely shaped to hold three and only three bases, thus establishing the triplet coding system. In addition, the ribosomes that establish the amino acid code names for making proteins are themselves made of 50 or more specific proteins. It takes specific proteins to establish the code for making specific proteins. Like batteries that can be used to start car engines that then recharge the batteries, so proteins can be used to code for production of proteins that can then "recharge" the coding proteins.[5-73] The ribosome is made up of two parts (Figure 5.13), called subunits. The larger of the two subunits is where the amino acids are added to the growing protein chain. The small subunit is where the mRNA binds and is decoded. Each of the subunits is made up of both protein and ribonucleic acid (RNA) components.[5-74] (This process is described in more detail in the section "How a Protein is Made—Step by Step" below.)

Ribosomes (Figures 5.1, 5.2, 5.6, 5.9, 5.10) are tiny granules in the cell that assemble the amino acids into proteins, as previously mentioned. Messenger RNA (mRNA), from the DNA in the cell nucleus, carries instructions to the ribosomes as to (1) what to do, (2) how much to do, and (3) when to stop and start. With the instructions received, more messages go out to other parts of the cell to send in various proportions of amino acids to those particular ribosomes. Within it, they are connected in their proper sequence to make one or more of the hundreds of different specialized proteins used in the body.[5-75] After ribosomes establish triplet codon names (start codon: AUG or GUG; stop codons: UGA, UAA, or UAG) for amino acids, the protein building blocks have no chemical way to recognize their code name on their own (A codon is defined after Figure 5.10a). The function of the transfer RNA (tRNA) molecules is to pick up amino acids and base pair them with their codons on the ribosome slots during protein translation (Figure 5.10).[5-76]

The base pairing of tRNA (transfer RNA) and mRNA (messenger RNA) triplets is based on interlocking shapes and ordinary chemical attraction, but the proper pairing of tRNAs with amino acids requires much more than ordinary chemistry. When "translating" DNA's instructions for making proteins, the activating enzymes need to unite specific tRNA/amino acid parts. Enzymes are proteins with special slots for selecting and holding other molecules for speedy reaction. Each activating enzyme has five slots: two for chemical coupling, one for energy (ATP—adenosine triphosphate), and, most importantly, two to establish a non-chemical three-base code name for each different amino acid group.[5-77]

How a Cell is Triggered to Make a Protein for a Specific Organ

Different chemicals in the blood affect different cells. Different cells have different cell receptors embedded in their cell membranes that faces on the outside of the cell. These cell receptors are mostly made of protein. The DNA (located in the nucleus of a eukaryotic cell) holds the genetic code for making proteins within a cell. Each organ has cells that code for genes for a particular organ. Only certain chemicals in the blood affect certain cells, depending on the organ. For example, chemicals for the kidney should only activate kidney cells to generate a protein for the kidney. The only genes that an organ should activate (turn on) are the genes that code for a specific organ, based on the chemicals embedded in that cell's membrane (See section "The Origin of the Cell Membrane"). A kidney cell has all the genetic information needed to make a protein for a kidney. A kidney cell is a kidney cell because it has the receptors for chemicals in the blood that will affect the kidney. The only part of the DNA that a kidney cell should ever activate (that is, the only genes that a kidney cell should ever turn on) are the genes that code for a kidney protein.

Different cells have receptors embedded in the cell membrane that are triggered by certain chemicals (for a particular organ) in the blood to make a protein. When specific chemicals (for a specific organ) in the blood bind to a specific cell receptor, this triggers the cell associated with the specific organ to begin the process of making a specific protein. Chemicals from the blood will signal the cell by binding to a specific receptor. When this occurs, it (the receptor) tells the DNA inside the nucleus of the cell (Figure 5.5) to begin the process of making a specific protein. (This is also how many drugs work. They either stimulate or block a receptor.)

In summary, each organ has cells that contain all the genetic information needed to make a protein for a particular organ. When an organ needs a protein, chemicals in the blood will signal the correct cell by binding to a specific receptor of a specific cell. When this occurs, it signals the DNA inside the nucleus of a cell to begin the process of making a specific protein for a specific organ. [5-77a] This process will be discussed next.

How a Protein is Made—Step by Step

The messenger RNA (mRNA) contains a sequence of bases which, read three at a time, code for the amino acids used to make protein chains. Each of the sets of three bases is known as a codon (Figure 5.10a). Translating the code into an actual protein chain is complicated due to the fact that individual amino acids will not interact with the messenger RNA (mRNA) chain. The amino acids have to be carried to the messenger RNA (mRNA) by another type of RNA known as transfer RNA (tRNA) (Figures 5.9, 5.10). This is controlled by a ribosome, another form of RNA (ribosomal RNA or rRNA). (The ribosome is illustrated in Figures 5.2, 5.3, 5.6, 5.9, 5.10)

The process first needs to find the starting point of the messenger RNA (mRNA) to find the correct instructions needed to build a protein. There is a length of RNA upstream of the start codon which is not actually used to build the protein chain. The system first needs to know where the code begins by finding the right AUG codon from all the ones which are probably strung out along the RNA to code for the amino acid methionine. (The start codons are AUG or GUG.) Ribosomes come in two parts, a small bit and a larger bit. The smaller bit is involved in finding the starting point. The ribosome attaches to the 5' end of the messenger RNA (mRNA—Figures 5.20, 5.21, 5.10) and moves along it (the mRNA) until it (the ribosome) comes to a particular pattern of bases that it can bind to. This pattern occurs just before the first occurrence of the AUG codon in the messenger RNA strand. In the example below the ribosome now has to build the protein chain starting with a methionine at the AUG codon it has just found.

Transfer RNA (tRNA) is responsible for carrying amino acids to the messenger RNA (mRNA) and then holding them there in a way that enables them to join together. Transfer RNA (tRNA) is a short bit of RNA containing about 80 or so bases. These are mostly the same bases as in messenger RNA (A, U, G and C, that is, adenine, uracil, guanine and cytosine), along with some modified bases as well. The amino acid becomes attached to the tRNA by forming an ester (chemical compound) between this—OH group and the—COOH group of the amino acid. This is carried out under the influence of an enzyme. (An ester is a compound produced by the reaction between an acid and an alcohol with the elimination of a molecule of water, as ethyl acetate or dimethyl sulfate.) The anti-codon (discussed shortly) attaches the transfer RNA (tRNA) with its amino acid to the right place on the messenger RNA (mRNA) molecule.

We will now use an example to explain how a protein is made. In Figure 5.11 below, the anti-codon (the rectangle) is for the amino acid methionine (UAC). The messenger RNA (mRNA) code (along the horizontal line - Figures 5.12 & 5.13) for amino acid methionine is AUG (Figures 5.12 & 5.13 first three letters along the horizontal line). The code in the tRNA anti-codon for amino acid methionine is UAC (Figures 5.11 & 5.13). That is exactly complementary to the codon AUG. The "A" (adenine) in the anti-codon pairs with the "U" (uracil) in the mRNA; and the "C" (cytosine) in the anti-codon pairs with the "G" (guanine) in the mRNA. The way a transfer RNA (tRNA) module picks up the right amino acid is by a process that is under control of enzymes that recognize the shapes of the various amino acid and tRNA molecules and make sure that they pair up properly. Figure 5.11 below depicts a transfer RNA (tRNA) with the amino acid methionine (Met) attached.

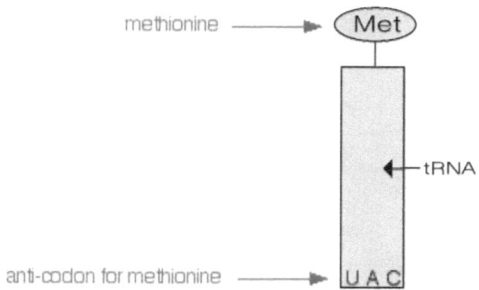

Figure 5.11

Translation is the process of turning the coded message in the messenger RNA (mRNA) into the final protein chain. Figure 5.12 below illustrates the lower, smaller part of a ribosome (depicted as a gray oval) with a mRNA (depicted as a horizontal line) attached to it at the AUG start codon, together with a small part of the RNA base sequence downstream of the start codon needed to make a protein chain. (Another start codon is GUG.) The bases upstream of the start codon are not needed any more once the ribosome has found the starting place.

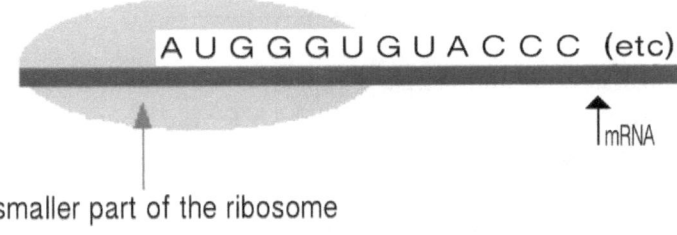

smaller part of the ribosome

Figure 5.12

Once the starting place is found, two things then happen. First, the transfer RNA (tRNA) (rectangle in Figure 5.13 below) carrying a methionine (UAC) attaches itself to the AUG codon by pairing its anti-codon bases with the complementary codon bases (AUG) on the messenger RNA (mRNA) (horizontal line in Figure 5.13 below). Then, the second, larger part of the ribosome (depicted as the larger gray oval) attaches to the tRNA as well (Figure 5.13 below).

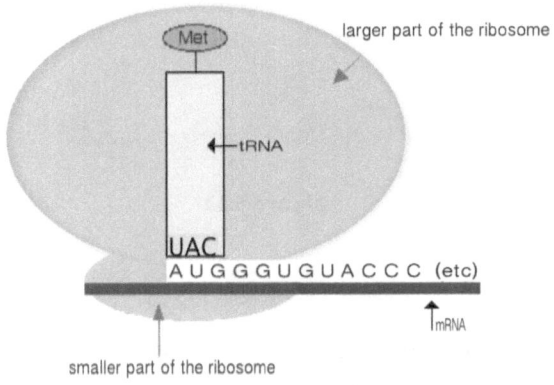

smaller part of the ribosome

Figure 5.13

In Figure 5.13 above, the large oval represents the upper part of the ribosome. The smaller oval underneath the larger one represents the lower part of the ribosome. The horizontal line going through the ribosome represents the mRNA. The codon (along the mRNA strand) for the amino acid methionine (Met) illustrated above is AUG. The rectangle represents the tRNA. The small oval with the letters "Met" above the tRNA rectangle indicates a methionine (Met) amino acid is attached to the tRNA. The anti-codon for amino acid methionine is UAC, as indicated at the base of the tRNA rectangle just above the mRNA (represented by the horizontal line).

After the larger part of the ribosome attaches to the AUG codon along mRNA strand, another transfer RNA (tRNA) (a rectangle) molecule with its attached amino acid now binds to the next codon (GGU) along the chain (the mRNA depicted as a horizontal line—Figure 5.14 below). The next codon on the messenger RNA (mRNA) (along the horizontal line) is GGU that codes for the amino acid glycine (Gly). The anti-codon (carried by the tRNA) is CCA. Remember that A (adenine) pairs with U (uracil), and G (guanine) pairs with C (cytosine).

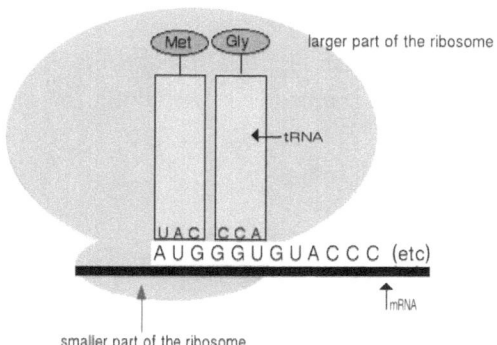

Figure 5.14

Next, the ribosome (depicted as a large and a small oval) moves along the messenger RNA (mRNA) (the horizontal line) chain to the next codon. At the same time a peptide bond is made between the two amino acids (illustrated by the line connecting the Met and Gly in Figure 5.15 below), and the first one (the amino acid methionine—Met) breaks away from its transfer RNA (tRNA) (the UAC rectangle). The transfer RNA (tRNA) (rectangle on the left in (Figure 5.15 below) molecule on the left (UAC) leaves the ribosome (the large oval and small oval under it) and goes off to pick up another methionine (Met) amino acid.

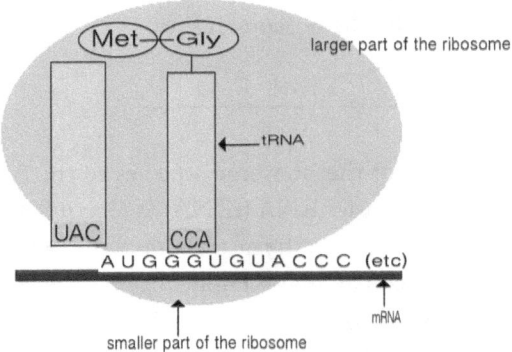

Figure 5.15

Now the process repeats. The next codon along the mRNA strand (depicted as a horizontal line in Figure 5.16 below) is the codon GUA which codes for valine (Val). For the codon GUA (along the mRNA strand), the anti-codon is CAU.

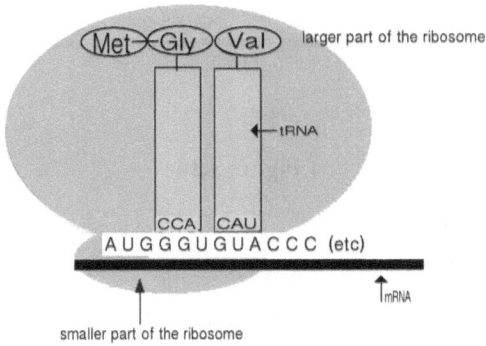

Figure 5.16

The ribosome (depicted as a large oval and a small oval under it) then moves forward one codon (3 nucleotide units symbolized as letters—GUA in this case), a new peptide bond is formed (Gly-Val Figure 5.17 below), and the transfer RNA (tRNA) on the left (CCA) breaks away to be used again later (Figure 5.17 below).

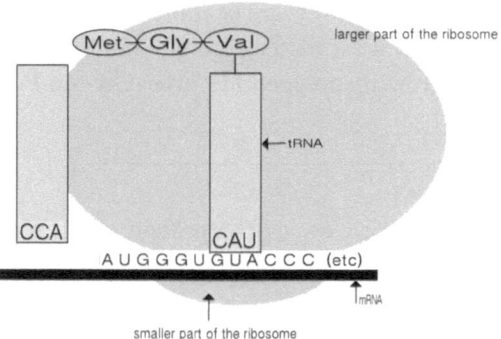

Figure 5.17

The next transfer RNA (tRNA) (anti-codon CAU) with its amino acid Val comes along above the mRNA strand (The codon for amino acid valine is GUA (Figure 5.17 above). The anti-codon for amino acid valine is CAU in our example.

Next, the tRNA amino acid anticodon GGG for Proline (Pro) comes along and joins with the codon CCC along the mRNA (Figure 5.18 below). The amino acid Proline is then added to the growing peptide chain.

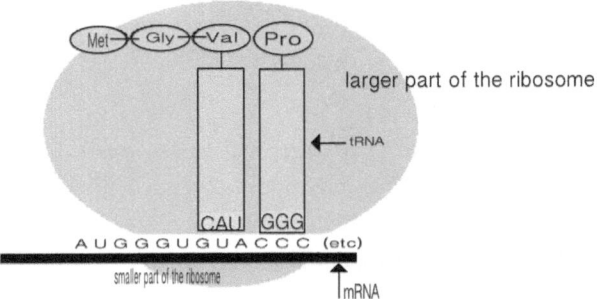

Figure 5.18

As mentioned previously, the chain of amino acids is called a "polypeptide," and when it is very long, it is called a "protein." In our example, Figure 5.18 above illustrates the growing polypeptide "Met-Gly-Val-Pro".

In the process described above, the protein chain produced up to this point (Met-Gly-Val-Pro) will continue to have additional amino acids added to it until the ribosome (the large gray oval and the smaller oval under it) comes to a stop codon along the mRNA molecule (the horizontal line). (There are three stop codons, UAA, UAG, and UGA. The stop codons serve as a signal that the end of the chain has been reached during protein synthesis. They do not code for any amino acids, and so the process will then come to a halt.) When all the correct amino acids are added to produce the correct protein and a stop codon is encountered, the process will stop. The completed protein will then be released from the ribosome, and will then fold itself up into its secondary and tertiary structures (Figure 5.6).[5-78] (See reference number [5-78a] in the Reference section for web link address to view video illustration of how a protein is produced inside a cell.)

When the ribosome first begins to translate an mRNA to form a protein, the ribosome determines whether the protein is to remain in the cytosol (Figure 5.2) or is to attach to the membranes of the endoplasmic reticulum (ER) forming "rough endoplasmic reticulum" (RER).[5-79] If a protein is meant to leave the cell, it is further modified in the endoplasmic reticulum (Figure 5.2), then it is shipped to the Golgi apparatus (Figure 5.2) where it may undergo further modification, or it could be marked for shipping to a specific location and packaged in vesicles for transport out of the cell. [5-80]

Protein Synthesis (How a Protein is Developed Inside a Cell)—continued

Creationist and Intelligent Design View

You cannot make proteins without DNA, but you cannot make DNA without enzymes, which are proteins. It is a chicken and egg situation. That a suitable enzyme should have cropped up by chance, even in a long period, is implausible, considering the complexity of such molecules. And there cannot have been a long time [in which to do it]." [5-81]

In translating the DNA instructions for making proteins, activating enzymes unite specific tRNA/amino acids pairs (tRNA = transfer RNA) (Figure 5.10). Enzymes are proteins with special slots for selecting and holding other molecules for speedy reaction. Each activating enzyme has five slots: two for chemical coupling (c, d), one for energy (ATP), and two to establish a non-chemical three-base "code name" for each different amino acid R-group (a, b).[5-82] The living cell requires at least 20 of these activating enzymes called "translases," one for each of the specific R-group/code name amino acids/rRNA (ribosomal RiboNucleic Acid) pairs to create a protein. The translases (100 specific active sites) would be: **(1)** worthless without ribosomes (50 proteins plus rRNA) to break the base-coded message of heredity into three-letter code names; **(2)** destructive without a continuously renewed supply of ATP (adenosine triphosphate) energy to keep the translases from tearing up the pairs they are supposed to form; and **(3)** vanishing if it weren't for having translases and other specific proteins to re-make the translase proteins that are continually and rapidly wearing out because of the destructive effects of time and chance on protein structures.[5-83]

After all the parts are in place, there is nothing "supernatural" or "mysterious" in the way cells make proteins. If the cells are: **(1)** continually supplied with the right kind of energy and raw materials, and **(2)** if all 75-plus of the RNA and protein molecules required for DNA-protein "translation" are **(a)** present in the right places **(b)** at the right times in the right amounts **(3)** with the right structure, then cells make proteins by using DNA's base series to line up amino acids. This is all accomplished at the rate of about two per second. It takes a living cell only about four minutes to develop an average protein (500 amino acids) according to DNA specifications.[5-84] The question is, "Could this process have developed by natural processes?" If so, how? (See section "The Origin of Biological Information

Stored in DNA and RNA" for further discussion.) The complexity of the DNA-protein relationship is very important. **First**, it takes specific proteins to make specific proteins. **Second**, among all the molecules that translate DNA into protein, there's not a single molecule in the living cell that's alive. A living cell is a collection of non-living molecules. A living cell is so complex that it has not been able to be created in a laboratory from raw chemicals alone.[5-85] (See section "Experiments Attempting to Synthesize Cells in the Laboratory" sub-section "Protocells created in a Lab" below for update.)

Intelligent Design View

The sequence of nucleotides in the mRNA (messenger RNA) is translated by the conventions of the genetic code into the amino sequence of a protein in the same way as a message in Morse Code can be translated into a sequence of letters by applying the translational conventions of Morse Code. Morse Code is a series of dots (.) and dashes (-) which make up a number or letter. For example,——equals M,.—equals A,—. equals N. If combined, the letters spell MAN. In precisely the same way, a sequence of nucleotides in an RNA molecule, such as AGU, CGA, UUG, ACA, can be translated into the amino acid sequence SER-ARG-LEU-THR by applying the following rules of the genetic code: Where AGU = the amino acid serine (SER); CGA = the amino acid orginine (ARG); UUG = the amino acid leucine (LEU); ACA = the amino acid threnine (THR). The nucleotide sequence in mRNA is read in successive non-overlapping triplets such that successive triplets of nucleotides in the mRNA specify successive amino acids in the protein. Every one of the sixty-four different nucleotide triplets which can be formed from the four nucleotides, A, U, G, C, has an exact meaning. Sixty-one triplets specify for amino acids. The remaining three, UAA, UAG, UGA, are used as punctuation signals that mean "stop," indicating the end of a particular message.

Two triplets have a double meaning and, depending on their position in the mRNA molecule and the surrounding nucleotide sequence, can also act as "Start" signals. These are the triplets AUG and GUG. AUG sometimes means to code for the amino acid methionine and at other times means "Start," while GUG may mean the amino acid valine or "Start."[5-86] After the translation of the mRNA (Figure 5.6), it (the mRNA) moves from the nucleus into the cytoplasm to the actual site of translation where the decoding of the message takes place. The translation of the mRNA molecule is carried out by a complex set of molecules that together constitute the translational apparatus.

An important component of the translational apparatus is a complex globular organelle, known as the ribosome (Figures 5.6, 5.9, 5.10), composed of an aggregate of some 50 proteins and three chains of RNA. The ribosome attaches itself to the mRNA at a special site on the mRNA known as the "ribosome binding site" (Figures 5.2, 5.3, 5.4, 5.6, 5.9, 5.10) which contains a "Start" triplet AUG or GUG. (Ribosomes are described above in sub-section "Ribosomes.") Like any other automatic decoding system, the translational system in the cell includes a set of elements (that converts one form of energy into another form) that relate each functional unit of the code. Each triplet in the RNA is matched up to the correct item in the translated message, that is, a particular amino acid. This key function is carried out by a special class of RNA molecules known as transfer RNA or tRNA (Figures 5.9 & 5.10). Each tRNA (transfer RNA) molecule consists of a short polymer of RNA some one hundred nucleotides long folded into a compact hairpin looped structure. Each tRNA can recognize a particular triplet in the mRNA (messenger RNA) as well as the appropriate amino acid specified according to the conventions of the code.

During the process of translation the mRNA passes through the ribosome (Figures 5.6, 5.9, 5.10) just as a magnetic tape passes the recording/reading head of a tape recorder. As each triplet reaches the reading head, it associates loosely with its appropriate tRNA that is also carrying the appropriate amino acid. Special proteins in the ribosome remove the amino acid from the tRNA and then the amino acid chain is gradually assembled, amino acid by amino acid, as successive tRNA bring their attached amino acids to the reading head of the ribosome (Figures 5.9 & 5.10). (Amino acids are like ingredients for making protein molecules and genes are like the recipe [instructions] for making a specific protein.[5-86a]) When the amino acid chain is completed, it is detached from the tRNA and folds automatically into its correct 3-D functional conformation.[5-87]

The mechanism of information storage is so efficient and so elegant that the mechanism of duplication of the DNA molecule may be the one and only perfect solution to the mechanism for information storage and duplication for self-replicating automation. The ribosome (Figures 5.6, 5.9, 5.10) serves as the self-replicating automation process. The ribosome is a collection of some fifty or so large molecules, mainly proteins, which fit tightly together. Altogether the ribosome consists of a highly organized structure of more than one million atoms which can synthesize any protein that it is instructed to make by the DNA, including the particular proteins which comprise its own structure. That is, the ribosome is capable of constructing itself. (Ribosomes are described above in Creationist and Intelligent Design View sub-section "Ribosomes".)

The protein synthetic apparatus can be designed to perform structural, logical, and catalytic functions. (See section "Protein Synthesis—How a Protein is Developed Inside a Cell.") The protein synthetic apparatus cannot only replicate itself, but, in addition, if given the correct information, it can also construct any other biochemical machine, no matter how complex, as long as its basic functional units are made up of proteins, which, because of the near infinite number of uses to which they can be put, gives it almost limitless potential.[5-88]

The Production of Proteins Within a Cell
(The Protein Factory)

Intelligent Design View

There are two types of nucleic acids, DNA and RNA. DNA is only found in the nucleus of the eukaryotic cell (Figures 5.6 & 5.9), equivalent to the head office of the factory analogy, and contains the master blueprints. (The DNA in a prokaryotic cell is located in the upper corner of the nucleoid, and then flows in the cytoplasm.) (A nucleoid is the central region in a prokaryotic cell, as a bacterium, that contains the chromosomes and that has no surrounding membrane.) RNA molecules perform the fundamental task of carrying the information stored in DNA to all the various parts of the cell where the manufacture of a particular protein is proceeding.

To explain this in terms of the factory analogy, the RNA molecules are photocopies of the master blueprint (DNA) that are carried to the factory floor (the cell being the factory) where the technicians and engineers convert the abstract information of the blueprint (RNA) into the concrete form of the machine (protein). In reality, the nucleic acid molecules are long chain-like molecules, rather than resembling blueprints as described in the analogy just mentioned.[5-89] Nucleic acids could be thought of as being analogous to magnetic storage devices often associated with computers. That is, nucleic acids could be thought of as playing the role of the library or memory bank that contains all the information necessary for the construction of all the various machines (proteins) on the factory floor. The nucleic acids could be thought of as a series of blueprints, each one containing the specification for the construction of a particular protein in the cell.[5-90]

DNA Replication

DNA replication is the process by which a double-stranded DNA molecule is copied to produce two identical DNA molecules. Replication is an essential process because, whenever a cell divides, the two new daughter cells must contain the same genetic information, or DNA, as the parent cell.[5-90-1]

Old Universe Progressive Creationist View

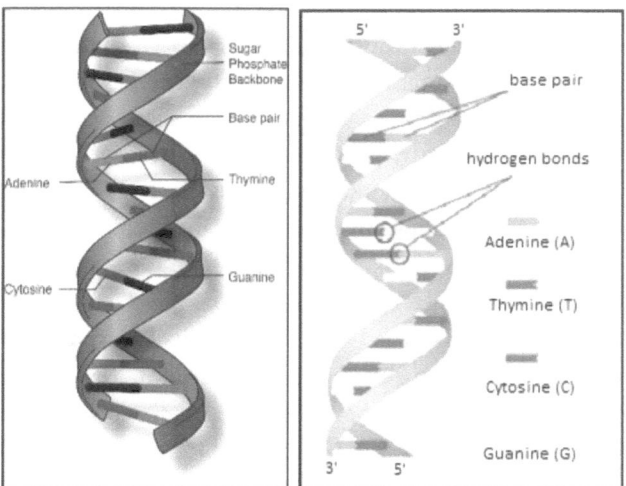

Figure 5.19 DNA Double Helix models

DNA consists of two polynucleotide chains aligned horizontally. The two strands are arranged parallel to another with the starting point of one strand in the polynucleotide duplex located next to the ending point of the other strand and vice versa. The paired polynucleotide chains twist around each other forming the DNA double helix. The polynucleotide chains are generated using four different nucleotides: adenosine (A), guanosine (G), cytidine (C), and thymidine (T). A special relationship exists between the nucleotide sequences of the two DNA strands. These sequences are considered complementary. When the DNA strand align, the A (adenosine) side chains of one strand always pair with T (thymidine) side chains from the other strand.

Likewise, the G (guanosine) side chains from one DNA strand always pair with C (cytidine) side chains from the other strand (Figure 5.19).

The DNA replication is quite extensive. The nucleotide sequences of the parent DNA molecule function as a template directing the assembly of the DNA strands of the two daughter molecules. It is a fairly straightforward process because after replication, each daughter DNA molecule contains one newly formed DNA strand and one strand from the parent molecule. Conceptually, template-directed, the DNA replication entails the separation of the parent DNA double helix into two single strands. According to the base-pairing rules, each strand serves as a template for the cell's process to follow as it forms a new DNA strand with a nucleotide sequence corresponding to the parent strand. Because each strand of the parent DNA molecule directs the production of a new DNA strand, two daughter molecules result. Each possesses an original strand from the parent molecule and a newly formed DNA strand produced by a template-directed synthetic process.

DNA replication begins at specific sites along the DNA double helix (Figure 5.19). Normally, prokaryotic cells have only a single origin of replication. More complex eukaryotic cells have multiple origins. The DNA double helix unwinds locally at the origin of replication to produce a duplication starting site (figures 5.20 & 5.21). The **replication starting site** expands in both directions from the origin during the course of replication. One the individual strands of the DNA double helix unwind and are exposed within the **replication starting site**, they are available to direct the production of the daughter strand. The site where the double helix continuously unwinds is the replication fork (Figure 5.21). Because DNA replication proceeds in both directions away from the origin, each **replication starting site** contains two replication forks.

DNA replication can proceed only in a single direction, from the top of the DNA strand to the bottom. Because the strands that form the DNA double helix align horizontally with the top of one strand that is put next to the bottom of the other strand, only one strand at each replication fork has the proper orientation (bottom-to-top) to direct the assembly of a new strand in the top-to-bottom direction. For this leading strand, the advancing replication proceeds rapidly and continuously in the direction of the advancing replication fork (Figure 5.21). DNA replication cannot proceed along the strand with the top-to-bottom orientation until the **replication starting site** expands enough to expose a sizable stretch of DNA. When this happens, DNA replication moves away from the advancing replication fork. It can proceed only a short distance along the top-to-bottom oriented strand before the replication process has to stop to wait for more of the parent DNA strand to unwind. After a sufficient length of the parent DNA template is exposed the second time, DNA replication can proceed again, but only briefly before it has to stop and wait for more DNA to become available. The process of discontinuous DNA replication takes place repeatedly until the entire strand is replicated. Each time DNA replication starts and stops, a small fragment of DNA is produced. These pieces of DNA (that actually comprise the daughter strand) are called Okazaki fragments (Okazaki fragments are defined below. Also see figures 5.20 & 5.21) . The discontinuously produced strand is the lagging strand, because DNA replication for this strand lags behind the more rapidly, continuously produced leading strand (Figure 5.20). The leading strand at one replication fork (Figure 5.21) is the lagging strand at the other replication fork because the replication forks at the two ends of the duplication starting site advance in opposite directions.[5-90a]

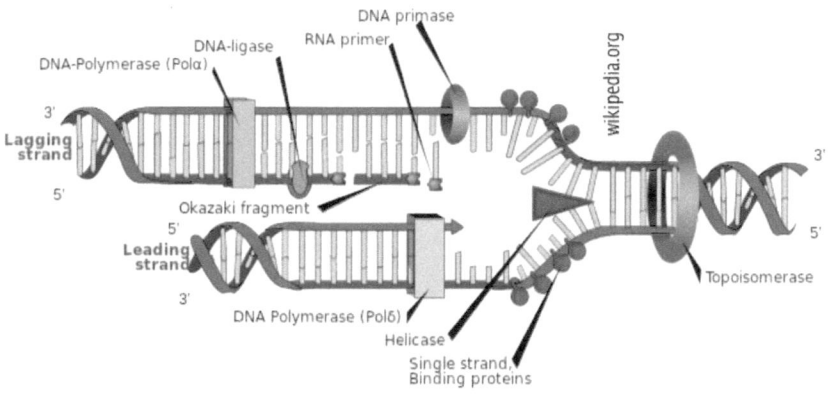

Figure 5.20 Unwinding of DNA

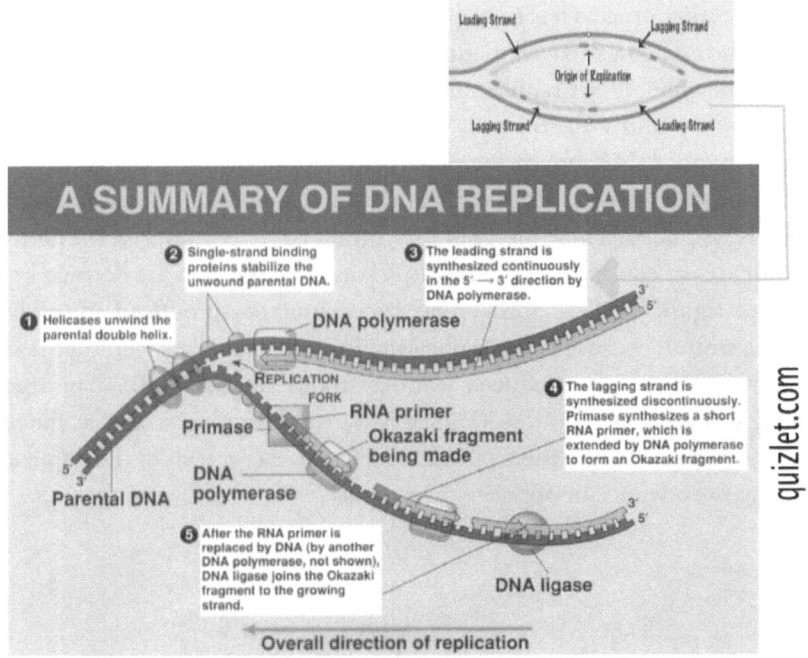

Figure 5.21 Okazaki fragment during DNA unwinding

A group of proteins is needed to carry out DNA replication. Once the origin recognition complex (which consists of several different proteins) identifies the replication origin, a protein called helicase (Figure 5.20) unwinds the DNA double helix to form the replication fork. The process of helix unwinding causes stress in the DNA helix downstream from the replication fork (Figure 5.21). Another protein, gyrase, relieves the stress, preventing the DNA molecule from over coiling. Single-strand binding proteins bind to the DNA strands exposed by the unwinding process. This association keeps the fragile DNA strands from breaking apart. Once the replication fork is established and stabilized, DNA replication can begin (Figure 5.21). Before the newly formed daughter strands can be produced, a small RNA primer must be made (figures 5.20 & 5.21). The protein that synthesizes new DNA by reading the parent DNA template strand (DNA polymerase) cannot start from scratch. It must be primed. A massive protein complex, the primosome, which consists of over fifteen different proteins, produces the RNA primer needed by DNA polymerase (Figure 5.20). Once primed, DNA polymerase will continuously produce DNA along the **leading strand** (See Figure 5.20 for illustration of leading and lagging strands). However, for the **lagging strand**, DNA polymerase can only generate DNA in spurs to produce Okazaki fragments (figures 5.20 & 5.21). Each time DNA polymerase generates an Okazaki fragment, the primose complex must produce a new RNA primer. After DNA replication is completed, the RNA primers are removed from the continuous DNA of the **leading strand** and the Okazaki fragments that make up the **lagging strand**. A protein called 3'-5' exonuclease removes the RNA primers. A different DNA polymerase fills in the gaps created by the removal of the RNA primers. Finally, a ligase protein connects all the Okazaki fragments biological together to form a continuous piece of DNA out of the lagging strand.[5-90a]

Topoisomerases (large ring in Figure 5.20) are enzymes that regulate the over-winding or under-winding of DNA. The winding problem of DNA arises due to the intertwined nature of its double helical structure. For example, during DNA replication (the process of producing two identical replicas of DNA from one original DNA molecule), DNA becomes over-wound ahead of a replication fork. If left unchecked, this tension would eventually halt DNA replication. (A similar event happens during transcription.) [5-63]

A **Okazaki fragment** is a relatively short fragment of DNA synthesized on the **lagging strand** (top strand portion in Figure 5.20 & bottom strand in Figure 5.21) during DNA replication (Figure 5.21). At the start of DNA replication, DNA unwinds (or unzips—Figure 5.20) and the two strands splits in two, forming two "prongs" which resemble a fork (thus, called replication fork). One of the strands called the **leading strand** goes from 5' to 3' (bottom strand portion in Figure 5.20 & top strand in Figure 5.21); the other strand called the **lagging strand** goes from a 3' to 5' (top strand portion in Figure 5.20 & bottom strand in Figure 5.21). Unlike the **leading strand** where DNA can be synthesized continuously, the **lagging strand** is synthesized discontinuously in the form of short fragments called Okazaki fragments that are later connected by chemical bonds (or links) that are formed by the sharing of electrons between atoms (that is, connected covalently) to form a continuous strand. This is because DNA synthesis can proceed only in one direction—the 5' to 3' direction (bottom strand portion in Figure 5.20—also see Figure 5.21).[5-64]

Okazaki fragments are short molecules of single-stranded DNA that are formed on the **lagging strand** during DNA replication (top strand portion in Figure 5.20). On the leading strand DNA replication proceeds continuously along the DNA molecule as the parent double-stranded DNA is unwound, but on the **lagging strand** the new DNA is made in installments, which are later joined together by a DNA ligase enzyme. This is because the enzymes that synthesize the new DNA can only work in one direction along the parent DNA molecule (the top strand portion in Figure 5.21). On the leading strand this route is continuous, but on the lagging strand it is discontinuous (that is, the lagging strand is terminated).[5-65]

Old Universe Progressive Creationist View - continued

Even though many details of the DNA replication process were left out, this brief description of DNA replication clearly illustrates its complexity and intricacies. It is difficult to comprehend how this biological system could have evolved a single time, let alone twice (the leading and lagging strands). There is no obvious reason for DNA replication to take place by a semiconservative, RNA primer-dependent, bidirectional process that depends on leading and lagging strands to produce DNA daughter molecules. Even if DNA replication could have evolved independently on two separate occasions, it is reasonable to expect that functionally distinct processes would emerge for bacteria and archaea/eukaryotes given their unique characteristics. But, they did not.

Many naturalistic origin-of-life researchers believe that because biochemical systems that comprise many intricately interlinked pieces, any particular full-blown system can only arise once. A great number of examples of molecular convergence (the tendency of unrelated animals and plants to independently evolve superficially similar characteristics under similar environmental conditions) that have already been discovered. It is very possible that many more will be discovered in the future. Each new instance of molecular convergence makes an evolutionary explanation for life less likely. The close analogy between biochemical systems and human designs logically compels one conclusion that life's most fundamental processes and structures reflect the design of a Creator.[5-90a]

The Origin of Biological Information Stored in DNA and RNA

Darwinist View

DNA replication (the process of duplicating a cell's genome - figures 5.20 & 5.21) is required every time a cell divides. Replication requires specialized proteins for carrying out the job. Creationists believe that the order of the chemical letters in DNA (A, G, C, T) (adenine, guanine, cytosine, thymine) is not determined by any known physical or chemical law, which means that the information in DNA cannot be explained by natural processes. The conclusion reached by creationists is that DNA must have been created. Scientists now know how new genetic information develops. A variety of well-known mutational mechanisms copy and modify the DNA letter sequence that makes up a gene. If the new sequence is advantageous to the organism, natural selection spreads the new gene through the population. (Natural selection is discussed in Chapter 9.) There are multiple examples where research groups have reconstructed the genes' origins.[5-91]

A great deal has been accomplished in the study of mutational processes that are involved in the origin of new genes. New genes can be created from an ancestral gene when a duplicate copy mutates and acquires a new function. This process becomes quite easier once a gene has been duplicated. This is because it increases the redundancy of the system. One gene in the pair can acquire a new function while the other copy continues to perform it original function. (See Chapter 9, Section "Mechanisms That Cause Changes in Organisms.") It has been determined that different genes can be formed by various mechanisms during evolutionary processes and that all genetic information was not always included in every organism's genome. There are various mechanisms by which new genetic information arises. They include: (1) exon shuffling (domain shuffling), (2) gene duplication, (3) retroposition, (4) mobile genetic elements (transposable elements), (5) lateral gene transfer, (6) gene fusion, (7) de novo gene origination.[5-92] (See Chapter 9 section "(6) Other Mechanisms That Cause Changes in Organisms" and Chapter 14, section "The Immune System" for further discussion of jumping genes and gene shuffling.)

The origin of DNA and RNA in the evolution of cells is currently unknown; however a fair amount of origin of life research has shown that the current simulations of primitive earth atmospheres, under certain circumstances, can give rise to the nitrogenous bases that are found in RNA. It is rather easy in the simulation experiments to produce adenine, and to produce cytosine, which are two of the bases. (The other two bases found in RNA are guanine and uracil. DNA contains thymine instead of uracil.) It is understood that this explanation does not completely answer the question as to how the complete RNA or DNA molecule evolved, but it does show that some of the building parts of it can be produced spontaneously in the laboratory under conditions that simulate the primitive earth.[5-93] (Chapter 6 discusses some of these experiments.)

"John Loves Mary" written in the sand on a beach

Intelligent Design advocates claim that biological information and living things contain abundant amounts of information. They conclude that the biological information found in DNA and RNA must come from a designer. Intelligent Design advocates use the illustration that if someone walks along the beach and sees something written in the sand such as "John loves Mary," that it is an example of information from which one can immediately conclude that it must have originated by a designer. They believe this proves the existence of an intelligent designer, a designer who thought of the message, coded it in the sand, and used symbols, symbolic language, in order to get that information across. They basically claim that all information must come from an intelligent cause, because there is information in DNA, and therefore, there must have been somebody there to write it. Upon careful analysis, it will be proven that this idea is false.

The first point is that the message "John loves Mary", on the beach, does not have the capacity to replicate as DNA does. It is never passed along in the process of reproduction as DNA is. It can never undergo genetic recombination as DNA can. It can never be subject to natural selection as the organisms and their characteristics coded for by DNA can. (Natural selection is discussed in Chapter 9.) That message is not part of a living organism, and the fact that messages in DNA are part of a living organism makes them entirely different.

The second point, however, that the analogy fails is something that any philosopher, any logician, would spot the mistake in logic immediately. When we look at the "John loves Mary" sentence, we know that a human being made that message because it is the kind of message that human beings make. We also know how that designer, the human being, made that message, probably by scratching a stick or other object into the sand to move the sand apart and create the message. And, **finally**, from our own ordinary experience, we have seen it happen. So we know the designer, we know the mechanism, and we have observed it happen in our own empirical experience. In the case of inferring a designer for DNA, the advocates of intelligent design do not meet those standards. They say, we can't tell who the designer is, we cannot know the mechanism, and we also do not know how the designer operated and we've never observed it. The conclusion therefore must be that the comparison between that kind of message and the kind of message in DNA fails even the most basic test of logic.

Scientific research has been done to determine whether or not there are natural explanations for new biological information. Research has determined that there are a number of mechanisms by which new genetic information is developed by the processes of evolution. By using the analogy that "John loves Mary" written on the beach, there are other mechanisms, other than some intelligence, that could have produced this intelligible phrase. There is no reason to assume that it had to be directly encoded by a designer. As mentioned earlier, some of these mechanisms include (**1**) exon shuffling, (**2**) gene duplication, (**3**) retroposition, (**4**) mobile genetic elements, (**5**) lateral gene transfer, (**6**) gene fusion and (**7**) de novo gene origination. Every one of these processes is a distinctly different molecular mechanism that results in the generation of new genetic information. None of these are hypothetical mechanisms. In every case the specific genes that were formed by these mechanisms were listed. There are a number of scientific studies that have been conducted that show how new genetic information originated by evolutionary processes. The conclusion is that intelligent design advocates are wrong in claiming that all of the information that is found in biological systems (DNA and RNA) must have been encoded by a creator/designer.[5-94]

Creationist and Intelligent Design View

Evolutionists claim that the illustration "John loves Mary" written on the beach used by intelligent design advocates to illustrate that some biological organisms are irreducibly complex to have originated by natural means, is faulty. Evolutionists claim that any philosopher, any logician, would spot the mistake in logic, because we know a human made that message, and probably made it with a stick, because we have seen such things happen in our own experience. The inference from the existence of designed objects in our world of experience to the conclusion of design in life is an example of an "inductive inference." In "inductive inference," one always infers from examples of what is known to examples of what is not known. The strength of the inference depends on similarities between the inferences in relevant properties. For example, in the Big Bang hypothesis, scientists extrapolated, or used "inductive reasoning" of their knowledge of explosions from our everyday world from things like fireworks and cannon balls and so on.

Scientists extrapolated, that is, they used "inductive reasoning," from their experience that the motion of objects away from each other indicates there was an explosion. They extrapolated from our common everyday experience to something that nobody had ever seen before, an entirely new idea, that the universe itself began in something like a giant explosion. Nonetheless, scientists were confident that this was a good method to use because they thought the relevant property, the parts moving rapidly away from each other, was what is understand from an explosion. And that is how science often reasons. In the same way, that if we saw such a message on the beach ("John loves Mary"), we could conclude that it had been designed, that is, that it did not appear on the beach by natural processes.

There are very complex functions performed within the cell. The cell was not understood in Darwin's day (1800s). And it is much better understood now. From this new information we can apply our knowledge of what we see in our everyday world to a different realm by the use of "inductive reasoning." Evolutionists admit that they have no idea (or explanation of) how DNA and RNA could have acquired all the information that is contained in them.[5-95]

There have been claims made by evolutionists that new genetic information can be generated by known processes, such as **gene duplication** and **exon shuffling**. **Gene duplication** is a process whereby a segment of DNA gets copied twice or gets duplicated and replicated so that where one gene was present before, a second copy of the exact same gene is now present in the genome of an organism. Or sometimes larger segments can be duplicated, so you can have multiple copies of multiple genes. **Gene duplication**, like photocopying, is just making another copy of the gene that originally existed. **Gene duplication** means that there is a copy of the old gene. There is nothing new. What occurred was that the same gene copied it twice. So it would be like photocopying a page. And now you have two pages, but it is just a copy of the first one, it is not something fundamentally or biologically new.

Once a gene has been duplicated, then perhaps one of those two copies can continue to perform the function that the single copy gene performed before the duplication, and the other one is sort of a spare copy. It is available to perhaps undergo mutation, and mutations accumulate changes, just as Darwinian Theory proposes. Perhaps it can go on to develop brand new properties.

Nobody disputes that random mutation and natural selection can make some small changes in pre-existing systems. The dispute is over whether random mutation and natural selection explains large complex functional systems. In the "John loves Mary" example, if discussing gene duplication, if we look at the spare copy of a gene, most likely the first copy continued to fulfill the function of conveying that information. Then, suppose a letter is changed. Suppose the final 'n' in the word "John" is changed to some other letter, like 'r' resulting in "Johr." This would not spell a name (or word) in the English language using our analogy, nor would it have any meaning in a cell as to how to make a protein. If this occurred in DNA or RNA, the instructions for how to develop a protein would be unintelligible. No protein would be made.

This is a kind of analogy that shows a loss of function in the message. In the terms of protein, the protein might no longer be functional. You might get some intelligible message. For example, if you deleted the 'r' and the 'y' from the end of "Mary", you might get "John loves Ma", or something like that. But you are not going to get anything radically different from that. So you are operating with the copy. The copy is operating with those same letters, the "John loves Mary", or some variation or deletions of that subset. A copy is a copy. It is essentially the same thing. And now the big problem that Darwinian processes face is, now what do you do? How do you generate a new complex function? And that is what occurs with **gene duplication**.

Exon shuffling is a little bit more involved than gene duplication. The gene for a protein can contain regions of DNA that actually code for regions of a protein interrupted by regions of DNA that do not code for regions of a protein. The regions that code for the part of the protein are called "exons." In cellular processes, similar to gene duplication and other processes, too, one can duplicate separate exons and sometimes transfer them to different places in the genome and other such processes. But to make it more understandable, that is, if we go back to the analogy of "John loves Mary", this may be more understood. In this sense, **exon shuffling** might be expected to generate something like, instead of "John loves Mary", perhaps "Mary loves John", or "John Mary loves", or something like that. But it is kind of a mixture of pre-existing properties, and nothing fundamentally new is being generated here. For example, one would not expect "Brad loves Jen" to generate from **exon shuffling** using the "John Loves Mary" beach example. **Gene duplication** and **exon shuffling** do have an impact on the concept of irreducible complexity. Scientists know all about the processes of **gene duplication** and **exon shuffling**. (Irreducible complexity is discussed in Chapter 14.)

In the blood-clotting cascade, many proteins look similar to each other, and they're often times pointed to as examples of **exon shuffling**. Nevertheless, experiments on the blood clotting cascade system do not explain how the blood clotting system might have arisen. The experiments that have been performed on the blood clotting system are sequence comparisons. And such information simply does not speak to the question of random mutation and natural selection being able to build complex new biochemical structures. (Natural selection and mutations are discussed in Chapter 9.) Similarly, those who are investigating the type III secretory system and the bacterial flagellum know all about "gene duplication" and "exon shuffling." And nonetheless, that information has not allowed them to explain the origin of either of those structures. (The blood clotting cascade, the type III secretory system, and the bacterial flagellum are discussed in Chapter 14.) Darwinists include those processes in their theory, but they do not explain where new complex systems come from. And it is an example of somebody accommodating this information to an existing theory rather than getting information that actually experimentally supports the theory. Random mutation along with natural selection can generate new information. (Natural selection is discussed in Chapter 9.) One needs to understand, however, that these are only small changes to pre-existing systems. There has been no demonstration to show that such processes can give rise to new complex systems. And there are many reasons to think that it would be extremely difficult to do so.

One researcher stated: *"It follows that if evolution by natural selection is to occur, functional proteins must form a continuous network which can be traversed by unit mutational steps without passing through nonfunctional intermediates."* 5-96a

Regarding two proteins binding to each other, scientists speak of unit mutational steps in terms of one of those interactions, maybe a plus charge and a minus charge or a hydrophobic group and another hydrophobic group. And to get two proteins to start change into something new and different with different properties, each one of those changes would have to be a beneficial one, or at least not cause any difficulties for the problem. And actually, seeing how that could happen is extremely difficult.

This same researcher concluded that: *"An increase in the number of different genes in a single organism presumably occurs by the duplication of an already existing gene followed by divergency."* 5-96a This is a standard scenario (which is standard in Darwinian thinking) that one has gene duplication and then divergence of the sequence of a gene, and that gives a brand new and complex protein. The word "presumably" is a presumption. That is, it may be true, and it may not true. But presumptions are not evidence. And so in order to support this idea, one needs more than the presumption that it occurs.

You have to be able to have a pathway that step by tiny step could lead from one functional protein to another. Another scientist states that: *"Given realistically low mutation rates, double mutants will be so rare that adaptation is essentially constrained to surveying, and substituting, one mutational step neighbors. Thus, if a double mutant sequence is favorable, but all single amino acid mutants are deleterious, adaptation will generally not proceed."* 5-96b This makes the point that, if you only need to change one little step, Darwinian evolution works fine. But if you need to change two things before you get to an improved function, the probability of Darwinian processes drops off dramatically. And if you need three things it drops off even more dramatically. And nonetheless even to get two proteins to stick together, multiple groups of proteins are involved. 5-96

As evolutionists would admit, it is unknown how DNA could have originated. It is also unknown how it is able to make copies of itself. There are two kinds of bases in the DNA code: purines (adenine and guanine) and pyrimidines (thymine and cytosine). (In RNA, uracil is present instead of thymine.) No one knows where these five chemicals came from. Scientists have figured out complicated ways in expensive laboratories to synthesize dead compounds of four of these five, using rare materials such as hydrogen cyanide or cyanoacetylene. (Thymine remains unsynthesizable.) Sugar can be made in the laboratory, but the phosphate group is extremely difficult. In the presence of calcium ions, found in abundance in oceans and rivers, the phosphate ion is precipitated out. In life forms enzymes catalyze the task, but it is unknown how enzyme action could occur outside of plants or animals.[5-97]

Naturalistic origin-of-life researchers have developed several important prebiotic compounds under a variety of laboratory conditions simulating those of early Earth. For example, several of the twenty amino acids used in cells to construct proteins have been synthesized in a simulated prebiotic soup. So, too, have the nucleotide bases: adenine, guanine, cytosine, and uracil. However, adequate prebiotic synthesis for numerous key biomolecules has yet to be achieved. Included in this list are the amino acids arginine, lysine, and histidine and several enzyme cofactors. All laboratory simulation experiments require a great deal of design and care on the part of the researchers to produce measurable quantities of either amino acids or nucleotide bases. Also, the yields are always small. Even though some progress has been made toward finding the right mixtures, the mere fact that laboratory experiments generated any preorganic compounds has been sufficient for most naturalistic researchers to conclude that a source for life's naturalistic origin has been established.[5-97a]

Scientists originally thought that each gene controls many different factors in the body, however, geneticists have discovered that each factor is controlled by many different genes. Because of this, it is therefore impossible, either for the DNA code to gradually "evolve," or to change. The DNA code had to be there "all at once," and once in place, that code could never change.[5-98] It was discovered that most characters, even simple ones, are regulated by many genes. For example, fourteen genes affect eye color in Drosophila. The mutation which suppresses 'purple eye' enhances 'hairy wing.' The mechanism is not understood. A single gene may influence several different characters. One of the lesser puzzles that must have baffled evolutionists who knew about it was the fact that some characters exist in more than one form, the best-known example being the four human blood groups. (This is known as polymorphism.) Why has natural selection not eliminated all but the most efficient of these blood types or why have genes for malaria resistance not become general in the population? (Natural selection is discussed in Chapter

9.) Until recently it was assumed that such polymorphisms were so rare that they could be neglected, and the matter was ignored. It has been estimated that as much as 30 per cent of all characters are polymorphic (that is, each character controlled several different factors, instead of merely one).[5-99]

It has been demonstrated that a control system in bacteria operates by switching off 'repressor' molecules, i.e., unmasking DNA at the correct 'line number' to read off the correct (polypeptide) subroutines. With a common type of bacteria, it has been suggested that 'sensor genes' react to an incoming stimulus and cause the production of RNA. This, in turn, activates a 'producer gene,' m-RNA is synthesized and the required protein eventually assembled as a ribosome. Many DNA base sequences may thus be involved, not in protein or RNA production, but in control over that production, that is, in switching the right sequences on or off at the right time. (See Chapter 13, section "Some DNA Portions are Used to Turn Genes On or Off" for further discussion of genes being turned on or off.)[5-100]

Creationist and Intelligent Design View

As mentioned earlier, cells are formed by cell division. (For further discussion of cell division, see sections "Types of Cell Division—Mitosis and Meiosis" and "Cell Division in Prokaryotes" above.) To have reproduction, there has to be cell division. And that presupposes the existence of information-rich DNA and proteins. The information needed for reproduction is what evolution is attempting to produce. In other words, for naturalistic evolution to occur, a self-replicating organism needs to exist, however, no such organism is possible until there is the information necessary in the DNA. Evolutionists believe that it is possible that DNA can replicate in a simple way and then later natural selection (discussed in Chapter 9) takes over. For example, some small viruses use RNA as their genetic material. (Viruses are discussed below.)

RNA molecules are simpler than DNA, and they can also store information and even replicate. Just to cite a couple of problems with replication having begun by RNA, the RNA molecule would need information to function, just as DNA would, which goes right back to the same problem of where the information came from. Also, for a single strand of RNA to replicate, there must be an identical RNA molecule close by. To have a reasonable chance of having two identical RNA molecules of the right length would require a library of ten billion billion billion billion billion billion RNA molecules. This effectively rules out any chance origin of a primitive replicating system.[5-101]

RNA should in theory be able to self-replicate without the help of proteins, however this is not seen in nature. There are RNA molecules that catalyze chemical reactions, a role usually carried out only by protein enzymes, these are called ribozymes. Ribozymes can facilitate the creation of both peptide bonds in proteins, and the bonds between phosphate and ribose in RNA. This means that under the right conditions some RNA sequences could both store information and implement the information.[5-100a]

Various self-organizational theories for the origin of information-bearing macromolecules have been developed. For example, scientists theorized that chemical attractions may have caused DNA's four-letter alphabet (A, G, C, T) to self-assemble or that the natural attractions between amino acids prompted them to link together by themselves to create a protein. (The letters A, G, C, T represent the bases adenine, guanine, cytosine, thymine.) Proteins are composed of a long line of amino acids, as discussed in section "Description of Proteins." DNA in life organisms is not merely a repetitive sequence. To convey information, irregularity in sequencing is needed. This irregular sequencing is what is used to convey information and what needs to be explained in DNA. The four letters of its alphabet (A, G, C, T) are highly irregular while at the same time conforming to a functional requirement, that is, the correct arrangement of amino acids to create a working protein. If there were only repeating characters in DNA, the assembly instructions would merely tell amino acids to assemble in the same way over and over again. The DNA would not be able to build all the many different protein molecules that are needed for a living cell to function.

The structure of DNA depends on certain bonds that are caused by chemical attractions. For instance, there are hydrogen bonds and bonds between the sugar and phosphate molecules that form the two twisting backbones of the DNA molecule. However, there is one place where there are no chemical bonds, and that is between the nucleotide bases, which are the chemical letters in the DNA's assembly instructions. In other words, the letters that spell out the text in the DNA message do not interact chemically with each other in any significant way. Also, the bases (letters) are interchangeable. Each base can attach with equal facility at any site along the DNA backbone in DNA, each individual base, or letter, is chemically bonded to the sugar-phosphate backbone of the molecule. This is how bases (letters) are attached to the DNA's structure. But there is no attraction or bonding between the individual letters. So, there is nothing chemically that forces the letters of

DNA into any particular sequence. The question is, could this sequencing have originated by random processes? If so, how?[5-101]

DNA and its Repair Mechanism

Creationist and Intelligent Design View

As part of the normal replication process for DNA, an enzyme travels down the DNA strand so that a copy strand of DNA can be produced (Figures 5.5 & 5.6). As the enzyme reads the sequence of molecules along the strand, and if an incorrect nucleotide is detected in the strand, there is a mechanism that uses other enzymes to cut out the bad nucleotide and insert the correct one, thus repairing the DNA. If the repair mechanism evolved first, what use is a repair mechanism if DNA has not evolved yet? If DNA evolved first, how would the DNA even know it would be better off with a repair mechanism? Can molecules think? DNA is not a stable chemical molecule, and without a repair mechanism, it would easily deteriorate by chemical oxidation and other processes. There is no mechanism to explain how DNA could exist for millions of years while the repair mechanism evolved.[5-102]

The Cell's Production Controls

The Cell's Computer-like Parity Code System to Detect Production Errors

Old Universe Progressive Creationist View

When adenine (A), guanine (G), thymine (T) (uracil (U) in RNA), and cytosine (C) (nucleobases that make up DNA and RNA) are incorporated into DNA, they instruct the double helix with a unique structural property that causes the information to behave like a parity code. None of the other nucleobases give DNA this special quality. Only the specific combination of A, G, T, and C does this. This is similar to computer software that utilizes parity codes to minimize errors in the transfer of information.

A parity code is a binary number 0 (no electronic pulse signal) or 1 (electronic pulse signal) that is assigned to numbers (0-9), alphabetic letters (A-Z, a-z) and special characters (such as !, @, #, $, %, &, etc.) to detect errors of the 7 or 8 bit code that represents the numbers, alphabet letters and special characters. The value of the parity bit is assigned either a 1 (electronic pulse, that is, bit is "on") or 0 (no electronic pulse, that is, bit is "off"), depending on whether the error-detection scheme is an even or an odd parity code system. If the error-detection scheme is an even parity code system, then the value of the parity bit is chosen so that the sum of the "on" (1) bits equals an even number. If the error-detection scheme is an odd parity code system, then the value of the parity bit is selected so that the sum of the "on" bits equals an odd number. For example, if the alphabetic capital letter A is represented by the 8 bit data unit 01000001 (the "on" bits are the 1s), if the even parity code system is used, then the parity bit would be 0 because when all the 8 bits that electronically represent the alphabetic letter A are added together, the result is an even number (0+1+0+0+0+0+0+1=2) being assigned to the beginning of the 8 bit data unit (001000001). If the alphabetic capital letter C is represented by the 8 data bit unit 01000011 (the "on" bits are the 1s), if the even parity code system is used, then the parity bit would be 1 because when all the 8 bits that electronically represent the alphabetic letter C are added together, the result is an odd number (0+1+0+0+0+0+1+1=3) being assigned to the beginning of the 8 bit data unit (101000011).

Every time the process within a cell transcribes a gene or replicates the DNA molecule, information is transmitted. Because transmission errors have disastrous consequences, error minimization (and consequently the DNA's parity code) is a critical structure in the cell's information systems. When the two DNA strands align, the adenine side chains of one strand always pair with thymine side chains from the other strand. Likewise, guanine always pairs with cytosine. When these side chains pair, they form crossbridges between the two DNA strands. The lengths of the A-to-T and G-to-C crossbridges are nearly identical. Adenine and guanine are both composed of two rings and thymine and cytosine are composed of one ring. Each crossbridge consists of three rings. When A (adenine) pairs with T (thymine), two hydrogen bonds dictate their interactions. Three hydrogen bonds accommodate the interaction between G (guanine) and C (cytosine). The specificity of the hydrogen-bonding interactions accounts for the A-to-T and G-to-C base-pairing rules. These base-pairing rules establish the complementary relationship between the nucleotide sequences of the two DNA strands. These complementary sequences play an important role in the transmission of information during DNA replication (figures 5.20 & 5.21). Likewise, the base-pairing rules plays a critical role when the process within the cell copies a gene. The nucleotide sequence of the resulting mRNA is complementary to the DNA sequence that stores the gene.

Once in a while base-pairing mistakes can occur. When A-to-T and G-to-C do not properly pair, the wrong information is transmitted. Quality control systems in the cell check for errors that might occur during DNA replication (figures 5.20 & 5.21) and transcription (figures 5.6 & 5.9). In addition to these quality control systems, another error-detection system code resides within the informational structure of DNA in the form of a parity code. Each hydrogen bond that links together A-to-T and G-to-C consists of donor chemical groups and acceptor groups. In the DNA informational system, if hydrogen bonds are considered analogous to an electrical pulse in binary number systems, then donor chemical groups can be assigned 1 and acceptor groups can be assigned a 0. For example, G would be assigned the bits 011 and C would be assigned bits 100. The parity bits correspond to the ring structure of the nucleobase.

If the nucleobase consists of a single ring, the parity bit is assigned a 1. When the nucleobase possesses two rings, the parity bit assumes a value of 0. The binary representation for G becomes 011, 0 and for C the binary representation would be 100, 1. (The binary depiction for these nucleobases is an even parity code system.) This system makes it easy to detect transmission errors. Relating the ring structure to the hydrogen bonding patterns between the nucleobases results in an optimal genetic alphabet. The even parity code found in DNA is identical to those used in computer hardware and software systems to check for errors when data is transmitted. It's as if an Intelligent Agent hand-selected the nucleobases A, G, T/U, and C to optimize DNA's structure so errors can be readily detected and minimized when any information is transmitted.[5-102a]

The Cell's Quality Control System to Detect Production Errors

Old Universe Progressive Creationist View

As part of the cell's protein production process, it has a well-designed protein development process that includes quality assurance procedures. Checkpoints occur at several critical stages during protein manufacture, including (1) tRNA and rRNA production, (2) mRNA production, (3) amino acid attachment to tRNA, (4) the movement of tRNA to the ribosome, and (5) the positioning of tRNA at the ribosome's acceptor site (Figure 5.10 – right side of the ribosome where the tRNA enters the ribosome). Once produced by the cell, molecules tRNA and rRNA remain stable, that is, they persist for a long period of time under normal growth conditions. The molecule mRNA, however, may not last as long as tRNA and rRNA. Even though tRNAs and rRNAs are produced at a relatively high successful rate, once and a while errors occur during the production process. If the errors are not detected, defective tRNAs and rRNAs will cause problems for the cell because these molecules are key factors in the biochemical process that manufactures proteins. If the process that makes the proteins within the cell does not work properly, the proteins either cannot be produced or will not be assembled correctly. The stability of these biomolecules further impairs the potential damage by defective tRNAs and rRNAs because, even if they are flawed, these molecules will persist in the cell. However, when flawed proteins are accidently made, the cell's process that detects faulty proteins eliminates them.

There is a strict quality control process governing the production of tRNAs and rRNAs in all cell types. When improperly made, the protein poly (A) polymerase adds several adenine nucleotides to the defective RNAs to form what is called a poly (A) tail. The addition of this poly (A) tail (polyadenylation) flags the faulty RNAs for destruction. If the three enzymes, poly (A) polymerase, RNAase R, PNPase, are inoperable, cell death inevitably occurs. In the process, defective tRNA molecules and rRNA fragments accumulate in the cell and the number of functional ribosomes decreases. Presumably, the defective rRNA molecules disrupt the assembly of working ribosomes.

Placing a quality assurance check at the point of rRNA and tRNA production is very beneficial to the cell. This ensures that the cell's functional abilities are in proper working order before protein production even begins. If this quality control system were not in place, the cell's protein production capabilities would become cluttered with inoperable production components. In prokaryotic cells, as well as eukaryotic cells, any tRNA and rRNA molecules that are faulty are targeted for destruction by polyadenylation. However, in eukaryotic cells, the poly (A) tail stabilizes mRNA and directs the splicing operations.

When the side chain of the DNA template is a C (cytidine), RNA polymerase adds a G (guanosine) to the growing mRNA strand (Figure 5.10). If the DNA side chain is a G (guanosine), RNA polymerase uses a C (cytidine) (because G and C always pair with each other – Figure 5.19). When the DNA side chain is a T (thymidine), RNA polymerase incorporates an A (adenosine) into the mRNA chain, and if the RNA polymerase encounters an A (adenosine), it slots in a U (uracil) (instead of a T). As mRNA moves along the DNA sense strand adding nucleotides to the mRNA molecule, it constantly checks to make sure the correct nucleotide has been added. If an error occurs and the wrong nucleotide becomes incorporated into the mRNA strand, the RNA polymerase removes the incorrect nucleotide, backs up, and repeats the combination step (this step is called proofreading). This quality control operation ensures that mRNA are accurately produced. (See sections "DNA and its Repair Mechanism" and "The Cell's Production Controls" for further discussion on detecting errors during RNA replication.)

In eukaryotes, newly formed mRNA undergoes several processing steps before it leaves the nucleus and makes its way to a ribosome. This processing includes adding a 7-methylguanine "cap" to one end of the mRNA and a poly A "tail" to the other end. Introns (noncoding intervening sequences within a gene) are removed and the remaining exons (the regions of a gene that contain information to make proteins) are spliced together. If a process within a cell makes errors in processing mRNA, so-called discard pathways remove flawed mRNA molecules. Once processed, mRNA migrates from the cell's nucleus through nuclear pores to the cytoplasm where translation occurs. Another quality assurance checkpoint prevents improperly spliced mRNA from exiting the cell nucleus (Figure 5.4). This quality control step is accomplished through binding and debinding of proteins to mRNA. When properly spliced, certain proteins that are part of the splicing procedure dissociate from mRNA. If errors occur in splicing, however, these proteins remain attached. After splicing is completed, other proteins bind to the fully processed mRNA. If not properly spliced, these proteins cannot bind to the defective mRNA. When it is associated with the wrong proteins, mRNA is not permitted to pass through the nuclear pore (Figure 5.4), which is how imperfectly processed mRNA is prevented from reaching the ribosome (figures 5.2, 5.3, 5.9, 5.10).

The complexity and intricacy of the posttranslational modifications that take place within the ER (Endoplasmic Reticulum – figures 5.2 & 5.3) make these processes susceptible to errors. It is not uncommon for proteins in the ER lumen to wind up misfolded or to be improperly assembled because of unbalanced subunit production. Quality control activities ensure that the proteins are properly produced and processed by the rough ER (Figure 5.2 #5 – Rough Endoplasmic Reticulum). It is believed that the quality assurance procedures of the ER to be the quintessential biochemical quality control system. Proteins in the ER lumen experience primary and secondary quality control checks. Primary quality control operations monitor general aspects of protein folding. Secondary quality control activities oversee posttranslational processing unique to specific proteins. One of the most remarkable features of the ER-quality assurance systems is the ability to discriminate between misfolded proteins and partially folded proteins that appear misfolded but are well on their way to adopting their intended three-dimensional architectures. If the quality control operations cannot efficiently make this distinction, it is devastating to the cell. As a matter of fact, some diseases have been linked to faulty quality control activities in the ER (Endoplasmic Reticulum – figures 5.2 & 5.3). When misfolded proteins escape detection, defective proteins accumulate in the cell. On the other hand, to mistakenly discard proteins in the process of being properly folded would be wasteful.

ER quality control systems use information contained within oligosaccharides (Figure 5.4b) as sensors to monitor the folding status of proteins. This process begins when the ER's processes attaches an oligosaccharide (Glc3 Man9 GlcNAc2) to newly made proteins after they've been manufactured by ribosomes and translocated into the lumen of the ER. Once inside the ER, two Glc units are then trimmed from the oligosaccharide to form Glc1 Man9 GlcNAc2 . This modified attachment signifies to the ER's process that it is time for chaperones to assist the protein with folding. Once completed, the remaining Glc residue is cleaved to generate the oligosaccharide Man9 GlcNAc2 . This attachment tells the ER's quality control system to examine the newly folded protein for any defects. If improperly folded, the ER's process reattaches Glc to the oligosaccharide and sends the protein back to the chaperones for another round of folding. (*An oligosaccharide (from the Greek ὀλίγος olígos, "a few", and σάκχαρ sácchar, "sugar") is a saccharide polymer containing a small number (typically three to ten) of monosaccharides (simple sugars). Oligosaccharides can have many functions including cell recognition and cell binding. For example, glycolipids have an important role in the immune response.* [5-102c])

Once the protein passes this stage of processing, the ER process removes a Man group to generate Man3 GlcNa2 . This maker triggers the ER process to send the protein to the Golgi apparatus (figures 5.2 & 5.3). If, however, the quality control system detects any evidence that proteins with the Man3 GlcNAc2 attachment are misfolded, it targets them for destruction. In other words, the quality control systems of the ER continually monitor the folding status of proteins as they're processed. If the structure of the bound oligosaccharide (Figure 5.4b) does not match the expected state of the protein, it triggers either a recycling step or a destruction sequence. If the ER's process determines that it is necessary to destroy the defective protein, the ER process shuttles the protein from the ER lumen to the cell's cytoplasm (figures 5.2 & 5.3). Once in the cytoplasm, the defective protein becomes coated with the protein ubiquitin and destroyed by the proteasome.

Only a designer who exercises thought and care could be so deliberate as to orchestrate effective quality control procedures for the operations found within the cell. The cell's quality assurance systems logically indicates that life's chemistry emanates from the work of a great Designer. As more and more research is accomplished as to how the cell functions, the evidence continues to mount that it, the cell, (and therefore all life) was designed and created by a Creator.[5-102b]

Viruses

Darwinist View

Viruses are simple submicroscopic parasites of plants, animals, and bacteria that often cause disease and that consist essentially of a core of RNA or DNA surrounded by a protein coat. A virus is a group of submicroscopic entities consisting of a single nucleic acid chain surrounded by a protein coat and capable of replication only within the cells of living organisms. That is, viruses are not able to replicate (reproduce) without a host cell, and are therefore not typically considered living organisms.[5-103]

A virus is an invasive biological agent that reproduces inside the cells of living hosts. When infected by a virus, a host cell is forced to produce many thousands of identical copies of the original virus, at an extraordinary rate. Unlike most living things, viruses do not have cells that divide; new viruses are assembled in the infected host cell. A virus consists of two or three parts: **(1)** genes, made from either DNA or RNA, long molecules that carry genetic information; **(2)** a protein coat that protects the genes; and **(3)** in some viruses, an envelope of fat that surrounds and protects them when they are not contained within a host cell. Viruses vary in shape from the simple helical and icosahedral to more complex structures. Viruses are about 1/100 the size of bacteria. The origins of viruses are unclear. Some may have evolved from plasmids (pieces of DNA that can move between cells) while others may have evolved from bacteria.[5-104]

Creationist View

Viruses have the DNA-protein (triplet base-R) coding relationship that suggests they were deliberately created. Viruses also have "docking proteins" that must attach to corresponding receptor proteins on a cell's membrane before they can enter the cell. No one will ever get a virus infection that first was not invited into the cells. This is why a virus will affect one organ system and not others, and why, for example, dogs get some viral diseases that do not affect humans, and vice versa. Some people may even be resistant to the AIDS virus because they do not have the receptor protein that is needed to get into their cells.

It is very possible that originally, the interlocking and docking and receptor proteins were designed to allow viruses to insert their DNA (or RNA) into only those cells in which gene transfer would be beneficial. In properly programmed receptor cells, some viruses can splice their DNA into the cell's genome, and the added (pre-existent, pre-programmed) genetic information multiplies along with the cell. In genetic engineering labs today, scientists use viruses as carriers and splicers of genetic information. It may be possible that viruses were originally designed as gene carriers, especially for bacteria, which are incredibly streamlined for genetic efficiency and rapid response to environmental stimuli. Then later came mutations affecting both viral docking and cellular receptor proteins. Mutated viruses inject genetic information into the wrong cells, where it causes havoc and disease.

Mutations only damage genes that previously had a beneficial function, and such damage may cause birth defects, viral and bacterial disease, and even loss of organ function. (Mutations are discussed in Chapter 9.) A machine such as a car or an airplane that breaks down does not prove that no intelligent design was involved in making the car or airplane originally. Imperfections in systems previously designed are a challenge to evolutionists, not to creationists. Evolutionists do not need examples of designed systems breaking down to prove evolution; they need examples of incomplete parts coming together to produce new and improved structures and functions. To date (2006), no such examples have been produced to illustrate this kind of evolutionary progress.[5-105]

Many of those who do not believe in creation question why a Creator would create pathogens. If animals and humans were not created with built-in, cell-based protective immune systems, all life would have quickly died off a long time ago. Many pollutants have entered the atmosphere by industry and business-related enterprises. Pathogens and parasites seem to have no other purpose than to cause disease. Research has discovered that in virtually all cases examined so far (as of 2012), a factor has been altered in the organism's environment or genetic makeup to cause it to change its food source, life cycle, natural environmental surroundings and/or symbiotic relationships. In the course of one or more of these factors being changed, the organism has become pathogenic and harmful in some way.

Since animals and humans were created with protective immune systems, we are able to survive. (See Chapter 14, section "The Immune System" for further discussion of the immune system.) These protective systems must have the ability to distinguish between good and harmful cells. When animals and humans produce new cells, markers are included so that the immune system can determine whether or not the item originated within the body or not. Any cell that does not have a self-make marker is attacked and destroyed. Items that are not self-made include bacteria, viruses, fungi, parasites and toxins. One's own cells may be included if infected inside by viruses or internally transformed by cancer. If so, portions of these non-self-proteins are carried from inside and placed on the cell surface. This marks its own cell for destruction.

Hundreds of millions of microscopic invaders that trigger defensive immune responses are called antigens. Once these cells mature, they not only recognize the cells that were self-made but will also express a receptor that will bind primarily to one type of antigen. Since the body makes over one billion different receptors, it may seem that any receptor that may be needed to destroy a pathogen has already been developed. The human body is not passively waiting for invading microbial antigens to determine which receptors will be made. Rather, the body has cells ready to meet any antigen that is detected. Elaborate systems shuffle segments, which allows a tremendous amount of combinations. Once receptor region determines its classification and allows it to interact with other cells. Another region is variable and will fit to some antigenic marker that initiates identification and the attack.

Then, at least 20 different chemical signals are sent throughout the body. Some attract helper cells while others close in at the site of infection. The activated cells start reproducing lineages of exact clones. Most clones are directed to undergo dramatic internal restructuring, thereby becoming cells able to make an antibody, which is an extremely important protein for fighting infection. These antibodies have regions that fit to antigens via similar genetic shuffling mechanisms as its ancestor cell. However, an intricate cyclical process ensures that the antigen-antibody fit gets progressively better by becoming even more specific. This is accomplished when the cell genes prescribing variable regions have genetically unstable areas guided by a mechanism allows hypermutation. A few new antibodies will fit to the antigen a little better than others. Then, after about six days of these mutation-selection cycles, thousands of cells will be making highly specific antibodies. Each of these cells manufactures antibodies at rates of about 2,000 per second. Then the antibodies may directly neutralize some threats, such as toxins, and destroys them. Some toxins, like snake bites, act so quickly that ready-make antibodies from another person or animal must be injected in victims to neutralize the toxin. Vaccines are compounds containing portions of dead or weakened antigens to deliberately expose people and, thereby, get their immune system primed. The immune system doesn't have a central dedicated organ. The system is dependent on every other body system. It is a complex system, and every area fights selflessly to rid the body of foreigners and to cleanse it of deteriorated cells. Only a Creator could account for such an elaborate system.[5-105-1]

Old Universe Progressive Creationist View

There are good and bad viruses. The vast majority of viruses do not cause disease if any kind. They are keeping the bacterial population in check.[5-105a] Coronaviruses have become very well known in recent years. Coronaviruses are a large family of viruses that have been around for a long time. Many of them can cause people to develop cold-like symptoms, such as coughing and sneezing. Before the SARS-CoV-2 (COVID-19) outbreak, coronaviruses were thought to cause only mild respiratory infections in people. Coronaviruses have all their genetic material in RNA (ribonucleic acid). When viruses infect someone, they attach to someone's cells. These viruses get inside the cells, and make copies of their (the virus's) RNA, which helps the virus to spread. If there's a copying mistake, the RNA gets changed (which is called a mutation). Some common coronaviruses include: MERS-CoV, a beta virus that causes Middle East respiratory syndrome (MERS), SARS-CoV, a beta virus that causes severe acute respiratory syndrome (SARS), and SARS-CoV-2, which causes COVID-19, also causes serious respiratory disorders. ("CO" stands for corona, "VI" is for virus, "D" is for disease, and "19" stands for the year 2019, the year in which the virus was first identified.) The COVID-19 coronavirus is one of several known to infect humans. This virus has probably been around for some time in animals. Sometimes, a virus in animals crosses over into people. So this virus isn't new to the world, but it is new to humans.[5-105c]

Most coronaviruses are normally found in bats. As a virus replicates (reproduces) without causing diseases in a bat, different mutations may occur in a virus being made. If the bat comes in contact with another animal, that particular mutation can set up an effective infection in a host bacteria that does not normally become infected. Then, once the virus has linked with a host bacteria, the host bacteria may then become mutated to allow it to be entered into the human population. (Viruses can also be created artificially and modified in a laboratory.[5-105b])

Bacteria are masters at reproduction and adaptation. Bacteria are able to take inorganic matter and turn them into amino acids and building blocks for other life forms. Bacteria would most likely cover every niche of our planet if viruses did not exist. There are more different viruses that can be numbered. Viruses can only reproduce inside a living cell and they outnumber bacteria by a factor of 10 or 15. Viruses keep bacterial populations in check so that other organisms can live and thrive on Earth. The balance of bacteria and viruses constantly needs to be kept in check. Bacteria are living cells. They are able to take nutrients from the environment around them. They can grow and reproduce. They can excrete waste. These are characteristics of living organisms. Viruses are not able to do any of these things. Viruses actually need a living cell in order for more viruses to be produced. Viruses are not alive and bacteria are alive. (Viruses are not considered living, although some naturalists would disagree.)

A bacteria (or bacterium for singular) can be killed by an antibiotic. However, it is very difficult to kill a virus because a virus is not a living organism and needs the resources of a host bacteria (which is a living cell) to survive. Therefore, the bacteria that supports the virus needs to be killed in order to stop the spread of the virus. The problem is that in most cases the bacteria that feeds a virus are also needed by other organisms. An antivirus medication is difficult but not impossible. Even though a virus by itself is not alive, it can be inactivated.

A virus does not mutate in order to cause harm to a bacteria. It is the result of a mutation within the virus. (Viruses are not alive, therefore, they cannot think or plan ahead.) The infected bacteria then provides resources for the virus to multiply. Contrary to what some naturalists may claim, viruses do not mutate in order to survive. Viruses are just genomes. They are stretches of DNA and RNA and are surrounded by proteins. They may or may not have an envelope of fat, made up of protein, around them. Viruses are made up of protein, nucleic acid, and may have an envelope of fat, made up of protein. A virus does not have any systems functioning until the virus gets inside a cell and unpacks. Once this occurs, then the cellular machinery picks up the viral genetic information and starts making more viruses. The host bacteria that has the virus sometimes makes mistakes in producing more cells, which are mutations in the genome. Those mistakes could be incorporated into new viruses, and as long as the mistakes do not prohibit the functioning process of the cell the mistakes will enter into the pool of viral sequences of a given virus. Only successful viruses (that is, those viruses that continue to exist) get picked up by a bacteria and reproduce. This does not mean that mutations of viruses cause any type of macro evolution. (A virus will remain a virus and a bacteria will remain a one-celled bacteria.)

Viruses were most likely created by God. They are very necessary to open up ecological space for other animals, including humans, to survive. Without viruses, Earth would be one giant ball of bacteria floating through space. Therefore, viruses are fundamentally essential for life on planet Earth, that is, for complex life such as plant, animal and human life, to exist. Viruses most likely were created because in the beginning viruses were helpful not harmful. The original purpose of viruses was to fine tune ecology throughout the Earth, not to cause harm to other living organisms. Viruses were actually extremely amazing entities.[5-105a]

How Different Traits are Obtained

Creationist View

Inside the DNA is the total of all the genetic possibilities for a given species. This is called the "gene pool" of genetic traits. It is also called the genome. That is all the traits a species can have; in contrast, the specific sub-code for an individual is the genotype, which is the code for all the possible inherited features a species could have. Genome is the total amount of genetic information in a species population. An individual carries genes on his or her chromosomes; the total of genetic instructions for that individual is the genotype. The genotype includes all the features a species could possibly have in its body, but what a species will actually have is called the phenotype. This is because there are many unexpressed or recessive characters in the genotype that do not show up in the phenotype. For example, you may have had both blue and brown eye color in your genotype from your ancestors, but your iris will normally only show one color. (This is the phenotype.) A fundamental concept of modern biology is the distinction between this genotype (the individual's code) and the phenotype (the physical body or expression of the code). But the genome applies to populations of a specific species. It is the sum total of all genotypes in a species.[5-106]

Within each species there is a range of possible changes that can be made through gene shuffling, within the gene pool of that species. That is why no two people look exactly alike. But this variation of range cannot cross the species barrier, as the DNA code will not allow it.[5-107] Non-mutant recombinations and reshufflings are built into the genes within the reproductive cells. (This is not to be confused with mutations that only cause damages to a gene.) Because of these normal reshufflings, offspring do not look exactly like their parents.[5-108] The genes and chromosomes, which determine inheritance, are in the DNA within each cell, and each species has its own unique DNA material. Potential variations based on the DNA material are already there, but the DNA cannot generate new structures. It is the complicated DNA code within each plant and animal type that erects the great wall that cannot be crossed.[5-109]

Experiments Attempting to Synthesize Cells in the Laboratory

Creationist View

> *"… it is not easy to see what replaced the flasks, pipettes, and stir bars of a chemistry lab during prebiotic evolution, let alone the hands of the chemist who performed the manipulations. (And yes, most of us are not comfortable with the idea of divine intervention.)"*[5-109a]

In 1970 it was proclaimed that Dr. J.P. Danielli had artificially produced a living cell. What he actually did was to carefully arrange a favorable environment of dishes with continuous warmth and food. Into one he placed several living cells obtained from living tissue. Carefully, he removed a part from a living cell and quickly put the same part from another living cell into the first one. Then he replaced another part from a different cell that was still whole and alive, and on it went, part after part. That was how Dr. Danielli "synthesized a living cell."[5-110]

In 2011, there has been a great deal of discussion regarding Dr. Craig Venter's experiments. Some proclaimed that Dr. Venter and his team created life in the laboratory, however this is not true. Although some claim that Dr. Venter and his team demonstrated how life could have originated by natural means, his experiments do no such thing. As amazing as their experiments were, they did not develop a living cell from chemicals (adenine, guanine, cytosine, thymine—in RNA uracil replaces thymine). They replaced DNA in an already existing cell. They succeeded in creating a new cell, however, they did not create life from non-life. Contrary to what some claim, the experiments conducted by Dr. Craig Venter and his associates, as astonishing as they are, did not prove or explain how life could have originated. His experiments were to help discover cures for diseases. Although Dr. Venter's experiments were very impressive, they did not begin with raw chemicals. Life was not developed from non-life from these experiments. DNA was replaced in a cell. Cells were altered to form new cells, thus creating new species, not life.[5-111]

In 2015, a team of Japanese researchers created protocells with the capacity to self-replicate continuously for multiple generations, mimicking the behavior of biological cells. These scientists formed vesicles that could be stimulated to grow and divide. They were also able to couple the replication of DNA in the vesicles' internal compartment to the growth and division process. These researchers even devised a way to replenish the daughter vesicles with nutrients after cell division took place so that these newly formed entities, in turn, could grow and divide. These scientists believe their work provides an evolutionary explanation for the origin of life.

To produce protocells with the capacity for multigenerational self-replication, the Japanese researchers had to devise a sophisticated process that required extremely knowledgeable and skilled scientists to execute the experiment under highly controlled conditions in the laboratory. For example, the researchers had to (1) carefully select the right types of phospholipids (PL) (the molecules that play a key role in forming cell membranes) to form stable vesicles. (2) They also designed and manufactured two synthetic, non-biological lipids with specialized properties. These lipids played a key role in coupling DNA replication to vesicle growth and division. One of these lipids, dubbed a catalyst (C), was added to the two PL species used to form the initial vesicle population. (3) The researchers had to carefully adjust the initial PL and C compositions to form stable vesicles that would then be capable of growing, dividing, and becoming tied to DNA replication. (4)

Next, the scientists encapsulated (a) nucleotides (dNTPs, the building blocks of DNA) (dNTP stands for deoxyribonucleotide triphosphate. Each dNTP is made up of a phosphate group, a deoxyribose sugar and a nitrogenous base. There are four different dNTPs and can be split into two groups: the purines and the pyrimidines.); (b) single-stranded DNA; (c) carefully designed DNA primers; and (d) a special type of DNA polymerase (an enzyme that replicates DNA) into the vesicles' interior compartment. This encapsulation process required the researchers to implement an exacting laboratory protocol that involved drying the lipids so that they formed a film and then rehydrating the film (with a solution containing the dNTPs, single-stranded DNA, DNA primers, and DNA polymerase), followed by incubation under a precise set of conditions. (5) Once the materials for DNA replication were encapsulated, the researchers carefully heated the vesicles, triggering DNA replication. Once produced by the replication process, the newly formed DNA molecules were automatically absorbed into the interior walls of the vesicles. This binding took place because the lipid catalyst (C) was designed to possess a positive charge, triggering its interaction with negatively charged DNA. (6) After cooling down the system, the researchers added another carefully designed synthetic lipid (V) to the vesicles. The C reacted with the V, converting it into a derivative material that caused the vesicles to grow and become destabilized. The destabilization caused the vesicles to fissure into two daughter vesicles. (7) In the next step, the researchers replenished the newly formed daughter vesicles with dNTPs so that the next round of DNA replication could take place. They accomplished this feat by encapsulating the dNTPs in vesicles dubbed conveyor vesicles. The lipid composition of these vesicles had to be carefully adjusted so that they possessed a negative charge on their surface. The negative charge made it possible for the conveyor vesicles to fuse with the daughter vesicles, which were designed to possess a positively charged surface. The fusion events were triggered by a dramatic change in solution pH, orchestrated by the researchers. (8) The researchers also had to carefully adjust the amount of the C in the conveyor vesicles so that when they fused with the daughter vesicles, the newly formed DNA would bind to the membrane wall of the daughter vesicles, allowing the cycle of growth and division to repeat.

The Japanese scientists' lab work is very impressive, to say the least. It illustrates the extent of how much the scientists know about a living cell. But does the system they developed explain how life originated from non-life? The Japanese scientists claim that a system such as the one they devised in the laboratory could have emerged near hydrothermal vents on early Earth (See section "Life May Have Originated Deep within the Ocean."). However, this research actually provides conclusive evidence that an intelligent agent must (in this case the Japanese researchers) play a crucial role in the transformation of nonliving material into a living cell. It is hard to envision how a system such as the one the Japanese scientists devised could have ever emerged on early Earth through unguided evolutionary processes. (See section "Could Life Have Begun By Natural Causes?") And in light of this work and other studies in synthetic biology, it is also difficult to imagine how anyone could conclude that life emerged via chemical evolution. The facts speak for themselves: work in synthetic biology affirms the case for intelligent design.[5-112]

In the next chapter (6) we will discuss the possibilities of how life could have originated by natural means, from a chemical perspective.

~~~

Chapter 6

How Did Life Begin on Earth?

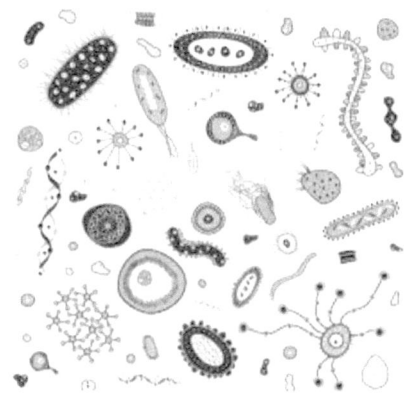

Early Life
(Single prokaryotic cell bacteria)

Introduction

In the previous chapter (Chapter 5) various theories were proposed to explain how life may have originated on Earth. Other topics discussed included how proteins are developed, how a cell functions, and the possibility of whether or not life could have developed by natural processes, including the concept of a "primeval soup." In this chapter (6) we will continue our discussion of how life could have originated on Earth, but will focus on the possibilities of how it could have actually happened from a chemical perspective rather than a biological one. We will especially look at the famous Stanley Miller Experiment of 1953, which was an attempt to re-create the origin of life in a laboratory.

Stanley Miller's 1953 Experiment

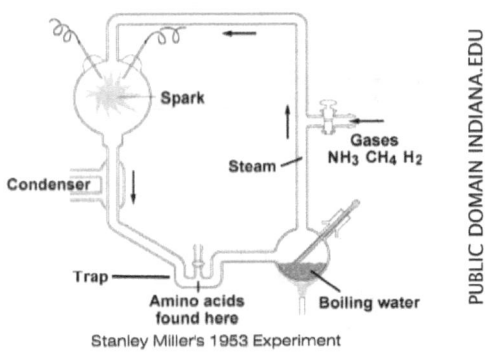

Stanley Miller's 1953 Experiment

Figure 6.1 Stanley Miller's 1953 Experiment

All of us who study the origin of life find that the more we look into it, the more we feel it is too complex to have evolved anywhere. We all believe as an article of faith that life evolved from dead matter on this planet. It is just that life's complexity is so great, it is hard for us to imagine that it did.

- Harold Clayton Urey [6-0]

Darwinist View

As mentioned in Chapter 5, it has been theorized that complex biological compounds somehow assembled by some chance out of an organic broth on the early Earth's surface. Once assembled, they were able to make copies of themselves. The first experiment to support this theory of life developing out of a primordial soup came from the famous 1953 experiment by Stanley Miller and Harold Urey, in which they made amino acids (the building blocks of proteins) by applying sparks to a test tube of **(1)** hydrogen, **(2)** methane, **(3)** ammonia, and **(4)** water. The idea was that if amino acids could come together out of raw ingredients, it could be possible that more complex molecules could form, if given enough time.[6-1]

The famous experiment performed by Harold Urey and Stanley Miller in the 1950s brought the question of the origin of life, and the "little warm pond," into the laboratory. Urey and Miller filled a flask with the gases they believed were present in the atmosphere of the ancient Earth, and suspended it over a small pool of water (Figure 6.1). They applied electrical sparks to the system, and observed that complex organic compounds, including amino acids, formed abundantly in the water. Amino acids are the most basic compounds of life on Earth. Their production in this experiment suggested that the beginnings of life could indeed have formed in appropriate settings on ancient Earth.

The Urey-Miller experiments have remained the standard explanation for the origin of life in school textbooks, but they leave many issues unresolved, and scientists have raised serious questions about nearly every aspect of the "little warm pond" model. For example, the atmosphere of ancient Earth may well have been composed primarily of carbon dioxide and nitrogen, which do not produce amino acids, rather than the more hospitable mixture of methane, hydrogen, and ammonia used by Urey and Miller. And even if amino acids were abundant in the ancient seas, it is not understood how they could have evolved into more complex forms, including the proteins that make up the genetic code of DNA that is found in all life forms today.[6-2]

In 1953, chemists Stanley Miller and Harold Urey believed their experiment simulated their concept of how life originated on Earth. They believed that life originated on Earth during a crucial instant around 3.9 billion years ago when a batch of simple inorganic molecules, zapped by a bolt of lightning (or maybe just the sun's warmth during a break in the clouds), fell together to form the prototypes for the complex organic compounds that life is made from.

In their lab, Miller and Urey shot flashes of lightning, in the form of cascades of sparks, through a flask containing an "ocean" of liquid water and an "atmosphere" of strongly reduced (that is, hydrogen-rich) gases, which included (1) methane (CH_4), (2) ammonia (NH_3), (3) hydrogen sulfide (H_2), and (4) water vapor (H_2O) (Figure 6.1). After a couple of days they checked the results. The experiment had formed all sorts of compounds, including large quantities of amino acids, the molecules that join to form proteins. This simple experiment seemed to demonstrate a vision Darwin had described a hundred years earlier, of life developing in some "warm little pond," (quoted in the beginning of Chapter 5) with all sorts of ammonia and phosphoric salts, light, heat, electricity, etc., present.

When Miller and Urey conducted their experiment it was thought that all of the planets had once shared a "primordial" atmosphere, which consisted of an atmosphere rich in methane and ammonia, believed to be left over from the formation of the planets. Because of their stronger gravity, the giant planets were believed to have retained this early atmosphere, while the atmospheres of Earth and the other, smaller planets had lost some of their lighter gases, hydrogen among them, to space. Urey therefore believed that an early Earth atmosphere, before its hydrogen had escaped and the life-driven process of photosynthesis had boosted its oxygen, would have been a lot like the atmosphere of the larger planets.[6-3]

Many of the compounds made in the Miller/Urey experiment are known to exist in outer space. If these organic compounds, such as amino acids, were not created in a reducing atmosphere here on Earth as Miller suggested, then where did they come from?[6-4]

The development of oxygenic photosynthesis on the Earth (about 3 billion years ago) and an oxygen-rich, non-reducing atmosphere can be traced through the formation of banded iron deposits, and later red beds of iron oxides. This was a necessary requirement for the development of aerobic cellular respiration, which is believed to have developed around 2 billion years ago. Simpler multi-cellular plants and animals began to appear in the oceans approximately a billion years ago. Shortly after these organisms developed, a great deal of life developed.[6-5]

As mentioned in Chapter 3, around 2 billion years ago, during the Proterozoic Eon (2500 mya-570 mya), there was a great deal of oxygen buildup occurring in the Earth's atmosphere, undoubtedly released by photosynthesis well back in the Archean Eon (3.8 bya-2.5 bya). The first advanced single-celled (prokaryotic cells), eukaryotes and multi-cellular life, roughly coincides with the start of the accumulation of free oxygen that may have been due to an increase in the oxidized nitrates that eukaryotes use, as opposed to cynobacteria (a phylum of bacteria that obtain their energy through photosynthesis).[6-6]

Oxygen is a by-product of photosynthesis, and oxygen produced by photosynthesizing bacteria built up in the atmosphere over hundreds of millions of years. This is why the geological deposits dating during this time are red-colored. Dissolved iron oxidized in the presence of free oxygen. The increase of oxygen in the atmosphere resulted in a drastic change in the environment. Many organisms could not live in the new oxygen rich environment, however, others were able to survive and adapt.[6-7]

Shortly after the Miller-Urey experiment was published, however, geologists revised their views regarding what they believed the Earth's volcanic emissions were, which in turn changed their views about the chemical makeup of an early Earth. The new theory indicated that neither methane nor ammonia spewed out of volcanoes. Instead, about 80 percent water vapor, 15 to 20 percent carbon dioxide, and traces of carbon monoxide and molecular hydrogen, were most likely present back when life first developed on the Earth. Scientists then came up with a much different picture of Earth's early atmosphere: an oxygen-rich mix of carbon dioxide, nitrogen, and water vapor. The catch is that oxygen, although an absolute necessity for multicellular, advanced life, is poison to pre-biotic synthesis.

Two competing theories have since developed. The discovery of microbes and other small organisms living in and around hydrothermal vents (underwater hot springs boiling from the ocean floor) has led to the idea that life may have started at the bottom of the sea (See Chapter 5, section "Life May Have Originated Deep within the Ocean."). Sharp differences in temperature and oxygen concentration at the boundaries around these vents make good catalysts for chemical reactions. The problem with this theory, however, is that the complex organic compounds likely to form life cannot remain stable for long at such high temperatures. Amino acids, instead of joining up, would tend to break down.

The other scenario states life first combined in the cold climates of outer space (See Chapter 5, section "Life May Have Originated Elsewhere in the Universe"), specifically, within the cold dark hearts of interstellar dust clouds. It has been theorized that long, complex organic molecules can be made when ionizing radiation leads to ion-molecule reactions. The intense cold prevents them from breaking down. These complex molecules are brought to Earth by incoming meteorites and comets. The problem is that most of a meteor is vaporized on impact with our atmosphere and therefore the survival potential for organisms would be very low. They would be burned to a crisp on the journey to Earth.[6-8] (See Chapter 5, Section "Amino Acids Found in Meteorite.")

There have been other doubts concerning Stanley Miller's experiment as well. For one thing, it is now believed that the early earth's atmosphere did not contain predominantly redundant molecules. Another objection is that this experiment required a tremendous amount of energy. While it is believed lightning storms were extremely common on the primitive Earth, they were not continuous as the Miller/Urey experiment portrayed. Thus it has been argued that while amino acids and other organic compounds may have been formed, they would not have been formed in the amounts which this experiment produced.[6-9]

Present-day life requires three types of molecules: **(1)** DNA (deoxyribonucleic acid), to store the genetic information that allows cells to replicate; **(2)** RNA (ribonucleic acid), which transfers that genetic information from the nucleus to the rest of the cell; and **(3)** the proteins that catalyze these reactions. Some believe that the answer of how life began on Earth may lie under the sea, in the volcanic activity that fires up super-hot hydrothermal vents (See Chapter 5, section "Life May Have Originated Deep within the Ocean.") Currently, the carbon released from the vents runs about 99 percent carbon dioxide, and about one percent methane, a slightly different mix than what comes from volcanoes on land. Currently, this is still only a theory that has not yet been proved or falsified.[6-8]

Creationist and Intelligent Design View

> *The notion that biological substances could arise from a purely natural process made scientists cheer and gave the clergy chills. But on reflection, less had happened than met the eye. Though the goo in Urey and Miller's beaker contained ingredients used by life, it did not come to life. It was just interesting goo. Now as then, nobody has any idea what makes chemicals start living. The origin of life is perhaps the leading unknown of contemporary science.*[6-10]

Stanley Miller's experiment was supposed to demonstrate that gases supposedly present in the early Earth's atmosphere could make the "building blocks of life" by themselves without any outside help. However, his outside, intelligent, help was necessary to save the molecules from their destructive chemical fate. Miller was able to produce amino acids (along with sugars and a few other things) that resulted from his 1953 experiment. The results, however, were wrong because Miller's Experiment had **(1)** the wrong starting materials, **(2)** used the wrong conditions, and **(3)** got the wrong results (chirality—discussed later in this chapter).[6-11]

Although Stanley Miller's experiment (Figure 6.1) demonstrates that organic compounds can be artificially produced in a laboratory, the resulting amino acids are not living things. Stanley Miller's experiment only demonstrates that in a laboratory setting with carefully designed and controlled conditions that intelligence was needed to develop the desired outcome.[6-12] Life was never formed from Stanley Miller's experiment. The product was amino acids, which are normal everyday chemicals that do not "live." There is still no known process that has ever converted amino acids into a life form of protein. Evolutionists know that amino acids do not live, but they call this experiment proof anyway because they claim amino acids are the building blocks of life. This claim suggests that if enough building blocks are present, life would result, but this conclusion is only an assumption and has never been demonstrated. Amino acids may be the building blocks of proteins, and proteins are necessary for life (as discussed in Chapter 5), but that does not mean that amino acids are the building blocks of life. At the end of the experiment, the products were found to contain a few amino acids. Miller succeeded in producing a few of the 20 amino acids found in living things. Since amino acids are the individual links of long chain polymers called proteins, they are important in our bodies.[6-13]

Scientists, including Stanley Miller and Harold Urey, were convinced that the laboratory evidence obtained from Miller's experiment was scientific proof that life could have been formed from chemicals by random chance natural processes.[6-14] However, if we look carefully at what has become known as Stanley Miller's experiment, we will see that it did not prove the evolution of life.[6-15] Instead of using carbon dioxide, nitrogen, and water vapor, which scientists believe were present in the early atmosphere, Miller used methane, ammonia, hydrogen, and water vapor. Without ammonia it would be impossible to synthesize the amino acids.[6-16]

The best hypothesis now is that there was very little hydrogen in the atmosphere because it would have escaped into space. Instead the atmosphere probably consisted of carbon dioxide, nitrogen, and water vapor. If Miller's experiment were re-done with these components, no amino acids would be created. If these elements were used, organic molecules would be created, however, what would develop would be Formaldehyde and Cyanide. These substances would fry proteins just from the fumes.[6-17] The two prime gases in the Miller spark chamber, Methane (CH_4) and ammonia (NH_3), could not have been present in large amounts. The ammonia would be dissolved in the oceans, and the methane would be found stuck to ancient (deep) sedimentary clays. If one attempts to simulate the origin of life using different starting materials, such as carbon monoxide and hydrogen cyanide, it would not be successful.[6-18]

The laboratory apparatus Stanley Miller used to produce amino acids consisted of two interconnected, chemical flasks (or bottles), arranged one above the other (Figure 6.1). The lower flask was heated and contained boiling water to produce water vapor (steam). The upper flask contained a mixture of gases including ammonia (NH_3), methane (CH_4), hydrogen (H_2) and water vapor (H_2O). (The upper flask had the presumed "primitive atmosphere," since it was known that if oxygen was present, the experiment would be a failure.)[6-19] A mixture of water, methane, ammonia, and hydrogen gases in the upper bottle was boiled, while a small electric spark continually ignited the chemical mixture that was supposed to be equivalent to a gigantic lightning bolt in the primitive environment. The lower bottle of water was kept boiling in order to keep the mixture in the upper bottle stirred up and circulating. The "primitive ocean" must have been pretty hot if this experiment was to simulate primitive natural conditions on the earth.[6-20]

The problem with this experiment is that the same electric spark that puts amino acids together also tears them apart, and sparks are much better at destroying amino acids than making them. This means that few, if any, amino acids would actually accumulate in the spark chamber. Miller, as well as other chemists, knew that an electric spark that puts amino acids together would also destroy them, so he set up his experiment to isolate the products that resulted away from the spark chamber by using a "trap" that would protect the amino acids from destruction by the conditions of their environment and also by the same electric spark that made them (Figure 6.1).[6-21]

Of course a trap would not be present in nature to protect any emerging life product (i.e, amino acids and other compounds). The trap guided and drained the product, essentially eliminating all natural effects, including its energy source (the electric spark). Without energy the product is not alive. If the product were taken out of the trap's place of safety and natural energy, the product could not live. Regardless, survival of the products would be difficult to comprehend. Even more unexplainable would be the probability of hundreds of them surviving and connecting in the proper position with hundreds of others to form just one protein. Evolutionists here face two problems. **First**, there could be no trap available on a primitive Earth. **Second**, a trap by itself would be fatal to any evolutionary scenario, for once the products are isolated in the trap, no further evolutionary progress is possible because no energy is available.[6-22]

After a week of the experiment, the fluid in the trap was chemically analyzed. It was found to have microscopic traces of a few L and D (right— and left-handed) nitrogen containing compounds called amino acids. As both L and D amino acids were formed by chemical action from Miller's experiment (as they always are when formed outside of living cells), it would be impossible for the amino acid mixture (the L and D mixture which formed in the trap) to be usable for life purposes. This is because only left-handed amino acids are used in animal and plant life. (L and D amino acids will be discussed in more detail in section "Left-handed and Right-handed Amino Acids".)[6-23] The only logical conclusion of the experiment was that an intelligently designed procedure and apparatus were needed to be designed for amino acids to form.[6-24]

Evolutionist View

Creationists do not believe life could have originated by natural processes because they believe there is no way developing life forms could become isolated to protect it from other harmful chemicals. This is simply not true. The ocean is not a big soup bowl in which free nucleotides have to find one another in the vast deep. Natural environments are full of "traps" for organic molecules – the surfaces of particles, to which organic molecules adhere and on which chemical reactions are catalyzed. Moreover, once the first nucleic acid molecules were formed, they didn't change into new, more complex sequences by chance. The ones that were more efficient in capturing organic molecules and replicating themselves more rapidly replaced the less efficient ones by natural selection.[6-24a]

Creationist View

Water

One of the early ideas in organic evolution was the "primordial soup." According to this view, at some point in Earth's history, the molten Earth cooled and oceans formed. The theory states as rain fell, chemicals in a hypothetical pool, warmed by the volcanic activity and energized by lightning, organized into proteins, lipids, and carbohydrates. These molecules then organized into cellular structures like proteins, DNA, and cell membranes (See Chapter 5 for discussion of proteins, DNA and cell membranes). The problem with this scenario is that in reality, it would not be possible. Proteins do not form from piles of amino acids, and DNA contains a specific code that must be copied from another strand of DNA. Proteins cannot form in water because the water breaks the bonds that hold the amino acids together—a process called hydrolysis.[6-25]

Evolutionists claim that amino acids formed out of seawater. But the seawater, needed to make the amino acids, would prevent them from forming into protein, lipids, nucleic acids and polysaccharides. Even if some protein could possibly form, the protein would hydrolyze with the abundant water and return back into the original amino acids. Those, in turn, would immediately break down into separate chemicals.[6-26]

It is commonly assumed today that life arose in the oceans. But even if this soup contained a decent concentration of amino acids, the chances of their forming spontaneously into long chains would seem remote. Other things being equal, a dilute hot soup would seem a most unlikely place for the first polypeptides to appear. The chances of forming tripeptides would be about one-hundredth that of forming dipeptides, and the probability of forming a polypeptide of only ten amino acid units would be something like $1/10^{20}$. The spontaneous formation of a polypeptide of the size of the smallest known proteins seems beyond all mathematical probability.[6-27] (See sections "The Improbability of Life Forming Naturally by Chemicals" and "Probabilities and Chance".)

It is clear that enzymes were not present in the primordial soup. Even if they were formed, they would not have lasted long since the primeval soup was by definition a conglomeration of nearly every conceivable chemical substance. There would have been innumerable enzyme inhibitors present to inhibit an enzyme as soon as it appeared. Thus, such molecules could not have formed; however, even with the assumption that they had formed, they could not have remained.[6-28] (Enzymes are various proteins, as pepsin and amylase, originating from living cells and capable of producing certain chemical changes in organic substances by catalytic action, as in digestion.)[6-29]

Even if chemical compounds and enzymes could survive the other problems, many organic products formed in the ocean would be removed and rendered inactive, as unfavorable chemical reactions would destroy such products. For example, fatty acids would combine with magnesium or calcium; and arginine (an amino acid), chlorophyll and porphyrins would be absorbed by clays. Many of the chemicals would react with other chemicals, to form non-biologically useful products. Sugars and amino acids, for example, are chemically incompatible when brought together.[6-30]

As mentioned in Chapter 5, the building blocks to produce a protein are amino acids. For DNA and RNA these building blocks are nucleotides, which are composed of purines, pyrimidines, sugars, and phosphoric acid. If amino acids are dissolved in water they do not spontaneously join together to make a protein. That would require an input of energy. If proteins are dissolved in water the chemical bonds between the amino acids slowly break apart, releasing energy the protein is said to hydrolyze. The same is true of DNA and RNA. To form a protein in a laboratory the chemist, after dissolving the required amino acids in a solvent, adds a chemical that contains high-energy bonds (referred to as a peptide reagent). The energy from this chemical is transferred to the amino acids. This provides the necessary energy to form the chemical bonds between the amino acids and releases H (hydrogen) and OH (oxygen and hydrogen) to form H_2O (water) (A water molecule contains one oxygen and two hydrogen atoms.). This only happens in a chemistry laboratory or in the cells of living organisms. It could never have taken place in a primitive ocean or anywhere on a primitive Earth. There would somehow need to be a steady input of appropriate energy. Destructive raw energy would not work. There would somehow also need to be a steady supply of the appropriate building blocks rather than useless material.

Water is one of the agents that damages DNA. If DNA somehow evolved on the Earth it (DNA) would have dissolved in the water. Ultraviolet light would also have destroyed DNA much faster than it could be produced. If it were not for DNA repair genes, DNA could not survive even in the protective environment of a cell (Chapter 5, Figures 5.1 & 5.2). How then could DNA survive when subjected to attack by all the chemical and other DNA-damaging agents that would exist on the hypothetical primitive Earth?[6-31] Life starting in water is also a problem since water tends to break the bonds of some amino acids and prevents them from forming chains. As mentioned earlier, Miller isolated the products with a trap (Figure 6.1) in order to avoid this destructive reaction.[6-32]

The gas phase reaction of H_2, NH_3, CH_4 and H_2O during the experiment has been shown to form alpha amino acids that are solids (that are soluble in water) and would fall into the ocean or on the ground when formed. There would be no further reaction on solid ground and would most likely dissolve (break down). If the amino acids would fall into the ocean, the concentration would dilute the formation of proteins, which would affect the composition of the amino acids. Since the formation of proteins from amino acids results in the elimination of water, then carrying out the reaction in the ocean (or small pond) or wherever water is in large excess, would be quite detrimental to the formation of proteins. This is because the reaction of amino acids to form proteins is a reversible reaction eliminating water and therefore, by the law of mass action, would be expected to reverse itself in the presence of a large amount of water.[6-33]

Oxygen

Evolutionists claim that life began in an oxygen-free atmosphere in order to explain why organic compounds such as amino acids, proteins, amine bases, etc. were not destroyed by oxidation over the long period of time necessary for life to begin. If this were true, then the same organic compounds would have been destroyed by ultraviolet radiation from the sun because without oxygen, there would not have been a protective ozone layer to protect them.[6-34] With oxygen in the air, the first amino acid would never have gotten started; without oxygen, it would have been wiped out by cosmic rays.[6-35]

Proteins are made up of amino acids. The amino acids that make up proteins cannot join together if oxygen is present. In other words, proteins could not have evolved from chance collisions of amino acids if the atmosphere contained oxygen. (As discussed in Chapter 5, section "Description of Proteins," proteins are very important to life, as all living matter is composed largely of proteins.) However, the chemistry of the earth's rocks, both on land and below ancient seas, show that the earth had oxygen before the earliest fossils formed. Even earlier, oxygen would have been produced by solar radiation breaking water vapor apart into oxygen and hydrogen. Some of the very light hydrogen would have escaped from the atmosphere into outer space, leaving behind oxygen.

To form proteins, amino acids must also be highly concentrated. However, the early oceans or atmosphere would have diluted amino acids to the point where the required collisions between them would rarely occur. Besides, amino acids do not naturally link up to form proteins. Instead, energy sources for forming proteins (the earth's heat, electrical discharges, or the sun's radiation) destroy the protein products thousands of times faster than they could have formed. The many attempts to show how life might have originated on earth have only demonstrated the futility of the effort, the immense complexity of even the simplest life, and the need for a vast intelligence to precede life.[6-36] (See section "Other Experiments Conducted to Determine How Life Could Have Originated on Earth.")

Stanley Miller did not include oxygen in his experiment because he knew oxygen would destroy the very molecules he was trying to produce. The presence of oxygen would make formation of amino acids impossible. It may be difficult to realize how "corrosive" oxygen is, since most modern living things depend on it. But oxygen is so valuable to life precisely because it's so chemically reactive, and aerobic living things today have systems to protect themselves against the harmful effects of oxygen, while using its chemical power to their advantage.[6-37] Research has shown that the early atmosphere did contain oxygen as proved by traces of oxidized iron and uranium in rocks estimated to be 3.5 billion years old. If oxygen was not present, life still could not have formed because of intense ultraviolet radiation that would have killed any molecules that happened to form by chance.[6-38]

The Earth's present atmosphere consists of 78% nitrogen (N_2), 21% molecular oxygen (O_2), and 1% of other gases, such as carbon dioxide CO_2), argon (Ar), and water vapor (H_2O). As mentioned previously, an atmosphere containing free oxygen would be fatal to all origin of life schemes. While oxygen is necessary for life, free oxygen would oxidize and thus destroy all organic molecules required for the origin of life. Thus, in spite of much evidence that the Earth has always had a significant quantity of free oxygen in the atmosphere, evolutionists will continue to claim that there was no oxygen in the Earth's early atmosphere.[6-39]

Because oxygen in the atmosphere would destroy all possibility of life arising by natural processes, evolutionists wrongly assumed the atmosphere had no oxygen. Scientists also assumed the atmosphere contained certain necessary ingredients, including (1) ammonia (NH_3), (2) nitrogen (N_2), (3) hydrogen (H_2), (4) water vapor (H_2O) and (5) methane (CH_4). However, it is well known that mixing these ingredients does not create life. Therefore, evolutionists theorized something else must be needed, such as a bolt of energy. The presence of oxygen would tend to destroy the organic compounds needed for life, but if oxygen were absent, the atmosphere would lack an ozone layer to shield the compounds from ultraviolet rays. Either way, oxygen would be a problem for evolution.[6-40] If there were no ozone layer the ultraviolet radiation would penetrate the atmosphere and would destroy the amino acids as soon as they were formed (if they were not somehow protected). So the dilemma facing the evolutionist can be summed up this way: amino acids would not form in an atmosphere with oxygen and amino acids would be destroyed in an atmosphere without oxygen.[6-41]

The present atmosphere—the air that we breathe—is composed of carbon dioxide (CO^2), nitrogen (N^2), oxygen (O^2), and water (H^20). The generally supposed primitive atmosphere would have had to have been composed of almost totally different chemicals: methane (CH^4), carbon monoxide (CO), carbon dioxide (CO^2), ammonia (NH^2), nitrogen (N^2), hydrogen (H^2), and water (H^20). Evolutionists then came up with the theory that when life first originated on earth, the entire world changed its atmosphere from reducing to oxidizing. A reducing atmosphere is an atmospheric condition in which oxidation is prevented by removal of oxygen and other oxidizing gases or vapors. It lacks free oxygen, and may contain such reactive gases as hydrogen and/or carbon monoxide that oxidize in the presence of oxygen, such as hydrogen sulfide. A reducing atmosphere is an atmosphere of hydrogen (or other constituent that willingly provides electrons) surrounding a chemical reaction or physical device; the effect is the reverse of an oxidizing atmosphere.[6-41a] An oxidizing atmosphere is a (planetary) atmosphere which oxidizes immersed (surface) compounds. It sometimes refers to an O^2-rich atmosphere. Oxidizing atmosphere refers to a gaseous atmosphere in which an oxidation reaction occurs, usually the oxidation of solids.[6-41b] Free oxygen is oxygen in its molecular forms, O^2 or O^3 (ozone), uncombined with other elements. Free oxygen is a requirement of all aerobic organisms and its copious presence in a planetary atmosphere is diagnostic of the presence of life.[6-41c] Evolutionists claim that photosynthesis caused the oxygen content in the Earth's atmosphere to increase. Evolutionists claim that the piece of amino acid that developed in the restless oceans was then able to form into part of a protein. But this possibility was proven false when a university study found that the plants could not suddenly have made all that oxygen, and oxygen had nowhere else to come from except through plants. However, the plants were not there at that time (since plants consist of living cells are also considered life), and whatever plants might have been there would all have died soon after, since they themselves need oxygen for their own cellular respiration.

In order to avoid the problem of mass action degradation of amino acids formed in seawater, someone else suggested that the amino acids were made in dry clays and rocks. (See Chapter 5, section "How Life Could have Originated on Earth" sub-section "Life May Have Originated from Concentrated Clays" and sub-section "Could Life Have Begun By Natural Causes?") But in that environment either the oxygen or ultraviolet light would immediately destroy those amino acids.

One scientist suggested that the amino acids were made on the edges of volcanoes, while a **second scientist** thought that dicyanimide (a compound not naturally occurring in nature) developed amino acids. A **third scientist** felt that phosphorus pentoxide in a jar of ether developed amino acids. A **fourth researcher** came up with the solution that the environment, in which all the amino acids developed, consisted of hydrogen cyanide.

It was later determined that none of these hypotheses would work. It was discovered that the **first scientist's** idea of the volcanic heat would ruin the amino acids as soon as they were formed. The **second scientist's** idea of Dicyanimide would not work because the original mixture in which the first amino acids were made had to have a more alkaline pH. The **third scientist's** idea of Phosphorus pentoxide is a novel compound that could not possibly be found in earth's primitive atmosphere. The **fourth research** worker's idea of the hydrogen cyanide would require an atmosphere of ammonia, which geological evidence shows never existed in our atmosphere.[6-42] (See section "Other Experiments Conducted to Determine How Life Could Have Originated on Earth.")

The Improbability of Life Forming Naturally by Chemicals

Old Universe Progressive Creationist View

Because Earth's first life must have possessed the full capability to produce all the chemical building blocks needed for cell activity (*autotrophism: 1) a plant capable of synthesizing its own food from simple organic substances or 2) an organism that is able to form nutritional organic substances from simple inorganic substances*), biochemical pathways for synthesizing amino acids, nucleotides, sugars, and fatty acids must have been in effect. In addition, metabolic pathways that degrade these compounds must have operated in Earth's first life. Finally, these life forms had reproductive potential and therefore DNA replication (Chapter 5, Section "DNA Replication" & figures 5.20 & 5.21) and cell replication processes (or its equivalent) must have been in place.

While some naturalistic origin-of-life scientists believe that messenger RNA, adapter RNA, ribosomes, and a diversity of specialized enzymes (the apparatus for the transcription of the genetic code) evolved slowly, through billions of years of evolution, this most likely is not true. The assemblage of prebiotics at most extended from the origin of the planet (4.566 billion years ago) until life's first appearance in the fossil record (3.5 billion years ago), which does not allow sufficient amount of time for life to have developed gradually, over billions of years. Earth's origin has been determined to be approximately 4.566 (plus or minus 0.002) billions year ago (Chapter 2). Fossils extend back 3.5 billion years. This time frame would seem to give nature over a billion years to abiogenesis (the emergence of life from nonlife). This, however, may not be true. This calculation provides the theoretical boundaries as to the time frame for life to develop. The question is, could life have developed naturally and gradually within this time frame of 1.066 billion years (4.566 – 3.500)?

It is very unlikely that life began to form shortly after the origin of the Earth 4.566 billion years ago. It has been determined that when the Earth first formed it was uninhabitable, that is, the conditions of the Earth were too hostile to sustain life. The extreme variability of the infant Sun's solar-type stars confirm this. It has been determined that between the formation of the Earth 4.566 billion years ago and 3.85 billion years ago, multiple giant bodies (such as asteroids) smashed into the planet causing much heat, enough to melt the surface. This means that no liquid water, solid rocks, conceivable life forms, or even basic prebiotic molecules could have survived anywhere on the boiling planet during such collision episodes. This may explain why no rocks older than 3.85 billion years old have been found. Regardless of whether the heavy bombardment of the Earth lasted 100 million years or less, when the Earth first became inhabitable, life appeared in a very short period of time. As soon as rocks formed, life appeared.

Some naturalistic origin-of-life scientists have calculated that it was possible that cyanobacteria might have evolved from a prebiotic soup within a 10-million-year time frame. They believe that life may have evolved from the following steps: 1.) formation of prebiotic compounds 2.) assembly of the first self-replicating molecules 3.) emergence of RNA molecules 4.) transition from an RNA system to a DNA-protein system 5.) emergence of starter proteins 6.) emergence of gene duplication and divergence. Two additional steps that are also necessary to sustain life that were not mentioned by the naturalistic origin-of-life scientists include: 7.) assembly of workable membrane (See Chapter 5, section "The Origin of the Cell Membrane") to protect and feed the RNA, DNA, and/ or protein molecules 8.) positioning of all the molecular components in the correct locations with respect to one another.

Given the complexity of the steps mentioned above, it is very doubtful that life could have developed within the time frame the naturalists estimate. All models for life's origin at some point involve the building and operation of RNA molecules. Thus the stability of nucleotide building blocks that comprise RNA molecules and the stability of the RNA molecules themselves provide an independent measure of the maximum time span for any naturalistic life origin scenario. At the time when life first formed on the Earth, the planet was very hot, somewhere between 158° and 194° Fahrenheit (70° and 90° Celsius), with no isolated cold spots. Nucleotide building blocks are known to fall apart quickly at warm temperatures in time periods ranging from nineteen days to twelve years. These extremely short survival rates for the four RNA nucleotide building blocks indicate that life's molecules would have to be assembled before any of the nucleotide building blocks decayed. Cold naturalistic origin-of-life possibilities were also calculated. At the freezing point of water, cytosine (one of the nucleotide building blocks) decomposes in less than 17,000 years. At any possible temperature the origin of life must take place quickly. It has been determined that since complex molecules tend to be fragile, it is very possible that unless life develops on a planet within a few thousand years, it will not develop at all. [6-42a]

Cytosine is one of the molecular components of nucleic acids (DNA and RNA) and important in the construction of DNA and RNA. Cytosine is a pyrimidine (*a pyrimidine could be any of a number of similar compounds having a basic structure that is derived from pyrimidine, including cytosine, thymine, and uracil, which are elements of nucleic acids*) and is a six-membered ring composed of four carbon atoms and two nitrogen atoms. Along with other ring compounds, such as adenine, guanine, thymine, and uracil, cytosine repeatedly extends from the chain-like backbone of DNA and RNA. The nitrogen-containing rings sequenced along DNA or RNA provide the chemical information that determines biochemical function.

Chemists have discovered two possible pathways that produce cytosine. One pathway involves a reaction between cyanoacetylene and cyanate, and the other pathway reaction begins with cyanoacetaldehyde and urea. These four compounds represent essential ingredients of early Earth's supposed prebiotic soup. Other scientists have demonstrated, however, that the two chemical pathways lack any relevance of their actually obtaining cytosine. They conclude that it is very unlikely that cyanoacetylene, cyanate, cyanoacetaldehyde, and urea existed in sufficient quantities on primordial Earth to effect the production of cytosine. Even if these compounds were sufficient amounts, interfering chemical reactions would have quickly consumed these compounds before cytosine could form. Cyanoacetylene rapidly reacts with ammonia, amines, thiols, and hydrogen cyanide. Cyanate undergoes rapid reaction with water. In the presence of water, cyanoacetaldehyde decomposes into acetonitrile and formate. When cytosine does form, it rapidly decomposes.

To date (2004), scientists have been unable to produce cytosine in a spark-discharge experiment, nor has cytosine been recovered from meteorites or extraterrestrial sources. Because meteorites (and other extraterrestrial sources) serve as a proxy for early Earth's chemistry, the absence of cytosine in these sources seem to confirm the conclusion that cytosine could not have existed in an early Earth. As with cytosine, it has been shown that adenine formation on early Earth (by currently recognized prebiotic routes) could not reasonably have occurred, for many of the same reasons. Other problems have also been discovered that would greatly decrease the possibility of cytosine developing during an early Earth. Nucleobases readily react with formaldehyde and acetaldehyde (compounds that most likely were present on early Earth) to form both small molecule derivatives and large intractable molecules. Even under mild conditions, these reactions take place so rapidly that they would preferentially occur at the expense of reactions that could lead to RNA. Thus, if nucleobases could form, competing reactions would likely consume them.

Ribose is a five-carbon sugar, like cytosine, that is an important component of RNA. (The closely related sugar deoxyribose takes ribose's place in DNA structure.) Ribose repeatedly alternates with phosphate to form RNA's backbone. This sugar also serves as the attachment point for cytosine, uracil, guanine, and adenine. Ribose production on early Earth stands as a central requirement for all origin-of-life scenarios that pass through RNA development. The only known plausible prebiotic pathway for ribose (and all sugars) is a reaction known as the Butlerow reaction (or the formose reaction). (The formose reaction is the formation of sugars from formaldehyde.) This reaction begins with the one-carbon compound formaldehyde, which readily forms in spark-discharge experiments.

In the presence of an inorganic catalyst (calcium hydroxide, calcium oxide, alumina clays, and so on), formaldehyde reacts with itself and resultant products to generate sugars containing two, three, four, five, six, or more carbon atoms. Although this pathway to ribose and other sugars exists, most scientists question its applicability to the origin-of-life scenario. Numerous side reactions dominate formose chemistry. As a consequence, this reaction yields over forty different sugar species with ribose as a minor component. If this reaction did operate on early Earth, it could never have yielded enough ribose to support an RNA structure. Laboratory formose reactions are free of contaminants that would likely be present on early Earth. Ammonia, amines, and amino acids, for example, react with formaldehyde and the products of the formose reaction. These side reactions would have consumed key reactants and prevented the formation of ribose and other sugars. As with cytosine, decomposition negatively affects ribose formation. Sugars decompose under alkaline and acidic conditions and are susceptible to oxidation. Even with a neutral pH range, sugars decompose. Deoxyribose, the sugar component of DNA, likewise possesses limited stability even under neutral conditions.[6-42b]

Hi-energy phosphate compounds are also essential to DNA and RNA. Phosphate chains, called polyphosphates, form a relatively unstable high-energy chemical structure in which the cells metabolic systems store energy. Although several plausible pathways to polyphosphates exist, researchers are not sure whether or not these chemical pathways could have even a remote possibility of occurring during an early Earth. For example, to produce polyphosphates from apatite and dihydrogen phosphate, water must be completely driven from the environment, which is an impossibility for phosphate minerals confined to rocks. Also, the high temperatures needed to form polyphosphates would in turn destroy any organic material.

The scenarios mentioned for the natural production of polyphosphates from high-energy chemicals (allegedly formed in spark-discharge reactions on early Earth) lacks chemical analysis. These reactions require unrealistically high levels of starting materials and produce low yields. Laboratory spark-discharge experiments performed under various chemical conditions failed to yield polyphosphates when phosphates were included in the reaction container. Even if there was a way on Earth to form polyphosphates, their availability for prebiotic reactions is unlikely because calcium ions drive polyphosphates to precipitate out of solutions. These ions would have been everywhere on early Earth. Given the extreme rarity (or nonexistence) of polyphosphate minerals on Earth today, the conclusion that primordial prebiotic polyphosphate synthesis could have taken place on early Earth seems impossible.[6-42c]

Scientists have yet to detect any of the biological five- and six- carbon sugars, such as ribose, deoxyribose, glucose, lactose, and fructose, in any nonbiological area of nature. The closest success was the discovery of extremely low levels of one three-carbon sugar in the Murchison and Murray (probably a piece of the Murchison) meteorites (See Chapter 5, section "Other Theories About How Life Could Have Originated on Earth," subsection "Amino Acids Found in Meteorite" for further discussion of the Murchison meteorite.). The absence of these homochiral sugars and homochiral amino acids in the nature causes difficulty in attempting to explain how life originated on this planet.[6-42d] Homochirality describes a geometric property of some materials that are composed of chiral units. Chiral refers to nonsuperimposable 3D forms that are mirror images of one another, as are left and right hands. A substance is said to be homochiral if all the constituent units are molecules of the same chiral form (enantiomer).[6-42e] (See section "Left-handed and Right-handed Amino Acids" below for further discussion of chirality).

Amino Acids

As mentioned earlier, the most interesting compound produced from Stanley Miller's 1953 experiment, were amino acids. Amino acids are compounds that are the simplest units out of which proteins can be assembled (as discussed in Chapter 5).[6-43] Amino acids are a group of chemicals that bond together to form proteins. Twenty amino acids join in different sequences to form the thousands of different types of proteins in the human body. Human cells have the metabolic capacity to manufacture 10 of the (20 or so) amino acids used to build proteins; the remaining 10 amino acids must be obtained from one's diet. All amino acids contain the elements carbon, hydrogen, oxygen and nitrogen. Two amino acids also contain sulfur.[6-44]

Left-handed and Right-handed Amino Acids

The Left (L) and Right (D) Amino Acid Molecules

The two molecules are identical in every way except shape. They are alike chemically, but different dimensionally. Each one is the mirror image of the other. One is like a left handed glove; the other a right handed one. But only the left shaped (L) amino acids are found in animal life. A typical amino acid in both forms is illustrated below.

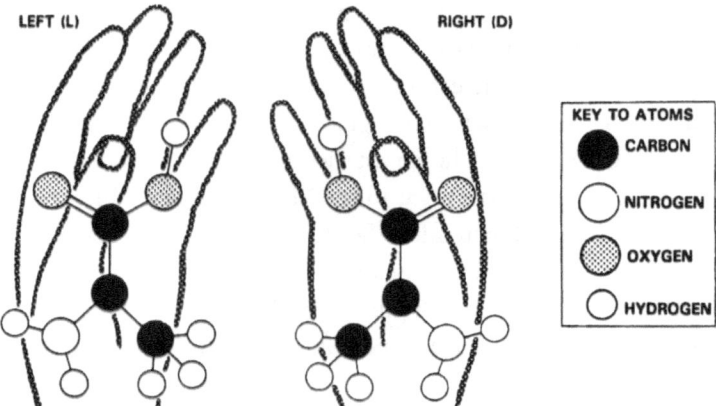

Figure 6.2 "Handedness" Illustration.[6-45]

Darwinist View

The chirality of DNA, RNA, and amino acids is utilized by all known life. As there is no functional advantage to right or left handed molecular chirality, the simplest explanation is that the type of amino acids used for life came about randomly in the early beginnings of life and passed on to all existing life through common descent.[6-46] Chirality is the characteristic of a structure (usually a molecule) that makes it impossible to superimpose it on its mirror image. Chirality is also called "handedness."[6-47] In chemistry, chirality usually refers to molecules. Two mirror images of a chiral molecule are called enantiomers or opticalisomers. Pairs of enantiomers are often designated as "right-handed" and "left-handed."[6-48] An enantiomer is either of a pair of optical isomers that are mirror images to each other.[6-49]

Standard alpha amino acids, except glycine, can exist in either of two optical isomers, called L or D amino acids. L and D amino acids mirror each other, called chirality. L-amino acids are in all amino acids found in proteins during translation in the ribosome and D-amino acids are in all amino acids found in some proteins produced by enzyme post-translational modifications after translation and translocation to the endoplasmic reticulum.[6-50]

The term chiral generally is used to describe an object that is non-superposible on its mirror image. Achiral (not chiral) objects are objects that are identical to their mirror image. Chirality is best described using human hands (Figure 6.2). The left hand is a non-superposable mirror image of the right hand. No matter how the two hands are arranged, it is impossible for all the major features of both hands to line up with each other. This difference in symmetry becomes obvious if someone attempts to shake the right hand of a person using his left hand, or in attempting to put a left-handed glove on a right hand.[6-51]

The origin of this homo-chirality in biology is the subject of much debate. Most scientists believe that Earth life's "choice" of chirality was purely random, and that if carbon-based life forms exist elsewhere in the universe, their chemistry could theoretically have opposite chirality. However, there is some suggestion that early amino acids could have formed in comet dust. In this case, circularly polarized radiation (which makes up 17% of stellar radiation) could have caused the selective destruction of one chirality of amino acids, leading to a selection bias that ultimately resulted in all life on Earth being homochiral.[6-52]

Creationist View

> *Many researchers have attempted to find plausible natural conditions under which L-amino acids would preferentially accumulate over their D-counterparts, but all such attempts have failed. Until this crucial problem is solved, no one can say that we have found a naturalistic explanation for the origin of life. Instead, these isomer preferences point to biochemical creation.*[6-53]

Although Miller and Urey formed amino acids in their experiments, all the amino acids that formed did not have chirality. The mixture of amino acids and other simple chemicals produced during the experiment is not correct for producing life. All known life uses amino acids that are exclusively of the "left-handed" form. The problem of having a mix of both "right-handed" and "left-handed" amino acids is that no known life can use any combination of both "right-handed" and "left-handed" amino acids (Figure 6.2). There is no natural process known that makes only left-handed amino acids, as only the left-handed (L) amino acids are found in animal life. Even if one "right-handed" amino acid were added to a chain of "left-handed" amino acids, this could destroy the entire chain. When amino acids are synthesized in the laboratory, there is always a 50% mixture of the two forms.[6-54] Just one long or right-handed amino acid inserted into a chain of short, left-handed amino acids would prevent the coiling and folding necessary for proper protein function. What Miller's experiment actually produced was a mixture of deadly poisons that would destroy any possibility for the chemical evolution of life.[6-55]

Laboratory experiments demonstrate that the presence of racemic (random mixtures of left and right handed molecules) mixtures of amino acids and sugars strongly inhibits the formation of amino acids and nucleotide chains. For example, the synthesis of an RNA or DNA strand that consists of right-handed ribose or deoxyribose is hindered by the presence of the same sugar in the left-handed configuration (and vice versa). One wrong-handed amino acid incorporated in a protein is enough to disrupt the folding of the protein, thereby blocking its capacity to function. Also, without homocharality, genetic material cannot copy itself. Two complementary strands of DNA cannot bind with each other into the crucial double helix structure unless all the nucleotides are of the same handedness (chirality).[6-55a]

The problem of chirality is that in our bodies, proteins and DNA possess a unique 3-dimensional shape, and it is because of this 3D shape that the biochemical processes within our bodies work as they do (See Chapter 13, Pseudogenes, "DNA Double Helix Strand" for further discussion of the 3 dimensional shape). It is chirality that provides the unique shape for proteins and DNA, and without chirality, the biochemical processes in our bodies would not do their job. In our body, every single amino acid of every protein is found with the same left-handed chirality. It is a well-accepted fact of chemistry that chirality cannot be created in chemical molecules by a random process. When a random chemical reaction is used to prepare molecules having chirality, there is an equal chance to prepare the left-handed isomer as well as the right-handed isomer.

It is a scientifically verifiable fact that a random chance process, which forms a chiral product, can only be a 50/50 mixture of the two optical isomers. There are no exceptions. There is also chirality in proteins and DNA. Proteins are compounds of amino acids and each one of the component amino acids exists as the "L" or left-handed optical isomer. Even though the "R" or right-handed optical isomers can be synthesized in the lab, this isomer does not exist in natural proteins. The DNA molecule is made up of billions of complicated chemical molecules called nucleotides, and these nucleotide molecules exist as the "R" or right-handed optical isomer. The "L" isomer of nucleotides can be prepared in the lab, but they do not exist in natural DNA. There is no way that a random chance process could have formed these proteins and DNA with their unique chirality. If proteins and DNA were formed by chance, each and every one of the components would be a 50/50 mixture of the two optical isomers. This is not what is seen in natural proteins or in natural DNA.

A random chance natural process cannot create proteins with thousands of "L" molecules and DNA with billions of "R" molecules. As nucleotide molecules come together to form the structure of DNA, they develop a twist that forms the double helix structure of DNA (See double helix model and accompanying explanation in Chapter 13, Pseudogenes, "DNA Double Helix Strand" for further discussion of the double helix). DNA develops a twist in the chain because each component contains chirality (handedness). It is this handedness that gives DNA the spiral shaped helical structure. If one molecule in the DNA structure had the wrong chirality, DNA would not exist in the double helix form, and DNA would not function properly. The entire replication process would be destroyed.

In order for DNA to have evolved, billions of molecules within our body would have had to have been generated with the "R" configuration all at the same time, without error. If it is impossible for one nucleotide to be formed with chirality, it is much less likely for billions of nucleotides to come together exactly at the same time, with all of them being formed with the same chirality. Without chirality, proteins and enzymes could not do their job and DNA could not function at all. (Proteins and enzymes are discussed in more detail in Chapter 5.) Without properly functioning proteins and DNA, there would be no life on this Earth. Chirality (handedness) is probably one of the best scientific evidences there is against random chance evolution. Chirality totally disproves the claim that life came from chemicals evolving on their own.[6-56]

Not only do the Miller-type experiments not produce the proper "handedness" of amino acids (left-handed amino acids only, instead of both-handed ones) (Figure 6.2), but the amino acids they produce are not the ones needed to produce life. Out of the hundreds of possible combinations, there are 20 essential amino acids. The synthesis of amino acids in laboratories only produces a few of the correct amino acids essential for life. It does, however, produce many non-essential, useless amino acids.

In considering Miller's 1953 experiment and subsequent experiments where amino acids were formed through applying heat to elements alleged to be in the primordial atmosphere, the following points need to be considered: **(1)** the amino acids were racemic (both D and L forms) and thus proteins formed from these would not support life; **(2)** the majority of amino acids formed by laboratory synthesis do not belong to the 20 amino acids that occur in natural protein molecules.[6-57]

To date (2004), the only possible physical mechanism known for driving racemic mixtures of amino acids and sugars into homochiral ones is an effect called "magnetochiral anisotropy," which links chirality and magnetism (A racemic mixture is one that has equal amounts of left-and right-handed enantiomers of a chiral molecule). So far, the two most probable sources of magnetochiral anisotropy are **(1)** photochemistry (light-induced chemistry) with circularly polarized light, and **(2)** parity violations in electroweak interactions (a particle physics phenomenon). Photochemistry with circularly polarized light, if sufficiently concentrated and turned to the appropriate wavelengths, can generate significant chirality. However, no natural circularly polarized light exists on Earth other than a tiny amount produced in the daylight sky. This amount falls far short of what would be needed to generate an excess of one configuration over its mirror image. Parity violations in electoweak interactions have never been observed in a natural realm. This effect is only seen in particle accelerator experiments. Other possible

Earth-based mechanisms as the source of homochirality yield also proved to be doubtful. The only other possible mechanism to obtain homochirality is circularly polarized ultraviolet (uv) light emanating from neutron stars and black holes. Circularly polarized uv light is the only proven means for generating significant homochirality. However, using 100 percent circularly polarized uv light, the most successful laboratory experiments yielded only a 20 percent excess of left-handed amino acids over right-handed ones. Experiments using polarized electrons and polarized positrons (anti-electrons) were even more doubtful. This problem is that this excess was achieved not by production, but by destruction.[6-57a]

Other Experiments Conducted to Determine How Life Could Have Originated on Earth

Origin of Life Experiments

Researcher(s)	Year	Reactants	Energy Source	Results	Probability
Miller	1953	CH_4, NH_3, H_2O, H_2	Electric discharge	Simple amino acids, organic compounds	unlikely
Abelson	1956	CO, CO_2, N_2, NH_3, H_2, H_2O	Electric discharge	Simple amino acids, HCN	unlikely
Groth and Weyssenhoff	1957	CH_4, NH_3, H_2O	Ultraviolet light (1470 - 1294 ?)	Simple amino acids (low yields)	under special conditions
Bahadur, et al.	1958	Formaldehyde, molybdenum oxide	Sunlight (photosynthesis)	Simple amino acids	possible
Pavolvskaya and Pasyrskii	1959	Formaldehyde, nitrates	High pressure Hg lamp (photolysis)	Simple amino acids	possible
Palm and Calvin	1962	CH_4, NH_3, H_2O	Electron irradiation	Glycine, alanine, aspartic acid	under special conditions
Harada and Fox	1964	CH_4, NH_3, H_2O	Thermal energy (900 - 1200s C)	14 of the "essential" amino acids of proteins	under special conditions
Oro	1968	CH_4, NH_3, H_2O	Plasma jet	Simple amino acids	unlikely
Bar-Nun et al.	1970	CH_4, NH_3, H_2O	Shock wave	Simple amino acids	under special conditions
Sagan and Khare	1971	CH_4, C_2H_6, NH_3, H_2O, H_2S	Ultraviolet light (>2000 ?)	Simple amino acids (low yields)	under special conditions
Yoshino et al.	1971	H_2, CO, NH_3, montmorilonite	Temperature of 700o C	Glycine, alanine, glutamic acid, serine, aspartic acid, leucine, lysine, arginine	unlikely
Lawless and Boynton	1973	CH_4, NH_3, H_2O	Thermal energy	Glycine, alanine, aspartic acid, ?-alanine, N-methyl-?-alanine, ?-amino-n-butyric acid.	under special conditions
Yanagawa et al.	1980	Various sugars, hydroxylamine, inorganic salts.	Temperature of 105o C	Glycine, alanine, serine, aspartic acid, glutamic acid	under special conditions
Kobayashi et al.	1992	CO, N_2, H_2O	Proton irradiation	Glycine, alanine, aspartic acid, ?-alanine, glutamic acid, threonine, ?-aminobutyric acid, serine	possible
Hanic, et al.	1998	CO_2, N_2, H_2O	Electric discharge	Several amino acids	possible

Table 6.1 The table above summarizes various experiments that have been conducted to determine how life could have originated on Earth.[6-58]

Probabilities and Chance

Creationist View

> *The probability of life originating from accident is comparable to the probability of the unabridged dictionary resulting from an explosion in a printing shop.*[6-59]

> *The chances that life just occurred are about as unlikely as a typhoon blowing through a junkyard and constructing a Boeing-747.* [6-60]

"Chance" does not cause anything. If someone flips a coin, someone else might say that there is a 50% chance that it will come up heads and a 50% chance that it will come up tails. But this is only an observation, not the cause for it to come up heads or tails.[6-61] Chance is not able to demonstrate a mechanism by which, for instance, the woodpecker could acquire a reversed claw. Chance cannot explain the processes by which an intricate organ like the ear was formed.[6-62]

Darwinist View

Chance is not an antonym (the opposite meaning) to the word "purpose." When we flip a coin, we trust that physical causes determines how it falls (heads or tails). It is only that we are ignorant of the exact physical forces operating on it, and so cannot predict how it will land. In general, we say that chance operates when physical causes can result in any of several outcomes, but we do not have sufficient knowledge to predict what that outcome will be in any particular case. At least some of what we call chance, then, is a name for our ignorance. Sometimes when we obtain additional information, we can make better predictions, that is, events will seem less random.[6-62a]

The Probability of Proteins Developing by Natural Causes

Darwinist View

Chemical elements interacting with other chemical elements produce complex products. These products in turn interact in complex ways. Creationists assume that protein molecules must take certain forms to produce life. They claim the odds are astronomical for proteins to simply form from amino acids. However, there are innumerable different proteins that promote biological activity. Taking this into account greatly decreases those odds.[6-63] The calculation of odds assumes that the protein molecule formed by chance. However, biochemistry is not chance, making the calculated odds meaningless. Biochemistry produces complex products, and the products themselves interact in complex ways. For example, complex organic molecules are observed to form in the conditions that exist in space, and it is possible that they played a role in the formation of the first life. The calculation of odds assumes that the protein molecule must take one certain form. However, there are innumerable possible proteins that promote biological activity. Any calculation of odds must take into account all possible molecules (not just proteins) that might function to promote life. The calculation of odds assumes the creation of life in its present form. The first life would have been very much simpler. The calculation of odds ignores the fact that innumerable trials would have been occurring simultaneously.[6-64]

When creationists speak of the impossibility of evolution, they claim that randomness cannot produce beneficial changes in an organism. They point out that a major belief of evolutionary theory is that the changes (mutations, genetic recombination, etc.) are purely random. Creationists claim that it is not possible for the beauty and organization of life to originate from randomness and chaos. Creationists then present an example such as this. This example involves a type of algae called "permutations" and "combinations." Suppose there is a particular protein found in an animal (proteins are long chains and the links in the chains are made up of amino acids.) Insulin and hemoglobins are examples of proteins. Suppose the protein polymer chain is 150 amino acids long (a small protein). Since there could be any of 20 amino acids that could fill that first site in the protein strand, the probability is 1/20 (read: one chance in 20), the second is 1/20, etc. The probability that this protein would randomly come together in the right sequence is 20 multiplied by itself 150 times, or 1/20 to the 150th power. Such a number would have over 150 digits in it making such a chance occurrence highly improbable. This argument by itself is completely valid. However, the important part left out of the creationists' argument is natural selection.

Neither randomness nor natural selection alone will cause evolution. (Natural selection is discussed in Chapter 9.) By the two working together, the probabilities become more realistic. Take the protein molecule example described above. Suppose that you had a die with 20 sides (each of the 20 amino acids). Then you roll the die until by chance you get the first amino acid. If you could keep it in its first position while you rolled the die for the second amino acid, the calculation changes dramatically. It now becomes 1/20 plus 1/20 etc. for the 150 amino acids. This produces a probability of 1/3,000 (the number 3,000 is derived by multiplying 150 by 20). This example has included both the principle of reproduction and selection with randomness. Because organisms can reproduce, they can duplicate and retain the organization they have already achieved. Each time they reproduce, if copying errors creep in (random mutations) we have modifications added by chance. Those modifications that are useful and help the organism survive are retained. If the change causes a disadvantage, the organism fails in the struggle for survival. The course of evolution does require tremendously low probabilities to occur. A chance event of 1/1,000,000 looks pretty unlikely to occur but if the event is repeated 1,000,000 times, it may occur. In the history of the Earth there have been billions of organisms alive at any given time for chance events to occur and billions of years for them to accumulate. With such a tremendous amount of time and events, the improbable becomes probable.[6-65]

Creationist View

Evolutionists claim that chance plays a big part in the development of a structure, characteristic, or functional result. The odds of getting all heads when simultaneously tossing 150 coins will have a chance of one in 2^{150}, or one in about 10^{45}. If it is virtually or absolutely impossible to get all heads for those 150 flipped coins, then how similar are the odds for getting all the steps and changes needed to get from one species to the next? If the odds look as improbable for creating the new species as it is to get all heads for the 150 coins, then both events might be considered equally impossible. The number 150 is used here because 150 is the number of amino acids contained in a small protein that was used in the Evolutionist View illustration above.

To calculate the chance of the whole series of steps occurring, the following information is needed: **(A)** What is the chance of getting a mutation? **(B)** What fraction of the mutations has a selective advantage (via natural selection)? **(C)** How many replications are there in each step of the chain of cumulative selection? **(D)** How many of those steps need to be completed to achieve a new species?

If a copying error occurs in one reproduction, it does not automatically become a typical step in the cumulative selection. To be a part of a typical step a mutation must: **(1)** Have a positive selective value, and **(2)** Add a little information to the genome. For the first of these parameters the mean mutation rate for animals is 10-10. It has been estimated that to get a new species would take about 500 steps. It has been determined that one small evolutionary step would comprise about 50 million births. We see, then, that 500 steps, that is, 500 unique mutations, will be required to get from one species to the next. It has been estimated that for each species alive today there are 1,000 species that are extinct. So, to get a new species there is a chance of 1 in 1,000. Another factor to consider in this chance is that species are known to change very little over time (which is called stasis). 1,000 times 1,000 = 1,000,000. Unless mutations survive long enough to spread throughout a population, then it's more likely that the mutation will disappear. Just because a mutation occurs is no guarantee it sticks over generations or time.[6-66] (Mutations are discussed in Chapter 9.) A simple cell requires at least a hundred different, specific protein molecules plus sugars, lipids, nucleic acids, etc. One lonely protein in an expanse of the 'soup' would be destroyed long before it met so many biological 'friends'. If such an unlikely combination did float together, how was it sufficiently separated from similar molecules whose presence would have fouled up the system?[6-67]

It would be extremely improbable for evolutionary processes to produce one protein. **(1)** There are 20 amino acids. **(2)** There are 300 amino acids in a specialized sequence in each medium protein. **(3)** There are billions upon billions of possible combinations of amino acids. **(4)** The right combination from among the 20 amino acids would have to be brought together in the right sequence in order to properly make one useable protein.[6-68] It has been determined that in the absence of any chemical competition with non-amino acids and nonbiologically relevant amino acids, the probability of undirected processes assembling a protein one hundred amino acids long is roughly one chance in 10^{191}.

There is a 50 percent chance of natural processes randomly selecting a left-handed amino acid (in nature, amino acids are produced 50 percent left-handed and 50 percent right-handed), a 50 percent chance of joining the two amino acids in the appropriate chemical bond, and roughly a 5 percent chance of selecting the right amino acid. Proteins in the cell typically consist of several hundred amino acids. This means that the likelihood of random chemical processes generating most proteins is far more remote. In effect, there is no chance that even a relatively small protein made up of a specified sequence could ever form by undirected processes. If we assume that all carbon on earth exists in the form of amino acids and that the amino acids are allowed to chemically react at the maximum possible rate of 10 $^{12/s}$ for one billion years (the greatest possible time between the cooling of the earth and the appearance of life), we must still conclude that it is highly improbable (~10^{-65}) that even one functional protein would be made.[6-68b] (See Chapter 5, section "Life Originated Too Early in Earth's History to Have Originated by Natural Causes" for further discussion.)

Evolutionist View

Creationists utilize the argument of probabilities in an attempt to disprove the view that life originated by natural causes. It is a fact that to the best of our current knowledge this probability is impossible to be estimated. We simply know too little about how abiogenesis (origin of life) occurred and the probability we should compute does in fact depend on the details of the scenario one considers, all of which are unknown. Without these details one cannot answer the question of whether or not it is possible for life to originate by natural means.[6-68c]

The Probability of DNA Developing by Natural Causes

Evolutionist View

Chance (unpredictable) and nonrandom (predictable) factors operate at the same time. The chance that an event will happen is altered by all sorts of factors. Creationists claim that the probability of life evolving from nonlife is vanishingly small. One of their arguments is the spontaneously formed nucleotides would be so dilute in the primitive ocean that they would have hardly any chance of aggregating into nucleic acids. But this ignores the fact that chemicals will accumulate in some places even if in the ocean as a whole they are greatly dispersed; or that organic compounds commonly adhere to surfaces, and so would be concentrated on the surfaces of sand grains or clay particles.

Consider the argument that the first DNA molecule could not have evolved by chance. Since each site on the molecule can be occupied by one of four nucleotide bases, the chance that a particular DNA molecule 1,000 nucleotides long would be formed is only 1 out of $4^{1,000}$ or about 1 out of 10 followed by 600 zeroes. This may be true if the first DNA molecule had to have any particular sequence of nucleotides. But any sequence of nucleotides would replicate itself. Any of a large number of mutational changes in the initial sequence could improve the stability of the molecule (its "survival") or improve its rate of replication, and cause subsequent evolution of the population of molecules. Any particular mutation that enhances the survival or the replication rate may occur with a very low probability, but the chance is much higher that one or another of the many possible mutations that have this effect will occur. Thus, a "new, improved" DNA molecule may be quite likely to evolve, but the one that actually evolves will be only one of the many improved DNA molecules that could have occurred.[6-68a]

Creationist View

Based on probability factors, any viable DNA strand having over 84 nucleotides cannot be the result of haphazard mutations. At that stage, the probabilities are 1 in 4.80×10^{60}. Mathematicians agree that any requisite number beyond 10^{60} has, statistically, a zero probability of occurrence. Any species known to us, including the smallest single-cell bacteria, have enormously larger numbers of nucleotides than 100 or 1000. In fact, single cell bacteria display about 3,000,000 nucleotides, aligned in a very specific sequence. This means, that there is no mathematical probability whatever for any known species to have been the product of a random occurrence— random mutations.[6-69]

The Probability of Cells Developing by Natural Causes

Intelligent Design View

According to Darwinian evolutionary theory, the initial elementary mutational changes, upon which natural selection acts, are entirely random. (Mutations and natural selection are discussed in Chapter 9.) Mutations are completely blind to whatever effect they may have on the function or structure of the organism in which they occur. That is, changes only occur by pure chance. Therefore, every adaptive advance, big or small, must have been found as a result of a purely random occurrence. Amino acids (which are needed to create proteins, which in turn are needed to create a cell) are small organic compounds consisting of about ten to twenty atoms. Each amino acid contains an amino (NH_2) and a carboxyl acid (COOH) group lined by a carbon atom, as well as a unique side chain. In most proteins the amino acid chain is between one hundred (100) and five hundred (500) amino acids long. Of all the various types of amino acids, living organisms use only twenty types of amino acids. In order for the correct amino acids to form, the correct chemical bases must be present.

Using the analogy of the English language, just as words must be spelled properly, the correct arrangement of chemicals (A, G, C, U) (adenine, guanine, cytosine, uracil) must be in place to create the correct RNA, which is utilized to create an amino acid. Not only must the correct letters (chemicals, bases) be present to create the correct amino acid, the correct syntax rules must be followed or no protein will be developed from the amino acids. The linear sequence of amino acids in a protein can be thought of as a sentence made up of long combinations of the twenty amino acid letters. Then, just as the proper sentence structure (using our English language analogy), consisting of subject, verb, and object must be followed, the correct amino acid sequences must be present to create a protein. Not only must the sentence structure be correct, but the sentence must make sense.

The number of all possible amino acid sequences (as with letter sequences in the English language analogy) is extremely large and specific sequences must obey certain rules like rules of grammar. Since these amino acid sequences are extremely large, it would be rare for them to come together in a meaningful sequence. Even short sentences just ten amino acids long only occur once by chance in about 10^{13} average-sized proteins. Unique sequences twenty amino acids long only occur once in about 10^{26} proteins. Unique sequences thirty amino acids long only occur once in about 10^{39} proteins.

It has been established that there are approximately 10^{40} possible proteins that could have ever existed on Earth since its formation. This means that, if protein functions reside in sequences any less probable than 10^{-40}, it becomes increasingly unlikely that any functional proteins could ever have been developed by chance on earth. Comprehending how a living cell could have gradually evolved through a sequence of simple proto-cells seems almost impossible. If the estimates above were anywhere near accurate, the possibility of life arising suddenly on earth by chance, is infinitely small. To get a cell by chance would require at least one hundred functional proteins to appear simultaneously in one place. That is one hundred simultaneous events each of an independent probability that could hardly be more than

10^{-20} giving a maximum combined probability of 10^{-2000}. [6-70] Sir Fred Hoyle, the famous British mathematician and astronomer, teamed up with Chandra Wickramasinghe in an analysis of the origin of life to determine the possibility of life beginning by chance. They mathematically determined that the likelihood that a single cell could originate in a primitive environment, given 4.6 billion years in which to do so, was one chance in 10^{40000}. That is one chance in 1 with 40 thousand zeros after it.[6-71] Hoyle and Wickramasinghe have estimated that the chance of life originating by natural causes (assuming functional proteins) has a probability of 10^{-2012}. No evolutionary biologist has ever produced any quantitative proof that the designs of nature are in fact within the reach of chance.[6-70]

In the next chapter (7) we will look at some specific fossils that may help shed light on whether or not species evolved from lower life forms.

~~~

Chapter 7

A Look at Some of the Transitional Fossil Evidence

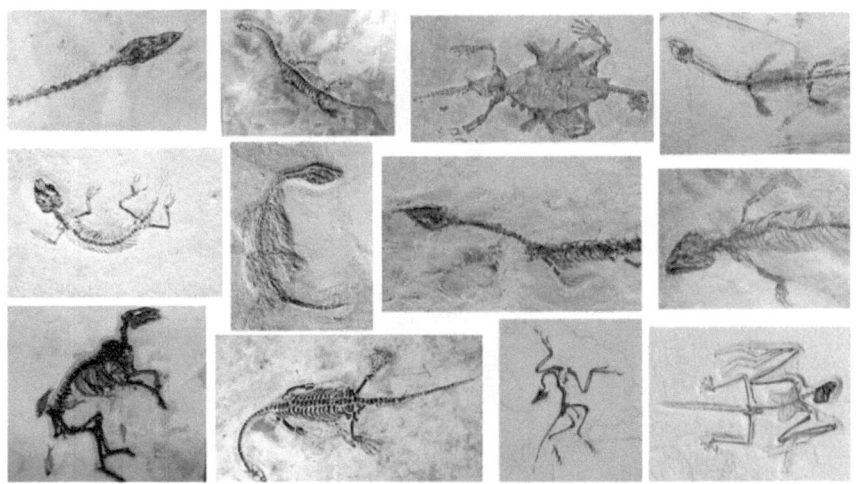

. . . the fossil record does not provide any support for the theory of evolution . . . [7-1]

There are no transitional forms between reptiles and mammals. [7-2]

If you [accept the idea] that birds evolved from reptiles, you should be able to find thousands of transitional fossils showing lizards growing wings and feathers . . . [however,] no transitional fossil has ever been found . . . [There is] no observable proof. Never in the history of

306

*paleontology has anyone found evidence of these strange creatures . . .
have you ever heard of an animal that's a cross between a reptile and
a bird?* [7-3]

*There can be no doubt that modern humans evolved, along with all
other life on our planet, and it takes a special kind of "faith," to even
attempt to deny this fact. The fact that "evolutionists," cannot fully
explain the origins of life as we understand it and did not "witness"
the event in no way detracts from this. The proof that life evolved is
overwhelming and there are no "missing links." Certainly we do not
have every single step of the process videotaped—but it is a far more
plausible answer than "God did it," which is no answer at all.* [7-4]

*How do you convince a creationist that a fossil is a transitional fossil?
Give up? It is a trick question. You cannot do it. There is no convincing
someone who has his mind made up already. But sometimes, it is even
worse. Sometimes, when you point out a fossil that falls into the middle
of a gap and is a superb morphological and chronological intermediate,
you are met with the response: "Well, now you have two gaps where you
only had one before! You are losing ground!"* [7-5]

Introduction

In this chapter (7), we will look at some of the fossil evidence to determine
whether or not species evolved from lower life forms. The various views
presented in this chapter regarding the origin of species are not exclusively
strict evolutionists and strict creationists. Some views are of those who believe
in variations of the two extremes. See "Various View Descriptions" section in
the back of this book for descriptions of the views presented in this chapter.

The fossil evidence that is said to represent an evolutionary progression
from one-celled bacteria to worm-like chordates to lancelets to fish was
discussed in Chapter 3, section "Evolution From Worm-like Chordates to Fish-
like Creatures." The fossil evidence will be further discussed in this chapter
as well as in chapters 8 and 16. This chapter (7) will discuss the evolution
of fish to amphibians, amphibians to reptiles, and reptiles to mammals and
birds. Chapter 8 will focus on the transition from reptile to the mammal jaw/
ear articulation. The fossil evidence leading up to human development will
be discussed in Chapter 16.

Questions for Creationists

Evolutionist View

Why would God create a fish with eight finger-like bones in its fins on the fifth day of creation, and then have it appear in the Earth's crust between ancient fish and the first amphibians with their seven to eight fingers? Did God do this to test our faith? Shouldn't the Bible (especially a literal reading of Genesis 1) and science be in agreement? And why would God on this same creation day (the fifth day) put tooth genes in both modern birds and toothless whales? Was this done to give the false impression that life has an evolutionary history?[7-6]

> *On the ordinary view of the independent creation of each being, we can only say that so it is; that it has pleased the Creator to construct all the animals and plants in each great class on a uniform plan; but this is not a scientific explanation.*
> • Charles Darwin The Origin of Species
> by means of Natural Selection or The Preservation
> of Favoured Races in the Struggle for Life,
> Chapter 14 - Mutual Affinities Of Organic Beings:
> Morphology—Embryology—Rudimentary Organs, 6th edition

Evolution From Fish to Amphibians

Evolutionist View

The first backboned land animals or "tetrapods" (the ancestors of amphibians, reptiles, birds and mammals, including humans) evolved from a group of fishes about 370 million years ago during the Devonian Period (408-360 mya) (The evolution of worm-like creatures to fish is discussed in Chapter 3, section "Evolution From Worm-like Chordates to Fish-like Creatures"). Even though scientists had discovered fossils of tetrapod-like fishes and fish-like tetrapods (vertebrates having four limbs or, as in the snake and whale, having had four-limbed ancestors) from this period, these were still rather different from each other and did not give a complete picture of the intermediate steps in the transition.[7-7]

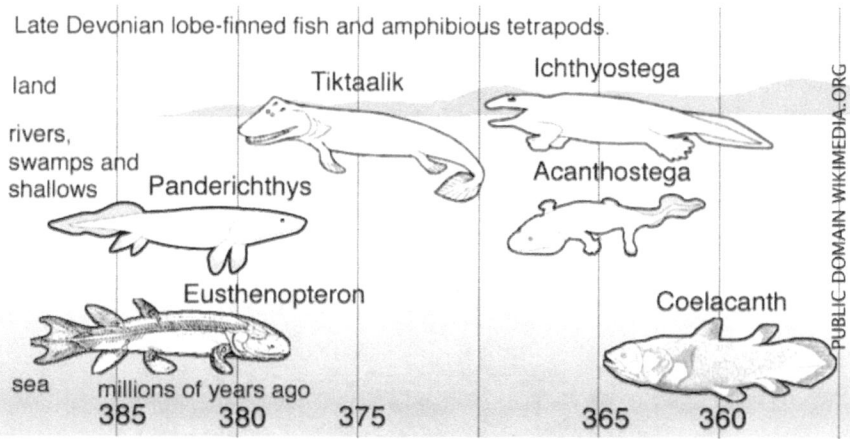

Lobe Finned Fish and Amphibians

Figure 7.1 Illustration of the progression of fish to land creatures.

Eusthenopteron is one of the oldest lobe-finned fish. Panderichthys is a prehistoric fish with a tetrapod-like head. Tiktaalik is a mixture of fish and tetrapod. Acanthostega and Ichthyostega are also intermediate between fish and amphibian. Coelacanth is an old lobe-finned fish.

- - - - - - - - - - - - - - - -

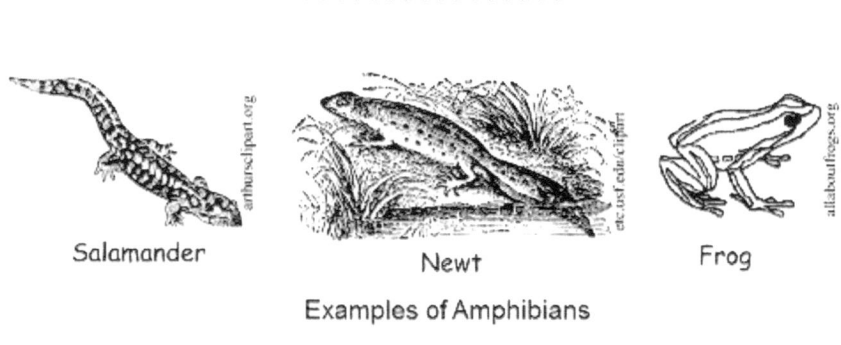

Examples of Amphibians

Examples of Amphibians

An amphibian is a cold-blooded vertebrate, including frogs and toads, newts and salamanders, and caecilians, the larvae being typically aquatic, breathing by gills, and the adults being typically semi-terrestrial, breathing by lungs and through the moist, glandular skin.[7-8] Amphibians are cold-blooded tetrapods (four-legged vertebrates) whose eggs lack a tough protective membrane around the embryo. Most amphibians have an aquatic stage where they spend part of their time, as well as a terrestrial stage where they spend part of their time. Many amphibians undergo a change from an aquatic larval stage in which they acquire oxygen from water and lack limbs, to a four-legged, air-breathing adult form adapted for living on the land. (See section "An Example of Fish Evolving into Amphibians: The Metamorphosis of Tadpoles to Frogs.")[7-9]

Artist conception of the first fish leaving the water.

The fish-amphibian transition took place in the water and eventually carried them to dry land. The very first amphibians seem to have developed legs and feet to travel around on bottom of the water, as some modern fish do, and not necessarily to walk on land. This aquatic-feet stage meant the fins did not have to change very quickly, the muscle structures for weight-bearing didn't have to be well developed and the muscle structures for movement didn't have to change at all. The idea of the aquatic feet stage is well supported by fragments of fossils found in the middle Upper Devonian strata (408-360 mya), and discoveries of late feet appearing in the Upper Devonian strata. Eventually amphibians moved onto the land. This involved the evolution of the attachment of the pelvis more firmly to the spine, and the separation of the shoulder from the skull. Since lungs are an ancient fish trait and were present already in some fish, it would not take too much of a transition for a fish to breathe out of water.

Amphibians most likely evolved from lobe-finned fish between 400 and 350 million years ago (mya). In contrast to numerous thin splints of bone in the flat fins of ray-finned fish (i.e., perch, trout), lobefins are fleshy and have large, distinctly defined bones. Many lobe-finned fish had both gills and lungs, as modern lungfish are evolutionary descendants of these fish. This allowed them to survive in shallow, swampy, and plant-filled waters, as well as in regions susceptible to droughts. Their limb-like fins gave them greater mobility, and eventually allowed them to go on land and to find new sources of food. Amphibians, after hatching, have gills and breathe water like a fish. After they develop into adults, they lose their gills and develop lungs to live on Land (See section "An Example of Fish Evolving into Amphibians: The Metamorphosis of Tadpoles to Frogs.").[7-10]

Fossils indicate many similarities between lobe-finned fish and the first amphibians. One of the earliest amphibians looks like a fish with legs. Both animals also had labrinthodont teeth (Greek "laburinthos: a maze; odontos; tooth) in common. The most common fossil evidence is the finger-like bones in the fins of some lobe-finned fish. There are eight bones in the fins of a lobe-finned fish known as "the fish with fingers." The first amphibians often had seven to eight digits, and obvious correlation with their origin from fingered fish. The fossil evidence shows that in the course of evolution, animals that began with eight digits, such as Acanthostega (Figure 7.1b), the number of digits was reduced to seven digits, then to six digits, and then to five digits (Figure 7.4). This is a good indication of transitional animals and is the reason why humans have five digits. (See section "Homology of Vertebrate Limbs.") About 300 million years ago, these bones were reduced to the five digits generally seen in modern four-legged animals and humans. It is easy to see how fins became limbs when the fossil record of lobed-finned fish and the first amphibians is considered along with the experimental process of developmental genes and molecules being placed in developing buds, and considering the fossil record of lobe-finned fish and the first amphibians, it is easy to see how fins evolved into limbs.[7-10] (See Chapter 10, Punctuated Equilibrium, section "Experiments Illustrating Punctuated Equilibrium" for further discussion of developing buds.)

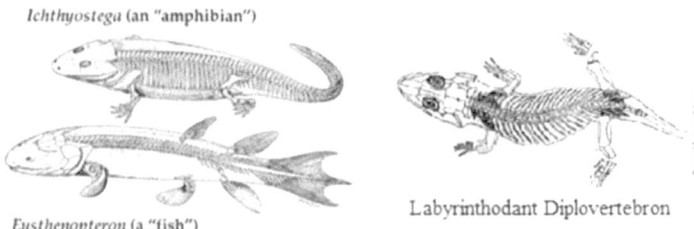

Figure 7.1a Eusthenopteron (a fish) (bottom left) compared to Ichthyostega (an amphibian) (top left) and Labyrinthodant Diplovertebron (an amphibian) (Right)

Comparing the Eusthenopteron (a lobed-fish) with the Ichthyostega (an amphibian), one can easily see the similarities (Figure 7.1a left). Eusthenopternon is a lobe-finned fish from the Upper (Late) Devonian Period (385 mya) and Ichthyostega is an amphibian from the Devonian Period (367-362 mya). Comparing Eusthenopteron with the Labyrinthodant Diplovertebron (Figure 7.1a right) from the Late Carboniferous Period (318-299 mya), the similarities are also very striking (The skeletal makeup of the Ichthyostega and the Diploverebron are almost identical). Although there are a considerable number of differences between the lobe-finned fish Eusthenopteron and the amphibian Diplovertebron, the skulls are very similar in structure (a tapering triangle, like a small-scale replica of a crocodile's skull), and many of the bones in the paired fins of the lobe-fin correspond with those of the legs of the amphibian.[7-10a]

Acanthostega

Acanthostega - Artist Conception

Figure 7.1b Acanthostega

Acanthostega (Figures 7.1 & 7.1b) was an early tetrapod (four legged) creature. It had well-defined fingers and toes (Figure 7.4), but it did not have wrists or ankles. Acanthostega had relatively long limb bones, but they were not able to support very much weight. Its hip also couldn't support much weight as it was weakly attached to the spine. Acanthostega had a long tail with a large bony fin, suitable for swimming. Acanthostega also had internal gills that include bony gill arches and post-branchial lamina and had small nostrils. Acanthostega possessed both fish-like and land animal-like characteristics. It had legs and feet rather than fins. Acanthostega had eight digits on the front leg and at least eight digits on the hind leg. It is now believed that Ichthyostega (Figure 7.1) also had more than five digits (Figure 7.4).[7-11]

Acanthostega is considered to be one of the first vertebrate animals with recognizable limbs. It appeared about 365 mya and is believed to be an intermediate between lobe-finned fishes and the first tetrapods (having four limbs) that were fully capable of walking on land. The front foot could not bend forward at the elbow and therefore it could not bear any weight for walking. Acanthostega was better suited for paddling than walking on land. It had lungs as well as gills. Its ribs were too short to provide any support to its chest cavity out of water. It had gills that were internal and were covered like those of fish, unlike external gills of some modern amphibians.[7-12]

Tiktaalik (Figure 7.1) is a fossil that was discovered in 2006 and has been estimated to be 375 million years old. Tiktaalik is technically a fish, complete with scales and gills, but it has the flattened head of a crocodile and unusual fins. Its fins have thin ray bones for paddling like most fish, but they also have sturdy interior bones that would have allowed Tiktaalik to prop itself up in shallow water and use its limbs for support as most four-legged animals do. It has a combination of features that show the evolutionary transition between swimming fish and their descendants, the four-legged vertebrates. Before Tiktaalik was discovered, paleontologists had studied many other extinct transitional organisms, such as Eusthenopteron and Acanthostega (Figure 7.1), which also provided clues about vertebrates migrating to land.[7-13] Tiktaalik roseae is an intermediate fossil that represents the transition of vertebrate life from water to land 375 million years ago. Tiktaalik roseae, discovered in Nunavut in 2004, is an ancient fish called a sarcopteryigian, or lobe-finned fish. Although it bears many similarities to fish like gills, scales, and fins, other key characteristics link Tiktaalik to land animals. While it did have fins, the bones inside the fins are homologous to the bones of the human hand and wrist, indicating it may have been able to bear weight. The animal also had a mobile neck and a strong ribcage, two critical traits that allowed four-legged (tetrapod) creatures to move onto land. Tiktaalik makes sense evolutionarily in the progression of other early tetrapods like the more aquatic Panderichthys and the clearly amphibious.[7-13a]

The pandericthyids (lobe-finned fishes) are related to early tetrapods (vertebrates having four limbs or, as in the snake and whale, having had four-limbed ancestors). The Tiktaalik roseae had an arm that is transitional between a fin and a limb.[7-14]

Creationist View

Evolutionists often use the rhipidistian fish Eusthenopteron and the amphibian Ichthyostega (Figure 7.1) to illustrate that fish evolved into amphibians. The problem is that they fail to show how fins could have changed into legs. Drawings often only show similarities between the jaws and skulls of these fossils, and ignore the supposed transition from fins to legs.

There is a lack of even a single photograph or drawing of a transitional fossil that supports the theory that fish developed legs and became amphibians. Some illustrations compare, for example, the pectoral fin skeleton of Eusthenopteron, the forelimb skeleton of Acanthostega, and the hindlimb skeleton of Ichthyostega. Eusthenopteron has four digits, Acanthostega has eight, and Ichthyostega has seven.

Evolutionists point out that the forelimb of Acanthostega had eight digits and the hindlimb of Ichthyostega had seven, which is not like the common pattern of five digits on the feet (or hands) of many amphibians, reptiles, birds and mammals (Figure 7.4). Evolutionists often describe some resemblances of the forelimb skeleton of Acanthostega (Figure 7.1b) to the pectoral fin skeleton of Eusthenopteron (Figure 7.1a). Paleontologists point out the similarities regarding the tetrapod limb bones with the fin-bones of lobe-finned fishes. The discovery of what is thought to be a fish-like gill (brachial) skeleton in Acanthostega, with grooves, are the same as those found in modern fishes (Figure 7.13). Evolutionists believe that Acanthostega retained fish-like internal gills that could have been used for aquatic respiration. Also, they believe that Acanthostega were similar to gill-breaching lungfish and that its legs with digits must have initially evolved for use in the water rather than walking on land. Evolutionary paleontologists claim that because this creature is a mosaic of fish and tetrapod-like characters, it is an indication that Acanthostega is a transitional creature. (Acanthostega was either a fish with legs or it was an amphibian with gills. There is no evidence that Acanthostega was a transitional creature. Without being able to examine the creature while it existed, it is difficult to determine what it was by only examining the bones or fossils of bones.)

Duck Billed Platypus

Australia's platypus is also considered a mosaic. It has milk glands and fur that classify it as a mammal, but it has a leathery egg, echo-location ability, a duckbill, webbed feet, poison spurs and other features that it shares in common with other animals, not only mammals, which make it difficult for evolutionists to classify it as a transitional animal. As evolutionists would admit, intermediates are very difficult to construct, as there is no evidence for them in the fossil record, although evolutionists do consider Archaeopteryx, also possessing a mosaic of traits, as an intermediate between reptiles and birds. (See section below "A look at the Archaeopteryx fossils" for discussion of Archaeopteryx.)

It is very difficult to determine the biological make-up of a creature based only on fossil evidence and can never be established with any degree of certainty (as will be discussed later with the coelacanth). This is because ninety-nine percent of the biology of any organism is in its soft tissue, which is not preserved in a skeletal fossil. The fossil record has so far revealed many types of fish, some of which have bones in their fin lobes, serving a useful purpose as in the coelacanth. The fossil record has also revealed many types of amphibians, including Ichthyostega and Acanthostega (Figure 7.1), in which the limb bones are firmly attached to the backbone and clearly designed for bearing the weight of the body in walking. Anything truly in-between these crucial fish and amphibian characteristics is not only hard to conceive, but has never been found. Acanthostega is not a transitional creature between fish and land animal but just another amphibian.[7-15]

Tikaalik (Figure 7.1) is another creature that evolutionists believe is an intermediary between fish and land animals. Although Tiktaalik was a fish, it had some features that were tetrapod-like. This means that, depending on one's point of view, Tiktaalik could be considered transitional to tetrapods (having four limbs). These features include: a lengthened snout (measured from the eyes to the tip of the skull), a mobile neck, overlapping ('imbricate') ribs and a pectoral girdle (shoulder girdle) that may have given it the ability to lift the front part of its body by its fins. On these grounds the animal is analyzed as being intermediate between the lobe-finned Panderichthys and the four-limbed Acanthostega and Ichthyostega (Figure 7.1). That is, Tiktaalik went on to evolve in two separate directions, Acanthostega on one branch and Ichthyostega on the other. Tiktaalik is not the only fish with a mobile neck. Mandageria, a fish closely related to Eusthenopteron and not thought to have been an ancestor of any tetrapod, also had a mobile neck. In evolutionist terms the feature must therefore be interpreted as 'convergent', i.e. it evolved in Mandageria, disappeared in Panderichthys, then re-appeared on another branch in Tiktaalik. Imbricate (overlapping) ribs do not occur in any other fish, nor in Acanthostega, but do occur in Ichthyostega.[7-16]

The coelacanth fish (Figure 7.1) was once thought to be a transitional fish because of the extra lobes growing from its fins, which were assumed to be its future land dwelling feet. This was when there were only fossils available to study.[7-17] For decades, scientists believed the coelacanth, a bony fish, was the link between fish and amphibians. Evolutionists claimed that it had been extinct for 70 million years. They said the fins on this fish were actually in the process of evolving into limbs (legs). Evolutionists imagined that the coelacanth fossil was in the early stages of growing limbs that could allow the creature to walk on land. Then in 1938 a fisherman caught a real, live coelacanth (or Latimeria) off the coast of South Africa.[7-18] The Coelacanth fish was believed to have been "extinct" since the Cretaceous Period (144-65 mya). It has not been found in the strata for the past 50 million years. The now-famous Coelacanth was a large fish known only from its fossil and allegedly extinct for 50 million years. It was thought to have been extinct until several specimens were found in the ocean. How could the Coelacanth have become extinct 50 million years ago, and then be found alive in 1938? In order to be declared "extinct" such a long time ago, the creature would obviously have had to have been found by paleontologists in older strata, and then not found at all in more recent strata. Why is the Coelacanth not in those more recent strata? This is clear-cut evidence that the sedimentary strata was the result of a rapid laying down of sediments during the Flood (discussed in Chapter 4), rather than the slow "inch a hundred years" deposition pattern theorized by the evolutionists.[7-19] The coelacanth was called "kombessa" by the Comorro Island natives.[7-20] With the discovery of a live coelacanth, the evolutionists' beliefs were proven wrong. Its fins turned out to be just fins, not little stumps in the process of becoming legs.[7-21] When scientists first looked at the fossil of the coelacanth, they imagined the fins turning into legs. Looking at a single fossil and coming to a conclusion about its past or future is pure guesswork. To really prove evolution through the fossil record, one would need a series of sequential fossils.[7-22]

An Example of Fish Evolving into Amphibians: The Metamorphosis of Tadpoles to Frogs

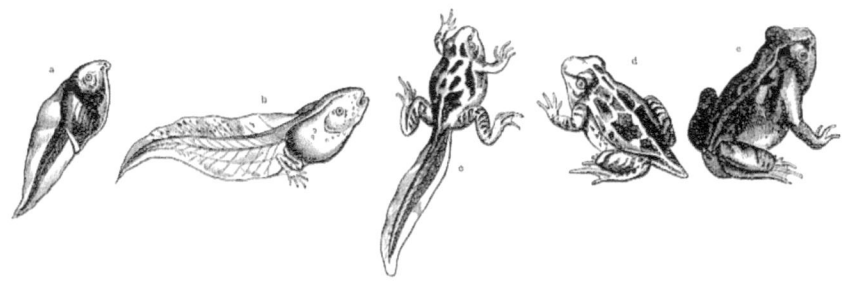

Figure 7.2 Tadpole to Frog

Evolutionist View

A tadpole changes into a totally different creature, a frog (Figure 7.2). This is similar to the metamorphosis of a caterpillar to a butterfly.

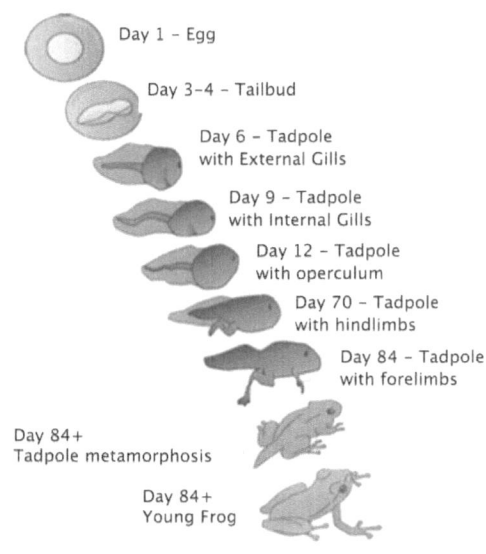

Day 1 – Egg

Day 3–4 – Tailbud

Day 6 – Tadpole with External Gills

Day 9 – Tadpole with Internal Gills

Day 12 – Tadpole with operculum

Day 70 – Tadpole with hindlimbs

Day 84 – Tadpole with forelimbs

Day 84+ Tadpole metamorphosis

Day 84+ Young Frog

Metamorphosis of Tadpoles to Frogs

Figure 7.3 Metamorphosis of Tadpoles to Frogs

A frog first lays eggs, which turn into tadpoles, which later develops into a frog. This process is very interesting. First, (Day 1) a frog lays eggs during the early spring in woodland ponds, among the leaves and sticks in the water. These eggs are large masses of a clear jellylike consistency enclosing hundreds of little black spheres about an eighth of an inch in diameter. Under a microscope one could see the small one-celled egg spheres divide into more and more numerous portions which are the daughter-cells, destined to form by their products the many varied tissues and organs of the developing larva and adult frog. (Days 3-4) After three or four days the egg changes from its globular form into an oval or elliptical mass (Figure 7.3), and from one end of this a small knob projects to become a flattened waving tail a few days later (Days 6-9). On the sides of the larger anterior portion shallow grooves make their appearance and soon break through from the throat or pharynx to the exterior as gill slits. Shortly afterward the little embryo wriggles out of its encasing coat of jelly, develops a mouth, and begins its independent existence as a small tadpole, with eyes, nasal and auditory organs, and all other parts that are necessary for a free life (Day 12). Thus the one-celled egg has transformed; and in doing this it has proved the possibility and the reality of organic reconstruction. [7-24] In Figure 7.3, Day 12 is labeled "Day 12 Tadpole with operculum." The word "operculum" is a lid or flap covering an aperture, such as the gill cover in some fishes or the horny shell cover in snails or other mollusks.[7-23]

The tadpole breathes by means of its gills, and it is at first entirely devoid of the lungs that the adult frog possesses and uses. The larval respiratory organs function as gills. They are truly like those of fishes, for the blood vessels that go to them are essentially the same as in the lower types and they are supported by simple skeletal rods like the gill-bars of fish (Figures 7.13 & 7.15). The tadpole feeds and grows during the months of its first summer, and hibernates the following winter; with the warmth of spring it revives and proceeds further along the course of its development (Day 70). Near the base of the tail two minute legs grow out from the back part of the body, and while these are enlarging two front legs appear a little behind the gills (Day 84). The transitioning tadpole now rises frequently to the surface where it takes small mouthfuls of air. Meanwhile great changes are occurring inside the body where the various systems of fish-like organs become remodeled into amphibian structures. A sac is formed from the wall of the esophagus out on land as a complete young frog (Day 84+). From this time on it breathes by means of its lungs instead of gills, even though it returns to the water to escape its predators, to seek its prey, and to hibernate in the mud of the lakebed during the winter, and this enlarges and divides to form the two simple lungs (Day 84+). (Lungs are discussed in more detail in section "Various Respiratory Systems.") The legs increase in size, the tail dwindles more and more, the gills close up, and soon the animal hops months. The tadpole is essentially a fish in its general structure and mode of life, even though its heritage is such that it can develop into a higher animal. When it does become a frog it proves beyond a doubt that there is no impassible barrier between fishes and amphibia. The connection between fish and amphibians is clear by the method and the sequence of a tadpole developing into a frog, a member of a higher class of vertebrates. This method is employed by developing frogs apparently because it follows the ancestral order of events and because the only way a frog knows how to become a frog is to develop from an egg first into a fish-like tadpole and then to alter itself as its ancestors did during their evolution in the past. It can be seen that in addition to the impressive fact of development itself, the mode of organic transformation is far more conclusive evidence of evolution, because it reveals an order of events which parallels the order established by comparative anatomy as the evolutionary sequence.[7-24]

Creationist View

Some claim that since tadpoles change into land-dwelling frogs with legs, this illustrates evolution. They believe that this metamorphosis is an example of evolution in action. This, however, is not true. Tadpoles are not fish. They may superficially look somewhat like a guppy with gills, but they are the offspring of fully functioning frogs, complete with all the genes for legs and the structures needed to use them. The tadpole is not yet fully grown, and in the incomplete stage has not acquired all the features present in the adult, but it is a juvenile frog nonetheless. A tadpole does, however, have all the genes needed for life in the water, as well as those genes needed to grow legs at the right time, then live on land, and eventually produce tadpoles (with gills and tails) which themselves become frogs (with lungs and limbs). Rather than a frog directly developing from an egg or in the womb, it first goes through an egg stage, followed by a fish-like tadpole stage where it can grow somewhat before fully developing into a frog. No new genetic information must be acquired by mutation as required by evolution. All the genetic information for an egg to develop into a tadpole and then into a frog are already present in the egg.

An invertebrate, like a jellyfish, a clam, or a worm, does not possess the genes necessary to construct a vertebral column, or to integrate it with all the other muscles, nerves, and organs needed by animals with a backbone, including fish, yet this transformation must have occurred if macroevolution is true. Similarly, a functioning fish does not possess the genes necessary to construct and utilize legs. The claim of creatures that go through metamorphosis proves evolution could also be made about a human fetus in its early stages. At one point a human fetus has no arms or legs (or eyes or lungs etc.) but it acquires them through genetically controlled growth prior to its birth. No evolutionary process is needed to transform a fertilized human embryo into a baby and then into an adult.

All the genes are present within the egg of a frog when it is laid. The conclusion is that neither growth nor metamorphosis demonstrates evolution. (See Chapter 11, sections "Embryology" and "What Really Occurs During the Stages of Human Development?" for further discussion of embryonic development.)[7-25]

Homology of Vertebrate Limbs

The similar framework of bones in the hand of a man, wing of a bat, fin of the porpoise, and leg of the horse—the same number of vertebrae forming the neck of the giraffe and of the elephant—and innumerable other such facts, at once explain themselves on the theory of descent with slow and slight successive modifications.

—Charles Darwin, Origin of Species,
Chapter 15—Recapitulation And Conclusion

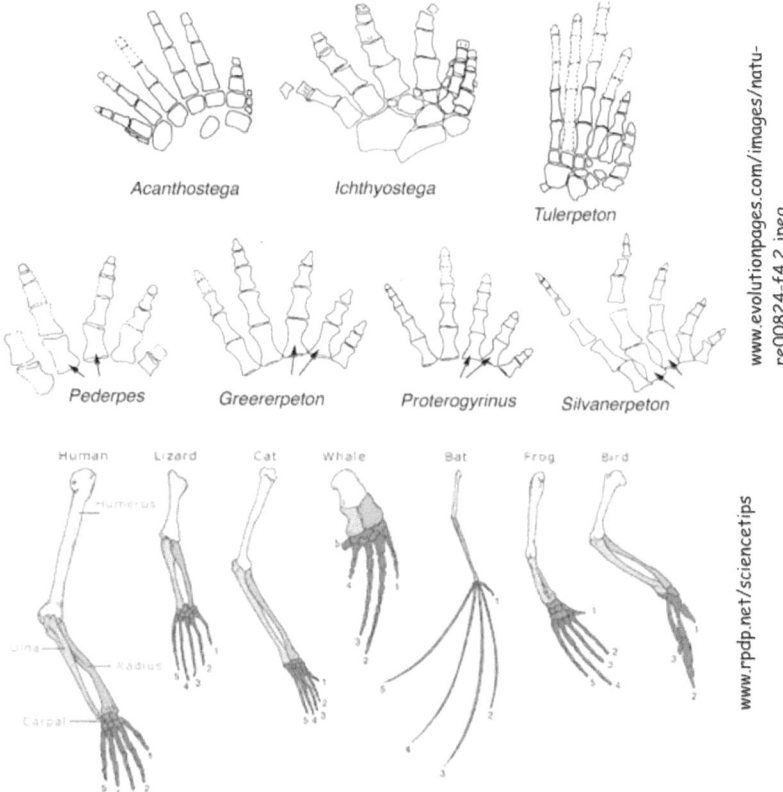

Forelimbs of Various Creatures

Figure 7.4 Hands and Forelimbs of Various Creatures

Evolutionist View

All animals with four limbs, called tetrapods, have five digits at the end of each limb (Figure 7.4 & Chapter 15, Figure 15.1). Even those animals with fewer than five digits in the adult animal, such as horses with hooves and the bats and birds with wings, their digits (fingers and toes) develop from an embryonic five-digit stage. There is no advantage having five digits. The development of pentadactyl (having five digits) creatures is due to an evolutionary past. All tetrapods descended from a common ancestor that had limbs with five digits. Over long periods of evolution, following that natural selection modified different pentadactyl creatures rather than starting over again to produce tetrapods with different numbers of digits. (Natural selection is discussed in Chapter 9.)

Creationists argue that species were created separately in their distinctive forms and did not descend from common ancestors. But the existence of the pentadactyl (having five digits on each hand or foot) limb indicates just the opposite. An ancestral tetrapod had five digits per limb, and all of its descendants did as well. The similarity is not restricted to the ends of the limbs: the bones of the arm, forearm, and hand of different vertebrates form a recognizable pattern (Figure 7.4), even though they have been adapted to different functions (some grasp, some climb, some fly, etc.) (Chapter 15, Figure 15.1). And aspects of the nerves, blood vessels, and other tissues in the limb reveal other homologous structures. Homologies are also seen in other structures, and can even be found biochemically, in the very genetic code that stores information for reproducing individuals. These molecular homologies provide some of the best evidence of a single common ancestor for all life on Earth.[7-26]

Creationist View

Evolutionists believe that humans evolved from lower life forms, such as a fish or an amphibian, perhaps a frog. Part of the evidence evolutionists use to prove that humans evolved from fish and amphibians is that humans, like fish and frogs, have five digits (Figure 7.4). If humans evolved from fish, then the human hand should develop the same as fins develop on a fish or hands on a frog. But in fact they do not. The frog hand develops differently than the human hand does. The human embryo develops a ridge at the limb tip, and then material between the digits dissolves (See Chapter 11, section "Embryonic Development" for further discussion). In frogs, the digits grow outward from buds. This fact strongly disproves the argument of common ancestry.[7-27] (See Chapter 15, section "Common Function or Common Ancestry?" for further discussion of common function and common ancestry.)

Evolutionists claim that the pentadactyl (five bone) limb pattern is found in the arm of a man, the wing of a bird, and the flipper of a whale, and this is held to indicate their common origin (Figure 7.4). If these various structures were transmitted by the same gene-complex, varied from time to time by mutations and acted upon by environmental selection, the theory would make good sense. Unfortunately for evolutionists, this is not the case. Totally different gene complexes in the different species produce homologous organs. The concept of homology in terms of similar genes handed on from a common ancestor has been disproved.[7-28] (See section "Scientific Research Casts Doubt That Birds Evolved From Reptiles (Dinosaurs)" for further discussion of the development of various limbs.)

Evolution From Amphibians to Reptiles

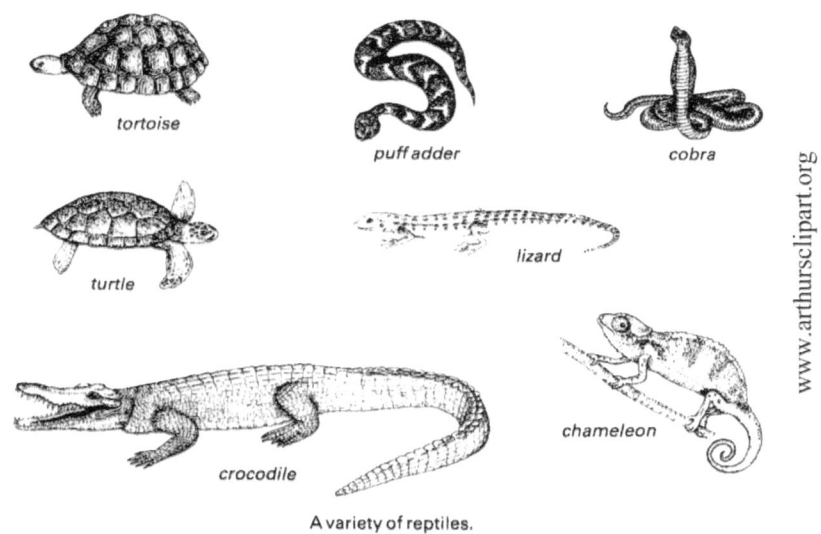

A variety of reptiles.

Examples of Reptiles

A reptile is a cold-blooded vertebrate, comprising of turtles, snakes, lizards, crocodilians, amphisbaenia (called amphisbaenians or worm lizards), tuatara, and various extinct members including the dinosaurs.[7-29] Cold-blooded vertebrates are animals that do not maintain thermal homeostasis; that is, they do not keep their core body temperature at a nearly constant level regardless of the temperature of the surrounding environment. Rather, cold-blooded animals have a variable body temperature, which reflects the environmental temperature. There are two characteristics that distinguished early reptiles from amphibians and enabled them to colonize terrestrial habitats more extensively than their ancestors, **(1)** scales and **(2)** the ability to lay hard-shelled eggs. Scales protect reptiles from abrasion and loss of body moisture. Hard-shelled eggs provide a protective environment in which embryos can develop.[7-30]

Evolutionist View

Reptiles arose from amphibians in the swamps about 310-320 million years ago during the Carboniferous period (360–3 mya). Increasing evolutionary pressure and the vast untouched niches of the land powered the evolutionary changes in amphibians to gradually become land based. Environmental selection (natural selection – Chapter 9) propelled the development of certain traits, such as a stronger skeletal structure, muscles, and more protective coating (scales) became more favorable. The evolution of lungs and legs are the main transitional steps towards reptiles, but the development of hard-shelled external eggs replacing the amphibious water bound eggs is the defining feature of the class Reptilia and is what allowed these amphibians to fully leave water. Another major difference from amphibians is the increased brain size, more specifically, the enlarged cerebrum and cerebellum. [7-31]

Intelligent Design View

Amphibians include creatures such as frogs and toads, salamanders and newts. A possible intermediate between amphibians and reptiles is a group of reptile-like amphibians, called the Seymouria. Seymouria is found in the lower Permian period (286-248 mya) and exhibits a combination of amphibian and reptilian characteristics. In terms of purely skeletal characteristics Seymouria would appear to be an intermediate, as it appears to be midway between amphibians and reptiles, however, there is a serious problem with this assumption. The major difference between amphibians and reptiles lies in their reproductive systems. Amphibians lay their eggs in water and their larvae undergo a complex metamorphosis (like a tadpole) before reaching the adult stage.

Reptiles develop inside of a hard shell-encased egg and are perfect replicas of the adult on first emerging. It is difficult to comprehend how the reptilian egg could have evolved gradually. The skeletal characteristics alone are insufficient for designating a particular organism or species as intermediate. Recently, a fossil of an immature form closely related to Seymouria has been found bearing laval gills (like a tadpole) which suggests that this group of amphibians were wholly amphibian in their reproductive system. Another difficulty with Seymouria being an ancestor to reptiles is that Seymouria appears too late in the fossil record to be an ancestor of the reptiles.[7-32] Seymouria fossils appear in the fossil record in the early Permian Period of

North America and Europe—approximately 280 to 270 mya (million years ago).[7-33] Reptiles first appear in fossil record in the Pennsylvanian Epoch (320-286 mya) stratum (Chapter 3, section "The Strata Layers of the Earth").

Darwinists believe that reptiles evolved from amphibians, but none explain how the major distinguishing adaptation of the reptiles, the amniotic egg, could have come about gradually as a result of a successive accumulation of small changes. The amniotic egg of the reptile is greatly more complex and very different to that of an amphibian. There are hardly two eggs in the whole animal kingdom that differ more fundamentally than reptile and amphibian eggs. Some of the main distinguishing features of the amniotic egg of the reptile are: **(1)** the tough impervious shell, **(2)** the two membranes, **(3)** the amnion which encloses a small sac in which the embryo floats, **(4)** the allantois in which the waste products formed during the development of the embryo accumulate, and **(5)** the yolk sac containing the food reserve in the form of the protein albumen. None of these features are found in the egg of any amphibian.[7-34]

Darwinists have been perplexed as to how the amphibian egg could have changed to produce a reptilian type egg. For this transition to have occurred, at least eight quite different changes needed to be combined to make the reptilian type amniotic egg possible. These changes include: **(1)** the formation of a tough impermeable shell; **(2)** the formation of the gelatinous egg white (albumen) and the secretion of a special acid to yield its water; **(3)** the excretion of nitrogenous waste in the form of water insoluble uric acid; **(4)** the formation of the amniotic cavity in which the embryo floats. This is surrounded by the amniotic membrane that is formed by an outgrowth of mesodermal tissue. Neither the amniotic cavity nor the membrane which surrounds it has anything similar in any amphibian; **(5)** the formation of the allantois from the future floor of the hind gut as a container for waste products and later to serve the function of a respiratory organ; **(6)** the development of a tooth or caruncle (fleshly outgrowth on the heads of certain birds) which the developed embryo can utilize to break out of the egg; **(7)** a quantity of yolk sufficient for the needs of the embryo until it is hatched; **(8)** changes in the urogenital system of the female permitting fertilization of the egg before the hardening of the shell.

The problem of the origin of the amniotic system is even more questionable considering that the basic problem is that amphibian eggs need water, whereas reptilian eggs have amniotic fluid, which eliminates their need for water. Some amphibian eggs have a tough gelatinous skin that will stand a certain degree of drought; other amphibians are live bearing. Certain amphibians are therefore quite independent of water for reproduction. The origin of the amniotic egg and the amphibian egg to the reptile type egg transition is just another of the major vertebrate divisions for which evolutionists have difficulty explaining. How the heart and aortic arches of an amphibian could have gradually converted to the reptilian and mammalian type raises many additional unexplainable questions.[7-35] (Also see Chapter 11, section "Embryonic Development" "Intelligent Design View," for further discussion of amphibian egg and reptilian egg.)

Creationist and Intelligent Design View

Evolutionists claim that amphibians evolved into reptiles, however, this is doubtful. The most important difference between amphibians and reptiles involves their reproductive systems. Amphibians lay their eggs in water and the larvae undergo a complex metamorphosis before becoming an adult. Reptiles do not go through this complex metamorphosis before becoming an adult. Evolutionists are not able to provide an explanation of how an amphibian could have developed into a reptile that has a different reproduction process. This may be due to the fact that soft parts do not fossilize well so there would be little way of knowing how the reproductive systems of amphibians could have evolved into reptile reproductive systems. Nevertheless, the fact remains that evolutionists are not able to explain how this transition could have occurred.[7-36]

Amphibian eggs possess an ideal structure for development in water. Reptiles, on the other hand, lay their eggs on land, and consequently their eggs are designed to survive there. The hard shell of the reptile egg, also known as an "amniotic egg," allows air in, but is impermeable to water. In this way, the water needed by the developing animal is kept inside the egg. If amphibian eggs were laid on land, they would immediately dry out, killing the embryo. This cannot be explained in terms of evolution, which asserts that reptiles evolved gradually from amphibians. That is because, for life to have begun on land, the amphibian egg must have changed into an amniotic one within the lifespan of a single generation. How such a process could have occurred by means of natural selection and mutation the mechanisms of evolution is difficult to comprehend. (Natural selection and mutations are discussed in Chapter 9.)

As mentioned previously, Seymouria is said to have a combination of amphibian and reptilian characteristics. Stephen Jay Gould, an evolutionist, admitted that Seymouria could not have evolved into a reptile. He said, *"Evolutionists at one time claimed that the Seymouria fossil . . . was a transitional form between amphibians and reptiles. According to this scenario, Seymouria was "the primitive ancestor of reptiles."* However, subsequent fossil discoveries showed that reptiles were living on the Earth some 30 million years before Seymouria. In the light of this, evolutionists had to put an end to their comments regarding Seymouria.[7-37] Reptiles first appear in fossil record in the Pennsylvanian Epoch (320-286 mya) stratum and Seymouria fossils appear in the fossil record in the early Permian Period (280 to 270 mya).

Currently, there is no satisfactory evidence that amphibians evolved into reptiles. As previously discussed, there are fossil amphibians called Seymouria that have some reptile-like skeletal characteristics, but they appear too late in the fossil record (approximately 280 to 270 mya) and recent evidence indicates that they were true amphibians. (Reptiles first appeared in the fossils during the Pennsylvania Epoch 320-286 mya.) The transition is in any case one which would be difficult to confirm with fossils, because the most important difference between amphibians and reptiles involves the unfossilized soft parts of their reproductive systems. As discussed previously, amphibians lay their eggs in water and the larvae undergo a complex metamorphosis before reaching the adult stage. Reptiles lay a hard shell-cased egg and the young are perfect replicas of adults on first emerging. No explanation exists for how an amphibian could have developed a reptilian mode of reproduction by Darwinian descent.[7-37a]

Evolution From Reptiles to Mammals

Examples of Mammals

Examples of Mammals

A mammal has the body more or less covered with hair, nourishing the young with milk from the mammary glands, and, (with the exception of the egg-laying monotremes,) giving birth to live young.[7-38] Like birds, mammals are endothermic or "warm-blooded," and have four-chambered hearts (See section below "Various Heart Types"). Mammals also have a diaphragm—a muscle below the rib cage that aids breathing. Some other vertebrates have a diaphragm, but mammals are the only vertebrates with a prehepatic ("pre" means "before" and "hepatic" means "liver" in Greek) diaphragm, that is, a diaphragm that is located in front of the liver. Mammals are also the only vertebrates with a single bone in the lower jaw.[7-39] (Reptile and mammal jaw/ear progression is discussed in more detail in Chapter 8.)

Evolutionist View

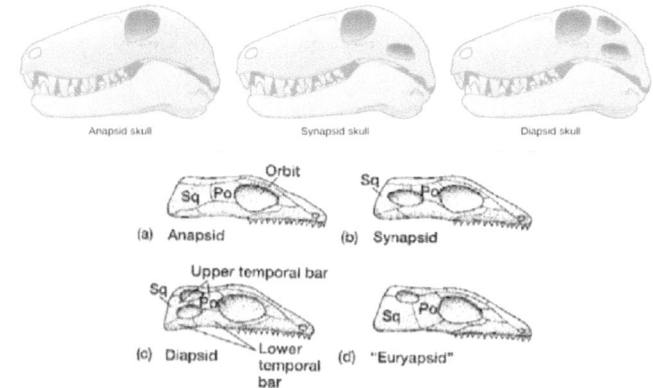

Figure 7.4a Skulls of Major Reptilian Groups

- - - - - - - - - - - - - - - - - - - -

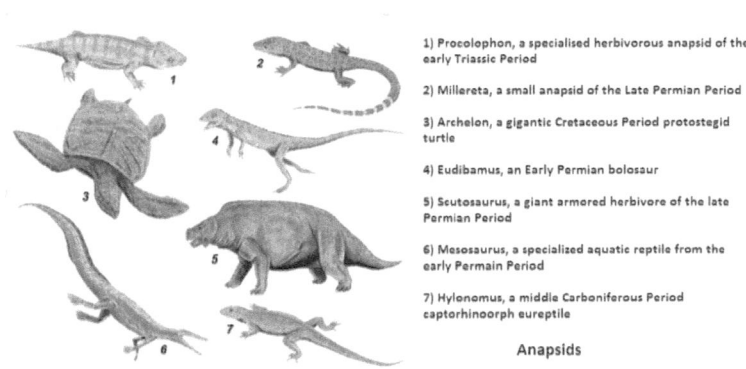

1) Procolophon, a specialised herbivorous anapsid of the early Triassic Period

2) Millereta, a small anapsid of the Late Permian Period

3) Archelon, a gigantic Cretaceous Period protostegid turtle

4) Eudibamus, an Early Permian bolosaur

5) Scutosaurus, a giant armored herbivore of the late Permian Period

6) Mesosaurus, a specialized aquatic reptile from the early Permain Period

7) Hylonomus, a middle Carboniferous Period captorhinoorph eureptile

Anapsids

Figure 7.4b Various Anapsids (artist perception)

Reptiles evolved into four major groups; **(1)** the anapsids (Figure 7.4b), which produced the turtles (Anapsids are amniotes whose skulls do not have openings near the temples. Amniotes existed approximately 340 mya during the Carboniferous period, that is, from the early Pennsylvanian Epoch (323–317 mya). They are vertebrates, comprising the reptiles, birds, and mammals, that have an amnion [membrane surrounding the fetus] during the embryonic stage),

Figure 7.4c Examples of Early Diapsids (artist perception)

(2) the diapsids (Figure 7.4c) which produced the dinosaurs, and an offshoot group (Diapsids [di meaning "two" and apsids meaning "arches"]). Diapsids existed about 300 mya during the late Carboniferous period, that is, from the late Pennsylvanian Epoch (ca. 303–290 mya) that are amniote tetrapods [vertebrates having four limbs or, as in the snake and whale, having had four-limbed ancestors] that developed two holes (temporal fenestra) in each side of their skulls). Descendants of Diapsids include the lizards, snakes, crocodiles, dinosaurs, pterosaurs, birds, and, in some classifications, the turtles.

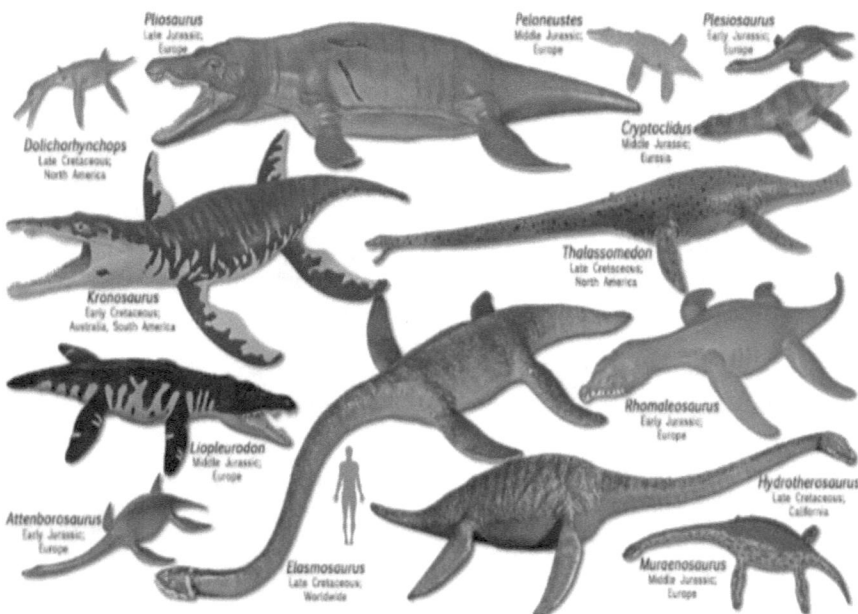

Figure 7.4d Examples of euryapsids (artist perception)

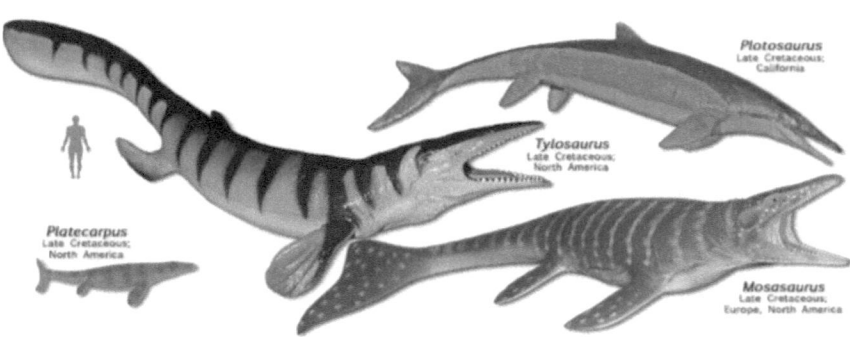

Figure 7.4e More Examples of euryapsids (artist perception)

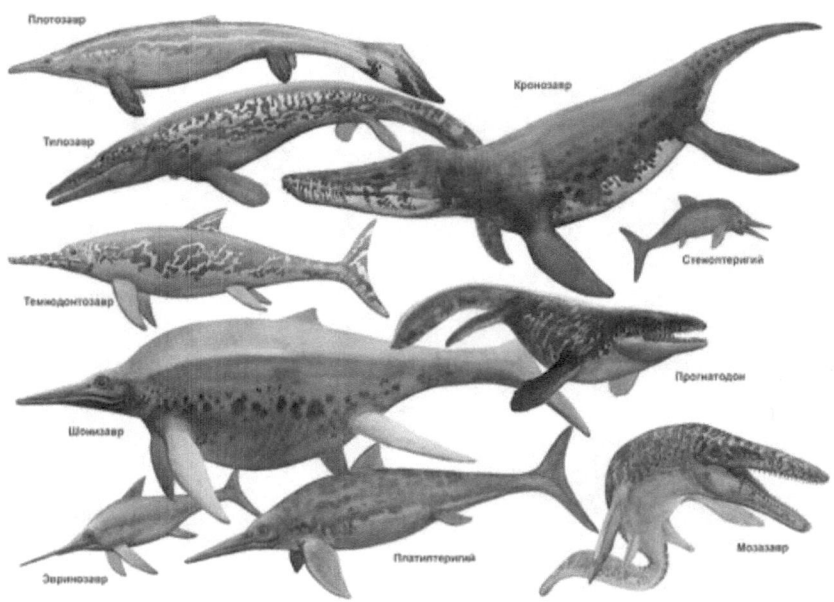

Figure 7.4f Examples of ichthyosaurs (artist perception)

(3) the euryapsids (A Mesozoic marine reptile of a group that have a single upper temporal opening in the skull, including the nothosaurs, plesiosaurs, and ichthyosaurs) (Figure 7.4d & Figure 7.4e), which produced the ichthyosaurs (Figure 7.4f). Euryapsids are reptiles that are distinguished by a single temporal fenestra, an opening behind the orbit [the cavity or socket of the skull in which the eye and its appendages are situated], under which the post-orbital and squamosal bones articulate (that is, these bones are united by forming a joint or joints). Euryapsids are believed to have evolved from diapsids (discussed above), having lost the lower temporal opening in the process. They were a very important part of the marine environment during the Mesozoic Period (250 – 65 mya). Euryapsida are different from Synapsida (Figure 7.4g), which also have a single opening behind the orbit (Figure 7.4a), by the placement of the fenestra (a small natural hole or opening, especially in a bone. The mammalian middle ear is linked by the fenestra ovalis to the vestibule of the inner ear, and by the fenestra rotunda to the cochlea.). Ichthyososaurs (ichthys meaning "fish" and sauros meaning "lizard") are fishlike marine reptiles of the extinct order Ichthyosauria.

Examples of Synapsids

Figure 7.4g Examples of Synapsids (mammal-like reptiles) (artist perception)

The final group, **(4)** the synapsids (Permian period, 299 to 251 mya) (Figure 7.4g), took a radically different path than the other groups and produced the therapsids (Carboniferous Period, from 359 million to 299 mya) (Figure 7.4h), which concentrated on osteo and physiological changes which eventually produced the mammals. Synapsids are reptiles with one temporal opening on each side of the skull (Figure 7.4a) (once called mammal-like reptiles), are now classified with the mammals. Therapsids are groups of mammal-like reptiles of the extinct order Therapsida. The group called the cynodontia (cyno meaning "dog" and dontia meaning "tooth") produced a lineage of forms intermediate between reptiles and mammals. (The earliest, most primitive synapsids are the Pelycosaurs. Pelycosaurs are discussed below in the next section, "Mammal-like Reptiles – The Synapsids" in "The First Group." Therapsids are discussed in "The Second Group" and Cynodonts are discussed in "The Third Group.")

Examples of Therapsids

Figure 7.4h Examples of Therapsids (Extinct) (artist perception)

There are several lines of evidence that support the conclusion that reptiles evolved into mammals. **First**, the progressive reduction in size and increase of mobility of the post-dentary bones is clearly seen in the cynodont fossil record (See Chapter 8, Figure 8.3). For instance, Dimetredon, the therocephalians, Thrinaxodon, Probainognathus and Morganucodont show the post-dentary bones in progressively more reduced form, and illustrate the step-wise transformation from the reptilian to the mammalian configuration. **Second**, the malleus articulates with the incus in exactly the same way as the articular articulates with the quadrate in advanced therapsids and the quadrate (incus) articulates with the stapes (See Chapter 8, Figure 8.1). **Third**, the ontogeny of the incus and the malleus reflects or recapitulates their reptilian derivation (see Chapter 11 for further discussion of recapitulation). When marsupials are still in the pouch, for instance, the malleus and the incus maintain the reptilian role of the articular and quadrate. (See Chapter 8 for discussion of these jaw/ ear bones.) Only when the young leave the pouch do these bones separate from the lower jaw and enter the middle ear.[7-40]

Mammal-like Reptiles – The Synapsids

Mammals evolved from reptiles many millions of years ago. The earliest mammalian fossils found are a group called the "Morganucodontids" such as Kuehneotherium, were found from the Triassic-Jurassic strata border almost 200 million years ago. Kuehneotherium had a mammalian jaw, gait, and tooth structure, and most likely was warm blooded. The origin of mammals can be traced back before the Morganucodontids, through a series of reptilian groups called mammal-like reptiles (Chapter 8). Mammal-like reptiles (the synapsids) is an informal name for three groups of animals that form a relatively complete record of fossils from the mid Carboniferous period to the end of the Triassic period (232-195 million years ago) when the earliest mammals appear (See Chapter 3 for further discussion of the strata periods). **The first group** of mammal-like reptiles were the pelycosaurs. (An image of pelycosaurs is in Chapter 3, Figure 3.1. It is the reptile with the large fin on its back). Archaeothyris, which is believed to have existed about 300 million years ago, is a good example of a pelycosaur. Archaeothyris was a lizard-like animal, approximately 50 cm (19.7 inches) long. It differed from the typical reptile in that there was an opening in the cheek region, on both sides of the face. Each opening is called a "temporal fenestra," which would have a muscle passing through it. The temporal fenestra is the defining characteristic of the mammal-like reptiles. Another animal which belonged to the pelycosaur group is the Dimetrodon (see Figure 7.4g).

The second group of mammal-like reptiles were the therapsids (see Figure 7.4h). Their temporal fenestrae (a small opening or perforation in a bone between the middle and inner ear) were larger than those of the pelycosaurs (an extinct group of large primitive reptiles that appeared during the Permian Period), and their legs were held in an upright position. **The third group** of mammal-like reptiles are the cynodonts, which first appeared in the late Permian period. Cynodont jaws are similar to mammal jaws and their teeth are multicuspid and differentiated down the jaw, similar to those of mammals. Mammalian teeth have a complex, multicuspid structure which differentiated down the jaw into canines, molars, and etc., whereas reptilian teeth form a relatively undifferentiated row and have a simpler structure. The mammalian jaw has cheek muscles which allow it to close with more power than the reptilian jaw. The reptilian jaw has its muscles at the back of the jaw, where it also articulates. It was from the cynodonts (small carnivorous reptiles) that the ancestors of the modern mammals such as the Kuehneotherium evolved. (See Chapter 8 for further discussion of reptilian and mammalian jaws.)

In population genetics, neo-Darwinism explains microevolution by changes in the frequencies of pre-existing variants. Most characteristics show variation and characteristics evolve as their structures are altered by natural selection. NeoDarwinism can explain how macroevolution can be explained without the need for extraordinary kinds of variation. Natural selection, on ordinary variation and mutation, is sufficient. (See Chapter 9 for further discussion of natural selection.) That is how the mammals evolved from the reptiles. Mammal-like reptiles were abundant during the Permian-Triassic period (286-213 mya). Mammal-like reptiles were the dominant group of large terrestrial vertebrates during this period. They were herbivorous and carnivorous. However, by the end of the Triassic period dinosaurs largely replaced the mammal-like reptiles. Mammals remained a minor group for almost 150 million years after their origin. Then they expanded rapidly 50-60 million years ago. Their expansion was slowed due to the presence of dinosaurs. But once the dinosaurs declined, mammals were able to rise more rapidly.[7-41]

The fossil record for the evolution of reptiles to mammals is quite complete and so gradual that their classification is difficult. In fact, the category of "mammal-like reptiles" emerged because scientists had trouble determining whether or not these animals were reptiles or mammals. In other words, they are clearly transitional creatures. The most distinguishing features in this evolutionary series appear in the teeth and jaws. (See Chapter 8 for a discussion of reptilian jaws transitioning into mammalian jaws.) As cold-blooded reptiles evolved into warm-blooded mammals, nutritional requirements increased.[7-42]

The evolution of teeth from reptiles to mammals passed through numerous transitional stages. Most reptiles have homodont teeth (same teeth). These are simple, cone-shaped, single-rooted teeth that are all about the same size and which are continuously replaced throughout life. As reptiles evolved into mammals, teeth began to lengthen at the corners of the mouth. Later mammal-like reptiles featured a reduction in the number of check teeth and the beginning of cusp (the points on teeth), resulting in a dentition that could slice tissue. Finally, mammals arose with distinct incisors, canines, premolars, and molars. They also developed only two sets of teeth during their lifetime—baby and adult dentitions. The back teeth were multi-rooted, wider from front-to-back, and had cusps that interlocked with those of the opposing jaw.[7-43]

There are many similarities between mammals and reptiles. They include: (1) Mammals and reptiles are two classes of the phylum Chordata. (2) The nerve cord of both mammals and reptiles are protected by a nerve cord. (3) Both mammals and reptiles have a sophisticated nervous system. (4) Both mammals and reptiles have bilateral symmetry. (5) Both mammals and reptiles are tetrapods, having four limbs. (6) Both mammals and reptiles breathe through lungs. (7) The respiratory system of both mammals and reptiles have a pharynx. (8) Both mammals and reptiles have a closed circulatory system with a heart. (9) Both mammals and reptiles have a complex exoskeleton made up of bones. (10) Both mammals and reptiles have well-developed sense organs. (11) The reproductive and excretory systems overlap in both mammals and reptiles. (12) Both mammals and reptiles undergo sexual reproduction as the major method. (13) Both mammals and reptiles are unisexual animals with internal fertilization [7-43a]

Intelligent Design View

Darwinists believe that intermediate groups existed between reptiles and mammals, such as the mammal-like reptiles, a group of extinct reptiles in which the morphology (structure) of the skull and jaw was very similar to that of a mammal. The possibility that the mammal-like reptiles were completely reptilian in terms of their anatomy and physiology cannot be excluded. (Morganucodontids such as Morganucodon and Kuehneotherium that were mentioned in the Evolutionist View above are discussed in Chapter 8, section "The Fossil Evidence" Creationist View.) The only evidence that exists regarding their soft tissue is their cranial endocasts, and these suggest that, as far as their central nervous systems were concerned, they were entirely reptilian. (An endocast is the internal cast of a hollow object, often referring to the cranial vault in the study of brain development in humans and other organisms.) Research has concluded that the mammal-like reptiles had brains typical of lower vertebrate size, since their endocasts were all very near the volume of these expected brain sizes. And since the endocasts present maximum limits on their brain sizes, the mammal-like reptiles could not have had brains that approached a mammalian size. The mammal-like reptiles were reptilian and not mammalian with respect to the evolution of their brains. There were a few mammalian features in the brains of the mammal-like reptiles. The forebrain, to the extent that its position is identifiable, was reptilian size and shape. This was not the case in the earliest known fossil mammal.

The earliest mammal for which there is reasonable evidence to support the view that mammals evolved from reptiles is the Triconodon of the Upper Jurassic period (161-146 mya), although it was apparently already at or near the level of living primitive mammals, such as the insectivores of the Cenozoic (Tertiary) Era—Early Paleocene Epoch (67-54 mya) or the Virginia opossum. The Virginia opossum possessed a larger brain than its reptilian ancestors of comparable body size. Also, many separate groups of mammal-like reptiles exhibited skeletal mammalian characteristics, yet only one group, the theromorphs [synapsids, pelycosaurs or therapsids], could have been the hypothetical ancestor of the mammals. The theromorphs [synapsids, pelycosaurs or therapsids], which are considered to be intermediate between the reptiles and the mammals, only possess a general resemblance to the mammals and not actually being ancestral to them. It is possible that extinct groups such as the mammal-like reptiles might have possessed features in their soft tissue that were completely different from any known reptile or mammal, which could eliminate them completely as potential mammalian ancestors, just as the discovery of the living coelacanth (Figure 7.1) revealed features in its soft tissue that were unexpected (as previously discussed). But, without the soft tissue to examine, any conclusions regarding mammal-like reptiles being the forerunners of mammals will only be sheer conjecture.[7-44]

Mammals have a number of unique features that are not found in any other group of organisms. They include: **1)** a hairy covering, each hair being a complex structure consisting of keratinized cuticle, **2)** a cortex and a central medulla mammary glands exhibiting alveoli surrounded by a network of myoepithelial cells responsive to the hormone oxytocin producing milk, a nutritious secretion containing fat globules and sugars **3)** specialized sweat glands in the skin **4)** a four-chambered heart with left ventricle delivering aerated blood to the aorta (Figure 7.5) **5)** unique shaped kidneys with a urinary tract filtering unit that removes waste matter from the blood and function to generate urine[7-45] (Unlike the kidneys of mammals and birds, reptile kidneys are unable to produce liquid urine more concentrated than their body fluid. This is because they lack a specialized structure called a "loop of Henle," which is present in the nephrons of birds and mammals.[7-46] Nephrons are numerous filtering units of the vertebrate kidney that remove waste matter from the blood.)[7-47] **6)** a large cerebral cortex with distinctive six layers of cells. The cerebral cortex is a complex outgrowth of neural tissue which forms the outer layer of the brain, which is the seat of all the higher mental functions and complex behavior patterns that is characteristic of mammals **7)** a diaphragm, a special muscle used by mammals for respiration (See Figure 7.13) **8)** three highly specialized ear ossicles consisting of a mallus, incus and stapes conducting vibrations across the middle ear (See Chapter 8) **9)** the organ of corti, a specialized organ for reception and analysis of sound. (See Chapter 8 for further discussion of the organ of corti.)

Each of the 9 characteristics listed above are unique to only mammals and essentially are in the same form.[7-48]

Creationist and Intelligent Design View

Evolutionists believe that mammals descended from reptiles. Some of the many differences between mammals and reptiles include: **1)** The basic structure of mammals is quite different than that of reptiles. **2)** Reptiles breathe in a totally different manner than mammals, for reptiles lack a diaphragm. **3)** Mammals primarily excrete urea, whereas reptiles excrete uric acid. **4)** Mammals have fur (although some, such as whales and elephants have relatively little); reptiles have scales. **5)** Mammals have much larger brains than reptiles have. **6)** Mammals maintain a constant body temperature, but reptiles do not. **7)** Mammals produce milk, but reptilian infants must get their nourishment from the egg. **8)** There are important vertebral differences between mammals and reptiles. **9)** Mammals have different blood. Their blood is nucleated (that is, the red blood cells have a nucleus that can undergo mitosis [mitosis is a type of cell division that results in two daughter cells each having the same number and kind of chromosomes as the parent nucleus]), whereas the blood of reptiles is un-nucleated. **10)** Mammals have three ear bones, whereas reptiles only have one. The inner ear of mammals is much more complex. (Chapter 8) **11)** Mammals have a palate separating the mouth from the nose cavity; reptiles lack it. **12)** Mammals consistently have a single dorsal aorta (their largest artery). Reptiles have two. How could one circulatory system change into a different one? (See "Various Heart Types" section for further discussion.) **13)** Mammals have a complex set of teeth, including temporary infantile ("milk") teeth. Reptiles have single peg-teeth.[7-49]

Any transition from a reptile to a mammal would require the development of completely new organ systems. Transforming the reproductive system, for example, is not just a question of changing where the eggs grow (whether inside or outside of the mother). It also requires the development of completely new organs like the placenta and mammary glands. It is doubtful whether natural selection and random mutations have the ability to produce such changes. (Natural selection & mutations are discussed in Chapter 9.) Many necessary anatomical changes would have to take place in a coordinated fashion. Transforming a reptile to a mammal requires the step-by-step conversion of many, separate physiological systems. It requires a coordinated change in the respiratory (Figure 7.13), circulatory (Figure 7.5), and reproductive systems, plus other changes as well. All of the intermediate organ systems must work, and in many cases, they must work together. Since vital organs are vital to the survival of an animal, every temporary loss of an organ could result in the death of the transitional animal forms.[7-50]

Evolutionists claim that there are multiple examples of reptile to mammal transitions. This would be evidence if these multiple fossils could be placed in a single line of descent that could conceivably lead from one particular reptile species to a particular early mammal descendent. Having multiple examples from different species only shows skeletal similarities but does not prove ancestry. Basically, it is difficult to determine whether a creature is a reptile or a mammal. The usual criterion to determine whether a creature is a reptile or a mammal is if its jaw contains several bones, one of which is the articular bone and it connects to the quadrate bone of the skull, then the fossil is considered to be a reptile. (See Chapter 8 for a discussion of the transition of the reptilian jaw/ear to a mammalian jaw/ear.) Just as amphibians and reptiles have different reproductive systems, reptiles and mammals also have vastly different reproductive systems. Reptiles lay eggs that hatch, whereas mammals give live births. Just as with the amphibian to reptile transition, since soft parts do not fossilize very well, there is no way for evolutionists to determine how a reptile reproductive system could transition into a mammal reproductive system.[7-51]

Various Heart Types

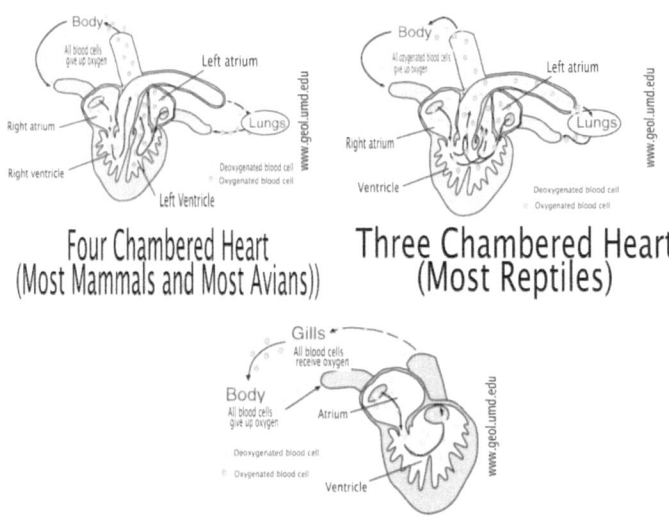

Figure 7.5 Various Heart Types

Intelligent Design View

Reptiles and mammals have different vascular and circulatory systems. Both reptiles and mammals have dual circulation, which means there are two pumping systems running simultaneously. The venous system pumps deoxygenated (spent) blood into the heart. The arterial system pumps the fresh blood out to the body. However, most reptiles have a three-chambered heart.

In a three-chambered heart (Figure 7.5), one chamber, called the atrium, receives deoxygenated blood from the venous system, while the other atrium receives oxygenated blood from the lungs. Both of these chambers empty into a third chamber, a single ventricle. The "spent" blood and the "fresh" blood get mixed together to some extent before getting pumped into the body.

A four-chambered heart (Figure 7.5) works differently. In a four-chambered heart, two chambers (one atrium, one ventricle) pump deoxygenated (spent) blood from the venous system to the lungs, where it is "recharged" with oxygen. The heart's other artium-ventricle combination pumps the oxygenated blood into the arterial system that carries the fresh blood out to the body. Because the three-chambered and four-chambered hearts work differently, they obviously have some different parts, and a different overall arrangement of those parts.[7-52]

Any transition of a reptile from a three-chambered heart to a mammal four-chambered heart (Figure 7.5) requires a series of coordinated physiological and anatomical changes. They include: **1)** lengthening and attaching the existing septum to create a new, separate ventricle chamber. **2)** replacing the forked abdominal aorta and two aortic arches with a single aorta. **3)** rerouting the pulmonary arteries and veins, and **4)** making various secondary structural changes to the walls and valves between chambers.

Some reptiles, crocodilians such as crocodiles, alligators, caimans, and gavials, do have four-chambered hearts, however, it is unlikely that these reptile four-chambered hearts could have evolved into mammals. There are three reasons for this. **First**, the crocodilians are not among the mammal-like reptiles that are thought to be ancestors to the first mammals. **Second**, even if crocodiles were ancestors to the mammals, there would need to be an explanation of how the crocodile's four-chambered heart first arose from the three-chambered heart of the typical reptile (since Darwinian common descent stipulates that all living things evolved from a former ancestor). In other words, an acceptable scenario would need to be developed to explain how a four-chambered heart could evolve step-by-step from a three-chambered heart while avoiding a non-functional (lethal, deadly) intermediate form. **Third**, even though the crocodile heart is a four-chambered heart (Figure 7.5), it is a special kind of four-chambered heart that is different from the mammal's four-chambered heart that is just as different from the reptile's three-chambered heart (Figure 7.5).[7-53]

It is very difficult to conceive mutations occurring early in the development of a creature could produce beneficial large-scale change. Even though some mutations do occur early in the development of some organisms, these mutations inevitably disrupt the orderly processes of organ system construction.[7-53] Evolutionists believe that fish evolved into amphibians that evolved into reptiles that evolved into mammals. The problem is that fish have a two-chambered heart; reptiles have a three-chambered heart; and mammals have a four-chambered heart (Figure 7.5). Since Darwinian evolution requires multiple, small changes that eventually result in large-scale improvements, any offspring with even minor changes to their heart would most likely not function as a fully developed heart. Any offspring with any significant change to their heart would have a weaker and less efficient heart and would therefore have a more difficult time of survival. Any three-chambered heart that had not yet evolved into a four-chambered heart, for example, would most likely not be as sufficient as either a three-chambered or four-chambered heart and would therefore be detrimental to the creature. It would most likely not survive long enough to produce offspring that would have a more improved, advanced heart. (See Chapter 11, section "Embryology" Creationist View for further discussion of the development of the heart.)

Evolutionist View

The various heart types appear to present a problem for evolution, however, this apparent dilemma can be explained by gradual evolution. It is very possible that systems like the four-chambered heart most likely arose as complete systems as the result of mutations that occurred early in the development of these organs. Because such mutations occur so early in development, they have the potential to produce larger-scale change and to affect the system as a whole, rather than just individual parts of the organ or system. Such mutations would therefore eliminate the need to coordinate the many minor modifications required by slow, gradual modifications.[7-54]

Evolution From Reptiles to Birds

CREDIT JANET M. RUTH/USFWS

Evolutionist View

A bird is any warm-blooded vertebrate of the class Aves, having a body covered with feathers, forelimbs modified into wings, scaly legs, a beak, and no teeth, and bearing young in a hard-shelled egg.[7-55] Primitive birds had a number of reptilian characteristics that indicate they evolved from dinosaurs. **First**, they had homodont (Greek: same tooth) teeth that are continually replaced throughout life, however, birds today do not have teeth. **Second**, the primitive birds had long tails similar to those of reptiles. Modern birds have short tails. **Third**, most primitive birds had three claws on each wing. These claws were the same as those found on the upper limbs of small, two-legged, meat-eating dinosaurs. Claws are rarely found on the wings of birds today. **Finally**, the first birds to evolve had a three-part lower limb similar to dinosaurs. The lower limbs included a femur, a tibia and fibula, and elongated upright foot bones. Both birds and dinosaurs walked on their toes. The three-part leg remains in modern birds, but the upright foot bones have fused into one bone in birds that exist today. This evidence indicates that the first birds were transitional creatures between dinosaurs and birds of today.[7-56] Similarities between birds and reptiles include: **1)** Skulls hinges on a single condyle **2)** Lower jaw is made of several bones **3)** Single middle ear bone **4)** Pneumatic bones **5)** Scales on legs **6)** Lack of skin glands **7)** Egg laying **8)** Nucleated red blood cells **9)** Ankle joint is intertarsal (bends forward) **10)** Uncinate process - overlapping tabs in the ribs **11)** embryonic development similarities.[7-56a]

As one looks through the various kinds of dinosaurs, it can be noted that the fourth and the fifth finger, first the fifth and then the fourth, become reduced and finally lost, until when animals like Allosaurus, Deinonicus, and Archaeopteryx are seen, they have only three fingers, and those are the first three fingers. The second finger is the longest, and that through time, these fingers and the hand bones become even longer and more graceful or slender. Those three fingers that are seen in Archaeopteryx at the end are still separate fingers, but in birds today, they are fused (Figure 7.4). (See section "A Look at the Archaeopteryx Fossils" for further discussion of Archaeopteryx.) If a turkey, for example, is dissected, the individual digit bones indicate that they later became fused. And this is because the bird is no longer using its hand for anything except flight. It is not using its fingers to pick up things or claw or scratch anymore. And early in the evolution of birds, when they began to only use these two limbs for flying and very little else, there was no further need to use these fingers for anything, and it made more sense to fuse them into position rather than use muscles to hold them there. And this is the evidence that we have of how these organs evolved.[7-57] (See Chapter 14, section "An Example of Irreducible Complexity—The Wing" for further discussion of the evolution of wings).

Dental features in birds today are an evolutionary vestige. (Vestiges are discussed in Chapter 12.) As mentioned previously, birds do not have teeth. However, a certain generic mutation in chickens results in the development of rudimentary teeth. But where did the genes to form teeth come from? The fossil record reveals that the first birds had teeth, and about 80 million years ago, they lost them. The reappearance of teeth in mutant chickens today indicates that though the genes for dental development were suppressed, they still remain intact.[7-58]

Scientific evidence confirms that birds today pass through the earliest stage of tooth development. They have the same dental initiation genes and molecules found in toothed animals. But this genetic dental development is suppressed in modern birds and teeth never form. Scientists have discovered ways to re-awaken these dormant genes. Combining skin tissue from the mouth of a bird with the inner jaw tissue from a toothed animal causes the bird tissue to produce enamel. Similarly, the placement of certain developmental molecules into the jaws of developing chicks can produce teeth.[7-59] (See Chapter 10, "Punctuated Equilibrium" for further discussion of these experiments.)

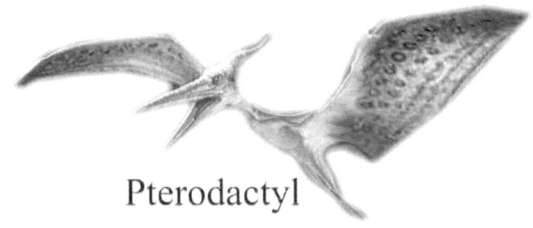

Pterodactyl

Pterosaurs are the earliest vertebrates known to have evolved powered flight. Birds are not believed to have evolved from pterosaurs, but rather from dinosaurs. The wings of Pterosaurs were formed by a membrane of skin, muscle, and other tissues stretching from the legs to an extremely lengthened fourth finger. Early species had long, fully toothed jaws and long tails, while later forms had a highly reduced tail, and some did not have teeth. Many had furry coats made up of hair-like filaments known as pycnofibres, which covered their bodies and parts of their wings. Pterosaurs are sometimes referred to as dinosaurs, but this is not correct. The term "dinosaur" refers to a certain group of reptiles with a unique upright stance (which includes birds), and therefore excludes the pterosaurs, as well as the various groups of extinct marine reptiles, such as ichthyosaurs, plesiosaurs, and mosasaurs. Pterosaurs are also sometimes referred to as "pterodactyls," however, this too is incorrect. "Pterodactyl" refers specifically to members of the genus Pterodactylus, and more broadly to members of the suborder Pterodactyloidea.[7-60] Like all pterosaurs, the wings of Pterodactylus were formed by a skin and muscle membrane stretching from its elongated fourth finger to its hind limbs. It was supported internally by collagen fibers and externally by keratinous ridges.[7-61]

The anatomy of pterosaurs was highly modified from their reptilian ancestors for the demands of flight. Pterosaur bones were hollow and air-filled, like the bones of birds. They had a keeled breastbone that was developed for the attachment of flight muscles and an enlarged brain that shows specialized features associated with flight. In some later pterosaurs, the backbone over the shoulders fused into a structure known as a notarium (the fused vertebra of the shoulder in birds and some pterosaurs that help brace the chest against the forces generated by the wings), which served to stiffen the torso during flight, and provide a stable support for the scapula (shoulder blade).[7-62]

Compsognathus

Compsognathus (artist depiction)

It is believed by many scientists that the dinosaur Compsognathus may have evolved into birds. The German forms of Compsognathus fossils are about the size of a chicken. The French version is about the size of a turkey. Some relatives of Compsognathus, such as Sinosauropteryx and Sinocalliopteryx, have been preserved with the remains of simple feathers covering the body like fur, causing some scientists to believe that Compsognathus might have been feathered in a similar way (in the section "The Origin of Wings and Feathers", see image above the caption "This fossil of Sinosauropteryx preserves evidence of hair-like feathers"). Therefore, some have concluded that Compsognathus had a covering of downy proto-feathers. However, no feathers or feather-like covering have been preserved with Compsognathus fossils, in contrast to Archaeopteryx (discussed below), which was found in the same sediments.[7-63]

wikipedia.org

Saurischian (artist depiction)

Avians (modern birds) are direct descendants of one group of theropod dinosaurs, descendants of saurischian dinosaurs. Saurischians ('lizard-hipped') are distinguished from the ornithischians ('bird-hipped') by retaining the ancestral configuration of bones in the hip. All carnivorous dinosaurs (theropods) are saurischians. Saurischians are distinguished from ornithischians by their three-pronged pelvic structure, with the pubis pointed forward. The ornithischians' pelvis is arranged with the pubis rotated backward. The ornithischian hip structure is somewhat similar to that of birds. This dinosaur was named saurischians "lizard-hipped" dinosaurs because they retained the ancestral hip anatomy also found in modern lizards. However it was eventually determined that the hip structure possessed by modern birds actually evolved independently from the "lizard-hipped" saurischians.[7-64]

Creationist View

There is no fossil evidence to suggest that a four-legged reptile ever evolved into a bird. For example, there is no fossil that has a limb in the process of becoming a wing. Evolutionists may label creatures such as Archaeopteryx as dinosaurs (reptiles), however, if a creature has feathers and wings then they are classified as birds. Birds might have evolved into different birds (microevolution) but dinosaurs (reptiles) did not evolve into birds (macroevolution). (See section "A Look at the Archaeopteryx Fossils" for further discussion of Archaeopteryx.) Some Darwinists have suggested that the small carnivorous dinosaur Compsognathus (discussed above) could be the transitional dinosaur between Archaeopteryx (discussed below) and the reptiles. Although Compsognathus appeared to have hollow bones, it is a poor candidate to be a transitional bird because it co-existed with Archaeopteryx (Figure 7.10). Moreover, dinosaurs are divided into two formal groups: lizard-hips and bird-hips. Modern evolutionists now believe that it is these lizard-hipped dinosaurs that evolved into birds, rather than birds evolving from bird-hipped dinosaurs. Since Compsognathus is a saurischian (lizard-hipped dinosaur depicted above), it is less likely that this creature could be a possible ancestor to the birds.[7-65]

The velociraptor fossil and other fossil finds considered to be dinosaur precursors to birds have turned out to be flightless birds similar to ostriches, such as *Protarchaeopteryx robusta* and *Caudipteryx zoui.* [7-65a] Just because some fossils have been designated as reptilian dinosaurs, it does not indicate that birds evolved from them.

Protoavis

Protoavis (Artist depiction)

Evolutionist View

Protoavis texensis is described as a fossil coming from a single animal, specifically a 35 centimeter (13.7795 inches) tall bird that lived in what is now Texas, USA, around 210 mya (million years ago). Although Protoavis existed before Archaeopteryx (155 – 150 mya), its skeletal structure is allegedly more bird-like (Archaeopteryx will be discussed later in section "A Look at the Archaeopteryx fossils."). Protoavis has been reconstructed as a carnivorous bird that had teeth on the tip of its jaws and eyes located at the front of the skull which suggests that they were nocturnal or were active during the twilight hours of the day. Reconstructions usually depict it with feathers, as the structures on the arm appear to be quill knobs, the attachment point for flight feathers found in some modern birds and non-avian dinosaurs. Some researchers have been inconclusive regarding whether or not these structures are actual quill knobs. This description of Protoavis assumes that Protoavis has been correctly interpreted as a bird. Many paleontologists doubt that Protoavis is a bird, or that all remains assigned to it even come from a single species, because of the circumstances of its discovery and unconvincing avian characteristics that are present in ancestral species in its fragmentary material. Much controversy remains over the animal, and there are many different interpretations of what Protoavis actually is.

Protoavis (meaning "first bird") is a problematic dinosaurian taxon known from fragmentary remains from Late Triassic Norian stage deposits near Post, Texas. (The Late Triassic Period extends from 220-213 mya.) When the remains were found at the Tecovas and Bull Canyon Formations in the Texas panhandle in 1984 (or 1973) in a sedimentary strata of a Triassic river delta, the fossils were a jumbled cache of disarticulated bones that may reflect an incident of mass mortality following a flash flood. When Protoavis was first announced, the fossils were described as being from a primitive bird which, if the identification is valid, would push back avian origins some 60-75 million years (210 – 150 = 60 mya).[7-64f]

Creationist View

Protoavis was discovered in 1973. Evolutionists find Protoavis puzzling, not because of the fossil remains, but because of other factors. Protoavis is dated to have existed approximately 210 mya, which is older than Archaeopteryx (155 – 150 mya) (Archaeopteryx will be discussed later in section "A Look at the Archaeopteryx fossils."). Archaeopteryx has been considered the oldest bird fossil for the past 100 years, since its first discovery in 1861. The other problem for evolutionists is that Protoavis is more bird-like than Archaeopteryx. Protoavis had a keel-shaped sternum and a shoulder girdle with supracoracoideus pulley system typical of modern flyers. (*Supracoracoideus is a muscle originating from the sternum (in birds) or the coracoid bone or plate (in reptiles and amphibians) and inserting on the humerus, serving (in birds) to elevate the wing in flight and (in reptiles and amphibians)* to support the body on the forelimb.) Some evolutionists reject the initial findings because it doesn't fit the theropod-to-bird-evolution.[7-64g]

Evolutionists believe that birds evolved from dinosaurs and that Archaeopteryx, considered the first known bird, was a reptile (dinosaur) that developed wings and became birdlike. Protoavis disproves this theory because it (Protoavis) existed 55 mya (210 – 155 = 55) before Archaeopteryx. Since dinosaurs first appear in the fossil record around 248 mya (see Chapter 3, section "The Strata Layers of the Earth"), this would mean that birds must have originated sometime before 210 mya, when Protoavis appeared, as Protoavis was a fully formed bird. This means that birds had to have originated within 38 mya (248 -210 = 38) of dinosaurs.

Coelophysis

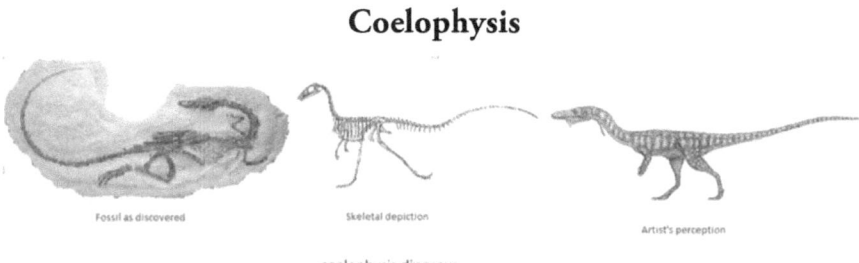

Fossil as discovered   Skeletal depiction   Artist's perception

coelophysis dinosaur

Coelophysis (C. Bauri) is an extinct genus of coelophysid theropod dinosaur that lived approximately 203 to 196 million years ago during the latter part of the Triassic Period. The fossil was found in the southwestern United States. It was a small, slenderly-built, ground-dwelling, bipedal carnivore, that could grow up to 3 meters (9.84 feet) long. Coelophysis is one of the earliest known dinosaur genera. Scattered material representing similar animals has been found all over the world in some Late Triassic and Early Jurassic formations. Coelophysis had a furcular (wishbone), the earliest known example of a dinosaur. [7-64a] (Theropods are carnivorus dinosaurs that had short (small) forelimbs and walked or ran on their hind legs.)[7-64b]

Coelophysis had hollow bones, hence the name Coelophysis. The Greek word "koilos" means "something hollow." The word "physis" refers to a form or structure. Due to its hollow bones, Coelophysis was a very light weight animal. It had 4 digits on each hand.[7-64c] An adult Coelophysis is about eight feet long, however, some species were smaller. The skeleton of this dinosaur is very similar to that of Archaeopteryx (Archaeopteryx is discussed below in section "A Look at the Archaeopteryx Fossils").[7-64d] Coelophysis apparently had superior vision, compared to most lizards, and ranked with that of modern birds of prey. The eyes appear to be the closest to those of eagles and hawks, with a high power of accommodation.[7-64e]

Caudipteryx

Evolutionist View

Caudipteryx (reconstructed skeleton)   **Caudipteryx (artist depiction)**

Caudipteryx is a genus of peacock-sized theropod dinosaurs that lived in the Aptian age of the early Cretaceous Period (144-65 mya). They were feathered and remarkably birdlike in their overall appearance. Mesozoic theropods were also very diverse in terms of skin texture and covering. Feathers or feather-like structures are attested in most lineages of theropods. Some palentologists believe that Caudipteryx was just a flightless bird, like a cretaceous ostrich. Many palentologists believe that Caudipteryx, like all maniraptorans, is a flightless bird, and that birds evolved from non-dinosaurian archosaurs. Many believe that Caudipteryx is a basal (primitive) member of the Oviraptoridae, and the oviraptorids are nonavian theropod dinosaurs. Some, however, came to a different conclusion. They found that the most birdlike features of oviraptorids actually place the whole clade within Aves itself, meaning that Caudipteryx is both an oviraptorid and a bird. In their analysis, birds evolved from more primitive theropods, and one lineage of birds became flightless, re-evolved some primitive features, and gave rise to the oviraptorids. Others, however, maintain that Caudipteryx was flightless. Others concluded that Caudipteryx is not a theropod dinosaur at all. They believe that Caudipteryx, like all maniraptorans, is a flightless bird, and that birds evolved from non-dinosaurian archosaurs. [7-65b]

Caudipteryx came later in the fossil record than archaeopteryx, which appeared from the Late Jurassic Period (213-144 mya) (161 million to 146 million years ago). Caudipteryx appears in the fossil record around the early Cretaceous Period (about 124.6 million years ago). Archaeopteryx is more bird-like than Caudipteryx, which appears to be flightless. Some paleontologists believe Caudipteryx are flightless birds that evolved from a flying ancestor, almost certainly Archaeopteryx.[7-65c] (Archaeopteryx is discussed in more detail below in section "A Look at the Archaeopteryx Fossils.") Caudipteryx is best considered a herbivorous flightless bird related to Protarchaeopteryx and Confusciusornis. It is more advanced than the significantly older Archaeopteryx. The existence of a flightless bird in the Early Cretaceous of China provides further evidence of the diversification of birds at this time, but contributes little to our understanding of the origin of birds.[7-65d]

Protopteryx

Evolutionist View

Protopteryx (artist depiction)

Protopteryx is an extinct genus of bird from the Cretaceous period (144-65 mya). The fossil is dated from 131 mya ago (million years ago). The name Protopteryx means "primitive feather." "proto-" meaning "the first of" and "-pteryx" meaning "feather" or "wing." The name comes from the fact that Protopteryx feathers are more primitive than those of modern birds, such as the two elongated tail feathers that lack barbs and rami. Protopteryx fossils show that they were roughly the same as a today's starling. The adult body length of Protopteryx was about 10 centimetres (3.93 inches), excluding the tail feathers. Protopteryx teeth were conical and unserrated, and some teeth had a resorption pit similar to those seen in Archaeopteryx (discussed in section "A Look at the Archaeopteryx Fossils" below). The body of Protopteryx was covered in 12 millimetres (0.47 inches). The barbs of the down feathers were laminar instead of hairlike and were frayed at the tips. The most distinctive feature of Protopteryx is that the tail consisted of two long feathers which only had barbs at their tips. Closer to the body, the long tail feathers were thin and needle-like. The only modern three types of feathers: down feathers, flight feathers, and long, ribbon-like tail feathers (see image illustration above). The body was mostly covered in feathers of about bird to share a feather type similar to Protopteryx is the red bird-of-paradise. The tail feathers also lack rami on the proximal end of the tail (A rami is an arm or branch of a bone, in particular those of the ischium and pubes or of the jawbone). Protopteryx was adapted for flying and had feathers with features similar to modern birds, as shown by its procoracoid, carina of the sternum, external tuberosity of the humerus, and deltoid crest, which suggest Protopteryx had a modern musculus supercoracoideus and pectoralis. Protopteryx also shares asymmetric wing flight feathers with flying birds, as well as Archaeopteryx and Confuciusornis. The tail feathers of Protopteryx lack of barbs and rami close to the body, suggesting a use outside of flight, such as display, thermoregulation, or sensory usage.[7-65e]

Oviraptor

Evolutionist View

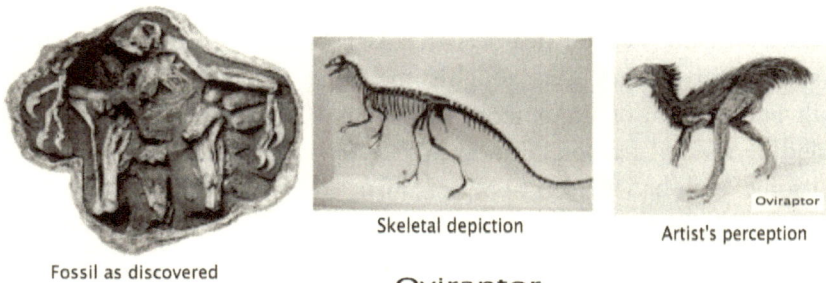

Fossil as discovered   Skeletal depiction   Artist's perception

Oviraptor

Figure 7.6 Oviraptor[7-66]

The Oviraptor dinosaur is a relative of the ostrich dinosaur. It has three clawed fingers on each hand (Figure 7.6).[7-67] Oviraptor is said to have existed 85-75 million years ago.[7-68] Oviraptor was one of the most birdlike of all dinosaurs, with a sharp, toothless beak and probably a coat of feathers. This theropod did not have wings, but it seems to have been close to the first flying birds. Oviraptor does not technically count as a true raptor, the breed of dinosaurs most famously represented by Deinonychus and, Velociraptor. [7-69]

Creationist View

Evolutionists claim that birds evolved from reptiles, such as Oviraptor. Oviraptor could not have been the predecessor of birds since Archaeopteryx (considered a transitional creature between reptiles and birds) is said to have already existed 155-150 million years ago. Oviraptor is said to have existed only 85-75 million years ago (Figure 7.10). (Archaeopteryx is discussed in section "A Look at the Archaeopteryx Fossils.") Just because Oviraptor may have had feathers does not indicate that birds evolved from dinosaurs. (Figure 7.10) The egg of a fish, which is supposed to have evolved first, is contained in nothing more elaborate than a blob of jelly. That is adequate for water-dwelling creatures, but this is not sufficient for land animals that need something more elaborate. The nucleus of the egg must be packaged in its own pool of liquid. The yolk, where the new life develops, is surrounded by the white part, which sustains the developing infant. The yolk is attached to the shell with a shock-absorbing suspension of elastic threads, while the shell that contains it all is a masterpiece of engineering. It has to be of exactly the right strength, hard enough to stay in one place when the bird sits upon it, but soft enough to let the chick peck its way out when the time comes. It has to be waterproof, and yet porous to air so that the chick can breathe. Evolutionists have no idea how the bird's egg could have evolved. All they say is that the reptile egg is similar to it and is supposed to have appeared first. This, of course, only throws the problem back one stage. They cannot say when the evolution of the egg might have happened.[7-70]

There are several differences between reptiles and birds. These differences include: **(1)** Reptiles have scales all over the body, whereas birds have scales on the legs and the rest of the skin is covered with fluffy feathers. **(2)** All the present-day reptiles are carnivore, but birds have many different types of food habits.[7-70f] **(3)** Birds are warm blooded whereas reptiles are cold blooded. (Fish and amphibians are also cold-blooded.) This means birds produce their own body heat rather than rely on the environment. Some reptiles, such as the leatherback turtle, can produce body heat but it isn't as highly modified a system as it is for birds. **(4)** Birds have a four chambered heart which, among the living reptiles, only crocodylians have. The other reptiles have three chambered hearts, with a poor separation between the ventricles. **(5)** Birds have no teeth, with a bill instead. Most, but not all, reptiles have teeth, except for turtles and a few others. **(6)** The tail bones of a bird have been reduced and compressed into a tiny nub called a pygostyle. Most reptiles have normal tail vertebrae, although some geckoes also have little nub tails.[7-70g] **(7)** Birds have a normal hard shelled egg whereas reptiles have a special type of egg.[7-70h]

Scientific Proof that Birds Evolved from Dinosaurs

Evolutionist View

Below is a diagram illustrating the evolution from reptile to bird:

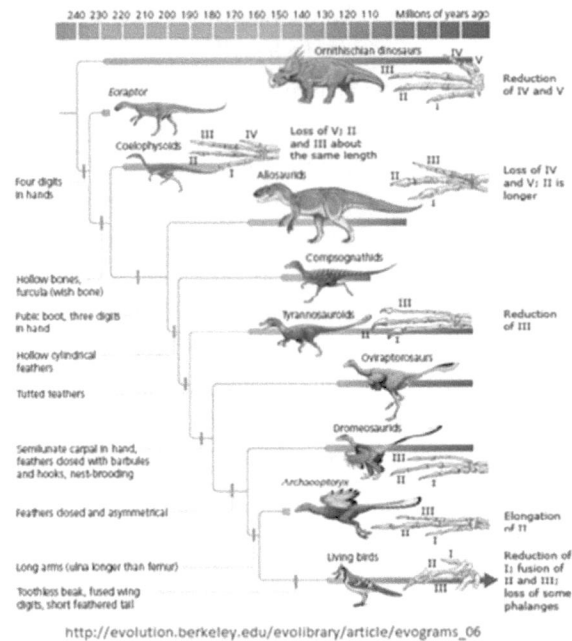

http://evolution.berkeley.edu/evolibrary/article/evograms_06

Figure 7.6a Evolution of the Modern Bird

The changes in the digits of the dinosaurs leading to birds indicates that birds evolved from dinosaurs. The first theropod dinosaurs had hands with small fifth and fourth digits and a long second digit. As illustrated above in Figure 7.6a, the theropod lineage that would eventually lead to birds, the fifth digit (e.g., as seen in Coelophysoids) and then the fourth (e.g., as seen in Allosaurids) were completely lost. The wrist bones underlying the first and second digits consolidated and took on a semicircular form that allowed the hand to rotate sideways against the forearm. This eventually allowed birds' wing joints to move in a way that creates thrust for flight.[7-70a]

All four-limbed creatures, including dinosaurs, evolved from an ancestor that had five digits at the end of its limbs. These became flippers, wings, hands or paws, and some or all of the digits disappeared altogether in some cases. Scientists believe birds evolved from a group of meat-eating dinosaurs called maniraptors including dinosaurs, evolved from an ancestor that had five digits at the end of its limbs. These became flippers, wings, hands or paws, and some or all of the digits disappeared some 150 million years ago, during the Jurassic period. Modern birds have three digits in each of their wings, which means two digits in the forelimbs of these dinosaurs would have had to be lost during evolution. Since birds evolved from dinosaurs, which had five digits, this means that two digits were lost during the evolution of birds. But which three digits survived and which ones were lost? Paleontologists and developmental biologists have disagreed heartily on this. A proposal made in 1999, called the "frame shift" hypothesis, explained a discrepancy in the evidence, but not everyone accepted it. The new study, conducted by researchers who transplanted cells from one part of a chick's body to another, adds to the support for the hypothesis. The results of their study appear in the Feb. 10, 2011 issue of the journal Science. If the digits are numbered so that digit 1 corresponds with our thumbs, digit 2 with our index fingers and so on, the fossil records shows that birds' wings evolved using digits 1, 2 and 3 of the dinosaur's forelimbs. However, in a bird embryo, the digits arise from the places on the limb bud associated with digits 2, 3 and 4. This conflict supported those who challenged whether birds were directly descended from dinosaurs.

In 1999, some researchers proposed that during development, digit 1 actually arose from the second position (where digit 2 should have arisen), and so on—a frame shift. Birds are dinosaurs, but developmentally the digits are 2, 3 and 4. In the new study, the researchers found evidence that the last digit of the wing does not correspond to the last digit of the foot. This supports the theory that the wing, unlike the foot, does not have a digit 4. Then the researchers mapped out digit development using cell-labeling techniques (enabling them to know where a certain cell ended up once it matured). (To determine the identity of the last digit in the chicks' "hands," the study used the development of the last digits in a five-digit mouse limb as a guiding model.) They found that by 3.5 days of embryonic development, a shift occurs, causing cells in the progenitor region for digit 4 to move forward and grow into digit 3. The same shift occurs for the digits that become 1 and 2. The study provides an excellent example of a shift in which one body part is transformed into another during embryonic development. The evolution of how the development of the limb changed during evolution of birds from their theropod ancestor demonstrates conclusively that birds did indeed evolve from dinosaurs. Some researchers who oppose the conclusion of this study are of the predominant scientific view that birds are descended from theropod dinosaurs (which include maniraptors). These opposing researchers, believe that birds and theropod dinosaurs share a common, earlier ancestor.[7-70b]

Scientific Research Casts Doubt That Birds Evolved From Reptiles (Dinosaurs)

Evolutionists with Critical View that Birds Originated from Dinosaurs

Not all biologists believe that birds are descendants of dinosaurs. All the evidence proposed by those who believe that birds originated from dinosaurs can also support the view that birds and dinosaurs had a common ancestor, and that they have diverged since that time. This group of scientists emphasize the differences between dinosaurs and birds, claiming that the differences are too great for the birds to have evolved from earlier dinosaurs.[7-70d]

Several recent studies, relying on completely different methods, contradict the view that birds evolved from dinosaurs. One study (Feduccia, A. 2001 BioMedNet News and Comment) examined the origins of the three bones of the fingers/wings and feet of the theropod dinosaurs and modern birds. The results clearly indicated that the wings of birds and hands of theropod dinosaurs are derived from different digits. Another study (Gibbons, A. 1997. Lung Fossils Suggest Dinos Breathed in Cold Blood Science 278: 1229. & Ruben, J.A., T.D. Jones, N.R. Geist, and W.J Hillenius. Lung structure and ventilation in theropod dinosaurs and early birds. Science 278: 1267.) showed that the theropod dinosaurs likely possessed a diaphragm and bellows-like septate lungs (similar to modern reptiles), which are not found in modern birds. The lack of adequate breathing structures in dinosaurs would suggest that they were incapable of supporting warm-blooded respiration and the aerobic requirements of flight.

The superficial resemblance between the bones of the theropod dinosaurs (the so-called "bird-like" dinosaurs) and Archaeopteryx (one of the most ancient birds - which lived roughly 150 millions years ago) has led scientists to hypothesize that birds are descendants of these dinosaurs. The bones of the wing of Archaeopteryx look very much like the bones of the theropod dinosaur, Deinonychus. It is true that the resemblance is so striking that it convinced a very large percentage of the scientific community that birds are descended from the theropod dinosaurs.

The theory that birds evolved from dinosaurs is in doubt now, since a recent study has demonstrated that the bones that make up the wings and feet of birds and the theropod dinosaurs are not derived from the same digits. One study indicates the digits of birds were not derived from the same dinosaur digits. The determination of what digits the feet and hands of theropod dinosaurs are derived from comes from the fossil record. All dinosaurs before the theropod had 5 digits on their feet and hands. The earliest known theropod dinosaur, Herrerasaurus, clearly demonstrates the presence of 5 digits in the hand, although the fourth and fifth digits are reduced in size. Other theropod dinosaurs show a similar loss of digits IV and V. The hands of Syntarsus/ Coelophysis, Plateosaurus englehardti, and Lesothosaurus diagnosticus, all show loss or reduction of digits IV and V, with Syntarsus/Coelophysis showing loss of digit V. Therefore, it seems clear that the theropod dinosaurs derive their hands from digits I, II, and III.

The paleontological evidence for the derivation of the wings and feet of birds is completely lacking, since there is no known predecessor of Archaeopteryx. Therefore, the fossil record cannot answer this question. A study of the embryology of reptiles and birds was done to determine from which digits the wings and feet of birds are derived. The authors of the study showed that during the development of the hand (or wing) of the bird, there was a transient appearance of digit V, which did not occur in the alligator and turtle, which develop all five digits. These results clearly indicate that the bird's wing is derived from digits II, III, and IV.

A comparison of the feet and wings of developing bird embryos showed that both demonstrate a transient appearance of digit V. In the feet, both digits I and V appear transiently before disappearing. These results clearly indicate that the bird's wing and feet are derived from digits II, III, and IV. The evidence is quite clear that the bird feet and wings could not have developed from digits I, II, and III of the theropod dinosaurs. The results of the recent studies show that the hands of the theropod dinosaurs are derived from digits I, II, and III, whereas the wings of birds, although they look alike in terms of structure, are derived from digits II, III, and IV. If birds were descended from the theropod dinosaurs, one would expect homologous structures to be derived from comparable regions. One could propose that bird wings were originally derived from digits I, II, and III, but later developed another fourth digit, while the first digit regressed. However, there is no fossil evidence that this ever happened (and would be extremely unlikely, since the bird wing was fully developed, even in Archaeopteryx). (Archaeopteryx is discussed in more detail below in section "A Look at the Archaeopteryx Fossils.")

Another study has revealed that the lungs of birds, mammals, and reptiles are vastly different in terms of morphology and function. The lungs of mammals and birds are far more efficient than those of reptiles, which allows for prolonged periods of intensive physical activity. The lungs of mammals consists of millions of alveoli (see Figure 7.13), which are highly vascularized air sacs. The degree of vascularization (the capillary blood supply surrounding the alveoli) and the large surface area allow for efficient exchange of oxygen and carbon dioxide, which allow for mammals' high metabolic rate. The lungs of reptiles are termed "septate," consisting of the equivalent of a large single alveolus divided by vascularized ingrowths, or septae. The bellows-like septate lung of the reptile is poorly vascularized, which prohibits endothermic ("warm-blooded") metabolism. The avian (bird) lung is also a septate lung, but consists of a series of extensive, highly vascularized air sacs, which extend into both the thoracic (chest) and abdominal cavities.

Both mammals and reptiles possess a diaphragm, the muscle separating the thoracic and abdominal cavities. The reptile breathes through contraction of the diaphragmatic muscles, which are attached to the pubis and the liver. This contraction pulls the liver further down into the abdominal cavity, therefore enlarging the thoracic cavity, which expands the lungs. The diaphragm of the reptile must completely separate the thoracic and abdominal cavities for this mechanism to work. Any gap in its integrity would result in the inability to induce a partial vacuum in the thoracic cavity, which would prevent the filling of the lungs. In contrast, the bird has no diaphragm and there is no separation of the thoracic and abdominal cavities. Breathing is accomplished through contraction of the muscles of the rib cage and pelvis.

The pelvic bones of modern perching birds and Archaeopteryx reveal that both probably assisted their breathing while perching by means of muscles attached between their pubis and tail. Through pelvic and tail movements, birds are able to assist their breathing by expanding or contracting their septate lungs. Although the design of the pelvic bones of modern birds and Archaeopteryx are profoundly different, they most likely served the same purpose - to assist breathing during perching. In contrast, the pelvic bones of the theropod dinosaurs look nothing like that of either modern birds or Archaeopteryx, but look very similar to that of modern reptiles, such as the crocodile. There is no way for the pubis of modern reptiles or the theropod dinosaurs to serve as an attachment point for suprapubic muscles to serve in assisting breathing during perching. Since there are no "intermediate" theropod which possesses a pelvic structure similar to Archaeopteryx, it seems unlikely that they could have given rise to Archaeopteryx. In addition, the fossil evidence clearly demonstrates that the theropod dinosaurs lack the avian jointed or hinged ribs and expansive sternum - all of which are necessary to maintain air flow in the avian lung. Therefore, it seems unlikely that the theropod dinosaurs could have given rise to modern birds either.

One study provided evidence in the soft tissue that theropods had the same kind of compartmentalization of lungs, liver, and intestines that is found in a crocodile. The fossil clearly shows the demarcation of the thoracic and abdominal cavities. A cross-section of a neonatal alligator showed a nearly identical pattern. The results indicate that the theropod dinosaurs almost certainly possessed a diaphragm, which separated the thoracic and abdominal cavities. This would make the theropod dinosaurs unlikely candidates as ancestors of modern birds, which possess no diaphragm.

It is virtually impossible for an animal that breathes by means of a diaphragm to evolve into an animal which breathes the way modern birds do, because the hypothetical intermediate creature would be severely hampered in its ability to breathe. The researchers concluded that: "Recently, the conventional view has held that birds are direct descendants of theropod dinosaurs. However, the apparently steadfast maintenance of hepatic-piston diaphragmatic lung ventilation in theropod throughout the Mesozoic Era (250 – 65 mya) poses a problem for such a relationship. The earliest stages in the derivation of the avian abdominal air sac system from a diaphragmatic-ventilating ancestor would have necessitated selection for a diaphragmatic hernia [or hole] in taxa transitional between theropod and birds. Such a debilitating condition would have immediately compromised the entire pulmonary ventilatory apparatus and seems unlikely to have been of any selective advantage."

Another study shows that the theropod dinosaurs did not possess the correct skeletal structure or lung structure to have evolved into birds. The evolution of theropods into birds would have required the introduction of a serious handicap (a hole in their diaphragm), which would have severely limited their ability to breathe.

There are other problems with the "birds are dinosaurs" theory. The theropod forelimb is much smaller (relative to body size) than that of Archaeopteryx. (Archaeopteryx is discussed in more detail below in section "A

Look at the Archaeopteryx Fossils.") The small "proto-wing" of the theropod is not very convincing, especially considering the rather hefty weight of these dinosaurs. The vast majority of the theropods lack the semilunate wrist bone, and have a large number of other wrist elements which have no homology to the bones of Archaeopteryx. In addition, in almost all theropods, nerve V1 exits the braincase out the side, along with several other nerves, whereas in birds, it exits out the front of the braincase, though its own hole.

There is also the problem that birds are warm blooded. The evidence for warm-blooded dinosaurs has been thoroughly dismantled lately and Archaeopteryx has been shown to be a bird in the modern sense, with fully developed elliptical wings similar to modern woodland birds, and asymmetric flight feathers that form individual airfoils, a flight scapula/coracoid arrangement, and a reserved hallux, found only in perching birds, but is not known in any dinosaur. There is also the minor problem that the vast majority of the theropods appeared after the appearance of Archaeopteryx. (See Figure 7.10. Archaeopteryx is discussed in more detail below in section "A Look at the Archaeopteryx Fossils.")

The conclusion reached from these studies indicate the problems with the evolution of birds from theropods is virtually incompatible with the evidence. Although the digital mismatch between birds and dinosaurs is anatomically the most serious problem, other versions of frame-shift hypotheses will be needed to explain such problems as the transformation of teeth and tooth replacement, the transformation of a dinosaurian septate, hepatic-piston breathing system to a bird flow-through lung, the complete abandonment of a balanced seesaw body plan to the avian model, and the reelongation of already foreshortened forelimbs, our present knowledge indicates that there are two requisites for flight origin: small size and high places. Also, it must be explained why these superficially birdlike theropods only occur in the fossil record 30 to 80 million years after the appearance of the earliest known bird, which is already well developed, and why Triassic theropods are devoid of birdlike features.[7-70c]

While those who hold to the view that birds evolved from dinosaurs emphasize the similarities that the Archaeopteryx skeleton has to dinosaurs, those who do not believe that birds evolved from dinosaurs claim that the Archaeopteryx skeleton has too many bird-like features to make it a coelurosaur. For instance, Archaeopteryx has a wishbone (furcula) and bird-like feet. This means that it is not merely a feathered dinosaur but something quite different. There are other differences between dinosaurs and birds:

Dinosaurs had serrated teeth, while birds have peg-like teeth. Bird feet have reversed toes used for perching in branches—something no dinosaurs has been seen to have. Meanwhile, dinosaurs had a characteristic joint in their lower jaws for grasping prey—something never found in birds.[7-70d]

The researchers said the newest findings are more consistent with birds having evolved separately from dinosaurs and developing their own unique characteristics, including feathers, wings and a unique lung and locomotion system. There are some similarities between birds and dinosaurs, and it is possible, they said, that birds and dinosaurs may have shared a common ancestor, such as the small, reptilian "theodonts," which may then have evolved on separate evolutionary paths into birds, crocodiles and dinosaurs. The lung structure and physiology of crocodiles, in fact, is much more similar to dinosaurs than it is to birds. It is possible that dinosaurs and birds may not have had a common ancestor somewhere in the distant past. However, it is quite possible that dinosaurs and birds are directly related, as is routinely found in evolution. It just seems pretty clear now from the latest research that birds were evolving all along on their own and did not descend directly from the theropod dinosaurs, which lived many millions of years later.[7-70e]

Creationist and Intelligent Design View

The theory that birds descended from reptiles (dinosaurs) is highly questionable since a recent study (that appeared in the journal, Science, Volume 278, no. 5338, dated 10/24/1997) has demonstrated that the bones that make up the wings and feet of birds and the theropod dinosaurs are not derived from the same digits (Figure 7.4).[7-71] This issue of Science journal (Science 24 October 1997: Vol. 278 no. 5338 pp. 666-668 DOI: 10.1126/science.278.5338.666) states (in part):

> *Theropod hands retain only digits I-II-III, so digits of the modern bird hand are often identified as I-II-III. [A Theropod consist of carnivorous dinosaurs that had short forelimbs and walked or ran on their hind legs.] Study of the developing manus [forelimb of a vertebrate, including the carpus and the forefoot or hand] and pes [foot or footlike part] in amniote embryos, including a variety of avian species, shows stereotyped patterns of cartilage condensations. A primary axis of cartilage condensation is visible in all species that runs through the humerus into digit IV. Comparison to serially homologous*

elements of the hindlimb indicates that the retained digits of the avian hand are II-III-IV . . .

The identity of digits in modern birds as I-II-III gained acceptance because the phalangeal formula [any of the bones forming the fingers or toes] of Archaeopteryx, an undisputed early avian, coincides with digits I-II-III of the generalized archosaur hand [2-3-4-5-3]. Phalangeal formulae are widely variant among many taxa, however, and individual specimens of Archaeopteryx have varying phalangeal formulae in the pes [foot] . . . the transition to modern birds from Archaeopteryx requires loss of additional phalanxes [Any of the long bones of the fingers or toes, numbering 14 for each hand or foot: two for the thumb or big toe, and three each for the other four digits.] to reach an avian phalangeal formula of 2,2,1 whether digits are numbered I-II-III or II-III-IV. The phylogenetic and ontogenetic flexibility of phalangeal number render this a dubious criterion for identifying the homology of digits between higher taxa . . . The developmental evidence of homology is problematic for the hypothesized theropod origin of birds. This conflict pivots on the significance awarded to different types of data in the identification of homology. Comparative ontogenetic data suggest that a conserved developmental program is causally involved in patterning the amniote limb. The identification of early embryonic topographic landmarks and the connectivity of cartilage precursors permits the identification of specific digits as they develop in the pentadactyl hand . . .

It is parsimonious [presumptuous] to assume that the theropod limb developed with a typical primary axis through distal carpal IV followed by typical development of the digital arch and digits III, II, and I, followed by a subsequent regression of the precursors of the fourth digit. Strong reduction of digit IV after its precocious appearance is seen in some lizards. The alternative would be an entirely new developmental program. It is unlikely that a shift between the typical amniote mode of development that generates digit IV through the primary axis, to a limb that develops digit III through a convergent primary axis, would maintain the pattern of cartilage condensation that is identical in avian, crocodilian, chelonian, and mammalian limbs, and the consistent patterns of gene expression between chicken and mouse limbs . . .

As the primary axis invariably gives rise distally to digit IV in amniotes, it serves as a consistent marker of digital identity and assigns the homologies of the reduced bird hand as digits II-III-IV. A variation of this pattern wherein the primary axis runs through digit III, would eliminate any phylogenetic significance from the morphological and molecular similarities in amniote limb development. If such a condition could be demonstrated, patterns of limb development would have to be decoupled from phylogeny, and this stereotypic pattern of development accepted as convergence . . .

The discrepancies that arise between different methods draw attention to a central problem of evolutionary biology, the distinction between homoplasy and synapomorphy. It remains an open question how heavily to weigh developmental characters in phylogenetic reconstruction. The inclusion of fossil characters is essential to our understanding of evolution. However, until we disqualify developmental patterns as a means of establishing homologies, the developmental patterns that identify avian digits as II-III-IV, cannot be ignored. [7-72]

The Origin of Wings with Feathers

Evolutionist view

Since the early 1990's, a great deal of fossils have been discovered in China that support the theory that birds evolved from small, two-legged, meat-eating dinosaurs (see section "Recent Discoveries of Ancient Bird-like Dinosaurs" for further discussion of recent discoveries). Nearly a dozen different species of these reptiles have bird-like features, including down (fluff) feathers and flight (vane) feathers. Chemical analysis reveals that the feathers are made of the same basic protein (keratin) found in reptilian scales and mammalian hair. However, feathered dinosaurs did not fly. Down feathers first evolved for insulation and temperature regulation, allowing these reptiles to become more active. Flight feathers then appeared on their arms and tails. By flapping vigorously, this innovation allowed for "wing" assisted running and jumping, similar to a partridge today. It also permitted feathered dinosaurs to parachute and glide from trees. Eventually, these feathers increased in length and number, and flying birds appeared.[7-73]

Small theropods related to Compsognathus (e.g., Sinosauropteryx) probably evolved the first feathers. (Compsognathus is also discussed in section above, "Compsognathus") These short, hair-like feathers grew on their heads, necks, and bodies and provided insulation. The feathers seem to have had different color patterns as well, although whether these were for display, camouflage, species recognition, or another function is difficult to tell.

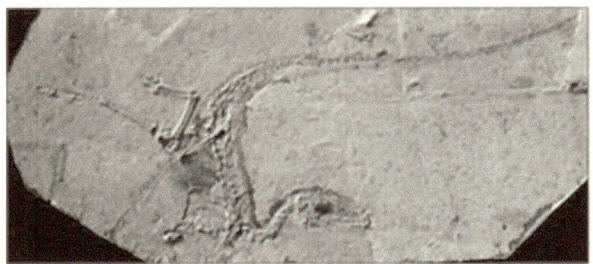

This fossil of Sinosauropteryx preserves evidence of hair-like feathers.

In theropods even more closely related to birds, like the oviraptorosaurs, there are several new types of feathers. One is branched and downy. Others have evolved a central stalk, with unstructured branches coming off it and its base. Still others (like the dromaeosaurids and Archaeopteryx) have a vane-like structure in which the barbs are well-organized and locked together by barbules (see Figure 7.7). This is identical to the feather structure of living birds.

Creationist View

Evolutionists point out that since feathers and scales are both made of the same protein, keratin, they assume that feathers evolved from scales. This, however, does not prove that reptiles evolved into birds. Keratin is a strong, lightweight, very useful protein that is found in many animals. It is in the dead outer skin layer, and in horns, hair, feathers, hooves, nails, claws, bills, etc.[7-74] Also, feather proteins (Q-keratins) are biochemically different from skin and scale proteins (A-keratins).[7-75] Feathers and scales attach to the skin differently and come from different genes on the chromosomes. There is no evolutionary relationship between the two.[7-76]

Feathers have been considered to be unique to birds for a long while. Certainly all living birds have feathers of some kind, while no living creature other than birds has been found to have a cutaneous appendage even remotely similar to a feather. Since most evolutionists believe that birds evolved from dinosaurs, or at least closely related to them, there has been an intense effort to find dinosaur fossils that show some suggestion of feathers or "protofeathers." Dinosaurs are reptiles, and so it is not surprising that fossil evidence has shown them to have a scaly skin typical of reptiles. For example, a recently discovered well-preserved specimen of Compsognathus (a small theropod dinosaur of the type believed to be most closely related to birds) showed unmistakable evidence of scales but no feathers.[7-77]

If birds evolved from dinosaurs or any other reptile, then feathers must have evolved from reptilian scales (or from beneath the scales). Since many evolutionists believe that feathers evolved from scales (while others believe feathers evolved beneath scales), they often claim that feathers are very similar to scales. Feathers have been described as a "horny outgrowth of skin peculiar to the bird but similar in structure and origin to the scales of fish and reptiles." In actual fact, feathers are profoundly different from scales in both their structure and growth. Feathers grow individually from tube-like follicles similar to hair follicles. Reptilian scales, on the other hand, are not individual follicular structures but rather comprise a continuous sheet on the surface of the body. Thus, while feathers grow and are shed individually (actually in symmetrically matched pairs), scales grow and are shed as an entire sheet of skin. The feather is made up of hundreds of barbs, each bearing hundreds of barbules interlocked with tiny hinged hooklets. This incredibly complex structure does not show the slightest resemblance to the relatively simple reptilian scale.[7-78]

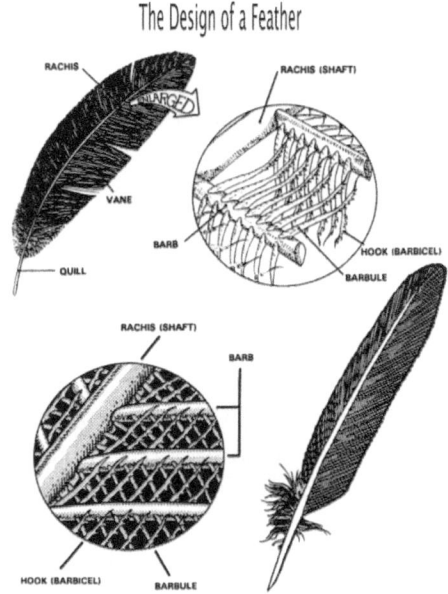

Figure 7.7 The Architecture of a Feather [7-79]

Feathers are unique to birds, although, according to mainstream evolutionary theory, some reptiles that had scales developed feathers that replaced the scales. (Some evolutionists believe feathers developed under the reptilian scales rather than directly from scales.) Out from the shaft of a feather are rows of barbs. Each barb has many barbules, and each barbule has hundreds of barbicels and hooklets (Figure 7.7). After a microscopic examination of one pigeon feather, it was revealed that it had several hundred thousand barbules and millions of barbicels and hooklets. The hooks hold all the parts of a feather together to make flat surfaces or vanes. [7-80]

A bird's flight feather is undoubtedly complex. A bird's feathers are extremely light and structurally sound, much more versatile than the rigid structure of an aircraft's wing—and far more readily repaired or replaced when damaged. The intricacy of the design that allows this can be appreciated by putting the feather under a microscope (Figure 7.7). It will seem that each parallel barb, slanting diagonally from the shaft, is not hair like, but appears as a miniature replica of the feather itself, with numerous smaller side branches, or barbules, that overlap those of the neighboring barbs in adhering to one pattern. These in turn have tiny projections called barbicels, many of which are equipped with minute hooks that neatly hold everything in place. The single pigeon feather under scrutiny may have several hundred thousand barbules and millions of barbicels and hooklets. Evolutionists attempt to explain how feathers could have evolved from scales of reptiles. They first describe the intricacy and perfection of a bird's feather. They then claim that these feathers developed from a loose, hanging, frayed scale. No such "scales" have been found anywhere in the fossil record. Loosely hanging and frayed scales would have meant they were no longer "equipped" to survive in their natural habitat and most likely would have been eliminated by natural selection.[7-81] (Natural selection is discussed in Chapter 9.)

Evolutionists are not adequately able to explain how feathers could have developed by natural means. The problem with the origin of feathers is that they had to be developed in advance so that, later on, birds could use them to fly. Evolutionists claim that feathers were originally in the form of down, and were used to keep animals warm. These primitive feathers were supposed to have served an evolutionary purpose at the time; only later did the first birds turn these downy-like feathers into wings. This argument is incorrect because wing feathers are entirely different from the downy variety, because they are provided with thousands of microscopic hook-like fasteners. These allow the wings to fold when not in use, and engage when the wing is stretched out ready for flight, thus turning the whole assembly into a continuous sheet to catch the wind. Evolutionists have no explanation for complex structures (such as feathers) which had to arise before there were any creatures that could utilize them. Their existence indicates that there must be an overriding purpose behind them. [7-82]

Recent Discoveries of Ancient Bird-like Dinosaurs

Figure 7.8 Sinornithosaurus (Artist Depiction)

Evolutionist description

The "four-winged" dinosaur Microraptor gui had long feathers on both its fore and hind limbs. These feathers are clearly visible in the fossil, about 120 million years old. It's not clear whether this animal could fly or only glide, but the rear "wings" almost certainly helped it land. In 1999, paleontologists discovered a dinosaur fossil (dromaeosasur) in northeast China (in the Yixian rock formation near Sihetun) named "Sinornithosaurus millenii," meaning "Chinese bird lizard of the millennium" (Figure 7.8) This dromaeosaurid dinosaur dates from about 124 to 125 million years ago (mya), from the Lower Cretaceous Period (late Barremian) (144-65 mya). Dinosaurs dominated the Earth during the Jurassic period (213-144 mya) (See Chapter 3 "The Strata Layers of the Earth") Sinornithosaurus (Figure 7.8) appears to have been covered with a coat of down-like fibers or early proto-feathers, as they are called. Sinornithosaurus is the fifth non bird-like feathered dinosaur type discovered. The features of the skull and shoulder are very similar to those of Archaeopteryx and other birds, which proves that the earliest dromaeosaurs were more like birds than the later dromaeosaurs were. Prior to the discovery of Sinornithosaurus other dinosaurs found in the same location were also covered with fibers. The features of these dinosaurs (dromaeosaur) may not have had scaly, reptile-like skin, but may have had a softer, downy coat. No fossil skin impressions of these dinosaurs had been found prior to the discovery of Sinornithosaurus.[7-83]

Sinornithosaurus clearly shows impressions of filamentous feathers, especially on the head and forelimbs. Its whole body was covered with long, thin feathers that were so small that they couldn't have helped it fly. Sinornithosaurus showed two features that indicate it had early feathers. **First**, several filaments were joined together into "tufts", like the structure of down feather. **Second**, a row of filaments (barbs) were joined together to a main shaft (rachis), making them similar in structure to those of normal bird feathers. However, they do not have the secondary branching and tiny little hooks (barbules) that modern feathers have, that allow the feathers of modern birds to form a distinct vane. The feathers covered the entire body in fossil NGMC—91, including the head in front of the eye, the neck, wing-like sprays on the arms, long feathers on the thighs, and an elongated-shaped fan on the tail like that of Archaeopteryx (See section "A Look at the Archaeopteryx Fossils" for discussion of Archaeopteryx fossil).[7-84]

Evolutionist View with Critical View of the Dinosaur to Bird Evolution

A fossil find from the Yixian formation in China revealed the theropod dinosaur called Sinosauropteryx, which was nicknamed the "feathered dinosaur." Subsequent studies have indicated that the feathers were probably "frayed collagenous fibers beneath the skin." (Anne Gibbons. 1997. Plucking the feathered dinosaur. Science 278: 1229) However, the remarkable preservation of the specimen reveals not only skin, but some of the internal organs as well.[7-84a]

Creationist View

The evolutionist descriptions of the Sinosauropteryx, Sinithosaurus and Sinornithosaurus millenii (Figure 7.8) fossils are debatable. The question is: Were the filamentous feathers, long, thin feathers-feathers, early feathers, really feathers, or were they long hairs? A filament is defined as "A fibril, fine fiber, or threadlike structure." This definition does not appear to describe feathers in the making, or something in the process of developing into a feather, but appears more like hairs. It is difficult to determine exactly what the covering was by merely examining a fossil.[7-85]

Sinosauropteryx is the earliest known feathered dinosaur. Although evolutionists claim many dinosaurs from Liaoning Province in northeastern China represent feathered dinosaurs, there is no evidence for this. Sinosauropteryx is quite non bird-like, as it has no structures that even resemble feathers. The protofeathers that are described for Sinosauropteryx and similar fossils are filamentous, interlaced structures, often referred to as dino-fuzz. Science now recognizes that these are actually connective tissue fibers (collagen) found in the deep dermal layer of the skin.[7-86]

In 2001, evolutionists made the claim that the fossil of the dinosaur Sinornithosaurus millennii ("Chinese bird lizard of the millennium") (Figure 7.8) showed evidence of rudimentary feathers, therefore providing evidence that Archaeopteryx's full-blown, modern feathers (Figure 7.9), did evolve from more primitive creatures. Contrary to this claim, it should be noted that the Sinornithosaurus fossil (Figure 7.8) is younger than that of Archaeopteryx. The Sinornithosaurus fossil is presumed to be 124 million years old, while the fossil of that of Archaeopteryx is thought to be 150 million years old. (See section below "A Look at the Archaeopteryx fossils," "Creationist View," Figure 7.10) If this is true, the feathers of the Archaeopteryx could not have evolved from the proto-feathers that did not appear until 26 million years later. The supposed proto-feathers (dinofuzz) found imprinted in the Sinornithosaurus fossil (Figure 7.8) are therefore more recent than fully developed, modern feathers. It may be possible that Sinornithosaurus existed prior to Archaeopteryx, however, since no older Sinornithosaurus fossils have been discovered, the fossil evidence does not currently support this. The only conclusion that can be reached is that dinofuzz could not possibly be a precursor of actual feathers. The so-called elementary feathers (or dinofuzz) that were found imprinted in the Sinornithosaurus fossil cannot be considered filaments, but are only fossilized impressions of filaments. This makes it virtually impossible to obtain an accurate analysis of what the impressions really are. It has been noted that the wispy hairlike structures are so different than modern bird feathers that it is doubtful that the filaments could even be related to feathers at all.[7-87]

A Look at the Archaeopteryx Fossils

PUBLIC DOMAIN IMAGES BING.COM

Figure 7.9 Archaeopteryx (Artist depiction)

Evolutionist View

Archaeopteryx (Figure 7.9) is the oldest-known fossil animal that is generally accepted as a bird. The eight known fossils are dated to be from the Late Jurassic Period (213-144 mya) (161 million to 146 million years ago).[7-88] All eight fossils were found in an area of southern Germany that is believed to have been much closer to the equator than it is now.[7-89] This area would have been tropical in nature as the fine-grained Jurassic strata limestone implies. Fossils in limestone are extremely well preserved. Several fossils from this area show clear impressions of feathers. Archaeopteryx is a good example of a transitional form between reptiles and birds. These fossils are well preserved. All the features, except feathers, are either reptilian or transitional. Its skeleton is definitely reptilian. It has reptilian teeth, claws on its wings, and a many-jointed tail (the joints fused as one in modern birds). The pelvic section is in three parts like reptiles, while in birds the pelvic bones and the sacro-vertebrae are all fused into one. Archaeopteryx, however, lacked a bird-like breastbone and strong breast muscles.[7-90] This kept Archaeopteryx from true flight and limited it to gliding. Archaeopteryx's sternum (breastbone) was made of cartilage rather than bone. This would have made flight difficult for Archaeopteryx, because cartilage's elasticity would have resulted in a lot of wasted energy. Modern birds avoid this problem by having bony sternums, which both Microraptor and Deinonychus also had. It has been stated that Archaeopteryx could go airborne, however, most likely it was not adept at flying. Another view is that Archaeopteryx was a true flyer. However, it was not as skilled as most modern birds. This is because its sternum was flat or slightly keeled.[7-91] Archaeopteryx is close to the ancestry of modern birds, and it shows most of the features one would expect in an ancestral bird. But, it may not be the direct ancestor of living birds, and it is not clear how much evolutionary divergence was already found in other birds at the time.[7-91b]

A recent study (March 2018) has concluded that Archaeopteryx most likely utilized an alternative flight style owing to major differences in the shape of its shoulder and arm bones, compared to modern birds. And crucially, it did not have a breast bone, which anchors the major flight muscles of living birds. Birds evolved from dinosaurs, but the pathway taken from terrible lizard to the graceful birds that are seen today has not yet been established, nor is it obvious how the capacity of flight first emerged in dinosaurs. This same study established that Archaeopteryx could actually fly, although its flying style was unlike anything seen today in modern birds. Archaeopteryx had the physical attributes and bone structure required for active flight, and that Archaeopteryx was meant to fly, although only in bursts over short distances. But because the now-extinct dinosaur Archaeopteryx featured a bone structure not seen in modern birds, paleontologists are not sure what Archaeopteryx's flying style actually looked like. It is quite possible that Archaeopteryx, being 150 million years old, is the oldest potentially free-flying dinosaur known. Unfortunately, scientists do not know if Archaeopteryx used its wings for passive gliding or powered, active flight. Another possibility is that Archaeopteryx was a ground-dwelling animal that did not fly at all, using its wings for something else, like snatching prey, leaping, or sexual displays.

It is difficult to infer physical capacities, movement styles, and behavior from fossilized etchings carved onto rock. Researchers concluded from their study, that Archaeopteryx exhibited a physical architecture and bone-structure consistent with the capacity for flight, and that these traits most closely matched living birds who use their wings to fly short distances in bursts, such as pheasants. The bones of Archaeopteryx plot closest to those of birds like pheasants that occasionally use active flight to cross barriers or dodge predators, but not to those of gliding and soaring forms such as many birds of prey and some seabirds that are optimized for enduring flight. Pheasants do not fly like eagles or albatrosses, but they are not completely grounded. When startled, pheasants burst into the sky at speeds reaching 38 to 48 mph (60 to 77 km/ hr), and when chased they can hit 60 mph (96 km/hr). It is not yet known for sure if Archaeopteryx was as capable in the air as the modern pheasant, but researchers are now a step closer to finding out. Much of Archaeopteryx's unique anatomy suggests it used a flapping motion and an aerial posture not seen in modern birds. As to what its flight style might have looked like, it is still not exactly known.[7-91a]

Archaeopteryx has the exact features one would expect of a transitional form. It has dinosaurian features including a tail, ribs without uncinate processes (a small part of the pancreas - a large gland behind the stomach which secretes digestive enzymes into the duodenum), small sternum (a long flat bone located in the central part of the chest), pubic peduncle (a stalk-like part by which an organ is attached to an animal's body), cervical ribs (A short, floating, rudimentary rib attached to the lowest neck vertebra on one or both sides), and an unfused backbone. It had intermediate features including small teeth, partially fused metatarsals (bones in an animal's hind limb, especially the bones of the foot), and downward facing pubic bones (the ventral and anterior of the three principal bones composing either half of the pelvis [the large bony structure near the base of the spine to which the hind limbs or legs are attached in humans and many other vertebrates]). Also, it had bird-like features including simple rear hinge joints, inturned femur heads, fused clavicles (collarbone) and feathers. Archaeopteryx had a tail and an unfused backbone and a pubic peduncle (a stalk-like part by which an organ is attached to an animal's body) and ribbed cervical ribs and lack uncinate processes (a small part of the pancreas [a large gland behind the stomach which secretes digestive enzymes into the duodenum]). Apart from Archaeopteryx, these features are normally only found in reptiles, not birds. Some creationists claim that Archaeopteryx is a forgery, which is not true. (See section "Could Archaeopteryx Fossils be Fakes?") A recent discovery is Unenlangia comahuensis, which is a link between Archaeopteryx and the Dromaeosauridae.[7-92]

Intelligent Design View

Archaeopteryx does possess certain skeletal reptilian features, such as teeth, a long tail, and claws on its wings. However, regarding flight, the most characteristic feature of birds, Archaeopteryx was already a true bird. On its wings there were flight feathers as fully developed as any modern bird, and recent research reported in 1979 suggests that it was as capable of powered flight as a modern bird. It is admitted that one could argue that Archaeopteryx does possess some reptilian features and thus could hint of reptilian ancestry, but surely hints do not provide a sufficient basis upon which to secure the concept that a reptile evolved into the Archaeopteryx, or that Archaeopteryx was the forerunner of the modern bird. There is no fossil record of a series of transitional forms from ordinary terrestrial reptiles leading up to gliding types until the bird type fossils are found (Compare Figure 7.6a with Figure 7.10).

Modern birds differ greatly from reptiles in many physiological and anatomical characteristics, particularly, for example, in their central nervous, cardiovascular (Figures 7.5) and respiratory systems (Figure 7.13). Because information about soft tissue is difficult to obtain from its skeletal remains, it is not known for sure to what extent Archaeopteryx was avian in regards to its major organ systems. The evidence available from the cranial endocast of Archaeopteryx appears that its brain was essentially avian in all important aspects.

The evidence indicates that Archaeopteryx exhibits a typical avian cerebral hemisphere and cerebellum (the part of the brain involved in balance and the coordination of fine motor activities), a part of the brain proportionally larger in birds than in any other class of vertebrates and generally considered an adaptation necessary for the control of the highly complex motor activities involved in powered flight. The possession of an essentially avian central nervous system lends further support to this idea, based on the basically modern form of its flight feathers and wing, that Archaeopteryx was as capable of powered flight as a typical modern bird. If Archaeopteryx was capable of powered flight, it could be assumed that it also possessed, of necessity, a fully avian heart (Figure 7.5), circulatory and respiratory system (Figure 7.13) to supply the vastly increased demand for oxygen that occurs during powered flight.[7-93]

Creationist View

Archaeopteryx gives evidence of being a regular bird in every way, except that it differs in certain features: **(1)** the lack of a sternum, **(2)** three digits on its wings, and **(3)** a reptile-like head, but there are explanations for all three points. **(a)—Lack of a sternum** (also known as the breastbone, is a long flat bone located in the central part of the chest). Archaeopteryx had no sternum, but although the wings of some birds today attach to the sternum, others attach to the furcula (also known as the wishbone, is a forked bone found in birds and some dinosaurs, and is formed by the fusion of the two clavicles [collarbone]). Archaeopteryx had a large furcula, so this would be no problem. **(b)—Digits on its wings**: Archaeopteryx had three digits on its "wings." Other dinosaurs have this also, but so do a few modern birds. This includes the hoatzin (Opisthocomus hoatzin), a South American bird, which has two wing claws in its juvenile stage. In addition, it is a poor flyer, with an amazingly small sternum, such as Archaeopteryx had. The touraco (Touraco corythaix), an African bird, has claws and the adult is also a poor flyer. The ostrich has three claws on each wing. Their claws appear even more reptilian than those of Archaeopteryx. **(c)—The shape of its skull**. It has been said that the skull of Archaeopteryx appears more like a reptile than a bird, but investigation by some researchers say it is shaped more like a bird.[7-94]

Bones of modern birds have been found in the same type of rock strata (the Jurassic Period - 213-144 mya) in which Archaeopteryx was found. (They have been found in eastern Colorado.) According to evolutionary theory, this cannot be, for millions of years ought to be required for Archaeopteryx to change into a regular bird. If it was alive at the same time as modern birds, how can it be their ancient ancestor? Birds have also been found in the Jurassic Period limestone beds of Utah. No bird bones of any type have been found below the late Jurassic Period, but within the Jurassic Period, they have been found in strata with Archaeopteryx, and now below it: Two crow-sized birds were discovered in the Triassic Dockum Formation in Texas (See Protoavis section). Because of the strata they were located in, those birds would, according to evolutionary theory, be 75 million years older than Archaeopteryx (Figure 7.10).[7-95]

In 1861, a feather was found from late Jurassic Period strata (213-144 mya). Soon after, in the same quarry, a fossil bird was found with the head and neck missing (The London Specimen – See Figure 7.11). The name Archaeopteryx ("ancient or early wing") had been given to the feather and so the same name was given to the bird. In 1877 (possibly 1874 or 1875), a second specimen was said to have been discovered close to the first, but this one had a neck and head (The Berlin Specimen – See Figure 7.11). In that head were 13 teeth in each jaw; the head itself had the elongated rounded shape of a lizard head. Including the feather fossil, there are six specimens of Archaeopteryx in the world. All six came from that same German limestone area. In addition to the single feather specimen and the first two more complete specimens, three other specimens are quite faint and difficult to use. It is almost impossible to tell what they are. Aside from the single feather specimen, the other specimens are located at London, Berlin, Maxburg, Teyler, and Eichstattall in Germany. They all came from the same general area. Only the first fossilized skeleton (the "London specimen" discovered in 1861) and the second one (the "Berlin specimen" discovered sometime between 1874 and 1877) are defined well enough to be usable (Figure 7.11). All six of those specimens were found in the Solnhofen Plattenkalk of Franconia, Germany, near the city of Eichstatt.[7-96] In April 1993, a seventh specimen, The Solnhofen-Aktien-Verein specimen (also known as the Munich Specimen) was found near Langenaltheim, Germany.[7-107a] This fossil is generally usable for research. The most recent Archaeopteryx fossil discovered was the Thermopolis Specimen, which was discovered in Bavaria, Germany, December 2, 2005.[7-113] This fossil has the best-preserved skull and feet of the twelve Archaeopteryx specimens found to date. The head, feet and the remains of a single wing of this fossil were preserved, although most of the neck and the lower jaw were not preserved.[7-113a] (For a complete list of Archaeopteryx fossils, see Reference # [7-96a].)

To believe that Archaeopteryx is a missing link between reptiles and birds, one must believe that scales evolved into feathers for flight or feathers grew out from under the scales or the offspring of reptiles grew feathers or feather-like structures instead of scales. Two basic theories have been developed by evolutionists to justify this claim. **(1)** One theory states that wings and feathers developed in tree-dwelling reptiles that jumped from treetops to escape their enemies or to pursue food. Thus, the feathers developed as a mechanism to help ease the animal's fall. **(2)** The second theory, on the other hand, views ancestral reptiles as ground dwellers that relied on speed for protection and for chasing prey. The development of feathered wings and tails helped to decrease wind resistance and also supplied lift to increase their speed.

Evolutionists believe that feathers gradually developed as air friction frayed the outer edges of reptilian scales. Eventually, in the course of millions of years, scales became more and more like feathers until the perfect feather emerged. Consider the meticulous engineering of feathers (Figure 7.7). The central shaft of a feather has a series of barbs projecting from each side at right angles. Rows of smaller barbules in turn protrude from both sides of the barbs. Tiny hooks, called barbicels, project downward from one side of the barbules and interlock with ridges on the opposite side of adjacent barbules. In some feathers there may be as many as a million barbules cooperating to bind the barbs into a complete feather, resistant to air penetration. In addition, the positioning of the feathers is controlled by a complex network of tendons that allow them to open like the slats of a blind when the wing is raised. As a result, wind resistance is greatly reduced on the upstroke. On the down stroke, the feathers close, providing resistance for efficient flight.[7-97] There are no clues in Archaeopteryx fossils that provide any hints as to how reptilian scales evolved into feathers. When feathers are found as fossils, they are fully developed and functional feathers. Feathers are quite complex structures, with little hooks and eyelets for zippering and unzippering them (Figure 7.7). Archaeopteryx not only had complete and complex feathers, but feathers of several different types, including the asymmetric feather characteristic of strong fliers.[7-98]

Archaeopteryx, which means "ancient wing" is said to have twenty-one specialized characteristics in common with particular kinds of dinosaurs. Multiple evidences demonstrate conclusively that Archaeopteryx is a full-fledged bird, not a missing link. Fossils of both Archaeopteryx and the kinds of dinosaurs Archaeopteryx supposedly descended from have all been found in a fine-grained German limestone formation said to be Late Jurassic Period (213-144 mya). The Jurassic period is said to have begun 190 million years ago, lasting 54 million years). (See section "Recent Discoveries of Ancient Bird-like Dinosaurs" "Creationist View.") Archaeopteryx is not likely to be a missing link, since birds and their alleged ancestral dinosaurs thrived during the same period of time. Other bird fossils were found in sediments classified by evolutionists as Late Triassic Period (248-213 mya), which is prior to the Jurassic Period. According to this hypothesis, these birds would have lived approximately 75 million years earlier than Archaeopteryx and, in fact, at the same time as the first dinosaurs. There is a possibility that Archaeopteryx may have existed prior to the time assigned to them, however, there is no fossil evidence to support this. This means that Archaeopteryx cannot have been the forerunner of birds since birds were already in existence.[7-99]

Evolutionists dated Archaeopteryx fossil to be about 150 million years old. (See section "Recent Discoveries of Ancient Bird-like Dinosaurs" "Creationist View") Fully formed bird fossils have been dated at 225 million years old (See section "Protoavis" for further discussion). If a reptile evolved into a bird 225 million years ago, how could the Archaeopteryx fossil, dated 75 million years after a bird had already evolved, be considered a transitional link between reptile and bird?[7-100]

Fossils of birds have been found in rocks dated in the Triassic Period (248-213 mya), which is a time in the fossil record that evolutionists believe preceded most dinosaurs (See Protoavis section for further discussion). Fossil remains claimed to be of two crow-sized birds 75 million years older than Archaeopteryx have been found [*Archaeopteryx is dated at 150 mya and the two crow-sized birds are dated at 225 mya (150 + 75 mya "older than Archaeopteryx" = 225 mya)*.]. The paleontologist who found the fossils, says they have advanced avian features. [7-100a]

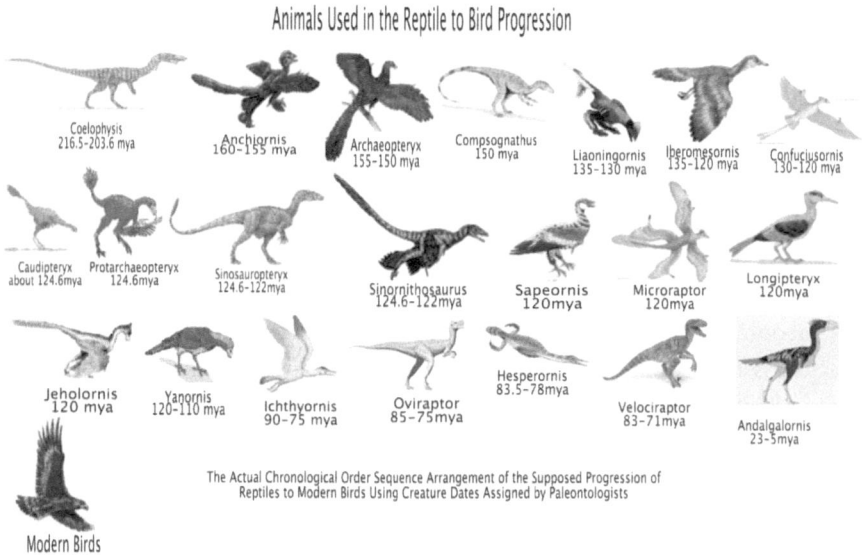

Animals Used in the Reptile to Bird Progression

Coelophysis 216.5-203.6 mya
Anchiornis 160-155 mya
Archaeopteryx 155-150 mya
Compsognathus 150 mya
Liaoningornis 135-130 mya
Iberomesornis 135-120 mya
Confuciusornis 130-120 mya

Caudipteryx about 124.6mya
Protarchaeopteryx 124.6mya
Sinosauropteryx 124.6-122mya
Sinornithosaurus 124.6-122mya
Sapeornis 120mya
Microraptor 120mya
Longipteryx 120mya

Jeholornis 120 mya
Yanornis 120-110 mya
Ichthyornis 90-75 mya
Oviraptor 85-75mya
Hesperornis 83.5-78mya
Velociraptor 83-71mya
Andalgalornis 23-5mya

Modern Birds

The Actual Chronological Order Sequence Arrangement of the Supposed Progression of Reptiles to Modern Birds Using Creature Dates Assigned by Paleontologists

Figure 7.10 Reptile to Bird Evolution

Any fossils discovered in the lower layers of the Earth cannot provide any "missing links" because fossil birds have already been found in lower layers.

By the evolutionist's own definition, a fossil qualifies as a missing link or transitional form in an evolutionary series if and only if it is found in both a morphologic series and a stratigraphic series, i.e., it must show graduation in structural features such as a "sceather" stage between scales and feathers (morphologic series), and these graduations must occur from lower to higher in a series of rock strata (stratigraphic series—Figure 7.10).[7-101]

The fact that Archaeopteryx has claws is no proof that this bird is a transitional creature between reptile and birds. There are three living birds with claws on their wings. Living birds with claws include: Hoatzin, Touraco, and Ostrich. The tail of Archaeopteryx is un-fused. This does not prove it evolved from a reptile because the penguin also has an un-fused tail vertebra.[7-102] Archaeopteryx has feathers, wings, and a beak. It has teeth in the bill, claws on the wings, no keel on the breastbone, an unfused backbone, and a long, bony tail. These are all characteristics that are normally associated with reptiles. The reptile-like features are not really as reptile-like as one might suppose. The ostrich, for example, has claws on its wings that are even more "reptile-like" than those of Archaeopteryx. Several birds, such as the hoatzin, does not have much of a keel. The penguin has unfused backbones and a bony tail. No living birds have socketed teeth, but some fossil birds do. Besides, some reptiles have teeth and some don't, so presence or absence of teeth is not particularly important in distinguishing the two groups.[7-103]

Evolutionists claim that reptiles evolved into birds. From examining the Archaeopteryx fossils, there are no clues as to how legs (or arms) evolved into wings. (See Chapter 14, section "Irreducible Complexity Explained.") When wings are found as fossils, they are completely developed, fully functional wings. That's true of Archaeopteryx, and it's also true of the flying insects, flying reptiles (pterodactyls), and the flying mammals (bats).[7-104] Archaeopteryx appears abruptly in the fossil record with wings and feathers that birds have today.[7-105] Evolutionists claim that since the Archaeopteryx lacks a keel, it was not able to fly. Actually, muscles for the power stroke in flight attach to the wishbone (furcula), and Archaeopteryx had an extremely strong furcular (wishbone). Some evolutionists (as mentioned above) believe that Archaeopteryx was a strong flier and the first bird, and not a missing link between reptiles and birds.[7-106] The initial Archaeopteryx fossil finds gave no evidence of a bony sternum (also known as the breastbone, is a long flat bone located in the central part of the chest), which led paleontologists to believe that Archaeopteryx could not fly or was a poor flyer. However, in April 1993, a seventh specimen, (The Solnhofen-Aktien-Verein specimen [7-107a]), was reported that included a bony sternum (breastbone). Thus, there is no further doubt that Archaeopteryx was suited for power flying as any modern bird.[7-107]

For reptiles to evolve into birds, the heart, lungs, brain, skeletal system, nervous system, body shape, and the reproductive system would all have to be drastically changed. (Circulatory—Figure 7.5; Respiratory—Figure 7.13) There is no evidence of any such change having taken place.[7-108]

Could Archaeopteryx Fossils be Fakes?

 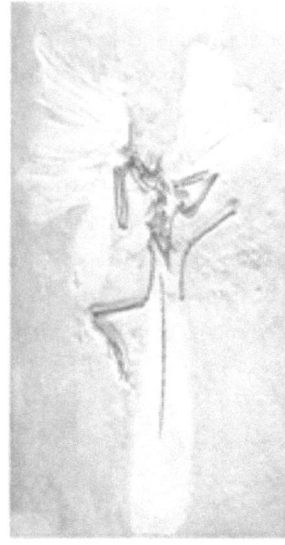

Archaeopteryx fossil
Berlin specimen

Archaeopteryx fossil
London specimen

Figure 7.11—Images of Berlin & London specimens

Creationist View

While some experts consider Archaeopteryx to have had a body like a bird, those who consider it a fake (some naturalists as well as some creationists) believe the fossilized body to be that of a reptile and that somebody took a reptile fossil and carefully added wings to it. Like the later Piltdown man (Chapter 16), Archaeopteryx seemed a perfect intermediate form. There are those (some naturalists as well as some creationists), however, who believe that Archaeopteryx was a carefully contrived fake. It would have been relatively easy to do. The nature of the hard limestone would make it easy to carefully engrave something on it. The head and body of Archaeopteryx is similar to that of a small coelurosaurian dinosaur, Compsognathus, and the flight feathers are exactly like those of modern birds. If the feathers were removed, the creature would appear to be only a small dinosaur. (See section "Evolution from Reptiles to Birds" for further discussion of Compsognathus.)

If the "London specimen" (discovered in 1861) (Figure 7.11) is examined, one will note that the flight feathers consist only of carefully drawn lines, and nothing else. It would be relatively easy for someone to take a genuine fossil of a Compsognathus, and carefully scratch those lines onto the surface of the smooth, durable limestone. All that would be needed would be a second fossil of a bird as a pattern to copy the markings from, and then inscribe its wing pattern onto the reptile specimen. All six of those specimens were found in the Solnhofen Plattenkalk of Franconia, Germany, near the city of Eichstatt. Nowhere else in the world have any Archaeopteryx specimens ever been discovered. Of all six of the specimens that have been discovered so far, only the London and Berlin specimens are usable (Figure 7.11); the rest are hardly recognizable as anything, that is, they are quite faint and difficult to use. Therefore, all the evidence pro and con must come from one or the other of these two specimens. (As an update to this, as mentioned previously, in April 1993, a seventh specimen, The Solnhofen-Aktien-Verein specimen, was discovered. There is not a great deal of information about this specimen, although it has been noted that the impression is very pronounced and has been classified as a new species, Archaeopteryx bavarica. It has been reported as possessing a small ossified sternum [breastbone], as well as feather impressions.[7-107a] For a complete list of Archaeopteryx fossils, see Reference # [7-96a].)

Ernst Haeckel (1834-1919) lived in Germany at the same time that these six specimens were found. He would have been in the prime of life at the time both the London & Berlin specimens were brought forth. Haeckel was the most rabid Darwinist advocate on the continent; and it is well-known that he was very active at the time the finds were made. He was continually seeking for new proofs of evolution, so he could use them in his lecture circuit meetings. It is known that Haeckel had unusual artistic ability that he put to work, producing pro-evolution frauds. He was fraudulently touching up and redrawing charts of ape skeletons and embryos so that they would appear to prove evolutionary theory. (See Chapter 11 for further discussion of Ernst Haeckel.)[7-109]

The feather imprints of the London Archaeopteryx specimen were forged. Evidence for this include:

* The feather impressions appear only on the slab, not on the counterslab.
* The surface texture is different between the feathered and unfeathered areas.

Slightly elevated "blobs" appear which are not always matched by depressions on the counterslab.

* The feathers show "double strike" impressions.
* Hairline cracks which pass through both bones and feathers could have formed by slight movements to the slab after the cement was in place.
* Under magnification, the limestone appears different in fossil and non-fossil areas of the specimen.
* Unknown material appears within the matrix in the fossil area.
* An x-ray chemical analysis showed chemical differences, including silicon, sulfur, and chlorine in the fossil area that were not present in the non-fossil area.

The above points indicate that the feather impressions were made by someone impressing feathers in a cement-like matrix that was added to the stone. Without the feathers, Archaeopteryx would be identified as the dinosaur Compsognathus, not as a transitional fossil (Compsognathus was discussed previously in this chapter). While Archaeopteryx had some traits in common with dinosaurs, it was clearly a bird.[7-110]

One creationist source does not consider Archaeopteryx to be a fake. They say that Archaeopteryx is "*far more useful to creationists as a real fossil, since based on uniformitarian dating methods, Archaeopteryx is older than all of the so-called feathered dinosaurs (Figure 7.10). It stands perched as an excellent example of the fact that evolutionists often ignore their own dating methods to produce transitional series.*" This same source states "*that none of cited sources here are creationist sources, so while some creationists have used this claim [that the feathers that appear on the Berlin and London Archaeopteryx fossils were forged onto the fossils], it is not a creationist claim."* [7-111]

Evolutionist View

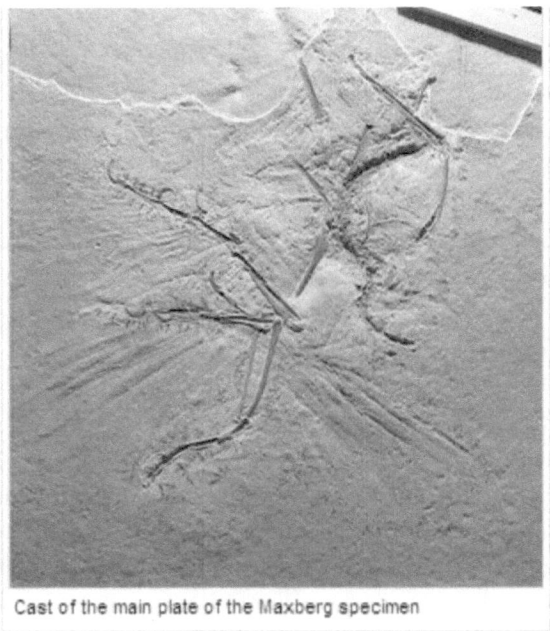

Cast of the main plate of the Maxberg specimen

Figure 7.12 Image of Maxberg Specimen (discovered in 1956)

Some claim the Archaeopteryx fossils are fakes. This is not true for the following reasons:

* There are nine other Archaeopteryx fossils discovered at different times and places under well-documented conditions (besides the London specimen that was the first labeled as Archaeopteryx). At least six of these also have unequivocal feathers. On the Maxburg specimen (Figure 7.12), the feathers continue under the bones and are overlain with dendrites that sometimes form within bedding planes, precluding the possibility of forgery. In addition, several other feathered dinosaurs have been discovered.
* Tiny fractures, infilled with calcite, extend through both feathers and bones, showing that they have the same source. They also match

perfectly from slab to counterslab, proving that the two fit together. These fractures are invisible to normal vision; a nineteenth-century forger would not even know they existed, much less be able to replicate them.

* The "double struck effect" on the counterslab is due to the fossilization method. Feather-degrading bacteria grew under the feathers, causing the sediments beneath to lithify, and so preserving a hardened feather impression. When the feathers decayed away, the sediments above pressed down to create a cast of the surface below. Evidence of this process, including lithified bacteria, is visible under high magnification and could not plausibly be forged. Other lack of detailed impressions results from the Archaeopteryx body resting on a flat surface without sinking into it much. The bulk of the fossil projected above the sea floor into the sediments that settled around and over it. When the shale split along the original seafloor surface, the upper part contained the bulk of the fossil, while the lower part showed only the impression which the body made on the sea floor. This pattern is typical of Solnhofen fossils.

* The difference in surface texture in the area of the fossils is due to the impression of the animal body.

* The elevated "blobs" are natural irregularities. There are none which don't have corresponding depressions on the counterslab. The two halves fit together well except where one surface has been destroyed by subsequent preparation.

* The double-strike impressions are not imprints; they are underlying feathers. A double-strike impression would be harder to forge than a single impression.

* The hairline cracks are infilled with calcite both in the original slab and in the area that was claimed to be cement. Plus, the cracks match between the slab and counterslab. None of this would be possible if the cracks formed after a cement layer were applied.

* Differences in appearance are due to different resolutions used in the SEM photography.

* The unknown materials are clearly not within the limestone matrix. The carbonate grains on top of them are simply dust.

* The chemical differences between the fossil and non-fossil areas are likely due to residues of preservatives applied to the fossil areas.[7-112]

The most recent Archaeopteryx fossil discovered was the Thermopolis Specimen, found about 2005.[7-113] The head, feet and the remains of a single wing were preserved, although most of the neck and the lower jaw were not preserved.[7-113a]

Various Respiratory Systems

We may cease marvelling at the embryo of an air-breathing mammal or bird having branchial slits and arteries running in loops, like those of a fish which has to breathe the air dissolved in water by the aid of well-developed branchiae.

Charles Darwin, Origin of Species,
Chapter 15—Recapitulation
And Conclusion

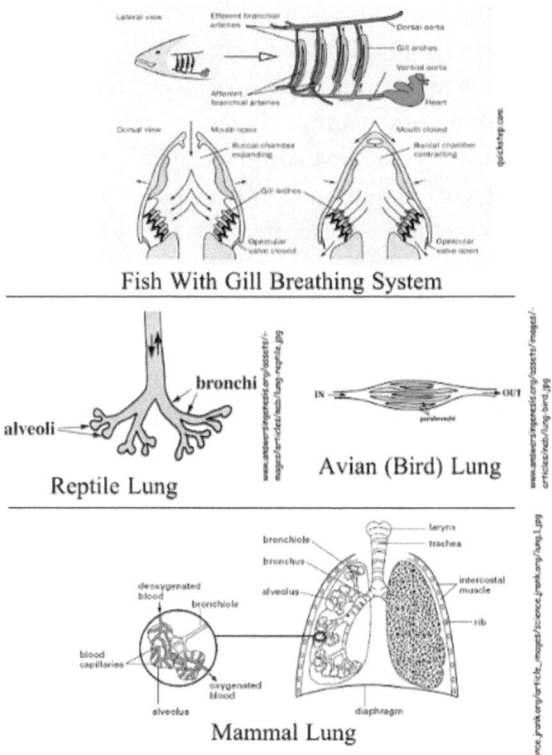

Figure 7.13 Various Respiratory Systems

Evolutionist View

In most fish respiration takes place through gills (Figure 7.13). Lungfish, however, possess one or two lungs. The labyrinth fish have developed a special organ that allows them to take advantage of the oxygen of the air, but is not a true lung. Both the lungs and the skin serve as respiratory organs in amphibians. The skin of these animals is highly vascularized and moist, with moisture maintained via secretion of mucus from specialized cells. (Vascularization refers to the development of extended blood vessels or other fluid-bearing vessels or ducts of a tissue—or—to supply an organ or tissue with blood vessels.) While the lungs are of primary importance to breathing control, the skin's unique properties aid rapid gas exchange when amphibians are submerged in oxygen-rich water.

The lungs of most frogs and other amphibians are simple balloon-like structures, with gas exchange limited to the outer surface area of the lung. Unlike mammals, which use a breathing system driven by negative pressure, amphibians employ positive pressure. The majority of salamander species are lungless salamanders, which respirate through their skin and tissues lining their mouth. The lungs of amphibians typically have a few narrow septa of soft tissue around the outer walls, increasing the respiratory surface area and giving the lung a honey-comb appearance. In some salamanders even these are lacking, and the lung has a smooth wall. In caecilians, as in snakes, only the right lung attains any size or development.

Reptilian lungs (Figure 7.13) are typically ventilated by a combination of expansion and contraction of the ribs via axial muscles and buccal pumping. Crocodilians also rely on the hepatic piston method, in which the liver is pulled back by a muscle anchored to the pubic bone (part of the pelvis), which in turn pulls the bottom of the lungs backward, expanding them. Turtles, which are unable to move their ribs, instead use their forelimbs and pectoral girdle to force air in and out of the lungs. The lung of most reptiles has a single bronchus running down the center, from which numerous branches reach out to individual pockets throughout the lungs. These pockets are similar to, but much larger and fewer in number than, mammalian alveoli, and give the lung a sponge-like texture. In tuataras, snakes, and some lizards, the lungs are simpler in structure, similar to that of typical amphibians.

Snakes and limbless lizards typically possess only the right lung as a major respiratory organ; the left lung is greatly reduced, or even absent.

Amphisbaenians, however, have the opposite arrangement, with a major left lung, and a reduced or absent right lung.

The lungs of mammals have a spongy and soft texture and are honeycombed with epithelium, having a much larger surface area in total than the outer surface area of the lung itself. The lungs of humans are a typical example of this type of lung (Figure 7.13). Breathing is largely driven by the muscular diaphragm at the bottom of the thorax. Contraction of the diaphragm pulls the bottom of the cavity in which the lung is enclosed downward, increasing volume and thus decreasing pressure, causing air to flow into the airways. Air enters through the oral and nasal cavities; it flows through the pharynx, then the larynx and into the trachea, which branches out into the main bronchi and then subsequent divisions. During normal breathing, expiration is passive and no muscles are contracted (the diaphragm relaxes). The rib cage itself is also able to expand and contract to some degree, through the action of other respiratory and accessory respiratory muscles. As a result, air is transported into or expelled out of the lungs. This type of lung is known as a "bellows lung" as it resembles a blacksmith's bellows.

Avian lungs (Figure 7.13 & 7.15) do not have alveoli as mammalian lungs do, they have Faveolar lungs. They contain millions of tiny passages known as para-bronchi, connected at both ends by the dorsobronchi. The airflow through the avian lung always travels in the same direction—posterior to anterior. This is in contrast to the mammalian system, in which the direction of airflow in the lung is tidal, reversing between inhalation and exhalation. By utilizing a unidirectional flow of air, avian lungs are able to extract a greater concentration of oxygen from inhaled air. Birds are thus equipped to fly at altitudes at which mammals would succumb to hypoxia. (Hypoxia refers to a condition whereby there is a deficiency in the amount of oxygen delivered to the body tissues.) This also allows them to sustain a higher metabolic rate than an equivalent weight mammal. The lungs of birds are relatively small, but are connected to 8-9 air sacs that extend through much of the body, and are in turn connected to air spaces within the bones. The air sacs are smooth-walled, and do not themselves contribute much to respiration, but they do help to maintain the airflow through the lungs as air is forced through them by the movement of the ribs and flight muscles. Birds have a posterior and an anterior sac, however, air is not stored in either the posterior or anterior sacs between respiration cycles (See Figure 7.15). Instead, air moves continuously from the posterior to the anterior of the lungs throughout respiration. This type of lung construction is called a circulatory lung, as distinct from the bellows lung possessed by other animals.[7-114]

The lungs of mammals are derived as a ventral outgrowth from the digestive tract in the early embryo. This relates humans to a group of lobe-finned fish from the Carboniferous period. The larynx is formed from the cartilages that made up the gill arches of human ancestors. The larynx is connected to the trachea which is basically a flexible tube held open at times by incomplete rings of cartilage. The trachea divides into the left and right bronchi that then enter the lungs and continue to divide forming narrower and narrower tubules called bronchioles. These tubules are surrounded by circular smooth muscle fibers. At the ends of the bronchioles are groups of alveoli or air-sacs. It is in the alveoli that gas exchange actually occurs (Figure 7.13). In humans, the gas exchange surfaces are the lungs. Humans do not breathe air from the atmosphere directly into the alveoli. The air passes into the bronchi and bronchioles but does not reach the alveoli. Oxygen must diffuse from the bronchioles into the alveoli and carbon dioxide must diffuse from the alveoli into the bronchioles. In fish, the gas exchange surfaces are the gills (Figure 7.13). Water is a denser medium than air and contains only a small quantity of dissolved oxygen per gallon. When a fish opens its mouth it sucks in water, the low pressure in the gill chambers will pull the gill flaps closed to prevent water being drawn back over the gill lamellae. The fish then closes its mouth and compresses the water in its pharynx by contracting it buccal muscles (Figure 7.13). The increased pressure forces the water over the gills, opens the gill flaps and the water flows out in a backwards direction away from the fishes' mouth. The gill lamellae are arranged as a series of flat plates sprouting from the gill arch. On their upper and lower surfaces there are many thin vertical flaps that contain blood capillaries. The blood flows through these capillaries in the opposite direction to the flow of water over the gills. This gives a highly efficient diffusion pathway, as the blood flows along and picks up oxygen from the water.[7-115]

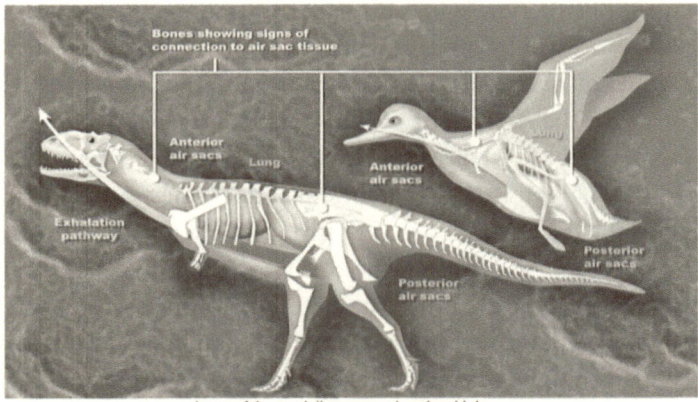

Lungs of theropod dinosaurs and modern birds

Figure 7.14 Large meat-eating dinosaurs appear to have had a complex system of air sacs similar to those found in modern birds, according to a recent study (Figure 7.14). The lungs of theropod dinosaurs (carnivores that walked on two legs and had birdlike feet) most likely pumped air into hollow sacs in their skeletons, such as is the case in birds. What was once formally considered unique to birds was present in some form in the ancestors of birds.[7-116]

Intelligent Design View

The evolution of the bird is said to have been derived from a reptile. The evolution of birds is far more complex than commonly portrayed. Birds possess unique adaptations that seem to be a problem for evolutionists to be able to explain. Adaptations such as the origin of the feather and flight and the origin of the avian lung and respiratory system (Figures 7.13 & 7.15), apparently, can only be associated with birds. In contrast to the avian lung, in all other vertebrates the air is drawn into the lungs through a system of branching tubes that finally terminate in tiny air sacs, or alveoli, so that during respiration the air is moved in and out through the same passage. In the case of birds, however, the major bronchi break down into tiny tubes that permeate the lung tissue. These so-called parabronchi eventually join up together again, forming a true circulatory system so that air flows in one direction through the lungs. This unidirectional flow of air is maintained during both inspiration and expiration by a complex system of interconnected air sacs in the bird's body that expand and contract in such a way so as to ensure a continuous delivery of air through the parabronchi. The existence of this air sac system in turn has necessitated a highly specialized and unique division of the body cavity of the bird into several compressible compartments.

Although air sacs occur in certain reptilian groups (Figure 7.14), the structure of the lung in birds and the overall functioning of the respiratory system is quite unique. No lung in any other vertebrate species is known to be anything like the avian system. Also, it is identical in all essential details in birds as diverse as humming birds, ostriches and hawks. Exactly how a vastly different respiratory system could have evolved gradually from the standard vertebrate design is extremely difficult to conceive, considering the maintenance of respiratory function is absolutely vital to the life of an organism to the extent that the slightest malfunction leads to death within minutes.

The avian lung (Figures 7.13 & 7.15) cannot function as an organ of respiration until the parabronchi system (which permeates it) and the air sac system (which insures that the parabronchi has adequate air supply) are both highly developed and able to function together in a perfectly integrated manner. The unique function and form of the avian lung necessitates a number of additional unique adaptations during avian development. **First**, the avian lung is fixed to the body wall and because of this cannot expand in volume. **Second**, because of the small diameter of the lung capillaries and the resulting high surface tension of any liquid within them, the avian lung cannot be inflated out of a collapsed state as it does in all other vertebrates after birth.

In birds, aeration of the lungs must occur gradually and starts three to four days before hatching with a filling of the main bronchi, air sacs and parabronchi with air. Only after the main air ducts are already filled with air does the final development of the lungs, and particularly the growth of the air capillary network, take place. The air capillaries are never collapsed, which is not the case of the alveoli of others vertebrates which do collapse. In birds, as the air capillaries grow into the lung tissue, the parabronchi are from the beginning open tubes filled either with air or fluid, which is later absorbed into the blood capillaries. Explaining how such an intricate and highly specialized system of correlated adaptations could have been achieved gradually through perfectly functional intermediates seems impossible.[7-117]

Intelligent Design View

All reptiles breathe through their lungs. Even though reptiles do not have a diaphragm muscle, reptiles have a diaphragm-type respiratory system and the act of breathing is accomplished by the reptile moving its throat or rib cage.[7-118] Most mammals, however, have two lungs and a diaphragm (Figure 7.13). The diaphragm pulls the lungs downward, which suck air in (much like a vacuum) to allow the mammal to breathe in oxygen, and the diaphragm then pushes upward on the lungs, to exhale (breathe out) carbon dioxide.[7-119] Birds require a respiratory cycle just like mammals in order to supply their muscles and tissues with oxygen and to remove built-up CO_2 (carbon dioxide) from the body. Unlike mammals, however, birds do not use just the lungs to breathe, in a simple in-and-out process. Birds require two respirations to pass oxygen all the way through the body, and use a more complicated system of air sacs to aid this process (Figures 7.13 & 7.15).[7-120]

Reptiles, from which birds are thought to have descended, as well as mammals, have a diaphragm breathing system. The avian (bird) breathing system, however, is totally different (Figures 7.13 & 7.15). For animals that have a diaphragm, breathing is a two-stage process: Stage one, air goes into the lung: stage two, air goes out of the lung (two directions). Reptile and mammal lungs are made up of millions of tiny balloons, called alveoli (Figure 7.13), which expand and contract as the animal breathes. But the lungs have no muscles. For mammals and some reptiles (as well as the ancestors of birds) lungs have no muscle, the lungs sit in an airtight sac or chamber (Figure 7.13). The bottom of this chamber is sealed with a large, thin muscle called the diaphragm. The dome-shaped diaphragm completely separates the chest cavity from the abdomen. When the breathing muscles force the diaphragm down, it changes the air pressure in the chest cavity, and the lungs fill with air. No known bird has a diaphragm, and a bird's lungs do not change size when it breathes. Instead, a bird has a network of air sacs that work together like bellows, drawing air in and through the bird's lungs.

Air sacs are unique to birds. A bird has two sets of air sacs, the posterior and the anterior (See Figure 7.15). Birds inhale and exhale by moving their ribs inward and outward, and by raising and lowering their very large sternum (wishbone). Also, unique to birds is the way their lungs function. Unlike reptile lungs (Figure 7.13), bird lungs have an opening at each end. Air flows through the bird's respiratory system, a one-way path, which provides an almost continuous supply of fresh, oxygen-rich air in the lungs. Also, in contrast to a reptile's two-stage breathing cycle (inhale-exhale), birds require four breathing stages to circulate air through their respiratory tracts (Figure 7.15) (inhale-exhale-inhale-exhale): in **stage 1** (first inhalation), some of the fresh air goes into the lungs, but most of it goes into the posterior air sacs. In **stage 2** (first exhalation), the air moves from the posterior air sacs into the lungs, where oxygen is exchanged for carbon dioxide. In **stage 3** (second inhalation), the air moves from the lungs to the anterior air sacs. In **stage 4** (second exhalation), the air passes up the bronchus and out of the bird's system.

It is very questionable as to how a reptile's diaphragm breathing system could transition into the birds' flow-through system. The two stage breathing apparatus of reptiles and mammals is an integrated system, made up of separate parts working together to perform a function. The four stage flow-through breathing apparatus of birds is also a functionally integrated system that performs the same function. These two systems, however, are different. These different systems perform the same function in different ways using different anatomical structures in different configurations. It is doubtful how such a transition could have taken place from a reptile to a bird breathing system.

There are four problems transforming a reptile/mammal balloon-like alveoli lung system into a bird airflow lung system (Figure 7.13). **First**, any credible scenario for the evolution from the reptilian breathing system to the avian breathing system needs to account for the gradual, step-by-step development of a new and unique structure: air sacs. Not just one, but nine air sacs in five variations: **(1)** abdominal, **(2)** caudal thoracic (air sacs located at or in the tail), **(3)** cranial thoracic (air sacs located in the skull), **(4)** cervical (neck), **(5)** and clavicular (collar bone). What would have evolved first, the new air sacs or the four-stage breathing cycle that uses the air sacs? If the air sacs came first, what was their advantageous (selectable) function before four-stage breathing arose? If the air sacs came later, how did four-stage breathing originate without them? **Second**, a viable explanation would also be needed to describe the transition from a two stage to a four stage breathing system. Was there once a three stage breathing system in between? What would it have looked like? Would there be two inhales for every exhale, or two exhales for every inhale? **Third**, the radical transformation of the lung also requires an explanation. What would the intermediate forms between the single opening (in-and-out) reptilian lung and the dual opening (flow through) avian lung look like? How would it happen in small, yet advantageous, steps? Can there even be a transition between a single opening and a dual-opening system? How would the balloon-like alveoli transform into the tube-like parabronchi? How would the lung maintain function? Would the lung transformation happen before or after the development of air sacs? Would it be before or after the four stage breathing cycle? **Fourth**, and finally, reptiles today do not have a diaphragm muscle. The reptiles thought to be the ancestors of birds most likely had a diaphragm breathing system. According to many evolutionary biologists, changing from a diaphragm lung system to a flow-through lung would require changing and increasing the musculature of the reptile's chest. At the same time, the diaphragm would need to gradually go away. This poses a fundamental problem, however. The earliest stages of this transformation would have required a hole or hernia in the reptile's diaphragm. This would have immediately caused problems for the entire breathing system and would lead to certain death for any animal obtaining this structure. If an animal cannot breathe, it will die.

Recently paleontologists have found a theropod dinosaur, Majingatholus atopus, with features such as caudal air sacs (air sacs located at or in the tail) that resemble the avian breathing system (Figure 7.14). If this finding is confirmed, it may show that some dinosaurs had a breathing system similar to what modern birds have. Some scientists believe this finding may help to solve the problem of the origin of the avian breathing system, since many evolutionary biologists believe birds evolved from theropod dinosaurs. Other scientists believe that even if this finding is confirmed, all it does is to relocate the problem. At some point, a novel mode of respiration arose. However, the ancestors of the theropod dinosaurs (the reptiles) did not possess this system, so that still leaves the questions of how and where such a system arose. It does not really matter whether the system first arose in birds or in theropod dinosaurs, it would still need to evolve from a group having a diaphragmatic, two-stage breathing system. Whether discussing a reptile-to-bird transition or in a reptile-to-theropod dinosaur transition, there is still the problem of maintaining continuous respiratory function during a series of anatomical transformations. Without adequate soft-tissue being preserved in fossils, it is difficult to determine how the reptile/ mammal lung could have transformed into a bird lung (Figures 7.13 & 7.15). The conclusion is that it appears physically impossible for a reptile to have evolved into a modern bird.[7-121]

Creationist View

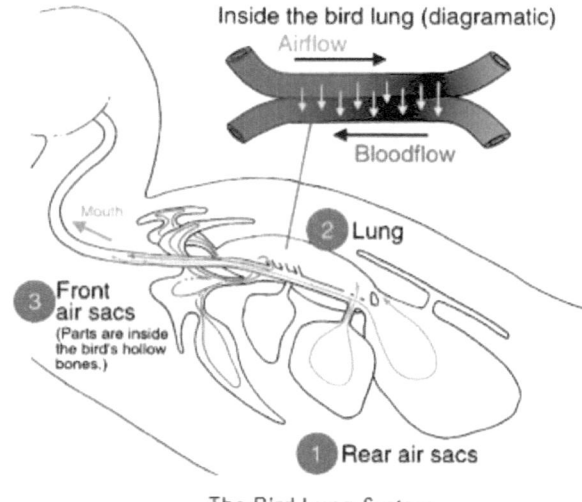

The Bird Lung System

Figure 7.15 The Bird Lung System

As a bird breathes, air moves into its rear air sacs (Figure 7.15 #1). These then expel the air into the lung (Figure 7.15 #2) and the air flows through the lung into the front air sacs (Figure 7.15 #3). The air is expelled by the front air sacs as the bird breathes out. The avian lung does not expand and contract as a reptile's or mammal's lung does. The blood that picks up oxygen from the lung flows in the opposite direction to the air so that blood with the lowest oxygen is exposed to air with the lowest oxygen. The blood with the highest oxygen is exposed to air with an even higher oxygen concentration. This ensures that, in every region of the circulation, the concentration of oxygen in the air is more than that of the blood with which it is in contact. This maximizes the efficiency of oxygen transfer from the air to the blood. This is known as counter-current exchange. Such very efficient lungs help birds to handle the energy demands of flight, especially at high altitudes. Bats do very well with the 'bellows' type of lung. This makes the evolutionist's argument for the origin of birds' lungs (i.e. that they 'needed' them) even more difficult to support.[7-122]

One of the most distinctive features of birds is their lungs (Figures 7.13 & 7.15). Bird lungs are small in size and nearly rigid, but they are, nevertheless, highly efficient to meet the high metabolic needs of flight. Bird respiration involves a unique "flow-through ventilation" into a set of nine interconnecting flexible air sacs sandwiched between muscles and under the skin. The air sacs contain few blood vessels and do not take in oxygen exchange, but rather function like bellows to move air through the lungs. The air sacs permit a unidirectional (one way) flow of air through the lungs resulting in higher oxygen content than is possible with the bidirectional (two way) airflow through the lungs of reptiles and mammals (Figure 7.13). The airflow moves through the same tubes at different times both into and out of the lungs of reptiles and mammals. This results in a mixture of oxygen-rich air with oxygen-depleted air. The unidirectional (one way) flow through bird lungs not only permits more oxygen to diffuse into the blood but also keeps the volume of air in the lungs nearly constant, a requirement for maintaining a level flight path.

If theropod dinosaurs are the ancestors of birds, one might expect to find evidence of an avian-type lung in such dinosaurs. While fossils generally do not preserve soft tissue such as lungs, a very fine theropod dinosaur fossil (Sinosauropteryx) has been found in which the outline of the visceral cavity has been well preserved. The evidence clearly indicates that this theropod had a lung and respiratory mechanics similar to that of a crocodile—not a bird. Specifically, there was evidence of a diaphragm-like muscle separating the lung from the liver; as seen in modern crocodiles (birds lack a diaphragm).[7-123] These observations suggest that this theropod (dinosaur) was similar to an ectothermic reptile, a reptile that regulates its body temperature, not an endothermic bird, a bird that has a chemical change due to an absorption of heat.[7-124] The lungs of reptiles (Figure 7.13) consist of millions of tiny air sacs; whereas, bird's lungs have tubes (Figures 7.13 & 7.15). The piecemeal evolution of bird's lungs from reptile's lungs seems virtually impossible. The survival of the hypothetical intermediate life forms possessing lungs, which consist of half tubes and half air sacs, seems impossible.[7-125]

The Evolution of Land Animals to Whales

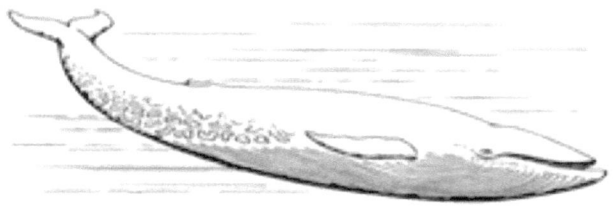

Whale

In North America the black bear was seen by Hearne swimming for hours with widely open mouth, thus catching, like a whale, insects in the water. Even in so extreme a case as this, if the supply of insects were constant, and if better adapted competitors did not already exist in the country, I can see no difficulty in a race of bears being rendered, by natural selection, more and more aquatic in their structure and habits, with larger and larger mouths, till a creature was produced as monstrous as a whale.[7-126]

—Charles Darwin
—On the Origin of Species or the Preservation of Favoured Races in the Struggle For Life
—First Edition 11/24/1859

Evolutionist View

The evolution of living cetacea, whales is very well established. The closest relatives are called basilosaurids. Then there are the protocetids, followed by a couple of forms from the Eocene Period called Ambulocetus (Ambulocetus natans – The scientific name Ambulocetus means "walking whale", and natans means "that swims") and Pakicetus (Figures 7.16 & 7.17), and eventually there came hippopotamus, which are the closest living relatives of whales, and outside of that are the early Eocene artiodactyls, or hoofed mammals, from which we have recognized certain characteristics that are shared between hippopotamus and whale. The skeletons of some fossils from the Eocene of hoofed mammals are members of the group artiodachtyl, the ones with the even toes. There are fossil remains of a partial skull and brain case of an animal called Pakicetus, which is closer to whales of today than hippos and the other Eocene Period artiodachtyls are. The oldest whales come from Pakistan, India, Egypt, which once was the edge of an ancient sea in the early part of the Tertiary period, fifty, sixty million years ago when the evolution of whales occurred. The brain cases and skulls of Pakicetus share features with whales that live today are not so much that it has a blow hole or flukes, but that it has an ear region with features that are only found in whales. It is concluded that Pakicetus share a common ancestor with the first whales. It is obviously a four-legged creature. Pakicetus walked around on the ground. It looks like a common quadruped, four-footed creature except for the ear region that indicates Pakicetus is related to whales. They are quadrupedal, that is, four-legged creatures. There are isotopes of oxygen, and oxygen comes in different kind of molecular forms, and the percentage of those forms varies between terrestrial and aquatic horizons, environments, so that when bones are found that are made with oxygen elements that contain this isotopic signal, it provides a clue as to whether these animals were primarily terrestrial or aquatic. The isotopes for Pakicetus demonstrates that it lived in a fresh water marine area. This evidence appears to indicate that this animal was spending at least part of its time in water, including brackish (briny or salty) or marine water.[7-127]

The first whales, such as Pakicetus (Figures 7.16 & 7.17), were typical land animals. They had long skulls and large carnivorous teeth. From the outside, they don't look much like whales at all. However, their skulls—particularly in the ear region, which is surrounded by a bony wall—strongly resemble those of living whales and are unlike those of any other mammal. [7-127a] Whales (cetaceans) are mammals, so they breathe air and give birth to live young. But unlike most other mammals, whales spend all of their time in the water. There are a series of fossils transitioning from land-based animals long ago to the ocean-dwelling whales of today. The transitional fossils that led to the development of sea mammals indicate that whales and dolphins evolved from land mammals with legs. In recent years, several transitional forms of whales with legs, both capable and incapable of terrestrial locomotion have been found. Sea cows (manatees and dugongs) are fully aquatic mammals with flippers for forelimbs and no hind limbs. It is believed that sea cows evolved from terrestrial ancestors with legs. A transitional fossil has been found in Jamaica of a sea cow with four legs.[7-128] In 2001, the discovery of Pezosiren portelli provided more clues as to land mammal to whale evolution. The fossil was discovered in Jamaica, and at 50 million years old represents the oldest member of the group Sirenia (sea cows). It is clear Pezosiren portelli

(Figure 7.15a) had 4 legs that it used to walk on land, but it also had heavy ribs that could indicate it lived part time in water, much like a hippopotamus. This species likely represents the transitional form of sea cows as it maintained the general body plan of a sea cow, but just without flippers. [7-128a]

Figure 7.15a Pezosiren portelli

Whales descended from land animals that entered the sea about 55 million years ago. It has been supposed by most biologists that whales evolved from a mammal that is amphibian in nature, living both on land and water.

Hippopotamus

hippopotamus

Biologists now believe that the whale could have evolved from the hippopotamus, which, although closely related to terrestrial mammals, is very aquatic for a land mammal. Whales have an excellent fossil record. Whales and their relatives, the dolphins and porpoises, are mammals. They are warm-blooded, produce live young whom they feed with milk, and have hair around their blowholes. Evidence from whale DNA, as well as vestigial traits like their rudimentary pelvis and hind legs, indicates that their ancestors lived on land. (See Chapter 12, Vestiges.)

Whales almost certainly could have evolved from a species of the artiodactyls, who are a group of mammals that have an even number of toes, such as camels and pigs. The fossil record for whale evolution includes about 24 transitional animals (Figure 7.17). As whales evolved, the nostrils gradually moved from the front of the face to the top of the head where they became a single or pair of blowholes. Early whales had shortened limbs and enlarged feet, which allowed them to paddle effectively in the water. Notably, some still had hooves. Over time, the bodies of ancient whales became elongated and more streamlined for an aquatic environment. Their tails also enlarged and evolved into the main power source of locomotion through the water. Consequently, the back legs of primitive whales were reduced dramatically. Today, all that is left in whales of their hips and hind legs is a tiny, non-functional, boney vestige of their evolutionary past. (See Chapter 12, Vestiges.)

The first whales also featured molars that resemble the land-dwelling mesonychids (an extinct order of carnivorous hoofed mammals. They have hooves on each toe). Their teeth had multiple roots and pointed cusps that were suited for meat eating. As these mammals began to live under water, chewing became very difficult and whales began to lose specialized teeth for grinding. In fact, the teeth of ancient whales are transitional. The teeth at the back of the jaw reflect their carnivorous past, while the front teeth are simple single cones like that of modern toothed whales. As the fetus develops in the womb, teeth appear in the jaws. These never develop full roots, nor do they attach to the jawbones. Most of these malformed teeth are rejected from the jaws before birth or soon afterward. Being mammals, these whales are nourished through the umbilical cord in the womb, and they drink milk from their mother's mammary glands once born. Like modern birds, whales still have tooth genes, which they inherited from their toothed evolutionary ancestors.[7-128] The scientific evidence for biological evolution is very convincing. Transitional fossils of whales definitely exist.[7-129]

The homologous structures in whales establish the fact that whales evolved from land mammals. One major homologous structure is the fin of a whale. All of the bones of a whale's fin match up to comparative bones in other mammals (Figure 7.4 and Chapter 15, Figure 15.1). This is evidence that whales, as mammals, share a common ancestor with other mammals. (See Chapter 15, section "Common Function or Common Ancestry?" for further discussion of common ancestry.) Another example of a homologous structure in whales is their inner ear bones. The inner ear bones of a whale are extremely similar to land mammals, but the one difference is that they (the inner ear bones) are fused together (in whales). Instead of several, detached ear bones, the ear bones of whales are attached, which helps in hearing under water. The hippopotamus (a land mammal) also has fused ear bones. [7-129a]

The evidence for the evolution of whales is extensive. A particularly good example is the transition from terrestrial reptiles called varanoids through the semi-aquatic aigalosaurs to the aquatic mosasaurs. It has been established that whales did not evolve from Mesonychids as paleontologists originally thought, but instead whales are related to hoofed mammals such as cows and hippopotamuses. Recently three transitional species were found including the Ambulocetus natans (Figures 7.16 & 7.17), which was a whale with four legs. Rodhocetus (Figure 7.16 & 7.17) had smaller hind legs but large enough for waddling like a seal. Basilosaurus, Protocetus and Indocetus all had hind legs that were too small to walk on. Prozeuglodon had tiny 15 cm legs on its 5m body. Eocetus had lost its hind legs entirely. [7-130]

It would not take much to transform a primitive mammal into a bat or a whale. Bats didn't evolve wings by inventing new structures: the wings are merely elongated fingers, with the same number of joints as in those of a hedgehog, and with an interdigital webbing grown out to the fingertips. The rest of a bat's skeleton is very similar to a shrew's. Whales are an even more striking case. Most whales, such as porpoises, are rather small. Their muscles and a thickened layer of fat give them a streamlined shape. The hind legs are reduced to vestigial pelvic bones. The front legs are flattened into paddles, with five digits (like primitive mammals); but the number of joints per digit is increased. The teeth are partly (in fossils) or entirely (in most modern whales) dedifferentiated, so they all have the same shape; and in modern (but not early fossil) species are increased in number – or else entirely lost, as in the blue whale. The most radical difference from other mammals consists largely of a forward extension of the jawbones out from under the nostrils, which are therefore situated on top of the head. In species such as the blue whale, the skin on the roof of the mouth is cornified like our calluses, and folded into sheets of baleen ("whalebone") that hang down into the mouth. (cornification is the formation of a horny layer of skin, or horny skin structures, as hair, nails, or scales, from squamous epithelial cells.) The only characteristics that are not mere modifications of primitive mammalian features are the baleen and the dorsal and tail fins, which are rigid folds of skin and fibrous tissue, like our ears. One of the most amazing aspects of evolution is how easy it is to account for major transformations through simple changes in developmental processes. Most of the differences among different kinds of mammals are quite simply accounted for by changes in the relative rates of growth of different parts of the body. Speed up the elongation of fingers to get a bat's wing; slow down the development of teeth or legs to reduce or eliminate them in whales; slow down the growth of the lateral toes and increase that of the middle one to get a horse's hoof. In more than 600 million years of animal evolution, there have only been a handful of truly novel features, such as wings of insects, that seem in our ignorance not to be mere modifications of something that came before.[7-130a]

The Evidence That Whales Evolved from Land Mammals

Evolutionist View

Multiple independent lines of evidence of whale origins demonstrate one of science's greatest tools for measuring the strength of a concept (theory). There are nine different types of evidence that prove whales evolved from terrestrial mammals. They are: **(1)** paleontological, **(2)** morphological, **(3)** molecular biological, **(4)** vestigial, **(5)** embryological, **(6)** geochemical, **(7)** paleoenvironmental, **(8)** paleobiogeographical, and **(9)** chronological. We will now take a look at each type of evidence in detail.

1. Paleontological evidence

The paleontological evidence shows changes in **(1)** the shape of the skull, **(2)** the shape of the teeth, **(3)** the position of the nostrils, **(4)** the size and structure of both the forelimbs and the hind-limbs, **(5)** the size and shape of the tail, and **(6)** the structure of the middle ear as it relates to directional hearing underwater and diving. It is obvious from the recorded history of the fossils that there has been increasing adaptation to life in the water, as lived by contemporary whales. We will now take a look at some transitional fossils.

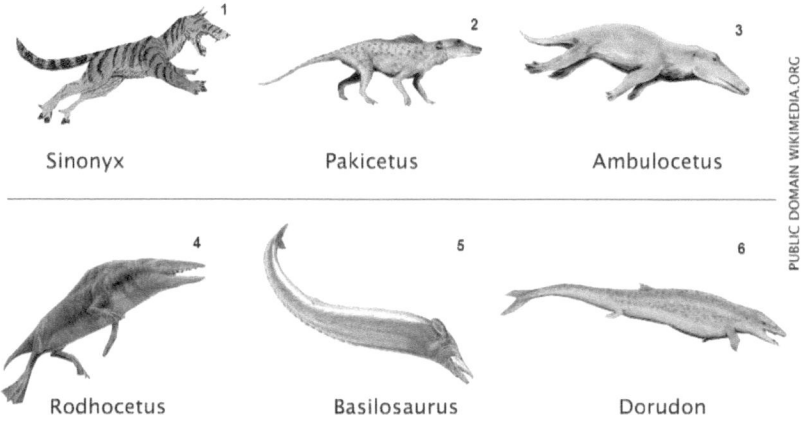

Land Mammal to Whale Evolution Sequence

Figure 7.16 Evolutionary sequence leading up to whales.

Sinonyx (Figure 7.16) was a wolf-like ungulate mesonychild (Figure 7.17) land mammal about the size of a wolf. It is a primitive hoofed animal, from the late Paleocene epoch (67-54 mya), about 60 million years ago. The characteristics that link Sinonyx to the whale include **(1)** an elongated muzzle, **(2)** an enlarged jugular foramen, or opening in—the skull base, and **(3)** a short basicranium. The tooth count of 44 was the primitive mammalian number. The molars were very narrow shearing teeth, especially in the lower jaw, but possessed multiple cusps. The elongation of the muzzle is often associated with hunting fish. All fish-hunting whales, as well as dolphins, have elongated muzzles.

Note: It was once believed that Sinonyx was the direct ancestor of whales and dolphins, but the discovery of well-preserved hind limbs of archaic whales as well as more recent DNA analysis now indicates that whales are more closely related to hippopotamids (as mentioned previously) and other artiodactyls (hoofed mammals having an even number of toes) than they are to mesonychids (extinct carnivorous mammals having hooves on each toe).

Pakicetus (Figures 7.16 & 7.17) is a land mammal and the oldest whale, is the first known archaeocete. It is from the early Eocene epoch (52 mya) of Pakistan. Although it is known only from fragmentary skull remains, those remains are very informative, and they are definitely intermediate between Sinonyx and later whales. This is especially true when examining the teeth. The upper and lower molars, which have multiple cusps, are still similar to those of Sinonyx, but the premolars have become simple triangular teeth composed of a single cusp with both front and back serrated. The teeth of later whales show even more simplification into simple serrated triangles, like those of carnivorous sharks, indicating that Pakicetus's teeth were adapted to hunting fish. A well-preserved cranium shows that Pakicetus was definitely a whale.[7-131]

Ambulocetus (Figures 7.16 & 7.17), which means 'walking whale,' was a whale with legs. The limbs were large and paddle-like. The hands and the feet were clearly being broadened and were apparently of some use to the animal in getting around in the water. Ambulocetus lived during the Eocene Period.[7-132] Ambulocetus is an amphibious long-legged mammal, dating from the early to middle Eocene epoch (50 mya). It was clearly a whale, but it also had functional legs and a skeleton that enabled it to maneuver on land as well. Ambulocetus had a large femur, but did not have the large attachment points for walking muscles, so most likely it walked only in the way that modern sea lions walk. Modern sea lions walk by rotating the hind feet forward and waddling along the ground with the assistance of their forefeet and spinal flexion. When walking, its huge front feet must have pointed laterally to a fair degree since, if they had pointed forward, they would have interfered with each other. The forelimbs were also intermediate in both structure and function. The ulna (the bone of the forearm on the side opposite to the thumb) and the radius (the bone of the forearm on the thumb side) were strong and capable of carrying the weight of the animal on land. The elbow was strong but it was inclined rearward, making possible rearward thrusts of the forearm for swimming. However, the wrists, unlike those of modern whales, were flexible.

The anatomy of Ambulocetus suggests that it swam by swaying its spine up and down, propelled by its back feet. Mammals using this method of swimming have back feet that are quite large. The toes of the back feet were hooves, which are unusual, thus indicating the hoofed ancestry of this animal. Ambulocetus' skull was quite whale-like. It had a long muzzle and teeth that were very similar to later archaeocetes. Although Ambulocetus apparently lacked a blowhole, the other skull features qualify Ambulocetus as a whale. The post-cranial features are clearly in transitional adaptation to the aquatic environment. Thus Ambulocetus is best described as an amphibious, sea-lion-sized fish-eater that was not yet totally disconnected from the land life of its ancestors.

Dalanistes (Figure 7.17) shares several features with Ambulocetus, and with its combination of terrestrial and amphibious adaptations, Dalanistes apparently is an intermediate species. Isotopic evidence indicates that Dalanistes may have had a marine diet.

Rodhocetus (Figures 7.16 & 7.17) is a long-legged aquatic mammal is dated in the middle Eocene epoch (46-47 mya). It is an ancient whale that evolved and became extinct at the end of the Eocene Epoch (40-34 mya). Rodhocetus had developed a powerful tail for swimming. The vertebrae along the spine also were unfused. This gave the spine more flexibility and allowed a more powerful thrust while swimming. Rodhocetus had shortened cervical vertebrae, heavy tail vertebrae, and large dorsal spines on the lumbar vertebrae for large tail and other axial muscle attachments. These features are all similar to those of modern whales. The pelvis of Rodhocetus was smaller than that of its predecessors, but it was still connected to the sacral vertebrae, meaning that Rodhocetus could still walk on land to some small degree.

Rodhocetus' skull was rather large compared to the rest of the skeleton, making it more whale-like. For the first time, the nostrils have moved back along the snout and are located above the canine teeth, showing blowhole evolution. Rodhocetus balochistanensis is believed to demonstrate a direct evolutionary link to artiodactyls, which are even-toed hoofed animals, modern examples of which include hippopotamuses, now believed to be the closest relative of the whales. The structure of the ankle bones of this species is specific only to artiodactyls. This matches studies of the genetic relations between whales and other animals. Because of these studies, previous fossil-based theories that whales were directly descended from mesonychids are no longer thought to be true. It is more likely that whales descended from the ancestors of hippos (as mentioned previously). Mesonychids are an extinct order of carnivorous hoofed mammals that somewhat resembled wolves, with hoofs instead of paws, and larger heads.[7-133]

Basilosaurus (basilosaurid) (Figures 7.16 & 7.17) is the next transitional step toward living whales. In basilosaurid, the nostrils moved farther up along the skull and the hind limbs were not just decoupled from the back bone, they were extremely reduced. There is a pulley shaped bone with a little hook off it. The ankle was still like the ankle of a terrestrial animal (a hoofed mammal) from which they evolved, even though this animal could no longer walk on land. There is a progression of features that became more and more whale-like from animals that are terrestrial and conventional land animals. These changes developed in areas such as the ear. The final creature is the living cetacean, which looks very much like the whales of today. They have almost completely lost the hind limbs.[7-134]

Basilosaurus is an aquatic mammal, lived during the latter part of the middle Eocene Epoch (45-35 mya). Its extremely long body length (about 15 meters or 49 feet) appears to be due to a feature unique even among whales. Its 67 vertebrae are so long compared to other whales of the time and to modern whales, that it probably represents a specialization that separates it from the lineage that eventually became modern whales. It also had a nearly complete pelvic section and set of hind limb bones. The limbs were too small for effective propulsion, less than 60 cm (1.9685 feet) long on this 15-meter-long (49.21 feet) animal, and the pelvic section was completely isolated from the spine so that weight-bearing was impossible. When the bones from this animal were reconstructed, it legs were placed external to the body. This indicates an important intermediate form in whale evolution. It appears that Basilosaurus spent most of its time in the water because of the changes developed in its skull. This animal had a large single nostril, further back than those of previous fossils. The movement from the forward extreme of the snout to a position nearer the top of the head is characteristic of only those mammals that live in marine or aquatic environments.

Dorudon (Figures 7.16 & 7.17) is another aquatic mammal, existed during the same time with Basilosaurus in the late Eocene Epic (about 40 mya) and probably represents the group most likely to be ancestors to modern whales. Dorudon had shorter vertebrae than Basilosaurus and was much smaller (about 4-5 meters or 13.1-16.4 feet in length). Dorudon did not seem to have the cranial anatomy necessary for sonar communication that whales of today have. Dorudon was a fully aquatic whale. In fact, it would not have been able to move around on land at all. Its size and lack of limbs that could support its weight made it become an aquatic mammal.

2. Morphological evidence

The common ancestry of whales and other animals with hooves becomes even more apparent when the morphological (structures of organisms) characteristics shared by the fossil whales and living animals with hooves, divided into the odd-toed perissodactyls and even-toed artiodactyls, is examined. For example, the anatomy of the foot of Basilosaurus links whales with artiodactyls, even-toed hoofed animals. Another example involves the incus, the anvil in the middle ear. The incus of Pakicetus, preserved in at least one specimen, is morphologically intermediate in all characters between the incus of modern whales and that of modern artiodactyls. Additionally, the joint between the malleus (hammer) and incus of most mammals is positioned at an angle between the middle and the front of the animal, while in modern whales and in hoofed animals, it is positioned at an angle between the side and the front. In the first fossil whale the joint is in a more forward position. Thus the joint has clearly rotated toward the middle from the ancestral condition in land mammals; Pakicetus provides a picture of the transition.

3. Molecular biological evidence

Molecular studies show that whales are more closely related to the hoofed animals than to all other mammals. Within this category of mammals, whales are most closely related to even-toed animals, or artiodactyls, even though there are differences in the details among the studies. By placing whales close to the artiodactyls, these molecular studies confirm the predictions made by evolutionary theory. This pattern of biochemical similarities must be present if the whales and the hoofed animals, especially the artiodactyls, share a close common ancestor. The fact that these similarities are present is therefore strong evidence for the common ancestry of whales and hoofed animals.

4. Vestigial evidence

The vestigial features of whales indicate two things. **(1)** They indicate that whales, like so many other organisms, have features that make no sense from a design perspective. These features serve no current function, they require energy to produce and maintain, and they may be harmful to the organism. **(2)** They also indicate that whales carry a piece of their evolutionary past with them, highlighting a history of a terrestrial ancestry. Modern whales often retain rod-like vestiges of pelvic bones, femora, and tibiae, all embedded within the musculature of their body walls. These bones are more pronounced in earlier species and less pronounced in later species. As the example of Basilosaurus shows, whales of intermediate age have intermediate-sized vestigial pelvis and rear limb bones (See Chapter 12 for further discussion of vestiges).

Whales also retain a number of vestigial structures in their organs of sensation. Modern whales have only vestigial olfactory nerves. Furthermore, in modern whales the auditory meatus (the exterior opening of the ear canal) is closed. In many, it is merely the size of a thin piece of string, about 1 mm in diameter, and often pinched off about midway. All whales have a number of small muscles devoted to nonexistent external ears, which are apparently a vestige of a time when they were able to move their ears, a behavior typically used by land animals for directional hearing. The diaphragm in whales is vestigial and has very little muscle. Whales use the outward movement of the ribs to fill their lungs with air. Finally, there are several occurrences of captured sperm whales with visible, protruding hind limbs. Similarly, dolphins have been spotted with tiny pelvic fins, although limb bones, as in those rare sperm whales, probably did not support their bodies. And some whales, such as belugas, possess rudimentary external ears, a feature that can serve no purpose in an animal with no external ear and that can reduce the animal's swimming efficiency by increasing hydrodynamic drag while swimming. (Vestiges are discussed in further detail in Chapter 12.)

5. Embryological evidence

Like the vestigial features, the embryological features also indicate two things. **First**, the whale embryo develops a number of features that it later abandons before it attains its final form. It makes no sense for a supernatural designer (Creator) to build structures only to abandon them or to destroy them. Would it not make more sense to have embryos attain their adult forms quickly and directly? It seems unreasonable for a supernatural designer (Creator) to send the embryo along such an unnecessary pathway. Evolution requires that new features be built on the foundation of previous features that would be modified or discarded later. **Second**, the embryology of the whale provides evidence for its land ancestry. As embryos, just as in adult whales, there are many discarded features that will be no longer used. Many whales, while still in the womb, begin to develop body hair. Yet no modern whales retain any body hair after birth, except for some snout hairs and hairs around their blowholes used as sensory bristles in a few species.

The fact that whales possess the genes for producing body hair shows that their ancestors had body hair. In other words, their ancestors were ordinary mammals. In many embryonic whales, external hind limb buds are visible for a time but then disappear as the whale grows larger. Also visible in the embryo are rudimentary external ear flaps, which disappear before birth, except in rare instances. And, in some whales, the olfactory lobes of the brain exist only in the fetus. The whale embryo starts off with its nostrils in the usual place for mammals, at the tip of the snout. But during development, the nostrils migrate to their final place at the top of the head to form the blowhole. (Embryology is discussed in more detail in Chapter 11.)

6. Geochemical evidence

The earliest whales lived in freshwater habitats, but the ancestors of modern whales moved into saltwater habitats and thus had to adapt to drinking salt water. Fresh water and salt water have somewhat different isotopic ratios of oxygen. The transition is recorded in the whales' skeletal remains, which are the teeth. Fossil teeth from the earliest whales have lower ratios of heavy oxygen to light oxygen, indicating that the animals drank fresh water. Later fossil whale teeth have higher ratios of heavy oxygen to light oxygen, indicating that they drank salt water. This reinforces the suggestion obtained from all other evidence that the ancestors of modern whales adapted from land creatures to freshwater creatures and then to saltwater animals.

7. Paleoenvironmental evidence

The sequence of whale fossils and their changes should also relate to changes observed in the fossil records of other organisms at the same time and in similar environments. The fossils of other organisms associated with the whale fossils indicate the environment that the whales lived in. The morphology (structure) of Sinonyx indicates that it was fully terrestrial. Its fossils are found associated with the fossils of other terrestrial animals. Pakicetus probably spent a lot of time in the water in search of food. Mammalians found with Pakicetus consists of rodents, bats, various artiodactyls (cattle, deer, etc.), perissodactyls (horses, etc.) and probiscideans (elephants, etc.), and even a primate (apes, chimpanzees, etc.). There are also aquatic animals such as snails, fish, turtles and crocodilians found in this same region. Moreover, the sediment associated with Pakicetus shows evidence of streaming or flowing, usually associated with soils that are carried by water. The paleoenvironmental evidence clearly shows that Pakicetus lived in the low-lying wet terrestrial environment, making occasional excursions into fresh water. Interestingly, both primary and permanent teeth of the animal are found in these sediments, supporting the idea that Pakicetus gave birth on the land. The sediments in which Ambulocetus was found contain leaf impressions as well as fossils of the turret-snail Turritella and other marine mollusks. Since these fossils were present with Ambulocetus fossil, it seems evident that Ambulocetus lived in what was once a shallow sea. Although leaves can be washed into the sea and fossilize there, marine mollusks would not be found on the land. Rodhocetus is found in green shales deposited in what is equivalent to the outer part of the continental shelf. Because green shales are associated with fairly low-oxygen bottom waters, Rodhocetus must have lived at a greater water depth than any previous whale. The fact that it is found in association with planktonic foraminiferans and other microfossils, agrees with this determination of water depth. Basilosaurus and Dorudon have been found in a variety of sediment types, indicating that they were wide-ranging and capable of living in deep as well as shallow water. From the paleoenvironmental evidence, as whales evolved, they made their way into deeper water and became less and less dependent on land.

8. Paleobiogeographic evidence

The range of Sinonyx is limited to central Asia. Specimens of Pakicetus have only been found in Pakistan; Ambulocetus and Rodhocetus seem to be similarly restricted. In contrast, Basilosaurus and Dorudon, representing the whales more adapted to living in the open sea, are found in a much wider area. Whales were initially found in a rather small geographic area and did not become distributed throughout the world until after they evolved into fully aquatic animals that were no longer tied to the land.

9. Chronological evidence

During the early Cenozoic Era (65-0 mya), mammals were able to diversify partly because of the vacuum left by the mass extinction of reptiles at the close of the Cretaceous Period (144-65 mya). Because the reptiles no longer predominated, there were new ways in which mammals could thrive. In the specific case of whales, as the swimming reptiles began to become extinct, the ocean became available for the expansion of the whales (See section "Evolution From Reptiles to Mammals" discussion of euryapsids and ichthyosaurs). Before the late Cretaceous Period reptilian extinctions, the Mesozoic Era (250-65 mya) marine reptiles such as the plesiosaurs, ichthyosaurs, mosasaurs, and marine crocodiles might well have feasted upon any mammal that strayed off shore in search of food. Once those predators became extinct, the evolution quickly produced mammals, including whales that thrived in the seas as they once did on land. The transition took some ten to fifteen million years to produce fully aquatic, deep-diving whales with directional underwater hearing.[7-135]

Whales Could Not Have Evolved from Land Mammals

Creationist View

Darwinian evolutionary theory states that gradual evolution by natural selection would require tremendous numbers of transitional forms for evolution to occur. (Natural selection is discussed in Chapter 9.) The gap between modern whales and land mammals is great. All known aquatic or semi-aquatic mammals such as seals, sea cows or otters are specialized representatives of distinct orders and none can possibly be ancestors to the present-day whales. In order to develop a scenario leading up to the existence of the whale, Darwinists usually assume various intermediate stages. These stages are thought to have begun with a shrew and developed gradually to otter-like then seal-like, and finally a presumed organism that eventually led to the modern whale. Even from the hypothetical whale ancestor stage one would need to assume many hypothetical primitive whales to bridge the huge gaps that separate the modern baleen whales and the toothed whales. It is impossible to accept that such a hypothetical sequence of species could have led up to the development of a whale. Such an assumption would be purely guess work and would go against Darwinian Theory and would defeat its major purpose of attempting to provide a natural explanation for evolution.

The differences in morphology (structure) between well-defined species, such as a rat to mouse or fox to dog, are extensive, if attempting to account for the modifications necessary for a land animal to evolve into a whale. This involves (1) forelimb changes, (2) the development of a tail fluke, (3) streamlining of the body, (4) reducing the hind-limbs, (5) alterations of the skull, (6) moving the nostrils to the top of the head, (7) changes to the trachea, as well as (8) developing specialized nipples for the young to feed underwater. These changes would most likely take hundreds, even thousands of transitional species, at least for a hypothetical land animal to evolve into a common ancestor and then into a modern whale. If the process is repeated to explain all of the gaps between different types of organisms and to connect all the unique and isolated groups such as whales, ichthyosaurs, pleisiosaurs, turtles, seals, and sea cows, it becomes obvious that the number of modifications necessary is inconceivably great.[7-136]

Creationist View

As is commonly known, the whale is not a fish, although it lives in the sea. The whale is a warm-blooded, air-breathing mammal, specially adapted for life in, and mostly under the water. Darwinists believe that somehow a whale must have evolved from an ordinary land-dwelling animal, which took to the sea and lost its legs. This appears to be feasible until of all the other changes that would need to take place before a land animal could become a whale are considered.

First, the body of a land mammal ends in a pelvis, which supports a relatively flimsy tail. This tail always moves from side to side, and is used mainly as a fly swish. A whale has no pelvis. Instead, it has an entirely different bony structure that supports a large flat tail, which moves up and down so that it can be used for propulsion under water. **Second**, land mammals have a skin full of sweat glands to keep them cool in the hot sunshine. The whale does not need these, but instead the skin of a whale is lined with a thick layer of blubber to keep it warm in cold water. Also, the whale's skin has a unique outer surface that helps to streamline the flow of water. **Third**, land mammals have eyes to be able to see on land, while the whale's eye is designed to be able to see under water. It is different than the eye of a mammal. **Fourth,** land animals communicate by air-borne noise, conveyed between their vocal systems and their ears. Whales, however, have an underwater system of communication that is entirely different. **Fifth**, land mammals are not capable of eating under water, as they would drown. Whales, however, do have the ability to eat under water. Many whales eat very small fish, which they catch in a kind of sieve of whalebone in their mouths, a device that is perfectly suited for this purpose. **Sixth**, land mammals give birth on land. If a mammal gave birth in the water, the infant would immediately drown. Also, there is no way for a mammal to be able to suckle its young in water. Whales, however, have no problem giving birth under water or feeding their young under water because they are built to an entirely different plan that enables them to bear and nurse their offspring in deep water. A land mammal that was in the process of becoming a whale would fall between two types of animals. The transitional animal would not be fitted for life on land or the sea and would have no hope of survival. The fossil evidence shows that all these changes presumably occurred within a time span of five to ten million years (Figure 7.17). This time frame is just a fraction of one per cent of the time that life is supposed to have been on earth.[7-137]

As mentioned above, Darwinists believe that whales evolved from some form of land mammal. However, there are many changes required for a whale to evolve from a land mammal. One of them is to get rid of its pelvis. This would tend to crush the reproductive orifice with propulsive tail movements. A shrinking pelvis, however, would not be able to support the hind limbs needed for walking. So the hypothetical transitional form would be unsuited to both land and sea, and hence be extremely vulnerable and most likely natural selection (Chapter 9) would have eliminated it. Also, the hind part of the body must twist on the fore part, so the tail's sideways movement can be converted to a vertical movement. Seals and dugongs are not anatomically intermediate between land mammals and whales. They have particular specializations of their own.

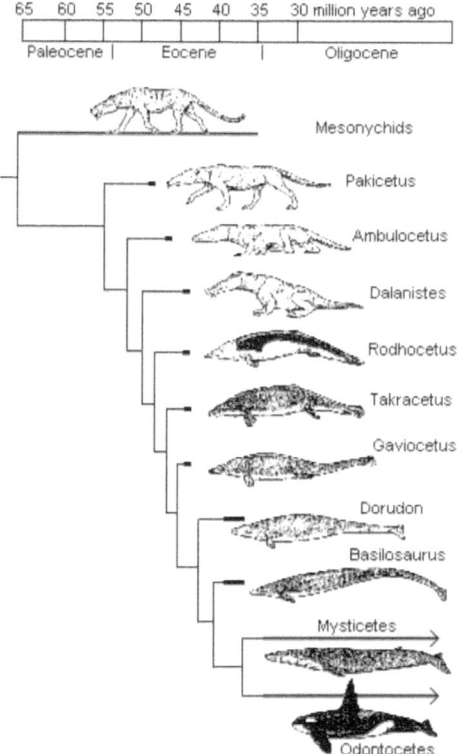

Figure 7.17 Evolutionary sequence of Whales[7-138]

The lowest whale fossils in the fossil record show they were completely aquatic from the first time they appeared. The supposed transitional fossils from mammal to whale are as follows: Mesonychid (55 million years ago); Ambulocetus (50 million years ago); Rodhocetus (46 million years ago); Prozeuglodon (40 million years ago). (Figure 7.17) Notice that there is a lack of time for the vast number of changes to occur by mutation and natural selection. (Natural selection and mutations are discussed in Chapter 9.) If a mutation results in a new gene, for this new gene to replace the old gene in a population, the individuals carrying the old gene must be eliminated, and this takes time. (If the old gene remains in the gene pool of the species, it may hinder the spread of the new gene. Perhaps the creatures with the new gene could somehow just become isolated from the creatures with the old gene, to allow the new gene to propagate through the offspring of the creatures with the new gene [see Speciation, Chapter 9].) Population genetics calculations suggest that in 5 million years (one million years longer than the alleged time between Ambulocetus and Rodhocetus), animals with generation lines of about ten years (typical of whales) could substitute no more than about 1,700 mutations. This is not nearly enough time to generate the new information that whales need for aquatic life, even assuming that all the hypothetical information-adding mutations required for this could somehow originate.

Many of the alleged transitional forms are based on fragmentary remains, which are therefore open to several interpretations, based on one's assumptions. Evolutionary bias means that such remains are often likely to be interpreted as transitional. But when more bones are discovered, then the fossils nearly always fit one type or another, and are no longer acceptable as transitional.[7-139] Evolutionists often point to vestigial hind legs near the pelvis. But these are found only in the Right Whale and upon closer inspection turn out to be strengthening bones to the genital wall.[7-140] (A Right Whale is a rare large (8 foot or 2.4 meters) whale with a massive head and jaw.) Many evolutionists support whale evolution by claiming that there are vestigial hind legs buried in their flesh. However, these so-called 'remnants' are not useless at all, but help strengthen the reproductive organs, as the bones are different in males and females. In reality, these four bones have special muscles attaching them to the abdomen to fortify the reproductive system. (Vestiges are discussed in Chapter 12.) As with the allegedly functionless limbs of Basilosaurus, one should not assume that ignorance of a function means there is no function. The predecessors of whales would have had to mutate in a beneficial manner to produce the above physiological adaptations. However, science shows that organisms don't survive a rapid rate of mutation. Also, essential design features would need to prevent the whales from hypothermia (abnormally low body temperature).

Mammals are warm-blooded creatures that function at a constant body temperature higher than fish, reptiles, or amphibians. Maintaining a core body temperature while being in an ocean of cold water would be a definite problem for the cetaceans during their evolution. If whales did evolve from land mammals, they did so at an unbelievable rate, accruing an amazing number of beneficial mutations and adaptations. The skeletal features would have needed to change radically, as well as the physiology.

Evolutionists claim that the Pakicetus is an evolutionary ancestor of whales. This claim is based solely on **(1)** the arrangement of cusps on the molar teeth, **(2)** a folding in a bone of the middle ear, and the **(3)** positioning of the ear within the skull. The fact that these features are not present in other land animals, but are seen in later Eocene Epoch (40-34 mya) is strong enough evidence for evolutionists to believe that these animals represent transitional forms in whale ancestry. But the reality remains that a small combination of characteristics in a land animal does not qualify it as an ancestor of whales, especially when compared to the vast majority of differences in the two species.

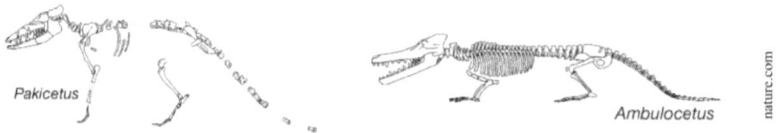

Figure 7.18 Skeletons of Pakicetus and Ambulocetus

The skeleton of the Pakicetus (Figure 7.18) is a four-footed animal, similar to that of the common wolf. It was found in a region full of iron ore, and containing fossils of such terrestrial creatures as snails, tortoises, and crocodiles. It is therefore obvious that this creature was part of a land stratum, not an aquatic one.[7-141]

Figure 7.19 Artist conception of Ambulocetus

Unlike drawings used by many evolutionists portraying Ambulocetus as having webbed feet and having swimming abilities (Figure 7.16 & 7.19), the skeleton sketch (Figure 7.18) is more similar to the actual skeleton of Ambulocetus. The legs are not fins. There are no imaginary webs between its toes. There is no reason for evolutionists to believe there was webbing between its toes, other than wishful thinking as a link to a more aquatic animal. It is obvious this animal used to walk on land because it had four legs, like all other mammals. It even had wide claws on its feet and paws on its hind legs. Ambulocetus had a backbone that ended at the pelvis. It obviously had powerful rear legs that extended from it, which is typical of all land mammals. This is an obvious difference from whales. In whales the backbone goes all the way down to the tail, there is no pelvic bone at all. Basilosaurus (a typical whale), believed to have lived some 10 million years after Ambulocetus (a typical land animal), has the anatomy of a whale. Basilosaurus would not have had sufficient time to make all of the changes necessary to have evolved from Ambulocetus (Figure 7.17).[7-142]

There are several problems with the theory that Basilosaurus and modern day whales evolved from Ambulocetus. The body shape of Basilosaurus is about ten times longer and different than Ambulocetus, as are the skull structure and the tooth shapes. Ambulocetus was 7 feet (2 meters) long and Basilosaurus was approximately 70 feet (21 meters) long.[7-143]

The evidence that is used for the claim that Ambulocetus was a walking whale is as follows: **(1)** A partial skeleton with a partial skull and lower jaw, **(2)** six cervical and upper part of the thorax vertebrae, **(3)** four ribs, **(4)** a fore arm with digits, and **(5)** a foot with digits, all found in one level. In another level five meters (16.4 feet) above the original level, a thorax vertebra, one vertebra near the tail and one femur were found. No scapula or pelvic girdle was found. Also, not enough of the tail vertebrae were recovered to determine the proper muscle attachment. Therefore, one is not able to determine for certain if Ambulocetus was a whale or even an ancestor to a whale. Evolutionists also claim that Pakicetus is a walking whale. It has been claimed that Pakicetus had an inner ear like a whale. The problem is that there were no other post cranial bones found to make any conclusive claim.[7-144] No species in the whale series could possibly be the ancestor of any other, because all of them possess characteristics they would first have to lose before evolving into a subsequent form. This is why the scientific literature typically shows each species of whale evolution branching off a supposed lineage (Figure 7.17).[7-145]

Evolutionist View

Because soft tissue do not fossilize, it is difficult to determine all the characteristics of a fossilized organism. The discovery of Ambulocetus natans provides additional support that ungulates (large mammals with hooves) evolved into whales. Creationists deny that ungulates evolved into whales. They claim that the skeletal arrangement of the creature is such that it could not undulate (move or go with a smooth up-and-down motion; to move with a winding or wavelike motion) its spine to produce a motion not unlike that of a whale's tail motion. Ambulocetus was found in geographically and chronologically just about where evolutionists hoped to find it. Examination of the configuration of basal ridges in fossilized reptile-mammal transitional form of skulls shows how changes developed gradually, even though the evolution of the soft, complex endothermic apparatus could not be directly observed.[7-145a]

Creationist View

The bones of Pezosiren portelli (Figure 7.15a), discussed in the Evolutionist View at the beginning of section "The Evolution of Land Animals to Whales", were put together by Daryl Domning. These bones came from three separate bone beds within the five-metre (16 feet) thick Guys Hill Member stratum of rock in Jamaica. (Note, in evolutionary thinking, those five metres (16 feet) would represent several million years' worth of accumulated sediments.) Bones of both land animals and ordinary sea cows have been found in the Guys Hill Member bone beds. Some of the bones that were utilized in the construction of Pezosiren portelli (Figure 7.15a) were found by themselves and some as partial skeletons. The stratum in question has yielded many hundreds of bones, including those of a rhinoceros (Hyrachyrus), a lizard, a crocodile (Charactosuchus kugleri), a turtle, sea cows (sirenians), and possibly a primate, along with lots of invertebrate marine fossils (mollusks, etc.). Some of the bones Domning used (the skull and ribs) have features typical of sea-cow bones, while others (the vertebrae, pelvis, and limb bones) have features typical of hoofed land animals. It seems quite likely that in constructing his 'legged sea-cow', Domning combined bones from different kinds of creatures.[7-145b]

Evolutionists claim that the whale is descended from land animals. They say this because it is warm-blooded and nurses its young with milk. But those are among the few things that whales have in common with mammals on the land. The whale has no neck to turn its head, and, because its eyeball is fixed, the whale must move its entire body to shift its line of sight. Its eyeball is ideal for seeing underwater, whereas land animals generally cannot do so. A special sclerotic coat protects its eye at great depths underwater. Whales produce excellent sonar. They have the ability to detect objects miles away through echolocation. Not only can they locate distant objects, but they can tell if they are neutral, friend, or enemy. According to the evolutionary theory of similarities, creatures with sonar and radar are all related; in other words, one is related to another. So the whale (which according to the theory that the whales evolved from land mammals) must be related to and therefore must have descended from the bat. The whale's nose and mouth are structured so that no water enters the body under the pressure of fast swimming or depth diving. Its forelimbs are jointless paddles or flippers; there is no fossil evidence that they evolved from animal arms. Except around the nose, the whale lacks the hair and fur that land animals have. Instead, it has thick layers of blubber to keep it warm. It has no sweat glands. The ears of a whale are designed remarkably differently than land animals. Sound is carried to the eardrums through a tube from a point beneath the surface near the eyes. It can hear other whales at a great distance.

The whale has special breathing equipment so that it can remain underwater for as long as two hours. A whale can withstand immense pressure while down at great depths that would crush any land animal that tried to go down there. The outer skin is marked with lines not found in land mammals. These lines help streamline water flow, giving it maximum speed for the least effort. In the mouth of the baleen whale are unique horny plates with fringed edges that permit it to strain out ocean water and catch tiny plant and animal plankton (the smallest creatures in the ocean) for food. The windpipe and gullet separate at about the same point in land animals, but in whales the two are located differently so the baby whale, as it nurses, will not get milk down its windpipe and choke. If a whale choked underwater, it would cough and that would carry enough water down its windpipe to kill it.[7-146]

Old Universe Progressive Creationist View

Recent scientific discoveries indicate that the first sea mammals appeared much earlier than paleontologists previously thought. Fossils of four extinct species of whales: Pakicetus (52 mya), Nalacetus (52 mya), Ambulocetus (50 mya), and Indocetus (48 mya) have been discovered (mya = million years ago). The dates assigned to these fossils indicate that sea mammals appeared before land mammals (large mammals [larger than rodents] which are presumed to have first appeared 66-23 mya during the Cenozoic Era in the Paleogene Period). These dates eliminate a naturalistic explanation for a newly found change in these whales' morphology. Phosphate isotopes in the teeth of these fossilized whales indicate a rapid transition from freshwater ingestion to saltwater ingestion. Geologists and anatomists discovered that Pakicetus and Nalacetus drank only freshwater. Ambulocetus drank freshwater at least through its formative years, probably all its life, and Indocetus drank saltwater only.

In just two to four million years (or less) whales' physiology changed radically. The transition from freshwater ingestion to saltwater ingestion requires completely different internal organs. The number and rapidity of "just right" mutations required to accomplish such a transition defies the limits set by molecular clocks (biomolecules for which mutation rates can be determined relatively easily).

Evolutionists who believe in punctuated equilibria (the alternative to gradual Darwinism discussed in Chapter 10) believe that dramatic genetic changes occurred in sudden jumps due to severe environmental stress. The period from 48 to 52 million years ago, however, appears to have been remarkably tranquil, far less stressful than such a scenario demands. For several decades now, evolutionists (those seeking a naturalistic explanation for the changes in life-forms over Earth's history) have pointed to "transitional forms" in the fossil record for proof that their explanation for life's history is correct. The fact that the bone structures of certain large land-dwelling whales, anciently saltwater-drinking whales, and modern whales exhibit an apparent progression persuades them that modern whales naturally evolved from land-dwelling mammals. Evolutionists often cite this progression as their best demonstration of Darwinian evolution. A recent discovery of ancient whale ankles, however, establishes that the ankles of all whale species are so distinct from those of mesonychids and artiodactyls, the only other suggested ancestor of whales, as to seriously call into question the descent of whales from land-dwelling mammals.

The whale is the least possible candidate to have evolved. No animal is a less efficient evolver than the whales. No animal has a higher probability for rapid extinction than the whales. Many factors severely limit their capacity for natural-process changes and greatly enhance their probability for rapid extinction. The ten most significant are: (1) Relatively small population levels (2) Long generation spans (the time between birth and the ability to give birth) (3) Low numbers of progeny produced per adult (4) High complexity of morphology and biochemistry (5) Enormous body sizes (6) Specialized food supplies (7) Relatively advanced cultural and social structures (8) High metabolic rates (9) Relatively small habitat size (for some species) (10) Relatively low ecological diversity.

These factors limit not only whales' capacity to change through natural selection and mutations but even their ability to adapt to environmental changes. A fundamental problem biologists observe (for well understood biochemical reasons) is that deleterious mutations vastly outnumbered beneficial mutations anywhere from 10,000 to 1 up to 10,000,000 to 1. Thus, it takes, for example, a very large population, a short generation time, and a small body size for a species to survive long enough to change through beneficial mutations before the onslaught of deleterious mutations and environmental stresses and changes (for example, declining rotation rate, increasing solar luminosity, changing chemical compositions of the atmosphere, changing biodeposits, supernova eruptions, asteroid and comet collisions, solar flaring, climate cycles, etc.) drive the species into extinction.

Crude mathematical models place the balance between a species capable of significant evolutionary advance and one doomed to eventual extinction at approximately a population size of one quadrillion individuals, a generation time of three months, and a body size of one centimeter. These conclusions are confirmed by field observations. Biologists directly observe significant evolutionary advance only for those species exceeding a quadrillion individuals with body sizes and generation times less than one centimeter and three months respectively.

As a consequence of the ten characteristics listed above, whales lack even the remotest possibility for significant evolutionary advance. Moreover, these ten characteristics imply that deleterious mutations and environmental changes and stresses will tend to drive any given whale species to extinction relatively rapidly. The same conclusions can be drawn for the so-called descent of horses (See section "The Evolution of the Horse" for further discussion of horse evolution). The same factors affecting whales also severely restrict horses' capacity to survive internal and external changes. Ecologists have observed several extinctions of horse and whale species during human history, but never a measurable change within a species, much less the appearance of a new one.[7-146a]

The Evolution of the Giraffe

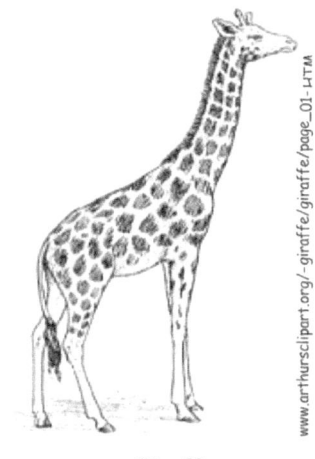

Giraffe

www.arthursclipart.org/~giraffe/giraffe/page_01-l.TM

Evolutionist View

Around 15 million years ago, antelope-like animals were roaming the dry grasslands of Africa. There was nothing very special about them, but some of their necks were a bit long. Within a mere 6 million years, they had evolved into animals that looked like modern giraffes, though the modern species only turned up around 1 million years ago. The tallest living land animal, a giraffe stands between 4.5 and 5 meters (14.763 to 16.4 feet) tall—and almost half that height is neck.[7-147]

Bohlinia (a deer-like ancestor of the modern giraffe)

The fossil record shows that giraffes evolved from a deer-like ancestor with a shorter neck. By about 1 million years ago, modern giraffes had appeared on the African savannah. Why did the long neck evolve? Until recently, the most popular theory involved finding food. Giraffe-like animals that were born with longer-than-normal necks were thought to have a big feeding advantage, since in times of scarcity, they could reach higher into trees to forage (search) for leaves. Longer-necked individuals were more successful at surviving, and passed their longneck genes onto their offspring. Over many generations, the modern giraffe evolved.

An average male giraffe's neck weighs 200 lbs. and can stretch 6 feet long. Giraffes fight over females by swinging their necks and heads like a medieval ball and chain. The longer and heavier the neck, the more momentum behind the often bone-shattering head slams. It is believed that it was the competition for mates that pushed the evolution of the giraffe's neck, with longer-necked animals more successful at reproducing. Female giraffes have many of the same genes, so their necks are long, too. But the females' necks stop growing in adolescence, while young males go on to add nearly 100 pounds of neck weight as they reach adulthood.[7-148] (See section "The Laryngeal Nerve—An Example of Poor Design" for further discussion of the giraffe's neck.)

Creationist View

> *So under nature with the nascent giraffe, the individuals which were the highest browsers, and were able during dearths to reach even an inch or two above the others, will often have been preserved. By this*

*process long-continued . . . combined no doubt in a most important manner with the **inherited effects of increased use of parts**, it seems to me almost certain that any ordinary hoofed quadruped might be converted into a giraffe.*
—Charles Darwin, Origin of the Species (1859), p. 202.

It is commonly thought that the long-necked giraffe obtained its long neck by natural selection (feeding competition hypothesis), however, it fails to explain, among other things, the size differences between males and females. (Natural selection is discussed in Chapter 9.) Giraffe cows are up to 1.5 meters (4.92 feet) shorter than the giraffe bulls, not to mention the offspring. The wide migration range of the giraffe and the low heights of the most common plants in their diet likewise argue against the dominant selection hypothesis.

Giraffes did not evolve from deer-like animals because: **1)** The fossil "links", which according to the theory should appear successively and replace each other, usually exist simultaneously for long periods of time. If giraffes with the genes for shorter necks remained in the gene pool, it would hinder (if not prevent) the giraffe from developing its longer neck, as the number of giraffes with the longer neck genes would always be less than those giraffes with the genes for the longer necks. **2)** Evolutionary derivations based on similarities rely on circular reasoning. **3)** The giraffe has eight cervical vertebrae. Although the 8th vertebra displays almost all the characteristics of a neck vertebra, as an exception to the rule the first rib pair is attached there. **4)** The origin of the long-necked giraffe by a macro-mutation is, due to the many synchronized structures, extremely improbable. **5)** Sexual selection also lacks a mutational basis and, what is more, is frequently in conflict with natural selection ("head clubbing" is probably "a consequence of a long neck and not a cause"). **6)** In contrast to the proposed naturalistic hypotheses, the intelligent design theory is basically testable. **7)** The long necked giraffes possibly all belong to the same basic type inasmuch as a gradual evolution from the short necked to the long necked giraffe is ruled out by the duplication of a neck vertebra and the loss of a thoracic vertebra. **8)** Chance mutations are principally not sufficient to explain the origin of the long-necked giraffe.[7-149]

The lungs supply the giraffe its necessary oxygen, but in a way that is unique to the giraffe. The giraffe's lungs are eight times the size of humans and its respiratory rate is about one third that of humans. Slower breathing is necessary in order to exchange the required large volume of air without causing windburn to the giraffe's twelve feet of rippled trachea. When taking in a fresh breath, the oxygen depleted previous breath cannot be 100% expelled. This problem is compounded for the giraffe by the twelve-foot trachea that will retain a larger volume of oxygen poor air than man can inhale in one breath. There must be, and is, enough lung volume to make this oxygen depleted air a small percentage of the total of each breath.[7-150]

The giraffe has the most powerful heart in the animal kingdom. This is due to the fact that it has double the normal blood pressure. This high blood pressure is required to pump blood all the way up to its brain. The giraffe's blood pressure is two or three times that of a healthy man and probably is the highest of any animal in the world. Because the giraffe has such a long neck (10-12 feet [3.048 – 3.6576 meters] in length), its heart must exert an immense force to pump blood through the carotid artery to the brain. The giraffe's heart is huge; it weighs 25 pounds [11.3398 kilograms], is 2 feet [0.6096 meters] long, and has walls up to 3 inches [0.0762 meters or 7.62 centimeters] thick. In contrast, the brain of any animal is a very delicate structure and is not able to stand high blood pressure.

What happens when the giraffe bends over to take a drink from a pond? High pressure is needed to get blood to the brain, yet that very pressure should destroy the brain when it lowers its head to the ground. Four carefully thought-out design factors nicely solve this problem: **(1)** The giraffe has in his jugular veins a series of one-way check valves. These immediately close as soon as the head is lowered. There is still a large amount of blood in the carotid artery; actually too much. **(2)** That extra blood is immediately shunted to a special spongy tissue, located near the brain and filled with small blood vessels, which absorbs it. In addition, **(3)** the cerebrospinal fluid, which bathes the brain and spinal column itself, produces a counter pressure to prevent rupture or capillary leakage. **(4)** the walls of the giraffe's arteries are thicker than those of any other mammal.[7-151]

The giraffe would pass out if it raised its head too quickly. Fortunately, the giraffe is equipped with special organs to allow it to lift its head without sustaining a brain injury. Without these special organs the giraffe would most likely faint if it raised its head too suddenly. The probability of a predator attacking would be greatly increased. If the giraffe evolved from an animal with a short neck, it would not have survived as his neck became longer and longer, if the blood flow (blood pressure) did not keep up with the needs and also if the oxygen levels did not adjust properly to the increasing neck length. How would random changes have been able to accomplish this, since changes are non-directional and serve no particular purpose? (That is, according to evolutionists, positive or negative mutational changes would occur naturally (randomly) prior to natural selection determining whether a change would be beneficial or not to a species in a particular environment. Multiple simultaneous positive changes would be needed to coordinate the increased blood pressure [which includes the heart muscle along with lungs becoming larger to accommodate the longer neck and increased blood flow] with longer neck lengths. See Chapter 9, "Natural Selection" and "Mutations" sections, for further discussion.) Would the evolving long-necked giraffe have been able to survive during its development?

As the giraffe bends his head down to get a drink of water from a stream, valves in the arteries in its neck begin to close. Blood beyond the last valve continues moving toward the brain. But instead of passing at high speed and pressing into the brain and damaging or destroying it, that last pump is shunted under the brain into a group of vessels similar to a sponge. The brain is preserved and the powerful surge of oxygenated blood gently expands this "sponge" beneath it. The giraffe's neck is designed in such a way that as he begins to raise his head, the arterial valves open. The "sponge" squeezes its oxygenated blood into the brain, the veins going down the neck contain some valves which close to help level out the blood pressure, and the giraffe can quickly be erect and running without passing out and allowing his predator to devour him. The giraffe was designed just like it is with all systems complete and ready for any emergency. There is no way the giraffe could have evolved its special features.[7-152]

Darwin cited the giraffe as an outstanding example of natural selection (See quotation at the beginning of this section.). (Natural selection is discussed in Chapter 9.) Supposedly, as a result of extended droughts, the supply of green leaves could be obtained only at the top of the trees, and therefore the shorter necked giraffe died off. The giraffes that grew longer necks survived. However, there is no evidence whatever in the fossil record or elsewhere that giraffes with short necks have ever existed. What would have happened to young giraffes with relatively short necks? Would they have starved to death if there were no leaves for them to reach?[7-153] Darwin did not realize that body characteristics in a creature's offspring are determined and programmed by DNA factors of the Genes or the Genetic material of the parents, and not by the stretching of the neck or any other bodily exercise.[7-154] Charles Darwin supported the idea of "use and disuse inheritance" developed by French biologist Jean-Baptiste Lamarck, though he (Charles Darwin) rejecting other aspects of Lamarck's theory.[7-154a] (See Darwin quote at the beginning of this section.)

Because a giraffe's neck is so long, there's a huge distance between a giraffe's heart and its brain. It needs auxiliary pumps (reinforced artery walls, by-pass and anti-pooling valves, etc.) to get blood to the brain so it will not faint when it raises its head up. And it needs pressure reducers so that when it bends its head down to take a drink, it will not blow its brains out. A long neck without these features would be deadly. It is therefore very unlikely that a giraffe could have gotten its long neck gradually by mutations and natural selection. The giraffe's neck is actually well designed.[7-155]

The Laryngeal Nerve—An Example of Poor Design

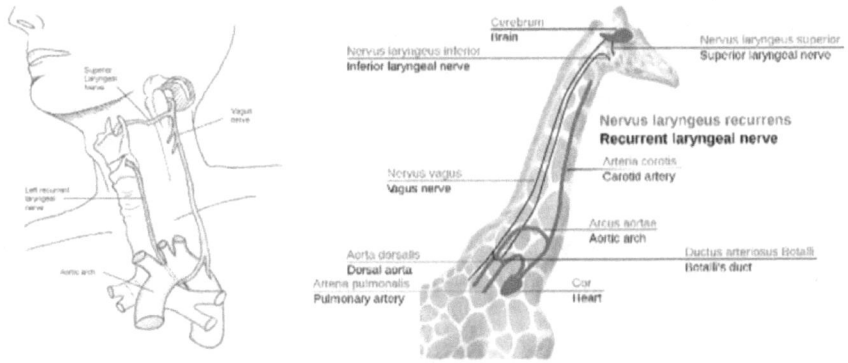

http://blog.eternalvigilance.me/wp-content/uploads/2013/04/giraffe-recurrent-laryngeal-nerve.png

The Laryngeal Nerve

Human (left)   Giraffe (right)

Figure 7.19a The Laryngeal Nerve – Human and Giraffe illustrations

Evolutionist View

There are many design flaws in mammals. This is due to evolutionary improvements. This is not, however, what would be expected if a creator designed mammals. One example is the extra-long laryngeal nerve. It is a branch of one of the cranial nerves, those nerves that lead directly from the brain rather than from the spinal cord. One of the cranial nerves, the vagus, has various branches, two of which go to the heart, and two on each side to the larynx (voice box in mammals). The laryngeal nerve goes to the larynx in an indirect route (Figure 7.19a). It goes right down into the chest, loops around one of the main arteries leaving the heart, and then goes back up the neck to the larynx. If the laryngeal nerve was designed, why did the designer make it go through such an indirect route?[7-156]

The laryngeal nerve goes from the brain to the larynx, which helps us to speak and swallow. The odd thing is that it is much longer that it needs to be. Rather than taking a direct route from the brain to the larynx, a distance of about one foot in humans, the nerve runs down into the chest, loops around the aorta and a ligament derived from an artery, and then travels back up to connect to the larynx (Figure 7.19a). The laryngeal nerve ends up being three feet long. The extra length makes it more prone to injury. For example, it might be damaged by a blow to the chest, making it hard to talk or swallow. This design only makes sense when viewed as being evolved. Like the mammalian aorta itself, it descends from those branchial arches of our fish-like ancestors. In the early fish-like embryos of all vertebrates, the nerve runs from the top to bottom alongside the blood vessel of the sixth branchial arch. It is a branch of the larger vagus nerve that travels along the back from the brain. And in adult fish, the nerve remains in that position, connecting the brain to the gills and helping them pump water.

During the evolution of humans, the blood vessel from the fifth arch disappeared, and the vessels from the fourth and sixth arches moved downward into the future torso so that they could become the aorta and a ligament connecting the aorta to the pulmonary artery. The laryngeal nerve, however, still behind the sixth arch, had to remain connected to the embryonic structures that become the larynx, structures that remained near the brain. As the future aorta evolved backward toward the heart, the laryngeal nerve was forced to evolve backward along with it. It would have been more efficient for the nerve to detour around the aorta, breaking and then re-forming itself on a more direct course, but natural selection could not have done that, for severing and rejoining a nerve is a step that reduces fitness. (Natural selection is discussed in Chapter 9.) To keep up with the backward evolution of the aorta, the laryngeal nerve had to become long. That evolutionary path is recapitulated during development, since as embryos humans begin with an ancestral fish-like pattern of nerves and blood vessels. (Recapitulation of evolutionary past during embryonic development is discussed in Chapter 11.) The end result is that the laryngeal nerve ends up being a bad design.[7-157] In giraffes the nerve takes a similar path, but it goes all the way down the long neck and back up again (Figure 7.19a). This unnecessarily long path causes the laryngeal nerve to go fifteen feet longer than if it had gone by a direct route.[7-158]

As the giraffe's neck slowly lengthened over evolutionary time, the cost of the detour gradually increased. The marginal cost of each millimeter of increase was slight. As the giraffe's neck began to approach its present length, the total cost of the detour might have begun to approach the point where (hypothetically) a mutant individual would survive better if its descending laryngeal nerve fibers broke themselves off from the vagus bundle and hopped across the tiny gap to the larynx. But the mutation needed to achieve this 'hop across' would have to have constituted a major change in embryonic development. Very probably, the necessary mutation would never happen. The recurrent laryngeal nerve in any mammal is good evidence against a designer. The giraffe illustration is an excellent example. The unnecessary long detour down the giraffe's neck and back up again is exactly the kind of thing one would expect from evolution by natural selection, and exactly the kind of thing that one would not expect from any kind of intelligent designer.[7-159]

Creationist View

The recurrent laryngeal nerve (RLN) controls the mammalian larynx (voice box) muscles. Evolutionists often claim that the loop of the recurrent laryngeal nerve (RLN) in mammals is proof of evolution. They claim that during the process of fish evolving into mammals the recurrent laryngeal nerve ended up with a long loop in mammals. Evolutionists claim that if the recurrent laryngeal nerve were designed, it was designed poorly. In mammals, this nerve does not have a direct route between brain and throat and instead descends into the chest, loops around the aorta near the heart, then returns to the larynx. That makes it seven times longer than it needs to be.

Although the laryngeal nerve does not take the shortest route to the larynx, this is also true for many other nerves. The optic nerves do not take the shortest route to the occipital lobe of the brain (the lobe near the back of the head), but rather cross over at the optic chiasm (where the two tracts cross over in the form of an "X") for reasons now known to be based on good design. The nerves from the right side of the brain go to the left side of the body (except for the right and left frontal branches off a facial nerve, which are supplied by both sides of the brain) also for good reasons. The left RLN has a different anatomical trajectory than one would first expect, and for very good reasons. In contrast to some of the claims made by evolutionists, the vagus nerve (the longest of the cranial nerves) travels from the neck down toward the heart, and then the recurrent laryngeal nerve branches off from the vagus just below the aorta (the largest artery in the body, originating from the left ventricle of the heart and extending down the abdomen). The RLN travels upward to serve several organs, some near where it branches off of the vagus nerve, and then travels back up to the larynx. This is the reason it is called the left recurrent laryngeal nerve. In contrast, the right laryngeal nerve loops around the subclavian artery just below the collarbone, and then travels up to the larynx. The fact that the longer left RLN works in perfect harmony with the right laryngeal nerve, disproves the faulty design claim. The most logical reason for the RLN design is due to limitations during embryo development. It has been noted that the recurrent laryngeal nerve that appears to be a poor design in adults is due to the necessary consequences of developmental dynamics rather than a result of evolution.

The embryo as a whole must be a fully functioning system in its specific environment during every second of its entire development. For this reason, adult anatomy can be understood only in the light of development. The human body begins as a sphere called a blastocyst (Chapter 11, Figures 11.3 & 11.4) and gradually becomes more elongated as it develops. Some structures, such as the carotid duct, are simply obliterated during development, and some are eliminated and replaced. Other structures, including the recurrent laryngeal nerve, move downward as development proceeds. The movement occurs because the neck's formation and the body's elongation during fetal development force the heart to descend from the cervical (neck) location down into the thoracic (chest) cavity. As a result, various arteries and other structures must be elongated as organs are moved in a way that allows them to remain functional throughout this entire developmental phase. The right RLN is carried downward because it is looped under the arch that develops into the right subclavian artery, and thus moves down with it as development proceeds.

The left laryngeal nerve recurs around the ligamentum arteriosum (a small ligament attached to the top surface of the pulmonary trunk and the bottom surface of the aortic arch) on the left side of the aortic arch. It likewise moves down as the thoracic cavity lengthens. The body must operate as a living, functional unit during this time, requiring ligaments and internal connections to secure various related structures together while also allowing for body and organ movement. For the laryngeal nerve, the ligamentum arteriosum functions like a pulley that lifts a heavy load to allow movement. As a result of the downward movement of the heart, the course of the recurrent laryngeal nerves becomes different on the right and left sides. These nerves cannot either be obliterated or replaced because many of them must function during every fetal development stage. It has been noted that no organ could exist that is functionless during its development, a situation that also applies to the nervous system. This movement appears designed to position the left RLN downward as the body elongates. The laryngeal branch splits up into other branches before entering the larynx at different levels. The various RLN branches serve several other organs with both motor and sensory branches, including the upper esophagus, the trachea, the inferior pharynx, and the cricopharyngeus muscle, the lowest horizontal band-like muscle of the throat just above the esophagus. The fact that the left RLN also gives off some fibers to the cardiac plexus is highly indicative of developmental constraints because the nerve must serve both the larynx (in the neck) and the heart (in the chest).

As mentioned previously, after looping around the aorta, the RLN travels back up to innervate the larynx. The upper and recurrent laryngeal nerves then innervate an area known as Galen's anastamosis. Other cases exist of one nerve splitting off early and providing direct innervations, and another taking what seems like a circuitous route. One example is the phrenic nerve that arises in the neck and descends to connect to the diaphragm. This is a necessary path, since the pericardium and diaphragm arise in the septum transversum (a thick mass of tissue that gives rise to parts of the thoracic diaphragm and the ventral mesentery of the foregut) in the neck area of the early embryo. It then migrates caudally (toward the tailbone) as the embryo enlarges by differential growth of the head and thorax areas, taking the nerve with it. The diaphragm cannot have evolved step-wise, since a partial diaphragm results in an imperfect chest-abdomen separation. Even a small defect results in herniation of the gut contents into the chest, which either compresses the lungs or results in strangulation of the gut. A complicated issue still being researched is how the incredibly complex nerve-muscle system, the component nerve fibers, and the laryngeal muscles arise from the neural crest (cells between the epidermis and the neural tube that develop into the brain and spinal cord) and dorsal somites (cells that develop into muscles and vertebrae) respectively in the early embryo, and then migrate anteriorly (towards the front of the body) into their final positions.

Without explaining the nerve structure's design system, function, and ultimate connections, alleging that the RLN is a poor design is a meaningless assertion. Thus, the claim that it has to loop up the distance from the ligamentum arteriosum for no reason is invalid. Some innervations to the larynx go directly to the larynx, including the sensory internal laryngeal nerve and the motor external laryngeal nerve. Other nerves, the left and right superior laryngeal nerves, branch off close to the larynx to provide this structure with direct innervation. The superior laryngeal nerve branches off of the vagus at a location called the ganglion nodosum and receives a nerve branch from the superior cervical ganglion (group of nerve cells near the neck) of the sympathetic nervous system (a branch of the autonomic nervous system).

In addition to the developmental reasons for the indirect route, certain benefits of overlapping sensory and motor innervations result when one of the nerves is slightly longer. One reason why laryngeal nerve branches are located both above and below the larynx (both branch off the vagus) is because this design allows some preservation of function if either one is interrupted. The redundant pathway also provides some backup in case of damage to one of the nerves. Knowledge of the laryngeal innervation will help to understand the necessity for the slightly longer route for a nerve, and a hint is provided from the fact that the two nerves regulate different vocal responses.

Several studies found that the existing path occupies a relatively safe position in a groove that allows it to be less prone to damage or injury than a more direct route. Arguing that the left RLN is poorly designed implies that if God designed the RLN, this God should have used different embryo developmental trajectories for all the structures involved to avoid looping the left RLN around the aorta. One who asserts that the RLN is a poor design assumes that a better design exists, a claim that cannot be asserted unless an alternative embryonic design from fertilized ovum to fetus, (including all the incalculable molecular gradients, triggers, cascades, and anatomical twists and tucks) can be proposed that documents an improved design. Lacking this information, the "poor design" claim uses evolution to fill in gaps in our knowledge. Furthermore, any alternative embryonic design pathway would likely result in its own unique set of constraints, also giving the false impression of poor design.

The left recurrent laryngeal nerve is not poorly designed, but rather is clear evidence of intelligent design. Much evidence exists that the present design results from developmental constraints. There are indications that this design serves to fine-tune laryngeal functions. The nerve serves to innervate other organs after it branches from the vagus on its way to the larynx. The design provides backup innervation to the larynx in case another nerve is damaged. No evidence exists that the design causes any disadvantage. The arguments presented by evolutionists are both incorrect and have discouraged research into the specific reasons for the existing design.[7-160]

The Evolution of the Horse

Evolution of the Horse

Figure 7.20 Evolution of the Horse

Evolutionist View

Of all animal lineages, the most famous general lineage of all is the horse sequences, which is called Perissodactyls (horses, tapirs, rhinos). It was the first such lineage to be discovered, in the late 1800's, and thus became the most famous. Some creationists wrongly believe that the horse sequence is questionable or outdated. As a matter of fact, the sequence has increased and has become more detailed and complete.[7-161] The horse family is a good example of a lineage that appears to be transitional to modern horses in a progressive straight line, however, this is not the case. Generally, there is no transition of horses that progresses in a straight line. At any one time there are multiple successful species, and only some continue to develop into modern horses. For example, Merychippus (Figure 7.20) evolves to almost nineteen new three-toed grazing horse species. These species traveled all over the world, although only one of these lines happened to lead to Equus (the modern horse).[7-162]

Some creationists claim the fossil record for horses from Eohippus (Hyracotherium) to Equus (modern horse) as insignificant (Figure 7.20). This is because only intermediate forms are known. Without the intermediate forms, there would be a large gap in the horse lineage. Eohippus was a very small species, only about eighteen inches long. Eohippus had four toes in the front and three toes in the back. The significance of the fossil record of horses becomes clearer when it is compared with that of the other members of the order Perissodactyla (odd-toed ungulates). The fossil record of the extinct titanotheres (also known as 'brontotheres,' are large, thick-skinned quadrupeds of extinct mammals that includes horses, rhinoceroses, and tapirs) is quite extensive, and the earliest representatives of this group are very similar to Eohippus. Also the earliest members of the tapirs and rhinos were very much like the Eohippus. Therefore, the different perissodactyl groups can be traced back to a group of very similar small ungulates.[7-163]

In the Paleocene period (67-54 mya) of western North America, a diverse group of dog-sized animals known as condylarths existed. The various species show transitions to such groups as the carnivores, the titanotheres, and the rhinoceroses (which at that time included both small running forms and amphibious species). One of the condylarths that appeared during the Eocene epic (54-34 mya) was Phenacodus, which had five toes with claws that were slightly developed into hooves. The central toe was also slightly enlarged. A close relative of Phenacodus was Hyracotherium, the Eocene eohippus or "dawn horse." Hyracotherium was much like Phenacodus, except that it had only four toes on the front foot and three on the rear, slightly longer foot bones, and a slight tendency for the cusps on the molar and premolar teeth to be united into crests. The differences between Phenacodus and Hyracotherium are equivalent to those that can often be seen within species. The number of toes, for example, varies within many species of living vertebrates.

Throughout the Eocene epic (54-34 mya), for more than 20 million years, most of the characteristics of Hyracotherium hardly changed, except for the teeth. The tendency of the cusps to form crests increased continually, so that the late Eocene epic (40-34 mya) form is given a different name, Epihippus. Then a slightly jump brings us to Mesohippus, in the early Ologocene epic (34-30 mya). Mesohippus is somewhat larger than Epihippus, has a longer face and longer legs, and the first toe (thumb) of the front foot is reduced to a vestigial nubbin. (Vestiges are discussed in Chapter 12.) Essentially, Mesohippus has three toes per foot, and its side toes are just as large as in Hyracotherium. The cusps of the teeth are joined into well developed crests, more suitable for grinding vegetation. Throughout the Oligocene epic (34-24 mya), Mesohippus changes gradually into Miohippus. It becomes larger, and an extra crest on the teeth that appears first as a variation within the Mesohippus population later becomes a typical feature of Miohippus.

It has not yet been determined why these changes occurred, but some suggestions have been proposed. The characteristics of Hyracotherium are those of a forest-dwelling animal that browsed on fairly soft foliage and scampered from thicket to thicket. As the Tertiary Period (Cenozoic Era 65 – 0 mya) progressed, the climate became drier, and grasslands replaced forests in much of North America. Because of the silica in the leaves, grass is difficult to chew and wears teeth down rapidly. It's likely that the increased ridges of the horses' teeth, and the greater height of the teeth, were adaptations to an increased diet of grass, as they are in certain other groups of mammals such as voles (field mice). In open country, moreover, many mammals escape predators by running long distances, rather than by springing quickly into thickets. The fusion of bones, enlargement of the central toe, and lengthening of the leg that happened in the evolution of the horses provided mechanical advantage for rapid running, as it has in other groups of mammals.

During the next 15 million years, the Miocene epic (24-5.3 mya), Miohippus grades into several distinct lines that diverged from each other. 1) Archaeophippus was a dwarfed version of Miohippus. 2) Anchitherium retained the three-toed condition and the small, simple teeth of Miohippus, but there are slight differences in the shape of the tooth crests. Anchitherium evolved into a larger form, Hypohippus, which became extinct in the early Pliocene epic (5.3 – 2.5 mya). 3) The third line that Miohippus gave rise to was Parahippus in the early Miocene epic (24 – 20 mya), a time during which grasslands became more widespread.

In the transition from Miohippus to Parahippus, several of the smaller crests on the teeth became enlarged, connecting the other crests into a complex series of ridges that were suitable for grinding grass. Moreover, there was a gradual increase in the height of the teeth, so that they could grow continually out of the gums as the tooth surface became worn down. This so-called "hyposodont" condition of the tooth was accompanied by the development of a cement-like substance on the tooth surface, between the ridges. All these changes, as well as an increase in body size and the length of the face, occurred very rapidly, so that Parahippus quickly became transformed into Merychippus. Merychippus had teeth well adapted for grazing grasses, but still had three toes. However, the side toes were large in some specimens of Merychippus and very small in others. By the end of the Miocene epic (5.3 mya) there was a great rise of species. Merychippus split into six different lines that varied in body size and the details of tooth structure. One of these was Pliohippus, which resembled one of the species of Merychippus in its distinctive pattern of tooth ridges and very small side toes. The later species of Pliohippus extended these trends even further, and became indistinguishable from the one-toed Equus, the modern horse.[7-163a]

Creationist View

Evolutionists portray the evolutionary horse tree as though it were fact, however, this view is highly debatable. Horses have great varieties within a single type of horse. There is no indication in the fossil record that the horse ever evolved. Paleontologists find bones of supposed ancestors and descendants in strata that they label by the same name. The conclusion is that these were all animals, whether they were horses or not. For example, scientists named some bones by the title of Phenacodus and claimed they were the ancestor of eohippus. Eventually, however, it was discovered that Phenacodus could not be the ancestor of eohippus because Phenacodus came too late in the fossil record (55 mya) to be an ancestor of eohippus (60-45 mya) (Hyracotherium), as Phenacodus existed the same time as eohippus and was also too large to fit into the early ancestral stages.[7-164]

The so called horse series is one of the most used "proofs" of evolution. The evidence purported by evolutionists is a familiar fossil series, progressing from the small, four-toed "dawn horse" or Eohippus (supposedly living 50 million years ago) to the large, one-toed "modern" horse, Equus. Shown in between, in order of increasing size, are usually such intermediates as Mesohippus, with three toes, and Merychippus, which had two of the toes smaller than the third. (There are actually over a dozen horse ancestors shown in a complete evolutionary "family tree.") Some diagrams also show a trend in tooth changes, supposedly evidence of evolution from browsing on bushes to grazing on grass. The horse series is non-existent.

Nowhere in the world can these fossils be found "correctly" sequenced in successive strata. In Oregon, three-toed and one-toed species are found in the same geological layer, and in South America, a one-toed species is found below its supposed three-toed ancestor. Three-toed horses are known today, in the American southwest, with all three toes of nearly equal size. Also, today's horses come in an amazingly wide range of sizes, so modern (large, one-toed) horses do not represent the last stage of a series.

Eohippus is a small four-toed "dawn horse" that does not look at all like a horse (Figure 7.20). The Hyracotherium, is identical to that of the hyrax or rock badger, a rabbit-like animal that lives today in the African bush. Eohippus (Hyracotherium) fossils have been found in surface strata together with two modern Equus species. These "distant evolutionary relatives," supposedly marking the beginning and end of the horse series, actually coexisted.[7-165]

The fossils for the horse series are not found in the proper time sequence as indicated by evolutionists, and the major types of horses appear abruptly, without transitions. There is a discrepancy in the skeletal development of this series. The Eohippus had 18 pairs of ribs. The Orohippus had only 15 pairs of ribs. Then Pliohippus jumped to 19 pairs of ribs. The Equus Scotti is back to having 18 pairs of ribs.[7-166] Evolutionists believe that the smaller horses evolved into the bigger ones, however, the dating method they use, the strata layer based on the geologic column, contradicts what they claim. Sometimes two of the so-called progressive forms were buried together. This means that they lived at the same time and therefore one could not be an ancestor to the other. In order for an animal to evolve into another species, one of the species needs to be millions of years older than the species that it supposed to have evolved into. The earliest horse fossils should appear in the oldest, lower rock strata, but this is not the case. Bones of the supposed older, more primitive horses have been found at or near the surface. Sometimes, bones of older horses are found next to modern horse fossils.[7-167]

The fossil record does not show the clear progression that evolutionists would have us believe. For example, in northeastern Oregon, the three-toed Neohipparion (16-5 mya) and one-toed Pliohippus (12-6 mya) were found in the same layer. This indicates that they were living at the same time, and thus provides no evidence that one evolved from the other.[7-168]

Preexisting Genetic Information

Creationist View

Different kinds of creatures were created with a great deal of genetic information. Natural selection (Chapter 9) can eliminate certain preexisting genetic information, by eliminating creatures not suited to a particular environment. Thus, many different varieties can be produced in different environments. Also, much of this genetic information may have been hidden in the original created kinds. (That is, the features coded for are not expressed in the offspring.) Scientists have also found that genetic information also had other controlling or regulatory genes that switch other genes 'on' or 'off.' That is, they control whether or not the information in a gene will be decoded, so the trait will be expressed in the creature. (See Chapter 13, section "Some DNA Portions are Used to Turn Genes On or Off.") This would enable very rapid changes, which are still changes involving already created information, not generation of new information. Note that these changes only occur within a kind. This would be an example of microevolution (discussed in Chapter 9).

Applying these concepts to the horse, the genetic information coding for extra toes is present, but is switched off in most modern horses. Sometimes a horse is born where the genes are switched on, and many horse fossils indicate that the genes had been switched on. This would explain, for example, why there are no transitional forms showing gradually smaller toe size. It is possible that body size and tooth shape were also controlled by regulatory genes. Scientists have found that a single protein, called BMP-4, prevents the gene that causes molars (back grinding teeth) to form, so incisors (cutting teeth) can grow instead. Without this protein no incisors will grow. This would explain the alleged horse evolutionary series as variation within the equine (horse) kind. The amount of variety within living horses, undoubtedly one kind, supports this.

Many evolutionists claim that the horse's splint bones in their legs are vestigial, that is, useless leftovers from its supposed evolutionary past. (See Chapter 12 for further discussion of vestiges.) Vestigial organs do not provide any evidence for evolutionary theory because it is impossible in principle to prove that an organ has no function. An organ could have a function that is presently unknown. The horse's splint bones serve several important functions. They strengthen the leg and foot bones, very important because of the enormous stress that galloping puts on the legs. They also provide attachment points for important muscles. Also, they form a protective groove that houses the suspensory ligament, a vital elastic brace that supports the horse's weight as it walks.[7-168]

The evidence of horse fossils indicates that the horse lineage definitely had biological change over time. Evolutionary scientists believe this indicates Darwinian evolution. Creation scientists, however, say that the evidence simply indicates changes within the horse basic type and that there is little evidence to suggest that horses developed from a non-horse ancestor. Since the type of change represented by the horse lineage can be explained by both evolution and creation theories, the horse series could be used effectively to support either evolution or creation.[7-169]

In the next chapter (8) we will continue our discussion of the fossil evidence, however, in the next chapter we will focus specifically on the progression of the reptilian jaw/ear to a mammalian jaw/ear.

~~~

Chapter 8

Transition of Reptile Jaw/
Ear to Mammal Jaw/Ear

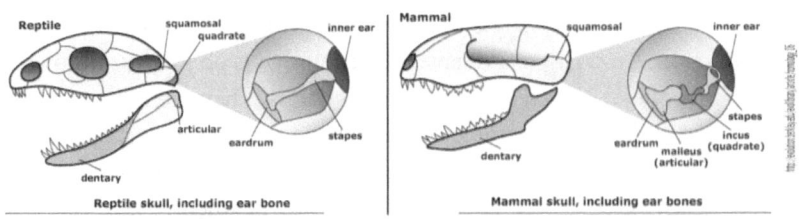

Reptile & Mammal Skulls, Jaws & Ears

Introduction

In Chapter 7, section "Evolution From Reptiles to Mammals," the possibility that reptiles evolved into mammals was discussed. In this chapter (8) we will continue our discussion of whether or not reptiles could have evolved into mammals, particularly looking at the evidence that pertains to the reptile ear/jaw transitioning into a mammal type ear/jaw.

Transition from Reptilian to Mammalian Type Jaw

Evolutionist View

It is commonly accepted among evolutionists that reptiles evolved into mammals. During this process, parts of the reptile jaw gradually became the ear bones of the mammal. This transition is an example of what creationists would otherwise call irreducible complexity (discussed in greater detail in Chapter 14). Creationists claim that it would be impossible for a reptilian jaw to evolve into a mammalian type jaw. The discussion that follows explains how this was accomplished by evolution. The development of the mammalian middle ear from a reptilian ear has baffled scientists for years.

The mammalian ear consists of three tiny bones called the **(1)** hammer (articular or malleus), **(2)** anvil (quadrate or incus), and **(3)** stirrup (stapes), which are joined together so that they relay signals from the eardrum to the inner ear (See Mammal ear Figures 8.1 & 8.2). If any one of these bones is removed, the middle ear would not function, thus it is said to be irreducibly complex. Yet there is an evolutionary explanation for how this complex system developed, that is, how jawbones from a reptile evolved into a mammalian ear. The fossil record shows that these three bones started out functioning as jawbones in reptiles. In transitional species between reptiles and mammals, the bones served dual functioning both in the jaw and in hearing. As other jaw bones evolved further, these **(1)** three bones became redundant in the jaw, **(2)** stopped assisting jaw motion, and **(3)** evolved further to support only hearing.[8-1]

As just mentioned, mammals have three tiny bones in their middle ears called **(1)** hammer (articular or malleus), **(2)** anvil (quadrate or incus), and **(3)** stirrup (or stapes) (Figures 8.1, 8.2). These bones transmit vibrations from the ear drum to the inner ear. Reptile ears work differently; instead of these bones, they have three extra bones in their lower jaw. As scientists look at transitional fossils leading from reptiles to mammals, they see an interesting transition in these bones. In the earliest fossils in this sequence (Figure 8.3), which are still mostly like reptiles and just a little like mammals, these three bones are still part of the jaw (Figures 8.2 & 8.3). In slightly later fossils, these three bones are farther back in the head and neck of the animal. The bones still play a function in chewing, but they are also located in the region of the body that reptiles use for hearing. In still later fossils, the other jawbones of the animals are further modified so that the three bones no longer play a role in chewing but still play a role in hearing. In later and later fossils in the sequence, the three bones become increasingly like the middle ear bones of mammals.[8-2]

The reptile skull on the left has five bones in the lower jaw (the dentary, articular, angular, surangular, and coronoid), and only one middle ear bone (the stapes). Reptile teeth all have the same basic shape.

The mammal skull on the left has only one lower jaw bone (the dentary) and three middle ear bones (the hammer, anvil, and stapes). Mammal teeth have different shapes.

The skeletal makeup of reptiles and mammals have two major differences that exist between them: **(1)** reptiles have five bones in the lower jaw (e.g. the dentary, articular, angular, surangular, and coronoid) (Figure 8.1), while mammals have only one bone in the lower jaw (the dentary), and **(2)** reptiles have only one middle ear bone known as the stapes, while mammals have three: hammer (articular or malleus), anvil (quadrate or incus), stirrup (stapes), collectively called the ossicles bones (Figure 8.1). During the evolution of the reptile into a mammal, two mammalian middle ear bones (the hammer and anvil, also known as malleus and incus) were developed from two reptilian jawbones. Thus there was a major evolutionary transition in which several reptilian jawbones, the quadrate, articular, and angular, were extensively reduced and modified gradually to form the modern mammalian middle ear. At the same time, the dentary bone, a part of the reptilian jaw, was expanded to form the major mammalian lower jawbone. During the course of this change, the bones that form the hinge joint of the jaw changed identity. The reptilian jaw joint is formed at the intersection of the quadrate and articular, whereas the mammalian jaw joint is formed at the intersection of the squamosal and dentary (Figure 8.1).[8-3]

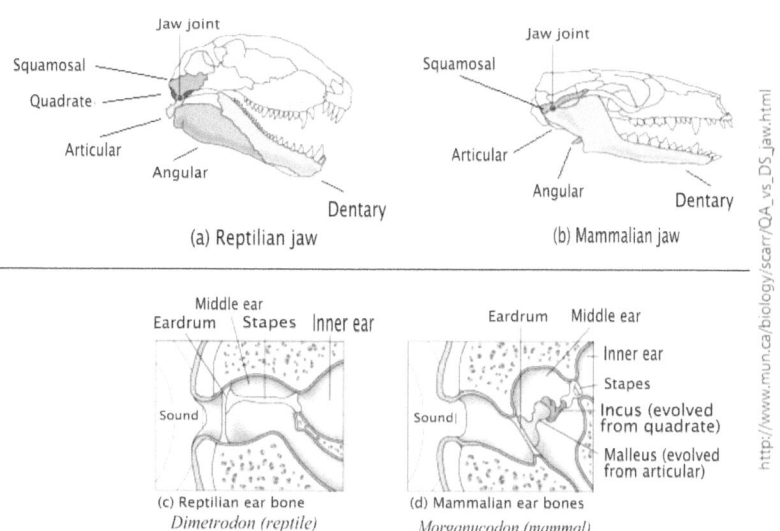

Evolution of the Mammalian Jaw

(a) Reptilian jaw

(b) Mammalian jaw

(c) Reptilian ear bone
Dimetrodon (reptile)

(d) Mammalian ear bones
Morganucodon (mammal)

http://www.mun.ca/biology/scarr/QA_vs_DS_jaw.html

Figure 8.1 illustrates reptile and mammal ear structures. The quadrate bone is also known as the incus or the anvil. The articular bone is also known as the malleus or the hammer. The stapes bone is also known as the stirrup.

Reptiles and mammals have different jaw and ear bone constructions. Reptiles have five bones in their jaw and one ear bone. Mammals have one bone in their jaw and three bones in their ear, as mentioned previously. The quadrate (incus or anvil) is directly connected to the stapes (stirrup), the articular (malleus or hammer) is connected to the quadrate (incus or anvil) and the angular (tympanic annulus) is connected to the articular (malleus or hammer) became free-floating and associated with the stapes (stirrup). In therapsids (mammal ancestors and their relatives), the lower jaw is made up of the dentary (the mandible in mammals) and a group of smaller "post-dentary" bones near the jaw joint (See Figure 8.2). As the dentary increased in size over millions of years, two of these post-dentary bones, the articular and angular, became increasingly reduced and the dentary eventually made direct contact with the upper jaw (see Figure 8.3). These post-dentary bones, even before their articular function was lost, probably transmitted sound vibrations to the stapes.[8-4]

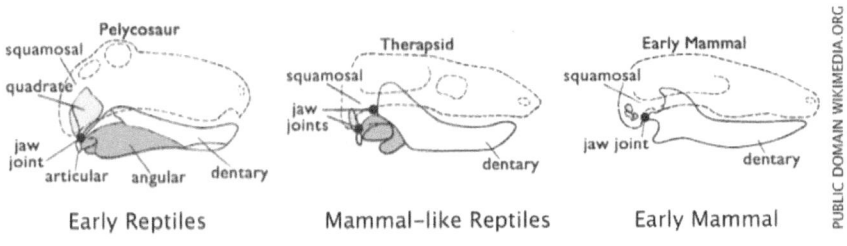

Figure 8.2—Progression of Reptilian Jaw/Ear to Mammalian Jaw/Ear[8-5]

Figure 8.2 show pelycosaur (early reptiles or non-mammalian amniote), therapsid (mammal-like reptile), and mammals jaws. These drawings illustrate the differences between reptilian jaws and ear-bone structures with those of mammals. The jaw joint is shown as a black dot (.), the quadrate (mammalian anvil or incus) is light gray, the articular (mammalian hammer or malleus) is shaded medium gray, and the angular (mammalian tympanic annulus) is shaded in dark gray. In the reptile, the jaw joint is located between the quadrate and the articular (with the angular close by), and in the mammal, the jaw joint is located between the squamosal above and the dentary below. In the reptile, the squamosal is located just above and contacting the quadrate. Advanced therapsids have two jaw joints: a reptile-like joint and a mammal-like joint.[8-6] (See Figure 8.3 for more detailed evolutionary progression.)

The single bone that makes up the mammalian lower jaw is the dentary. At its posterior end is an articular (condyloid) process, which articulates with a bone called the squamosal in the upper jaw. In therapsids (mammal-like reptiles), and other vertebrates with jaws, the lower jaw is made up of a number of bones, including the dentary and a series of additional bones, which in therapsids are concentrated in the rear half of the jaw and are collectively sometimes called the "postdentary" bones. The jaw joint is between one of these, the articular (this is different from the articular process of the dentary) and the quadrate of the upper jaw.[8-7]

Note: It has been suggested that the therapsid (mammal-like reptiles) drawing in Figure 8.2 is not totally accurate. It is claimed that the dentary/squamosal contact is actually much closer to the quadrate/articular contact. The two joints are lateral and medial to one another, not anterior-posterior.[8-8]

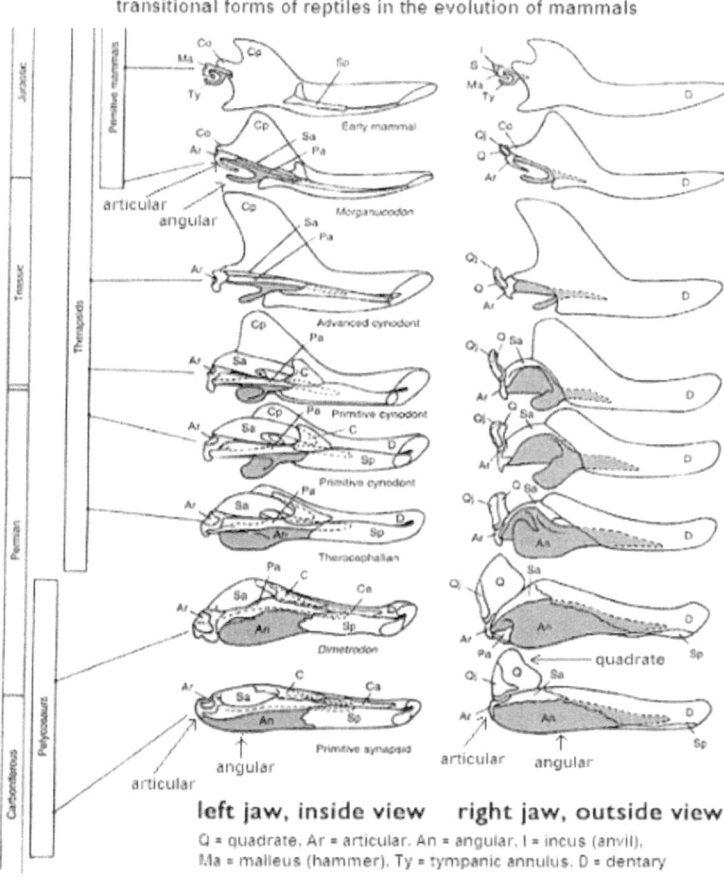

A comparison of the jawbones and ear-bones of several transitional forms of reptiles in the evolution of mammals

left jaw, inside view   right jaw, outside view

Q = quadrate, Ar = articular, An = angular, I = incus (anvil),
Ma = malleus (hammer), Ty = tympanic annulus, D = dentary

Figure 8.3—A comparison of the jawbones and ear-bones of several transitional forms of reptiles in the evolution of mammals. Approximate strata periods of the various therapsids are indicated at the far left (more recent on top). For clarity, the teeth are not shown, and the squamosal upper jawbone is omitted (it replaces the quadrate in the mammalian jaw joint, and forms part of the jaw joint in advanced cynodonts and Morganucodon).[8-9]

The reptile to mammal record reveals different stages in the evolution of jaws (Figures 8.2 & 8.3). Reptiles have a lower jaw made up of numerous bones, and the jaw joint connects the articular and quadrate bones. In contrast, the bottom jaw in mammals is only one bone and the joint is between two entirely different bones, the dentary and squamosal. As the reptilian jaw evolved, the dentary bone increased in size and the other bones reduced. Most remarkably, some mammal-like reptiles had two jaw joints—a reptilian joint and a mammalian joint. These are clearly transitional fossils, since they feature distinct characteristics from both classes of animal. Eventually, the dentary bone became the lower jaw in mammals, and the articular and quadrate bones separated from the jaws and reduced even further in size to become tiny ear bones (the incus and malleus, respectively.)[8-10] A pattern definitely exists in the fossil record with regard to tooth and jaw evolution from reptiles to mammals. Evolutionary developmental biology helps clear up how these transitions occurred. In dental development, there is an incredibly complex series of molecular interactions between the tissues that produce the two basic components of a tooth: enamel and dentine. Different types of teeth feature different combinations of embryological molecules during development.[8-11]

Scientists state that the evolution of mammals can be seen clearly by a study of the therapsids (synapsid mammal-like reptiles considered to be ancestors of the mammals) that have double jaw joints (See Chapter 7, Figure 7.4g & Figure 7.4h). Although there are many transitional features, paleontologists focus on the jaw joints: the mammalian jaw is formed by the dentary and squamosal bones while the reptile jaw is made of the quadrate and articular bones, bones which form parts of the ear in mammals (Figures 8.1, 8.2 & 8.3). Regarding the transitional forms being able to chew, the evidence shows that no jaws were "unhinged" at any point during the transition. Diarthrognathus (literally "double jaw joint") has a full set of both reptilian and mammalian jaw joints, so it can be safely assumed that none of these animals starved to death during transitional changes.[8-12]

The sinodun (cynodonts) is an ancient relative of mammals. The jaw joint in this animal is formed by two bones, one is called the articulator (mammalian hammer or malleus), in the lower jaw. The quadrate (mammalian anvil or incus) and the articular (mammalian hammer or malleus) are the two bones that in all other animals except mammals make up the jaw.

Another ancient relative of mammals is a creature called probanigmasis (Probainoganthus), which is also not a mammal. It is a little closer in relation to mammals than the sinodun is. This animal not only has the articulation between the quadrate (mammalian anvil or incus) and the articular (mammalian anvil or incus) in the upper and lower jaws, but also there is an articulation between the bone in the lower jaw called the dentary and the squamosal in the skull, and this is located in the area where the dentary and the squamosal would meet right next to the quadrate (mammalian anvil or incus) and the articular (mammalian hammer or malleus).

Sinodun (cynodonts) and probanigmasis (Probainoganthus) actually have what is called a dual jaw joint of two pairs of bones that are actually articulating next to each other on the upper and lower sides of the skull. Morogenucidaun (Morganucodon or Morganucodontans) is another animal that is slightly closer to mammals that also has a dual jaw joint of the two bones. A typical possum has changed this articulation found in more ancient ancestors so that only the dentary and the squamosal bones are connected. In the possum, the quadrate (mammalian anvil or incus) and the articular (mammalian hammer or malleus) are no longer part of the jaw joint.

The transition originally began with the quadrate articular joint only. Then, both the quadrate articular and the dentary squamosal joints became present, then only the dentary squamosal joint was present. This is the way that scientists understand this transition to have taken place. In early mammals, there is the stapes bone.

The stapes is a long structure with a big hole in the middle. The malleus (articular or hammer) is an ear bone that connects to the eardrum in the inner ear. There is also a bone called the Cochlea that is shaped like a little snail that has been in animals ever since they came out on land. The quadrate (mammalian anvil or incus) and the articular (mammalian hammer or malleus) are two bones that previously made up the jaw joint, but in mammals, now make up part of the ear bone.

In mammals, there is the stirrup shaped bone (stapes) next to a bone which is the anvil (quadrate or incus), and the bone next to it which is the malleus, or hammer. The malleus (articular or hammer) and the incus (quadrate or mammalian anvil), or the hammer and the anvil, are actually the quadrate (mammalian anvil or incus) and the articular (mammalian hammer or malleus) that used to be in the jaw joint, and now are hooked up to the stapes (stirrup) in the ear. They always were connected to the stapes (stirrup), but now are moved so that the hammer, or the articular, moved into the skull rather than being part of the lower jaw. The transition of the jaw/ear bones in reptiles to mammals is clearly evidenced embryologically as well as in the fossil record. The articular bone (mammalian hammer or malleus) moved from the lower jaw in reptiles and became part of the ear in mammals.[8-13]

How Hearing and Chewing Could Have Been
Maintained During the Transition

Evolutionist View

As multiple transitional fossils demonstrate, the bones that transfer sound in the reptilian and mammalian ear were in contact with each other throughout the evolution of this transition. In reptiles, the stapes contacts the quadrate, which in turn contacts the articular (Figure 8.1). In mammals, the stapes (stirrup) contacts the incus (anvil or quadrate), which in turn contacts the malleus (hammer or articular) (Figure 8.1). Since the quadrate evolved into the incus (anvil), and the articular evolved into the malleus (hammer), these three bones were in constant contact during this evolutionary change. Furthermore, a functional jaw joint was maintained by redundancy. Several of the intermediate fossils have both a reptilian jaw joint (from the quadrate and articular) and a mammalian jaw joint (from the dentary and squamosal). Several late cynodonts and Morganucodon clearly have a double-jointed jaw. The reptilian-style jaw joint was then freed to evolve into a new specialized function in the middle ear.

Some modern species of snakes have a double-jointed jaw involving different bones, so such a mechanical arrangement is certainly possible and functional. In reptiles, the eardrum is connected to the inner ear by a single bone, the stapes or stirrup, while the upper and lower jaws contain several bones not found in mammals. During the transitional period of the evolution of mammals, one lower and one upper jaw bone (the articular and quadrate) lost their purpose in the jaw joint and were put to new use in the middle ear, connecting to the stapes and forming a chain of three bones—hammer (articular or malleus), anvil (quadrate or incus), stirrup (stapes)—(collectively called the ossicles) which amplify sounds and allow more acute hearing.[8-14]

The Articular and Quadrate bones

Evolutionist View

Mammals use two bones for hearing while all other vertebrates (tetrapod—four footed) use these for chewing. The earliest vertebrates had a jaw joint composed of the articular and the quadrate (Figure 8.1). All reptilian vertebrates use this system including lizards, crocodilians, dinosaurs, their descendants, (the birds) and therapsids; so the only ossicle in their middle ear is the stapes (Figure 8.1).

The discovery of Morganucodon and other fossils shows definite examples of how reptiles evolved into mammals. The transition in function between the quadrate and articular bones (in reptiles) is evident. In Morganucodon, the quadrate (anvil) and the articular (hammer) serve as mammalian-style ear bones and reptilian jawbones simultaneously. In fact, even in modern reptiles the quadrate and articular (Figure 8.1) serve to transmit sound to the stapes and the inner ear. The transition, then, is a process where the ear bones, initially located in the lower jaw, became specialized in function by eventually detaching from the lower jaw and moving closer to the inner ear. [8-15]

The Fossil Evidence

Evolutionist View

Changes to reptiles occurred over millions of years. Changes of the articular-quadrate to the dentary-squamosal jaw articulation have occurred through the pelycosauromorpha to Tetraceratops to therapsids such as Probainoganthus and Diathrognathus and Thrinaxodon to Morganucodon. Probainoganthus is unquestionably transitional between reptiles and mammals. This is true whether or not it is in the direct line leading to mammals. It is much more difficult to establish exact descent than it is to establish transitional status. This is why paleontologists cannot be certain as to the exact common ancestor of the mammals, but quite certain that it was a mammal-like reptile. Creationists tend to ignore this distinction.

Evolution is not like a staircase of uniform steps, it is actually more like a bush than a tree. For simplicity, a single "line" is chosen for ease of discussion purposes. Also, not all transitional forms are presented. Tetraceratops, for instance, fits in nicely between the Dimetrodon (see Chapter 7, Figure 7.4g) and the Theracophalia, but is not included in the data. The data presented are meant to demonstrate transition, not give a complete account of it. The transition, however, has been amply documented. Notice that a particular species will not be intermediate in all respects and we do not expect to find a fossil that is exactly half-reptile and half-mammal. The cynodont Cynognathus, had a very large dentary. In this respect it was rather mammalian, however, it had a reptilian jaw articulation. This means that different characteristics change at differing rates. This is known as mosaic evolution.

Evolutionary theory makes no claim that all characteristics evolve at the same rate. When comparing major groups, it is important to choose appropriate characteristics. Jaw articulation is appropriate because it accurately distinguishes reptiles from mammals. This is also true of dentary size, temporal fenestra (an opening) size and coronoid (beak-like projection) depth. The number of teeth and incisor/canine height is less appropriate and so has been omitted. This is because these characteristics are sensitively correlated with the way of life.

Creationists believe that the evidence does not indicate a reptilian-mammalian transition. They claim that a transitional stage between reptiles and mammals is impossible because they could not imagine a functional intermediate jaw-joint. Probainognathus and Diarthrognathus have a double jaw-joint that neatly bridges the reptiles and mammals. Also, the dentary increasingly took up more of the lower jaw and the teeth became differentiated as reptiles evolved towards the mammalian jaw type (Figure 8.2). An example of a primitive mammal-like reptile is the Trochosaursus. They have simple back teeth, smaller brain case and shorter dentary compared to Thrinaxodon. Thrinaxodon was more advanced with complex back teeth, larger brain case and longer dentary. Some creationists totally ignore that Probainognathus ever existed, or that it had a double jaw articulation. They also ignore the fact that the dentary bone and temporal fenestra of Procynosucus were medium sized. At first glance, it appears that marsupials look like any other mammal. Some, however, are born with a double jaw joint. The joint shifts to the typical mammalian condition before they begin chewing. It is questionable why these mammals would have this feature unless it had been inherited from an ancestor.

Biologists predicted from such evidence that transitional forms between reptiles and mammals, the mammal-like reptiles, would have this double jaw joint. Their prediction can be considered as a reconstructed sequence based on comparing existing organisms. Later, palaeontologists discovered Diarthrognathus in the fossil record. Diarthrognathus did have this double jaw articulation. By this find, biologists discovered the original sequence and found that it was almost identical to their reconstruction. Scientists are very confident that mammals evolved from mammal-like reptiles such as a diathrognathus-like or probainognathus-like form, even if they can never resolve the exact ancestry. In this instance, they are very confident that they are mostly correct about the past even though they may never be perfectly correct. Comparataive anatomists were initially puzzled by the origin of the hammer and stirrup bones in the middle ear (Figure 8.1). By examining mammalian embryos the mystery is solved. The hammer forms from the backmost piece of a cartilage structure that, in reptiles, remains part of the lower jaw. In the mammals it has lost its function in jaw articulation, became reduced and was then changed to serve a new function that of amplifying sound in the ear.[8-16]

Megazostrodon is considered one of the earliest mammal-type creatures discovered. Megazostrodon had a mammalian jaw joint. The tooth-bearing bone formed a new jaw joint alongside the other one. Both jaw joints continued working together, although the former jaw-joint bones became looser and smaller. The inner-ear casing below the brain was so large it became the biggest feature there. Megazostrodon lived during the earliest part of the Jurassic period (213-144 mya), 210 million years ago in southern Africa. At the same time in the temperate Northern Hemisphere, Morganucodon lived in what is now England and China. At that time, all these areas were moderate in temperature and were parts of the same vast super-continent.[8-17]

Intelligent Design View

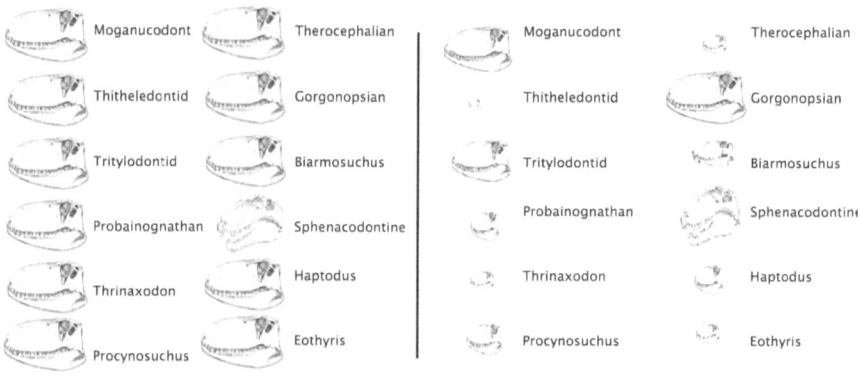

Sequences of Mammal-like reptiles, as typically presented in textbooks.

Sequences of Mammal-like reptiles, shown to scale.

Figure 8.4 Sequence of Mammal-like reptile progression toward mammalian species. **Left two columns illustration** – Typical textbook progression illustration. **Right two columns illustration** – Progression illustration with more realistic skull size proportions.[8-18a] Both illustrations begin with the oldest mammal-like reptile, Eothyris, on the lower right column, and progressing up to the top of the column and continuing at the bottom of the column to the left and progressing up to Morganucodont, on the upper left column. Morganucodont being the first species classified as a mammal, although the lower-jaw seems very much like that of a reptile.

Evolutionists often list a series of mammal-like reptiles to illustrate their progression toward mammals. The following sequence of mammal-like reptiles is often used by evolutionists to represent this transition: Eothyris, Haptodus, sphenacodontine, Biarmosuchus, gorgonopsian, therocephalian, Procynosuchus, Thrinaxodon, probainognathan, tritylodontid, thitheledontid, morganucodont (morganucodont being the earliest mammal, similar to a modern shrew) (Figure 8.4 Left illustration). The problem with this arrangement is that the actual sizes of these fossils do not follow in the order of their supposed transitional progression. If their skulls were arranged side-by-side with their actual skull sizes it would be more difficult to see their progression from one to the other (Figure 8.4 Right illustration). For example, thitheledontid is the smallest fossil, followed by Eothyris, probainognathan, Thrinaxodon and therocephalian, which are approximately all the same size. Procynosuchus, Haptodus, Biarmosuchus, tritylodontid, morganucodont, sphenacodontine and gorgonopsian become progressively larger. Gorgonopsian, although the largest of the series, appears somewhere in the middle of the transitional list. Another problem with this supposed fossil sequence is that the different skeletons supposedly representing a transitional sequence from reptiles to mammals were not found close together geologically. Some supposed ancestors and descendants were found in widely separated layers of sedimentary rock, representing tens of millions of years of geologic time.[8-18]

There is little evidence to support the view that mammals evolved from reptiles. Not only is there a lack of intermediate fossils, but it is hard to see how it could possibly have happened. The main question is: how could mammals evolve their jaw and ear from a reptile jaw? All reptiles have a lower jaw made up of at least four separate bones on each side, and a single bone in each ear. In every known mammal, either alive or extinct, the opposite is true. Mammals have a one-piece jawbone and three bones in the ear. All these bones fossilize readily, yet there is not a single fossil species with two bones in the ear or with two or three bones in the jaw.[8-19]

Creationist View

Evolutionists claim that mammals evolved from reptiles. They say there are five evidences that this evolution occurred and can be seen in therapsids and early mammal Morganucodon. These evolutionist evidences are: (1) connection of the limbs, (2) mobility of the head, (3) fusing of the palate, (4) musculature of the jaw, and (5) migration of certain bones from the jaw to the middle ear. In considering these claims, it is important to remember that soft tissues, such as those in the circulatory and reproductive systems, are not preserved in the fossils. Also, evolutionists portray a structural or morphological series rather than a lineage. It is impossible to tell which of the numerous therapsid (see Chapter 7, Figure 7.4h) species represented in the fossil record were in fact the ancestors of mammals. If evolution were true, there would be actual transitional fossils rather than a multitude of fossils with mammal-like features. If several therapsid (see Chapter 7, Figure 7.4h) linages evolved into mammals, it is doubtful that they all independently developed with a mammal ear.[8-20]

Now let's take a look at the two creatures that evolutionists believe represent the most definitive transitional forms between reptiles and mammals in more detail. They are Morganucodon and Kuehneotherium. (The fossils of Kuehneotherium are limited to teeth, dental fragments, and mandible (lower jaw or jawbone) fragments.[8-20a]) These are the creatures that evolutionists claim possessed the mammal-type jaw-joint side by side with the reptile-type jaw-joint. Articulation of the jaw with the skull is indirect, with the articular (one of the bones of the jaw) articulating with the quadrate bone of the skull, a bone not found in mammals. Another fundamental difference between reptiles and mammals is the fact that all reptiles, living or fossil, have a single bone in the ear, a rod-like bone known as the columella (stapes), which connects the tympanum (eardrum) to the inner ear (Figure 8.1). Morganucodon and Kuehneotherium each possessed a full complement of the reptilian bones in its lower jaw (Figure 8.3 – Morganucodon is the second skull from the top). Furthermore, there was no reduction in the functional importance of the reptilian (quadrate-articular) jaw-joint, even though these creatures are supposed to be intermediates between reptiles and mammals. Allegedly they possessed a mammalian (squamosal-dentary) jaw-joint in addition to the reptilian jaw-joint (Figure 8.1). There is no doubt that Morganucodon had a powerful standard reptilian type jaw-joint.

Almost all of the available material related to Morganucodon consists of bones that were scattered about and almost all of the individual bones consist of fragments. A fragment of a jaw was recovered with the quadrate bone still in contact with the articular bone, leaving no doubt about the existence of a reptilian jaw-joint in this creature. It is debatable whether Morganucodon and Kuehneotherium have, in addition to this reptilian jaw-joint, a point of contact between the dentary and squamosal (Figure 8.1). It is questionable as to whether this indicates the appearance of a formation of a mammalian type jaw-joint. (If this is true, then all the jaw and ear images in Figure 8.3 are reptilian, except for the images at the top labeled "Early Mammal.") The evidence is extremely fragmentary and no fossils are available showing the dentary in actual contact with the squamosal of the skull. In fact, not even a single intact lower jaw is available, all such specimens being reconstructed from fragments. The evidence for a squamosal-dentary joint consists of an alleged condyle on the dentary. (A condyle is a rounded area at the end of a bone forming a ball and socket joint with the hollow part (termed the fossa) of another bone.) In mammals there is a very prominent condyle on the posterior end of the dentary that articulates to the squamosal bone of the skull. The squamosal contains a fossa for the reception of the condyle and the contact forms the jaw-joint.

The dentary in Morganucodon and Kuehneotherium extends sufficiently to make contact with the squamosal and the alleged point of contact on the dentary is called the condyle. Whether the dentary bone of these creatures actually made contact with the squamosal can only be implied. But if there had been a real contact between the dentary and squamosal, it is doubtful that this constituted as a mammalian jaw-joint which existed alongside the reptilian jaw-joint. It must be emphasized that these creatures had a fully developed, powerful reptilian jaw-joint. The anatomy required for such a jaw-joint, including the arrangement and mode of attachment of musculature, must be quite different from that required for a mammalian jaw-joint. How then could a powerful, fully functional reptilian jaw-joint be accommodated along with a mammalian jaw-joint?

Evolutionists believe that as the bones in the reptilian jaw (except for the dentary) gradually became relieved of their function in the jaw they were now free either to evolve out of existence or to assume some new function. Evolutionists believe the quadrate and articular bones of the jaw of the reptile became free. These bones were firmly attached to the dentary in the Morganucodon fossil. Evolutionists claim that somehow these bones worked their way into the middle ear of the mammal to eventually become the incus (anvil) and malleus (hammer), respectively. The problems associated with such a proposed process are vastly greater than merely imagining how two bones precisely shaped to perform in a powerfully effective jaw-joint of a reptilian jaw could detach themselves, then make their way into the middle ear of a mammal, reshape themselves into the malleus and incus in a mammalian ear. Remember these bones are precisely engineered to function with a remodeled stapes in a vastly different auditory apparatus, while at the same time the creature continued to chew and to hear.

There are also other factors that need to be considered in the supposed transition of the reptilian jaw to a mammalian jaw. The possibility of this transition actually occurring seems to be rather remote when we consider the fact that the essential organ of hearing in the mammal, called the organ of Corti, is not possessed by a single reptile. Nor is there any evidence that would provide even a hint of where this organ came from. The organ of Corti is an extremely complicated organ. It is not found in reptiles. There is no possible structure in the reptile from which it could have been derived. According to evolutionary theory, all evolutionary changes occur as the result of mistakes during the reproduction of genes. These mistakes are called mutations (discussed in chapter 9), and each change brought about by such mutations which survived must be superior to preceding forms.[8-21] If evolution is true, one must believe that a series of thousands and thousands of random changes occurred gradually that caused the organ of Corti to function in an ear. And at the same time two bones from the jaw (quadrate, articular) migrated to the ear (incus, malleus) and were redesigned. Furthermore, each intermediate stage had to remain fully functional during this transition.[8-22] (The organ of Corti (or "spiral organ"), found only in mammals, is part of the cochlea of the inner ear and is provided with hair cells or auditory sensory cells.)[8-23]

In the next chapter (9) we will discuss the various components of evolution that explain what causes changes in species.

~~~

Chapter 9

The Components of Evolutionary Theory: Natural Selection, Adaptation, Speciation, Mutations, and Genetic Drift

Everywhere we look in nature, we see animals that seem beautifully designed to fit their environment, whether that environment be the physical circumstances of life, like temperature and humidity, or the other organisms—competitors, predators, and prey—that every species must deal with. It is no surprise that early naturalists believed that animals were the product of celestial design, created by God to do their jobs.[9-1]

We see dogs in great variety, even some extinct species, but no half dog/ half something else.[9-2]

Introduction

In this chapter (9) we will look at the following components of evolutionary theory: (**1**) natural selection, (**2**) adaptation, (**3**) speciation, (**4**) mutations, (**5**) genetic drift, and (**6**) other mechanisms that are said to cause changes in organisms.

Evolution Defined

Evolution is a process of change over time that characterizes the natural history of life on this planet. There are basically three core elements to evolutionary theory. The **first core** element of evolution is the observation that life really has changed over time, that the life of the past is different or was different from the life of the present, and that the natural history of this planet is characterized by a process of change over time. The **second core** element of evolution is the principle of common descent, and that is the notion that living things are united by a core of common ancestry, that living things trace their ancestry back to common ancestors that gave rise to the many forms of life that exist today. The **third core** element of evolution is the process that drove the change through time from common ancestors and common descent, driven by forces and principles and actions that are observable in the world today. There are many individual forces that drive the changes. Grouped together, these forces and processes are called natural selection. (Natural selection will be discussed below.)[9-3]

Macroevolution Versus Microevolution

Many so called "creationists" attempt to deny evolution based on the fact that there is no "evidence" for "macro evolution," only micro-evolution. This is nonsense. There is no such thing as macro evolution, where one species turns into another species in one step, and is just a case of using semantics to deny the truth. There is only micro-evolution—millions of tiny changes occurring over thousands (or millions) of years, until such times as two parts of the same original population are unable to interbreed and a new species is born. The Bible does not offer an explanation of the origins of human life. Science does.[9-4]

Creationist view

There are two kinds of evolution: microevolution and macroevolution. Microevolution represents small changes within a species as it adapts to its particular environment. These are cyclical, in-species variations. Macroevolution is the process by which fish develop legs, reptiles evolve into birds and apes evolve into humans. Evolutionists believe that if microevolution is true, then macroevolution must also be true.[9-5]

Intelligent Design View

Macroevolution cannot be explained in terms of micro-evolutionary processes, or any other currently known mechanisms, even though the acceptance of macroevolution resulting from multiple microevolutions appears logical. Since a certain degree of evolution has been shown to occur, any degree of evolution is possible. There is obviously an enormous difference between the evolution of a color change in a moth's wing and the evolution of an organ like the human brain, and the differences among the fruit flies of Hawaii. Comparing these minor differences of fruit flies, for example, to the differences between a mouse and an elephant, or the difference between an octopus and a bee, is extensive.[9-6]

(1) Natural Selection

Natural selection acts only by the preservation and accumulation of small inherited modifications, each profitable to the preserved being; and as modern geology has almost banished such views as the excavation of a great valley by a single diluvial wave [relating to a flood or floods, especially the biblical Flood], so will natural selection banish the belief of the continued creation of new organic beings, or of any great and sudden modification in their structure.

• Charles Darwin The Origin of Species
by means of Natural Selection or The Preservation
of Favoured Races in the Struggle for Life,
Chapter 4 - Natural Selection, 6th edition

Consequently natural selection would have had different materials or variations to work on, in order to arrive at the same functional result; and the structures thus acquired would almost necessarily have differed. On the hypothesis of separate acts of creation the whole case remains unintelligible.

• Charles Darwin The Origin of Species
by means of Natural Selection or The Preservation
of Favoured Races in the Struggle for Life,
Chapter 6 - Difficulties Of The Theory, 6th edition

... I think that ... light has been thrown on several facts, which on the belief of independent acts of creation are utterly obscure. We have seen that species at any one period are not indefinitely variable, and are not linked together by a multitude of intermediate gradations, partly because the process of natural selection is always very slow, and at any one time acts only on a few forms; and partly because the very process of natural selection implies the continual supplanting and extinction of preceding and intermediate gradations.

• Charles Darwin The Origin of Species
by means of Natural Selection or The Preservation
of Favoured Races in the Struggle for Life,
Chapter 6 - Difficulties Of The Theory, 6th edition

Why, on the theory of Creation, should there be so much variety and so little real novelty? Why should all the parts and organs of many independent beings, each supposed to have been separately created for its own proper place in nature, be so commonly linked together by graduated steps? Why should not Nature take a sudden leap from structure to structure? On the theory of natural selection, we can clearly understand why she should not; for natural selection acts only by taking advantage of slight successive variations; she can never take a great and sudden leap, but must advance by the short and sure, though slow steps.

• Charles Darwin The Origin of Species
by means of Natural Selection or The Preservation
of Favoured Races in the Struggle for Life,
Chapter 6 - Difficulties Of The Theory, 6th edition

The one mistake is that some believe Natural Selection is random, which underlies much of the skeptical backlash against evolution. [Random] Chance cannot explain life. Design [creation] is as bad an explanation as chance because it raises bigger questions than it answers. Evolution by natural selection is the only workable theory ever proposed that is capable of explaining life, and it does so brilliantly.[9-7]

The theory of evolution teaches that a species may become a different species over time, that is, develop new traits that never existed before. However, it has never been explained that the origin of the truly new

traits needed to produce a truly new kind of organism, something more
than just a variation of some existing kind. There are many other
logical limits to extrapolation from natural selection to evolution,
but the simplest is this: natural selection cannot explain the origin
of traits.[9-8]

Evolutionist View

Natural selection is the gradual, non-random process by which biological traits become either more or less common in a population as a function of differential reproduction of their bearers. It is a key mechanism of evolution. Natural selection acts on the phenotype, or the observable characteristics of an organism, but the genetic (heritable) basis of any phenotype that gives a reproductive advantage will become more common in a population (allele frequency). Over time, this process can result in populations that specialize for particular ecological niches and may eventually result in the emergence of new species. Natural Selection is most often defined to operate on heritable traits, because these are the traits that directly affect evolution. However, natural selection is "blind" in the sense that changes in phenotype (physical and behavioral characteristics) can give a reproductive advantage regardless of whether or not the trait is heritable (non-heritable traits can be the result of environmental factors or the life experience of the organism). Traits that cause greater reproductive success of an organism are said to be selected for, whereas those that reduce success are selected against. Selection for a trait may also result in the selection of other correlated traits that do not themselves directly influence reproductive advantage.[9-9]

The main component of evolution is natural selection, although there are other mechanisms that contribute as well. Natural selection occurs naturally in that nature selects those individuals with the characteristics that prove to have a higher probability of survival and enables them to increase in population over a period of time.[9-10] Natural selection improves what was before. It produces a creature that is more fit, but not the fittest.[9-11] It is non-random but directs evolution toward improvement. Natural selection is a process of how the genes that produce better adaptations become more frequent over time. This is done by eliminating those creatures that are not able to survive their environment.[9-12]

Natural selection can be subdivided into two categories: **ecological selection** (generally referred to as natural selection) and **sexual selection**.

Ecological selection occurs when creatures that survive and reproduce increase the frequency of their genes in the gene pool over those that do not survive (thus causing more creatures of their offspring than other similar creatures with different traits, such as color of skin or hair). Sexual selection occurs when creatures that are more attractive to the opposite sex because of their features reproduce more and thus increase the frequency of those features in the gene pool.[9-13] Sexual selection is a mode of natural selection in which members of one biological sex choose mates of the other sex to mate with and compete with members of the same sex for access to members of the opposite sex. This form of selection means that some individuals have better reproductive success than others within a population, either because they are more attractive or prefer more attractive partners to produce offspring. For example, in the breeding season, sexual selection in frogs occurs with the male frogs gathering at the water's edge and making their mating calls by croaking. The females then arrive and choose the males with the deepest croaks and best territories. In general, males benefit from frequent mating and monopolizing access to a group of fertile females. Females can have a limited number of offspring and maximize the return on the energy they invest in reproduction.[9-13a]

Sexual selection comes about from differences in mating success and natural selection (ecological selection) is due to variance in all other fitness components (that will be discussed shortly). Natural and sexual selection are different than neutral processes that cause organic evolution, such as genetic drift (Genetic drift is discussed later in this chapter.). A great deal of evidence has been obtained that verifies the concept of sexual selection, and has established the fact that both females and males are in fact choosy and that the outcome of reproductive competition is determined by an individuals' phenotype. (*Phenotype is the observable properties of an organism that are produced by the interaction of the genotype and the environment.*) To date (2011), it is not clear whether or not traditional sexual selection is a major factor in driving trait evolution. (See section "Sexual Selection" sub-section "How Did Sexual Reproduction Originate and is it Beneficial to a Species?" below.)[9-13b]

Ecological Selection (A process of Natural Selection)

Evolutionist View

As defined previously, natural selection (ecological selection) is "the process whereby organisms that are better adapted to their environment tend to survive and produce more offspring." Natural selection is the differential survival and reproduction of individuals due to differences in phenotype (the composite of the organism's observable characteristics or traits). It is a key mechanism of evolution, the change in the heritable traits characteristic of a population over generations. Charles Darwin popularized the term "natural selection", contrasting it with artificial selection (see sub-section "Limitations of Artificial Breeding (Artificial Selection or Selective Breeding)"), which in his view is intentional, whereas natural selection is not. Natural selection is further defined as "the process by which forms of life having traits that better enable them to adapt to specific environmental pressures, as predators, changes in climate, or competition for food or mates, will tend to survive and reproduce in greater numbers than others of their kind, thus ensuring the perpetuation of those favorable traits in succeeding generations."[9-13c]

As an example of natural selection, all the young giraffes of any one generation would vary with respect to the length of the neck. Those with longer necks would have a slight advantage over other giraffes in the extended sphere of their grazing territory. Those giraffes with the shorter necks would be weaker, as they would have less access to food and therefore would not live as long to produce as much offspring as the giraffes with the longer necks. The giraffes with the longer necks, being better nourished than the others with shorter necks, would be stronger and so they would be more able to escape from their flesh-eating predators, like the lion. For the reason that their variation would be congenital (due to an abnormal inherited trait or condition that is present at birth, as a result of either heredity or environmental influences) and therefore already transmissible, their offspring would vary according to the advanced condition, and further selection of the longer necked individuals would lead to the modern result.[9-14] (The origin of the modern giraffe is discussed in Chapter 7.)

Creationist View

After the fossil record (discussed in chapters 3, 7, 8, 16), the second supporting component of evolution is natural selection. Nature selects those organisms that are best fit to survive the struggle for existence. In that way evolution will inevitably work. Natural selection is the major force driving evolutionary change.[9-15]

Natural selection decreases genetic information of a species due to the selection of creatures according to their traits that are more advantageous to their environment than those whose traits are less conducive of their environment. The genes of those creatures with less advantageous traits would not be as plentiful as those creatures with more advantageous traits. Those creatures with less advantageous traits may even die out. Natural selection allows organisms to survive better in a given environment by selecting those creatures better suited for their environment to survive. Natural selection cannot increase or provide new genetic information, that is, natural selection cannot cause new traits to develop in a creature, it can only select traits that are already present.[9-16]

Many believe that natural selection means "survival of the fittest." This is not necessarily true. While a species may be very healthy, it may not be the best fit for the environment in which it is living. Natural selection may be better defined as "members of a population that leave the most offspring to the next generation." Natural selection seems to promote short-term survival at the expense of long-term extinction. The fossil evidence indicates that those species that were generalized were the long-term survivors rather than the specialized forms that natural selection generated to use for short-term advantage. Natural selection cannot plan ahead. Natural selection focuses on that certain trait combinations that will win the immediate struggle for survival, becoming the fittest, no matter what that does to the future of the species.[9-17]

Natural selection is a blind process that does not plan, has no purpose, and cannot look ahead toward goals.[9-18] Natural selection cannot select what is not there. If the trait is not already in the genes it cannot be selected for use or adaptation. Selecting which trait will be used is not evolution, for a trait needs to exist before it can be selected.[9-19]

An Example of Natural Selection—Peppered Moths

LIGHT FORM
OF THE PEPPERED MOTH

DARK FORM
OF THE PEPPERED MOTH

The Evolution Cruncher

Examples of White and Black form Peppered Moths

Evolutionist View

In microevolution small changes in species, caused by the mechanisms of evolution, accumulate over a few decades or centuries. This allows species to adapt to an environment and sometimes to split into two or more species. One example of microevolution is the peppered moth (species Biston betularia) in Great Britain. In 1848 nearly all peppered moths had white bodies with small black spots, but a few had black bodies. By 1895, 98 percent of moths in Manchester, England, had black bodies, and only two percent had the original white body color. What caused the change between 1848 and 1895 is that industrial coal plants in the Manchester area spewed soot into the air, and the coal soot settled on trees. This condition caused the tree bark and the lichen on the trees to become a darker color. Before the accumulation of soot the white moths were well camouflaged when they landed on the tree trunks. After the tree trunks became darker, the small population of black-bodied moths suddenly had the advantage. They were less likely to be eaten by birds and more likely to survive and lay eggs. In other words, darker moths had more reproductive success then the white moths. Microevolution had caused the species to change. Since the late 1800s pollution controls have reduced the amount of coal soot, and the original white-bodied moths have become numerous again. Within the past decade or so a few people have disputed whether peppered moths are a good example of microevolution. But scientists who have actually studied peppered moths as part of their research say that, while the story is a little more complex than is presented in textbooks, peppered moths are a good example of microevolution.[9-20]

The dark form of moth was first recorded in 1848 near Manchester. It is possible that no dark moths existed before the 1800s, however, it is also possible that the dark moths already existed in the population at the time of the industrial revolution (1850s), and natural selection simply increased their frequency; but we can use the case as an example as if there were no dark moths when the environment changed. The industrial revolution imposed on the moths an environment they had never encountered before.[9-21]

Creationist View

Evolutionists commonly cite the case of the peppered moth (Biston betularia) of England as an example of present-day Darwinian evolution. Peppered moths have always existed in light, intermediate, and dark-colored varieties. Before the advance of the industrial revolution, the tree trunks were light and light-colored moths were camouflaged; whereas the dark-colored moths were easily spotted and eaten by birds. Consequently, the dark-colored moths constituted a very minor proportion of the total population. As the industrial revolution progressed, however, and pollution increased, the tree trunks became darker and within 45 years the situation was reversed. In the Manchester vicinity, for example, 95 percent of the moths were of the dark-colored variety. This occurrence could hardly be considered evolution. This process did not produce anything new. It did not result in increased complexity and organization. The dark-colored moths had always existed. The air pollution simply caused a shift in populations of the dark versus light-colored moths. Although no evolutionary change occurred in these moths, the case of the peppered moths does illustrate the principle of natural selection.[9-22]

For many years, the prime example of natural selection has been the peppered moths. On trees with light bark, the dark moths stood out. The light moths on light colored bark trees would be camouflaged. On trees with dark bark, the light moths stood out. The dark moths on dark colored bark trees would be camouflaged. Presumably, that's the way birds saw them too, back in the 1850's when the moth experiment was done. While the tree barks were still light colored, birds ate mostly dark moths, and light moths made up over 98 percent of the population. But then pollution killed the lichen on the trees, revealing the dark color of the bark. As a result, the dark moths were more camouflaged than the light colored moths. Thus, the dark colored moths had a better chance of surviving and leaving more offspring to grow into dark moths in succeeding generations. The population then shifted. The "dark environment" just naturally selected the dark moths as more likely to survive and reproduce. By the 1950's, the population was over 98 percent dark moths, proving the definition of natural selection.[9-23]

There are obviously all kinds of changes through time that are not evolution, so evolution must be only a particular kind of change through time. Natural selection certainly produces change in populations, but is it the evolutionary kind of change? Take a look at the peppered moth example between the dark and light colored moths. What did we start with? There were dark and light varieties of the peppered moth, species Biston betularia. After 100 years of natural selection, what resulted were dark and light varieties of the peppered moth, species Biston betularia. The moths themselves didn't change; there were always dark colored moths and always light colored moths from the earliest observations. The only thing that changed was the percentage of moths in the two categories caused by natural selection.[9-24]

In spite of what might be claimed, natural selection has been observed to produce only variation within kind, that is, merely shifts in populations, for example, to moths with greater percentages of darker moths, to flies resistant to DDT (Dichloro-Diphenyl-Trichloroethane), or to bacteria resistant to antibiotics. Modern evolutionists believe, however, that such small changes plus vast amounts of time could lead to huge changes (macro-evolution) from one kind to another.[9-25] One might have thought that the dark-colored peppered moth was the most fit to survive in a polluted forest because it was the most camouflaged. But what if the extra production of melanin interfered with, say, sex hormone production and made the dark-colored moths sterile? Obviously, the superior camouflage would not make such a moth the most fit to survive. Evolutionists think the camouflage helped, of course, but the dark moths were really determined to be the most fit to survive because a greater percentage of their offspring survived in polluted forests than the percentage for any other color form.[9-26]

Natural selection does not produce a better moth; the dark moth was already present. Natural selection really operates as the "great eliminator" or "terminator." Natural selection did not produce a new and improved moth; the dark moth was already present. Pollution made the light form of moth less camouflaged, and so presumably natural selection eliminated more light than dark moths. Had natural selection "gone to completion" and totally eliminated the light moth, the peppered moth (Biston betularia) might now be well on the road to extinction, since reduction in pollution has now made the light moth more camouflaged again.[9-27] Natural selection will only promote increased death of less camouflaged moths; it did not produce either dark or light color moths. Mutations are supposed to produce new traits for natural selection to select, but known mutations are either neutral (having no effect) or harmful, producing defects, disease, and disease organisms. (Mutations will be discussed in greater detail later.) Perhaps the most important role of natural selection is acting as a brake, slowing down the accumulation of harmful mutations, eliminating or reducing genetic decay by only permitting those more fit to survive (survival of the fittest).[9-28]

The peppered moth experiments beautifully demonstrate natural selection or survival of the fittest in action, but they do not show evolution in progress, for however the populations may alter in their content of light, intermediate, or dark forms, all the moths remain from beginning to end Biston betularia.[9-29]

Because of dominant and recessive genes (Mendelian genetics), the moth continued to produce both light and dark offspring for thousands of years, while the birds kept eating the dark varieties. Yet dark ones continued to be born. This is proof of the stability of the species, which is exactly the opposite of evolutionary proof. In recent years, industrial pollution laws are making the air cleaner, and the lighter moths are again becoming more common. This is not evolution, but simply a color change back and forth within a species. This is an excellent demonstration of the function of camouflage, but, since it begins and ends with moths and no new species is formed, it is quite irrelevant as evidence for evolution.[9-30]

Studies on industrial melanism have confirmed the hypothesis that natural selection takes place in nature. The dark and light moth observation is the story of the black mutant of the common peppered moth which increases in numbers in the vicinity of industrial centers and decreases, being more easily exposed to predators, in rural areas. The neo-Darwinists claim that this is natural selection, that is, evolution, is actually going on. It is true that natural selection takes place in nature; however, the color of moths or snails or mice is clearly controlled by visibility to predators, but this is not evolution. Changes in living creatures do not come about by random chance mutations. Minor changes in species come about by genes being turned on or off, such as in the case of the color of a moth.[9-31] The changes are variation within species, not evolution across species. It is a re-assortment of the DNA and genes, but nothing more. The changes within living things are the result of naturally reshuffled genes. If the changes had been the result of mutations, the result would have been weakened stock whose offspring would tend eventually to become sterile or die out.[9-32] (Genes being turned on or off is discussed in Chapter 13, section "Some DNA Portions are Used to Turn Genes On or Off." Gene shuffling is discussed in Chapter 5, section "How Different Traits are Obtained.")

All scientists agree that elimination of the unfit is a major effect of natural selection. Eliminating defects to repair an unfit species may keep it alive, but it will never turn an unfit species into one that is more fit. Natural selection may eliminate the weaker species that may reduce breeding more weak species, but it will never create a more superior species than what is already in existence.[9-33] Scientists have found in reality that natural selection deals only with the number of a species, not the change of the species to another. It has to do with the survival of the species rather than the beginning of a new species. Natural selection only preserves existing genetic information (DNA); it doesn't create genetic material that would allow an animal's offspring to sprout a new organ, limb or other anatomical feature.[9-34]

Natural Selection and Darwin's Galapagos Finches

Darwin's Finch Sketches

Charles Darwin's Sketches of Finches

Evolutionist View

When Charles Darwin was a young man, he sailed on the ship "Beagle" (December 27, 1831-October 2, 1836), not expecting that this trip would eventually lead him to begin a lifelong quest searching for evolutionary evidence. The ship landed on San Cristobal and remained there for some time, originally stopping to find some food. During Darwin's stay in the Galapagos Islands (Floreana, Isabela, San Cristóbal, and Santiago) he noticed that the characteristics of the creatures that roam, fly and swim around these islands often were very unique from those seen elsewhere. Not only were many

Galapagos species distinct from those on the mainland, but that between islands many species of similar features were so perfectly adapted for their environment. Among those that struck Darwin so greatly were the finches, with such varying diets as cactus and seeds, fruits and blood. Darwin would later base some of his views from the thought that these finches were all descendants of the same lineage. Because Darwin had not yet formulated his theory of evolution at the time of his voyage, he did not record which finches appeared on which islands.

Much of the study on the Galapagos Islands included the various varieties of finches that lived there. Altogether there are fourteen species of Darwin's finches that live in the Galapagos islands. Many of these species have different body sizes and beak shapes. During times of drought, the only seeds that the finches have to feed on are large and hard. The finches with the larger stronger beaks (finches #1 & #2 above) are able to break the shells, whereas the finches with the smaller, weaker beaks (finches #3 & #4 above) are not able to do so. It is therefore most likely that the finches with the larger beaks would have a better chance of survival, via natural selection, than the finches with the smaller beaks.

The beak size of the Galapagos finches shows a continuous variation of a large class of characterizations to study. Simple characterizations often have distinct variation, however, many of the characterizations of species are like beak size in the Galapagos finches that vary continuously, and every individual finch differs slightly from every other finch. There are no distinct categories of beak size within a species. It is not known exactly which genes produce any given beak size. Characterizations like beak size are probably controlled by a large number of genes, each of small effect. It is possible that beak size is controlled by a single pair of alleles (a variant of a single gene), with one allele dominant to the other. In this case the population of finches would contain two categories of individuals, possibly controlled by two alleles each. The conclusion (first developed by Gregor Mendel in 1865 and confirmed in 1910 by East, Nilsson-Ehle and others) is that multifactorial inheritance (a characterization being influenced by many genes) can generate a continuous frequency distribution.[9-35]

Creationist View

Darwin made the voyage of the Beagle the most important sea voyage in the history of science. His most striking discovery came in a group of islands in the Pacific, about 650 miles west of Ecuador, called the Galapagos Islands. What attracted Darwin's attention during the five-week stay on the islands was the variety of finches on the islands. They are known as Darwin's finches to this day. He found the birds divided into at least fourteen different species, distinguished from one another mainly by differences in the size and shape of their bills. These particular species did not exist anywhere else in the world, but they resembled an apparently close relative on the South American mainland.[9-36] On the Galapagos Islands in the western Pacific, Charles Darwin found 13 (some say 14, 17, or 19) varieties of a dark brown finch. This finch series was closely related to finches found in South America. After having returned to England in 1838, he decided that those finches he had earlier seen were probably 13 different species. Darwin therefore concluded that this discovery was proof of evolution. He believed that one species had crossed over the species barrier and produced still other species. But a close examination of the 13, or so, species reveals they are only variations of one brown finch. In legs, feet, bodies, coloration, eyes, and internal organs, they are all essentially the same. The only differences are variations in beaks, body sizes, and food gathering habits. The collection of 3,700 specimens of the 13, or so, varieties of Darwin finch, were later studied by others who determined that the Galapogos finches consisted only of subspecies of a single finch.[9-37] Because they are almost identical in every way, except for very slight differences in body and beak size, and different food habits, it is debatable whether they should be classified as separate species or simply variations of the same species. No one can know for certain whether there were 13, 14, 17, or 19 subspecies, since the birds look so much alike. Yet Charles Darwin chose to classify them into four different genera.[9-38]

Darwin, knowing nothing of modern genetics and the boundary imposed by DNA to changes across basic types, imagined that perhaps these birds were all different types and evolution across types had occurred. The Galapagos Island finches (often called 'Darwin finches') do look just about alike. They are sub-species of a single parent species that, at some earlier time, reached the island from South America. (If hummingbirds can fly across the Gulf of Mexico, finches ought to be able to be borne by storms to the Galapagos Islands.)[9-39] When Charles Darwin looked at the finches on the Galapagos Islands, he noticed variations in beak sizes. Darwin thought that the harder seed during the dry period was causing the beaks of finches to grow stouter from use. As is now known, beak sizes do not change as a result of "use and disuse," as Lamarck thought and as Darwin stated, "inherited effects of the increased use of parts" (The Origin of Species, 6th ed, London 1902, p 278.) In reality, what occurred was natural selection. A long-term drought in the islands was causing the seed cases to harden. The finches with the stronger, heavier, beak allele in the genome were favored and the finches with the lighter, less durable, allele were not. (Alleles are discussed in section "Genetic Drift.") The finches with the stronger beaks became more dominant because they were able to eat more seeds and were more able to survive to pass on the stronger beak alleles to their offspring. The stronger beak finches were not the result of 'use and disuse of parts.'

The allele for heavier beaks was already in the genome and was just brought out as a result of the environment (hard seeds), as those finches with the stronger beaks thrived. When the rains came back, the finches with the lighter beaks had more efficient beaks for eating than the finches with the heavier beaks. Therefore, the number of finches with heavy beaks decreased. This is an example of Natural Selection. There was no change in the genome of the finch and no new species developed as a result of these changes. The genome within finches produced a variety by recombination of the alleles and causing the phenotype to display its given types. Phenotype is the appearance of an organism resulting from the interaction of the genotype and the environment.[9-40]

Natural Selection, Mutation and Resistance to Drugs and Poisons

Evolutionist View

Features that make bacteria ideal for studies of evolution in the lab are their huge population sizes and short generation times. The chance of a mutation producing antibiotic resistance is high. And those bacteria that are resistant to a drug will be those that survive (natural selection), leaving behind genetically identical offspring that are also drug-resistant. Eventually the effectiveness of the drug diminishes, and once again there is a medical problem.[9-41]

It is generally believed that drug resistance occurs because somehow the patients themselves change in a way that makes the drug less effective. But this is wrong: resistance comes from evolution of the microbe, not habituation of patients to the drugs. When penicillin was introduced in the early 1940s, penicillin was a miracle drug, especially effective at curing infections caused by the bacteria Staphylococcus aureus ("staph"). In 1941, the drug could wipe out every strain of staph in the world. Seventy years later, more than 95 percent of staph strains are resistant to penicillin. What happened was that mutations occurred in individual bacteria that gave them the ability to destroy the drug, and of course these mutations spread worldwide.[9-42]

Viruses, the smallest form of life capable of change, have also evolved resistance to antiviral drugs, notably AZT (azidothymidine), designed to prevent the HIV (Human immunodeficiency virus) virus from replicating in an infected body. (Viruses are discussed in Chapter 5.) Evolution even occurs within the body of a single parent, since the virus mutates at a furious pace, eventually producing resistance and rendering AZT ineffective.[9-43]

Intelligent Design View

Bacterial antibiotic resistance works in two ways. **The first** is an enzyme defense system. For example, penicillin kills a bacterium by "gumming up" a molecular machine responsible for synthesizing the bacterium's cell wall. However, bacteria that are resistant to penicillin have an enzyme called "penicillinase" that chemically cuts the antibiotic and causes it to become harmless. If there is penicillin in the bloodstream, bacteria that don't produce penicillinase will die. Bacterial cells that can produce penicillinase survive and reproduce, passing their DNA on to their descendants. As a result, the next generation of bacteria is also resistant to the antibiotic. **The second** way that bacteria become resistant to some antibiotics is through mutation. Bacteria multiply very quickly. During reproduction, some of their offspring have genes that have been altered by a "copying error" in the DNA. In some cases, this copying error replaces one amino acid in a bacterial protein with a different amino acid. Usually a mistake of this kind would be harmful to a bacterium, but occasionally this actually helps it survive. Antiobiotic agents attack critical functions of bacteria, such as genetic information processing. (This includes processes such as DNA replication, RNA synthesis, and protein synthesis.) Specific molecular machines made of many protein components perform these functions.

Antibiotics poison bacteria by homing in on a protein component of one of these essential molecular machines. The antibiotic fits, hand-in-glove, into a vulnerable spot in the structure of a key protein. When an antibiotic latches on to this "target site" called an "active site," it prevents the machine component (the protein) from working properly. The bacterium either fails to grow or dies outright. However, sometimes a DNA mutation changes the shape of the active site on the target protein. If the mutation changes the shape of the protein enough to prevent the antibiotic from binding to it, but not so much that it destroys the protein's function, then the antibiotic no longer fits, and cannot latch on. The result is that a single base-pair mutation has given the bacterium a competitive advantage. After the non-resistant cells (those lacking the altered protein) die, the mutant bacterium reproduces to form a large population of antibiotic-resistant bacteria. In a few generations, an antibiotic-resistant strain arises. The development of antibiotic resistance in bacteria is a powerful example of random mutations providing a source of new variation. Mutations produce new genetic information upon which natural selection can act (Mutations are discussed later in this chapter.) The result is a new, fitter strain of bacteria.[9-44]

Creationist View

Evolutionists have conducted extensive research to determine why some bacteria become resistant to some insecticides and drugs. It was discovered that strains of bacteria resistant to penicillin, aureomycin, or chloromycetin appeared when these drugs were given for various diseases. The idea that there could be beneficial mutations was proven wrong when it was discovered that those mutations did not arise because of exposure to antibiotics, but instead occurred spontaneously at a constant rate, regardless of whether or not antibiotics were present. Because those resistant strains were mutants, they were always weaker and soon died out from natural causes other than the antibiotics. Doses of antibiotic kill off the natural strain, and the mutated form takes over. Then when the antibiotic treatment is stopped, the natural strain increases and the resistant strain soon die out. This is because, as a mutated form, it was never strong.[9-45]

Certain strains of bacteria, induced with penicillin, and flies, induced with DDT (Dichloro Diphenyl Trichloroethane - a synthetic insecticide that is highly toxic toward a wide variety of insects as a contact poison that apparently exerts its effect by disorganizing the nervous system), seemed to become resistant to these chemicals. These resistant bacteria and flies already existed and it only seemed that the fittest were surviving.[9-46] Evolutionists believe that resistant strains of bacteria are a marvelous demonstration of evolution. Yet this is not true. No change across the species barrier has occurred. Just as creatures in nature respond to predators with new defenses, viruses and insects also show an amazing resilience.[9-47]

Bacteria that produce mutated bacteria will be resistant to antibiotics, thus natural selection will cause the mutated bacteria to survive while the regular bacteria will be killed through the antibiotics. One commonly cited evidence for evolution is the development of resistance in bacteria against antibiotics. In the 1940s, penicillin killed many types of disease-causing bacteria. However, it is not that effective today. The development of resistance against antibiotics is supposedly direct evidence for evolution. However, it is not true that this is evidence of evolution. Before the development of penicillin, some bacteria species were already resistant but many were not. After penicillin was used against the bacteria, the non-resistant bacteria were killed. The resistant bacteria survived and reproduced to produce more resistant bacteria. The immunity example is then due to natural selection, not mutation. The population increase of resistant bacteria is not evolution. It is natural selection at work. A new species of bacteria did not evolve. No mutation or genetic or DNA change resulted from the use of penicillin. The bacteria are still the same species as they were before. What happened was only the removal by natural selection of the non-resistant bacteria.[9-48]

Antibiotic Resistance of Bacteria is Not An Example of Evolution in Action

Creationist View

The ability of certain bacteria to develop resistance to antibiotics, which are otherwise useful in speeding recovery from some illnesses, has been proclaimed as a textbook example of evolution in action. Evolutionary scientists have studied these bacteria with the hope that they will reveal secrets as to how molecules-to-man evolution could have happened. Evolutionists claim bacteria that develop a resistance to antibiotics illustrates evolution in action. Antibiotics are natural substances secreted by bacteria and fungi to kill other bacteria that are competing for limited nutrients. (The antibiotics used to treat people today are typically derivatives of these natural products.) Scientists are dismayed to discover that some bacteria have become resistant to antibiotics through various alterations, or mutations, in their DNA. These bacteria grow in an environment filled with sick people who have poor immune systems and where antibiotics have eliminated competing bacteria that are not resistant. Bacteria that are resistant to modern antibiotics have even been found in the frozen bodies of people who died long before those antibiotics were discovered or synthesized.

Antibiotics were first discovered in 1928. These discoveries eventually led to the large-scale production of penicillin from the mold Penicillium notatum in the 1940s. As early as the late 1940s resistant strains of bacteria began to appear. Currently, it is estimated that more than 70% of the bacteria that cause hospital acquired infections are resistant to at least one of the antibiotics used to treat them. Antibiotic resistance continues to expand for a multitude of reasons, including (1) over prescribing of antibiotics by physicians, (2) non completion of prescribed antibiotic treatments by patients, (3) use of antibiotics in animals as growth enhancers (primarily by the food industry), (4) increased international travel, and (5) poor hospital hygiene.

Bacteria can gain resistance through two primary ways: (1) By mutation, and (2) By using a built-in design feature to swap DNA (called horizontal gene transfer), bacteria share resistance genes. An antibiotic kills a bacterial cell by simply disrupting a critical function. Antibiotic resistance of bacteria only leads to a loss of functional systems. Evolution requires a gain of functional systems for bacteria to evolve into man. The antibiotic binds to a protein so that the protein cannot function properly. The normal protein is usually involved in copying the DNA, making proteins, or making the bacterial cell wall, all important functions for the bacteria to grow and reproduce. If the bacteria have a mutation in the DNA that code for one of those proteins, the antibiotic cannot bind to the altered protein; and the mutant bacteria survive. In the presence of antibiotics, the process of natural selection will occur, favoring the survival and reproduction of the mutant bacteria. (The mutant bacteria are better able to survive in the presence of the antibiotic and will continue to cause illness in the patient.) Although the mutant bacteria can survive, the change has come at a cost. The altered protein is less efficient in performing its normal function, making the bacteria less fit in an environment without antibiotics. Typically, the non-mutant bacteria are better able to compete for resources and reproduce faster than the mutant form.[9-49]

An Example of How Bacteria Becomes Resistant

Anthrax is a serious infectious disease caused by gram-positive, rod-shaped bacteria known as Bacillus anthracis. Anthrax can be found naturally in soil and commonly affects domestic and wild animals around the world. People (as well as animals) can become infected with anthrax when spores get into the body, either by injecting or simply coming in contact with the bacteria. When the spores become active, the bacteria can multiply, spread out in the body, produce toxins (poisons), and cause severe illness. This can happen when people (as well as animals) breathe in spores, eat food or drink water that is contaminated with spores, or get spores in a cut or scrape in the skin.[9-49a]

Ciprofloxacin (Cipro) is an antibiotic that has been administered to those who have become infected with the Anthrax bacteria. Ciprofloxacin belongs to a family of antibiotics known as quinolones, which bind to a bacterial protein called gyrase, decreasing the ability of the Anthrax bacteria to reproduce. This allows the body's natural immune defenses to overtake the infectious bacteria as they are reproducing at a slower rate. Quinolone-resistant bacteria have mutations in the genes encoding the gyrase protein. The mutant bacteria survive because the Cipro cannot bind to the altered gyrase. This comes at a cost as quinolone-resistant bacteria reproduce more slowly. Resistance to this family of antibiotics is becoming a major problem with one type of bacteria that causes food poisoning. These bacteria increased its resistance to quinolones 10-fold in just five years. Bacteria can also become antibiotic resistant by gaining mutated DNA from other bacteria. Unlike humans, bacteria can swap DNA. But this still is not an example of evolution in action. No new DNA is generated (a requirement for molecules-to-man evolution), it is just moved around. This mechanism of exchanging DNA is necessary for bacteria to survive in extreme or rapidly changing environments like a hospital.

The mechanisms of mutation and natural selection aid bacteria populations in becoming resistant to antibiotics. However, mutation and natural selection also result in bacteria with defective proteins that have lost their normal functions. Evolution requires a gain of functional systems for bacteria to evolve into man, functioning arms, eyeballs, and a brain, to name a few. Mutation and natural selection, thought to be the driving forces of evolution, only lead to a loss of functional systems. Therefore, antibiotic resistance of bacteria is not an example of evolution in action but rather is a variation within a bacterial kind.[9-49]

Intelligent Design View

It is commonly agreed that minor mutations can result in some advantages to some organisms, however, it is doubtful whether such mutations can result in new living forms. There are limits as to how far a mutation can go in an organism. For example, antibiotic resistance indicates organisms have the capacity for minor change, but not the extensive change stipulated by the Darwinian Theory. In the case of penicillin resistance, when penicillin is present in the bloodstream, a bacterial strain that already has a gene coding for penicillinase will have a significant survival advantage over a strain that does not. Bacterial cells either have a penicillinase gene, or they do not. They do not develop such a gene when penicillin is introduced. In this case, the enzyme defense system is not an indication that mutations are capable of producing new forms of life.

Micro evolutionary change is evident when point mutations result when bacterial cells become resistant to antibiotics (Point mutations are the result of what happens when a single chemical "letter" (A, C, G, T—adenine, cytosine, guanine, thymine) in the DNA sequence is changed as a result of heat, chemicals, or radiation.). Such change is small scale. It is doubtful that mutations like those that cause antibiotic resistance can go on to produce major (macroevolutionary) changes in organisms. Instead, recent discoveries about how bacteria acquire antibiotic resistance actually show that there are limits to the amount (and kind) of change that mutations can produce. When an antibiotic binds to a critical "active site" of a target protein in an essential molecular machine, such as a polymerase or a ribosome (discussed in Chapter 5), it impairs the bacterial cell's ability to copy or process its own genetic information. As a result, one of two things happens. Either **(1)** the cell loses the ability to replicate itself, or **(2)** it loses the ability to make essential proteins to keep the cell alive. Either way, without these vital functions, it is not long before the bacterium dies. Also, mutations can cause bacteria to become resistant to antibiotics. Antibiotics aim at "target" proteins in one of the bacterium's essential molecular machines. However, mutations sometimes produce a change in the shape of the active site of the target protein. Because of this change in shape, the antibiotics no longer recognize the protein as a target. And because the antibiotic doesn't recognize the protein, the antibiotic does not bind to the protein. As a result, the antibiotic no longer interferes with the machine's function. When this happens, an antibiotic-resistant strain of bacteria originates. The negative to this process is that the very same mutation usually hampers the molecular machine's ability to function. It changes a critically important protein. The bacterium is faced with the lesser of two evils. Either (1) it accepts a severe handicap that fools the antibiotic, or (2) it will die from the antibiotic.

There is another problem for the bacterium. If more mutations of this type occur, they will inflict additional damage on vital systems. The cell cannot endure an unlimited number of mutation-induced changes at these critical active sites. At some point, the cell's information processing system will be damaged so badly that it stops functioning altogether. For this reason, multiple mutations at active sites inevitably do more harm than good. This helps to explain that once antibiotics are removed from the environment, the original (non-resistant) strain "out-competes" the resistant strain, which dies off within a few generations. This occurs because a mutation gives some bacteria a resistance to antibiotics. However, that very same mutation also impairs that strain's ability to perform other vital functions like information processing. When the environment returns to normal, the impaired mutant strain is less fit in the struggle for survival.

When bacteria acquire antibiotic resistance by mutations at active sites, the new mutant strain of bacteria pays a price for its short-term competitive advantage. And because mutations at these critical active sites come with a fitness cost, creationists argue that additional mutations of the same kind are most likely to destroy essential functions than to produce fundamentally new forms of life. This implies that there are limits to the amount of change that such mutations can produce. Mutations sometimes affect non-active sites of proteins as well. Although these mutations do occur, they're unhelpful as agents of change. They produce one of two possible outcomes: they either (1) have no effect whatsoever on the structure or function of the protein, or (2) they interfere with the protein and destroy its function. Either way, a mutation at a non-active site isn't likely to produce beneficial large-scale change either.

There is no possibility that resistance-producing mutations could eventually produce a new form of life. In every case where mutations lead to antibiotic resistance, resistance results from small changes to a single protein molecule. For this reason, it is doubtful that the kind of mutations that produce antibiotic resistance can ever produce fundamentally new forms of life, no matter how many times the same molecule is altered. Evolutionists claim that many mutations of many different types of proteins are necessary to produce a major biological change. This is not possible because mutation-induced antibiotic resistance provides no support for the idea that mutations produce major biological change. The mutations that cause antibiotic resistance only change a small site on the surface of a relatively large protein molecule and that these mutations do not alter the overall structure of the protein. Since the kinds of mutation that produce antibiotic resistance do not change the structure of the protein components of the organism, they will not fundamentally change the organization of the organism or the organism as a whole.[9-50]

Creationist View

One way that Staphylococcus bacteria become resistant to penicillin is by a mutation that disables a control gene for production of penicillinase, an enzyme that destroys penicillin. When Staphylococcus bacteria have this mutation, the bacterium over-produces this enzyme, which means it is resistant to huge amounts of penicillin. In the wild, however, this mutant bacterium is less fit, because it wastes resources by producing unnecessary penicillinase. In this case, the mutation causes information loss, even though it might be considered beneficial. This is not an example of evolution (new species) because no new information is generated when the Staphylococcus bacteria becomes mutated.[9-51]

Evolutionists claim that bacteria that have become resistant to antibiotics, due to natural selection selecting resistant strains, is proof that new kinds of genetic information has been developed. What has really occurred is that in many cases some bacteria already had the genes for resistance to the antibiotics. Some bacteria that was frozen before scientists developed antibiotics was thawed and shown to already be antibiotic-resistant. When antibiotics are applied to a population of bacteria, those lacking resistance are killed, and any genetic information they carry is eliminated. The surviving bacteria carry less information, but they are all resistant.

In other cases, antibiotic resistance is the result of a mutation, but in all known cases, this mutation destroyed information. In this case, the loss of information was actually helpful in keeping the bacteria alive. One example of bacteria being resistant to antibiotics, are those resistant to the antibiotic penicillin. As previously mentioned, bacteria normally produce an enzyme, penicillinase, which destroys penicillin. The amount of penicillinase is controlled by a gene. There is normally enough produced to handle any penicillin encountered in the wild, but the bacterium is overwhelmed by the amount given to patients. A mutation disabling this controlling gene results in much more penicillinase being produced. This enables the bacterium to resist the antibiotic. But normally this mutant would be less fit as it wastes resources by producing unnecessary penicillinase. Another example of acquired antibiotic resistance is the transfer of pieces of genetic material, called plasmids, between bacteria, even between those of different species. These various types of bacteria have never obtained or produced new information.[9-52]

An Example of Natural Selection is Antibiotic Resistance

Creationist View

An example of natural selection is that of antibiotic resistance in bacteria. Such natural selection is commonly portrayed as evolution in action, but in this case, natural selection works in conjunction with mutation rather than designed variation. Antibiotics are natural products produced by fungi and bacteria, and the antibiotics we use today are typically derivatives of those. Because of this relationship, it is not surprising that some bacteria would have resistance to certain antibiotics; they must do so to be competitive in their environment. As previously mentioned, antibiotic-resistant bacteria have a mutation in the DNA that codes for that protein. The antibiotic then cannot bind to the mutated protein produced from the mutated DNA, and thus the antibiotic-resistant bacteria live. If the antibiotic-resistant bacteria are grown with the non-mutant bacteria in the environment without antibiotics, the non-mutant bacteria will live whereas the mutant antibiotic-resistant bacteria will die because the mutant bacteria is not able to obtain proper nutrients. This is because the mutant bacteria produce a mutant protein that does not allow them to compete with other bacteria for necessary nutrients. Mutant bacteria will die because they are not able to obtain the proper nutrients to remain alive.[9-53]

An Example of Natural Selection and Bacteria

Creationist View

Helicobacter pylori is a spiral-shaped gram negative bacterium that can live in the stomach and in the duodenum which is the section of intestine below the stomach. It is the most common cause of ulcers of the stomach and duodenum.[9-54] Regular H. pylori does not have a mutation and is not antibiotic-resistant and has the ability to produce an enzyme. Antibiotic-resistant H. pylori bacteria have a mutation that results in the loss of information to produce an enzyme. (An enzyme is any of various proteins, as pepsin, originating from living cells and capable of producing certain chemical changes in organic substances by catalytic action, as in digestion.) This enzyme normally converts an antibiotic to a poison, which causes death to the regular bacteria H. pylori. But when the antibiotics are applied to the mutant H. pylori, these bacteria (the mutant H. pylori) can live but cannot produce an enzyme while the normal bacteria (H. pylori) are killed by the antibiotics. By natural selection the mutant bacteria that lost information will survive and pass this trait to their offspring. The mutant bacteria (H. pylori) produces no enzyme but continues to reproduce offspring that resist antibiotics used to treat it. The normal bacteria H. pylori died because antibiotics were applied to them. A bacterium can get antibiotic resistance by gaining the mutated DNA from another bacterium. Bacteria can swap DNA, unlike humans. It is important to note that this is still not considered a gain of genetic information since the information already exists and that while the mutated DNA may be new to a particular bacterium, it is not new overall.

As discussed previously, natural selection cannot increase or provide new genetic information, that is, natural selection cannot cause new traits to develop in a creature, it can only select traits that are already present. **(1)** Through mutations, genetic information is lost. **(2)** The antibiotic resistant bacteria only survive well in an environment with antibiotics; they are less able to survive in the wild. It is important to keep in mind that the gain of antibiotic resistance is not an example of a beneficial mutation but rather a beneficial outcome of a mutation in a given environment. These types of mutations are rare in other organisms as offspring are more limited in number, therefore, there is a greater need to preserve genetic integrity. **(3)** A particular mutation in a bacterial population was selected for natural selection. **(4)** As a result of natural selection Helicobacter pylori (mutated) is still H. pylori. No evolution has taken place to change it into something else. It's still the same bacteria with some variation. Antibiotic resistance in bacteria is an example of natural selection, not evolution. [9-55]

Limitations of Artificial Breeding
(Artificial Selection or Selective Breeding)

... pigeons with feathered feet have skin between their outer toes; pigeons with short beaks have small feet, and those with long beaks [have] large feet. Hence, if man goes on selecting [breeding], and thus augmenting, any peculiarity, he [the breeder] will almost certainly modify unintentionally other parts of the structure, owing to the mysterious laws of correlation.

• Charles Darwin The Origin of Species
by means of Natural Selection or The Preservation
of Favoured Races in the Struggle for Life,
Chapter 1 - Variation Under Domestication, 6th edition

Slow though the process of selection may be, if feeble man can do much by artificial selection [breeding], I can see no limit to the amount of change, to the beauty and complexity of the coadaptations between all organic beings, one with another and with their physical conditions of life, which may have been effected in the long course of time through nature's power of selection, that is by the survival of the fittest.

• Charles Darwin The Origin of Species
by means of Natural Selection or The Preservation
of Favoured Races in the Struggle for Life,
Chapter 4 - Natural Selection, 6th edition

Evolutionist View

The term "natural selection" was popularized by Charles Darwin, who intended it to be compared with artificial selection, what we now call "selective breeding." Natural selection is an important process (though not the only process) by which evolution takes place within a population of organisms. As opposed to artificial selection, in which humans favor specific traits, in natural selection the environment acts as a sift through which only certain variations can pass.[9-56]

Darwin and other people were impressed at how much plant and animal breeders could influence the ultimate characteristics by selecting individuals from a breeding population. Plant breeders have done the same thing for years.

This was the methodology that was utilized when all sorts of beneficial strains of plants were developed by artificial selection. Darwin believed that the same process performed by animal breeders also happens in nature. Darwin pointed out that there is a struggle for existence, and not all organisms are able to pass their genes on to the next generation. Those that do the best in that struggle for existence, and it is not just a struggle to survive, it is a struggle to find mates, to reproduce, and to raise those offspring. So in many respects things that are very cooperative are important in this struggle. Darwin called this process 'natural selection.'

Darwin realized that those organisms that had the characteristics that suited them best in that struggle, were the ones that were going to maintain their characteristics in the next generation, and he realized that is pretty much what plant and animal breeders do, and therefore over time the average characteristics of a population could change in one direction or another and they could change quite dramatically. Darwin did not understand exactly how these changes occurred because during the 1800s genetics was unknown. The entire process depends scientifically on what that mechanism of inheritance is. Darwin did not know anything about the mechanism of inheritance. There was no way that he could have known. Nobody knew it at the time. When modern genetics came into being, the work of Gregor Mendel was rediscovered.[9-57]

Creationist View

In Darwin's day, pigeon breeding was very popular, and he too took up pigeon breeding. Yet despite all the variations in tails and feathers that was the result of breeding, all the pigeons Darwin observed remained descendants of the common, ordinary rock pigeon. These changes represent cyclical change in gene frequencies but no new genetic information. As a result of this breeding, Darwin surmised what must have occurred in the distant past. He reasoned that if the common rock pigeon can be so greatly changed within a few years at the hand of a breeder, he envisioned what changes could have taken place in nature over thousands, even millions, of years. Darwin came to the conclusion that given enough time, chance would be virtually unlimited, and the pigeon might even be transformed into a completely different kind of bird. The problem is that Darwin, nor anyone else, had actually witnessed evolution occurring. And to make reasonable assumptions and hypotheses, one must have good evidence for believing that the process could continue at a steady rate. This is the major flaw of Darwin's theory. Centuries of experiments have actually shown that the change produced by breeding does not continue at a steady rate from generation to generation. Instead, change is rapid at first, then levels off, and eventually reaches a limit that breeders cannot cross.

There is a reason why there are limits as to the extent of changes that could develop in a species. The reason is once all the genes for a particular trait have been selected, breeding can go no further. Breeding shuffles and selects among existing genes in the gene pool, combining and recombining them. But because breeding does not create new genes, no new genes are introduced into the genome. A bird cannot be bred to grow fur. A mouse cannot be bred to grow feathers. A pig cannot grow wings. As breeders continue the selection pressure on a species, the organism grows weaker until it finally becomes sterile and dies out.

There is a natural barrier that no amount of breeding is able to cross. When an organism is no longer subject to selective pressure, it tends to revert to its original type. If left to themselves, the offspring of the fancy pigeons that Darwin observed during his breeding of pigeons will revert to the wild rock pigeon. Darwin was mistaken in his conclusions. The more recent Neo-Darwinism claims that mutations cause changes in species that account for the variety of species that now exists. The problem is that most mutations are harmful, often lethal, to an organism, so that if mutations were to accumulate, the result would more likely be devolution (become less advanced of having the ability to survive) than evolution. Neo-Darwinists believe that some mutations will somehow be beneficial. And since the evolution of a single new organ or structure may require many thousands of mutations, Neo-Darwinists believe that vast numbers of these rare beneficial mutations will occur in a single organism. In reality, the possibility of this actually occurring to produce a better, improved, species is extremely rare, if not impossible. (Mutations are discussed in more detail below.)[9-58]

In total contrast to natural selection is "artificial selection" or "selective breeding." Evolutionists point to the results of selective breeding as an example of what natural selection accomplishes. But there is a vast difference between them.

Several points should be kept in mind, among which are these: **(1)** The results of breeding never cross the species line; they are always improvements within a species. **(2)** There is a limit to how much change can be made. Beyond that limit, no further changes can be made. The wall imposed by the genetic code cannot be penetrated. **(3)** "Improvements" through breeding may improve certain qualities, but others will be weakened. The original was generally stronger and more vigorous than the "improved" varieties. **(4)** After being left alone for a time, the improved varieties will slip back toward the original pattern. **(5)** The very fact of success in breeding points out that intelligent minds caused it, by careful pre-planning, purposive activities, and continual observation at each step. It is just that: "selective breeding." The evolutionist's "natural selection" is totally different: it involves no intelligence, no planning, no design, and no purpose.[9-59]

Artificial selection, practiced by breeders of agricultural plants and domesticated animals, has commonly been used as a model of the action of natural selection. However, Natural selection has no purpose. For any given generation, natural selection is a consequence of the differences between individuals with respect to their capacity to produce progeny (offspring). Artificial Selection, in contrast, is a purposeful process. It has a goal that can be visualized. Natural selection can and does take place in domesticated and laboratory organisms, and in mankind, under all sorts of natural and artificial conditions. Artificial selection is man-made, however. Natural selection has no selector, it is a self-generated outcome of interactions between organisms and their environments.[9-60]

Intelligent Design View

It is true that both human breeders and nature can produce some changes in populations. Artificial selection is the intentional breeding of animals that have certain traits with the goal of enhancing those traits. This may be an example of microevolution. No artificially bred animal has ever become a different animal. Similarly, natural selection cannot develop different animals. There is a limit as to what artificial and natural selection can do. With artificial breeding, a change in an animal can only go so far before the change becomes detrimental to the animal. Darwin's theory states that the unguided force of natural selection is supposed to be able to do what the intelligent breeder can do. However, even a process of careful, intentional selection encounters limits that neither time nor the efforts of human breeders can overcome. Therefore, natural selection has only limited capabilities.[9-61]

Sexual Selection (A process of Natural Selection)

Evolutionist View

How Did Sexual Reproduction Originate and is it Beneficial to a Species?

Sex is a costly process that should have been lost quickly from populations. Yet, most eukaryotes (multi-celled animals) are sexual. So, what selective benefits does sex provide that outweigh these significant costs and how did the sexual reproduction process develop? Sex might have evolved as a parasitic adaptation among genetic elements to allow them to spread to other genetic lineages. Unfortunately, the unique origin of sex, coupled with the fact that it occurred in the distant past under selective conditions about which one can only guess, and which might have changed dramatically since, makes testing theories for the evolutionary origins of sex extremely difficult. (For more information on the how sex evolved, see Reference # [9-61f].) Evolutionary biologists might be limited to possible theories and might never have a conclusive answer as to why sex evolved in the first place. Regardless of whether a selective explanation for the initial spread of sex across the tree of life is necessary, it is required for its continued maintenance against significant evolutionary costs. Also, the selective forces maintaining sex must still be operating, and operating in a diversity of species. It is possible that sex is simply a mechanism for repairing DNA damage, in particular double-stranded DNA damage. DNA damage is a problem for all organisms, and so selection based on repair would have some common functionality required to explain the widespread nature of sex. Moreover, many of the enzymes involved in recombination do have roles in DNA repair and probably did evolve initially for that function, only later being utilized for sex. However, this may not be the reason why sex has continued. There are organisms that never have sex, but do not apparently suffer from catastrophic DNA damage.

The main result of sex is that genetic material from two individuals is mixed together into a single individual, leading to the production of offspring that are genetically distinct from either parent. The genetic mixing that results from sex is, however, not guaranteed to increase the heritable genetic variation for fitness, which is the ultimate determinant of the rate of adaptation. Even under conditions where sex was beneficial, the benefits were rarely substantial enough to outweigh the two-fold costs of producing males (that do not actually give birth). Sex appears to be beneficial because sexual groups are more evolutionarily successful (they have a longer evolutionary lifespan) than asexual groups (that is, species that do not need to have two parents to produce offspring). Generally speaking, sex produces higher quality genotypes (*The genotype of an organism is the chemical composition of its DNA, which gives rise to the phenotype, or observable traits of an organism.*), and the genes for sex thrive on the high-fitness genotypes that they create. The evolutionary success of sex is down to the diversity that it creates. In a complex world in which environments are constantly changing, competitors, parasites and prey are constantly evolving and mutation is continually eroding adaptation, the differences produced by the sexual cycle provide an important evolutionary advantage. (Adaptations and mutations are discussed later in this chapter.) This advantage favors genes for sex and recombination within populations, and can also have profound implications for the evolutionary lifespan of populations and species.[9-61a]

Creationist View

How and why did the two sexes originate? Sexual reproduction is only possible when both sexes have fully functional reproductive organs at the same time. By definition, an evolutionary process is not directed by some purposeful strategic plan. How is it then possible that such different and complex organs, which fit one another in every morphological and physiological detail, could have evolved suddenly? [9-61b]

The origin and subsequent maintenance of sex and recombination is a phenomena not easily explained by Darwinian evolution. Evolutionary mechanisms such as natural selection are not able to explain why organisms should evolve from asexual reproduction (reproduction not needing a mate) in favor of the more costly and inefficient sexual reproduction. Most single-celled organisms (prokaryotes, such as a bacterium) reproduce asexually that is, they do not need to have two parents to produce offspring. Asexual reproduction is the formation of new individuals from cells of only one parent, without gamete formation or fertilization by another member of the species. If life on earth is derived entirely from these single-celled creatures, then why was this simple, yet efficient, method of asexual reproduction not continued (as life progressed and as new species evolved) in favor of sexual reproduction? From the perspective of evolutionary biology, sex is without question an inefficient way to reproduce.[9-61c]

Evolutionists claim that amoebas evolved into intermediate organisms, which then further evolved into amphibians, reptiles, mammals, and, eventually, humans (See chapters 3, 7, 8, 15, 16). (*An amoebas is a type of cell or unicellular [one celled] organism which has the ability to alter its shape, primarily by extending and retracting pseudopods. Amoebas do not form a single taxonomic group; instead, they are found in every major lineage of eukaryotic organisms.*) Yet, evolutionists never explain biologically exactly when or how independent male and female sexes originated. (Evolutionists only show examples of various forms of reproduction and claim the more developed methods evolved from simpler life forms.) Somewhere along this evolutionary path (according to evolutionists), both males and females were required in order to ensure the procreation that was necessary to further the existence of a particular species. But how is this explained? Evolutionists have no explanation as to where males and females actually came from, nor do they have an explanation as to the evolutionary origin of sex. How could nature evolve a female member of a species that produces eggs and is internally equipped to nourish a growing embryo, while at the same time evolving a male member that produces motile sperm cells? Also, how is it that these gametes (eggs and sperm) evolved so that they each contain half the normal chromosome number of somatic (body) cells? (Somatic cells reproduce via the process of mitosis, which maintains the species' standard chromosome number; gametes are produced via the process of meiosis, which halves that number. These processes will be discussed later under sub-sections "Mitosis & Meiosis" and "Types of Cell Division—Mitosis and Meiosis".)

The evolution of sex (and its accompanying reproductive capability) remains unknown because no matter how many theories evolutionists develop, they still must first explain the origin of the first fully functional female and the first fully functional male that is necessary to begin the process. Evolutionists admit that the origin of the sexual process remains one of the most difficult problems in biology. Sex occurs in all major groups of life. But why is this the case? When (and how) did sexual reproduction evolve? The difficulty is that sexual reproduction creates complexity of the genome and the need for a separate mechanism for producing gametes. The metabolic cost of maintaining this system is great, as there is the need of providing organs that are specialized for sexual reproduction. Sexual reproduction requires organisms first to produce, and then maintain, gametes (reproductive cells, that is, sperm and eggs). Also, various kinds of incompatibility factors (such as the blood Rh factor between mother and child) pass along more costs (some of which can be life threatening) that are automatically inherent in this expensive means of reproduction.

Many single-celled organisms (prokaryotes, such as a bacterium) reproduce asexually, that is, they do not need to have two parents to produce offspring. If all multi-celled animals (and humans) descended from these single-celled creatures, then why did the simple, yet efficient, method of asexual reproduction develop sexual reproduction? Asexual reproduction is the formation of new individuals from cells of only one parent, without gamete formation or fertilization by another member of the species. Asexual reproduction therefore does not require one egg-producing parent and one sperm-producing parent. A single parent is all that is required. Sporulation (the formation of spores) is one method of asexual reproduction among protozoa and certain plants. A spore is a reproductive cell that produces a new organism without fertilization. In certain lower forms of animals, and in yeasts, budding is a common form of asexual reproduction as a small protuberance on the surface of the parent cell increases in size until a wall forms to separate the new individual (the bud) from the parent. One lower form of animal, the hydra, is a minute freshwater coelenterate with a tubular body and a ring of tentacles around the mouth. (*A coelenterate is an aquatic invertebrate animal of a phylum that includes jellyfishes, corals, and sea anemones. They are distinguished by having a tube- or cup-shaped body and a single opening ringed with tentacles. A coelenterate is also called cnidarian.*) Regeneration is another form of asexual reproduction that allows organisms (e.g. starfish and salamanders) to replace injured or lost parts.

Some naturalists theorize that sexual reproduction provides the best defense against the rapidly reproducing, infectious species that threaten the existence of organisms. The diversity in the species that results from combining different gene pools favors the survival of those that are sexually reproduced over those that by cloning inherit repetitive genetic similarity. The evidence does indicate that species go for vast periods of time without changing much. It has been stated that bacteria evolved in such a way that ultimately it (bacteria) would be responsible for sexual reproduction. Yet if that is the case, why, then, have the bacteria themselves remained virtually unchanged (from an evolutionary viewpoint) for billions of years of Earth history?

In an asexual organism, any mutation that occurs in one generation will be passed on automatically to the next. (Mutations will be discussed in section "(4) Mutations" below.) DNA can be damaged in at least two ways. **First,** ionizing radiation or mutagenic chemicals can alter the genetic code. Or, **second**, a mutation can occur via errors during the replication process itself. Most mutations are harmful. Eventually, the quality deteriorates severely. As asexual organisms continue to accumulate mutations, they could eventually become both unable to reproduce and unviable, neither of which would be at all helpful to evolution. Sexual reproduction allows most plants and animals to create offspring with good copies of two genes via crossover and would therefore help eliminate the accumulation of mutations, since mutations, although they might still be passed on from one generation to the next, would not necessarily be expressed in the next generation. This is because a mutation must appear in the genes of both parents before it is expressed in the offspring. Some believe that sexual reproduction makes it easier for an evolving organism to get rid of harmful changes. That should certainly be the case if there is more than one genetic change and if their combined effect on the fitness of the evolving organisms is greater than the sum of their individual changes acting separately. [9-61d]

Mitosis & Meiosis

There is more to the problem in determining the origin of sex, however, than just rare, beneficial mutations, along with the harmful, deleterious mutations. There is the added problem related to the two different types of cell division (mitosis and meiosis). During mitosis, all of the chromosomes are copied and passed on from the parent cell to the daughter cells. Meiosis (from the Greek meaning to split), on the other hand, occurs only in sex cells (i.e., eggs and sperm); during this type of replication, only half of the chromosomal material is copied and passed on. The claim that meiosis allegedly has evolved the ability to halve the chromosome number (but only for gametes), and that it actually can provide unlimited new material, seems unbelievable. The critical nature of meiosis to life as we know it has been acknowledged, even by evolutionists.

The sexual process involves a great deal of complexity, including the complexity involved in reproducing the information carried within the DNA. Besides the difficulties associated with the sheer rarity of beneficial mutations and the harmful deleterious mutations, there is the added complexity related to the two different types of cell division (mitosis and meiosis). During mitosis, all of the chromosomes are copied and passed on from the parent cell to the daughter cells. Meiosis, however, occurs only in sex cells (i.e. sperm and eggs).

During the latter type of replication (meiosis), only half of the chromosomal material is copied and passed on to the subsequent generation. Meiosis results in the production of completely new combinations of the parental genes, all of them uniquely different genotypes. These, in turn, produce unique phenotypes, providing unlimited new material for the process of natural selection. (*Phenotype is the observable properties of an organism that are produced by the interaction of the genotype and the environment.*) It is because meiosis has allegedly evolved the ability to halve the chromosome count, but only for gametes, and that it can actually provide unlimited new material, which make the meiotic process so incredible. The mechanism of meiosis is critical for sexual reproduction. Evolutionary biologists have not been able to provide an adequate explanation as to how somatic cells reproduce by mitosis (thereby maintaining the species' standard chromosome number in each cell), while gametes are produced by meiosis, wherein that chromosome number is halved so that, at the union of the male and female gametes during reproduction, the standard number is restored. (See Chapter 5, section "Description of a Cell - Continued" sub-section "Types of Cell Division—Mitosis and Meiosis" further discussion of this topic.)

The origin of sex remains a mystery for those who believe in a purely materialistic view of reality, not to mention the origin of the incredibly complex meiotic process that makes sex possible, or the intricate development of the embryo (which is itself an engineering marvel). At conception, the chromosomes inherited from the sperm are paired with the chromosomes inherited from the egg to give the new organism its full chromosomal complement. Evolutionists believe that undirected occurrences brought about this amazingly interdependent process of **(1)** halving the genetic information; and **(2)** recombining it through sexual reproduction. Not only is such a sophisticated mechanism required for the production of a sperm or egg cell via meiosis, but another equally intricate process also joins the genetic information during fertilization in order to produce the zygote, which will later become the embryo. It is highly unlikely that purely materialistic processes, governed by the laws of chance and natural selection, could have produced such a mechanism.[9-61c] (See Chapter 11 for further discussion of embryonic development.) Evolutionists have not been able to propose an adequate cause or mechanism to explain how opposite sexes could have developed even one time let alone for millions of species. This is why it is very possible that the existence of sexual reproduction in mankind, as well as in numerous animals, may be the best argument against evolution. [9-61e]

(2) Adaptations

Evolutionist View

> *The failure of many plants and animals to adapt, which is well over*
> *99 percent of the species that ever lived . . . poses an enormous problem*
> *for theories of intelligent design (ID) [creationists]. It doesn't seem so*
> *very intelligent to design millions of species that are destined to go*
> *extinct, and then replace them with other, similar species, most of*
> *which will also vanish* [9-62]

Adaptation may be defined as is any alteration in the structure or function of an organism or any of its parts that results from natural selection and by which the organism becomes better fitted to survive and multiply in its environment. Adaptation is a form or structure modified to fit a changed environment. Another definition of adaptation is the ability of a species to survive in a particular ecological niche, especially because of alterations of form or behavior brought about through natural selection.[9-63]

Natural selection can create complex adaptations, sometimes in surprisingly little time. Adaptations are changes living things develop to better survive in their environment. Three things are involved in creating an adaptation by natural selection. **First**, the starting population has to be variable. A creature must show some difference, otherwise this trait cannot evolve. **Second**, some proportion of that variation has to come from changes in the form of genes, that is, the variation has to have some genetic basis, called heritability. If there were no genetic difference between the same species there would be no evolutionary change. The **third** and last aspect of natural selection is that the genetic variation must not inhibit an individual's probability of leaving offspring.[9-64] The theory of natural selection explains how every adaptation evolved, step-by-step, from traits that preceded it. This includes not just body form and color, but the molecular features that underlie everything.[9-65] An adaptation must evolve by increasing the reproductive output of its possessor. For it is reproduction, not survival, that determines which genes make it to the next generation and cause evolution. A gene that kills a creature after reproductive age incurs no evolutionary disadvantage. It will remain in the gene pool. The accumulation of such genes by natural selection, in fact, is widely thought to explain why creatures deteriorate in so many ways as they reach old age.[9-66] Adaptation always increases the fitness of the individual, not necessarily of the group or the species. The idea that natural selection acts "for the good of the species" is not true. Evolution can produce features that, while helping an individual may harm the species as a whole.[9-67]

Creationist View

Evolutionists believe that all adaptations begin with time and chance, that is, with random changes in DNA and hereditary traits called mutations. In evolutionary theory, those chance mutations that suit an organism better to its environment are preserved by the process called natural selection.[9-68] There is no convincing evidence or argument that fitness or natural selection can lead to adaptation, but there is evidence and logic that adaptation can lead to natural selection.[9-69]

(3) Speciation

As the amount of modification which animals of all kinds undergo partly depends on the lapse of time, and as the islands which are separated from each other, or from the mainland, by shallow channels, are more likely to have been continuously united within a recent period than the islands separated by deeper channels, we can understand how it is that a relation exists between the depth of the sea separating two mammalian faunas, and the degree of their affinity, a relation which is quite inexplicable on the theory of independent acts of creation.

• Charles Darwin The Origin of Species
by means of Natural Selection or The Preservation
of Favoured Races in the Struggle for Life,
Chapter 13 - Geographical Distribution, 6th edition

Evolutionist View

Speciation is the formation of new species as a result of geographic physiological, anatomical, or behavioral factors that prevent previously interbreeding populations from breeding with each other.[9-70] Speciation occurs as a breeding population splits into two, and they go their separately evolving ways. Among sexually reproducing species, speciation is said to have occurred when the two gene pools have separated so far that they can no longer interbreed. Speciation begins by accident. When separation has reached the stage where there is no interbreeding even within a geographical barrier, this causes the origin of a new species.[9-71] Speciation is the evolution of different groups (species) that cannot interbreed, that is, groups that can't exchange genes.[9-72] The diversity of nature encompasses millions of species, each with its own unique set of traits. And all of this diversity came from a single ancient ancestor. To explain biodiversity, then, one would have to do more than explain how new traits originate. One must also explain how new species originate. If speciation did not originate then there would be no biodiversity at all. There would only be a single, long-evolved descendant of that very first species. The explanation to the origin of species is based on how species are defined. Species is defined as a group of individuals that resemble one another more than they resemble members of other groups.[9-73] Species is further defined as a group of interbreeding natural populations that are reproductively isolated from other such groups. The term "Reproductively isolated" simply means that members of different species have traits (differences in appearance, behavior, or physiology) that prevent them from successfully interbreeding, while members of the same species can interbreed readily.[9-74]

The origin of species, as described in general terms by Charles Darwin, is the process whereby species descended from other species. That is, the family tree of life is a branching one, which means that more than one modern species can be traced back to the same ancestor. For example, lions and tigers are now members of different species, but they both sprung from a single ancestral species. This ancestral species may have been the same as one of the two modern species, or it may have been a third modern species, or it may be a species that became extinct. Similarly, humans and chimps now clearly belong to different species, but their ancestors of a few million years ago belonged to one single species. (Human evolution is discussed in chapters 15 & 16.) Speciation is the process by which a single species becomes two species, one of which may be the same as the original single one.

The reason speciation was previously thought to be a difficult problem is that all members of the single would-be ancestral species are capable of interbreeding with one another. Therefore, every time a new daughter species begins, the new species is in danger of being lost due to interbreeding with their ancestor. For example, would-be ancestors of the lions and the would-be ancestors of the tigers would fail to split apart if they kept interbreeding with one another and would therefore stay similar to one another. There would be no problem of interbreeding if the ancestral lions and the ancestral tigers happen to be in different parts of the world where they could not interbreed with each other. Different parts of the world do not necessarily need to be different continents. Geographical separation could be any barrier that would prevent interbreeding with the ancestor. Some barriers include: different sides of a desert, a mountain range, a river, or even a motorway. Barriers can also apply to animals separated by no barrier other than great distances.

If the two separated species were ever to come in contact with each other after multiple of generations (and changes), they would no longer be able to interbreed, as they became two distinct different species. Natural selection would then be said to finalize the process of reproductive isolation that began with the chance intervention of a barrier. Speciation is then said to be complete. In reality, if the two species were ever to re-intermingle with each other, the two species would not co-exist very long because they would compete with each other (such as competing for food) and possibility one would be driven to extinction. It is also possible that the two species would not have the same way of life, which would mean that they could co-exist with each other.[9-75]

An Example of Speciation

Evolutionist View

Suppose, for example, that an ancestral species of a flowering plant was split into two portions by a geographic barrier, like a mountain range. The species may, for example, have been scattered over the mountains by birds. Let's take a look at one population of a flowering plant living in a place having a lot of hummingbirds but only a few bees. In this area, the flowers will evolve to attract hummingbirds as pollinators. Typically the flowers would become red, a color that the birds find attractive, produce abundant nectar that would reward birds, and have deep tubes to accommodate hummingbird's long bills and tongues. The population on the other side of the mountain may find its pollinator reversed, having few hummingbirds but many bees. There the flowers would evolve to attract bees. They may become pink, a color bees favor, and evolve shallow nectar tubes with less nectar. This is because bees have short tongues and do not require a large nectar reward. The flowers would also be flatter and would have petals that form a landing platform. Eventually, the two populations of flowering plants will differ in the form of their flowers and amount of their nectar, and each will be specialized for pollination by only a single type of animal, in the case here, either hummingbirds or bees. If the geographic barrier disappeared, and the newly different populations of flowering plants found themselves back in the same area, an area containing both bees and hummingbirds, they would now be reproductively isolated: each type of flower would be served by a different pollinator, so their genes would not mix via cross-pollination. The flowers would become two different species and would no longer be able to exchange genes, that is, interbreed. Their gene pools would be restricted to those plants that were able to share pollen and genes.[9-76]

Creationist View

If evolution is constantly producing species, why are the species not changing? [9-77]

Unfortunately, biologists have at times classified certain plants and animals as "species," when they are not species, but subspecies. True species cannot cross breed with other species, although some sub-species do not interbreed (some dogs will not interbreed but they are still dogs). True species cannot cross breed, not for mechanical reasons, but because they have different basic DNA code. Monkeys cannot breed with owls and produce young, etc. [9-78]

Evolutionists believe that two or more species that come from one kind indicates evolution. Some would use the example of flies on certain islands that used to interbreed but no longer do so. They've become separate species, which some would claim proves evolution. Any real evolution (macro-evolution) requires an expansion of the gene pool, the addition of new genes (genomes) with new information for new traits as life is supposed to move from simple beginnings to more varied and complex forms. Suppose there are islands where varieties of flies previously traded genes, but no longer interbreed. This is not evidence of evolution. Each variety resulting from reproductive isolation has a smaller gene pool than the original and a restricted ability to explore new environments with new trait combinations or to meet changes in its own environment. The long-term result would be extinction of varieties of flies rather than evolution.[9-79]

Ring Species

Evolutionist View

Ring species provides very conclusive evidence that speciation occurs and thus provides additional evidence of evolution occurs in the natural world. A ring species is defined as: *A biological species consisting of overlapping subgroups each of which can interbreed with the next but which cannot freely interbreed when taken as a whole.* In biology, a ring species is a connected series of neighboring populations, each of which can interbreed with closely sited related populations, but for which there exists at least two "end" populations in the series, which are too distantly related to interbreed, though genetically connected, "end" populations may co-exist in the same region thus closing a "ring" (See figures 9.1a, 9.1b). Ring species provide important evidence of evolution in that they illustrate what happens over time as populations genetically diverge, and are special because they represent in living populations what normally happens over time between long deceased ancestor populations and living populations, in which the intermediates have become extinct. Formally, the issue is that interfertile "able to interbreed" is not a transitive relation – if A can breed with B, and B can breed with C, it does not follow that A can breed with C – and thus does not define an equivalence relation. A ring species is a species that exhibits a counterexample to transitivity.[9-79a]

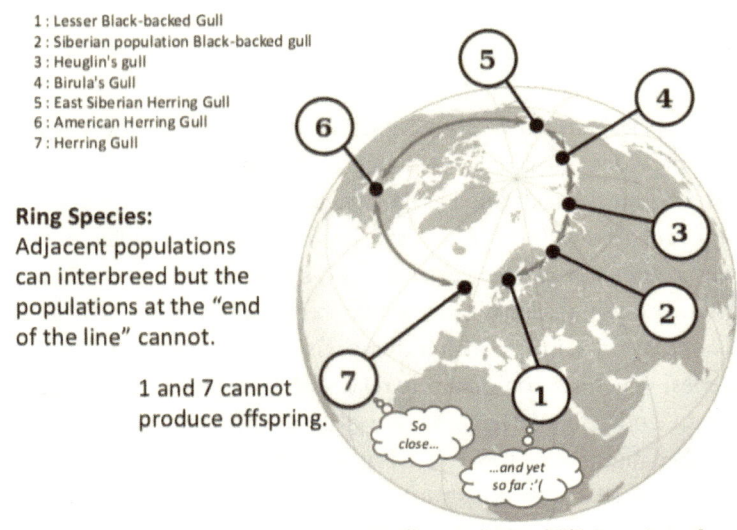

Figure 9.1a Ring Species Illustration of Gulls in Northern Hemisphere

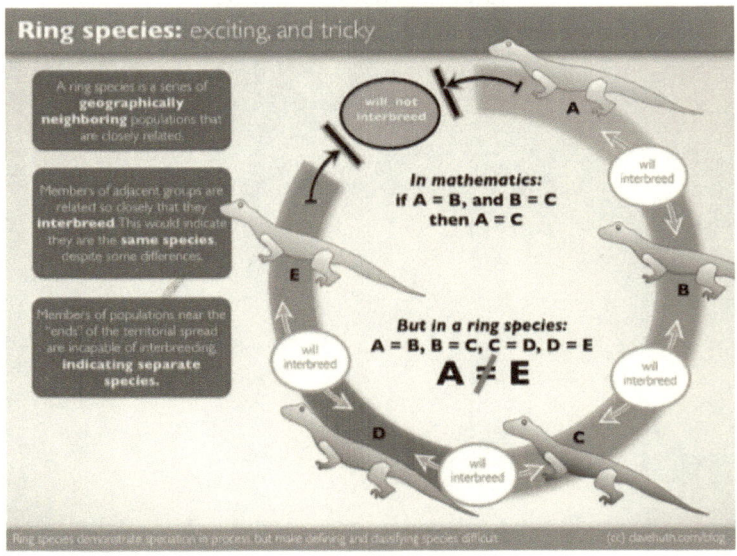

Figure 9.1b Ring Species Illustration

A ring species is a connected series of neighboring populations that can interbreed with relatively closely related populations, but for which there exists at least two "end" populations in the series that are too distantly related to interbreed. Often such non-breeding-though-genetically-connected populations co-exist in the same region thus creating a "ring". Ring species provide important evidence of evolution in that they illustrate what happens over time as populations genetically diverge, and are special because they represent in living populations what normally happens over time between long deceased ancestor populations and living populations. If any of the populations intermediate between the two ends of the ring were gone they would not be a continuous line of reproduction and each side would be a different species. [9-79b]

Creationist View

The fascinating phenomenon known as 'ring species' is sometimes quite incorrectly used to 'prove' evolution. The classic example is as follows. In Britain, the herring gull is clearly a different species from the lesser black-backed gull. Not only can they be easily told apart, but apparently they never interbreed, even though they may inhabit the same areas. By the usual biological definition, they are therefore technically different species. Yet when the ends of the ring meet, the two do not interbreed and so are for all intents and purposes separate species. However, as you go westward around the top half of the globe to North America and study the herring gull population (Figure 9.1a), an interesting fact emerges. The gulls become more like black-backed gulls, and less like herring gulls, even though they can still interbreed with herring gulls from Britain. Now go still further via Alaska and then into Siberia (see map below Figure 9.1c). The further west you go, the more each successive population becomes less like a herring gull and more like the black-backed. As you travel west via the route shown by the gray band at the top of the globe (Figure 9.1c), each successive population of herring gull seems more like the black-backed gull.

Figure 9.1c Ring Species Illustration of Gulls in Northern Hemisphere

At every step along the way, each population is able to interbreed with those you studied just before you moved further west. Therefore, you are never technically dealing with separate species. Until, that is, you continue your journey into Europe and back to Britain, where you find that the lesser black-backed gulls there are actually the other end of a ring that started out as herring gulls. At every stage around the ring, the birds are sufficiently similar to their neighbors to interbreed with them. Yet when the ends of the ring meet, the two do not interbreed and so are for all intents and purposes separate species.

It is clear from such examples that species are not fixed and unchanging, and that two apparently different species may in fact be genetically related. New species (as man defines them) can form. The herring gull and the lesser black-backed gull could not have been initially created as two separate groups reproducing only after their kind, or else they would not be joined by a chain of interbreeding intermediates. There are also observations of other wild populations from which a reasonable person must infer that certain very similar species did indeed share the same ancestor, even though there is no complete 'ring'. Changes do appear, however, but what sort are they? Many have been misled into thinking this is evidence for evolution. The wolf, the dingo and the coyote are all regarded as separate species. However, they (perhaps along with several other species) almost certainly 'split off' from an original pair. A 'mongrel' dog population can be 'split' into separate sub-groups, the varieties of domestic dog (breeders can isolate portions of the total information into populations which do not contain some other portions of that information).

This sort of variation does not add any new information. On the contrary, it is genetically downhill. It involves a reduction of the information in each of the descendant populations compared to the ancestral one. Thus, a population of pampered lap-dogs has less genetic information/variability, from which nature or man can select further changes, than the more 'wild' population before evolution selection took place. But is it conceivable that such change (which is obviously limited by the amount of information already present in the original kind) can extend to full, complete formation of separate species without any new information arising, without any new genes? (In other words, since evolution means lots of new, useful genes arising with time, can you have new species without any real evolution, that is, without an increase in genetic information?) An ancestral species can split into other species within the limits of the information already present in that kind. In the example just discussed, there is no reason to believe that the differences between the two gull species are the result of any new, more complex, functional genetic information not already present in an ancestral, interbreeding gull population. Because there is no evidence of any such information-adding change, it is misleading to say this gives evidence of evolution, of even a little bit of the sort of change required for a fish to eventually evolve into a human.[9-79c]

Ring Species is defined as a species with a series of connected populations which through time spread around a geographic barrier (Figure 9.1d) and where neighboring populations are able to interbreed but the distant populations that meet after the barrier are unable to interbreed.

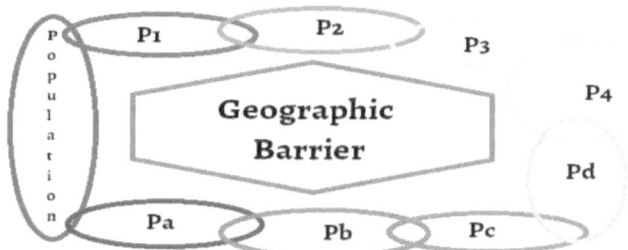

Figure 9.1d Ring Species Illustration

By definition, there are several basic characteristics that must be matched by a plant or animal to qualify as a Ring Species. The three most prominent characteristics (Figure 9.1d) are: **1)** a series of connected populations encircling a geographic barrier (P1, P2, P3, P4 and Pa, Pb, Pc, Pd) **2)** continuous gene flow along both branches of the ring and **3)** the end populations cannot interbreed where they come together (P4 and Pd). Evolutionists have made the assumption that Ring Species like the salamanders and the gulls illustrate in the spatial dimension something that must happen in nature. Most of the time, speciation is considered to occur because of reproductive isolation through time. However, within Ring Species, the possibility of speciation without such isolation would be a great find toward showing Darwinian style evolution as it would create a situation where a single species could become two species, due to divergent populations, even with some connected genetic flow. If a ring were truly found, the two ends that are unable to interbreed would act like two separate species. Yet, the entire ring, from one end to the other, is able to breed and would therefore be considered one species.

Through the past century, several suggested candidates for Ring Species have been suggested. The most notable are the: **1)** Ensatina Salamanders surrounding the Central Valley of California **2)** the Larus Gulls near the Arctic Circle (Figure 9.1a), **3)** the Greenish Warbler surrounding the Tibeten Plateau, **4)** the Crimson Parrot in Australia, and **5)** the Caribbean Slipper Spurge in Central America. Within the last decade, these species have been studied, including genetically, and the results were not as expected. The major downfall of the Ring Species concept is that the end populations, which by definition cannot interbreed, have been found to interbreed even in the wild, typically with fertile offspring. Therefore the species is only showing some variations (as varieties or subspecies) based on the surrounding environment. It remains the same species.

The changes that are actually observed are in choosing a mating partner rather than successful fertilization not occurring when attempted because of genetic dissimilarity. Furthermore, some suggested Ring Species merge back together, rather than diverge, when they meet, the opposite of what is needed for evolution to occur. It can also be noted that the gene flow which is expected to continue around the ring has usually not been found. Instead, most of the groups have sharp breaks in the genetics which show distinct populations rather than gradual change. Within the Larus gulls, the opposite happens and far too much hybridization occurs with other species not originally included within the ring and make the formation of a ring impossible. Ring Species do not show evolution and does not hold up to scrutiny.[9-79d]

(4) Mutations

It must not be forgotten that mutation is the ultimate source of all genetic variation found in natural populations and the only new material available for natural selection to work upon.[9-86]

Evolutionist View

Mutations are the occasional changes that occur in the DNA of genes during replication. Mutations are changes, but they are not mistakes. These changes can cause changes in the genes that can either be harmful or beneficial or have no effect at all on the organism. Mutations provide the essential variation on which life depends. Without mutations, life could not have adapted and survived on Earth.[9-81]

Mutations occur because of copy errors that happen during DNA replication. (DNA replication is the biological process of producing two identical replicas of DNA from one original DNA molecule - see Chapter 5, figures 5.20 & 5.21 and Chapter 5, section "DNA Replication.") Genetic variation occurs because of random mutations that occur at a certain rate in all organisms. Mutations are permanent and transmissible. Mutations result if: **1)** copying errors occur in the genetic material during cell division; **2)** there is exposure to radiation, chemicals, or viruses. In multicellular organisms, mutations can be passed on to the offspring, or can lead to malfunction (that may also cause cancer) or death of a cell.[9-82]

Mutations occur regardless of whether they would be useful to the individual or not. Mutations are simply errors in DNA replication (Chapter 5, figures 5.20 & 5.21). Most of them are harmful or neutral, but a few can turn out to be useful. The useful ones are the raw material for evolution.[9-83] Mutations come as small changes, gradually over time. There might be some point where the intermediate steps threaten to be fatal or are lost before becoming part of a final adaptation. Most mutations are recognized to be harmful and too many mutations cause damage rather than enhanced benefits. Therefore mutation rates need to be slow to make evolution work on the genetic level. Mutations can affect either small or large sections of both genes and chromosomes. New genetic information may come about by gene duplication or by the exchange of nucleotides. Mutations can either have a great impact on an organism or none at all. Genetic research has demonstrated that even slight genetic changes can alter the size, shape, and growth rate of an organism. Mutations have resulted in an organism's survival rate such as in antibiotic resistant bacteria.[9-84] (Antibiotic resistant bacteria are discussed in section "Natural Selection", sub-section "Natural Selection and Bacteria.")

Experts have proven that a newly mutated gene would not automatically form a new species. Nor would it automatically replace the preexisting form of the gene, and so transform the species. Replacement of one gene by a mutation form of the gene, they said, could happen in two ways. **(1)** The mutation could enable its possessors to survive or **(2)** reproduce more effectively than the old forms; if so, it would increase by natural selection, just as Darwin had said. The new characteristic that evolved in this way would ordinarily be considered an improved adaptation. One expert pointed out, however, that not all genetic changes in species need to be adaptive. A new mutation might be no better or worse than the preexisting gene – it might simply be "neutral." In small populations such a mutation could replace the previous gene purely by chance – a process he called random genetic drift.[9-84a]

By far the most important way in which chance influences evolution is in the process of mutation. Mutation is, ultimately, the source of new genetic variations, and without genetic variation there cannot be genetic change. Mutation is therefore necessary for evolution. But it is not sufficient. A new mutation exists at first in just one individual of the species, and then in the individual's offspring. Therefore it is carried by just a few individuals at any time, unless something makes those individuals reproduce more than others, so that the mutation becomes more common. That "something" is either genetic drift or natural selection. Evolutionary change requires at least two ingredients – **(1)** mutation and **(2)** genetic drift or natural selection. Creationists often accuse biologists of attributing all of evolution to chance, and this would be true if mutation were the whole story. But natural selection is part of the process also, which is not chance. Natural selection shapes order out of mutational chaos.

At the molecular level, there are many kinds of mutations. Perhaps the most common type is a change in one or more of the nucleotides in a ladder-like arrangement (Chapter 5, Figure 5.19). The exact sequence of four kinds of nucleotides determines the exact sequence of twenty kinds of amino acids in a protein that is produced by the DNA. The amino acid sequence of the protein determines its biochemical function, which in turn affects the development, form, and physiology of the plant or animal. Some mutations can therefore have profound effects. They can alter the structure of a critical protein so much that the organism becomes severely distorted and may not survive. Other mutations may cause changes in the protein that don't affect its function at all. Such mutations are adaptively neutral – they are neither better nor worse than the original form of the genes. Still other mutations are decidedly advantageous. Between these extremes, there is a complete spectrum of effects, and this is probably where most mutations fall.

Many mutations in fruit flies, plants, and other organisms are known that increase or decrease the activity of enzymes and proteins to a greater or lesser degree. At the level of the whole organism, most mutations have slight effects. In fruit flies, for example, most mutations cause slight increases or decreases in (1) the rate of growth, (2) body size, (3) the length of wings or legs, (4) the number of bristles, (5) the ability to detoxify DDT and other poisons, (6) and so on. Some mutations, have drastic effects, like the vestigial fruit fly that reduces the wings to little nubbins. Because such mutations are easy to measure, they have been studied extensively: but the vast majority of mutations have much more subtle effects. It is possible to make a population of fruit flies that is genetically completely homogeneous, and to measure new genetic variation in the population as it comes into existence. If you look for the appearance of very deleterious or lethal mutations, you will find them: but if you look for new variation in the number of bristles or any other feature, you will find that as well. Such variation increases from generation to generation, as subtle new mutations come into existence.

Most biologists believe that mutations occur simply because organisms can't prevent them. From generation to generation, DNA molecules replicate themselves by separating the two sides of the DNA "ladder" (Chapter 5, Figures 5.20, 5.21); enzymes then insert each of the four kinds of nucleotides into the right positions, so that the two identical ladders are formed from the original. Sometimes the wrong nucleotide is inserted into a particular position. There are "repair enzymes" that help to correct such mistakes, but they don't correct them all. (See Chapter 5, sections "DNA and its Repair Mechanism", "The Cell's Production Controls", "The Cell's Quality Control System to Detect Production Errors" and Chapter 9, section "Mutations and the Genetic Code" below for further discussion.) Such mistakes presumably occur because chemical reactions don't happen with perfect precision every time. In fact, a variety of chemicals such as caffeine can increase the mutation rate, as can various forms of radiation.[9-84b]

Because most mutations have only a slight effect on each of the characteristics that they alter, almost all characteristics display what is called "continuous variation." There is a complete spectrum of height and hair color, for example, in many human populations. Typically there are dozens of genes that affect a characteristic, each existing in different forms that alter the characteristic very slightly. Suppose for the sake of example that three groups, A, B, and C affect height. Every individual has two representatives of each gene. Let the average height of a plant with genotype aabbcc be 50 inches. Then if a capital letter represents a mutation that adds an inch, a plant that is AaBbCc will be 53 inches. Now suppose each such mutation has just come into existence, so that it is very rare in the population. When the genes are shuffled about during sexual reproduction, very few individuals will inherit more than one upper-case mutation, so most of the plants will be about 50 or 51 inches tall. It is possible for a 56-inch (AABBCC) plant to be produced, but this will be extremely uncommon as long as the upper-case (mutated) genes are rare.

If, however, 51-inch plants, which carry either the mutation A or B or C, survive somewhat better than 50-inch plants (via natural selection), each of these mutations will have been distilled or concentrated in the population, so that they become more common. When these plants mate with each other, some of their offspring are now likely to inherit two or more upper-case (mutated) genes and be 52 inches or taller. The same process, if repeated again and again, would make the upper-case (mutated) genes so common that a 56-inch offspring would become likely rather than unlikely. The population will then have evolved – and quite quickly, too – from an average height of 50 inches to an average of nearly 56, well beyond the range of variation in the original population. This hypothetical mode of inheritance – many genes, each adding or subtracting a little bit – is exactly the way most characteristics are inherited. The selection process just described has enabled geneticists to **(1)** alter the milk production of cows, **(2)** the yield of corn, **(3)** the resistance of crops to diseases, **(4)** and almost every conceivable characteristic of the fruit fly: wing length, mating behavior, sensitivity to temperature, rate of genetic recombination. It doesn't matter whether the selection is imposed by a geneticist who wants to breed taller corn, or by competition in nature that favors plants able to shade out their competitors; if the mutations are there, and if the selective pressure in favor of them is strong enough, the population will evolve.[9-84c]

Evolution occurs due to biological changes in species. Macro evolution occurs because there is no limit as to the number of changes that could occur. Let's illustrate this process. There exists all types of phenotypes (The phenotype is the set of observable characteristics or traits of an organism), and are arranged next to one another to the amount of difference between their genotypes (*The genotype of an organism is its complete set of genetic material. Genotype also refers to the alleles or variants an individual carries in a particular gene or genetic location. The number of alleles an individual can have in a specific gene depends on the number of copies of each chromosome found in that species, also referred to as ploidy.*) Now take a functional point A and a functional point B that are separated by a few hundred thousand points in this space. The probability of a macro-mutation, changing A to B is astronomically small. Likewise, the probability of A changing stepwise to B through a series of random events, is astronomically small. Creationists think that evolution takes place in a logical manner, but it does not. The rate of mutation in organisms is one mutation per locus per 10^5 to 10^6 gametes is sufficient to give natural selection an immense amount of variation to work with at any functional starting point. It is probable that one mutant of A will be a functional point closer to B. if this descendant is selected for, it will have a large number of descendants, whose gametes will undergo the same rate of mutation, making it extremely likely that another functional point closer to B will be produced. Without natural selection, the process would be random, and a connection from A to B would require an extremely improbable sequence of events to occur. But if thousands, hundreds of thousands, or millions of mutants are created at each locus, the selective pressures choose those forms that are closer to B, then it is virtually inevitable that B will arise. This scheme is somewhat simplistic, because it assumes a constant environment and constant selective forces, making the result almost teleological in character, which is not an accurate account of evolutionary processes. In actuality, the environment fluctuates, so selective pressures vary constantly, and lifeforms must constantly change to keep up. In a fluctuating environment, it is less probable that a path will be traced to B. But this is hardly a problem, because evolution is not teleological. Nothing in evolution says that a multiple-celled animal had to arise gradually from bacteria. But even in a fluctuating environment, it is clear that we could expect something to very structurally different and reproductively isolated from A eventually arise. Nothing in evolution claims that the history of life had to occur exactly the way it did, but evolution is by far the best, and certainly a plausible, explanation of why life developed as it did. The predictive power of evolution at the macro-level is low, but the explanatory power of evolution is immense.[9-84d]

Creationist View

> *The process of mutation is the only known source of the raw materials*
> *of genetic variability, and hence of evolution. The mutants which*
> *arise are, with rare exceptions, deleterious to their carriers, at least in*
> *the environments which the species normally encounters.*[9-85]
>
> —Theodosious Dobzhansky

> *Mutations modify or destroy already existing information, but they*
> *never create new information. Mutations never create, for example,*
> *an entirely new biological organ such as an eye or an ear. Herein lies*
> *the error of Neo-Darwinism, which teaches that fundamentally new*
> *information is created by mutations.*[9-80]

Modern evolutionary theory, from the mid-twentieth century (1950s) onward, is based on the idea that mutations plus natural selection, plus time can produce changes in all living creatures and have been responsible for all the astounding faculties and complicated organs that we see in plants and animals. Since DNA in the cell is the blueprint of the form that life will take, it seems reasonable to assume that if the blueprint could be changed, the life form might greatly improve.[9-87] Mutations are always random and never purposeful or directed. This has repeatedly been observed in actual experimentation with mutations.[9-88] Mutations tend to have a widespread effect on the genes.[9-89] Mutations are random "accidents" that cause slight changes in the DNA of a creature.[9-90] They occur with no goal and no direction. Mutations are usually damaging. Therefore, a huge number of reproductions are necessary to create the possibility of a beneficial mutation, and a tremendous number of beneficial mutations would be necessary for something beneficial to result. In the process, natural selection would destroy the weak or inadequate organisms. A developing organism would need to develop quickly and become useful; otherwise it would be eliminated by natural selection.[9-91]

Mutations may change regulatory genes in DNA, however, the chances of the DNA molecule, crucial to all life, evolving by natural processes is very remote. Without an outside controlling intelligence of some kind, it is virtually impossible for positive changes to occur. Natural selection is not capable of developing anything. It only controls what already exists. DNA is a super-molecule which stores coded hereditary information. It consists of two long "chains" of chemical "building blocks" paired together. In humans, the strands of DNA are almost 2 yards long (approx. 1.82 meters), yet less than a trillionth of an inch thick (approx. 0.0000254 microns). Many scientists are convinced that cells containing such a complex code and such intricate chemistry could never have come into being by undirected chemistry. No matter how chemicals are mixed, they do not create DNA spirals or any intelligent code whatsoever. Only DNA reproduces DNA.[9-92]

Mutations are random and nearly always represent a disadvantage from the point of view of the struggle for existence. Therefore they would most likely be quickly eliminated by natural selection (discussed previously).[9-93] It has been noted that once in a while, at random (about once in ten million times during cell division) the genes make a copying mistake. These mistakes are known as mutations, which are mostly harmful. They lead to a weakened plant, or a sick or deformed creature. They do not persist within the species, because they are eliminated by natural selection.[9-94]

Nearly all mutations are harmful. Some mutations are neutral, that is, they are not an improvement nor are they detrimental. In most instances, mutations weaken or damage the organism in some way so that it would not long survive.[9-95] (If a mutated gamete appears in a reproductive cell of an organism and the organism were able to produce offspring, then if the mutation appears in the genotype and the phenotype (the observable properties) of the offspring, the offspring could inherit the mutation and it could be weak or deformed.) Copying errors normally produce negative results. [9-96] If copying errors were to accumulate; a species, instead of improving, would eventually degenerate and die. But geneticists have discovered a self-correcting system. (See Chapter 5, section "DNA and its Repair Mechanism").[9-97] Modern evolutionary theory states that the diversity of life we see around us is a result of small successive changes over billions of years. These changes come about as a result of genetic mutations that are selected or discarded based upon their ability to increase biological fitness. Only mutations that help the organism survive in their particular environment are kept by natural selection. Neutral mutations may or may not stay around to turn into beneficial mutations later on, but are not selected until they are beneficial.[9-98]

Neither natural selection nor mutations introduce any new genetic data into the organism's DNA. Natural selection only selects out the disfigured, weak, or unfit individuals of a population for extinction. It cannot produce new species, new genetic information, or new organs and thus cannot make anything evolve. Mutations have never been observed to have any useful effect despite thousands of experiments. Even if a mutation occurs in an organism, it has to occur in the reproductive cells of the organism for a change to occur in its offspring. For example, a human eye altered by the effects of radiation or other causes will not be passed onto subsequent generations.[9-99]

> *It would be difficult to give any rational explanation of the affinities*
> *of the blind cave-animals to the other inhabitants of the two continents*
> *on the ordinary view of their independent creation.*
> • Charles Darwin The Origin of Species
> by means of Natural Selection or The Preservation
> of Favoured Races in the Struggle for Life,
> Chapter 5 - Laws Of Variation, 6th edition

A mutation usually results in the loss of information or just a copy of information. Mutations never add any new information to the genome. This would appear to indicate that mutations never cause added complexity to a genome. Research has indicated, however, that there are some beneficial mutations for certain environments. For example, blind cave catfish are the result of the mutation that lost the information of an eye. This mutation of the eye, which was useless in a dark cave and often caused disease and injury, actually helped the catfish to survive in the cave. The mutated eye in the catfish genome did not have any new information added to it. Actually, the eye genes were lost to the genome. The result is that if a blind catfish ever left the cave, it would not survive because natural selection, which worked well for the blind catfish in the cave, would cause the blind catfish to be killed very quickly outside of the cave. The blind catfish would have a disadvantage outside of the dark cave, for predators would quickly be able to take advantage of the catfish being blind, as the blind catfish would not be able to detect predators very easily. The fact is that only the genome of the catfish was utilized. At no point was any new information added to change the catfish genome to something else.[9-100] What caused catfish to become blind within a few generations? Research has determined that water-conductivity sensors on fish eggs signal the inhibition of a specific developmental protein. Epigenetic controls shut down eye genes during development in such a way that enables eye traits to reappear in future generations. [9-100a] Whether catfish are born with the ability to see or not depends on an identifiable selective activity in fish embryos. In a true selection between "on" or "off" mechanisms regulated by logic-based information, "control switches" for eye development were turned off in the blind fish. [9-100b]

Genomes are like rubber bands that can be stretched out so far but then they will always revert back to the original when released. One only needs to look at the many fruit fly experiments that used radiation to accelerate the copying errors of DNA to try to produce another species. These experiments only resulted in fruit flies with missing parts, or dead flies, or flies that were too crippled to produce offspring. No experiment ever produced any other type of fly or any other creature. This is because mutations cause a loss of information rather than new information being added to the genome of the fruit fly.[9-100] (See sections "Hox Genes" and "Experiments with Fruit Flies" for discussion of experiments conducted on fruit flies.)

Mutations generally produce one of three types of changes within genes or chromosomes: (1) an alteration of DNA letter sequence in the genes, (2) gross changes in chromosomes (inversion, translocation), or (3) a change in the number of chromosomes (polyploidy, haploidy). But whatever the cause, the result is a change in genetic information. For mutations to be considered the major source of change that produces new species mutations must: (1) Occur very frequently. (2) Be consistently beneficial. (3) Cause a dramatic enough change (involving, actually, millions of specific, purposeful changes) so that one species will be transformed into another. Small changes will only damage or destroy the organism.[9-101]

Genes can be affected by radiation, x-rays, atomic bombs, ultraviolet light, and certain chemicals, for they can produce mutations. Mutations can make tiny changes within genes. A mutation is a change in a hereditary determiner, which is a DNA molecule inside a gene (Chapter 5 Figure 5.5). Genes, and the millions of DNA molecules within them, are very complicated. If such a change actually occurs, there will be a corresponding change somewhere in the organism, and in its descendants. If the mutation does not kill the organism, it will weaken it. But the mutation will not change one species into another. Mutations are only able to produce changes within the species. They never change one kind of plant or animal into another kind.[9-102]

Harmful mutations happen constantly. Without repair mechanisms, life would be very short and might not even get started because mutations often lead to disease, deformity, or death. The earliest, "simple" creatures in the evolutionist's primeval soup (discussed in chapters 5 and 6) or tree of life would have needed a sophisticated repair system. But the mechanisms not only remove harmful mutations from DNA, they would also remove mutations that evolutionists believe build new parts. (See Chapter 5, section "DNA and its Repair Mechanism.") Evolutionists have difficulty explaining the evolution of mechanisms that prevent evolution from reversing, all the way back to the very origin of life.

Only mutations in the reproductive (germ) cells (Meiosis occurs in reproductive cells) of an animal or plant would be passed on. Mutations in the eye or skin of an animal would not matter. Mutations in DNA happen fairly often, but most are repaired or destroyed by mechanisms in animals and plants. All known mutations in animal and plant germ cells are neutral, harmful, or fatal. Yet in spite of this fact evolutionists believe that millions of beneficial mutations built every type of creature that ever existed.[9-103] (See Chapter 5, section "Types of Cell Division—Mitosis and Meiosis" for further discussion of meiosis.)

The mathematical possibility of even two major mutations occurring is very unlikely (See Chapter 6, section "Probabilities and Chance."), yet millions of coordinated, beneficial mutations would be required to produce a new species. Variable mutations with the major morphological or physiological effects are exceedingly rare and usually infertile; the chance of two identical rare mutations arising in sufficient proximity to produce offspring seems too small to consider as a significant evolutionary event.[9-104]

In actual cells the mistakes made when DNA is copied are far fewer than commonly thought, and mistakes like point mutations occur on average only once in each complete copying of the whole 200,000 chains. (*A point mutation (also called gene mutation) is a mutation that changes only one small area or one nucleotide in a gene. A nucleotide is a compound consisting of a nucleotide combined with a phosphate group and forming the basic constituent of DNA and RNA. A nucleotide is a group of molecules that, when linked together, form the building blocks of DNA or RNA: in DNA the group comprises a phosphate group, the bases adenine, cytosine, guanine, and thymine, and a pentose sugar; in RNA the thymine base is replaced by uracil (Chapter 5).* [See Chapter 5, Figures 5.5 & 5.19, and image in the beginning of Chapter 13, "DNA Double Helix Strand"]) So instead of obtaining large numbers of natural mutations for natural selection to act upon, the copying of DNA seems to be remarkably accurate, which is detrimental to the modern form of the Darwinian Theory (Neo-Darwinism).[9-105]

Scientists performed experiments by taking x-rays on plant genes. They were very surprised that mutations were not the source of many varieties of flowers. These variations were caused by genetic factors unrelated to mutations. Flower and plant varieties are often very positive and quite beneficial, and it was hoped that they were caused by mutations. But this was not the case. In fact, it was found that x-rays were generally not very effective in inducing variations in plants. (Even if mutations had been the cause of the many varieties of flowers, for example, those varieties would still involve only changes within kinds and not across kinds.) As with animal life, so with plants; it was found that most mutations resulted in harmful effects and semi-sterile life forms. Many of the plant mutations involved splitting and reattaching chromosomes, and most were found to be lethal.[9-106]

When scientists began experimenting with plants, they noticed some very interesting results. They discovered that irradiated "budding eyes" of roses would dramatically increase mutational production in roses. Fifty irradiated budding eyes could yield more mutations than a million rose plants in a field in a lifetime of reproductions. Now much faster, more thorough work on plant mutations could be obtained. Plant research intensified. But then it was discovered that, although some plant mutations showed usefulness, most were worthless or fatal. Of the useful ones (change in petal number, loss of color, etc.), all of the plants having these mutational changes were weaker than their non-radiated parents. In the end, all of the useful mutated plants failed commercially since they were not healthy enough to survive under varying garden conditions. In every instance, even the best of the mutated plant forms were significantly weaker, or had a reduced fertility. The only exceptions were those few that could be given special care throughout their lifetime, such as certain sheltered, in-house ornamental plants.[9-107]

Mutations are said to be the cause of changes in genes, which eventually cause changes that would appear in the offspring of a creature. According to the theory, mutations somehow had to make those structural changes and produce a different species. Some minor finishing touches may have been said to be accomplished by natural selection (by selecting creatures with certain traits, thus limiting the gene pool), but the change would have had to be done by mutations.[9-108] Another reason why mutations are so detrimental has only recently been discovered. Geneticists discovered the answer in the genes. Instead of a certain characteristic being controlled by a certain gene, it is now known that each gene affects many characteristics, and many genes affect each characteristic (See last paragraph in section "(6) Other Mechanisms That Cause Changes in Organisms" at the end of this chapter that discusses an experiment of a mutated chicken gene). We have here a complicated interweaving of genetic-characteristic relationships never before imagined possible. If a delicate system, such as the genes, are changed by mutations, interlocking havoc results. Even assuming mutations could produce those complex structures such as feathers, birds would have wings on their stomachs, where they could not use them, or the wings would be upside down, without light-weight feathers, and under, or oversize. Most animals would have no eyes, some would have one, and those that had any eyes would have them in strange places, such as under their arm-pits or on the soles of their feet. The very randomness of mutations alone would annihilate any value they might otherwise provide.[9-109]

Evolutionists often say that given enough time, anything could happen. Evolutionists claim that it may take 5 billion years for mutations to do the job. But 5 billion years is, in seconds, only 1 with 17 zeros (1×10^{17}) after it. And the whole universe only contains 1×10^{80} atomic particles. So there is no possible way that all the universe and all time past could produce such odds as 1×10^{30001} And this is the estimate of the odds that evolutionists say it would take to produce just one horse by evolution: 1 with 3,000 zeros after it. (Horse evolution is discussed in Chapter 7.)

Evolution requires millions of beneficial mutations all working closely together to produce delicate living systems full of fine-tuned structures, organs, hormones, etc. And all those mutations would have to be non random and intelligently planned for anything beneficial to result. In no other way could they accomplish the needed task. Random, non-directional, mutations could not produce positive changes and there is no other way that life forms could develop by means of mutations.[9-110]

Experienced geneticists are well aware of the fact that the traits contained within the genes are closely interlocked with one another. That which affects one trait will affect many others. They work together. Because of this, all the traits would have to be there together, instantly, in order for a new species to form. Scientists describe the problem as follows: Each mutation occurring alone would be wiped out before it could be combined with the others. They are all interdependent.[9-111]

The probabilities that a mutation will survive or eventually spread in the course of evolution tend to vary inversely with the extent of its somatic effects. (*A somatic cell is one of the cells that take part in the formation of the body, becoming differentiated into the various tissues, organs, etc. A somatic cell is any cell other than a germ cell, which is the sexual reproductive cell at any stage from the primordial cell to the mature gamete.*) (*A gamete is the male or female reproductive cell that contains half the genetic material of the organism.*[9-111a]) Most mutations with large effects are lethal at an early stage for the individual in which they occur and hence have zero probability of spreading. Mutations with small effects do have some probability of spreading. As a rule those changes that are small and have a lesser effect have a better chance of survival than those changes that are dramatic and have a greater effect on the creature.[9-112]

Mutational changes are impossible as a means of constructive, beneficial change in an organism. Most, if not all, of our characteristics are polygenetic, that is, they are under the control of not one but a number of genes. For instance, eye color in Drosophila (the fruit fly) is under the control of 15 genes. The desirable alteration of a certain characteristic, if that is possible at all, most likely would require changes in more than one particular gene. Precisely coordinated changes in several genes would probably be required. If all these problems could be solved, which seems incredible, one insuperable difficulty would yet remain. In each gene there are thousands of nucleotides, but only four different kinds of bases: adenine (A), cytosine (C), guanine (G), and thymine (T). (Neucleotides are discussed in Chapter 5.) In a gene of 10,000 nucleotides, there would be, on the average, 2,500 of each of the four different kinds of bases.[9-113]

Sickle Cell Anemia and Evolution

Creationist View

Evolutionists point to sickle-cell anemia as an example of beneficial evolutionary change through mutation. Those with sickle-cell anemia are less likely to contract malaria. A long time ago, a mutation occurred in someone in Africa. In this instance, the shape of the red blood cells was changed, from its normal flattened shape, to a quarter-moon shape. Instead of causing death in one or two generations, this mutation passed into the race and became a recessive factor. The problem was that, although the blood of a person with sickle-cell anemia does not properly absorb food and oxygen, that person will be less likely to acquire malaria from the bite of an anopheles mosquito. As a result, the sickle-cell anemia factor has become widespread in Africa.[9-114]

As mentioned above, Sickle-cell anemia is often given as an example of a favorable mutation, because people carrying sickle-cell hemoglobin in their red blood cells are resistant to malaria. The gene (sickle-cell hemoglobin) will automatically be selected (natural selection) when the death rate from malaria is high, but evolutionists themselves admit that short time advantages, all that natural selection favors, can produce negative results detrimental to long-term survival.[9-115]

Sickle-cell anemia is said to be beneficial because it helps Africans resist malaria. It is doubtful if anyone who understood the consequences of sickle-cell anemia would be willing to acquire this disease in order to lessen his or her chances of getting malaria. This is because, all aside from malaria, sickle-cell anemia itself terribly weakens one's system and shortens life.[9-116] Mutations and natural selection do produce changes, but not beneficial ones. Mutations are no real help in explaining the origin of species, but mutations are great for explaining the origin of disease, disease organisms, and birth defects. Molecules-to-man evolution is all about phenomenal expansion of genetic information. It would take thousands of information adding mutations to change "simple cells" into invertebrates, vertebrates, and mankind. If there were any scientific merit at all to mutation selection as a mechanism for evolution, an evolutionist should be able to provide multiple examples. The problem with evolution is with the fundamental nature of information itself. The information in a book, for example, cannot be reduced to, nor derived from, the properties of the ink and paper used to write it. Similarly, the information in the genetic code cannot be reduced to, nor derived from, the properties of matter or the mistakes of mutations.

Evolutionists believe that mutations are continually producing new and different genes, although this view is highly debatable. Currently, there is no known evidence whereby mutations add genetic information to a genome. Genes of the same kind, like those for straight and curly hair or those for yellow and green seeds, are called alleles. There are over 300 alleles of the hemoglobin gene. By concept and definition, alleles are just variants of a given gene, producing variation in a given trait. Mutations produce only alleles, which mean they can produce only variation within a kind, not changes from one kind to another (evolution). Genes of the same kind can be defined as segments of DNA that occupy corresponding positions (loci, sing, locus) on homologous chromosomes. Homologous chromosomes are pairs that look alike, but come from two different parents, so their genetic content is similar but not identical. They pair up and then separate in the kind of cell division (meiosis) required for sexual production. Genes that pair up in meiotic cell division, therefore, can be identified as genes of the same kind. Genes of the same kind are also turned on and off by the same gene regulators. (Meiosis is discussed in Chapter 5, sub-section "Types of Cell Division—Mitosis and meiosis." Genes being turned on and off are discussed in Chapter 13, section "Some DNA Portions are Used to Turn Genes On or Off.")

Mutations, random changes in the genetic code, do produce "new genes" not present, but the so-called "new genes" are still found at the same locus, still pair the same way in meiosis, and are still turned on and off by the same regulators, so they are really only genes of the same kind as the original, and represent only variation within kind (usually harmful variation in the case of mutations).

The terms "new genes" or "different genes" can have two radically different meanings. Geneticists normally call genes of the same kind "alleles." The genes for tongue rolling and non-rolling are "different genes" in one sense, but only variations of the same kind of gene, affecting the same trait, found in corresponding positions (loci) on homologous chromosomes, pairing up in meiosis, and turned on and off by the same regulators. They are not different genes in the sense that genes for tongue rolling, and genes for making sickle cell hemoglobin are. Similarly, the sickle cell gene is a "new gene" in the sense that it was not present at creation, but it is only a new (and harmful) version of a pre-existing gene, (1) one that occupies the same chromosomal position, (2) pairs the same way, and (3) is turned on and off by the same regulators as the gene for making normal hemoglobin. (See Chapter 13, section "Some DNA Portions are Used to Turn Genes On or Off.") The gene for sickle cell hemoglobin differs in base sequence at only one position out of several hundred in the normal gene for making hemoglobin, which is just variation within kind or allelic variation. (See Chapter 15, section "The Evolution of Hemoglobin" for discussion of hemoglobin.)[9-117]

Evolutionists believe that mutations are the causes of change while creationists do not believe that mutations could account for the changes that are observed in nature. Research has indicated that differences within major animal groups (kinds) are due to genetic variation that is already present within various kinds of animals and plants. That is, the genes that determine the characteristics of a creature are either dormant (turned off) or active (turned on). (See Chapter 13, section "Some DNA Portions are Used to Turn Genes On or Off.") It has been observed that there is enough variation within two creatures to account for multiple variations. There is enough variation within human beings, for example, to produce all the variations among human beings. Research has determined that 6 or 7 times in 100, the pair of genes for a given trait differ, like the genes for free or attached ear lobes, or for rolling or not rolling the tongue. It has been determined that this 6.7 percent variety of two parents could produce 10^{2017} children before they would run out of variation and have to produce an identical twin.

Geneticists call the shuffling of pre-existing genes "recombination." The genetic pool remains the same however, within each individual the genes that are turned on are unique and different. This means that individuals within a group are different but the genetic pool remains the same. The human gene pool, for example, is no bigger and no different now than the gene pool that was present at creation. As people multiplied, the genetic variability built right into the first created human being came to visible expression. The variation in size, color, form, function, etc., is present in the ancestors of all the various kinds (groups) of plants and animals. (Human evolution is discussed in chapters 15 & 16.)

According to the creation concept, each kind began with a large gene pool present. As descendants of these created kinds become isolated (speciation), each average-looking (generalized) type would tend to break up into a variety of more specialized descendants adapted to different environments. Thus, the created ancestors of dogs, for example, have produced such varieties in nature as wolves, coyotes, and jackals. Varieties within a created kind have the same genes (genons), but different percentages of various alleles. Differences from average allele percentages can come to expression quickly in small populations (a process called "genetic drift" that will be discussed later in section "(5) Genetic Drift.")

Many varieties of plants and animals retain the ability to reproduce and trade genes, despite differences in appearance. Varieties of one kind may also lose the ability to interbreed with others of their kind. For example, fruit flies multiplying through Central and South America have split up into many subgroups. Since these subgroups no longer interbreed, each can be called a separate species (See section "(3) Speciation" for further discussion). The fact that some species of fruit flies have lost their ability to interbreed is not a proof of evolution. Some define evolution as "a change in gene frequency." Using this definition, the fly example mentioned above would prove evolution, but it would also prove creation, since varying the amounts of already existing genes is what creation is all about.[9-118]

Mutations and the Genetic Code

Old Universe Progressive Creationist View

Research has established that the genetic code has methods to minimize errors during the protein production process (Chapter 5). The failure of the genetic code to transmit and translate information accurately can be devastating to the cell. Generally, mutations could have a disastrous effect on the cell's capability to function. A mutation refers to any change that takes place in the DNA nucleotide sequence. Several different types of changes to DNA sequences can occur with substitution mutations being the most frequent. As a result of these mutations, one or more nucleotides in the DNA strand is replaced by one or more nucleotides. When substitutions occur, they alter the codon (Chapter 5, Figure 5.10a) that houses the substituted nucleotide. And if the codon changes, then the amino acid specified by that codon also changes, altering the amino acid sequence of the polypeptide chain (Chapter 5, Figure 5.10) specified by the mutated gene. This mutation can then lead to a distorted chemical and physical profile along the polypeptide chain. If the substituted amino acid has dramatically different physiochemical properties from the native amino acid, then the polypeptide folds improperly. An improperly folded protein has reduced or even lost function. Mutations can be harmful to cell abilities because they hold the potential to significantly and negatively impact protein structure and function.

The genetic code's redundancy is not haphazard but carefully thought out. It has been established that deliberate rules were set up to protect the cell from the harmful effects of substitution mutations. For example, six codons encode the amino acid leucine (Leu). If at a particular amino acid position in a polypeptide, Leu is encoded by 5' CUU, substitution mutations in the 3' position from U to C, A or G produce three new codons – 5' CUC, 5' CUA, and 5' CUG, respectively, all of which code for the amino acid Leu (leucine). (See Chapter 5, figures 5.20 & 5.21 for illustrations regarding 5' to 3' Leading Strands and 3' to 5' Lagging Strands.) The net effect leaves the amino acid sequence of the polypeptide unchanged. And, the cell successfully avoids the negative effects of a substitution mutation. Also, a change of C in the 5' position to a U generates a new codon, 5' UUU, which specifies phenylalanine, an amino acid with physical and chemical properties similar to Leu. Changing C to an A or a G produces codons that code for the amino acid isoleucine (Ile) (AUU) and amino acid valine (Val) (GUU), respectively.

These two amino acids possess chemical and physical properties similar to the amino acid leucine (Leu). All considered, it appears as if the genetic code has been constructed to minimize the errors that could result from substitution mutations.

The number of possible genetic codes based on one or two nucleotide codons is far fewer than for codes based on coding triplets. The idea that a simpler code that codons consisted of one or two nucleotides makes code evolution much more likely from a naturalistic perspective. One complicating factor, however, for these proposals arises from the fact that simpler genetic codes cannot specify twenty different amino acids. Instead, they are limited to sixteen different amino acids at the most. For such a scenario to be possible, it would mean that the first life-forms had to make use of proteins that consisted of no more than sixteen different amino acids. There are some proteins found in nature that are produced with only thirteen amino acids (such as ferredoxins). Therefore, it appears at first that the genetic code found in nature could have developed from a simpler code. The problem is that proteins like ferredoxins are not typical. Most proteins require all twenty amino acids. This requirement, along with the conclusion that life in its most minimal form needs several hundred proteins, makes these types of theories for code evolution doubtful. The optimal nature of the genetic code and the difficulty explaining the genetic code's origin from an evolutionary perspective together indicate that an Intelligent Designer programmed the genetic code, and therefore, life.[9-118a]

Limitations of Mutations

Creationist View

Darwinists believe that evolution is due to the combined effect of mutation and natural selection. Mutations are the result of 'copying errors' in the genes. When a plant or animal reproduces, the new generation is usually almost exactly like the parents. The genes are extremely complex chemical substances in the germ cells, and they contain a sort of blueprint of the parents that is passed on to the next generation. If something goes slightly wrong when the genes duplicate themselves, the result may cause a minor defect in an offspring. According to Darwinists, such mutations give natural selection something to work on. Occasionally a mutant offspring is better equipped for survival than its normal siblings. When that happens, the normal variety may die out locally, while the unusual one takes its place. And according to Darwinists, the outcome may be a new species. The only problem is that neither mutation nor natural selection works the way Darwinists claim they do. Mutations have been extensively studied in the laboratory and in nature, from bacteria to plants and animals. The result of the research has revealed that mutations do not take succeeding generations further and further from their starting point. Instead, the changes do not extend much beyond the original. There are invisible but firmly fixed boundaries that mutations can never cross.[9-119]

Mutations do not appear to bring progressive changes. Genes seem to allow changes to occur within certain narrow limits, and to prevent those limits from being crossed. Mutations very easily produce new varieties within a species. They might occasionally produce a new, though similar, species; but despite enormous efforts by experimenters and breeders, mutations seem unable to produce entirely new forms of life.[9-120] Genes are passed on from generation to generation. To be passed on they are copied. Like in any other copying process, the copying of genes can result in mistakes, called mutations. Genes contain the information code that determines what the offspring is going to be. Within this code various characteristic details for the individual offspring are also determined. Half of this information comes from the female and half from the male. In humans, the DNA from the parents is two six-foot long strands. This DNA is cut up into 46 pieces called chromosomes. Genes are little sentences along these DNA chromosome strands (Chapter 5, Figure 5.5). The information in these genes, which determines for example whether one will have green or blue eyes, is not passed on like a copier, but is copied one letter at a time. The total information in the DNA and all the coded letters would fill volumes and volumes of books. In a matter of twenty minutes, all the information, all the letters and sentences that determine what we are going to be, is copied from the 46 chromosomes in our cells for reproduction. Every once in a while a typo mistake occurs in this process which results in a mutation of that gene.

Our bodies have filtering mechanisms that filter out many of these mistakes. (See Chapter 5, "DNA and its Repair Mechanism.") Sometimes, despite this, a mistake gets by. If this happens in the sperm of the male or egg of the female the mistake is passed on permanently from generation to generation. The reason why we don't see this accumulation of defects in ourselves more greatly than we do is because it will only show up if both parents have this same defect in the same gene. If one does not, the good gene is produced rather than the defective one.[9-121]

According to scientists, the minimum number of mutations necessary to bring about the simplest new structure in an organism is five. However, these five mutations must be of the right kind and must be functionally related. This means that not just any five mutations will do the job. The odds of five mutations taking place functionally related at the same time in a single organism by chance are astronomically impossible. Scientists believe only one mutation takes place in every 100,000 gene replications. The probability of five non-harmful mutations taking place functionally related in a single cell is 1 in 10^{40}. If one hundred trillion, 10^{14}, bacteria were produced every second for five billion years, 10^{17} seconds, the resulting population would be only 1/1,000,000,000 of what was needed.

This is just the first step. Next, these five mutations must somehow be brought together to be integrated and function in concert with each other. Next, this integrated function must provide some advantage or be scattered within the population due to interbreeding. With these odds it is not possible to explain, through these means, how complex life forms came into being, given any amount of time. For example: the wing of the fruit fly alone involves 30-40 genes. This means that given the age of the universe, as evolutionists define it, through random selection you might produce one wing of a fruit fly. That is trillions of years away from a fully formed fruit fly let alone an animal or human. Evolution is not the answer to how we got here.[9-122]

The Possibilities of a Species Evolving Into New Species— Not as Remote as Creationists Claim

Evolutionist View

Once mutations introduce variation into the genetic makeup of species, then chance plays much less a role. The beneficial variations will be quickly selected in a logical, predictable manner (See section, "Natural Selection"). For example, flowers whose color variation attracts hummingbirds will more likely be pollinated than other flowers and this will result in that color variation becoming more common in that species of flower. There are chance occurrences that also influence evolution. For example, the extinction of dinosaurs may have been partly caused by a meteorite striking the earth. A hurricane might blow a few individual members of a species onto a remote island, separating them from the rest of the population. A new species may develop as a result of the limited gene pool in which the species would have to choose from (See Section, "Speciation"). Most evolution, however, is directed by natural selection instead of pure chance processes.[9-123] (See Section, "Natural Selection.")

The Odds of Mutations Resulting in a Positive Change

Creationist View

Mutations are rare. They normally occur on an average of perhaps once in every ten million duplications of a DNA molecule. But even assuming that all mutations are beneficial, in order for evolution to begin to occur in even a small way, it would be necessary to have a series of closely related and interlocking mutations, not just one. All of these changes would need to occur at the same time in the same organism. The odds of getting two mutations that are in some slight manner related to one another is the result of two separate mutations: ten million times ten million, or a hundred trillion. That is a 1 followed by 14 zeros (in scientific notation written as 1×10^{14}).[9-124] Any two mutations might produce no more than a wavy edge on a bent wing of a fly or a honey bee. But it is still a fly or a honey bee; the creature has not changed from one species to another. That's a long way from producing a truly new structure and certainly a long way from changing a fly into some new kind of organism. More related mutations would be needed for a species to change into a different species.

The odds of getting three mutations in a row are one in a billion trillion (1×10^{21}). The odds here are tremendous. The ocean isn't big enough to hold enough bacteria to make it likely to find a bacterium with three simultaneous or sequential related mutations. Four mutations that were simultaneous or sequentially related would be 1 with 28 zeros after it (1×10^{28}). But all the earth could not hold enough organisms to make that possibility come true. And four mutations together do not even begin to produce real evolution. Millions upon millions harmonious, beneficial characteristics would be needed to transform one species into another. But all those simultaneous mutations would have to be beneficial, whereas in real life mutations very rarely occur, and they are almost always harmful.[9-124]

It would take many more than four simultaneous mutations to change a fish into an amphibian or an amphibian into a reptile or a reptile into a mammal or a dinosaur into a bird or a land animal into a whale. It was at this level (just four related mutations) that microbiologists gave up on the idea that mutations in asexual lines could explain why some bacteria are resistant to four different antibiotics at the same time.

The odds against the mutation explanation were simply too great, so evolutionists began to look for another mechanism to explain it. Researchers found that bacteria were resistant to antibiotics, even before commercial antibiotics were developed (See sections "Natural Selection, Mutation and Resistance to Drugs and Poison" and "Antibiotic Resistance of Bacteria is Not An Example of Evolution in Action"). Resistant bacteria were even found in the bodies of explorers frozen more than a century before medical antibiotic use. Resistant forms of bacteria did not become resistant by mutations. Genetic variability was already built into the bacteria. Certain bacteria have little rings of DNA, called plasmids that they trade around among themselves and they pass on their resistance to antibiotics in that way. It wasn't mutation and asexual reproduction at all; it was just ordinary recombination of genes and variation within a kind. Bacteria can be made antibiotic resistant by mutation, but such forms only cripple. The mutation typically damages a growth factor, so that the mutated bacteria that were crippled can scarcely survive outside the lab or hospital. The antibiotic resistance carried by plasmids results from enzymes produced to break down the antibiotic. Such bacteria do not have their growth crippled by mutation. Their resistance is by design.[9-125]

Evolutionists claim that we cannot see evolution taking place because it happens too slowly. A human generation takes about 20 years from birth to parenthood. Evolutionists say it took tens of thousands of generations to form man from a common ancestor with the ape, from populations of only hundreds or thousands. (Human evolution is discussed in chapters 15 & 16.) We do not have these problems with bacteria. A new generation of bacteria grows in as short as 12 minutes or up to 24 hours or more, depending on the type of bacteria and the environment, but typically 20 minutes to a few hours.

There are more bacteria in the world than there are grains of sand on all of the beaches of the world (and many grains of sand are covered with bacteria). They exist in just about any environment: hot, cold, dry, wet, high pressure, low pressure, small groups, large colonies, isolated, much food, little food, much oxygen, no oxygen, in toxic chemicals, etc. There is much variation in bacteria. There are many mutations. In fact, evolutionists say that smaller organisms have a faster mutation rate than larger ones. But they never turn into anything new. They always remain bacteria. Fruit flies are much more complex than already complex single-cell bacteria. Scientists like to study fruit flies because a generation (from egg to adult) takes only 9 days. In the lab, fruit flies have been studied under every conceivable condition. There is much variation in fruit flies. There are many mutations. But they never turn into anything new. They always remain fruit flies.[9-126]

Experiments with Fruit Flies

Creationist View

Scientists have performed multiple experiments to study the results of mutations. As discussed previously, the common fruit fly is often used in these experiments, and for good reason. Since fruit flies reach sexual maturity in only five days, the effects of mutations can be observed over several generations. Using chemicals or radiation to induce mutations, scientists have produced flies with purple eyes or white eyes; flies with oversized wings or shriveled wings or even no wings; fly larvae with patchy bristles on their backs or larvae with so many bristles that they resemble hedgehogs. But even through all this experimentation, evolutionary theory has not advanced at all. Nothing has ever emerged from the multiple experiments except odd forms of fruit flies. The experiments have never produced a new type of insect.

Mutations alter the details in existing structures, such as the eye color or wing size, however they do not lead to the creation of new structures. Fruit flies have remained fruit flies. Like breeding, genetic mutations produce only minor, limited change. Also, the minor changes observed do not accumulate to create major changes, which is the major part of Darwinism. The conclusion is that mutations are not the source of endless, limitless change required by evolutionary theory. Whether performed in breeding experiments or laboratory experiments, the outcome is always the same: Change in living things remains strictly limited to variations. No new complex structures emerge as a result of mutations.[9-127]

So far, no one has been successful in changing fruit flies into different species. In fact, the multiplied millions of mutations induced by countless exposure to x-rays or other radiation did not in one instance even improve the fruit fly species, much less change it into another one. All that was accomplished was the production of deformed flies.[9-128] The fruit fly has long been the favorite object of mutation experiments because of its fast gestation period (twelve days). X-rays have been used to increase the mutation rate in the fruit fly by 15,000 percent. All in all, scientists have been able to catalyze the fruit fly evolutionary process such that what has been seen to occur in Drosophila is the equivalent of many millions of years of normal mutations and evolution. Even with this tremendous speedup of mutations, scientists have never been able to come up with anything other than another fruit fly. Most important, what all these experiments demonstrate is that the fruit fly can vary within certain upper and lower limits but will never go beyond them.[9-129]

Since the 1920s scientists have been experimenting with the common fruit fly (Drosophila), trying to prove that all life on planet earth is the result of a series of random beneficial mutations. Evolutionists believe that the almost endless variety and complexity of plants and animals 'evolved' from an ancient pool of 'primordial soup.' They believe this is possible by millions and billions of accidental beneficial mutations. The evidence is overwhelming that such accidental mutations either make the gene worse, or at best, no better than the original. Because fruit flies reproduce many generations in a very short time, scientists picked them for the experiment hoping to compress thousands of years of evolution into a few years of lab work. After 80 years and millions of generations of fruit flies subjected to x-rays and chemicals which cause mutations, all they have been able to produce are more of the same: fruit flies. These mutated fruit flies have all been no better or stronger and many have been weaker. All the changes eventually reached limits that, when approached, the strains of the fruit flies grew progressively weaker and died. And when the mutated strains were allowed to breed for several generations, they gradually changed back to the original form. One experiment produced fruit flies without eyes. Yet, after a few life cycles, flies with eyes began to appear. Some kind of genetic repair mechanism took over and blocked any possibility of evolution.[9-130]

Hox Genes

Evolutionist View

The particular genes turned on at a particular moment in the nucleus will determine which proteins are present in the cell and which chemical reactions take place. Often these proteins or their products can in turn also influence which genes are turned on or off in the nucleus and which additional proteins are manufactured. (See Chapter 13, section "Some DNA Portions are Used to Turn Genes On or Off.") This complex cascade of signals is what ultimately determines how an individual organism develops as an embryo, the specific traits the individual ends up with, and how that organism ultimately functions as an adult.

Vertebrates, including humans, have a complex of genes called the "homeotic complex," which is similar in DNA base sequence to bithorax, which controls the development in fruit flies, fish, and mammals. Homeotic (Hox) genes were discovered to be master control genes. These genes share an identical 180-DNA base segment called the "homeobox." Their protein products all share an identical region called the "homeodomain," which allows specific proteins to bind to DNA, and thus influence the activity of particular developmental genes. Mammals have thirteen Hox genes and fruit flies have eight. Hox genes control the specific identity of each segment in a developing animal embryo by directing the fate of the cells in which the genes are active. Therefore, even the smallest of imperfections in a Hox gene can have catastrophic implications for later development. Because most mutations of Hox genes are deadly to a developing embryo, these genes cannot evolve rapidly. As a result, the genes that control patterning in species from insects to fish to mammals are very similar, in both DNA sequence and function, to the genes originally discovered in fruit flies.

It has been discovered that gene families, if altered by mutations, cause fruit flies to develop debilitating pathways that often prove to be deadly. Some mutations observed in laboratories include apterous (flies without wings), vestigial (flies with shriveled, useless wings), white (flies with the white eye mutation), ebony (flies with dark-colored bodies), and notch (flies with a notch at the tip of each wing). All of these gene families found in the fruit flies are also found in one form or another in humans. They are essential to the control of development in fruit flies, but many are also active in adult human cells, regulating whether a cell will become cancerous. When particular genes in these gene families are altered, they can cause mistakes early during human embryological development (See Chapter 11 for further discussion of embryology). Alternatively, if the mutation in the gene is subtle, the consequence of the mutation might not be apparent until adulthood when affected cells contribute to tumor formation. In other words, many developmental regulatory genes are now understood to have strong links to human cancers.

In fruit flies, the notch gene codes for a specific protein that ends up embedded in the cell membrane of the embryonic cells that give rise to wing tissue. The notch gene controls normal wing development through a series of complex cell-to-cell interactions and cascading molecular signals. An altered notch gene, however, produces an altered notch protein, which in turn produces an altered trait, the notched wing, in the adult fly. Even after nearly one billion years of evolution, the notch gene is still recognizable in humans, yet it has duplicated and evolved. Humans now have four of these notch genes—all different—and therefore four different notch proteins. These protein products of the human notch genes are from 11 percent to 64 percent identical to the protein coded for by the fruit fly notch gene.

In humans, notch gene mutations cause cancer. For example, one of these notch genes, Notch 1, is linked to a form of leukemia (cancer of the white blood cells). Another notch gene, Notch 3, is active in the smooth muscle development of vein and artery cells in adult humans. Hyperactivity of the gene, however, has been strongly linked to lung tumor formation and subsequent human lung carcinoma. The presence of notch genes, like Hox genes, in both humans and fruit flies (and every other species) is more evidence of evolution through a common ancestor.

Considering the vast fossil record and the diversity in animal forms that have appeared, it is understandable that the conclusion might be that new genes must have evolved to create the new body types that are observed. In this view, individuals with those new traits (and the apparently new genes that code for them) deemed "most fit" by the current environment were then selected to produce more offspring than individuals without the new traits and genes. This hypothesis, however, has proven to be incorrect. Instead of new genes appearing over time, the same genes were being used in different ways. In other words, they provided more evidence that evolution works by modifying preexisting genes to take on new roles required by changing environments rather than by new genes evolving.

Research of the species in the phylum Arthropoda (including spiders, centipedes, lobsters, and fruit flies) has revealed that they share all ten Hox genes with a distant cousin, the Onychophorans. Onychophorans are small organisms (most of them are no bigger than a human hand) found in the tropics and subtropics. They resemble caterpillars and they have many jointed appendages like arthropods (discussed in Chapter 3). It is interesting how these very different groups of animals use the same genes to control their embryonic development. During the development of the embryos of these animals, a subset of the Hox genes are turned on (actively making their protein products) in some segments of the embryo and turned off in other segments. Which genes are on and which are off in each segment have changed over evolutionary time, resulting in different arrangements of appendages and segments, therefore the appearance of very different body forms. For example, in arthropods, the segments in which number eight and nine Hox genes are turned on, or expressed, have legs. When gene expression is turned off in these segments and replaced by the expression of the number seven Hox gene, another type of appendage, maxillipeds, mouthparts modified for handling food, develop instead. (Genes being turned on and off are discussed in more detail in Chapter 13, section "Some DNA Portions are Used to Turn Genes On or Off.")

It is now understood the implications of this versatility of when and where master control genes are expressed (turned on) and goes far beyond body segments and includes hearts, fingers, eyes, and other organs. These genes are so similar among distantly related animals that they can be transplanted from one organism to another without losing functionality. For example, it has been demonstrated that the master control gene that initiates eye development in the mouse, called Pax 6, can be inserted in the fruit fly genome to stimulate the fly's own series of instructions for eye development. Pax 6 is now known to initiate eye development in organisms as different sea squirts and humans. By itself, the presence and function of a gene like Pax 6 among animals with eyes is evidence of common ancestry. The fact that this gene is also found in animals without eyes is evidence that through descent with modifications, genes can take on novel functions and result in a unique structure like the eye. (See Chapter 14, section "The Eye" for further discussion of the eye.)[9-131]

Intelligent Design View

Evolutionists believe that hox genes can provide a significant source of major variation in living organisms. Hox genes are "master regulator" genes that turn other genes in the cell on and off during the embryonic developmental process. Hox genes determine when other genes in the cell will transmit their instructions for building proteins. Because hox genes are so important for coordinating the activities of the cell, some researchers think that mutations in these genes can cause large-scale changes in the structure of an organism. Many evolutionists think that influencing the hox genes through mutation can provide a new source of beneficial variation. This, however, is questionable. Hox genes are so influential; no experimental mutations so far (as of 2007) in hox genes have proven helpful to the organism.

Living organisms depend on many interrelated systems and genes. Living organisms require a great many coordinated changes to transform one system into another without losing function during the transitional steps. The more the individual parts of a system depend on each other, the harder it is to change any one part without destroying the function of the organism as a whole. Since hox genes affect so many genes and systems, it seems unlikely that they could be mutated without damaging the way some of the genes are switched 'on' or 'off'. (See Chapter 13, section "Some DNA Portions are Used to Turn Genes On or Off.") The question whether hox genes will be able to provide the source of variation that natural selection needs, is debatable. The question is whether mutations in hox genes will always harm the organism in which they occur.

Higher-level instructions, for building tissues, organs, and body types, are not stored only in DNA. That means DNA could be mutated and there still would not be a new body plan. Even the mutation of hox genes would not produce a new body plan.[9-132]

Recent research has found that genes may not do as much as scientists previously thought. Gene sequences in DNA provide the assembly instructions for building proteins. These instructions, found in the DNA of every plant and every animal, have been compared to precisely sequenced strings of code. Some scientists, however, state that while the genes in DNA carry assembly instructions for building proteins out of amino acids, they do not carry the assembly instructions for building organs out of proteins, or for building whole creatures out of organs and other body parts. In other words, many biologists (especially embryologists) are beginning to think that genes do not by themselves carry the instructions for building a whole organism or animal. They believe that animals are made of many body parts and organs arranged in very specific ways. Body parts and organs are made of special kinds of tissues. Each type of tissue is made of different types of cells. Each different cell type has many proteins, including some that may be unique to that particular cell type.

According to scientists, current research shows that DNA is actually closer to the bottom of the organizational ladder than the organs and body parts. This obviously poses a big problem for evolutionists. Even if all the information is needed to produce an organism, additional information is still needed to arrange all the components into functioning body parts. Many scientists now believe that this is true. An organism needs genetic information to build proteins. It also needs higher-level assembly instructions to arrange tissues and organs into body plans. Scientists are not entirely sure where these higher-level assembly instructions are stored. However, these instructions are clearly necessary, and many scientists now doubt that they are stored in DNA alone. Evolutionary scientists believe that new biological form arises when natural selection acts on randomly occurring mutations and variations in DNA. But new research seems to say that mutations in DNA assembly instructions will produce, at best, a new protein. Higher-level instructions, for building tissues, organs, and body types, are not stored only in DNA. That means DNA could be mutated and there still would not be a new body plan. Even the mutation of hox genes would not produce a new body plan (Much research still needs to be completed in this area before any conclusion should be reached).

Some developmental biologists now think that two other cellular features, (1) the cytoskeleton and the (2) cell membrane (See Chapter 5, section "The Origin of the Cell Membrane" that discusses the origin of the cell membrane.), store structural information that affects how the embryo develops, but there is much that is still unknown. If DNA doesn't control development, then something else must. Whatever this something else is, is currently unknown. It is also unknown whether this something else can be altered by mutation, which would provide variations on which natural selection could act. Currently, there are no answers to these questions.[9-133]

Creationist View

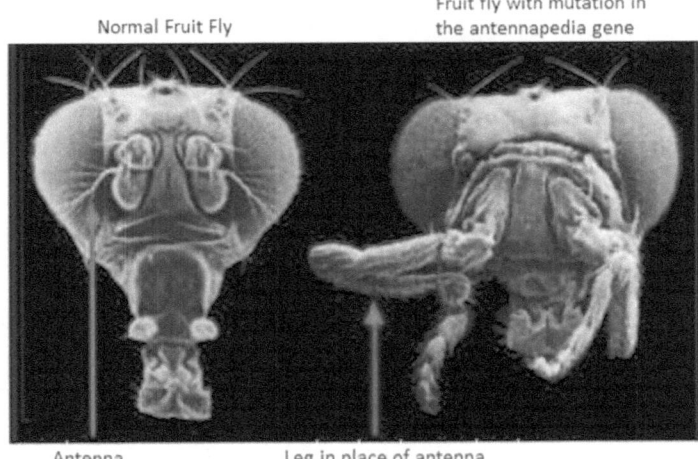

http://z3.invisionfree.com/bogleech/ar/t2096.htm

Figure 9.2 Normal Fruit Fly and Fruit Fly with Mutation in the Antennapedia Gene

Evolutionists claim that the homeobox (Hox) family of developmentregulating genes in animals can have complex effects. They note that Hox genes direct where legs, wings, antennae, and body segments should grow. In their classic example, evolutionists point out that in fruit flies, the mutation called Antennapedia causes legs to sprout where antennae should grow (see Figure 9.2 above). Evolutionists claim that this is an example of evolution. It may be an example of the result of a mutation, that is, the change in a switch in a Hox gene, but it is not an example of a new species. No new information was ever developed as a result of this mutation. A mutation in the hox gene results in already-existing information being switched on in the wrong place. The hox gene merely moved legs to the wrong place. It did not produce any of the information that actually constructs the legs, which in ants and bees include a complex mechanical and hydraulic mechanism that enables these insects to stick to surfaces. These abnormal limbs are not functional. Their existence demonstrates the genetic mistakes can produce complex structures. Such deformities would be a hindrance to survival and most likely natural selection would eliminate organisms with this mutation.[9-134]

(5) Genetic Drift

Evolutionist View

Genetic drift or allelic drift is the process of change in the genetic composition of a population of living things due to chance and random events rather than by natural selection. An allele is one of several different codes a gene might have. An organism has two copies of most genes, one from its father, and one from its mother. If the two copies are identical, the organism is said to have a single allele for that gene. If the two copies are different, it's said to have two alleles for that gene. If an entire species had descended from a single pair of animals in the recent past, the genetic diversity within the species should be limited. Genetic tests of many individuals in that species would find that most genes had at most four alleles, two from the original father and two from the original mother. (A few genes might have five or six alleles because of mutation.) Closely related species often share many of the same alleles, while more distantly related species share fewer of the same alleles.[9-135]

The alleles in offspring are a random sample of those of the parents, and chance has a role in determining whether a given individual survives and reproduces. The frequency of occurrence of allele change is only a fraction of the gene replications that share a particular form. Genetic drift occurs when the number of alleles selected in each generation fluctuates. As a result of random chance, there is a possibility that certain alleles will be selected over others. If one specific allele is not selected at all during a generation, that allele will become lost.[9-136] In other words, genetic drift is random change in the frequency of genes over time.

Creationist View

Genetic drift is changes in small groups of sub-species that have become separated from the rest of their species. Oddities in their DNA code factors became more prominent, yet there was no change in species.[9-137] Genetic drift and natural selection, over time, cause populations to be very specialized and adapted to their particular environment, but exhibiting so little variation as to be unable to adapt to other environments. The net effect of genetic drift is that populations become adapted to a particular environment, but unable to adapt to others. And as the population becomes more isolated in its environment, genetic drift increases continually, until the population is almost completely well adapted to one environment, but unreceptive to change. The evolutionary claim that populations increase in diversity over time is debatable. Populations decrease in diversity over time, because genetic drift removes changes from populations more quickly than mutation can add to it.[9-138] Genetic drift is a force working against natural selection, for it tends to preserve or destroy genes without distinction, whether favorable, neutral or unfavorable.[9-139]

(6) Other Mechanisms That Cause Changes in Organisms

Under domestication we see much variability, caused, or at least excited, by changed conditions of life; but often in so obscure a manner, that we are tempted to consider the variations as spontaneous. Variability is governed by many complex laws, by correlated growth, compensation, the increased use and disuse of parts, and the definite action of the surrounding conditions.

• Charles Darwin The Origin of Species
by means of Natural Selection or The Preservation
of Favoured Races in the Struggle for Life,
Chapter 15 - Recapitulation And Conclusion, 6th edition

Variability is governed by many unknown laws, of which correlated growth is probably the most important. Something, but how much we do not know, may be attributed to the definite action of the conditions of life. Some, perhaps a great, effect may be attributed to the increased use or disuse of parts. The final result is thus rendered infinitely complex.

• Charles Darwin The Origin of Species
by means of Natural Selection or The Preservation
of Favoured Races in the Struggle for Life,
Chapter 1 - Variation Under Domestication, 6th edition

Creationist View

Biologists have discovered multiple mechanisms that can cause radical changes in the amount of DNA possessed by an organism. Research has determined that there are a number of mechanisms by which new genetic information is developed by the processes of evolution. As mentioned in Chapter 5, section "The Origin of Biological Information Stored in DNA and RNA," some of these mechanisms include (1) exon shuffling, (2) gene duplication, (3) retroposition, (4) mobile genetic elements, (5) lateral gene transfer, (6) gene fusion and (7) de novo gene origination. Gene duplication, polyploidy, insertions, etc., do not help explain evolution, however. Gene duplication represents an increase in the amount of DNA, but not an increase in the amount of functional genetic information. These mechanisms create nothing new. For macroevolution to occur, new genes for developing feathers on reptiles, for example, are needed. In plants, but not in animals, possibly with rare exceptions, the doubling of all the chromosomes may result in an individual that can no longer interbreed with the parent type. This is called "polyploidy." Although this may technically be called a new species, because of the reproductive isolation, no new information has been produced, only repetitious doubling of existing information. Duplication of a single chromosome is normally harmful, as in Down's syndrome. Insertions are a very efficient way of completely destroying the functionality of existing genes. These mutational changes are actually examples of loss of specificity, which means they involved a loss of information, rather than an increase in information, as evolutionists claim.

The evolutionist's "gene duplication idea" is that no existing gene may be doubled, and one copy does its normal work while the other copy is redundant and non-expressed. Therefore, it is free to mutate free of selection pressure, that is, it will have the possibility of survival. However, such "neutral" mutations are powerless to produce new genuine information. Evolutionists point out that natural selection (discussed earlier) is the only possible naturalistic explanation for the immense design in nature. They claim that random changes produce a new function, and then this redundant gene becomes expressed somehow and is fine-tuned under the natural selective processes. This idea relies on a change copying event, genes somehow being switched off, randomly mutating to something approximating a new function, then being switched on again so natural selection can tune it. (See Chapter 13, section "Some DNA Portions are Used to Turn Genes On or Off" for further discussion of genes being switched on and off. See Chapter 5, section "The Origin of Biological Information Stored in DNA and RNA" Creationist View for further discussion of gene duplication.)

Mutations do not occur in just the duplicated gene, they also occur throughout the genome. Consequently, all the harmful mutations in the rest of the genome have to be eliminated by the death of the unfit. Selective mutations in the target duplicate gene are extremely rare, perhaps representing only 1 part in 30,000 of the genome of an animal. The larger the genome, the bigger the problem, because the larger the genome, the lower the mutation rate that the creature can sustain without causing death. As a result, it takes even longer for any mutation to occur, let alone a desirable one, in the duplicated gene. There just has not been enough time for such a naturalistic process to account for the amount of genetic information that we see in living things. Evolutionists have recognized that the information space possible in just one gene is so huge that random changes without some guiding force (intelligence) could never come up with a new function. There could never be enough experiments (mutating generations of organisms) to find anything useful by such a process. An average gene of 1,000 base pairs represents 4^{10000} possibilities, that is, 10^{602}. Such a neutral process cannot possibly find any sequence with specificity (usefulness), even allowing for the fact that more than just one sequence may be functional to some extent.[9-140] (See Chapter 5, section "The Origin of Biological Information Stored in DNA and RNA," Creationist View for further discussion on the possibilities of gene changes causing positive changes in species.)

The effects of genes on development are often varied. For example, in the house mouse, nearly every coat-color gene has some effect on the body size. Out of seventeen x-ray induced eye color mutations in the fruit fly Drosophila melanogester, fourteen affected the shape of the sex organs of the female, a characteristic that one would have thought was quite unrelated to eye color. Almost every gene that has been studied (as of 1970) in higher organisms has been found to effect more than one organ system, a multiple effect that is known as pleiotropy (one gene being responsible for or affecting more than one phenotypic characteristic). (*Phenotype is the appearance of an organism that results from the interaction of the genotype and the environment. Genotype is the genetic makeup of an organism or group of organisms with reference to a single trait, set of traits, or an entire complex of traits.*) Not only are most genes in higher organisms pleiotropic in their influence on development but, as is clear from a wide variety of studies of mutational patterns in different species, the pleiotropic effects are invariably species specific.

When an experiment of one particular gene in a domestic chicken was done, there were multiple effects. A mutation of this gene caused developmental abnormalities in a variety of systems. This one gene obviously is involved in the development of some structures unique to birds, such as air sacs, downy feathers, lungs, and kidneys. The mutated gene resulted in **(1)** either the wings did not develop at all or they only formed small stumps; **(2)** the hind limbs reached full length although two or more digits were often fused; **(3)** the downy cover remained underdeveloped; **(4)** the lungs and air sacs were missing although the trachea and extrapulmonary bronchi were normal; **(5)** the ureter (the duct which carries the urine from the kidney to the bladder) did not grow and failed in the development of the kidney. The only conclusion that can be reached from this experiment is that non-homologous genes are involved to some extent in the specification of homologous structures.[9-141]

In the next chapter (10) we will continue our discussion of how various changes could occur in species by taking a look at the concept of Punctuated Equilibrium.

~~~

Chapter 10

Punctuated Equilibrium

Introduction

In Chapter 9, we discussed various components that may explain how changes could occur in species. In this chapter (10) we will discuss the theory of Punctuated Equilibrium, which states that species exist for long periods of time with no change followed by periods of rapid change.

> *There are, however, some who still think that species have suddenly given birth, through quite unexplained means, to new and totally different forms. But, as I have attempted to show, weighty evidence can be opposed to the admission of great and abrupt modifications. Under a scientific point of view, and as leading to further investigation, but little advantage is gained by believing that new forms are suddenly developed in an inexplicable manner from old and widely different forms, over the old belief in the creation of species from the dust of the earth.*
>
> • Charles Darwin The Origin of Species
> by means of Natural Selection or The Preservation
> of Favoured Races in the Struggle for Life,
> Chapter 15 - Recapitulation And Conclusion, 6th edition

Could Positive Changes Occur Quickly in Species?

Evolutionist View

Stephen Jay Gould, co-originator of the theory of Punctuated Equilibrium, was a paleontologist and an evolutionary biologist, who once stated that "*The fossil record with its abrupt transitions offers no support for gradual change. Macroevolution proceeds by the rare success of these hopeful monsters, not by continuous small changes within the populations.*"[10-1] Stephen Jay Gould described his theory of punctuated equilibrium this way: "*Thus, our model of 'punctuated equilibria' holds that evolution is concentrated in events of speciation and is an infrequent event punctuating the stasis of large populations that do not alter in fundamental ways during the millions of years that they endure.*"[10-2]

> *Darwinism is a theory of cumulative processes so slow that they take between thousands and millions of decades to complete.*
> —Richard Dawkins[10-3]

The fossil evidence has revealed, in some cases, that evolution is gradual, however, in most cases, transitional forms have not been found. In 1972, evolutionary scientists Stephen Jay Gould and Niles Eldredge proposed another explanation, which they called "punctuated equilibrium." That is, species are generally stable, changing little for millions of years. This slow pace of genetic change is "punctuated" by a rapid burst of change that results in a new species and that leaves few fossils, as there would not have been an abundance of intermediate species to fossilize.[10-4] When Niles Eldredge and Stephen Jay Gould first proposed their theory of punctuated equilibria in 1972, they proposed that the fossil record may very well be imperfect. They believed that it is possible that supposed gaps in the fossil record are a true reflection of what really happened rather than the possibility of there being an imperfect fossil record. Eldredge and Gould suggested that evolution really did in some sense occur in sudden bursts, that is, punctuating long periods of 'stasis', when no evolutionary change took place in a given lineage. It is conceivable that there were never really any intermediates, that is, it is conceivable that large evolutionary changes took place in a single generation. For example, a son might be born so different from his father that he would be classified as a different species from his father. He would be a mutant individual, and the mutation would be such a large one that we should refer to it as a macromutation.

It is doubtful, however, that major changes could occur within a single generation. This view can be easily rejected because if a new species really did arise in a single mutation step (within a single generation), members of the new species might have a hard time finding mates. It is also difficult to believe that major changes could occur within a single generation because it is commonly believed that evolutionary changes are advances in complexity of design. The eye is a good example of this. It is inconceivable that a parent with no eye could produce offspring that have fully functional eyes, complete with variable focus lens, iris diaphragm, retina with millions of three-color photocells, all with nerves correctly connected up in the brain to provide the offspring with correct, binocular, stereoscopic color vision. (The eye is discussed in Chapter 14.)

The concept proposed by Eldredge and Gould did not really deviate from Darwin's theory that proposes changes occur slowly. The only difference that Eldredge and Gould proposed is that all the gradual changes are compressed into brief bursts, rather than having changes occur all the time. Eldredge and Gould emphasized that most of the gradual change goes on in geographical areas away from the areas where most fossils are dug up. They still believed that changes occur gradually. The only difference is that they believed that evolution (still undeniably gradualistic evolution) occurs rapidly during relatively brief bursts of activity.

Evolutionary changes could be so slight that they might not be detectable by human observers. Since the rate of evolutionary change is so slow and very minor, it would not be noticed during an ordinary human lifetime. It would appear that no evolution was going on at all. Nevertheless, they would be evolving, but very slowly. Therefore, the proper way to view the concept of punctuationism is that it is gradualistic, but with long periods of "stasis" (evolutionary stagnation—inactivity) punctuating brief episodes of rapid gradual change.[10-5] Rapid changes build up in the smaller gene pool. This creates new species that leave behind few fossils since fossilization is so rare anyway. The low number of fossils may be evidence of rapid change, rather than missing evidence for gradual evolution.[10-7] In biology, a species is one of the basic units of biological classification and a taxonomic rank. A species is often defined as a group of organisms capable of interbreeding and producing fertile offspring.[10-6] While searching the fossil records for evidence of punctuated equilibrium, scientists have found one group of coral-like sea organisms in particular, called bryozoan that shows evidence of punctuated equilibrium. The well-preserved fossil record of bryozoans shows that one species first appeared about 140 million years ago and remained unchanged for its first 40 million years. Then there was an explosion of diversification, followed by another period of stability for vast amounts of time. Studies in population genetics have shown that small changes can accrue quickly in small populations. Evolutionary developmental biology (Evo Devo) is revealing new mechanisms that regulate the appearance of small genetic changes in ways that can have a large effect on the physical look of an organism as opposed to its genetic makeup.[10-8] (See Chapter 9, section "Hox Genes" for further discussion of hox genes influencing changes in species).

Intelligent Design View

The theory known as punctuated equilibrium, developed by Niles Eldredge and Stephen Jay Gould, states that evolution occurs in episodes, in a process occurring in spurts and starts interspaced with long periods of no change. According to this view, new species originate rapidly in small isolated groups. During the active period, as new species emerge the population undergoes rapid morphological change after which it spreads over a wide geographical area and undergoes little further change.

While punctuated equilibrium is an explanation for the gaps between species, it is doubtful if it can be extended to explain the larger gaps. The gaps that separate species: dog to fox, rat to mouse, etc, are trivial compared to a primitive terrestrial mammal and a whale or a primitive terrestrial reptile and an Ichthyosaur. Even these relatively major gaps are trivial compared to those that separate major phyla (groups) such as mollusks and arthropods (discussed in Chapter 3). Such major gaps could not have been closed through one or two transitional species occupying isolated geographical areas. There would have to have been hundreds, or thousands, or possibly millions of intermediate species that would be necessary for the resulting changes to have occurred. This seems highly improbable given the isolation and the small populations of animals.[10-9]

Creationist View

The theory of punctuated equilibrium states that small groups of animals that are ready to "evolve" break away from the pack, become separated from the main population, isolate themselves in remote areas, change into new and unrelated species and then disappear forever without leaving fossil or any other evidence.[10-10] Punctuated equilibrium views species populations as systems that display recurrent patterns of evolution. Rather than the smooth, gradual change imagined by Darwin now known as 'gradualism', Gould and Eldredge suggested species tend to remain stable, changing little over long periods of time. (The system is then in 'stasis' or 'equilibrium.') Eventually, that stability is 'punctuated' by an episode of rapid change.[10-11] The concept of punctuated equilibrium states that two creatures simply give birth to an entirely different creature, however, the theory does not explain the vehicle by which such a thing could happen, or where this descendant would get its mate.[10-12]

Punctuated Equilibrium has been proposed as an explanation for the incomplete fossil record. However, punctuated equilibrium is merely another form of gradual change, there is no evidence that even heightened mutation rates could produce the amount of novel body plans found in the Cambrian strata. The mutation rates would have to be amazingly fast in order to account for the fossil record. (Mutations are discussed in Chapter 9).[10-13] Although macro-mutations of many varieties produce drastic changes, the vast majority would be incapable of survival, let alone show the marks of increased complexity. If structural gene mutations are inadequate because they do not produce significant changes, then regulatory and developmental mutations appear even less useful as they have a greater likelihood of non-adaptive or even destructive consequences. At present, it is doubtful, whether mutations, great or small, are capable of producing limitless biological change. [10-14] There is no evidence that punctuated equilibrium ever occurred in the natural world. No one has ever witnessed it occurring either in the past or in the present. For multiple "good" mutations to occur simultaneously in order to produce the genes in a creature so that its offspring will emerge with irreducible complex systems intact is more improbable than gradual evolution.[10-15]

Experiments Illustrating Punctuated Equilibrium

Evolutionist View

Creationists view Punctuated Equilibrium as being an impossibility and only appears to be an answer by evolutionists for the insufficient fossil evidence. Creationists argue that too many genetic changes, all occurring at the same time, would be required to account for the non-gradualistic pattern in the fossils. Creationists believe the chance of this happening through natural processes is highly improbable, if not impossible. However, this line of thinking was a pre-1950s understanding of genetics in which one gene was responsible for one trait. This mindset is now out of date. In the new emerging field of evolutionary developmental biology (Evo Devo), experimental studies reveal that manipulating only one developmental gene or molecule could cause significant changes in the anatomy of an organism. For example, the fins of fish and limbs of amphibians, reptiles, birds, and mammals begin as buds at the side of the body. As these grow, similar developmental genes and molecules appear sequentially, but they are expressed in differing combinations between animals, and this leads to different types of fin/limb anatomy. (Limbs are discussed in Chapter 7, section "Homology of Vertebrate Limbs.")

Simple experiments placing these molecules in a developing bud can alter the final number of bones in a limb and change their shapes dramatically. Therefore, a minor genetic modification in the release of just one developmental molecule can produce different concentrations and combinations of these molecules, resulting in a major anatomical change. With this process in mind, and considering the fossil record of lobe-finned fish and the first amphibians, it would be easy to see how fins evolved into limbs. (See Chapter 7, section "Homology of Vertebrate Limbs.")[10-16]

Experiments reveal that the placement of specific developmental molecules in the front of the mouth can change the expected simple incisor tooth into a complex multi-cusped molar. In other words, small genetic changes in the production and release of embryological molecules can easily produce the many different types of teeth seen in the fossil record. Similar developmental mechanisms can lead to dramatic changes in the shape and number of bones in the lower jaw. Experimental limbs reveal that bone structure is very flexible. Evolution development can also explain the appearance of a new joint in mammal-like reptiles. (See Chapter 8 for more details of transitional jaw joints.)

Basic embryological programs, like the instructions for teeth, are made up of groups of genes that are arranged in a sequence. These genes can be turned on or off at different times during development, and they can even be initiated at different places in a developing organism. For example, scientists have caused eyes to develop on legs and wings. Thus, it would take only a small genetic change to switch on or turn off the gene program for a joint between two adjacent jaw bones, as seen in the origin of the mammalian jaw joint (See Chapter 7, section "The Evolution of the Horse," Chapter 9, section "Hox Genes," and Chapter 13, section "Some DNA Portions are Used to Turn

Genes On or Off" for further discussion of genes being turned on or off and see Chapter 7, section "Evolution From Reptiles to Mammals" and Chapter 8 for reptile to mammal evolution).[10-17]

In the next chapter (11) we will discuss embryology and the Theory of Recapitulation.

~~~

Chapter 11

Embryology and The Theory of Recapitulation

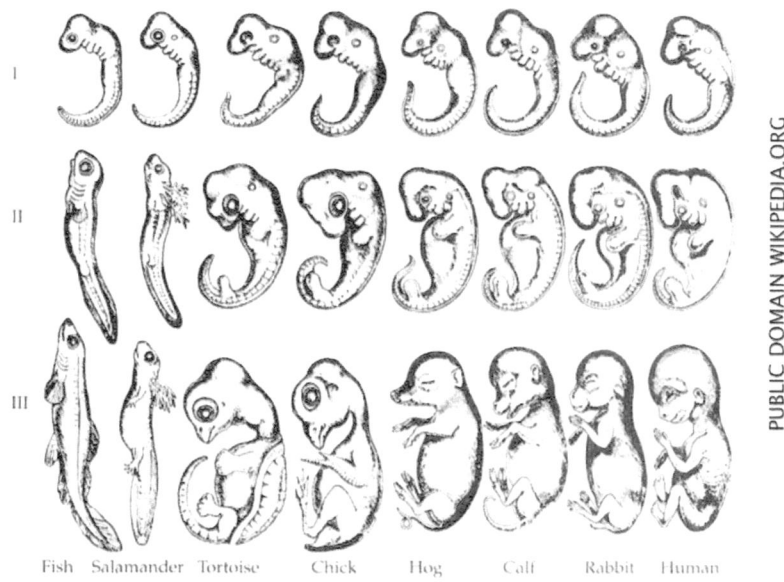

Fish   Salamander   Tortoise   Chick   Hog   Calf   Rabbit   Human

Ernst Haeckel's Embryo Drawings

Figure 11.1 The drawings above are Ernst Haeckel's twenty-four embryos that he used to illustrate the different stages of development. The different species are arranged in columns with the different stages of development from row I to row III. The first two rows labeled I & II show similarities.

Introduction

In this chapter (11) we will discuss whether or not vertebrate embryos go through a process called recapitulation, whereby during development the embryo goes through various stages that reflect their evolutionary past. Recapitulation is a theory that states during embryonic development, an organism passes through various stages of structural change that repeat its ancestral lineage.[11-1]

Ernst Haeckel's Biogenetic Law

Evolutionist View

Ernst Haeckel, a German evolutionist and Darwin's contemporary, formulated a "biogenetic law" in 1866, famously summarized as "Ontogeny recapitulates phylogeny." This means that the development of an organism simply replays its evolutionary history. But this notion is true in only a limited sense. Embryonic stages don't look like the adult forms of their ancestors, as Haeckel claimed, but like the embryonic forms of ancestors. Human fetuses, for example, never resemble adult fish or reptiles, but in certain ways they do resemble embryonic fish and reptiles. Not every feature of an ancestor's embryo appears in its descendants. Haeckel's law has been disregarded not only because it wasn't strictly true, but also because Haeckel was accused, largely unjustly, of falsifying some drawings of early embryos to make them look more similar than they really are. Yet the concept should not be discarded just because it was not totally accurate.[11-2] Haeckel thought only the final stages of development could be altered during the process by evolution, but this is not the case. All developmental stages can be modified during evolution, though the developmental stage may be more limited than other stages. Embryos still show a form of recapitulation: features that arose earlier in evolution often appear earlier in development. This makes sense only if species have an evolutionary history.[11-3]

Some creationists wrongly believe that because Ernst Haeckel's "Biogenetic Law" is false, therefore embryology can no longer provide evidence for evolution. Neither modern evolutionary theory nor modern developmental biology is based on Haeckel's observations and theories. The ideas of Ernst

Haeckel greatly influenced the history of embryology in the 1800s. Haeckel's ideas were then superseded by those of Karl Ernst von Baer, his predecessor.

Von Baer suggested that the embryonic stages of an individual should resemble the embryonic stages of other closely related organisms, rather than resembling its adult ancestors. Evolutionists no longer promote either Von Baer's Laws or Haeckel's Biogenetic Law, as both of these theories have been proven to fail as scientific laws. Haeckel thought only the final stages of development could be altered appreciably by evolution, but it has been known to be false for nearly a century. All developmental stages can be modified during evolution, though the developmental stage may be more constrained than others.[11-4]

Intelligent Design View

It is not true that embryos are really similar in their earliest stages of development, or the adult stage of development that Haeckel claimed. Even the embryos of closely related animals, such as chickens and ducks, display specific differences very early in development. There are noticeable differences in the second day of development.[11-5]

Ernst Haeckel's Drawings

Evolutionist View

> . . . from the many slight successive variations having supervened in the several species at a not early age, and having been inherited at a corresponding age, the young will have been but little modified, and they will still resemble each other much more closely than do the adults . . .
>
> —Charles Darwin, Origin of Species, Chapter 14

> . . . being inherited at a corresponding not early period of life, we clearly see why the embryos of mammals, birds, reptiles, and fishes should be so closely similar, and so unlike the adult forms.
>
> —Charles Darwin, Origin of Species, Chapter 15

It is admitted that Haeckel exaggerated the similarities in very early embryos of different species (Figure 11.1). Comparing his drawings with actual photographs of embryos, however, shows strong and undeniable similarities (Figure 11.2). The embryos of reptiles, birds, and mammals all resemble one another much more than the adult forms do, exactly as Darwin noted in Origin of Species. The similarities of evolutionary embryonic development are evident.[11-6]

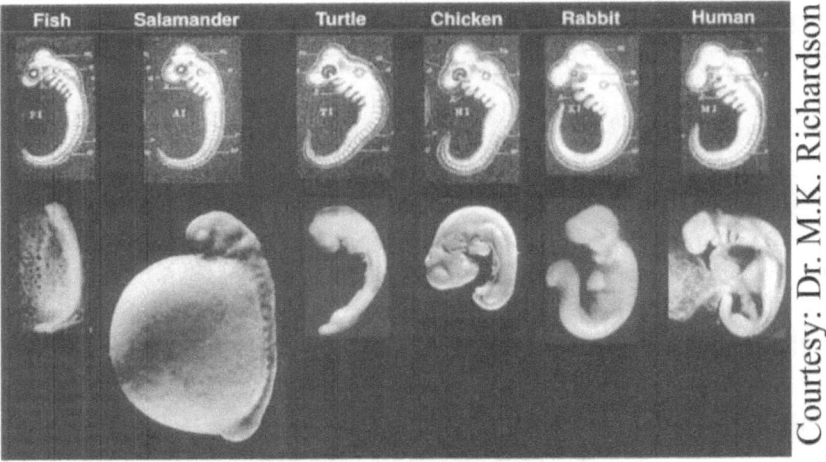

Comparison of Ernst Haeckel's Drawings
With Actual Photos by Dr. Michael Richardson

Figure 11.2 Comparison of Ernst Haeckel's Drawings with Actual Photos
The top row is Haeckel's images of embryos.
The bottom row is actual photos of the species.

Dr. Michael Richardson, an embryologist, in a critique of Haeckel's work, upon re-examining all of his drawings, wrote that basically Haeckel's theory regarding vertebrate embryos passing through similar stages during development was correct and that the development of vertebrate embryos does indeed reflect a relationship of their evolutionary past.[11-7] Dr. Richardson stated that variations in the late stage embryo are often indicated by modifications of early embryonic development. He also stated that evolution has produced a number of changes in the embryonic stages of vertebrates, such as: **(1)** Differences in body size; **(2)** Differences in body plan (for example, the presence or absence of paired limb buds); **(3)** changes in the number of units in repeating series such as a member of a paired segments of arches near the area around the pharynx; (4) changes in the pattern of growth of different fields; (5) changes in the timing of development of different fields during evolution.[11-8] Haeckel's embryo drawings are very simplified versions of true embryonic development, and he admitted they were. Haeckel, in his admission, pointed out that all diagrammatic figures are inaccurate, in that they are simplified. It may be possible that Haeckel was stating that diagrams are simplifications of more complex data.[11-9]

Creationist View

The theory of Haeckel's embryos supposedly going through stages of evolution during development assumes that living embryos re-experience the evolutionary process that their pseudo-ancestors underwent. In a series of drawings that came to be known as "Haeckel's embryos," (Figure 11.1) German biologist Ernst Haeckel attempted to show that embryos are all the same at their early stages. Haeckel not only utilized deceptive data but also used doctored drawings to deceive his followers. His dishonesty was so blatant that he was charged with fraud by five professors and convicted by a university court at Jena. His forgeries were made public with the 1911 publication of "Haeckel's Frauds and Forgeries."

The drawings Haeckel labeled as "first stage" are actually somewhere in the middle of development. Haeckel realized that first stage embryos look nothing alike, so he chose the "worm like" stage that comes later. But even as mid-stage drawings, he made them appear to be more similar than they really are. Actual photographs of a large number of different embryos show that embryos of different kinds are very different. The only way for Haeckel to have drawn them looking so similar was to have cheated. Therefore, the idea of extensive embryonic similarities is outdated and based on fraud.

Many evolutionists still believe that embryo developments go through similar stages as Haeckel's faked drawings illustrate. Haeckel also apparently chose his examples so they would look similar and therefore support his Biogenetic Law. In Haeckel's drawings, he shows four different mammals. The issue is that they are all placental mammals. It was not very scientific of Haeckel to use only data that supported his hypothesis, and ignore the data that disagreed with his theory.

It appears that Haeckel began with similar embryo drawings and changed them so that the embryos would resemble one another. He added fake organs to embryos, removed organs from others and illustrated embryos of very different sizes as being the same in scale. The clefts that Haeckel depicted as "gills" in the human embryo had in fact nothing to do with gills at all. They were actually the middle ear canal and the beginnings of the parathyroid and thymus glands (See Figure 11.6). The embryos did not in fact resemble one another at all. Haeckel also falsified the scale of some embryos to exaggerate similarities among species, even when there was a difference of ten times in size. Haeckel included "gill-slits" in every drawing in the first two rows, possibly to promote what he believed, that all embryos begin as fish. In the third row of his drawing, only the fish and salamander were drawn as having "gill-slits."

In the 1990s the British embryologist Dr. Michael Richardson and his colleagues compared Haeckel's embryo drawings with actual embryos under the microscope. Although Dr. Richardson and his colleagues are evolutionists, they concluded that there is no resemblance of Haeckel's drawings to actual vertebrate embryos. After their study was completed, Dr. Richardson and his colleagues published genuine photographs of embryos in the August 1997 issue of the journal Anatomy and Embryology (Figure 11.2).[11-10] The September 5, 1997 issue of Science magazine entitled "Haeckel's Embryos: Fraud Rediscovered," mentions that Haeckel's Embryos, being part of supposed evidence of evolution, states that the whole scientific world agrees that Haeckel's embryo drawings had been a fraud. The article also mentions that in addition to Haeckel adding or omitting features, Dr. Richardson and his colleagues reported that Haeckel falsified the scale of the embryos to exaggerate similarities among species, even when an embryo was ten times the size of an embryo being compared with. Dr. Richardson and his colleagues also mention that Haeckel did not provide scientific names, stages or sources of the specimens illustrated.

Dr. Richardson and his colleagues stated that even closely related embryos such as those of fish vary quite a bit in their appearance and developmental pathway. Dr. Richardson stated that the problem of comparing embryos is that embryonic stages are subjected to shifts in developmental timing during evolution so that different organs develop at different times in different species. The developmental timing variation makes it impossible to define a specific stage at which all vertebrate embryos have the same combination of the first stages of organ development. In summary, there is no one stage where embryos of all species have identical structures, which makes comparisons difficult.[11-11] In the March 2000 issue of Natural History magazine, evolutionist and paleontologist Stephen Jay Gould said that he had been aware of Haeckel's fraud for a long time but remained silent about the issue, but eventually he realized the fraud needed to be exposed.[11-12]

Regarding variations in the late stage embryo being indicated by modifications of early embryonic development, there does not appear to be a way to either confirm or falsify the hypothesis that mutations in genes that regulate embryonic development might provide whatever is needed to produce genetic changes in the final creature. Creatures that appear to start out looking similar during the embryonic stages are much different as adults. Some believe that simple changes in genetics regulating development could induce an embryo to develop in an unusual manner. Future studies in embryology might shed some light on this. Currently, however, there is not enough information in this area to provide any verifiable information. Even if further research succeeds in altering a genetic program of a fish embryo so that it develops into an amphibian, this would not prove that amphibians evolved or could have evolved from fish. This is because evolutionary changes are claimed to be random, mutational changes. A random change would cause a deformity in an organ, rather than a positive change.[11-13]

Embryology

In the higher Vertebrata the branchiae have wholly disappeared—but in the embryo the slits on the sides of the neck and the loop-like course of the arteries still mark their former position. But it is conceivable that the now utterly lost branchiae might have been gradually worked in by natural selection for some distinct purpose . . .

—Charles Darwin, Origin of Species, 1859,
Chapter 6—Difficulties Of The Theory

Every evolutionist will admit that the five great vertebrate classes, namely, mammals, birds, reptiles, amphibians, and fishes, are descended from some one prototype; for they have much in common, especially during their embryonic state.

—Charles Darwin, The Descent of Man,
1882, p. 158

Evolutionist View

Embryology is the study and growth of the early development of an organism. Evolution theory states that species that are more closely related should go through similar embryonic development, which they do. The early embryos of all vertebrates resemble each other. Embryos of creatures that breathe air go through a gill slit stage, complete with aortic arches and a two-chambered heart, like those of a fish. Creationists have no valid explanation as to why this would occur.[11-14]

In the past, biologists studied both embryology (how an animal develops) and comparative anatomy (the similarities and differences in the structure of different animals). Their studies revealed many peculiarities that at the time did not make sense. For example, all vertebrates begin development in the same way, looking, rather like an embryonic fish (Figure 11.1). As development continues, different species begin to appear, but in peculiar ways. Some blood vessels, nerves, and organs that were present in the embryos of all species at the start suddenly disappear, while in others they go through strange changes, even moving to a different location. Eventually, development ends in the very different adult forms of fish, amphibians, reptiles, birds, and mammals.[11-15]

The fishy fetuses of all vertebrates are limb-less and have a fish-like tail. The fish-like feature is a series of five to seven pouches, separated by grooves along each side of the embryo near its undeveloped head. These pouches are called the "branchial arches." Each "arch" contains tissues that develop into nerves, blood vessels, muscles, and bone or cartilage. As fish and shark embryos develop, the first arch becomes the jaw and the rest become gill structures: the clefts between the pouches open up to become the gill slits. In fish and sharks the development of gills from the embryonic arches appears more straightforward. These embryonic features simply enlarge without much change to become the adult breathing organ. However in other vertebrates that don't have gills as adults, these arches develop differently. In mammals, for example, they form the (1) three tiny bones of the middle ear, (2) the Eustachian tube, (3) the carotid artery, (4) the tonsils, (5) the larynx, and (6) the cranial nerves (Figure 11.6).

Human blood vessels go through very strange contortions during embryonic development. In fish and sharks, the embryonic pattern of vessels continues without much change into the adult system. But as other vertebrates develop, the vessels move around, and some of them disappear. Mammals, including humans, are left with only three main vessels from the original six.

As human embryonic develop proceeds, the changes resemble an evolutionary sequence. The human fish-like circulatory system develops into one that is similar to that of embryonic amphibians. In amphibians, the embryonic vessels grow directly into adult vessels, but humans continue to change into a circulatory system that resembles that of embryonic reptiles. In reptiles, this system develops directly into the adult one. Human development, however, continues to change until it becomes a true mammalian circulatory system, complete with carotid, pulmonary, and dorsal arteries.[11-16]

These patterns raise a lot of questions: **(1)** Why do vertebrates, which are completely different from one another, all begin development looking like a fish embryo? **(2)** Why do the same embryonic structures used to form heads and faces for mammals become the gills of fish? **(3)** Why do vertebrate embryos go through such a contorted sequence of changes in the circulatory system? **(4)** Why don't human embryos, or lizard embryos, begin development with an adult-like circulatory system, rather than making a lot of changes in what developed previously? And **(5)** why does the human sequence of development copy the order of our ancestors, fish to amphibian to reptile to mammal?

The "recapitulation" of an evolutionary sequence is also noted in human kidneys. During development, the human embryo actually forms three different types of kidneys, one after the other, with the first two discarded before the final kidney appears. The transitory embryonic kidneys are similar to those that are found in species that evolved before humans in the fossil record, jawless fish and reptiles, respectively.

Each vertebrate develops in a series of stages. The sequence of these stages appears to follow the evolutionary sequence of its ancestors. For example, a lizard's embryonic structure resembles an embryonic fish, then later it looks like an embryonic amphibian, and finally becomes an embryonic reptile. Mammals go through the same evolutionary sequence with the additional stage of an embryonic mammal.[11-17]

Creationist View

The concept of embryos passing through various stages of evolution was originally called the 'biogenic law' by Haeckel and is often stated as 'ontogeny recapitulates phylogeny.' This interpretation of embryological sequences will not stand close examination, however. This concept has been discredited by almost every modern scientist, although, the idea still has a prominent place in evolutionary thinking.[11-18] If the developing embryo is supposed to reenact the sequence in the evolutionary history of the race, why are so few stages included? Why should we find some of them appearing in the wrong order? Why should we not find thousands of steps instead of only a few? Why does the embryo go through some steps that could not possibly have been included in the history of the animal? How can such stages as the egg, tea, pupa, and adult of a butterfly be explained? Why do some parts of an embryo show recapitulation and other parts never show it?[11-19] Fetal structures cannot represent adult ones, and mutations are supposed to modify all stages. According to the theory of recapitulation ('ontogeny recapitulates phylogeny'), the development of the individual recapitulates phylogeny, the development of the race. In this form the theory runs into so many difficulties it clearly cannot be true. An immediate problem is presented by the fetal membranes, the umbilical cord, and other fetal structures that cannot represent adult structures of any period. Furthermore, mutations have been shown to modify all stages of development, not just the final stages.[11-20]

According to the theory of recapitulation, the embryo-like parts of the adult repeat each stage of what its ancestors were shaped like. But another quality found in embryos that is not to be found in their supposed "ancestors," is that embryos will have two types of organs, while their supposed "ancestors" only had one. **First**, there are the organs that will not function until after the infant is born. Such an organ would be the lungs. For this reason people only develop one set of lungs in their lifetime. **Second**, there are the organs that have a special function in pre-birth as well as afterward. Such organs frequently change form two or three times. Two examples are the heart and kidneys. If recapitulation was correct, such multi-changing hearts and kidneys should also be found in adult mice and minnows. But this never occurs in the adult form of animal life.[11-21]

Evolutionists claim the development of the heart in humans illustrates an evolutionary past. Supposedly, the heart passes through a worm, fish, frog, and reptile stage before reaching its final form. It is true that at one step, the heart in the human embryo has one chamber (as in the worm), two chambers (as in the fish), three chambers (as in the frog), and four chambers with the connection of the two sides (as in the reptile). But it should be noted that the heart in human beings starts out with two chambers which fuse into one for a time. This sequence actually reverses the stages of supposed evolution. There are reasons for each step. The 'reptile stage' is necessary to churn the blood around the lungs until after birth. Since oxygen is received from the placenta before birth there is no use in sending a large supply of blood to the lungs when it is not needed.[11-22]

If the human embryo really did recapitulate its assumed evolutionary ancestry, the human embryonic heart should first have one chamber, then change it into two, then three, and finally four chambers. For that is the arrangement of hearts in our supposed ancestors according to evolutionary theory. But instead of this, the human heart first begins as a two-chambered organ, which later in fetal development fused into a single chamber. This single chamber later, before birth, changed into the four-chambered heart humans now have. So, the actual sequence of heart chambers in a human fetus is 2-1-4, instead of the one required by recapitulation: 1-2-3-4.[11-23] (Various heart types are discussed in Chapter 7, section "Various Heart Types".)

If embryos repeat past ages of history in their development; what about the embryos of the invertebrates? Why do they not "recapitulate'" also? Invertebrate embryos are so varied that it is impossible for those who believe in recapitulation to use them to illustrate evolutionary development. The invertebrates have been examined for evidences of recapitulation also, but they have not been of much help. Their eggs often hatch into larvae that are so unlike the adult that they may not be recognized as belonging together without breeding experiments. These larval stages are a kind of continuation of the embryo and have been included in the search for clues of taxonomic and phylogenetic relationships.[11-24]

Embryonic Development

Intelligent Design View

The evolution of living creatures would be more convincing if the comparison of homology (similar organ or structure changed by nature to perform different functions) among living organisms along with embryological and genetic comparisons showed homologous (similar) patterns of development, however, this is not the case. The first division of the egg cell to the blastula stage in amphibians, reptiles, and mammals is different (Figures 11.3 & 11.4 below). It is clear that neither the blastula (a solid sphere of cells formed during an early stage of embryonic development in animals) itself, nor the sequence of events which lead to its formation, are identical in amphibian, reptile, or mammal. The differences become even more apparent in the next major phase in embryo formation, which is gastrulation.

Gastrulation is a phase early in the embryonic development of most animals, during which the single-layered blastula is reorganized into a three-layered structure known as the gastrula (Figure 11.3 below). These three germ layers are known as the ectoderm, mesoderm, and endoderm. Gastrulation involves a complex sequence of relative cell movements whereby the cells of the blastula rearrange themselves, eventually resulting in the transformation of the blastula into the intricate folded form of the early embryo, or gastrula, which consists of the three basic germ cell layers. These basic germ cell layers are (1) the ectoderm, which becomes the skin and the nervous system, (2) the mesoderm, which becomes connective tissue and (3) the endoderm, which becomes the lining of the alimentary tract as well as to the liver and pancreas. Most scientists believe that the gastrulation and the gastrula are similar in all vertebrates, yet the way the gastrula is formed, and particularly the positions in the blastula of the cells which become the germ layers and their migration patterns during gastrulation, are unique in the different vertebrate classes. In some ways the egg cell, blastula and gastrula stages (Figure 11.3) in the different vertebrate classes are so different that, if the body plans were not as similar as adult vertebrates, they would most likely not have been classified as belonging to the same phylum (group). Because of the great differences of the early stages of embryo-genesis in the different vertebrate classes, organs and structures considered homologous in adult vertebrates cannot be traced back to homologous cells or regions in the earliest stages of embryogenesis. In other words, homologous structures are arrived at by different routes.

Even after gastrulation (a process, as invagination, by which a blastula or other form of embryo is converted into a gastrula—Figure 11.3), the sites from which homologous structures are derived are different in different vertebrate classes. Structures as obviously homologous as the vertebrate alimentary canal are formed from quite different embryological sites in different vertebrate classes. The alimentary canal is formed from the roof of the embryonic gut cavity in the sharks, from the floor in the lamprey, from roof and floor in frogs, and from the lower layer of the embryonic disc, which is called the blastoderm in birds and reptiles. Another class of organs considered strictly homologous are the vertebrate forelimbs, yet they generally develop from different body segments in different vertebrate species. The forelimbs develop from different trunk segments. (See Chapter 7, section "Homology of Vertebrate Limbs" "Creationist View" for further discussion of forelimb development.) Also, the position of the occipital (the back of the head or skull) relative to body segmentation varies widely in different vertebrate species.

The development of the vertebrate kidney provides another problem to the evolutionary assumption that homologous organs are developed from homologous embryonic tissues. In fish and amphibians the kidney is derived from an embryonic organ known as the mesonephros, while in reptiles and mammals the mesonephros degenerates towards the end of embryonic life and plays no role in the formation of the adult kidney, which is formed instead from a discrete spherical mass of mesodermal tissue, the metanephros, which develops quite independently from the mesonephros. Even the ureter, the duct that carries the urine from the kidney to the bladder, is formed in a completely different manner in reptiles and mammals from the equivalent duct in amphibians. Another problem for Darwinian evolutionists to explain is the development of the two unique membranes, the amniotic and allantoic, which surround the growing embryo in reptiles, birds, and mammals. These membranes are considered to be strictly homologous in all the vertebrate groups in which they occur, but in mammals the processes which lead to their formation and the cells from which they are derived differ completely from those in reptiles and birds.[11-25] (Also see Chapter 7, A Look at Some of the Transitional Fossil Evidence, section "Evolution from Amphibian to Reptile", Intelligent Design View, for further discussion of amphibian egg and the reptilian egg.)

What Really Occurs During the Stages of Human Development?

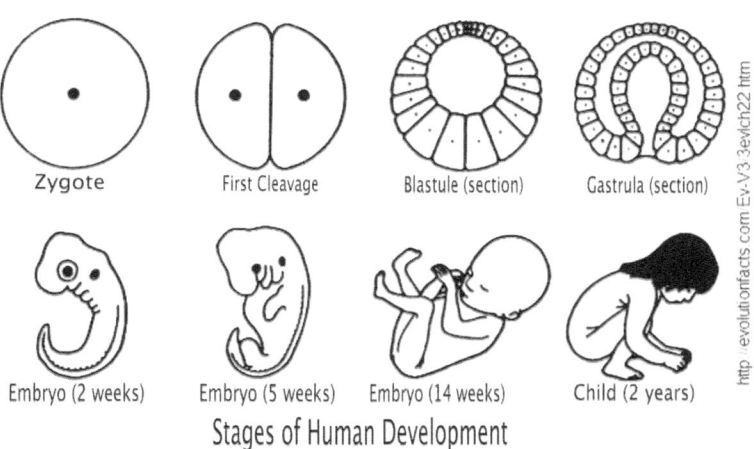

Stages of Human Development

Figure 11.3 Illustration of the various stages of human development from conception.

A zygote is the initial cell formed when two gamete cells are joined by means of sexual reproduction. In multicellular organisms, it is the earliest developmental stage of the embryo. In single-celled organisms, the zygote divides to produce offspring, usually through mitosis, the process of cell division.[11-26] (In Figure 11.4 below, the first stage labeled "fertilization—1 day" is called a zygote, depicted in Figure 11.3 above.) In embryology, cleavage (Figure 11.3) is the division of cells in the early embryo (Figure 11.3). The zygotes (Figure 11.3) of many species undergo rapid cell cycles with no significant growth, producing a cluster of cells the same size as the original zygote. The different cells derived from cleavage are called blastomeres and form a compact mass called the morula. Cleavage ends with the formation of the blastula (Figure 11.3).[11-27] A blastocyst is the blastula of the mammalian embryo, consisting of **(1)** an inner cell mass, **(2)** a cavity, and **(3)** an outer layer, the trophoblast. It is also called the blastosphere, which is the blastula of mammals. (The blastocyst is indicated as appearing on the fifth day in Figure 11.4 below.) The blastula consists of a sphere of cells (trophoblast) enclosing an inner mass of cells and a fluid-filled cavity (blastocoel).[11-28]

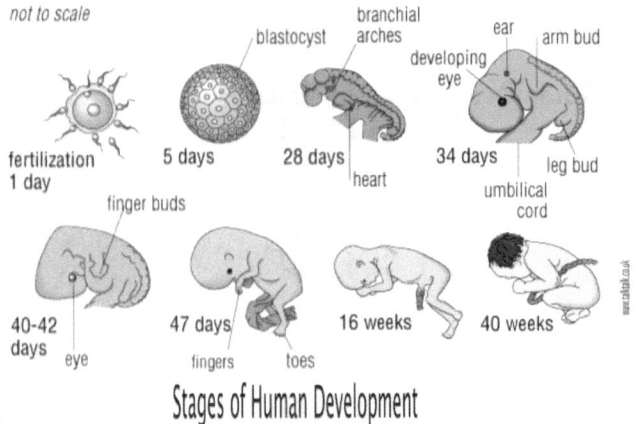

Stages of Human Development

Figure 11.4 Illustration the stages of the human development from conception at 1, 5, 28, 34, 40-42, 47 days, 112 days (3 ½-4 months), and 280 days (9 months).(See reference number [11-28a] in the Reference section for web link address to view video illustration of human development from day 15 to day 28. See reference number [11-28b] for web link address to view animated illustration of development from conception to birth.)

- - - - - - - - - - - - - - -

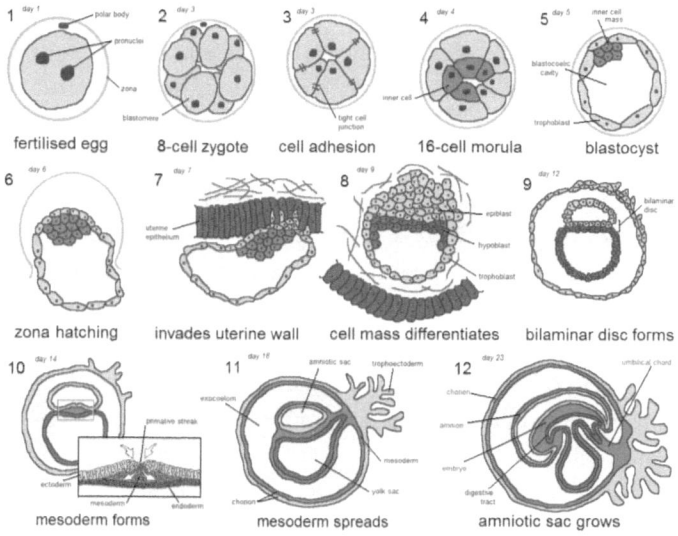

Figure 11.4a Human Embryo Development

(1) **Day 1** - Fertilization (and mitosis) occurs in the Uterine Tube (Fallopian tube) (1 celled Zygote). (2) **Day 2** – The 1 celled Zygote becomes a 2 celled Zygote that becomes 4 celled Zygote embryo (cleavage) (mitosis continues). The Zygote will then increase to 8 cells. The Zygote moves along Uterine Tube towards the Uterus until Blastula forms in the uterus on day 5 or day 6. (3) **Day 3** - Zygote increases the numbers of cells from 8 to 16 cells. (4) **Day 4** – 16 Celled Zygote (morula – inner cells develop on day 3 or 4). (5) **Day 5** to (6) **Day 6** – Zygote becomes a Blastocyst and continues to move towards the uterus. (7) **Day 7** – after Zygote has over 100 cells, it is called a Blastula. (The blastula is the phase that comes after the morula (solid ball phase), but before the blastocyst. It is a group of cells arranged in a hollow sphere, containing fluid. The blastocyst is likewise a ball of cells filled with fluid, but with a new addition: an inner cell mass, called the embryoblast. The embryoblast eventually forms the embryo. The outer cell layer is called the trophoblast and eventually contributes to the placenta. [11-28c]) (Trophoblast appears on the outer layer of the blastula). The Blastula invades the uterine wall. (8) **Day 8** to **Day 9** – Blastula implants onto the uterine wall. (9) **Day 10 to Day 13** - Blastula becomes Gastrula. (During gastrulation, the blastula folds upon itself to form the three layers of cells. Each of these layers is called a germ layer and each germ layer differentiates into different organ systems. The three germs layers are (1) the endoderm, (2) the ectoderm, and (3) the

mesoderm. The endoderm becomes the columnar cells found in the digestive system and many internal organs. The ectoderm becomes the nervous system and the epidermis. The mesoderm becomes the muscle cells and connective tissue in the body.[11-28d]) **(10) Day 14** to **Day 17** - Mesoderm (what becomes muscle and connective tissue) forms. **(11) Day 18** to **Day 19** - Mesoderm spreads. **Day 20** to **Day 22** – embryo develops. **(12) Day 23** – amniotic sac (protects the fetus) grows. **Day 24** to approximately **Day 270** (9 months) - embryo (after the 77th day, the embryo is called a fetus) continues to grow until birth, at which time the fetus becomes an infant. (An embryo is the early stage of human development in which organs are critical body structures are formed. An embryo is termed a fetus beginning in the 11th week of pregnancy, which is the 9th week of development after fertilization of the egg.[11-28e]) (Time spans are approximate.)

Creationist View

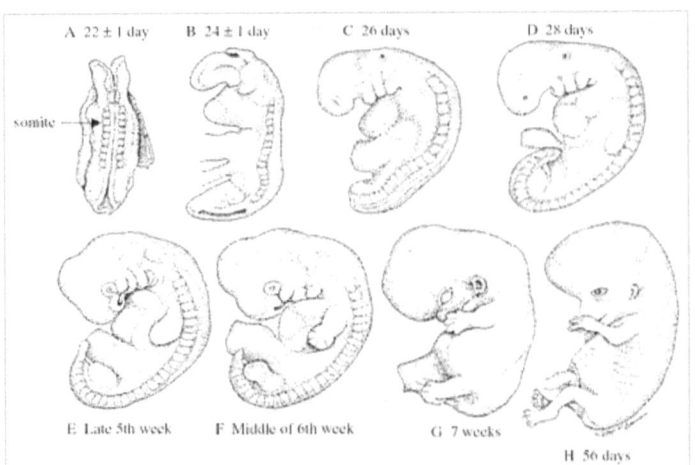

Figure 11.5 Development of a human embryo at (A) 22, (B) 24, (C) 26, (D) 28, (E) 32, (F) 39, (G) 49, and (H) 56 days.

The concept of "recapitulation" is based on the fact that there are similarities among embryos of people, animals, reptiles, birds, and fish. It is true that similarities do indeed exist. Babies, before they are born, look quite a bit alike during the first few weeks. This includes human babies, raccoon babies, robin babies, lizard babies, and goldfish babies. They all begin as little round balls. Then, gradually arms, legs, eyes, and all the other parts begin appearing (Figures 11.4 & 11.5). At one stage, there is just a big eye with skin over it and little flippers. Each part of every embryo has a definite purpose. But when animals are just beginning to form, and while they are very, very small, there is only one ideal way for them to develop. Literally thousands of parts are developing inside something that is extremely small. There are simply too many extremely tiny organs clustered in one near-microscopic object. When creatures are that tiny, there are only a very few ideal ways for them to be shaped, in order to develop efficiently.[11-29]

Evolutionists claim that human embryos have organs that are left-overs from ancestors. For example, gill slits like a fish. Evolutionists ask: *"What good are fish gills in your body?"* They then reason: *"Such organs are useless to people, so they must be "vestiges" from our ancestors. Since earlier ancestors of humans needed those organs, but humans do not, that proves that we are descended from those lower forms of life."* So human embryos are said to repeat or "recapitulate" various stages of their ancestors (such as the fish stage), and this recapitulation is declared to be evidence of evolution.

The chicken sac is the so-called "yolk sac" (Figure 11.6) in your body. In a baby chick, the yolk sac is the source of nourishment that it will live on until it hatches. This is because the chick embryo is in an eggshell and has no connection with its mother. But in a baby human being, this little piece of bulging flesh has no relation to a chick yolk sac, except for the shape. It is a small nodule attached to the bottom of the human embryo, even before it develops feet. A very tiny human being is connected to its mother and receives nourishment from her; therefore it does not need a yolk sac. But it does need a means of making its own blood until its bones are developed. For, although nourishment passes from the mother to the embryo, blood does not. That tiny human being must make its own. We humans make our blood in the marrow of our bones. Embryos are only beginning to form their bones and the marrow within them, so they cannot make blood in their bones and, for a time, need another organ elsewhere to fulfill that function.

The first blood in the human body came from that very tiny sack-like organ, long before a human baby is born (Figure 11.6). If the sack-like organ is removed from an embryo prematurely, death would immediately follow. When further developed, the blood within the human body is made within the bones, but during development of the human embryo a different source of blood is needed (which is the sack-like organ). The problem before birth is that it takes blood to make the bones that will make the blood. Even though the sack-like organ in a human embryo looks like a "tail," the sack-like organ later becomes the lower part of the spinal column in the child and adult. The reason why the spinal column is larger in proportion to the rest of the embryo is because the spinal column is full of very complicated bones, and the total length of the spine starts out longer in proportion to the body than it later will be. There are such complicated bones in your spine that it needs to start out larger and longer in relation to the body. Later, the trunk grows bigger as internal organs develop. Scientists have recently discovered that another reason the spine is at first longer than the body, is because the muscles and limbs do not develop until they are stimulated by the spinal nerve. So the spine must grow and mature enough that it can send out the proper signals for muscles, limbs, and internal organs to begin their growth. For this reason, the spine at first is bigger than the limbs, but later the arms and legs become largest.

Evolutionists claim that fish gills prove that humans are descended from fish. The theory, that people have gill slits when they are embryos, is false. In the embryo there are, for a time, three small folds to be seen in the front of its throat. These three bubble outward slightly from the neck. Upon careful examination of these folds, there are no gills to extract oxygen out of water, and no gill slits (no openings) of any kind. These are not gill slits. There are no slits and no gills. More recent careful research has disclosed that the upper fold contains the apparatus that will later develop into the middle ear canals, the middle fold will later become the parathyroid, and then the bottom fold will grow into the thymus gland (Figure 11.6). [11-30] Organs in mammals do not develop in the same order as they do in the smaller creatures. In the earliest fishes, (from which mammals, including humans, were claimed to have evolved from,) there are teeth but no tongue. But in the mammalian embryos, the tongue develops before the teeth. [11-31]

Evolutionists promote theoretical evidence for gill slits, yolk sac, and tail (just to name a few) in a human embryo (Figure 11.6). In the early stage of development the human fetus has folds or creases that resemble those found in a fish embryo. As these folds and creases develop, however, the resemblance ends. In the fish, the folds develop into gills. In humans, the folds develop into the glands and structures in the ear and neck areas (Figure 11.6). If humans were related to fish, the gills would evolve into the lungs, trachea, and mouth. Similarly, the embryonic human "tail" is really the development of the coccyx (tail bone), which is a fully human feature. The so-called yolk sac in humans is not a source of nourishment as in a bird egg, but is the source of the human embryo's first blood cells (Figure 11.6). Everything about the human embryo is totally unique and human. (See Chapter 12, Vestiges, for further discussion of the coccyx.)[11-32]

As discussed previously in sections "Ernst Haeckel's Biogenetic Law" and "Ernst Haeckel's Drawings", Haeckel theorized that during its development in its mother's womb, the human embryo first displayed the characteristics of a **(1)** fish, and then those of a **(2)** reptile, and **(3)** finally those of a human. It has since been proven that this theory is incorrect. It is now known that the "gills" that supposedly appear in the early stages of the human embryo are the initial phases of the middle-ear canal, parathyroid, and thymus (Figure 11.6). That part of the human embryo that was similar to the "egg yolk pouch" (Figure 11.6) is really a pouch that produces blood for the human infant. The part that was identified as a "tail" (Figure 11.6) by Haeckel and his followers is the backbone, which resembles a tail only because it takes shape before the legs do.[11-33]

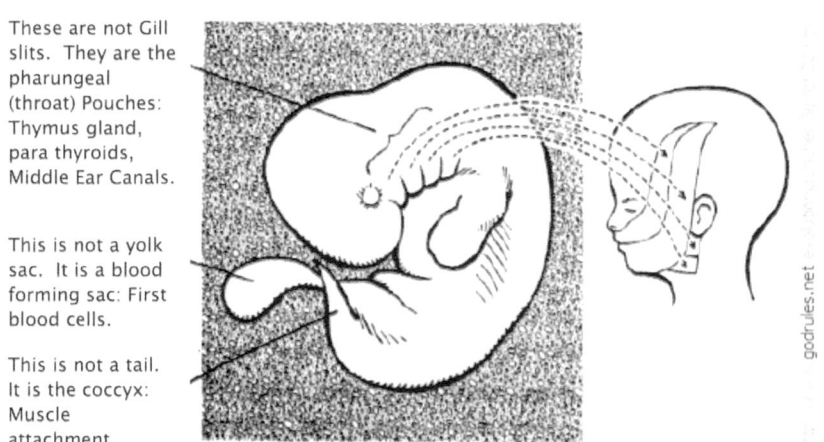

These are not Gill slits. They are the pharungeal (throat) Pouches: Thymus gland, para thyroids, Middle Ear Canals.

This is not a yolk sac. It is a blood forming sac: First blood cells.

This is not a tail. It is the coccyx: Muscle attachment.

Figure 11.6 Development of the Human Embryo.[11-34]

Gill Slits—A one-month human embryo has wrinkles in the skin where the "throat pouches" grow out. Occasionally, one of these pouches will break through, and a child will be born with a small hole in the neck. If these pouches were literally gills there would be blood vessels all around them, for absorbing oxygen from water, as a gill does. But there is no such structure in humans of any age. Humans do not have the DNA instructions for forming gills. The gill slits are actually the throat (or pharyngeal) grooves and pouches (Figure 11.6). They are not mistakes in human development (or leftovers of gill slits) but become essential parts of human anatomy. The first pouches form the palatine tonsils that help fight disease. The middle ear canals come from the second pouches, and the parathyroid and thymus glands come from the third and fourth (pouches). The thymus prepares T cells, the immune cells necessary for fighting all diseases. Without the parathyroids, we would be unable to regulate calcium balance and could not even survive. Another pouch, thought to be vestigial by evolutionists until just recently, becomes a gland that assists in calcium balance.

Yolk Sac—In chickens, the yolk contains much of the food that the chick depends on before hatching. But humans develop differently. They are attached to their mothers, and are nourished directly from their mothers. The so-called "yolk sac" is the source of the human embryo's first blood cells, and death would result without it. In the adult, the blood cells form inside the bone marrow (Figure 11.6). The blood cells are very sensitive to radiation damage, and bone would offer the most protection. Blood is needed in order to form the bone marrow that later is going to form blood. The main building materials for making it (the blood) are DNA (deoxyribonucleic acid) and protein. Because the yolk sac is located outside the embryo, it is easily discarded after it has served its temporary, but vital, function.

Human Tails—The nervous system of humans begins as an open system along the back. As it develops it forms ridges and then closes. It closes in the middle first then advances toward the ends. Sometimes it doesn't close at the bottom. This produces a serious birth defect called "spina bifida" (discussed in the next section). Spina bifida results when during the development stage of an embryo, the nervous system starts stretched out open on the back. Then during its development, it rises up in ridges and rolls shut. It starts to zipper shut in the middle first, then it zippers toward either end. Sometimes it closes beyond the end (Figure 11.6). Then the baby will be born with a fatty tumor that evolutionists call a tail. In reality it is skin and a little fatty tissue that the doctor cuts off. It is nothing like a true tail that has muscle, bones, and nerve, so cutting it off is not complicated.[11-35] To summarize, every stage in the development of an embryo plays a crucial role in embryonic growth. There are no redundant vestiges of former evolutionary phases.

Human Tails are Not the Result of One Who Has Spina Bifida

Evolutionist View

Some creationists claim that spina bifida occurs when the end of the tailbone closes beyond the end. They claim that when a baby is born with spina bifida, it has a fatty tumor that evolutionists call a tail. This is false. There is a difference between a true tail and something from spina bifida. True tails in reported cases lack cartilage and vertebrae, but they are still similar to mammal tails. True tails are when the tail in the fetal stages does not go away like it is supposed to. These creationists claim that spina bifida is simply skin and a little fatty tissue that a doctor cuts off. They claim that spina bifida is not like having a true tail that has muscle, bones, and nerve. This is also not true. They confuse pseudo-tails with true tails. True tails have nothing to do with spina bifida. Some creationists further claim that evolutionists believe that humans have a "tailbone" (also called the sacrum and coccyx) and that it is an indication that humans evolved from apes. This claim is also incorrect. These creationists also falsely claim that the human tailbone is not useless. They claim that the sacrum and coccyx are among the most important bones in the whole body. They claim that the human tailbone forms a crucial point of muscle attachment that is required for humans to have an upright posture and is useful for defecation. This too is false. The muscles are typically muscles once used for tail function and are not specifically used to keep upright posture. People with functional true tails can wag their tails because that is what that bone and muscle group was originally for.

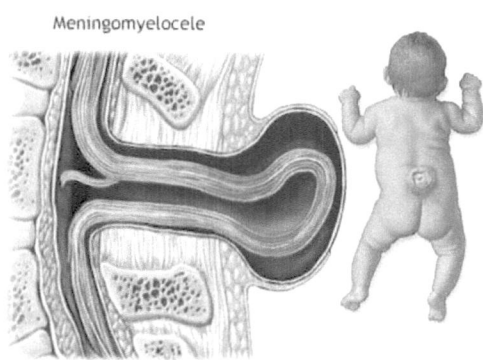

Meningomyelocele

An Example of Spina Bifida

Figure 11.7—Spina Bifida[11-36]

Myelomeningocele is the most common type of spina bifida. It is a neural tube defect in which the bones of the spine do not completely form, resulting in an incomplete spinal canal. This causes the spinal cord and meninges (the tissues covering the spinal cord) to stick out of the child's back (Figure 11.7). That is what most cases of spina bifida look like. But there is a difference between spina bifida versus someone that is born with a true tail. Most pseudo-tails, or true tail cases have nothing to do with spina bifida. A true tail will project out from the very bottom of the spinal column, being an extension of the coccyx or tailbone (Figure 11.8). A tail caused by spina bifida will be attached somewhere higher on the back in the Limbaugh region (Figure 11.7).[11-37] A true human tail is a rare occurrence and is defined as a caudal, vestigial, midline protrusion with skin covering connective tissue, muscle, vessels and nerves. A true case of human tail in a child, which is a very rare occurrence, has been reported (Figure 11.8). Humans have a tailbone (the coccyx) attached to the pelvis, in the same place which other mammals have tails. The tailbone is formed of fused vertebrae, usually four, at the bottom of the vertebral column. It doesn't protrude externally, but retains an anatomical purpose: providing an attachment for muscles like the gluteus maximus.[11-38]

The coccyx, or tailbone, is the remnant of a lost tail. All mammals have a tail at one point in their development. In humans, it is present for a period of 4 weeks, during stages 14 to 22 of human embryogenesis. This tail is most prominent in human embryos 31-35 days old (Figure 11.5). The tailbone, located at the end of the spine, has lost its original function in assisting balance and mobility, though it still serves some secondary functions, such as being an attachment point for muscles, which explains why it has not degraded further. In rare cases a short tail can persist after birth.[11-39]

The true, or persistent, vestigial tail of humans arises from the most distal remnant of the embryonic tail. It contains adipose and connective tissue, central bundles of striated muscle, blood vessels, and nerves and is covered by skin. Bone, cartilage, notochord, and spinal cord are lacking. The true tail arises by retention of structures found normally in fetal development. It may be as long as 13 cm (5.12 inches), can move and contract, and occurs twice as often in males as in females. A true tail is easily removed surgically, without residual effects. It is rarely familial (that is, it rarely occurs in more members of a family than expected by chance alone). Pseudo-tails are varied lesions having in common a lumbosacral protrusion and a superficial resemblance to persistent vestigial tails. The most frequent cause of a pseudo-tail was an anomalous prolongation of the coccygeal vertebrae. Additional lesions included two lipomas, and one each of teratoma, chondromegaly, glioma, and a thin, elongated parasitic fetus.[11-40] A distinction between a true and pseudo tail in lumbo-sacral region is important since treatment and prognosis are different. The occurrence of true tail in a neonate (newborn infant child from birth to end of first year of life), is rare.

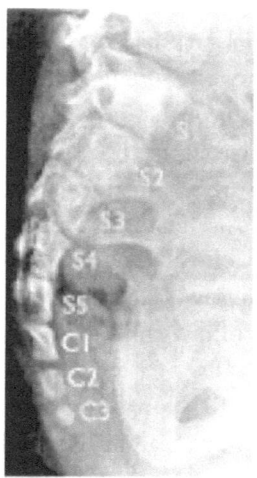

Figure 11.8 X-ray image of an atavistic tail found in a six-year old girl.[11-41]

Figure 11.8 is a radiogram of the sacral region of a six-year old girl with an atavistic tail. The tail was perfectly midline and protruded to form the lower back as a soft appendage. (The five normal sacral vertebrae are indicated as S1-S5 in Figure 11.8. The three coccygeal tail vertebrae are indicated as C1-C3 in Figure 11.8.) The entire coccyx (usually three or four tiny fused vertebrae) is normally the same size as the fifth sacral vertebrae.[11-41] Humans actually have the tail genes, the same genes we find in Chimps and other mammals with tails.[11-42] Scientists have actually discovered the tail genes inside the human genome (the Wnt-3a and Cdx1 genes). Humans contain both the gene to develop tails along with apoptosis (programmed cell death) that plays a significant role in removing the tail while humans are still in the embryo form. The tail genes are retained from distance ancestors to humans and apoptosis was adapted later during the course of our ancestors' evolution.[11-43]

Creationist View

The development of a "tail" is indeed contributed to a condition known as spina bifida, contrary to what some evolutionists claim. Spina bifida is a complex birth defect that affects the development of a baby's spinal cord, spine and developing brain. In utero, the baby's spine fails to close in the first few months of its fetal life. In the most common forms, there is an abnormal development of the backbones, spinal cord, surrounding nerves and/or the fluid-filled sac that surrounds the spinal cord. The abnormality can develop anywhere along the spine, and can cause a portion of the spinal cord and surrounding structures to develop outside, instead of inside, the baby's body. Scientists do not know for sure what causes spina bifida, but they think that genetic, nutritional and environmental factors may all play a role. Researchers believe that a deficiency of folic acid (a B vitamin) in the mother may be a strong contributor. And having one child with spina bifida increases the chance of having a second child with the same condition.[11-44]

There are two types of tails involved in a discussion of human tails, the non-bony tail and the bony tail. **Non-bony tail**—A baby's "tail", like nearly all cases of human "tails", is not a real tail. And it is not evidence of evolution. It does not have any bones in it, doesn't have fur like a monkey's tail, and does not have a nerve cord. A baby's tail does not have anything to do with the idea that humans and monkeys may be related. It is just skin and fatty tissue, and can easily be cut off. **Bony tail**—The second type of "tail", a rarer type, is one that has bone in it. A few evolutionists claim that this is evidence that humans evolved from creatures with tails. Even if every human had a bony tail a few inches long, it could not possibly show that humans have evolved from monkeys. That is because the existence of something is not evidence of its origin.[11-45] (See Chapter 12, section "Human Vestigial Structures" Creationist View Point #2 for further discussion of the human tailbone.)

In the next chapter (12) we will look at some vestiges to help determine whether or not they provide evidence for the ancestry of species.

～～～

Chapter 12

Vestiges

Ostrich

Creationism cannot explain why a designer would give useless or unneeded organs and structures to living things.[12-1]

Rudimentary organs . . . are either quite useless, such as teeth which never cut through the gums, or almost useless, such as the wings of an ostrich, which serve merely as sails.

> • Charles Darwin, Origin of Species,
> Chapter 14—Mutual Affinities of Organic Beings:
> Morphology—Embryology—Rudimentary Organs

On the view of descent with modification, we may conclude that the existence of organs in a rudimentary, imperfect, and useless condition, or quite aborted, far from presenting a strange difficulty, as they assuredly do on the old doctrine of creation, might even have been anticipated in accordance with the views here explained.

• Charles Darwin The Origin of Species
by means of Natural Selection or The Preservation
of Favoured Races in the Struggle for Life,
Chapter 14 - Mutual Affinities Of Organic Beings:
Morphology—Embryology—Rudimentary Organs, 6th edition

Introduction

In this chapter (12) we will look at some organisms that appear to serve no function at all or have some limited function not originally intended. We will also discuss whether or not these vestige organisms are in fact useless, and also whether or not these organisms, claimed to be useless, indicate an evolutionary past. Only a few examples will be discussed, as the same concepts apply to any organ considered to be vestigial.

Evolutionist View

And why did God – sorry, the Intelligent Designer – give whales a vestigial pelvis, and the flightless kiwi bird tiny, nonfunctional wings? Why do we carry around in our DNA useless genes that are functional in similar species? Did the Designer decide to make the world look as though life had evolved? What a joke! And the Designer doesn't seem all that intelligent, either. He must have been asleep at the wheel when he designed our appendix, back, and prostate gland.[12-1a]

... if animals were specially created, why do they have "parts that are quite useless and often in fact positively detrimental to them?" and why should a human possess a system of functionless muscles for his external ear, a useless hairy covering before birth, and a worse than useless vermiform appendix? No advocate of the theory of special creation has ever been able to give a satisfactory answer to these questions.[12-1b]

Vestigial organs are said to be remnants from the evolutionary past. Since evolution is a bottom up designer, it can only modify existing structures. Examples of vestigial organs include the wings of flightless birds, the limb girdles of snakes, the appendix and the ear muscles of humans, and the scale leaves of parasitic flowering plants. Another example of vestiges is the pelvic girdle of humpback whales. All whales and dolphins have vestigial pelvic girdles. These structures are no longer useful for locomotion, and are not attached to the vertebral column, as they are in fossil ancestors of whales. (See Chapter 7, section "The Evidence That Whales Evolved from Land Mammals" for further discussion of whale vestiges and whale evolution.)

Useless Body Parts—Leftovers From Ancestors

Evolutionist View

A vestigial structure is defined as a part or organ that was well developed in ancestral forms, but the size and structure of which have diminished until it no longer has its original function. In order for a vestigial structure to be considered genuine the part in question must serve no contemporary useful purpose. A vestige may also be defined as a reduced and simple structure compared to the same complex structure in ancestral organisms. Some of the best evidence for evolution is the various nonfunctional (or simple) vestigial traits. Vestigial traits, if functional, complete relatively simple, or unessential functions. They use structures that were clearly once used for other complex purposes. Many vestigial organs have no function. Ostrich wings are an example of a vestige with minimal function. The vestigial wings of ostriches may be used for relatively simple functions, such as balance during running, or for protection, however, they are not used for the original purpose, which was flight. They are useless as wings.[12-2]

The human body is full of relics of antiquity. This characterization is justified by those vestigial and rudimentary structures that represent organs of value to human ancestors among the lower animals, though they play a less active part at the present time in human economy. A few structures will be discussed. As compared with those of the apes, the human wisdom teeth are degenerate; in the gorilla they are cut at the same time as the other molars and in the lower human races (pre-human hominids) come through the gums in early youth, while in the more advanced Caucasian races the wisdom teeth are cut only in later life or not at all. The reduced appendix of man, a source of much ill health, is another structure that is a counterpart of a relatively larger and useful part of the digestive tract in the lower primates and other animals. Furthermore, the human tail is a reality, not a fiction.

Occasionally, an individual is born with a tail that may reach a length in later life of eight or ten inches; such structures are, of course, abnormal. (See Chapter 11, section "Human Tails are Not the Result of One Who Has Spina Bifida" for further discussion of human tails.) But in every normal human being there is a series of little bones at the lower end of the vertebral column, constituting the coccyx, and this is just where the abbreviated tail of the ape and the still longer prehensile tail of the monkey arises from the body. Unless the coccyx is a tail, what can it be? And if it does not represent a reduced counterpart of the tails of other mammals, what does it represent?[12-3]

Creationist View

Vestigial structures, as defined by evolutionists, are: body structures that have lost their original function in a present-day organism but were probably useful to an ancestor. Evolutionists say that these structures are evidence for evolution as they show structural change over time. It is debatable as to whether apparent vestigial structures have truly changed their function.

Human Vestigial Structures

Evolutionist View

1). Male nipples. Men have nipples because females need them, and the overall architecture of the human body is more efficiently developed in the uterus from a single developmental structure. **2). Coccyx.** The human tailbone is all that remains from our common ancestor's tails, which were used for grasping branches and maintaining balance. (See Chapter 11, section "Human Tails are Not the Result of One Who Has Spina Bifida," "Evolutionist View" for further discussion of human tails.) **3). Appendix.** This muscular tube connected to the large intestine was once used for digesting cellulose in our largely vegetarian diet before we humans became meat eaters. The human appendix is a classic example of a vestigial organ. The appendix was used by ancient ancestors to help with the digestion of tough roots. Since tough roots have not been eaten for tens of thousands of years, the appendix has lost its original function.[12-4]

Creationist View

Evolutionists claim that evolution is not a perfect process. As environmental changes select against certain structures, others are retained, sometimes persisting even if they are not used. A structure that seems to have no function in one species, yet is homologous to a functional organ in another species, is termed vestigial.[12-5]

In response to **point #1** above, evolutionists claim that vestigial organs in humans include: **(1)** vestiges of the reproductive structures, **(2)** nipples in men, **(3)** vestiges (in the female) of the Wolffian duct. (*This duct, also called the Archinephric Duct [or the mesonephric Duct, or the Leydig's Duct, or the Nephric Duct], is one of a pair of embryogenic tubules that carry urine from primitive or embryonic kidneys to the exterior or to a primitive bladder in mammals, including humans during embryogenesis. It is an organ that [evolutionists claim] persists in the female chiefly as part of a vestigial organ and in the male as the duct system leaving the testis and including the epididymis, vas deferens, seminal vesicle, and ejaculatory duct.*), and **(4)** vestiges (in the male) of the Mullerian ducts. (*These ducts, also called the Paramesonephric ducts, are paired ducts of the embryo that run down the lateral sides of the urogenital ridge and terminate at the sinus tubercle in the primitive urogenital sinus. In the female, they will develop to form the uterine tubes, uterus, cervix, and the upper one-third begins of the vagina. In the male, they are lost.*) These structures, however, clearly reflect the embryonic development of a sexually dimorphic organism which its development in a sexually indifferent condition with structures characteristic of both sexes. They certainly do not reflect phylogenetic development. No one supposes males evolved from females or vice versa.[12-6] Males have nipples because of the common plan followed during early embryonic development. Embryos start out producing features common to male and female.[12-7]

In response to **point #2** above, evolutionists claim that humans have a "tail bone" (also called the sacrum and coccyx), and that it is an indication that humans evolved from apes, however, this is highly debatable. The human tailbone is not useless. In one sense, the sacrum and coccyx are among the most important bones in the whole body. They form a crucial point of muscle attachment required for our distinctive upright posture (and also for defecation). Therefore, far from being a useless evolutionary leftover, the "tail bone" is quite important in human development. It is true that the end of the spine sticks out noticeably in a one-month embryo (Chapter 11, Figures 11.5 & 11.6), but that is because muscles and limbs do not develop until stimulated by the spine. As the legs develop, they surround and envelop the "tail bone," and it ends up inside the body.[12-8] The coccyx, the coccyges vertebrae, is the bottom of the spine in humans. Scientists have found that important muscles (the levator and coccyges) attach to those bones. Without those muscles, the human pelvic organs would collapse, that is, would fall down. Without them the human could not have the ability to excrete solid waste, nor could the human have the ability to walk or sit upright.[12-9] (The coccyx is discussed in more detail below in the section "A Look at Some Other Organisms that are Claimed to be Vestiges") (See Chapter 11, section "Human Tails are Not the Result of One Who Has Spina Bifida," "Creationist View" and Chapter 11 on embryonic development, section "What Really Occurs During the Stages of Human Development?" for further discussion of the coccyx.)

In response to **point #3** above, recent studies have determined that the appendix contributes to human immune function. The human appendix, long considered only an accessory rudimentary organ, is now considered to play an important function in the immune system, especially in early childhood. Today medical researchers recognize that the appendix is clearly a functional organ in humans and therefore cannot be considered as vestigial.[12-10] The appendix is part of the immune system, strategically located at the entrance of the almost sterile ileum from the colon with its normally high bacterial content.[12-11] Science has discovered that the appendix is not useless in humans. It helps protect humans from gastrointestinal problems in the lower ascending colon. The appendix is now known to be an important part of what is called the "reticulo-enabthelial system" of the body. Like the tonsils, the appendix fights infection. The tonsils and appendix are not functionless, as evolutionists claim. The human alimentary canal is a long tube leading from mouth to anus. Near each opening, there is an organ called the alimentary canal, whose function is to protect the entire gastrointestinal tract from pathogenic invasion while an infant. The appendix is crucial during the first months, and tonsils are crucial during the first several years. In later years, one does not have as urgent a need for either tonsils or appendix as one did as a small child. Both tonsils and appendix are now believed to guard against Hodgkin's disease.[12-12] The appendix is the special body structure pointed to by evolutionists as a prime example of a vestigial organ which is no longer used. If that is true, then we should be able to trace our ancestors through it in a direct line. Which other animals have an appendix? Here they are: apes, rabbits, wombats, and opossums. Four are totally different from each other. Which one descended from which?[12-13] Apes possess an appendix, whereas their less immediate relatives, the lower apes, do not; but it appears again among the still lower mammals such as the opossum. Evolutionists cannot explain this.[12-14] (The tonsils and the appendix are discussed in more detail below in the section "A Look at Some Other Organisms that are Claimed to be Vestiges")

A Look at Some Other Organisms that are Claimed to be Vestiges

Creationist View

In the early 1900s, over 100 organs in the human body were declared vestigial organs. Several recent biology textbook authors still claim there are about 100 vestigial organs in the human body, but they only list five or six. It has been said that as our knowledge has increased, the list of vestigial organs has decreased.[12-15] Medical research has found uses for every one of the claimed supposed vestigial organs.[12-16] If humans evolved from lower forms of life, evidence of organs or structures that functioned in our evolutionary past, but not in the present, should be seen in our bodies. For this reason, Darwin and other early evolutionists looked for examples of organs that were useful in lower level animals but not in more advanced animals. All of the examples of vestigial organs cited by Darwin and others have now been shown to be functional, and many, such as the thymus gland, are now known to be critically important.[12-17]

Vestigial organs

It has been said that vestigial organs are a major proof for evolution because nothing else can explain the existence of many useless organs.[12-18] However, it will be illustrated that the assumed vestigial organs are not useless or relics of past ancestral species but are useful and beneficial.

The most common definition of a vestigial (or rudimentary) organ is: "Structures in living creatures, including humans, that have no useful function but which represent the remains of organs that once had some use." Vestigial organs has more recently been described as: "A vessel that has lost its original function in the course of evolution, and is usually much reduced in size. A remnant of a structure that functioned in a previous stage of a species' evolution."[12-19]

The penguin flippers

Evolutionists believe that the penguin was once able to fly but have lost their ability to fly because they eventually did not need to fly and subsequently lost their ability to fly. (This is Lamarckian' view – use and disuse of organs.) There is no fossil evidence that penguins were once able to fly. If penguins lost their ability to fly due to non-use, the species would have died out. Mutations that reduced efficient wing use would have impeded survival, and therefore would have been selected against survival. Any mutations that aided swimming would be positively selected to survive. This would, theoretically, result in the improvement of both skills (flying and swimming), not the loss of one. The paddle-like flippers do not function as wings, but function quite well to enable penguins to swim. Penguin flippers are not vestigial, nor is there evidence (fossil or otherwise) that penguins were once able to fly.[12-20]

The human appendix

The appendix has been regarded for a long time as being a vestigial organ. The appendix is perhaps the best example of a vestigial organ, mainly because it was determined to be useless in humans. Because the appendix was not known to be of any useful value in humans, the appendix was often cited as one of the strongest evidences of evolution and disproves creationism. Evolutionists believe that human ancestors [apes] ate a much coarser diet and is not needed in humans. It is now known that the human appendix is useful in many ways. In a healthy human body, there are more bacteria than human cells. Most of these bacteria are beneficial and serve several functions such as to help digest food. Sometimes the intestinal bacteria are purged, such as by antibiotics or diseases such as cholera or amoebic dysentery. One of the appendix's jobs is to reboot the digestive system when this occurs. The very features of the human appendix that caused people to believe it was useless or worse (it sometimes becomes inflamed and needs to be removed) turn out to be important for its function. For example, the small worm-like shape of the appendix restricts access, thus protecting the bacteria inside of it, enabling it to function as a 'safe haven' for good bacteria to thrive.

Tissue studies have determined that the human appendix wall is thickened by an extensive development of lymphoid tissue which forms an almost continuous layer of large and small lymphatic nodules. Because the appendix is situated near the junction of the small intestine and the colon, the appendix has early access to antigens entering the cecum. Its location also supports the evidence that it protects the almost sterile ileum part of the intestine from infection in the cecum region where the colon begins. The large intestine has billions of bacteria, and if they travel into the ileum, E. coli or other infections could result. Antibodies are often manufactured close to the organs where they are used, such as along the entire gut lining. This explains why the human appendix has masses of lymphatic tissue in its walls and seems to provide a local defense against infection from microorganisms in the colon. The small intestine is sterile, but the large intestine contains an enormous amount of bacteria that can back-up into the small intestine at the cecum through the ileocecal sphincter.

The appendix has also been determined to help in fighting the effects of post-radiation infection. Death after radiation overdose is usually not directly caused by the radiation, but by subsequent infections due to an impaired immune system. If radiation damage reaches a certain level, antibody production in the spleen is temporarily impaired. After radiation, lymphoid cells (such as those that might exist in the appendix if it has been shielded) migrate to the impaired spleen. There they manufacture antibodies until the spleen can recover enough to take over this role again. Lack of an appendix may significantly increase the danger of problems occurring after radiation exposure. Studies have indicated that the appendix may also play a role in cancer prevention. In a study of several hundred leukemia, Hodgkin's disease, colon cancer and ovarian cancer patients it was discovered that over 80% of those who had these diseases had their appendix removed prior to contracting the disease.[12-21]

Tonsils

In the 1930s over half of all children had their tonsils removed partly because it was felt they were useless organs that only caused problems later. It was reasoned that it was best to remove them when young rather than wait.[12-22] The major reason why tonsils were removed was to prevent throat infection. When the throat is irritated for any reason, the tonsils also tend to become inflamed in response to the nearby throat infection.[12-23] After it was determined that the tonsils are an important part in the immune system, the body's first line of defense against bacteria, the number of tonsillectomies dropped drastically. By 1971, the frequency was 14.6 per 1,000 children, and it is currently (as of 2019) only 0.53 per 1,000 children. Today tonsillectomies are usually not done except to treat severe problems.[12-24]

Gill slits

Most Darwinists once believed that as the human embryo developed, it passed through most of its major past evolutionary stages, proving that humans evolved from fish. Many Darwinists still believe that the human embryo still passes through the major evolutionary stages. Although Darwinists have dropped most of the historically accepted ontogeny stages, they sometimes claim that a few evolution stages are revealed in embryological development. [The phrase: "ontogeny (individual development) recapitulates (repeats) phylogeny (evolutionary descent)" later became known as "recapitulation" or the "biogenetic law."] One of the most commonly cited examples is the fish stage. The gill-slit argument is used in an attempt to claim that the biogenic law is still valid. Darwinists claim that all mammal embryos pass through a stage in which they have gills like a fish, illustrating that mammals are descended from fishlike ancestors. In reality, these views are not true. The claim that mammals have gill slits is important because it is the major evidence for the fish stage evolutionists have claimed exist in human embryo in early developmental stages.[12-25] It has been determined that pharyngeal clefts of vertebrate embryos are neither gills nor slits.[12-26] Not a single evolutionary transitional, atavistic, or vestigial organ exists in any stage of embryological development.[12-27]

The so-called gill slits are neither slits nor gills, but actually are endoderm that includes epithelial tissue located in the neck region of the embryo which forms a set of alternating pouches and ridges which are also called grooves, folds, or creases. Although they superficially resemble the structure in fish that develops into gill slits, the supposed human "gill slits" are in the neck and throat area; in fish the "gill slits" are located on the side of the head adjacent to the neck area. In fish, the gill slits are literally slits that form openings to allow water in and out of the internal gills that remove oxygen from the water. The gill-slit region in humans does not contain even partly developing slits or gills, and has no respiratory function. Human pharyngeal pouches (called gill-slits by many evolutionists) develop into structures that become part of the face, the ear cavities, bones of the middle ear, muscles of mastication and facial expression, the lower jaw, certain neck parts, and the thymus, thyroid, and parathyroid glands. (See Chapter 11, Figure 11.6, for illustration.) As has been discussed, the term "gill-slits" is inaccurate, outdated, and very misleading and should not be used when discussing the pharyngeal pouches, thymus and thyroid areas. The term "pharyngeal folds" (or "pharyngeal pouches") should be used instead because it is accurate and is not misleading.[12-28] (See Chapter 11, section "What Really Occurs During the Stages of Human Development?", Creationist View, for further discussion of gill slits.)

The spleen

The spleen was also once thought to be a useless, rudimentary organ in humans until research has determined that the spleen is very functional and necessary. The spleen, located next to the pancreas, which is next to the liver, is now known to help locate and fight infections and remove damaged or worn-out red blood cells. If the spleen is removed, another organ will usually take over some of its functions. Before birth, and for a short time afterwards, the spleen produces various types of blood cells. In adulthood, the spleen usually does not do this. This is why some Darwinists have labelled the spleen as being vestigial. The spleen contains ten times as much monocytes as the blood. This is important for those who have a heart attack, as one who has had a heart attack requires healing that depends on monocytes. The spleen also produces opsins, useful in marking antigens to help the immune system. The spleen produces properdin convertase, a protein family that activates other proteins to do their job of protecting the body. The spleen also produces tuftsin, a tetra-peptide that has an immune-stimulatory effect, useful for overall health. One who does not have a spleen are more likely to die of heart disease, pneumonia, and other diseases. The spleen, once regarded as a useless vestige by many evolutionists, has now been documented to have numerous important roles in the body, including immunological, blood storage and quality control functions. The useless organ view has hindered looking for the actual function of numerous organs and structures, and the spleen is just one more example.[12-29]

The yolk sac

The yolk sac has long been regarded as a vestigial organ and has commonly been used as evidence for Darwinism. It is now known that the yolk sac serves several critical functions, including the formation of blood cells, sex cells, and a network of vessels that provides the embryo with both nutrients and serves as a waste-disposal system. Since the yolk sac does not contain yolk, it is now properly referred to as an umbilical vesicle. Part of the membrane derived from the yolk sac becomes incorporated into the umbilical cord, and the remainder lies in the cavity between the chorion and the amnion near the placenta. The allantois forms during the third week as a tube extending from the early yolk sac into the connecting stalk of the embryo. It, too, forms blood cells and gives rise to the umbilical arteries and vein.

The yolk sac is useful during organogenesis. (Organogenesis is the phase of embryonic development that starts at the end of gastrulation and continues until birth. See Chapter 11, section "What Really Occurs During the Stages of Human Development?" for further discussion of embryonic development.) As development progresses, the dorsal part of the yolk sac is incorporated into the developing embryo and gives rise to the epithelium of the trachea, bronchi, lungs and digestive tract. During the development of the body's organs and before the placental circulation, the yolk sac is the primary source of exchange between the embryo and the mother. The yolk sac has nutritive, endocrine, metabolic, immunologic, secretory, and hemopoietic functions that serve in the transport of critical substances from the mother to the fetus at a time when the yolk sac is the sole or main maternofetal transport system. The yolk sac forms early in development because of its critical importance to the embryo. The yolk sac is now known to have several critical functions in the developing embryo, including organ formation, gonocyte production, and hematopoiesis. As the organism develops, the yolk sac's roles are taken over by other structures in the developing embryo. Not long after the bone marrow develops, the yolk sac almost totally disappears. The proper term used today for the structure is not yolk sac but "umbilical vesicle." This term describes it quite well because it is a vesicle extending from the umbilical cord away from the embryo and toward the placenta.[12-30] (See Chapter 11, section "What Really Occurs During the Stages of Human Development?", Creationist View, for further discussion of the yolk sac.)

Outer Ear

Charles Darwin, and the majority of evolutionists, believe that since they believe that humans evolved from a lower form, such as a primate that could move its outer ear (pins or auricle) to help direct sound waves into the ear canal, Charles Darwin considered the human outer ear as being a functionless remnant of the much larger outer ear that our supposed evolutionary ancestor possessed. Evolutionists believe that the outer ear in humans became greatly reduced in size and has become ineffective as a funnel for concentrating sound waves. If this claim is true, it would be evidence of the loss of function, which would be devolution, not evolution. In contrast to the evolutionist's claim, the external ear is a well-designed structure that effectively collects sound in the frequency ranges most important to humans. It also directs these frequencies toward the external auditory meatus, commonly called the ear canal. Since humans are able to move their necks easily, there is no need for the outer ear to have the ability to move.[12-31]

Ear Muscles

Evolutionists believe that the three ear muscles, the attrahens, the retrahens, and the attollens auriculum, are vestigial, useless leftovers from our early ancestors. They claim these ear muscles have become degenerated in humans. The question is, why would this trait be selected against in humans?[12-32]

Darwin's Point

The "Darwin's Point" (also known as "Darwin's bump", "Darwin's tubercle", or "Woolner's tubercle") has been proclaimed by Charles Darwin and many evolutionists as being a leading proof that humans evolved from lower life forms. One major problem is that our supposed closest ancestors, the chimps and most other higher apes, lack a pointed ear as well as a Darwin's Point. Neither gorillas, orangutans, gibbons nor dwarf chimps called bonobos or most large primates have pointed ears. The few primates with a pointed ear include macaques and yellow baboons. The trait can be bilateral, present on both ear auricles, or asymmetrical, present on one ear only. There seems to be no relation between the presence or absence of Darwin's Point and evolution; it seems, instead, to be part of the tremendous variation found everywhere in the natural world.[12-33]

The coccyx

All lower primates have tails and the human coccyx is interpreted by Darwinists as a rudimentary tail left over from our distant evolutionary past. A major problem with the conclusion that the coccyx is evidence of evolution is that our nearest relatives do not have tails. All of the major primates including chimpanzees, gorillas, orangutans, bonobos, gibbons or the lesser apes such as siamangs lack tails. Only a few of the over 100 types of monkeys and apes, such as the spider monkeys, have tails. The primates that have tails tend to be the small cat-like lemurs, tarsiers and loris. The major function of the coccyx is an attachment site for the series of interconnected muscle fibers and tissues that support the bladder neck, urethra, uterus, rectum, and a set of structures that form a bowl-shaped muscular floor collectively called the pelvic diaphragm. The coccyx muscle system expands and contracts during urination and bowel movements and also distends (swells or bloats) to help enlarge the birth canal during childbirth.

True tails, believed by Darwinists to be remnants of embryological tails, contain mesenchymal tissue (adipose, muscle, connective, blood vessels, cutis and nerve tissue). Instead of curving toward the midline of the body as occurs in normal development, the coccyx bones in true tails curves away from the body midline. So, instead of a true tail, what actually often occurs is an abnormal development of the coccyx. Of the few cases that have been declared as true tails, the tails lack bone, cartilage, notochord, and spinal cord elements. Of the few cases that had a true tail, very few could actually move. The evidence indicates that human tails, both true tails and pseudo tails, are all birth abnormalities.[12-34] (See Chapter 11, section "Human Tails are Not the Result of One Who Has Spina Bifida" for further discussion of the coccyx.)

Editor's Note: Only the most popular organs that are considered to be vestiges by evolutionists are included here, although there are many other organs and traits in humans that have been determined by evolutionists to be useless. To include the vast majority of organs considered to be vestiges would require a great deal of additional pages. Please refer to the resources listed in the References section for Chapter 12 for further study.

In the next chapter (13) we will continue our investigation of whether or not different species are related to other species (and thereby evolved from related ancestors) by taking a look at pseudogenes to determine if there is any evidence that will provide clues regarding ancestry.

~~~

Chapter 13

Pseudogenes

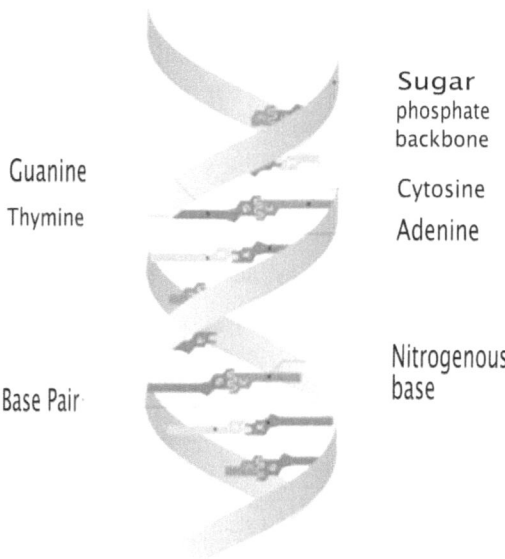

DNA Double Helix Strand

The DNA (Deoxyribonucleic acid) double helix model illustrates how DNA and base pairs are arranged on a chromosome. It is called a "double helix" because there are two intertwining spiral coils that are attached by molecules (base pairs), similar to rungs or steps on a ladder. The chromosomes are located inside the nuclei of every living cell (Chapter 5). Sections of a DNA strand are called genes (Chapter 5, Figure 5.5 lower left, 5.19).

DNA is helix shaped because the DNA contains an extreme amount of code and must fit inside the chromosome (Chapter 5, Figure 5.5 lower left, Figure 5.19)[13-1] A molecule consists of four nucleotide units: **(1)** one containing adenine, **(2)** one guanine, **(3)** one cytosine, and **(4)** one either thymine (in DNA) or uracil (in RNA). The sides of the double helix consist of alternating deoxyribose sugars and phosphates (base pairs) (Illustration above and Chapter 5, Figure 5.19). Adenine and thymine base pairs are attached to the DNA sugar phosphate backbone (the curled double-helix) like rungs on a ladder. Every other rung consists of guanine and cytosine.[13-2] (Adenine, thymine, guanine, and cytosine are discussed in Chapter 5.)

Introduction

In Chapter 5 we discussed how amino acids, DNA, genes, chromosomes and cells are used in living things. In Chapter 9 we discussed how mutations develop and what results from mutations. In this chapter (13) we will discuss the concept that some genes are said to have been neutralized (became useless) by mutations; that is, they are no longer able to create proteins. We will also discuss the evidence that provides clues that these alleged useless pseudogenes are evidence that indicates various species arose from former, different species.

Genes and Pseudogenes

Evolutionist View

A gene is the molecular unit of heredity of a living organism. Genes comprise some sections of deoxyribonucleic acids (DNA) and ribonucleic acids (RNA) that code for a polypeptide or for an RNA chain that has a function in the organism, although there still are controversies about what plays the role of the genetic material. Living beings depend on genes, as they specify all proteins and functional RNA chains. Genes hold the information to build and maintain an organism's cells and pass genetic traits to offspring. All organisms have many genes corresponding to various biological traits, some of which are immediately visible, such as eye color or number of limbs, and some of which are not visible, such as blood type, increased risk for specific diseases, or the thousands of basic biochemical processes that comprise life.[13-3] (Chromosomes, DNA, RNA and genes are discussed in more detail in Chapter 5.)

Genes are the basic physical units of heredity. Genes are segments of a special molecule called DNA (deoxyribonucleic acid) that serve as a sort of instruction manual for cells. Genes reveal a great deal about the history of an organism and are therefore very important in the study of evolutionary biology. DNA tells each cell how to manufacture the chemicals it needs. In the analogy that every organism can be thought of as a building, the genes would be like the blueprint for the building. Genes explain how the building is to be built. Each gene is like one page of the blueprint describing one particular detail of the organism (Chapter 5).[13-4]

Pseudogenes are broken or nonfunctional genes. A pseudogene looks like an ordinary gene but has one or more defects that prevent the body from using it to make molecules. It is like a page of a blueprint with a big "X" through it indicating that the carpenters should ignore it. An ordinary gene can turn into a pseudogene if a mutation occurs that stops it from functioning. When a species no longer needs a particular gene in order to survive, that gene can mutate into a pseudogene that can remain in the DNA of the species without harming the species. It is questionable that two different animals would have the same pseudogenes in the same place if both creatures were created. Animals with the same pseudogenes indicate they share similar ancestors.[13-5]

Pseudogenes no longer perform their original function, however, the organism itself may still retain that function if it still has the functional counterpart of the pseudogene. The lines of evidence that support the belief that most pseudogenes are nonfunctional are: **First**, the presence or absence of most specific pseudogenes has no measurable effect on a creature as a whole. **Second**, there are good genetic arguments indicating pseudogenes have little, if any, function. Pseudogenes have complex sequences highly similar or identical to those required for the proper function of other enzymatic or structural proteins. Normal genes are actively transcribed and translated into proteins, whereas pseudogenes are not. Pseudogenes cannot perform the functions of the proteins they encode. **Third**, if a pseudogene has little or no function, then most mutations in the pseudogene will have only minor functional consequences, and many mutations will not be weeded out by purifying selection. **Fourth**, it is understood how redundant pseudogenes are developed, and scientists have observed the development of new redundant pseudogenes in the lab and in the wild. Redundant pseudogenes originate by gene duplication and subsequent mutation. Gene duplication rarely creates pseudogenes. These creations would not only be rare but random as well. It is important to note that any duplicated DNA is inherited, including pseudogenes. Thus, finding the same pseudogene in the same chromosomal location in two species is strong evidence of common ancestry.[13-6]

Evidence of common ancestry is provided by molecular examples of DNA sequences known as pseudogenes. Pseudogenes are very closely related to functional, protein-coding genes. The similarity involves both the primary DNA sequence and often the specific chromosomal location of the genes. Normal genes are transcribed into mRNA (messenger RNA), which is in turn actively translated into functional proteins. (Chapter 5, section "How a Protein is Made—Step by Step" discusses mRNA.) Pseudogenes have faulty regulatory sequences that prevent the gene from being transcribed into mRNA, or they have internal stop codons (UGA, UAA, UAG) that keep the functional protein from being made. In this sense, pseudogenes are molecular examples of vestigial structures, discussed in the previous chapter.[13-7]

Pseudogenes and the Production of Vitamin C

Evolutionist View

A human pseudogene of note is GLO, so named because in other species it produces an enzyme called L-gulono-y-lactone oxidase. This enzyme is used in making vitamin C (ascorbic acid) from the simple sugar glucose. Almost all mammals have the ability to make vitamin C, except for primates (which includes humans), fruit bats, and guinea pigs. In these species, vitamin C is obtained directly from food sources. Normal diets provide the necessary amounts of vitamin C. The reason why primates and these few other mammals do not make their own vitamin C is because they do not need to.[13-8] The process for making vitamin C from glucose involves four steps, each developed as the product of a different gene. Primates (which includes humans) and guinea pigs still have active genes for the first three steps, but the last step, which requires the GLO enzyme, does not take place. The GLO in this instance has mutated and no longer functions. It has become a pseudogene. GLO does not work because a single nucleotide in the gene's DNA sequence is missing. This is exactly the same nucleotide missing in other primates. This indicates that the mutation that prevents primates from making vitamin C was passed from the ancestor of all primates onto its descendants.[13-9]

The fact that humans (and other primates) are missing the ability to manufacture vitamin-C is due to their ability to obtain sufficient vitamin C in our diet of fruit. As mentioned above, the reason for the deficiency in humans (and other primates) is that they carry defective copies of the gene that codes for an enzyme that is involved in producing vitamin C. These defective genes, called pseudogenes, could also be called "vestigial genes." (Vestiges are discussed in Chapter 12.) The evolutionary explanation is that the ancestors of humans had functional copies of the gene. (Chimpanzee genes and human genes are discussed in Chapter 15.) Mutations caused the genes to become defective, but these individuals were not discontinued by natural selection because their diets contained sufficient vitamin C, which allowed primates (which includes humans) to continue to exist. (Mutations and natural selection are discussed in Chapter 9.) It is very unlikely that God would have directly created non-functional genes in humans and other primates to give the impression that the one was descended from the other. This is an indication that primates are the ancestors of humans (See Chapter 15 for further discussion).[13-10]

Questions for Creationists

Evolutionist View

If God created humans out of nothing, why would God put a functional gene that produces vitamin C in most animals, but reuse the mutated gene found in chimpanzees for humans? Why would God make 600 of our 1,000 genes for scent receptors defective?[13-11] If the common function theory were true, it is hard to explain why humans, chimps, and other apes would have a non-functional pseudogene in the same location that other mammals have a functional vitamin C gene. But if common ancestry is true, the existence and location of the vitamin C pseudogene makes sense. [13-12] (Common function and common ancestry are discussed in Chapter 15.)

Aren't Junk Genes Just Useless Leftovers from Our Ancestors?

Evolutionist View

For years, scientists had no explanation for why so much of the human DNA doesn't code for proteins. These non-coding parts were dismissed with the term "junk DNA". But gradually this position has begun to look less plausible, for many reasons. One of the biggest surprises occurred in 2001, when the Human Genome Project completed their study of the human genome sequence. The surprise that over 98 per cent of the DNA in a human cell is junk. That is, these portions of DNA don't code for any proteins. The release of the human genome sequence of DNA in human cells by the Human Genome Project in 2001 allowed scientists to know where all the genes for humans are positioned relative to one another, and their sequences. (The genome is the entire sequence of DNA in human cells.) This has made it much faster and cheaper to find the mutations, even in rare diseases.

Previously, scientists thought of proteins as the final end points required for life, but they will never be properly produced and coordinated without the junk. The whole organization only works when all the components are in place. And so it is with the human genome. It has been determined that the (so-called) junk genes have some use. It has been determined that Junk genes prevents our DNA from unravelling and becoming damaged. Other structural regions of junk DNA act as anchor points when chromosomes are shared equally between different daughter cells during cell division. Other junk genes act as insulation regions, restricting gene expression to specific regions of chromosomes. But a great deal of our junk DNA is not simply structural. It doesn't code for proteins, but it does code for a different type of molecule, called RNA. A large class of this junk DNA forms factories in the cell, helping to produce proteins. Researchers are only just beginning to unravel the subtleties and interconnections in the vast networks of junk DNA. Recent research has determined that some genetic diseases are caused by mutations in junk DNA.[13-12a] Mutations in junk DNA has been the cause of diseases, such as Idiopathic Pulmonary Fibrosis, Roberts Syndrome and Cornelia de Lange Syndrome, Brittle Bone Disease, Myotonic Dystrophy, Amyotrophic Lateral Sclerosis (ALS) (also known as Motor Neuron Disease or Lou Gehrig's Disease), Kaposi's Sarcoma, Hutchinson-Gilford Progeria, Human Cartilage-Hair Hypoplasia, as well as many other diseases.[13-12a-1]

The human genome is constantly bombarded by potentially damaging stimuli in our environment. From ultraviolet radiation in sunlight to carcinogens in food, or emission of radon gas from granite rocks, we humans have always been attacked by potential threats to our genomic integrity. If only one in 50 of our base pairs is important for protein sequence because the other 49 base pairs are simply junk, then there's only a one in 50 chance that a damaging stimulus that hits a DNA molecule will actually strike an important region. Is insulation the only role of junk DNA? Also, where did all this insulating material come from in the first place? [13-12b]

When considering how the human cells work, almost every scientist over the last sixty years (1955 - 2015) has been focused on the impacts of proteins. Scientists had known for many years that there were stretches of DNA that didn't code for proteins. But hardly anyone anticipated how important these regions would prove to be, nor that they would provide the explanation for certain genetic diseases. The bits of junk between the parts of a gene that code for amino acids were originally considered to be nothing but nonsense or rubbish. They were referred to as "junk" or "garbage DNA", and pretty much dismissed as irrelevant and useless. The term «junk" is used to denote any DNA that doesn't code for protein. If these regions were completely nonsensical we would expect them to have changed randomly over time, but they haven't. This suggests that the normal repeats have some function.[13-12c]

Through research, it has been determined that protein-coding genes aren't formed from one continuous sequence of DNA. They are constructed in a modular fashion, with protein-coding regions interrupted by stretches of junk. In general, human genes are much longer than the genes in fruit flies or the microscopic worm called C. elegans, which are very common model systems in genetic studies. But human proteins are usually about the same size as the equivalent proteins in the fly or the worm. It's the junk interruptions in the human genes that are very big, not the bits that code for protein. In humans, these intervening sequences are often ten times as long as in simpler organisms, and some can be tens of thousands of base pairs in length. This creates a big problem when analyzing genes in human sequences. Even within one gene there's just a small region that codes for protein, embedded in a huge stretch of junk.[13-12d]

The human genome contains an extraordinary amount of DNA that doesn't code for proteins. Ninety-eight per cent of our genetic material doesn't act as the template for those all-important molecules believed to carry out the key functions of a cell or an organism. Why do humans have so much junk? Maybe the junk has no function or biological significance. It can be a mistake to assume that because something is present, it has a reason to be there. Some scientists speculated back in 2001 that this might also be true of most of the junk DNA in the human genome, but this view was proven to be incorrect. It was thought that the non-coding regions of the human genome might be "simply parasitic", selfish DNA elements that use the genome as a convenient host. But this isn't necessarily a logical prediction. Just because something has no apparent function in a specific organism, it doesn't mean it is irrelevant in all species.

Because evolution is usually building from a relatively limited selection of components there is a tendency for features to be co-opted for new functions. So, junk DNA could easily have roles in other organisms, especially ones that are more complex. It is also worth bearing in mind that there is a functional cost for a cell in containing so much junk DNA. If less than 2 per cent of the DNA is important, why would evolution maintain the other 98 per cent if it is simply functionless junk? As we have already acknowledged, the greatest evidence in favor of evolution of species lies in all those things we are stuck with because of our forebears (such as vestiges discussed in Chapter 12). But using huge amounts of resources to reproduce "useless" base pairs for every one that performs a function seems like taking redundancy a bit far.[13-12e]

Some of the most striking data suggested that over 75 per cent of the genome was copied into RNA at some point in some cells. This was quite remarkable. No one had ever anticipated that nearly three-quarters of the junk DNA in the human cells would actually be used to make RNA. When researchers compared protein-coding coding messenger RNAs with long non-coding RNAs they found a major difference in the patterns of expression. In the fifteen cell lines they studied, protein-coding messenger RNAs were much more likely to be expressed in all cell lines than the long non-coding RNAs. The conclusion they reached from this finding was that long non-coding RNAs are critically important in regulating cell fate.[13-12f]

If 80 per cent of the human genome has function, one could predict that there should be a significant degree of similarity between the human genome and at the very least the genomes of other mammals. The problem is that only about 5 per cent of the human genome is well-preserved across the mammalian class, and the well-preserved regions are highly biased towards the protein-coding entities. In order to address this apparent inconsistency, researchers speculated that the regulatory regions have evolved very recently, and mainly in primates. Using data from a large-scale study of DNA sequence variation in different human populations, they concluded that the regulatory regions have relatively low diversity in humans, whereas the diversity is much higher in areas that have no activity at all.

It has been argued that protein-coding sequences are highly well-preserved in evolution because a particular protein is often used in more than one tissue or cell type. If the protein changed in sequence, the altered protein might function better in a particular tissue. But that same change might have a really damaging effect in another tissue that relies on the same protein. This acts as an evolutionary pressure that maintains protein sequence. But regulatory RNAs, which don't code for proteins, tend to be more tissue-specific. Therefore they are under less evolutionary pressure because only one tissue relies on a regulatory RNA, and possibly only during certain periods of life or in response to certain environmental changes. This has removed the evolutionary brakes on the regulatory RNAs and allowed us humans to diverge from our mammalian cousins in these regions. But across human populations, there has been pressure from evolution to maintain the optimal sequence for these regulatory RNAs.[13-12g]

Junk Genes May Not Be Junk

Evolutionist View

> *Who would have thought that the majority of the genome would be copied or transcribed into RNA and that it would in fact be functional? Only a few years ago the scientific community believed that less than 5% of the genome was actually functional and the rest was non-functional evolutionary remnants . . . Who would have predicted this besides creationists and intelligent design theorists?[13-13]*

> *Creationists and intelligent design theorists have been claiming for many years that the concept of "Junk DNA" (as well as vestigial structures) was not entirely correct . . . only now are mainstream scientists finally starting to realize the significant errors in their long-cherished beliefs when it comes to the ill-conceived notion of junk DNA, an idea . . . based on ardently held evolutionary presuppositions that blinded mainstream science and prevented them from searching out the hidden treasures of so-called "junk DNA". . .When are*

scientists going to start realizing that the creationist paradigm does indeed have very good predictive scientific value when it comes to accurately understanding and investigating the physical world and universe?[13-13]

Recent findings and conclusions have verified what intelligent design theorists have been claiming for a long time, that pseudogenes have important functions and therefore are not really "pseudo" (or useless non-functional) after all.[13-13]

It was because of the evolutionary bias that these non-coding regions of DNA were assumed to be junk . . . and therefore overlooked and unrecognized as key informational components in the genome . . . such findings actually support the predictions of intelligent design theory while disproving long-held evolutionary assumptions.[13-13]

The failure to recognize the full implications of this particularly the possibility that the intervening noncoding sequences may be transmitting parallel information in the form of RNA molecules may well go down as one of the biggest mistakes in the history of molecular biology . . . [13-13]

Pseudogenes (junk genes) are DNA sequences that resemble functional genes but are generally thought to have no purpose. Many scientists think that pseudogenes are nothing more than discarded genetic fossils of a bygone era when they did have some sort of important function. Of course, it logically follows that similar pseudogenes that are shared by different species give evidence of common ancestry and even potential times of divergence. For example, the eta-globin pseudogene, which is found in both humans and chimps, has been used as an argument for the common ancestry of the two species. The first pseudogene was reported in 1977. Since that time, a large number of these genes have been reported and described in humans and many other species. There are two types of pseudogenes known as "processed" and "unprocessed" pseudogenes. Processed genes are found on different chromosomes from their functional counterparts. They lack introns and certain regulator genes, often terminate in adenine series, and are flanked by direct repeats (which are associated with movable genetic elements). They may be complete or incomplete copies of genes or mixtures of several genes. They are believed to have occurred through a 3-step process: **(1)** Copying DNA into RNA, **(2)** editing the introns to make mRNA, and **(3)** then turning the code in the mRNA back into DNA through a reverse transcription process. This process is thought to have created the "L1 family of pseudogenes." Other theories include retroviruses as means of pseudogene transport between different organisms.

Unprocessed pseudogenes are usually found in clusters of similar functional sequences on the same chromosome. They usually have introns and associated regulatory sequences. Their expression is usually prevented by a "misplaced" stop codon or codons (stop codons: UGA, UAA, UAG). There may be other changes from the "original" as the result of deletions, insertions, and point mutations. Some form of mRNA may or may not be produced depending on the damage to the gene. Many of these are believed to have arisen by gene duplication, which produced an extra copy of the gene. The extra copy could then accumulate mutations without harming the organism since it would still have a completely functional original copy. (The evolutionary gene duplication hypothesis suggests that over time, random mutations may produce a new gene with new functions by using this gene duplicate while maintaining the original gene function). (See Chapter 5, section "The Origin of Biological Information Stored in DNA and RNA" Evolutionist View and Chapter 9, section "Mechanisms That Cause Changes in Organisms" Creationist View for further discussion of duplicate genes).

It is felt by many, especially evolutionary biologists, that shared pseudogenes (which have no function in any form in different species) are examples of common ancestry. Comparison of DNA sequences from humans, chimps, and other mammals shows a great number of shared pseudogenes. Perhaps the best-known example of a shared pseudogene is the eta-globin gene. The eta gene is located on chromosome 11 in humans and is fourth in a series of 6 beta globin genes (five are functional). It has no start codon (AUG or GUG) and it has several stop codons (UGA, UAA, UAG). So obviously, no mRNA is made and therefore no protein. Humans, chimps, and gorillas have the same number of beta-globin genes arranged in the same sequence. The exon sequences within these genes are also similar, as are the exons of the eta gene. It is thought that the eta-globin gene originated by a duplication of the gamma-A-globin gene because of the high similarity of the sequences. Also, both genes are present in primates.

The very fact that (so-called) pseudogenes are still present and recognizable after tens of millions of years without any beneficial function just does not seem to make sense. Certainly, without some beneficial function, natural selection would not have maintained their sequences for such long periods of time. (Natural selection is discussed in Chapter 9.) There is in fact a cost to maintaining (the alleged or so-called) non-functional DNA. It takes energy to replicate and maintain DNA that does not pay for its keep, although this cost might seem small over the short term. Even an extremely small cost compounded over the course of millions of generations starts to turn into a significant disadvantage. So, the fact that pseudogenes have any recognizable gene-like structure at all suggests that they do in fact serve some kind of purpose.

The persistence of (so-called) pseudogenes is in itself evidence for their activity. This is a serious problem for evolution, as it is expected that natural selection would remove this type of DNA if it were useless, since DNA manufactured by the cell is energetically costly. Because of the lack of selective pressure on this neutral DNA, one would expect that old pseudogenes would be scrambled beyond recognition as a result of accumulated random mutations. Moreover, a removal mechanism for neutral DNA is now known. Typically when people say that the human genome contains 27,000 genes or so, they are referring to genes that code for proteins. But even though that number is still tentative, estimates range from 20,000 to 40,000. It seems to confirm that there is no clear correspondence between the complexity of a species and the number of genes in its genome. Fruit flies have fewer coding genes than roundworms, and rice plants have more than humans. The amount of non-coding DNA, however, does seem to scale with complexity.

An increasingly number of evolutionary biologists now realize that there is a large collection of genes that are clearly functional even though they do not code for any protein but produce only RNA. The term 'gene' has always been somewhat loosely defined; these RNA-only genes muddle its meaning further. To avoid confusion, evolutionary biologists tend not to talk about genes anymore; they just refer to any segment that is transcribed to RNA as a 'transcriptional unit.' Based on detailed scans of the mouse genome for all such elements, it has been estimated that there will be 70,000 to 100,000. Easily half of these could be non-coding. If that is right, then for every DNA sequence that generates a protein, another works solely through active forms of RNA, forms that are not simply intermediate blueprints for proteins but, rather, directly alter the behavior of cells.

Given this, it is not known if all of what are currently thought of, as pseudogenes have absolutely no function. In fact, some pseudogenes are believed to function as sources of information for producing genetic diversity. It is thought that partial pseudogenes are copied into functional genes during genetic recombination, producing variants of the functional gene. This phenomenon has been reported many times to include various immunoglobulins within mice and birds, mouse histone genes, horse globin genes, and human beta-globin genes. It is not known if this could be a possible role for the eta-globin gene as well. However, the fact that the eta-globin pseudogene is located between the fetal and adult genes suggests that it might play a role in gene switching (there seems to be some preliminary evidence to this effect, although the eta gene sequence's part in this is still unknown).

Regarding information, it all seems like the protein coding genes are actually very simplistic (on the level of bricks and mortar for building a house) and that the real informational complexity and functionality lies in the non-coding (so-call pseudogene) portion of the genome (the blueprint for directing where to put the bricks and mortar for building the house). This portion of the genome directs when and where the protein building blocks are placed. It is therefore vitally important to the overall structure and ultimate function of the resulting creature. There is evidence that Alu (a SINE "Short Interspersed Nuclear Element) sequences are involved in gene regulation, such as in enhancing and silencing gene activity, or can act as a receptor-binding site. This is surely a precedent for the functionality of other types of pseudogenes. Around 1998, a molecular biologist started advancing what seemed like an odd way to explain Alu's unusual affinity for genes. This molecular biologist suggested Alu sequences resided near genes because they are not really sequences, but are rather useful sequences involved with a mechanism that helps cells repair themselves. With the entire genome map in front of them, showing so many instances of Alu sequences around genes, scientists are beginning to take these ideas seriously. It looks pretty convincing according to other molecular scientists.

A team of molecular geneticists discovered in 2001 two areas where the same SINEs inserted independently: Vertebrate retrotransposons have been used extensively for phylogenetic analyses and studies of molecular evolution. (*A phylogenetic character is referring to the evolutionary development and diversification of a species or group of organisms, or of a particular feature of an organism.*) Information can be obtained from specific inserts either by comparing sequence differences that have accumulated over time in orthologous copies of that insert or by determining the presence or absence of that specific element at a particular site. The presence of specific copies has been deemed to be an essentially homoplasy-free phylogenetic character because the probability of multiple independent insertions into any one site has been believed to be nil. *(Homoplasy is the correspondence between organs or structures in different organisms that are acquired as a result of evolutionary convergence or of parallel evolution. A phylogenetic character is defined above.)* Two areas for SINE insertion within mys-9 have been identified and at each area it has been found that two independent SINE insertions have occurred at identical sites. These results have major repercussions for phylogenetic analyses based on SINE insertions, indicating the need for caution when one concludes that the existence of a SINE at a specific locus in multiple individuals is indicative of common ancestry. Although independent insertions at the same locus may be rare, SINE insertions are not homoplasy-free phylogenetic markers.

Researchers assumed that because humans, mice, and rats look so different, there would be differences in the genome. They did see the expected differences in the shared genes from the assumed 'common ancestor', but they were surprised to find long stretches of shared non-coding "junk" DNA that were exactly the same in humans and rodents. There were about five hundred stretches of DNA in the human genome that had not changed at all in the millions and millions of years that separated the human from the mouse and the rat. This was a great surprise. It is very unusual to have such an amount of conservation continually over such a long stretch of DNA. Many of these stretches of DNA, called "ultraconserved" regions, do not appear to code for protein, so they might have been dismissed as junk if they had not shown up in so many different species. Researchers confirmed that negative selection is three times stronger in these regions than it is for non-synonymous changes in coding regions. Researchers say that it is a mystery that molecular mechanisms would place virtually every base in a segment of size up to 1 kilobase (i.e., 1000 bp) under this level of negative selection. That is 500 regions of DNA up to 1000 bp that are identical between rats and humans, up to 500,000 identical genetic sites in DNA. What is also surprising is that these same regions largely matched up with chicken, dog and fish sequences as well; but are absent from sea squirt and fruit flies. Note that the last supposed common ancestor for all of these creatures was thought to live some 400 million years ago. Of course, it is only logical to assume that if nature has gone to so much trouble to preserve these ultra conserved regions over all these years, then they must be more important than just 'junk.'

Researchers now believe that the most likely scenario is that they control the activity of indispensable genes and embryo development. From what we know about the rate at which DNA changes from generation to generation, the chance of finding even one stretch of DNA in the human genome that is unchanged between humans and mice and rats over these hundred million years is less than one divided by ten followed by 22 zeros. It is a tiny, tiny fraction. It is virtually impossible that this would happen by chance. Scientists, however, still believe very strongly that humans and rats do in fact share a common ancestor that lived a hundred million years ago or so. The idea that perhaps humans and rats might have actually been individually created, deliberately, is not considered. This discovery suggests that the genomes of both humans and rats must be doing something other than coding for proteins, but the purpose of these ultraconserved regions remains a mystery. Scientists have concluded that other bits of 'junk' DNA will turn out not to be junk. One argument to consider is that various pseudogenes in different species often have certain shared "mistakes" that must have originated in a common ancestor. However, there is some evidence that nucleotide changes may not be completely random in certain gene locations.

It is interesting that among the many various substitution mutations in the "GLO" pseudogene that many, though not all, would be shared, to include a single deletion mutation that is shared by all primates (when compared to the rat). The general argument is, if not for common descent why would the sequences of human, chimpanzee, gorilla and orangutan reveal a single nucleotide deletion at position 97 in the coding region of Exon X? The odds are slim that out of 165 base pairs the same one would be mutated in all these primates by random chance. This is overwhelming evidence of common evolutionary ancestry. This would indeed seem to be the case at first approximation. However, in 2003, the complete sequence of the guinea pig GLO pseudogene (which is thought to have evolved independently) was compared to that of humans. To the surprise of the researchers, they reported many shared mutations (deletions and substitutions) were present in both humans and guinea pigs.

It must be pointed out that humans and guinea pigs are thought to have diverged at the time of the common ancestor with rodents. Therefore, a mutational difference between a guinea pig and a rat should not be shared by humans with better than random odds. But, this was not what was observed. Many mutational differences were shared by humans, including the one at position 97. According to researchers, this indicated some form of non-random bias that was independent of common descent or evolutionary ancestry. The probability of the same substitutions in both humans and guinea pigs occurring at the observed number of positions was calculated, by researchers to be 1.84 x 10-12, which is consistent with mutational areas. What is interesting here is that the mutational areas found in guinea pigs and humans exactly match the mutations that set humans and primates apart from the rat.

What makes mutational areas so prone to change is in the chemical nature of the mutational region. The type of molecular bonds, their stability or instability, or other molecular interactions may lend themselves to specific nucleotide pair switches, especially given certain environmental changes. No one really knows for sure except to say that mutational areas do exist. So, given that they do exist, similar genes should be expected to function in similar ways and this includes having similar "mutational areas" and/or "shared mistakes." In any case, it is interesting to note that there are no such examples of "shared errors" between mammals and other groups of animals (although there are plenty of common "errors" that are shared by widely divergent mammalian groups). There are no examples of 'shared errors' that link mammals to other branches of the genealogic tree of life on earth. Therefore, the evolutionary relationships between distant branches on the evolutionary genealogic tree must rest on other evidence besides 'shared errors.' Of course the argument used to explain this fact is that mammals split off from other groups of animals over 200 million years ago. Given this amount of time, random mutations would have obliterated any trace of common genetic errors.

The question remains, however, as to why are some identifiable genetic errors are maintained as long as they are if they are in fact functionless? Also, "processed pseudogenes" are very similar to "movable genetic elements" which are often transmitted from animal to animal by viruses. Certain interspecies pseudogenes of this type might in fact share a common ancestor while the various types of animals themselves, that harbor certain of these genetic sequences, may not be related through common descent so much as they are partially related through common infection. In any case, there are really no "foolproof" genetic markers of common decent. All of the ones proposed so far to be foolproof have been shown to have significant flaws. The prediction that pseudogenes, transposons (SINEs and LINEs) and other shared mutational mistakes are conclusive evidence for common descent has not held up over recent years.

The old notions that pseudogenes and other forms of shared "junk" DNA give clear evidence of common ancestry over common functional need, will have to be discarded. Certainly if organisms share similar environments and have similar morphologic (*the form and structure of organisms without consideration of function*) appearances and needs, one should not be surprised to find similar functional genetic elements shared between such creatures. Such sequences cannot be used to clearly establish evolutionary trees and to estimate divergent times since such beneficial sequences would be maintained over time via natural selection without any significant changes. (Natural selection is discussed in Chapter 9.) The similarities and differences would not be based so much on evolutionary changes over the time since a shared common ancestor as they would be the result of similarities and differences in functional needs that have always been there, maintained by the forces of natural selection, since these creatures came to be. It is unknown what further research will find in regards to the current understanding of genetics.

What was once labeled as junk because it was not understood may, in fact, turn out to be the very basis of human complexity. In between the genes and the sequences known to regulate their activity are long, tedious stretches that appear to do nothing. The term for them is "junk" (useless) DNA, reflecting the presumption that they are merely remnants from our evolutionary past and have no biological function. The genome turns out to be a highly complex, interwoven machine with very few inactive stretches, the researchers report. Genes, it transpires, are just one of many types of DNA sequences that have a functional role. And "junk" DNA turns out to have an essential role in regulating the protein-making business.

It is becoming more and more clear that the key functional differences between living things, like humans and apes, are not so much found in protein-coding genes, but in the non-coding regions of DNA once thought to be functionless "junk-DNA," evolutionary remnants of past mistakes that are shared between various creatures. This notion is starting to be obvious with more and more discoveries that show that many of these same regions are not just functional; they carry the vast majority of the genetic information. The "genes" that were once thought to be so important for genetic function are turning out to be equivalent to the most low-level basic building blocks within the genome, like bricks and mortar. Surprisingly, it is the non-coding regions of DNA control what is done with these building blocks that determine what kind of protein is to be built.[13-13] (See Chapter 5 for discussion of the purpose of proteins.)

So-called Pseudogenes Are Not Useless Remnants

Creationist View

Researchers identify pseudogenes by comparing similar sequences in the genome, the genetic content of heredity in DNA or RNA, to functional genes within an organism. To find a pseudogene for vitamin C in the human genome, a study of both a human genome and the genome of another organism with the ability to produce vitamin C would have to be made. In 1994, scientists found a DNA sequence in humans that resembled the rat gene that codes for the enzyme (L-gulono-y-lactone) that sets in motion the last step of vitamin C production. Humans are missing this final enzyme necessary in the final stage of producing vitamin C, however, they do have all of the other enzymes necessary to convert glucose into vitamin C. Claiming that only the last enzyme is missing, implies to the untrained individual that there is a biochemical pathway (organic substances that could convert to another organic substance) that is useless. In reality, the biochemical pathway that results in vitamin C production in rats also leads to the production of five-carbon sugars in virtually all animals.

Human pseudogenes have been catalogued by the thousands, but in spite of the similarities to functional genes, the exact role of pseudogene sequences in the genome is basically not known by any scientist. It is debatable that pseudogenes are remnants of once functioning genes that now waste space in the DNA. It is possible that these regions of DNA actually perform a function in both humans and animals. As science continues to make advances in DNA sequences the role of these genomes may yet be discovered. Considering what we now know it seems unreasonable to consider any function of a portion of DNA as an evolutionary leftover. More research needs to done (For an update, continue reading below. Also, see the section "Junk Genes May Not Be Junk" above). Currently, there are indications that some so-called pseudogenes in humans are actually functional.[13-14]

Many evolutionary scientists claim that certain DNA segments are the dead, useless remains of genes that many generations ago did code for proteins. Recent experiments, however, show that many pseudogenes are not useless. When certain pseudogenes are turned off, the organism suffers either fatal or injurious consequences. (See section "Some DNA Portions are Used to Turn Genes On or Off.") Geneticists now realize that these pseudogenes somehow protect the protein-coding genes from breakdown or malfunction.

Other so-called pseudogenes, they have found, actually encode for functional proteins. Still other pseudogenes were also misidentified and later found by researchers to encode for the construction of molecules at first were wrongly considered to serve no purpose.[13-15]

Since the 1980s, scientists referred to DNA that does not code for the manufacture of proteins as "junk" DNA. Recently, evolution-minded scientists are beginning to realize these so-called junk DNA are not useless remnants after all. (See section "Junk Genes May Not Be Junk" above). This so-called junk is now known to serve several life-critical functions. Since scientists only use the protein coding portions of genes when comparing human and chimpanzee sequences, they do not include the so-called junk DNA sequences in their comparisons. (The non-coding regions of the DNA are not included in chimpanzee and human DNA comparisons made by evolutionists, as mentioned in Chapter 15, section "Comparison of Human and Chimp DNA," Creationist View: "One of the main problems with a comparative evolutionary analysis between human and chimp DNA is that some of the most critical DNA sequence is often omitted from the scope of the analysis . . . . Evolutionary scientists have used only the coding portion of a gene for comparative analyses." See Chapter 15, section "Comparison of Human and Chimp DNA" for discussion of human and chimpanzee DNA comparisons.)

Geneticists have discovered five kinds of non-protein-coding DNA: (1) pseudogenes, (2) SINES, (3) LINES, (4) endogenous retroviruses, and (5) LTRs. The term **pseudogenes** comes from the assumption that certain DNA segments are the dead, useless remains of genes that many generations ago did code for proteins. **SINES** is an acronym for "short interspersed nuclear elements." Emerging research shows that these DNA elements serve at least two distinct purposes. Some help protect the cell when it experiences stress. Others help regulate the expression of the protein-coding genes. **LINES** is an acronym for "long interspersed nuclear elements." Recent findings show that some LINES play a central role in X-chromosome inactivation. When such inactivation fails, serious genetic disorders result. Another discovered LINES function is to turn off one of the two protein-coding genes inherited from an individual's parents. (See section "Some DNA Portions are Used to Turn Genes On or Off.") Evolutionists once presumed that all **endogenous retroviruses** were the product of retroviral infections. They hypothesized that retroviral DNA becomes incorporated into the host's genome. New research, however, shows that many endogenous retroviruses protect the organism from retroviral infections by disrupting the life cycle of invading retroviruses. Others function as protein-coding genes. **LTRs**, an acronym for "long terminal repeats", were once thought to originate from **endogenous retroviruses**. Recent studies show that several LTRs play crucial roles in protecting organisms from retroviral attacks. Other research demonstrates that some LTRs help regulate the expression of certain protein-coding genes. Rather than being junk, non-protein-coding DNA serves many amazing life-beneficial purposes. These purposes would never have been discovered and understood if geneticists had continued to study only the protein-coding DNA.[13-16]

Whenever evolutionists previously discovered a new section of DNA that had no known function, they assumed it was junk DNA, believing it was a leftover of evolution. This view, however, is now being reconsidered as new evidence indicates otherwise. For example, the DNA of organisms more complex than bacteria contains regions called exons that code for proteins, and non-coding regions called introns.[13-17]

It has been discovered that any 'single' gene (in the sense of a single continuously read passage of DNA text), is not all stored in one place. (This concept is similar to how files are stored on a computer disk. As a file grows, each successive save will cause parts of the file to be stored in different locations in the disk with pointers indicating where the next segment is located.) The genes along the chromosome are fragments called exons, separated by portions of 'nonsense' genes called introns. Any one 'gene' in the functional sense, is in fact split up into a sequence of fragments (exons) separated by meaningless introns. It appears that each exon ends with a pointer indicating where the code continues. A complete gene is then made up of a whole series of exons, which are actually strung together only when they are eventually read during translation that translates the genes into proteins.[13-18] (This translation process is discussed in Chapter 5, section "How a Protein is Made—Step by Step.") During the translation process, the introns are removed and the exons are spliced together to form the mRNA (messenger RNA) that is finally decoded to form a protein (See Chapter 5). This also requires elaborate machinery called a "spliceosome." This assembles on the intron, chops it out at the right place, and joins the exons together. This must be in the right direction and place, because it makes a huge difference if the exon is joined even one letter off. It is questionable as to why more complex organisms should evolve such elaborate machinery to splice the introns if they are really useless.

More likely, natural selection (discussed in Chapter 9) would favor organisms that did not have to waste resources processing a genome filled with 98 percent of junk (introns). And there have been many uses for junk DNA discovered, such as the overall genome structure and regulation of genes, and enablement of rapid diversification, if and when needed. Damage to (so-called 'meaningless') introns can be disastrous. One example was deleting four letters in the center of an intron, preventing the spliceosome from binding to it, resulting in the intron being included. Mutations in introns interfere with imprinting, the process by which only certain maternal or paternal genes are expressed, not both. Expression of both genes results in a variety of diseases and cancers. It has also been noted that the non-coding DNA regions, or rather their non-coding RNA negatives, are important for a complicated genetic network. These interact with each other, the DNA, mRNA, and the proteins. It is believed that the introns function as nodes, linking points in a network. The introns provide many extra connections, what in computer terminology would be called multi-tasking and parallel processing. (Introns are also discussed in Chapter 15.) This could control the order in which genes are switched on and off. (See section "Some DNA Portions are Used to Turn Genes On or Off.") This means that rewiring the network could produce a tremendous variety of multicellular life. In conclusion, when DNA is compared between different life forms, all regions need to be compared, not just those that are used for production of proteins.[13-17] (See Chapter 5 section "The Origin of Biological Information Stored in DNA and RNA" for further discussion of DNA.)

Looking at the guinea pig and the simian primates it is obvious they do not share the same ancestry even though their GLO is alike. If the pseudogene 'shared mistake' argument is taken to its logical conclusion, then humans are more closely related to guinea pigs than to primates because of the unexpected degree of similarity between guinea pig GLO pseudogene and those of the higher primates, including humans. Even if organic evolution is accepted as fact, evolutionary ancestry is just not the answer to the mystery of why guinea pig and human GLO is so similar.[13-19]

Old Universe Progressive Creationist View

Many evolutionary biologists believe "junk" DNA is one of the most powerful pieces of evidence for biological evolution. They believe that junk DNA results when undirected biochemical processes and random molecular and physical events transform a functional DNA segment into a useless molecular artifact. This segment remains part of an organism's genome only because of its attachment to functional DNA. Evolutionary biologists consider pseudogenes to be the dead remains of once functional genes. They believe that severe mutations destroyed the capacity of the cell's processes to "read" and process the information in these genes. They believe that pseudogenes possess tell-tale signs that allow molecular biologists to recognize them as genes, although the pseudogenes are nonfunctional.

Junk DNA at one time represented an insurmountable challenge to the biochemical intelligent design argument and appeared to make an ironclad case for evolution. Now, recent advances suggest otherwise. Much to the surprise of scientists, junk DNA has function. Based on the characteristics possessed by pseudogenes, few molecular biologists would have ever thought junk DNA plays any role in the cell's operation. However, several recent studies unexpectedly identified functions for both duplicated and processed pseudogenes. Some duplicated pseudogenes help regulate the expression of their corresponding genes. And, many processed pseudogenes code for functional proteins. Biologists will soon establish functional roles for endogenous retroviruses (discussed below) and their compositional elements. Recent advances indicate that this class of noncoding DNA regulates gene expression and helps the cell prevent retroviral infection by disrupting the assembly of retroviruses after they take over the cell's processes.

Endogenous retroviruses are viruses that are permanently incorporated into the host organism's genome. Like all viruses, retroviruses consist of protein capsules that house genetic material (either DNA or RNA). Viruses infect organisms by invading specific cell types of their hosts. (See Chapter 5, section "Viruses" for further discussion of viruses.) Once viruses attach to the target cell's surface, they inject their genetic material into the healthy cell. Then the viral genetic material exploits the cell's processes to produce more viral genetic material and proteins, which combine forming new viral particles. When the newly formed viruses escape from the host cell, the infection cycle repeats. Instead of DNA, RNA is the genetic material used by retroviruses. After it is injected into the host cell, reverse transcriptase uses the retroviral RNA to make DNA. This newly made DNA can then direct the production of new retroviral particles. (HIV, the virus responsible for AIDS, is a retrovirus.) The DNA copy of the retroviral genetic material can become incorporated into the host cell's genome. If the retroviral DNA suffers severe mutations, the retrovirus becomes disabled. When this happens, the retrovirus DNA presumably remains nonfunctional in the host genome and is referred to as an endogenous retrovirus.

SINEs and LINES are two types of noncoding DNA that are known as transposable elements, that is, pieces of DNA that jump around the genome or transpose. In the process of moving around, transposable elements direct the cell's process to make additional copies and consequently increase the number of these elements. SINEs (short interspersed nuclear elements) and LINEs (long interspersed nuclear elements) belong to a class of transposable elements called retroposons. Molecular biologists believe SINEs and LINEs duplicate and move around the genome through an RNA intermediate and the work of reverse transcriptase. As with pseudogenes and endogenous retroviruses, molecular biologists now recognize that the SINE DNA found in the genomes of a wide range of organisms play an important part in regulating gene expression and offers protection when the cell becomes distressed. Researchers have also identified another potential function of SINEs - regulation of gene expression during the course of development.

SINEs possess regions that the cell's process methylaters (attaches the methyl chemical functional group). This process turns genes off. Depending on the tissue type, SINEs display varying patterns of DNA methylation. These diverse patterns implicate SINEs in the differential gene expression that occurs during development. Much in the same way that scientists acknowledges function for SINE DNA, molecular biologists now recognize that LINEs critically regulate expression. For example, researchers have identified a central role for LINE DNA in X chromosome inactivation. This inactivation occurs in healthy females to compensate for duplicate genes found on the two X chromosomes. (Females have two X chromosomes. Males have an X and a Y chromosome.) The inactivation of one set of X chromosomal genes ensures proper levels of gene expression in females. If X chromosome inactivation does not occur, genetic disorders result. Scientists now believe LINE DNA controls monoallelic gene expression, a situation in which only one of the two genes inherited from both parents is used. This process completely "turns off" the other gene.

Many molecular biologists now believe that even though they cannot directly identify functional roles for many classes of noncoding DNA, these DNA segments must be functional because so many distantly related or unrelated organisms share them. Evolutionary biologists reason that these noncoding DNA classes must serve in a critical capacity. If they did not, mutations would readily occur in them. These noncoding DNA segments must have resisted change, otherwise mutations would have rendered them nonfunctional, and that would have been harmful to the organism.[13-19a]

Some DNA Portions are Used to Turn Genes On or Off

Evolutionist View

In biology, genes tell that something is to appear in an organism, such as a leopard is to have spots, but does not stipulate exactly where each spot is to be located. A gene is a specific segment of deoxyribonucleic acid (DNA) that indirectly provides the instructions for the specific proteins that a cell manufactures. The DNA does this from inside the nucleus of a cell by transcribing instructions using a similar molecule, ribonucleic acid (RNA) (Chapter 5, Figures 5.3, 5.4, 5.5, 5.6). The specific sequence of the building blocks (adenine, guanine, thymine, and cytosine) that make up DNA determines that specific sequence for the building block of the RNA that leaves the nucleus. (In RNA, uracil is present instead of thymine.) The RNA instructions are used to build specific proteins, many of which will perform specific tasks, by way of chemical reactions for the cell, giving each cell a unique functional identity (DNA, RNA, proteins and genes are discussed in Chapter 5).[13-20] (The process of building proteins is discussed in Chapter 5, section "How a Protein is Made—Step by Step.")

Intelligent Design View

DNA does not consist entirely of genes containing encoded messages for the specification of proteins. A great portion of DNA is involved in control purposes, that is, switching off and on different genes at different times and in different cells. (See Chapter 7, section "Evolution of the Horse" sub-section "Preexisting Genetic Information" for further discussion of genes being turned on and off.) Biological design is far more complex than once believed. It has been discovered that many sequences of DNA that perform the crucial control functions related to information retrieval are situated not adjacent to the genes which they control, but actually are embedded within the genes themselves. This complex biological coding system is too complex to have ever been developed by random changes driven by natural selection.[13-21]

(See Chapter 9 section "Hox Genes" for additional discussion of changes in genes and their effects during the development of an embryo.)

In the next chapter (14) we will look at some organisms that are said by some to be too complex to have originated by natural causes.

~~~

Chapter 14

Biological Complexity and
Intelligent Design

*IN crossing a heath [an area of open uncultivated land - vegetation
dominated by dwarf shrubs of the heath family], suppose I pitched my
foot against a stone, and were asked how the stone came to be there;
I might possibly answer, that, for any thing I knew to the contrary,
it had lain there for ever: nor would it perhaps be very easy to show
the absurdity of this answer. But suppose I had found a watch upon
the ground, and it should be inquired how the watch happened to be
in that place; I should hardly think of the answer which I had before
given, that, for anything I knew, the watch might have always been
there. Yet why should not this answer serve for the watch as well as
for the stone?*

*Why is it not as admissible in the second case, as in the first? For
this reason, and for no other, viz. that, when we come to inspect
the watch, we perceive (what we could not discover in the stone)
that its several parts are framed and put together for a purpose, e.g.
that they are so formed and adjusted as to produce motion, and that
motion so regulated as to point out the hour of the day . . . This
mechanism being observed (it requires indeed an examination of the
instrument, and perhaps some previous knowledge of the subject, to
perceive and understand it; but being once, as we have said, observed
and understood), the inference, we think, is inevitable, that the watch
must have had a maker: that there must have existed, at sometime,*

and at some place or other, an artificer or artificers who formed it for the purpose which we find it actually to answer; who comprehended its construction, and designed its use.

—William Paley D. D.
Natural Theology; or, Evidences of the Existence and Attributes of the Deity, Collected From the Appearances of Nature. Twelfth Edition. 1809[14-1]

If it could be demonstrated that any complex organ existed, which could not possibly have been formed by numerous, successive, slight modifications, my theory would absolutely break down. But I can find out no such case. No doubt many organs exist of which we do not know the transitional grades, more especially if we look to much-isolated species, around which, according to the theory, there has been much extinction. Or again, if we take an organ common to all the members of a class, for in this latter case the organ must have been originally formed at a remote period, since which all the many members of the class have been developed; and in order to discover the early transitional grades through which the organ has passed, we should have to look to very ancient ancestral forms, long since become extinct. We should be extremely cautious in concluding that an organ could not have been formed by transitional gradations of some kind. Numerous cases could be given among the lower animals of the same organ performing at the same time wholly distinct functions; thus in the larva of the dragon-fly and in the fish Cobites the alimentary canal respires, digests, and excretes. In the Hydra, the animal may be turned inside out, and the exterior surface will then digest and the stomach respire. In such cases natural selection might specialise, if any advantage were thus gained, the whole or part of an organ, which had previously performed two functions, for one function alone, and thus by insensible steps greatly change its nature. Many plants are known which regularly produce at the same time differently constructed flowers; and if such plants were to produce one kind alone, a great change would be effected with comparative suddenness in the character of the species. It is, however, probable that the two sorts of flowers borne by the same plant were originally differentiated by finely graduated steps, which may still be followed in some few cases.

Again, two distinct organs, or the same organ under two very different forms, may simultaneously perform in the same individual the same function, and this is an extremely important means of transition: to give one instance—there are fish with gills or branchiae that breathe the air dissolved in the water, at the same time that they breathe free air in their swim-bladders, this latter organ being divided by highly vascular partitions and having a ductus pneumaticus for the supply of air.

To give another instance from the vegetable kingdom: plants climb by three distinct means, by spirally twining, by clasping a support with their sensitive tendrils, and by the emission of aerial rootlets; these three means are usually found in distinct groups, but some few species exhibit two of the means, or even all three, combined in the same individual. In all such cases one of the two organs might readily be modified and perfected so as to perform all the work, being aided during the progress of modification by the other organ; and then this other organ might be modified for some other and quite distinct purpose, or be wholly obliterated . . .

—Charles Darwin
On the Origin of Species by Means of
Natural Selection, or the Preservation of
Favoured Races in the Struggle for Life.
Chapter VI, Difficulties of the Theory
of Natural Selection.
First Edition. 1859 [14-2]

Introduction

In Chapters 5 and 6 we discussed what amino acids, genes, DNA, proteins, and cells are. We also looked at the theories and possibility of how cells could have originated by natural means. In Chapter 9 we discussed whether mutations and natural selection could produce new species. In this chapter (14) we will discuss whether or not some organisms are too complex to have developed (or self-assembled) by natural processes.

The various views presented in this chapter regarding the origin of various species and organisms are not exclusively of strict evolutionists and strict creationists. Some views are of those who believe in variations of the two extremes. See "Various View Descriptions" section in the back of this book for descriptions of the views presented in this chapter.

Irreducible Complexity Explained

Creationist View

The concept of irreducible complexity of organs is based on the belief that some organs are too complex to have evolved by natural processes. (The cell is an example of an organism that is too complex to have originated by natural causes. Chapter 5 discusses the cell in detail.) Since irreducible complex systems are extremely complicated, it is doubtful they evolved by natural means. Evolutionists believe that by numerous, successive, slight modifications in subsequent generations of offspring, changes are made in an organism. Thus, the only way for evolutionists to explain complex organisms is by means of an indirect Darwinian pathway in which structures and functions evolve simultaneously. Evolutionists believe that one way structures and functions could develop is for parts (having one function) previously targeted for other systems could break free and become part of a different organism and perform a new function. It is as though pieces from a car, bicycle, motorboat, and train were suitably recombined to form an airplane.

If we look at the bacterial flagellum (discussed later in this chapter), an irreducible complex system, we can see how impossible it would be for this system to be developed piecemeal. All of the pieces within the system have a specific purpose and function. Multiple protein parts from separate functioning systems would need to break away, then reunite in a completely different formation, all at the same time. Imagine a cell with the components necessary to form bacterial flagellum yet currently serving other functions. It seems impossible that these would just cease their original function, reintegrate spontaneously to form the flagellum system. Not only would each part have to reassemble, each part, like the pieces of a puzzle, would have to **(1)** be the right size, **(2)** be the right shape, and **(3)** perform the right function. All spontaneous changes necessary to form an irreducible complex system seem impossible because they require multiple coordinated changes from multiple functioning systems to bring about an irreducibly complex system.

According to the Darwinian Theory, the changes take place gradually, in increments, over time. Darwinists claim that a gradual increase in complexity in which new parts that enhance function are added and eventually become indispensable. The logic is very simple. Some part (some organism), say 'A', initially performs some function, but possibly not very well. Another part, say 'B', later becomes added to 'A'. If 'B' proves to either help or improve 'A', then natural selection may allow it to continue. This new part 'B' at this point is not essential; it merely helps or improves the function of 'A'. Then, later, 'A' (which could eventually include more or less parts) may change in such a way that 'B' now becomes indispensable. This process continues as further parts become integrated into the system. Eventually, many parts become essential for the organism to function. For this to occur the functions of each part during the stages of evolving would be different from the final function because the final function is exhibited by an irreducibly complex system, and as such, cannot be part of any simpler system. The problem is that there is no evidence that irreducibly complex biochemical systems like the bacterial flagellum came about by this method of (1) add a component, (2) make it indispensable, (3) add another component, (4) make it indispensable, etc.[14-3]

Intelligent Design and Creationist View

For an irreducibly complex system (organism) to develop through natural processes, it would require that all components from the irreducible core be in place simultaneously for basic function to be achieved. It follows that if natural selection (discussed in Chapter 9) is going to select for the function of an irreducibly complex system, it has to produce the irreducible core all at once or not at all. The problem is that complex systems are not simple. The irreducibly complex biochemical systems are protein machines that consist of numerous distinct proteins each of which is indispensable to the machine's basic function. Darwinism is based on a gradual evolutionary process that piece-by-piece builds complexity and function. The problem is that if a function does not exist, it will not be selectable by natural selection until the irreducibly complex system is already in existence.[14-4]

The concept of Intelligent Design specifies that a single system (organism) composed of several well matched, interacting parts that contribute to the basic function, wherein the removal of any one of the parts causes the system to effectively cease functioning. An irreducibly complex system cannot be produced directly by slight, successive modifications of a pre-cursor system, because any pre-cursor to an irreducibly complex system that is missing a part is by definition non-functional. In a multi-part system, all the parts are necessary for it to function. A system cannot obtain five or more parts all at one time (that may or may not perform a function) and then gradually build up to a more complex system because there is no new function possible until the last part is in place. This is why evolution cannot produce such complex systems. And since natural selection can only choose systems that are already working, then if a biological system cannot be produced gradually, it would have to arise as an integrated unit, that is, all at once, for natural selection to have anything to act upon.[14-5] (Natural selection discussed in Chapter 9.)

Four irreducibly complex systems have been currently determined to be irreducibly complex. They are: (1) the bacterial flagellum, (2) the cilium, (3) the blood-clotting cascade, and (4) the immune system.[14-6] (Each system will be discussed in this chapter.)

Darwinist View

Biology is the study of complicated things that give the appearance of having been designed for a purpose.

—Richard Dawkins[14-7]

. . . our own existence presented the greatest of all mysteries, but that it is a mystery no longer because . . . [Charles] Darwin and [Alfred Russel] Wallace solved it . . .

—Richard Dawkins[14-8]

Every complex structure develops through small mutational steps. This is completely in agreement with natural selection. Every step must be beneficial to the organism in order to have a reasonable chance to be selected. The step-by-step process and the beneficial changes of intermediate steps are essential to the Neo-Darwinist explanation of the adaptations of living organisms. There are two aspects of the "irreducible complex" theory that must be considered. (1) The state of a system and (2) the origin of the system are important to understanding Neo-Darwinism. To call something 'irreducible complex' is a description of the state of the system. It means that all parts of that system are necessary for the functioning of the whole system. The origin of a complex system is completely different than the state of a complex system. With this in mind, the production of these irreducible complex systems is not too difficult to understand.[14-9]

Intelligent design advocates identify some biological machines and some structures that they regard as irreducibly complex systems. They are as follows: (1) the bacterial flagellum, (2) the blood-clotting cascade, (3) the eukaryotic cilium, and (4) the immune system.

The basis of the biochemical argument from irreducible complexity made by Intelligent Design advocates is that the individual parts of that machine have no function of their own. And because these individual organisms have no function of their own, they cannot be produced by natural selection and, therefore, the reason a complex biological system cannot develop from individual parts is because the individual parts have no function of their own prior to their being part of a complex system.

Intelligent Design advocates claim that all the parts are needed for a particular complex system to function. But their claims are false. Contrary to the claims of Intelligent Design advocates, these complex machines come from pre-existing machines that had functions of their own, that is, the individual parts performed different functions prior to their becoming part of a more complex system.[14-10]

An Example of Irreducible Complexity—The Wing

Creationist View

Regarding the Darwinian Theory, someone may point out that any variation has to be of immediate value to its possessor if it is going to have a better chance of survival than its fellow creatures. Of what 'survival value' are the first dim beginnings of forelimbs starting to flap about feebly and nakedly prior to the final development of a wing?[14-11]

In our illustration of irreducible complexity, we will use the wing. The wing is not necessarily the best example, but is useful in illustrating the point. An adaptation of a skeletal structure for flight without feathers (or adapted 'hand' without adapted 'shoulder') would simply not make sense. The mutated 'hand' (what a wing is presumed to have evolved from) would, in reality, be a disadvantage until it could be used for flight or some other useful function, which would take millions of years to evolve. If a mutation had neither advantage nor disadvantage (such as in the supposed stages during the evolution of feathers) it would not emerge in a new species simply because natural selection (Chapter 9) would eliminate it as non-functional or even detrimental. Unless feathers have some alternate function, such as warmth, it would seem senseless for the tremendous number of other neutral mutations (subsystems of the wing) necessary for flight, to randomly emerge, since these alone would not produce the goal and benefit of flight. For one to believe the wing evolved, many 'small' mutations in specific order and timing would be necessary. The difficulty here would be that these small mutations are not as small as they appear. (Mutations are discussed in Chapter 9.) Even with billions of years of random mutational change, the probability of these small, neutral mutations remaining is minimal. As the same type of creature continues to produce offspring, the new small mutations would not be dominant, and would therefore not appear in every new offspring, which in turn may gradually be eliminated altogether as the species evolves. Even if a new organ all of a sudden appeared in a creature, if it were not proven to be beneficial, it would be a disadvantage to the creature and would subsequently lead to its elimination by natural selection.[14-12] (Sudden changes are discussed in Chapter 10.)

Darwinist View

Creationists claim that since the development of wings from four legged animals would be gradual, an intermittent form would be a part leg/part wing and natural selection (Chapter 9) would eliminate the species, assuming the partly formed wing was useless and even detrimental to the survival of the creature. Darwinists believe that each successive stage of wing development would need to be functional, even the probable stumpy little partial wings that would not be aerodynamically capable for flight. It is very possible that although an organ may not have been originally formed for some specific purpose, it may have been useful for a different purpose while it is in the process of developing into the wing. For example, someone could make a machine for some specific purpose with used old wheels, springs, and pulleys. The parts could be slightly altered for the machine to function as planned. In nature every creature has parts that have been slightly modified to perform different purposes. This re-using of parts for different purposes in different organisms is called "exaptation," in which an organism that originally evolved for one purpose is used for a different purpose. The developing stages in wing evolution would have had uses other than for aerodynamic flight. Half wings were not poorly developed wings, they would have been well developed for some other purpose, perhaps thermo-regulating devices. The first feathers in the fossil record, for example, are hair-like and resemble the insulating down of modern bird chicks (See Chapter 7 section "The Origin of Wings with Feathers" and Chapter 7 section "Recent Discoveries of Ancient Bird-like Dinosaurs"). Since modern birds probably descended from bipedal therapod dinosaurs, wings with feathers could have been utilized for regulating heat. This could have been accomplished by holding their wings and feathers close to the body to retain heat and then stretching them out to release heat.[14-13]

Intelligent Design View

To explain the evolution of a bat, for example, one would need to imagine a succession of small mammalian species in which the fingers are gradually lengthened (Chapter 7, Figure 7.4). This would result in the loss of normal forelimb function before the necessary development of wings and specialized muscle to sustain powered flight had been attained. Would a primitive wing, far less efficient than a modern bat's, allowing only very restricted movement through the air be of such selective advantage that an organism would sacrifice it forelimbs in its favor? If the wings of bats really did evolve gradually from gliding organs then at some point during their evolution the creature was at a disadvantage (because at one point the species that previously had forelimbs and had offspring that developed the mutated forelimbs, the offspring would not be able to function normally as its ancestor did, but would also not be able to glide or fly) but natural selection chose it regardless of its limited usefulness, even though the mutated offspring would no longer be able to hunt normally, nor would it be able to properly escape predators efficiently.[14-14] (Natural selection is discussed in Chapter 9.)

Creationist View

Evolutionists claim that the bat evolved from a small, mouse-like creature whose forelimbs (the "front toes") developed into wings by gradual steps. They believe that the "front toes" grew longer and skin grew between them, eventually forming the bat wing. During this evolutionary process it is obvious that as skin began to grow between the "front toes" the animal could no longer run as it previously could. The problem with the evolutionists' claims is that during the intermediate stages of transition from a four-legged animal to a creature with wings, at some point the forelimbs would be useless because the toes with skin between them could no longer function properly as feet. During this developmental stage, the forelimbs would not yet be long enough to function as wings. During these hypothetical transitional stages, the intermediate creatures would have limbs too long for running but too short for flying. Natural selection (Chapter 9) would eventually cause this intermediate creature to become extinct, as it would not be able to either run or fly sufficiently to escape predators, such as owls, coyotes, foxes, cats, weasels, snakes, etc. There is no conceivable pathway for bat wings to be formed in gradual stages. This conclusion is confirmed by the fossil record, as no transitional fossils have been found that resemble intermediary creatures between mice and bats. The first time bats appear in the fossil record, they are already fully formed and virtually identical to modern bats.[14-15] (See Chapter 7, section "Evolution From Reptiles to Birds.")

Both Evolution and Intelligent Design Could be True

Theistic Darwinist View

Consider the possibility of basic parts of a cell being created with the ability to self-assemble. If this were true, both creationists' and evolutionists' views would be true. Creationists would be correct in stating that the Creator designed and created the basic parts of all living things. Evolutionists would be correct in claiming that simple, basic parts self-assembled into more complex organisms. It may be possible, for example, that an engineer (Designer) could have designed multiple parts to be adaptable for various uses. The parts could be simple organisms (proteins) such as: (1) two gears, forward and reverse, water-cooled, proton motive force, (2) a stator, (3) a rotor, (4) a u-joint, (5) a drive shaft, (6) a propeller, for the manufacture of a flagellum or proteins such as (1) alpha-tubulin, (2) beta-tubulin, (3) dynein, (4) nexin, (5) spoke protein, and (6) a central bridge; to be used for example, to develop a Cilium (discussed later in this chapter). As various parts come in contact with each other, some become attached (self assembled) and become functional. The parts came together by themselves; however, an engineer (Designer) designed the parts so that they would become adaptable for various purposes.[14-16] Some believe that irreducible complex structures were actually present in the first living cell and then became utilized by organisms as they evolved. However, due to the effect of mutational drift of genes on complex structures, it is highly unlikely that irreducibly complex structures were ever present in the first living cell.

Some Actual Examples

We will now look at some specific organisms and discuss the possibilities of how they could have developed by natural processes.

Flagellum

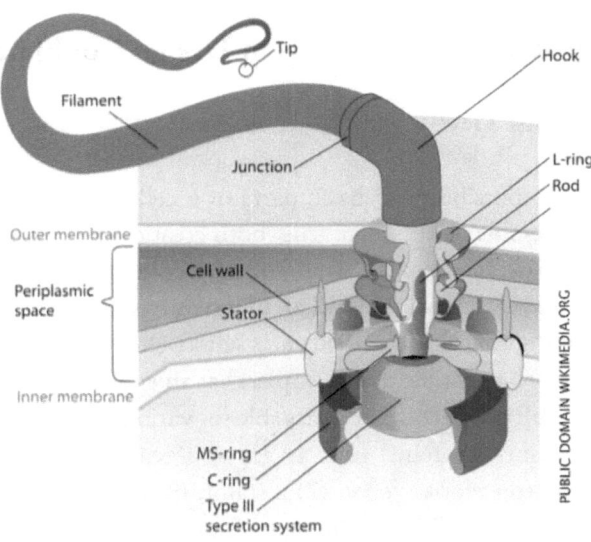

Flagellum

Figure 14.1 Flagellum

Bacterial Flagellum is Irreducibly Complex

Intelligent Design View

The bacterial flagellum (Figure 14.1) is a long, hair-like filament embedded on the outside wall of a prokaryotic cell (Chapter 5, Figure 5.1 bacteria) membrane (Chapter 5, Figures 5.4a, 5.4b and Figure 14.1a). The external filament consists of a single type of protein, called flagellin. The flagellin filament is the paddle surface that contacts the liquid during swimming. At the end of the flagellin filament near the surface of the cell, there is a bulge in the thickness of the flagellum (Figures 14.1 & 14.1a "Hook"). It is here that the filament attaches to the rotor drive. The attachment material is comprised of something called "hook protein." The filament of a bacterial flagellum, unlike a cilium, contains no motor protein; if it is broken off, the filament just floats stiffly in the water. Therefore the motor that rotates the filament-propeller must be located somewhere else. Experiments have demonstrated that the motor is located at the base of the flagellum (Figure 14.1), where electron microscopy shows several ring structures occur (Figure 14.1a C, L, M, P & S rings).[14-17]

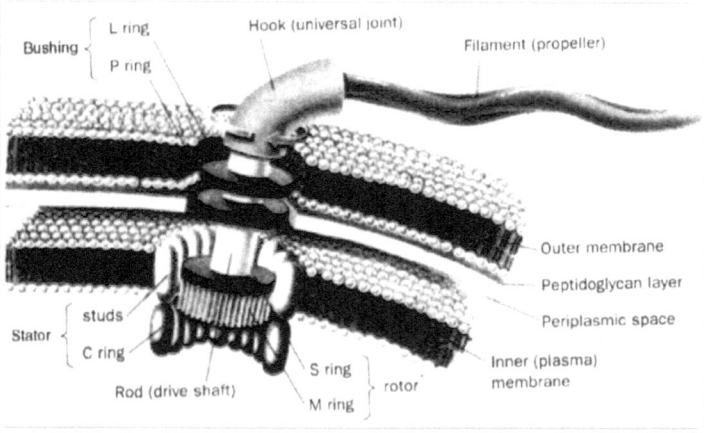

Figure 14.1a Flagellum

The parts and three-dimensional structure of the flagella motor embedded on the outside walls of prokaryotic cells is a marvel of engineering on a miniaturized scale. It is a rotary engine capable of 100,000 rpm, hardwired into a sensory mechanism so that it gets feedback from its surrounding environment. The flagella motor consists of: **(1)** two gears, forward and reverse, water-cooled, proton motive force, **(2)** a stator, **(3)** a rotor, **(4)** a u-joint, **(5)** a drive shaft, and **(6)** a propeller. These components function as their names indicate. The flagella motor has about 40 different parts that are necessary for it to work (Figures 14.1, 14.1a). If any of those parts are missing the machine simply will not function. In evolutionary terms you have to be able to explain how natural processes built these systems gradually, when there was no function (or no useful function), until you have all those parts in place. The bacterial flagellum not only requires 40 parts to function but like many complex machines, it also requires a precise sequence of assembly. Other complex molecular machines perform the assemblies of these molecular machines within the cell. There are tens of thousands of such functions being performed within the cell. There are numerous examples of irreducible complexity throughout the entire cell.[14-18]

Darwinists claim that the flagellum, a tail-like structure that mainly propels single prokaryotic cells (bacteria) (see Chapter 5, Figure 5.1), arose by natural processes, including natural selection. (Natural selection is discussed in Chapter 9.) For natural selection to select a cell that has a change, a cell must first change by mutation or some other process, such as **(1)** exon shuffling (domain shuffling), **(2)** gene duplication, **(3)** retroposition, **(4)** mobile genetic elements (transposable elements), **(5)** lateral gene transfer, **(6)** gene fusion, **(7)** de novo gene origination.

There are enormous difficulties that actual molecules would encounter in the cell for the flagellum to arise. It is questionable that an irreducibly complex molecular system could arise from individual functional precursors. For example, for individual proteins that perhaps existed in the cell already, there are six different proteins, and the complex molecular machine now is, as supposed by Darwinists, to be a combination of all six proteins with a new function that the system has that the individual parts previously did not have. Proteins are the components of molecular machines. Certain proteins stick to other proteins. If they did not stick together they would just float away and not be very useful for a complex system. It is known how proteins stick to other proteins, however, it is very difficult to explain why certain proteins stick to other proteins. Darwinists claim that complex systems arose by obtaining various preexisting components that performed different functions before becoming part of the complex molecular system.

To use mechanical machines as an analogy, if you put a part in a place different from where it usually is, that often times breaks the machine. If you took the propeller of an outboard motor and you put it on top instead of down by the rotor, then the machine would not function. The same is true of molecular machines. If an independent component (protein) attempts to become part of a more complex molecule, the system ceases to function. The new component (protein) would cause the system to malfunction and will not function at all.[14-19]

The Flagellum, Type III Secretion System, and Irreducible Complexity

Darwinist View

The flagellum is a minuscule object consisting of an internal molecular motor system far superior to any mechanical design man has conceived. For this reason, it is recognized by creationists as an excellent example of irreducible complexity. The flagellum is indeed spectacular but it is not irreducible. The molecular motor of the flagella consists of 40 complex proteins. Creationists and Intelligent Design advocates believe the absence of just one of these proteins would cause the functioning of the flagella to cease all together. To the average person, this may seem irreducibly complex, but according to fairly recent scientific research, this is not the case. In many ways, the most popular example for irreducible complexity has been the bacterial flagellum. The well-matched parts of the bacterial flagellum ion-driven rotary engine pose a seemingly impossible challenge to Darwinian evolution. Once again, however, upon a close examination of this remarkable biochemical machine a solution becomes possible.

There is more than one type of "bacterial flagellum." Flagella found in the archae-bacteria are clearly not irreducibly complex. Recent research has shown that the flagellar proteins of these organisms are closely related to a group of cell surface proteins known as the Class IV pilins. Since these proteins have a well-defined function that is not related to mobility, the archael flagella fail the test of irreducible complexity. Recently the flagella of eubacteria were discovered to be closely related to a non-motile cell membrane complex known as the Type III secretory apparatus. When biologists recently uncovered the structure of the basal body of the flagella, they concluded that the structure is very similar to the Type III Secretion System (commonly abbreviated as TTSS) (See Figure 14.1 base of flagellum). TTSS is a needle-like structure used by pathogenic germs to inject toxins into a living eukaryotic cell. The base of the needle has 10 elements in common with the flagellum, but is missing the complete 40 proteins needed to allow the flagellum to function. As a result, scientists have concluded the structure of TTSS disproves the claim made by Intelligent Design advocates that taking proteins away from the flagellum would cease its functionality.

These complexes (TTSS) play a deadly role in the cytotoxic (cell-killing) activities of bacteria such as Yersinia pestis, the bacterium that causes bubonic plague. When these bacteria infect an organism, bacteria cells bind to host cells, and then pump toxins directly through the secretory apparatus into the host cytoplasm. Efforts to understand the deadly effects of these bacteria on their hosts led to molecular studies of the proteins in the Type III apparatus, and it quickly became apparent that at least 10 of them are homologous to proteins which form part of the base of the bacterial flagellum (Figure 14.1). This means that a portion of the whip-like bacterial flagellum functions as the "syringe" that makes up the Type III secretory apparatus. In other words, a subset of the proteins of the flagellum is fully functional in a completely different context, not motility, but the deadly delivery of toxins to a host cell. This observation disproves the central claim of the biochemical argument from design, namely, that a subset of the parts of an irreducibly complex structure must be, "by definition nonfunctional." There are 10 proteins from the flagellum which are missing not just one part but more than 40, and yet they are fully-functional in the Type III apparatus.[14-20] Since the bacterial flagellum consists of more than forty proteins and enzymes, creationists and Intelligent Design advocates claim that it is extremely unlikely that so many elements could be assembled by chance. In reality, cell structures are continually reusing parts for other functions. The proteins or combinations of proteins are not as unique as some believe. The proteins of flagellum also function differently in other areas of the cell. A small portion of those same proteins (or similar proteins) is found in other bacteria in the Type III apparatus.

Creationists and Intelligent Design advocates rely on an Intelligent Design whenever a scientific explanation is unknown. This kind of thinking prevents further research. Although science has not yet been able to explain how some irreducible complex organisms developed, there is no reason to assume some type of intelligence did it. Further research needs to be done.

As mentioned above, a subset of the proteins from the flagellum does have a selectable function in Type III secretion. However, a more general statement can be made about many of the components of the eubacterial flagellum.

Proteins that make up the flagellum itself are closely related to a variety of cell surface proteins, including the pilins found in a variety of bacteria. A portion of the flagellum functions as an ion channel, and ion channels are found in all bacterial cell membranes (See Chapter 5, section "The Origin of the Cell Membrane" that discusses the origin of the cell membrane). All bacteria possess a membrane protein complex known as the ATP synthase that uses ion movements to produce ATP. The synthase uses the energy of ion movements to produce rotary motion. In short, at least four key elements of the eubacterial flagellum have other selectable functions in the cell that are unrelated to motility. Even the eubacterial flagellum consists of components that have other selectable functions. The four components are: **(1)** The flagellins are cell surface proteins like the pilins. **(2)** The ion channels are present in all cells. **(3)** The rotary motor is present in all cells. **(4)** The protein secretion through the membrane is found in all cells. These facts demonstrate that the one system most widely cited as the premier example of irreducible complexity contains individual parts that have selectable functions. What this means, in scientific terms, is that the hypothesis of irreducible complexity is falsified.

One might claim that since a detailed, step-by-step explanation of the evolution of the flagellum has not been provided, that the development of irreducible complex organisms cannot be explained. This lack of explanation would therefore appear to indicate that there exists organisms that are too complex to have self-assembled without any intelligence. Actually, it is true that currently no evolutionary explanation of the eubacterial flagellum has been written, and because no explanation has currently been provided, one might assume the argument of there being irreducible complex organisms must be correct. Actually, the argument would be true if an evolutionary explanation of the eubacterial flagellum would be necessary to disprove the idea of there being irreducible complex organisms.[14-21]

Intelligent design advocates claim that a biological system is irreducibly complex because unless all the parts are assembled, the individual parts could not have existed prior to combining and forming a more complex system. This is because, as they claim, the individual parts would not have had a useful function prior to combining into a more complex structure and therefore would have been eliminated by natural selection. (Natural selection is discussed in Chapter 9.) Therefore, scientists should be able to take the bacterial flagellum, for example, break its parts down, and discover that none of the parts are good for anything except when they are all assembled in a flagellum, if the intelligent design advocates' claims are true. If evolutionary theory holds true, however, then this claim would be falsified. If the intelligent design advocates' claims are true then when we look at these irreducibly complex structures, we ought to be able to find parts of those systems that actually do have useful functions within them that are found nowhere else in any other organism.

If 30 parts (proteins) are removed from the bacterial flagellum, only 10 proteins will remain that span the inner and outer membrane. Many of these bacteria are surrounded by two membranes. These 10 remaining parts are located at the very base of the flagellum near one of the cellular membranes. These 10 remaining parts are known to microbiologists as the Type III Secretory System (Figure 14.1—see the base of flagellum). The Type III Secretory System is a little molecular syringe that some of the nastiest bacteria in all of nature have. Yrsinia pestis, for example, which is the organism that causes bubonic plague, is a type III secretor. And what it does is, it gets inside one's body, crawls up alongside, and uses a syringe to inject poisons into a human cell. The connection between this and the flagellum is that the 10 proteins in the Type III Secretory System are almost a precise match for the corresponding 10 proteins in the base of the bacterial flagellum. A subset of those proteins has an entirely different function, a beneficial function, not for humans, but for the bacterium, and a function that can and is favored by natural selection.[14-22]

Intelligent Design View

The closest thing that evolutionary biologists have been able to find as a possible evolutionary precursor to the bacterial flagellum is what is known as a Type III Secretory System (TTSS) (See Figure 14.1 base of flagellum). The TTSS is a type of pump that enables certain pathogenic bacteria to inject virus-type proteins into host organisms. One bacterium possessing the TTSS is Yersinia pestis, the organism responsible for the black plague that during the Fourteenth Century killed a third of the population of Europe. The TTSS was the delivery system by which Yersinia pestis inflicted its massive destruction of human life. The ten or so proteins that go into the construction of the TTSS are similar (homologous) to proteins found in the bacterial flagellum

(Figure 14.1). Also, the TTSS corresponds roughly to the part of the flagellum used in the construction of its filament, the long whip-like tail. However it is not simply a matter of substituting the TTSS for the corresponding part of the bacterial flagellum to have a functioning flagellum. This is because the proteins in the TTSS are not adapted to the proteins of the bacterial flagellum, the resulting monstrosity would be nonfunctional. Despite such difficulties relating the TTSS to the bacterial flagellum, suppose the TTSS is treated as a subsystem of the flagellum. As a subsystem, it performs a function distinct from the flagellum. Finding a subsystem of a functional system that performs some other function is hardly an argument for the original system evolving from that other system. Indeed, multipart, tightly integrated functional systems almost invariably contain multipart subsystems that could serve some different function.

The TTSS does represent one possible step in the indirect Darwinian evolution of the bacterial flagellum. This, however, would still not provide a solution as to how the bacterial flagellum could have evolved. What is needed is a complete evolutionary path and not merely a possible scenario. There is another problem here as well. The whole point of mentioning the TTSS was to speculate it as an evolutionary predecessor to the bacterial flagellum. The best current evidence put forward by evolutionary biologists, however, points to the TTSS as evolving from the flagellum and not vice versa. The bacterial flagellum is a motility structure for propelling a bacterium through its watery environment. Water has been around since the origin of life. Indeed, evolutionary biologists believe that the bacterial flagellum is 2 to 3 billion years old. But the TTSS is a delivery system for animal and plant pathogens. Its function therefore depends on existence of multicellular organisms. Accordingly, the TTSS could only have been around since the rise of multicellular organisms, which evolutionary biologists place around 600 million years ago. It follows that the TTSS does not explain the evolution of the flagellum.

The bacterial flagellum could explain the evolution of the TTSS, however, even that is not quite right. The TTSS is, after all, much simpler than the flagellum. The TTSS contains ten or so proteins that are homologous to proteins in the flagellum. The flagellum requires an additional thirty or forty proteins, which are unique. Evolution needs to explain the emergence of complexity from simplicity. But if the TTSS evolved from the flagellum, then all that has been done was to explain the simpler in terms of the more complex, but have not explained how the more complex flagellum originated.

Despite these difficulties, Darwinists continue to speculate the TTSS as an evolutionary predecessor to the bacterial flagellum. Some of them even go so far as to speculate a few intermediate structures by which the TTSS is supposed to have evolved into bacterial flagellum.

To justify how the flagellum could have evolved, Darwinists need to show that each step in it is reasonably likely to follow from the previous one. This requires being able to assess the probability of transitioning from one step to the next. And this in turn presupposes that the biological structures at each step are described in sufficient detail so that it is possible to assess the probabilities of transitioning between steps. Evolutionary biologists never provide any details to explain their models. The steps in these models are so unspecific and absent of detail that these questions are unanswerable. To actually test such models requires being able to evaluate the likelihood of transitioning from one step in the model to the next. Yet because the intermediate systems described at the various transitional steps are so lacking in detail, as they are hypothetical, they do not, as far as we know, currently exist in nature, they are not available in any laboratory, and researchers for now have no experimental procedures for generating them in the laboratory, the models offer no way to carry out this evaluation.[14-23]

Old Universe Progressive Creationist View

Evolutionary biologists claim that irreducibly complex systems did not originate all at once but could have emerged in a stepwise fashion. According to this view, the biomolecules of irreducibly complex biochemical systems originally played other roles in the cell and were later recruited (or incorporated) or co-opted, one-by-one, to be part of transitional systems that eventually led to the irreducibly complex systems of contemporary biochemical operations. Some evolutionary biologists claim that the bacterial flagellum evolved from the type III secretion apparatus through the process of co-option. Pathogenic bacteria use the type III secretion system to export proteins into the cells of the host organism. The molecular architecture of the type III secretion system closely resembles the part of the flagellum embedded in the bacterial cell envelope. Evolutionists believe that the first flagellum originated from the merger of the type III secretion apparatus and a filamentous protein system. Presumably, both structures provided the microbe with prior services, neither of which had anything to do with motility (the ability of an organism or fluid to move). The claims made by evolutionists seems possible at first glance, however, these claims have little, if any, factual support. Invariably, the naturalistic scenarios proposed to account for the origin of irreducibly complex systems are highly speculative and lack any type of detailed convincing explanations. This problem is clearly the case for all the evolutionary explanations offered to account for the origin of bacterial flagellum. In addition to the speculative nature of the evolutionary explanations for the origin of the bacterial flagellum is the problem of which came first; the flagellum or the type III secretion system. Evolutionists realize that there is insufficient evidence to determine whether (1) the type III secretion apparatus originated first or (2) the bacterial flagellum came first, or (3) both biological systems evolved from the same precursor system. If evolutionary analyses indicate that the type III secretion system emerged from the flagellum, it thoroughly weakens the co-option explanation. On the other hand, the type III secretion system, the proposed evolutionary stepping stone to the bacterial flagellum, is an irreducibly complex structure.

It appears that the type III secretion system is the result of intelligent design. Further evidence for intelligent design comes from the recognition that highly similar flagellar systems appear to have emerged independently (from an evolutionary perspective) in bacteria and archaea. It makes sense that an Intelligent Designer designed the type III secretion system to be used for more than one purpose, just as a human designer might design a piece of machinery to be adaptable for more than one use.

Evolutionary biologists have the burden of proof to support their claims. If irreducible complex systems do, indeed, have an evolutionary origin, evolutionists must provide a detailed explanation for each step in the sequence of molecular events that caused the origin of the type III secretion system as well as the entire bacterial flagellum. Also, evolutionists must demonstrate that this proposed sequence of steps could have happened in the available time and with the resources available at its disposal. It is not enough to merely propose a chronology of events that "may have happened" or "most likely took place." Virtually every evolutionary explanation for irreducible complex biological systems is proposed with these types of qualifiers. Evolutionists need to provide adequate evidence and detailed explanations to support their claims. [14-23a]

An Example of a Bacterium Evolving

Darwinist View

Escherichia coli (E. coli) is a good example of a bacterium that uses the flagellum proteins to form a different sort of mechanism. (See Chapter 5, Figure 5.1—Prokaryotic Cell (structure of a bacterium) and Figure 14.1—Flagellum.) In E. coli, the flagellum is built with a kind of pump that squirts out proteins. Scientists believe that E. coli flagellum may have begun like the pump and syringe of other pathogens. The syringe then developed a long needle and a base with a flexible hook. Finally, it linked with another kind of pump. This gave the flagellum a spinning movement where the needle became the propeller. E. coli has a tail structure but also carries remnants of genes for a second type of tail.[14-24]

Researchers have conducted an experiment on an E. coli bacterium that was forced to adapt or perish. The experiment was conducted to determine how a bacterium would react under adverse conditions. The experiment was to address the question: Would the organism find a way to survive or would it die? During the experiment it was observed that the E. coli bacterium developed a new way of making molecular bolts called "disulfide bonds," which are stiffening struts in proteins that also help the proteins fold into their proper, functional, three-dimensional shapes. This new method restarted the bacterium's motor and enabled it to move toward food before it starved to death. A particular strain of mutant bacteria was developed in a laboratory that lost its ability to make disulfide bonds. These disulfide bonds are essential for the bacteria's ability for its propeller-like swimming motor, the flagellum, to work. These mutated bacteria were then no longer able to move, as their flagellum lost its ability to propel the bacteria. The mutant bacterium has the same flagellum that Intelligent Design theorists often use to support their theory of irreducible complexity.

Another experiment was conducted to see how these same mutated bacteria would react if alterations were made in the DNA that encodes the thioredoxin. The researchers wanted to see if an altered version of thioredoxin could be coerced to make disulfide bonds for other proteins in the bacteria. They made random alterations in the DNA encoding thioredoxin in the mutated bacteria, in a process similar to natural selection, that is, to simulate natural selection. (Natural selection is discussed in Chapter 9). The researchers then subjected thousands of these mutated non-swimming bacteria to a test to see if they would survive. The researchers placed the mutated non-swimming bacteria on a dish of food where, once they had exhausted the food they could reach, they either had to repair the broken motor in their flagellum or starve.

What happened was that a mutant bacterium, carrying only two amino acid changes in thioredoxin, restored its ability to move. The motor in the mutated bacteria's flagellum had been repaired by the altered thioredoxin. The bacteria in the experiment were able to use the altered thioredoxin protein, which normally destroys disulfide bonds, to make the disulfide bonds instead. The altered thioredoxin was able to carry out disulfide bond formation in numerous other bacterial proteins all by itself, without relying on any of the components of the natural disulfide bond pathway. The mutant bacteria managed, with the altered thioredoxin protein, to solve the problem and swim away from starvation. The naturally occurring enzymes involved in disulfide bond formation are a biological pathway whose main features are the same from bacteria to man. Normally, Intelligent Design advocates would claim that the researchers acted as an intelligent designer. However, in this case, the researchers were acting as the force of natural selection because they made random changes to the bacteria. This experiment provides strong support for evolution.[14-25]

Creationist View

Escherichia coli (E. coli) are a small bacterium that has been studied quite a bit. This little bacterium is one of the smallest living creatures known, and yet it carries genetic information within its cells that is so complicated that after studying and applying mathematics to this bacterium, it was decided that the DNA code within this single bacterium could not have been formed by chance, even if given 5 billion years to evolve. In spite of such a scientific fact, evolutionists teach that all of life, including that single bacterium and every other living thing in our world, originated out of nothing and evolved by random chance within the last 5 billion years.[14-26]

We will now take a look at some examples of how bacteria adapt to their environment without the need to declare that the changes are examples of evolution. (Chapters 5 & 6 discuss the origins of life.)

E. coli develops ability to process citrate

Creationist View

Over several decades, naturalistic scientists conducted extensive and intensive evolutionary experiments on bacteria. They tracked changes in over 40,000 generations of E. coli. Although these bacteria normally cannot import citrate (salt or ester of citric acid) in the presence of oxygen, after 30,000 generations, some did. This raised many questions. Then, many experiments were conducted in an effort to answer the question: Did the bacteria invent a new mechanism to import citrate? If so, then how? Some were claiming that this confirms evolution in action, but what if the bacteria were designed to modify themselves? Naturalistic scientists wanted to determine whether or not the strain of E. coli that acquired the new ability to import citrate, called Cit+, was able to construct new functional, biochemical machinery by chance-base mutations. To verify that this new trait represents the kind of evolution that reflects Darwinian evolution, naturalistic researchers needed to determine exactly what occurred for bacteria to develop this new trait. (Prior to this new trait, none of the other observed changes that researchers had tracked showed that the bacteria evolved into a different basic kind.) After much research, it was determined that even though the bacteria developed a new trait, they were all still E. coli. And so far, none of those changes specified any new functional coding. (The research was summarized in Nature magazine. The article summary states in part: *"At least three distinct clades coexisted for more than 10,000 generations before its emergence. The Cit(+) trait originated in one clade by a tandem duplication that captured an aerobically expressed promoter for the expression of a previously silent citrate transporter. The clades varied in their propensity to evolve this novel trait, although genotypes able to do so existed in all three clades, implying that multiple potentiating mutations arose during the population's history."* [14-26a])[14-27]

As mentioned previously, E. coli are able to metabolize (process) citrate. It already had that ability in non-oxygen environments. Intelligent Design researchers stated that, after 30,000 generations, one of their lines of cells has developed the ability to utilize citrate as a food source in the presence of oxygen. E. coli in the wild cannot do that. Now, wild E. coli already has a number of enzymes that normally use citrate and can digest it. The wild bacterium, however, lacks an enzyme called a "citrate permease" which can transport citrate from outside the cell through the cell's membrane into its interior. So all the bacterium needed to do to use citrate was to find a way to get it into the cell. The rest of the machinery for its metabolism was already there. Some evolutionary researchers have observed that the only known barrier to aerobic growth on citrate is its inability to transport citrate under conditions when oxygen was present. Researchers in the past several decades have also identified mutant E. coli that could use citrate as a food source. In one instance a protein coded by a gene called citT, which normally transports citrate in the absence of oxygen, was over expressed (constitutively expressed). The over expressed protein allowed E. coli to grow on citrate in the presence of oxygen. It seems likely that other mutant genes will turn out to be either this gene (citT) or another of the bacterium's citrate-using genes that were tweaked somewhat to allow it to transport citrate in the presence of oxygen.

Scientists speculate that there are three possible explanations for the ability of E-coli to digest citrate: (1) A rare single mutation (for example, more than one point mutation in the right places in only one generation). (2) A rare chromosome inversion (The gene string gets inserted backwards.) (3) The accumulation of several mutations in a sequence. (A point mutation (also called gene mutation) is a mutation that changes only one small area or one nucleotide in a gene.)

Researchers provided two possibilities of how two mutations could eventually develop the ability to transport citrate into the cell for digestion, including the physiological mechanism that would be involved. (1) One mechanism is Epitasis (the effects of one gene that is modified by one or several other genes), whereby the functional expression of the mutation that finally yielded the Cit+ phenotype requires interaction with one of more mutations that evolved earlier. (2) A second possibility is that the physical production of the mutation that produced the Cit+ phenotype requires some previous mutation that allows the final sequence to be generated. For example, the insertion of a mobile genetic element creates new sequences at its junctures, and one of these new sequences might then undergo a mutation that generates a final sequence that could not have occurred without the insertion.

Scientists have speculated about the mechanisms of how the E coli bacteria could develop the ability to digest citrate. (1) One possibility is that the Cit+ lineage activated a 'cryptic' (preexisting but dormant) transporter, that is, some once-functional gene that has been silenced by mutation accumulation. (2) A second and more likely possibility is that an existing transporter has been utilized for citrate transport when oxygen was present. This transporter may have previously transported citrate under anoxic conditions or, alternatively, it may have transported another substrate in the presence of oxygen. The evolved changes might involve gene regulation, protein structure, or both. Some scientists prefer the second possibility because they say that the first possibility seems unlikely because the Cit—phenotype is characteristic of the entire species, one that is very diverse and therefore very old. One would expect a cryptic gene to be degraded beyond recovery.[14-28] Regardless of which option may explain the ability to digest citrate, there is no evidence that suggests mutations caused a positive change in the bacteria. No new traits were formed as a result of mutations and these changes did not result in a new species.

E. coli evolves into a Cit+ strain

Creationist View

One Intelligent Design scientist reviewed 12 new phenotypes, which are outward expressions of genetic coding that previous experiments conducted on E. coli had revealed. This Intelligent Design scientist categorized the known genetics producing each new bacterial phenotype as either losing, shuffling, or gaining what is called "functional coded elements," which include genes and gene promoters. All the known changes in the bacteria were either a loss or reorganization of pre-existing functional coded elements. None of the new phenotypes came from a gain of functional coded elements, and yet this is what Darwinian evolution theory requires. The question is: Did E. coli evolve into a Cit+ strain by mutations that constructed new and functional coded elements to its DNA? If so, it would be the first in recorded biological history. If not, then it would be just another loss or modification of a pre-existing piece. (Mutations are discussed in Chapter 9.)

In one experiment, the bacteria (both Cit+ and wild-type) already possessed a gene named citT. It encodes a protein that transports a range of citrate-like chemicals. The results showed that the bacteria made extra copies of the gene citT and a neighboring sequence, a process called "gene amplification." More copies of the gene should translate to higher amounts of the transporter protein that it encodes. With enough transporters, the bacteria could access enough citrate. The problem is that oxygen deactivates the gene citT and having many copies of a gene that is turned off is not very useful. But the bacteria solved this problem of deactivating the gene citT when the amplification event also moved the gene sequence to a different place in the bacterial chromosome, where a different but pre-existing promoter could regulate it. Unlike the original one, it appears that the new promoter does not have an "oxygen off" switching mode. Instead, it allowed expression of citT in the presence of oxygen so that the bacteria successfully imported enough citrate to grow. So, the bacteria solved the problem of accessing an alternative food source by generating extra copies of the critical gene and by placing those copies under the control of an appropriate promoter. This does not resemble natural, undirected Darwinian evolution. This mechanism did not develop new functional coded elements. Instead, it merely modified pre-existing elements. Therefore, not only did the Cit+ bacteria not evolve as evolutionary theory stipulates, but they demonstrated how ingenious these preexisting DNA rearrangement mechanisms actually are.[14-29]

E. coli processes lactose

Creationist View

It was known previously that bacteria could digest different types of sugars, including the most common kind, called glucose, as well as another, much less common sugar, called lactose, which is found in milk. When bacteria were grown in the presence of glucose, they could not use lactose. Only in the absence of glucose and the presence of lactose could they digest the milk sugar. When glucose was missing, the bacteria made proteins that could pull lactose into the cell and metabolize it, but when no lactose was around, the bacteria did not make those proteins. (Metabolism is the sum of the physical and chemical processes in an organism by which its material substance is produced, maintained, and destroyed, and by which energy is made available.) This made great biological sense, since in normal conditions the bacterium would waste energy if it manufactured proteins that could metabolize only a rarely encountered sugar. The interesting question was, how did the bacteria "know" when to switch on the genes for making the proteins?

Some evolutionary researchers have discovered a defective mutant bacterium that made lactose-using proteins all the time, even in the absence of lactose. It was lacking a control mechanism. These scientists reasoned that the bacteria contained another, hidden protein, which they called a "repressor." They surmised that the repressor would ordinarily bind to a specific sequence of DNA near the genes that generated the lactose-using proteins, switching them off. In the presence of lactose, the milk sugar would bind to the repressor itself, changing the protein's shape enough to make it fall off the DNA, switching back on the previously blocked genes. These evolutionary researchers correctly concluded that the mutant bacteria had a broken repressor. This research indicates that there really is not much new information being presented, at least in terms of explaining how evolution can proceed past adaptation to create new information. (See Chapter 7 section "The Evolution of the Horse" "Creationist View," Chapter 9 section "Hox Genes," and Chapter 13, section "Some DNA Portions are Used to Turn Genes On or Off" for further discussion of genes being turned on and off.)

Starvation is frequently a state that bacteria have to endure. In order to survive, they need to have special adaptive capabilities. Both adaptive mutation and antagonistic pleiotropy come into play. *(Antagonistic pleiotropy is when an existing system is traded for an altered phenotype that is better suited to survive the specific stressful environment. Pleiotropy is the phenomenon of one gene being responsible for or affecting more than one phenotypic characteristic. Pleiotropy is a condition in which single gene affects many characters or single gene affects many phenotypes. Pleiotropy occurs when one gene influences two or more seemingly unrelated phenotypic traits.)* A number of research experiments using E-coli and other bacteria have been conducted to determine how bacteria would react under certain conditions. Each of the mutant strains had an antagonistic pleiotropy characteristic. An existing system was traded for an altered phenotype that was better suited to survive the specific stressful environment. Regulation was reduced to enable over expression. DNA repair and DNA polymerase fidelity were reduced to enable increased mutation rates (increasing the probability of a "beneficial" mutation). A gene was inactivated by a process that concurrently activates a silent gene. Such trade-offs provide a temporary benefit to the bacterium, thus increasing its chances of surviving specific starvation conditions. However, these mutations do not account for the origin of the silenced genes, as their prior existence is essential for the mutation to be beneficial.

Bacteria have remarkable capabilities to mutate and to survive as is seen with antibiotic resistance. But these capabilities do not apply to multi-celled organisms as a whole nor do they provide a genetic mechanism that accounts for the origin of biological systems or functions. Rather, they require the prior existence of the targeted cellular systems. It must be pointed out that all of the known mechanisms have a limited ability to make new cellular structures. (1) Point mutation, (2) deletion, (3) insertion, (4) gene duplication, (5) transposition, (6) genome duplication, (7) self-organization, (8) self-engineering, or any other processes currently known have not been able to explain how bacteria develop new cellular structures. Despite the evolutionary experiments that provide insight into the adaption of bacteria to their food supply, these experiments have failed to demonstrate the mechanism for evolution.

Natural selection is active in nature and mutations explain the adaptation of bacteria to their environment. (Natural selection and mutations are discussed in Chapter 9.) Bacteria are a different domain of life and they have remarkable capabilities, but these mechanisms would largely not apply to higher organisms. The adaptability of bacteria to starvation does not begin to reveal a mechanism for macroevolution. Currently, there is no mechanism known to explain life or its macroevolutionary path.[14-30]

Cilium

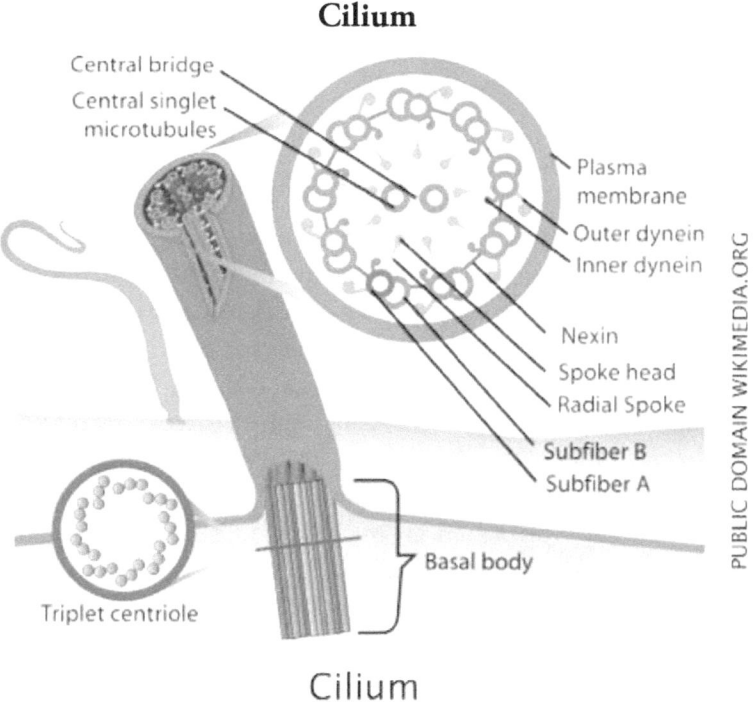

Cilium

Figure 14.2 Cilium

The Cilium and Irreducible Complexity

Creationist View

Cilia (Figure 14.2) are hair-like organelles on the surfaces of many animal and lower plant eukaryotic cells (Chapter 5, Figure 5.3) that serve (1) to move fluid over the cell's surface or (2) to "row" single cells through a fluid. In humans, for example, epithelial cells lining the respiratory tract each have about 200 cilia that beat in synchrony to sweep mucus towards the throat for elimination.[14-31] There are two types of cilium. One is the eukaryotic motile cilium and the other is the eukaryotic flagellum. Structurally, they are both identical. The only difference is that eukaryotic cilia are short, hair like appendages extending from the surface of a living cell. Eukaryotic flagellum are long, threadlike appendages on the surface of a living cell. Eukaryotic cilia has slender, microscopic, short hair-like structure whereas eukaryotic flagella have long hair-like filamentous cytoplasmic complex structure. The biophysical force of the eukaryotic flagellum that causes motion in the base of the structure is based on the use of ATP (adenosine triphosphate) produced in the mitochondria, a completely different type of system than that used in bacteria. The bacterial flagellum uses a proton pump system.[14-30a]

A cilium consists of a membrane-coated bundle of fibers called an axoneme. An axoneme contains a ring of 9 double microtubules surrounding two central single microtubules. Each outer doublet consists of a ring of 13 filaments (subfiber 'A') fused to an assembly of 10 filaments (subfiber 'B'). The filaments of the microtubules are composed of two proteins called alpha and beta tubulin. The 11 microtubules forming an axoneme are held together by three types of connectors: (1) subfibers 'A' are joined to the central microtubules by radial spokes; (2) adjacent outer doublets are joined by linkers that consist of a highly elastic protein called nexin; and (3) the central microtubules are joined by a connecting bridge. Finally, every subfiber 'A' bears two arms, an inner arm and an outer arm, both containing the protein dynein.

Experiments have indicated that ciliary motion results from the chemically-powered "walking" of the dynein arms on one microtubule up the adjacent subfiber 'B' of a second microtubule so that the two microtubules slide past each other. However, the protein cross-links between microtubules in an intact cilium prevent neighboring microtubules from sliding past each other by more than a short distance. These cross-links, therefore, convert the dynein-induced sliding motion to a bending motion of the entire axoneme.

Cilia are composed of at least a half dozen proteins: **(1)** alpha-tubulin, **(2)** beta-tubulin, **(3)** dynein, **(4)** nexin, **(5)** spoke protein, and **(6)** a central bridge protein. These combine to perform one task, ciliary motion, and all of these proteins must be present for the cilium to function. If the tubulins are absent, then there are no filaments to slide. If the dynein is missing, then the cilium remains rigid and motionless. If nexin or the other connecting proteins are missing, then the axoneme falls apart when the filaments slide. The cilium, as it is made up, must have the **(1)** sliding filaments, **(2)** connecting proteins, and **(3)** motor proteins for function to occur. In the absence of any one of those components, the apparatus is useless. It is believed by some scientists (Intelligent Design advocates) that if an organism is too complex to have evolved from simpler organisms, then it is said to be "Irreducible Complex", that is, too complex to have self-assembled. The majority of evolutionary scientists (Darwinists), however, disagree with this view and believe that every organism is capable of having evolved. It may be that technology has not yet advanced enough to be able to explain how some complex organisms evolved.[14-31]

In addition to the bacterium flagellum and the cilium, there are also other irreducible complex organic systems, such as **(1)** the DNA and RNA information storage systems, **(2)** the sliding clamp, and **(3)** the gated portal that are irreducible complex. The DNA and RNA systems are very sophisticated information storage systems. DNA and RNA are information systems that store digital information. DNA is an information system that is optimized to wipe out point mutations, rather than resist them, which are necessary for that selection to function. (A point mutation (also called gene mutation) is a mutation that changes only one small area or one nucleotide in a gene.) The sliding clamp is a DNA polymerase (Chapter 5, Figures 5.20 & 5.21) that is the copying mechanism for DNA replication. (DNA replication is the biological process of producing two identical replicas of DNA from one original DNA molecule - See Chapter 5, section "DNA Replication.") The question is: How could these systems have originated by natural processes? There are also the macro-molecular machines that are highly complex, with ordered parts that are considered irreducibly complex.

The clamp protein has been found in E. coli. The clamp protein is a more primitive organism. It is a beta subunit of E. coli polymerase. The clamp protein forms a ring around this double helix of DNA. Its origin is unknown. If the protein sequences that form the structure between E. coli and yeast are compared, they would not seem to be similar. One might think that replicating DNA would be one highly basic process if the prokaryotic cell

(Chapter 5, Figure 5.1) eventually evolved into the eukaryotic cell (Chapter 5, Figure 5.2) from a common ancestor, but instead there is a protein that has a similar structure, forming the same function, but completely different amino acid sequences. This eliminates the hypothesis that the prokaryotic cell eventually evolved into the eukaryotic cell. The gated portal (an informational transport system) is found in the nucleus of a eukaryotic organism (Chapter 5, Figure 5.4). The gated portal is a gate that allows the material that needs to come from the nucleus to the outside or from the outside back into the nucleus. Evolution is not able to explain the origin of this system or any of the systems above just described.[14-32]

Intelligent Design View

Every cilium that has been examined to date (1985) has been found to possess essentially the same basic structure. There is no hint anywhere of any structure halfway to the complex molecular organization of these fascinating micro-hairs through which their evolution might have occurred. [14-33]

Darwinist View

One of the structures that can be used to investigate irreducibly complex systems is the eukaryotic cilium (Figure 14.2), an intricate whip-like structure that produces movement in cells as diverse as green algae and human sperm. The argument developed by intelligent design advocates regarding the removal of any one of the parts of an irreducibly complex system effectively causes the system to stop working, is simply not true. The cilium provides us with a perfect opportunity to test that assertion. If it is correct, then scientists should be unable to find examples of functional cilia anywhere in nature that lack the cilium's basic parts. Contrary to those who believe in intelligent design, there are examples of functional cilia that have been found that do not have all the basic parts. Nature presents many examples of fully functional cilia that are missing key parts. One of the most compelling is the eel sperm flagellum, which lacks at least three important parts normally found in the cilium: (1) the central doublet, (2) central spokes, and (3) the dynein outer arm.[14-34] Cilia and flagella have a very similar internal arrangement of tubes.[14-35]

It is also known that a wide variety of systems having the ability to move do exist that are missing parts of this supposedly irreducibly complex structure. Also, biologists have known for years that each of the major components of the cilium, including proteins (1) tubulin, (2) dynein, and (3) actin, all have distinct functions elsewhere in the cell that are not related to ciliary motion. Those who believe in irreducible complexity claim that the parts of an irreducibly complex structure have no function on their own. The individual parts of the cilium, including (1) tubulin, (2) the motor protein dynein, and (3) the contractile protein actin are fully functional elsewhere in the cell. What this means, of course, is that a selectable function that has an independent different function exists for each of the major parts of the cilium, and therefore the argument that the cilium is an irreducible complex system is false.[14-36]

Intelligent Design View

The way the cilia functions is that there are nine pairs of microtubules, which are long, thin, flexible rods, which encircle two single microtubules. The outer microtubules are connected to each other by what are called "nexin linkers." And each microtubule has a motor protein called dynein. (Nexin linkers and dynein are illustrated in Figure 14.2.) The motor protein attaches to one microtubule and has an arm that reaches over, grabs the other one, and pushes it down. So the two rods start to slide lengthwise with respect to each other. As they start to slide, the "nexin linkers," which were originally like loose rope, get stretched and become taut. As the dynein pushes farther and farther, it starts to bend the apparatus; then it pushes the other way and bends it back. This is how the rowing motion of the cilium is obtained.

There is more to the cilium than has been described above. These three parts: (1) the rods (microtubules), (2) linkers (nexin), and (3) motors (dynein), are necessary to convert a sliding motion into a bending motion so the cilium can move. If it were not for the linkers, everything would fall apart when the sliding motion began. If it were not for the motor protein, it would not move at all. If it were not for the rods, there would be nothing to move. The cilium only has motion when all the parts are working together. None of the parts are able to do this independently. For natural processes to do this one would have to imagine how this could develop gradually, which no one has yet been able to explain. Some Darwinists have pointed out that there are examples of other cilia that do not have some of the parts that creationists claim are essential. Darwinists claim there are many forms of cilia lacking one or more of the components supposedly essential to the function of the apparatus. If science could explain how a series of less complex structures that progress from one to another in order to create a functional cilia, then the concept of the cilia being irreducibly complex would be proven wrong. However, to date, science has not been able to demonstrate how functional cilia could have self-assembled.

Those who deny that any irreducible complex organisms exist make the claim that one of the several microtubules could be removed and the cilium would still function. Advocates of irreducible complexity do not dispute this claim. In reality, the cilum still have the same basic components: (1) microtubules (flexible rods), (2) nexin (linker), and (3) dynein (motor protein), and it still functions, even if some parts are missing. The cilium has some redundant components. One of the microtubules may be taken away and it will still function, though maybe not as well. Evolution does not start with the completed cilium and take parts away; it has to build things up from the bottom. And all cilia have the three critical components that have been mentioned (rods, linkers, and motors). There have been experiments where scientists have removed one of the three components and the cilium does not work.[14-37]

Coagulation Cascade Blood Clotting System

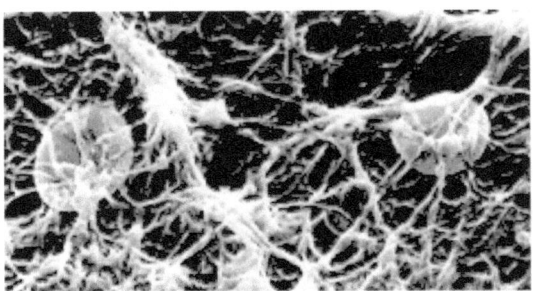

Blood Clot

Creationist View

A good illustration of irreducible complex organisms is the blood clotting system. It is a relatively simple process necessary for animal and human life. This system is necessary because it provides the ability for blood to clot to seal a wound and prevent an injured animal or person from bleeding to death. Yet the only way this intricate system works is when many complicated chemical substances interact. If only one ingredient is missing or does not function in the right way, as in the genetic blood disorder hemophilia, the process fails, and the victim bleeds to death.

How can complex substances appear at just the right time in the right proportions and mix properly to clot blood and prevent death? Either they function flawlessly or clotting does not work at all. Medical science is aware of the problems of blood clotting at the wrong time. Blood clots that cut off the flow of oxygen to the brain are a leading cause of strokes and often result in paralysis or death. With blood clots, either everything works perfectly or the likely outcome is death. For evolution to have led to this complicated process, multiple mutations of just the right kind had to converge simultaneously or the mutations would be useless. Evolutionists can offer no realistic explanation of how this is possible.[14-38]

Intelligent Design View

The critical process with blood clotting is not so much the clot itself. A blood clot is just a blob that blocks the flow of blood. The real critical process is the regulation of the system. If a clot is made in the wrong place, say, the brain or lung, death will occur. If a clot is made twenty minutes after all the blood has drained from the body, death will occur. If the blood clot is not confined to the cut, the entire blood system might solidify, and death will result. If a clot does not cover the entire length of the cut, death will result. To create a perfectly balanced blood-clotting system, clusters of protein components have to be inserted all at once. That rules out a gradualistic Darwinian approach and fits the hypothesis of an intelligent designer.[14-39] When scientists tried to come up with a step-by-step scenario of how blood-clotting could have developed, they were not able to do so. There is no meaningful scientific explanation of what could have caused the development of the blood-clotting cascade.[14-40]

The explanations that some people attempt are only theories. Science is supposed to complete experiments to show something is true. No one has ever performed experiments to show how blood-clotting could have developed. No one has been able to show how a duplicated gene can develop some new function where it starts to make a new and irreducibly complex pathway, that is, an organic substance that could convert to another organic substance.[14-41] One of the most important parts of this whole blood clotting machine is the ability it has to keep the clotting localized to the area of the wound and to stop the clotting cascade. Most heart attacks and strokes are caused by blood clots lodging in the wrong places.[14-42]

The Blood Clotting Cascade Process

Creationist View

Blood clotting is a biological system where a number of circulating proteins and blood cells combine to form a clot that plugs a site of injury thereby reducing blood loss and also reducing the risk of infection. Blood clotting involves over a dozen different proteins like thrombine, fibrinogen, accelerin and many more. Some of these proteins are involved in forming the clot. Others are responsible for regulating clot-formation. Regulating proteins are needed because there should only be clots forming at the site of a wound, not in the middle of flowing arteries. Other proteins take care of removing the clot once it is no longer needed. Biochemical research clearly shows that there are many factors involved in the process of blood clotting, and none of them should be missing for the process to succeed. If one factor fails, the whole system of clot-formation fails. So again this leaves us with the question: How could this system have been evolved from a 'simple' to a 'complex' form if it only functions as a whole?[14-43]

What Happens When We Cut Ourselves?

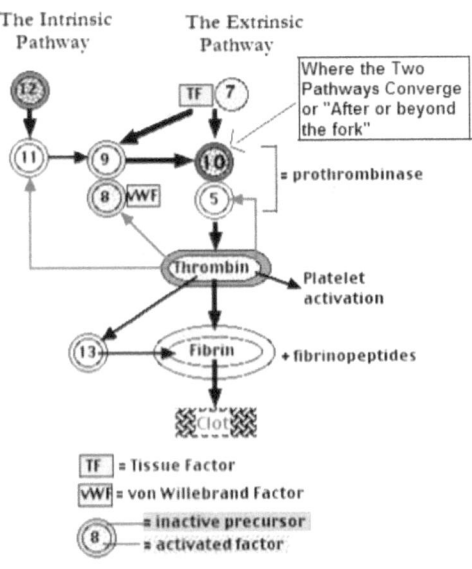

How blood clots in vertebrates

Figure 14.3 How blood clots in vertebrates.

(The circled numbers in the diagram above represent a specific Factor number. They are not step numbers.)

Land-dwelling vertebrates have two processes that can initiate blood clotting. A very rapid process so-called the "extrinsic pathway" (Figure 14.3—Extrinsic Pathway begins at upper right Factor 7) and a slower but larger "intrinsic pathway" (Figure 14.3—Intrinsic Pathway begins at upper left Factor 12). **The Extrinsic Pathway**—Damaged cells display a surface protein called tissue factor (TF). Tissue factor binds to activated Factor 7 (Figure 14.3). The TF-7 heterodimer is a protease with two substrates: Factor 9 and Factor 10. **The Intrinsic Pathway**—Factor 12 (also called the Hageman factor) circulates in the blood. If blood escapes into tissue spaces (e.g., as a result of an injury), contact with collagens in the tissue space activates Factor 12. Activated Factor 12 is a serine protease that activates Factor 11 that, in turn, activates Factor 9 that, in turn, activates Factor 10 (Figure 14.3). **The Two Pathways Converge**—Factor 10—produced by either or, more likely, both pathways—binds and activates Factor 5. This heterodimer is called prothrombinase because it is a protease that converts prothrombin (also known as Factor II) to thrombin. (Protease is a group of enzymes that catalyze the hydrolytic degradation of proteins or polypeptides to smaller amino acid polymers. Enzymes are a group of complex proteins or conjugated proteins that are produced by living cells and act as catalysts in specific biochemical reactions.) Thrombin has several different activities. Two of them are: **(1)** proteolytic cleavage of fibrinogen (aka "Factor I") to form soluble molecules of fibrin and **(2)** a collection of small fibrinopeptides activation of Factor 13 that forms covalent bonds between the soluble fibrin molecules converting them into an insoluble meshwork—the clot. (Thrombin and activated Factors 10 ("Xa") and 11 ("XIa") are also serine proteases.)[14-44]

How The Blood Clotting Cascade Works in Most Land-dwelling Vertebrates

1. A cut occurs and Hageman Factor (Factor 12 or XII) sticks to the surface of cells near the wound (Factor 12—See Figure 14.3—Intrinsic Pathway). Bound Hageman Factor reacts with another enzyme (proteins that increase or decrease the rates of chemical reactions) called HMK (high molecular weight kininogen) to produce Activated Hageman (XIIa).

2. Pre-Kallikrein reacts with Activated Hageman (XIIa) to produce Kallikrein.

3. Hageman Factor (Factor 12 or XII) also reacts with HMK and Kallikrein to form Activated Hageman.

4. Plasma Thromboplastin Antecedent (PTA) (XI), an enzyme, reacts with Activated Hageman (XIIa) and HMK to produce Activated PTA (XIa).

5. Christmas Factor (IX) reacts with Activated PTA (XIa) and Convertin to produce Activated Christmas Factor (IXa).

6. Antihemophilic Factor is activated by Thrombin (IIa) to produce Activated Antihemophilic Factor.

7. Stuart Factor (X) reacts with Activated Christmas Factor (IXa) and Activated Antihemophilic Factor to produce Activated Stuart Factor (Xa).

8. Proconvertin (VII) is activated by Activated Hageman Factor (XIIa) to produce Convertin.

9. When a cut occurs, Tissue Factor, which is only found outside of cells, is brought in near the wound where it reacts with Convertin and Stuart Factor (X) to produce Activated Stuart Factor (Xa).

10. Proaccelerin is activated by Thrombin (IIa) to produce Accelerin.

11a. GLU-Prothrombin reacts with Prothrombin Enzyme and Vitamin K to produce GLA-Prothrombin. Note that Prothrombin (Factor II) cannot be activated in the GLU form so it must be formed into the GLA form. In this process ten amino acids must be changed from glutamate to gama carboxy glutamate.

11b. GLS-Prothrombin is then able to bind to Calcium. This allows GLA-Prothrombin to stick to surfaces of cells. Only intact modified Calcium-Prothrombin Complex can bind to the cell membrane and be cleaved by Activated Stuart (Xa) and Accerlerin to produce Thrombin (Factor IIa).

12. Prothrombin-Ca (bound to cell surface) is activated by Activated Stuart (Xa) to produce Thrombin (Factor IIa).

13. Prothrombin (Factor II) also reacts with Activated Stuart and Accelerin to produce Thrombin (Factor IIa). (Step 13 is much faster than step 12.)

14. Fibrinogen (Factor I) is activated by Thrombin (Factor IIa) to produce Fibrin. Threads of Fibrin are the final clot. However, it

would be more effective if the Fibrin threads could form more cross links with each other.

15. FSF (Fibrin Stabilizing Factor) is activated by Thrombin (Factor IIa) to form Activated FSF.

16. When Fibrin reacts with Activated FSF many more cross ties are made with other Fibrin filaments to form a more effective clot.

This irreducibly complex system of blood clotting must have a way to remove the clot once the wound has healed. This is accomplished by the following process:

17a. A blood protein, Plasminogen, is activated by +—Pa to produce Plasmin. This acts like tiny chemical scissors that cut up the Fibrin filaments of the clot.

17b. The rate at which the clot is broken up is controlled by another blood protein named Alpha 2 Antiplasm, which in turn inactivates Plasmin.

18. Antithrombin inactivates Activated Christmas (IXa), Activated Stuart (Xa) and Thrombin (Factor IIa).

19. Protein C is activated by Thrombin to produce Activated Protein C.

20. Activated Protein C inactivates Accelerin and Activated Antihemophilic.

21. Finally, Thrombomodulin, which lines the inside of the blood vessels of Land-dwelling vertebrates, prevents Thrombin (Factor IIa) from activating Fibrinogen (Factor I).[14-45]

In order to have a controlled clotting mechanism you have to have two proteins for each step. Both the pro enzyme (non-activated) and its activator are required. If a protein appeared in one step with nothing to do, then mutation and natural selection would tend to eliminate it. (Natural selection and mutations are discussed in Chapter 9.) To prevent this from happening, evolutionists are forced to imagine large clusters of proteins evolving all at once.[14-46] Evolutionists are not able to explain how the blood clotting cascade system could have developed by natural processes. They only offer theories and examples of some animals that have varied complex blood clotting systems but cannot explain how any of them could have evolved by natural processes. What is needed is a step-by-step explanation of how this system could have evolved.[14-47]

Darwinist View

Scientific research is beginning to give possible (and testable) scenarios for how supposed irreducible complex systems, such as blood clotting system and bacterial flagellum (discussed previously) could have evolved. For example, the blood-clotting pathway (an organic substance that could convert to another organic substance) of vertebrates involves a sequence of events that begins when one protein sticks to another in the vicinity of an open wound. That sets off a complicated cascade reaction, sixteen steps long, each involving an interaction between a different pair of proteins and culminating in the formation of the clot itself. (See section above "What Happens When We Cut Ourselves" for the sixteen steps.) Altogether more than twenty proteins are involved.[14-48]

It is not yet known for sure, but scientists have evidence that the system could have been built up in an adaptive way from simpler precursors. Through duplication, many of the blood-clotting proteins are made by related genes. Once they develop, such duplicated genes can then evolve along separate pathways so that they eventually perform separate functions, as they now do in blood clotting. And we know that other proteins and enzymes in the pathway had different functions in groups that evolved before vertebrates. For example, a key protein in the clotting pathway is called fibrinogen, which is dissolved in blood plasma. In the last step of blood clotting, an enzyme cuts this protein, and the shorter proteins (called fibrins) stick together and become insoluble, forming the final clot. Since fibrinogen occurs in all vertebrates as a blood-clotting protein, it presumably evolved from a protein that had a different function in ancestral invertebrates, which were around earlier but lacked a clotting pathway. It is hopeful that an evolutionary explanation will be developed within the near future. There must have been an ancestral protein from which fibrigen evolved by natural means.[14-49]

Around 1990, it was discovered that a protein in a sea cucumber (an echinoderm) led to blood clotting. (See Chapter 3, "Examples of Mollusks" drawing for an example of a sea cucumber.) Sea cucumbers branched off from the vertebrate lineage at least 500 million years ago, yet they have a protein that, while clearly related to blood-clotting proteins of vertebrates, is not used to clot blood. This means that the common ancestor of sea cucumbers and vertebrates had a gene that was later implemented in vertebrates for a new function, precisely as evolution predicts. Since then, scientists have worked out a possible and adaptive sequence for the evolution of the entire blood-clotting cascade from parts of precursor proteins. All of these precursors are found in invertebrates, where they have other, non-clotting functions, and were evolutionarily implemented by vertebrates into a working clotting system. And the evolution of the bacterial flagellum (discussed previously), although not yet fully understood, is also known to involve many proteins implemented from other biochemical pathways, organic substances that could convert to another organic substance.[14-50]

Even though evolutionists have a realistic evolutionary idea that is supported by gene sequences in living organisms, it cannot be said with any certainty how the clotting system evolved. That would be making far too much of science's limited ability to reconstruct the details of the past. Nonetheless, there is little doubt that enough is known to develop a plausible and scientifically valid scenario for how it might have evolved. And that scenario makes specific predictions that can be tested and verified against the evidence.[14-51] (See Chapter 15, section "The Evolution of Hemoglobin" for discussion of hemoglobin, which aids in the process of forming blood clots.) By assuming the cascade blood clotting system is irreducibly complex, proponents of irreducible complexity must also assume that all the components of the coagulation process have always been necessary for the system to function, thus ruling out natural selection. However, natural selection (discussed in Chapter 9) can lead to complex biochemical systems forming from simpler systems, and has been widely documented to do so. It has also been noted that natural selection can combine multiple functioning systems to perform a new function. It should be noted that one of the clotting factors that has been listed by intelligent designer advocates as a necessity, was later found to be absent in whales (the absent clotting factor is Hageman factor 12; it is now a pseudogene in the whale genome), dolphins and the puffer fish, demonstrating the blood-clotting cascade was not irreducibly complex. (Pseudogenes are discussed in Chapter 13.) It has been shown multiple times that simpler aspects of the coagulation system have been noted in lesser complex organisms with simpler structures. Based on these observations, there is no reason to assume that the blood clotting system is irreducibly complex. It is actually readily explained by natural causes.[14-52]

Is the Blood-clotting Cascade Irreducibly Complex?

Darwinist View

When we cut ourselves, we do not just bleed and bleed and bleed and bleed, but rather that cut eventually seals with a blood clot. That is even more important on the inside of the body, because a bruise is actually a result of broken blood vessels. If they did not close with a clot, one would bleed out internally and die. Blood clotting is an enormously complicated biochemical process. If one were to view a red blood cell while in the process of forming a clot under an electron microscope, one would notice the action of this pathway produces a cross-link protein known as fibrin, which produces a mesh-work which actually stabilizes the clot and helps blood to stop flowing. Creationists and Intelligent Design advocates claim that such interactive systems are very strong arguments for intelligent design and are virtually impossible to explain in terms of Darwinian evolution. Creationists claim that only when all the components of the system are present and in good working order does the system function properly. Some of the clotting proteins share discrete regions of their sequences with other proteins. The question is, does that mean that they derive from one another? It may. But consider that even if this were the case, all of the proteins need to be present simultaneously for the blood clotting system to function. Creationists and Intelligent Design advocates argue this is an example of design because it is a multi-part system, and all of the parts have to be put together, presumably by a creator/designer before the system will work. A standard, simple and straightforward scientific test of the claim that 'all parts must be present,' is to eliminate one of the parts and see if the blood will clot. If it will not clot anymore, the claim might be right. If it will clot, the claim could be wrong.

Whales and dolphins do not have Factor 12 (Figure 14.3), however their blood clots when necessary. Whale Hageman Factor 12 is basically now a pseudogene in the whale genome. That is why it is not produced. (Pseudogenes are discussed in Chapter 13.) Even with the missing factor, whale blood clots just fine. So the scientific prediction made by creationists turns out to be wrong. Another point to be made here is that three parts of the cascade can be removed and the system will still work. The three parts known as the "contact phase system" include (1) Factor 12, (2) Factor 11, and also (3) the factor that catalyzes the conversion of Factor 12 to the active form (see Figure 14.3). These three parts that were eliminated are missing in a vertebrate known as the puffer fish. Even though the puffer fish and sea squirt genomes are missing three parts of the blood-clotting system their blood clots perfectly well. So the prediction by creationists and Intelligent Design Advocates has, in fact, been invalidated, that is, refuted by the scientific evidence. It was refuted by the scientific evidence in 1969 that was confirmed by genome studies of the whale, and it has been further refuted by Jiang and Doolittle's study of the contact phase system.[14-53]

Intelligent Design View

There is some experimental evidence that complex biochemical systems can arise by Darwinian processes. The first one is the "lac operon" and the second one is called the "blood-clotting cascade." If one takes an electron micrograph and views some red blood cells caught in a mesh work of a protein called fibrin, they will notice it forms a blood clot. During a minor cut the blood flow slows down, stops, and heals over. It does not seem like much at all, but thorough investigation over the past 40 to 50 years has shown that the blood clotting system is a very intricate biochemical system. In blood clotting, the material that forms the clot cannot, of course, be in its solid clotted form during the normal life of an animal. The material that forms the clot exists as something called fibrinogen, which is actually a soluble pre-cursor to the clot material. It floats around in the bloodstream during normal times. But when a cut occurs, fibrinogen is transformed into something called fibrin. Fibrin is formed when another protein comes along and cuts off a small piece of fibrinogen, a specific piece that exposes a sticky area (sticky in the sense that those two proteins have complimentary surfaces). This protein exposes a sticky site on the surface of the fibrinogen, which allows the many copies of fibrinogen, now turned into fibrin, to aggregate and stick to each other, forming the blood clot.

The component that cuts fibrinogen and activates it is another protein called thrombin. If thrombin were always active, it would cause all the blood to clot. That would congeal the blood and kill the animal. Therefore, thrombin itself is an inactive form called prothrombin. It has to be activated when a cut occurs. And that is the responsibility of another protein. That protein also exists in an inactive form. The activation of that is the responsibility of another protein. So within the blood there is what is called a blood-clotting cascade. One component acts on the next that acts on the next that acts on the next and so on. That is why the process is called a cascade. The blood-clotting cascade actually has two branches. There is one labeled the 'intrinsic pathway' and the other is labeled the 'extrinsic pathway' (Figures 14.3, 14.4, 14.6). So there are actually two branches to this blood-clotting cascade.

Darwinists who claim that the blood clotting system evolved point to DNA sequencing studies of something called a puffer fish. The entire DNA of its genome was sequenced, and scientists looked for genes that might code for the first couple components of the intrinsic pathway. However, they were not found. Darwinists claim that since three components (factors XI, XII, and XIIa) were missing in the puffer fish, they must have evolved later from other life forms. Nonetheless, puffer fish have a functioning clotting system. And so Darwinists argue that this is evidence against irreducible complexity. The problem with their claim is that the control pathway for blood clotting splits in two (that is, has two branches—Figure 14.5). Potentially then, there are two possible ways to trigger clotting. Regarding the two possible ways to trigger clotting (leaving aside the system before the fork (Figures 14.4, 14.5, 14.6) in the pathway where some details are less well-known) the blood clotting system fits the definition of irreducible complexity. The components of the system beyond the fork (Figure 14.4 & 4.5) in the pathway are fibrinogen, prothrombin, Stuart factor, and proaccelerin (Figures 14.3, 14.4, 14.6). Regarding the extrinsic pathway, the pathway can be activated by either one of two directions.

Now let's focus on the parts that were close to the common point after the fork (Figures 14.3, 14.4, 14.5, 14.6). If you concentrate on those components, a number of those components are ones that have been experimentally knocked out, such as fibrinogen, prothrombin, and tissue factor. If you knock out those components, the blood-clotting cascade is broken. This idea of how the blood clotting works in humans would be similar to a light having two switches, and the blood clotting system would be the light. The extrinsic and intrinsic pathways would be two separate switches to turn on the system. If one switch becomes broken, the other system could still be used. The sequences of many components of the blood-clotting cascade have been understood for quite some time now. Even though there is a clear understanding of how these sequences work, Darwinists are unable to provide a clear pathway of how these complex systems evolved. Experts in this area of study assert that molecular sequence data simply cannot tell what an ancestral state was. They believe fossil evidence is required. Since there is no explanation as to how the blood clotting system could have evolved, the conclusion is that the blood clotting cascade is irreducibly complex.[14-54]

A Closer Look at the Details of the Blood-clotting Cascade

Intelligent Design View

Darwinists claim that since the blood-clotting cascades in whales and dolphins lack Factor XII (also called the Hageman factor 12), and the blood-clotting cascade in puffer fish lacks factors XI, XII, and XIIa, that this is an indication that the blood-clotting cascade in humans evolved. Darwinists do not mention that there are two ways for blood to clot in humans. Generally speaking, in land-dwelling vertebrates, there are two different pathways by which the blood-clotting cascade can be initiated. They are the "intrinsic" pathway, and the "extrinsic" pathway (Figures 14.3 & 14.4, 14.6). (There can be some crossover between the two pathways.) The final stages of the blood-clotting cascade take place after either pathway reaches Factor X, also called the Stuart factor. These final stages of the cascade are what is called "beyond the fork" or "after the fork." Figure 14.4 below illustrates the area "Extrinsic Pathway" (boxed in) on the left:

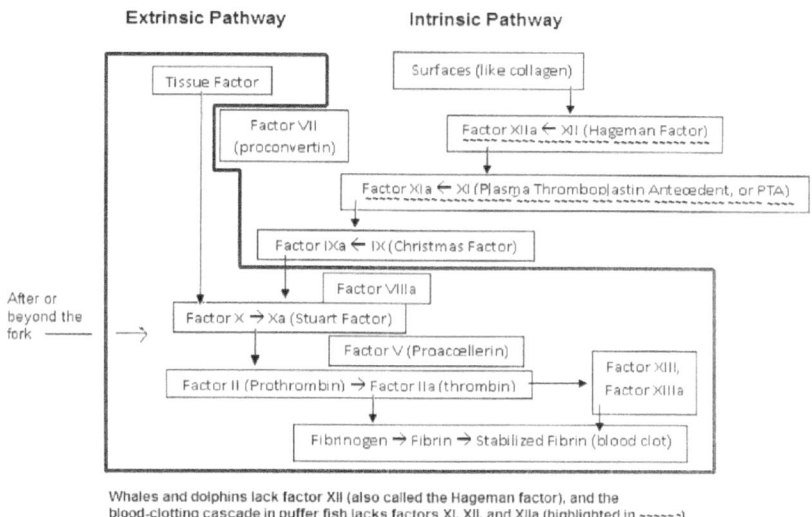

Whales and dolphins lack factor XII (also called the Hageman factor), and the blood-clotting cascade in puffer fish lacks factors XI, XII, and XIIa (highlighted in ------)

Figure 14.4 Diagram illustrating how blood clots in vertebrates

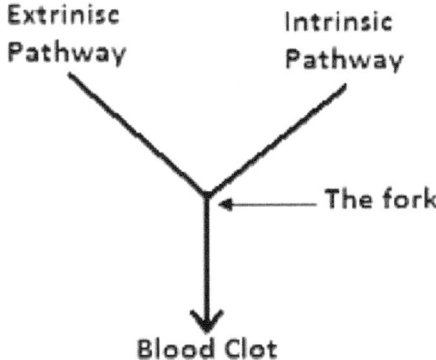

Figure 14.5 The fork in the path

Figure 14.5 is a rough sketch of the intrinsic and extrinsic pathways of the blood-clotting cascade. This figure contains a very rough description of how the blood-clotting cascade can be initiated by either the extrinsic or intrinsic pathway before it forms a final clot. Figure 14.4 (above) and Figure 14.6 (below) contain full descriptions of the land-dwelling vertebrate blood-clotting cascade.

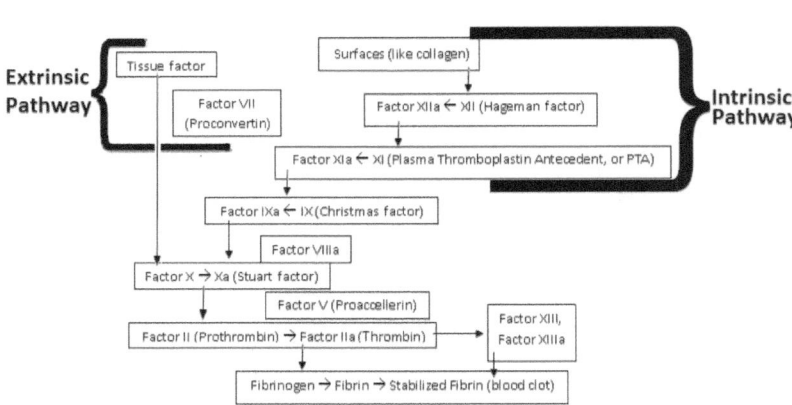

Figure 14.6 The blood-clotting cascade of land-dwelling vertebrates with both intrinsic and extrinsic pathways labeled.

The cascade "after the fork" (Figures 14.3, 14.4, 14.5), where the Intrinsic and Extrinsic pathways converge (or fork—Figures 14.3, 14.4, 14.5 & 14.6), is the area of discussion. Intelligent design advocates focus on this area but do not discuss other areas because adequate experimental tests had not yet been performed to demonstrate irreducible complexity with respect to some of the other factors, particularly those involved in the intrinsic pathway. It is true that some vertebrates (like dolphins or jawed fish) lack certain components involved in the intrinsic pathway (Factors XI, XII, and XIIA) found in land-dwelling vertebrates (Figure 14.4). Darwinists, however, fail to mention that land-dwelling vertebrates, jawed-fish, and water-dwelling mammals like dolphins and whales still have the extrinsic pathway intact, as well as everything after the point where the intrinsic and extrinsic pathways combine in land-dwelling vertebrates. In other words, dolphins and jawed fish still have the factors in the blood-clotting cascade that Intelligent Design advocates consider irreducibly complex (i.e. those "after the fork"—Figure 14.5). Dolphins and jawed fish have the factors on the blood-clotting cascade's extrinsic pathway. The only factors that dolphins and jawed fish appear to be missing are the portions on the intrinsic pathway (Factors XI, XII, and XIIA—Figure 14.4).

The blood-clotting cascade pathway of jawed fish has an important difference from that of land-dwelling vertebrates because fish do not have an intrinsic pathway found in land-dwelling vertebrates. That does not mean that the rest of the cascade is not irreducibly complex, for both jawed fish and land-dwelling vertebrates may have a core system of parts that is irreducibly complex. The fact that both jawed vertebrates and land-dwelling vertebrates contain all of the components in the extrinsic pathway, and everything "after the fork," (Figures 14.3, 14.4, 14.5) that indicates there is an irreducible core that comprises the components in the extrinsic pathway (Figure 14.4). Perhaps the intrinsic pathway of land-dwelling vertebrates (Figure 14.6) is not part of this irreducible core. If that is the case, this does not refute irreducible complexity for the rest of the system; it just shows that those components in the intrinsic pathway are not indispensable to the system. The fact that land-dwelling vertebrates have an intrinsic pathway does not cancel out the existence of an irreducible core. To reiterate, the extrinsic pathway, as illustrated in Figure 14.6, contains all of the components that are part of the irreducible core.

It is true that dolphins (like jawed fish) lack Factor XII (Hageman factor) in their blood-clotting cascade (Figures 14.3 & 14.4). Darwinists also claim that dolphins are believed to have descended from land-dwelling vertebrates (which have Factor XII) however, their blood-clotting cascade is like that of jawed fish, which lack Factor XII. This implies that there may be a functional limitation on water-dwelling vertebrates to have a different blood-clotting activation pathway than what land-dwelling vertebrates have. For some reason, water-dwelling vertebrates may not need the Intrinsic Pathway.

Darwinists claim that the dolphin's lack of Factor XII as evidence of convergent evolution, but one might also see it as evidence of a functional constraint or a case of common design. The fact that jawed fish lack Factor XII is not necessarily evidence that their blood-clotting cascade was a "primitive evolutionary precursor" to the land-dwelling vertebrate blood-clotting cascade, but evidence of a functional constraint for water-dwelling vertebrates, a constraint that is confirmed in that dolphins also lack Factor XII. This is an interesting issue that will require further research to sort out.[14-55]

The Immune System

Intelligent Design View

All animals and some plants have extremely complicated immune systems to protect them. Yet they had to have those immune patterns—dating back to the beginning. If not, they would never have survived long enough to develop them. Each immune system can identify bacteria, viruses, and toxins, and recognize whether each is safe or harmful. Each system has a complete, complex pattern for organizing a variety of organisms to eliminate such problems as soon as possible. In fact, each invasion is indelibly remembered by the system, so they can better protect the body the next time. The immune system could not slowly evolve; it had to completely be there to begin with.[14-56]

Is the Immune System Irreducible Complex?

Anything [any explanation] is good enough for the evolutionist, and nothing [no explanation] is good enough for the creationist. [14-57]

Darwinist View

Creationists and Intelligent Design advocates claim that AIDS (Acquired Immune Deficiency Syndrome) virus, Type 3 (III) secretory system (discussed previously in this chapter), and anthrax, are systems that are irreducibly complex. Creationists and Intelligent Design advocates believe that the immune system is also irreducibly complex. Why would a creator create harmful systems and then need to create an immune system to combat the harmful systems? Why wouldn't a creator just not create the harmful systems, then there would be no need for an immune system to defend organisms against pathogens?[14-58]

The immune system is a system of our body that is widely distributed. We have cells from our immune system sort of engaging in patrol, floating throughout the blood stream and the tissues. And it is a system that enables our bodies to identify, defend against, and to repel foreign invaders. Chicken pox is a virus, and when it invades the human body, the immune system recognizes the code proteins on the virus, makes cells that can continue to recognize it, and produces proteins called antibodies that will bind to the surface of the virus. After a week of suffering, the person affected would be permanently immune to the chicken box. (Viruses are discussed in Chapter 5.) Vaccinations are designed to stimulate our immune system from diseases far worse than chicken pox such as polio, diptheria and whooping cough in an effort to make sure no one suffers from these diseases. The essential protein, researchers call immunoglobulin, is more commonly called an antibody. These are the essential molecules of the immune system. If one gets the chicken pox virus, that foreign particle-binding site on the chicken pox antibody will bind to the surface of the virus. Then another one (a chicken pox antibody) will bind to another site (a chicken pox virus). And gradually, the virus will be cross-linked into a mesh world, which the immune system recognizes, eliminates from the circulation, and destroys it. The chicken pox molecule is made up of four parts, called polypeptides. There are a number of other such diseases besides chicken pox that people may be vaccinated against, such as polio, diphtheria, measles, and many others. The antibodies in a vaccinated person's body against polio differ from the antibodies one would have received against diptheria in the variable regions. Diptheria have a different shape than poliovirus because the viruses or the bacteria have different molecules on the surface. The amazing aspect of the immune system is that it can produce an antibody that will attach to, stick to, identify, and destroy just about anything. One of the most important things in our immune system is the ability, basically, to produce antibodies against any conceivable molecule that might get inside our body.

In the 1980s it was determined exactly how antibodies had the ability to produce such diversity. There is a system in the genes of cells in the immune system known as a VDJ recombination system (Variable, Diverse, and Joining gene segments of vertebrates). At a certain point in development, parts of DNA, in a variety of genes, are literally shuffled. They are tossed from one side to another, and they are rearranged to form a final gene. Now some elements of this shuffling are random, but it is in that random shuffling that our immune system develops the ability to produce an antibody to just about anything. That shuffling is the main reason why the immune system works. If anything goes wrong with this process, the individual in which it goes wrong loses the ability to make diverse antibodies and they get very sick. If they encounter foreign organisms, they will be in great danger. In 1994, a number of scientists, speculated that this process, which is called VDJ recombination (Variable, Diverse, and Joining gene segments of vertebrates), might actually have evolved from a system known as transposition, a system in which genes jump around.[14-59] V(D)J recombination is a mechanism of genetic recombination in the early stages of immunoglobulin (Ig) and T cell receptors (TCR) production of the immune system. VDJ recombination nearly randomly combines Variable, Diverse, and Joining gene segments of vertebrates. Due to its randomness in choosing different genes, it is able to diversely encode proteins to match antigens from bacteria, viruses, pollens, parasites, and dysfunctional cells, such as tumor cells.[14-60] The gene shuffling system could have been part of retrotransposons and had a DNA rearrangement function in a previous function. It is possible that the ancestors of these genes, they are called RAG (recombination activating genes) genes, may have been horizontally transferred into a metazoan multi-cellular animal lineage at a recent point in evolution.[14-61]

There are very strong biochemical similarities found between the gene shuffling process, the V(D)J recombination, and the way in which retroviruses shuffle their DNA. Investigators noticed there were biochemical similarities between the way the genes are shuffled in the immune system and the way that retroviruses go into other cells. The cutting and transposing enzymes that are normally used for these transposable genetic elements can be replaced by the RAG enzymes, which do the cutting and pasting in the immune system. The RAG (Recombination Activating genes) enzymes have been shown to cause transposition in mammalian cells. What this means is, not only can they shuffle the immune system pieces of DNA, they can shuffle other pieces of DNA as well. The VDJ recombinase was shown to cause transposition, that is, shuffle DNA around, found in mammalian cells, including human cells. The significant similarity between the transib transpases and RAG core, the common structure of these transpases and others, as well as the similar size of these basically catalyzed by these enzymes, directly support the 25-year-old hypothesis of a transposon related origin of the VDJ machinery. Researchers also point out there have been other hypotheses that have been considered. Previously, the RAG transposon hypothesis was open to challenge by alternative models of convergent evolution. (Convergent evolution is the process in which organisms that are not closely related independently evolve similar features.)

Because there were no known transpases similar to the gene shuffling ones, it could be argued that our gene shuffling enzymes, the RAG1 independently developed some transposon-like properties rather than deriving them from a transposable element encoded transpases. These arguments can now be put to rest. It has now been established that this system developed from evolution. The immune system came from a transposable element system. Each element of the transposon hypothesis has been confirmed. When the enzymes that do this gene shuffling are actually put to an analysis to see how closely related they are, that is, to see if they themselves match the evolutionary predicted tree, they match that tree perfectly.[14-62]

Details Pertinent to the Immune System Discussion

Darwinist View

A Transposon is a segment of DNA that is capable of independently replicating itself and inserting the copy into a new position within the same or another chromosome or plasmid.[14-63] Figure 14.7 below illustrates which species contain Transib transposons, RAG1, RAG2, and RAG1-Like proteins in Eukaryotes cells:

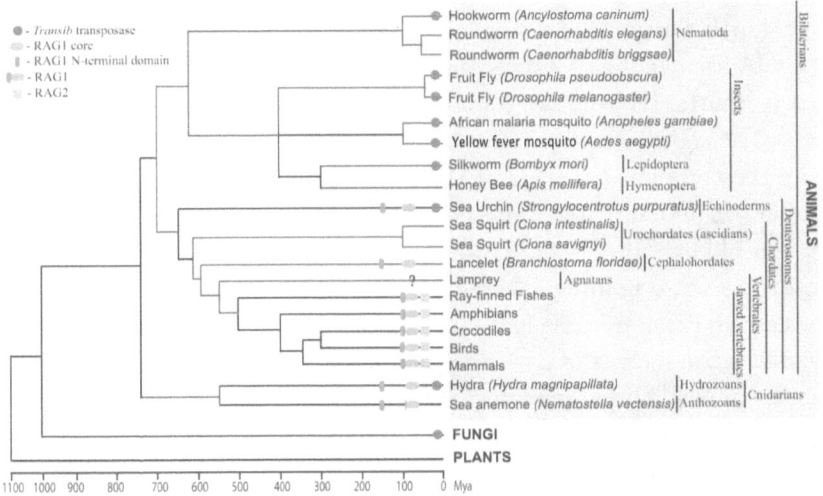

Figure 14.7 RAG1 Core and V(D)J Recombination Signal Sequences Were Derived from Translib Transposons[14-64]

Figure 14.7 above illustrates that hookworms, roundworms, fruit flies, African malaria mosquito, yellow fever mosquito, and silkworm all have Transib transposase. Vertebrates, such as ray-finned fishes, amphibians, crocodiles (reptile), birds (avian) and humans (mammal) have RAG1 and RAG2 (RAG = Recombination Activating Genes). The sea urchin and hydra (small, fresh-water animal) have RAG1 N-Terminal domain, RAG1, and Transib transposase. Since the sea urchin and the hydra possess RAG1 N-Terminal domain, RAG1 and Transib transposase, perhaps they will provide the link between worms, fruit flies, mosquitoes, and ray-finned fishes, amphibians, crocodiles, birds, mammals.

Transposase is an enzyme involved in the movement of a DNA fragment from one site in the genome to another. An enzyme is a protein that catalyzes chemical reactions of other substances without itself being destroyed or altered upon completion of the reactions. Enzymes are divided into six main groups: **(1)** oxidoreductases, **(2)** transferases, **(3)** hydrolases, **(4)** lyases, **(5)** isomerases, and **(6)** ligases.[14-65]

Researchers have documented some of the changes that occurred during the emergence and development of the antibody-based immune system (AIS). These researchers have also speculated as to how these changes may have come about. The origin of the AIS only appears to be sudden, whereas in reality it represents a culmination of a long preparatory phase characterized by gradual accumulation of small changes over an extended period. Researchers also have examples of such changes for each of the three levels constituting the system: organs, cells, and molecules. The emergence of the AIS must have involved deployment of a number of existing signaling cascades, which evolved earlier to serve other functions. The integration of these cascades into the AIS was presumably accomplished by the insertion of a few regulatory molecules into the pathways. These molecules themselves may have served other functions before being utilized by the AIS. The evolution of the AIS itself must have begun long before the divergence of agnathans and gnathostomes from their common ancestor. The evolution consisted initially of changes unrelated to immune responses that were selected to serve other functions. The different functions may have been unrelated to one another, but ultimately, a combination of these functions arose by chance, which presented the potential for the development, in a number of small steps, of a qualitatively new system. The actualization of the potential required integration of the different functions merged into one whole system. The necessary steps for this integration were undertaken in the gnathostome lineage, whereas the agnathans evolved in a different direction. Once the critical steps were accomplished in the gnathostome lineage, the integration created the illusion of a sudden, explosive change. However, in reality, the entire process was gradual, consisting of accumulation of small changes over an extended period.[14-66]

Intelligent Design View

Evolutionists claim the immune system evolved, however, they have not been able to provide a detailed, testable explanation as to how the immune system could have developed by natural means. They only provide assumptions that may or may not be true. Scientific research conducted by evolutionists has concluded that the adaptive immune system appears to have arisen suddenly, within a relatively short time interval. This would be a clear indication that it did not evolve, since evolutionary theory states that adaptations have to originate by numerous successive slight modifications very gradually, over a great period of time. It also acknowledges that scientists have no idea how the immune system could have evolved. The purpose of the immune system is to defend an organism against pathogens (disease-producing agents, such as a virus or bacterium or other microorganism). The purpose of the immune system is to combat viruses and to defend organisms, not to destroy pathogens, per se.[14-67]

The Eye

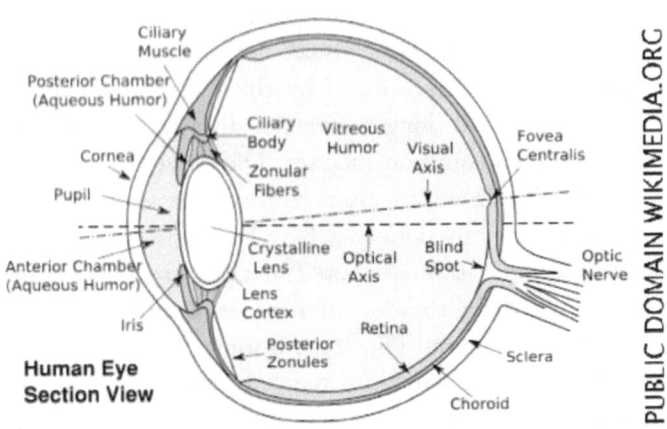

Figure 14.8 The Human Eye

Darwinist View

To suppose that the eye with all its inimitable contrivances for adjusting the focus to different distances, for admitting different amounts of light, and for the correction of spherical and chromatic aberration, could have been formed by natural selection, seems, I freely confess, absurd in the highest degree . . . Reason tells me, that if numerous gradations from a simple and imperfect eye to one complex and perfect can be shown to exist, each grade being useful to its possessor, as is certainly the case; if further, the eye ever varies and the variations be inherited, as is likewise certainly the case and if such variations should be useful to any animal under changing conditions of life, then the difficulty of believing that a perfect and complex eye could be formed by natural selection, though insuperable by our imagination, should not be considered as subversive of the theory. How a nerve comes to be sensitive to light, hardly concerns us more than how life itself originated; but I may remark that, as some of the lowest organisms in which nerves cannot be detected, are capable of perceiving light, it does not seem impossible that certain sensitive elements in their sarcode [protoplasm, specifically protozoans[14- 67a]] should become aggregated and developed into nerves, endowed with this special sensibility.

• Charles Darwin, Origin of Species,
Chapter VI, Difficulties of the Theory

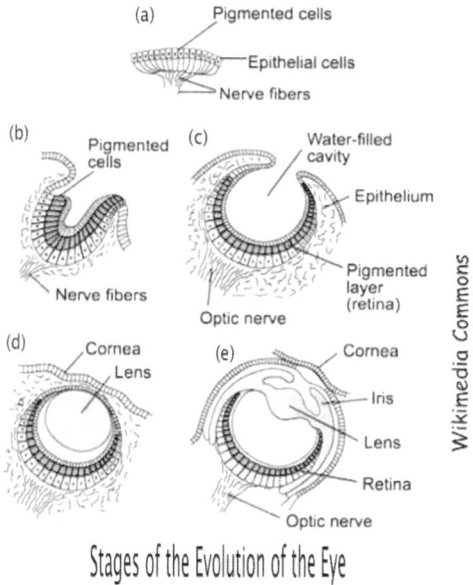

Stages of the Evolution of the Eye

Figure 14.9 Stages in the Evolution of the Eye The Stages of the Evolution of the Eye is a hypothetical scheme for the evolution of chambered (vitreous humor fluid filled) eyes.[14-68]

The eye (Figure 14.8) is a common example used by creationists of a supposedly irreducibly complex structure, due to its many elaborate and interlocking parts, seemingly all dependent upon one another. However, the same problems we found with the irreducibility of blood coagulation, as discussed previously, are just as applicable to the eye. Research has revealed that the earliest eyes were photoreceptor cells, which are specialized cells consisting of two molecules and a membrane. The first molecule is known as the opsin, which is a light sensitive protein surrounding the chromophore pigment, a molecule which distinguishes colors. This is another perfect example of how complex structures stem from simplistic beginnings. The eye is no different than any number of examples.

The original eye was not nearly as advanced as it is today. At most, the original eye most likely could only distinguish the intensity of light with a faint sense of direction. While the complex optical system originated as a multi-cellular patch, the patch deepened into a depressed cup (cavity—Figure 14.9), allowing for a slightly better perception of intensity and direction of surrounding light. A rapid period of evolution for a more complex eye appears to have occurred in the Cambrian Explosion with radical improvements in image processing and detection of light direction. (See Chapter 3 section "A Closer Look at the Cambrian Explosion"). During this period of rapid evolution, as indicated by the fossil record, the eye was one of many complex organs to have quickly evolved to include much more complex structures.[14-69]

It is well known that the fundamental mechanisms of eye function evolved long before the Cambrian Explosion. From the molecular evidence what happened was not that eyes instantly evolved, but that the evolution of body armor gradually increased from the pre-Cambrian Period through the Cambrian Explosion, making the organization of eyes visible in the fossil record.[14-69a]

The primitive nautilus eye functions very similar to a pinhole camera. The "pinhole camera" eye was developed as the pit deepened into a cup, then a chamber (or cavity—Figure 14.9). By reducing the size of the opening, the organism achieved true imaging, allowing for fine directional sensing and even some shape-sensing. These eyes lacked a cornea or lens. They had poor resolution and dim imaging, but were still, for the purpose of vision, a major improvement over the early eye patches. It is likely that a key reason eyes specialize in detecting a specific, narrow range of wavelengths in the visible spectrum is because the earliest species to develop photosensitivity were aquatic. Only two specific ranges of electromagnetic radiation can travel through water, the most significant of which is visible light. The adult box jellyfish possesses a simple eyespot, but in addition has true eyes, complete with lens, iris and retina. The box jellyfish does not contain a brain, which indicates that a brain is not necessary for eyes to evolve.

The development of the lens in camera-type eyes probably followed a different progression. The transparent cells over a pinhole eye's aperture split into two layers, with liquid in between. The liquid originally served as a circulatory fluid for oxygen, nutrients, wastes, and immune functions, allowing greater total thickness and higher mechanical protection. In addition, multiple interfaces between solids and liquids increase optical power, allowing wider viewing angles and greater imaging resolution. Again, the division of layers may have originated with the shedding of skin, with intracellular fluid filling in some of the depth of the area. Note that this optical layout has not been found because fossilization rarely preserves soft tissues. And even if soft tissues were preserved, the remains would have dried up, or they would have become damaged by sediment layers, causing the soft tissue layers to fuse together.

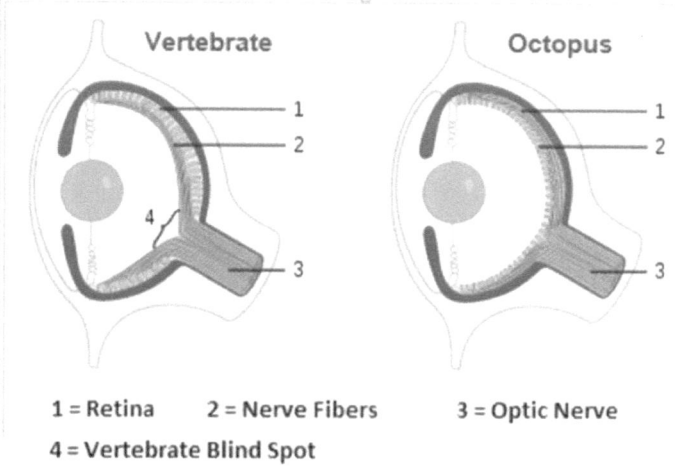

1 = Retina   2 = Nerve Fibers   3 = Optic Nerve
4 = Vertebrate Blind Spot

In vertebrate eyes, the nerve fibers route *before* the retina, blocking some light and creating a blind spot where the fibers pass through the retina. In cephalopod eyes, the nerve fibers route *behind* the retina, and do not block light or disrupt the retina. 4 denotes the vertebrate blind spot. In vertebrates, 1 denotes the retina and 2 the nerve fibers, whereas in cephalopods, it is opposite. 3 is the optic nerve.

Figure 14.9a Vertebrate Eye Compared to an Octopus Eye

The evolutionary history is often most observable in imperfect design. The eyes of vertebrates (including humans) are an example of this. The vertebrate eye is both upside down and backwards (Figure 14.9a). Photons of light must travel through the cornea, the lens, the aquaeous fluid, blood vessels, ganglion cells, amacrine cells, horizontal cells, and bipolar cells before reaching the light sensitive rods and cones. There the light can be converted to neural impulses that are sent to the back of the brain for processing. The rods are more sensitive to light than the cones, which is why colors are so difficult to see in dim lighting. But the cones can resolve more detail than the rods. The reason for this difference in resolving power between rods and cones is that only one cone cell is connected to the next nerve cell in the relay, but several rod cells are connected to the next nerve relay cell. This reduction in efficiency may be countered, as it is in humans, by the formation of a reflective layer of tissue, called the tapetum behind the retina (Figure 14.9b). This tapetum is in the eye of many vertebrate animals. It reflects visible light back through the retina, increasing the light available to the photoreceptors. This improves vision in low-light conditions, but can cause the perceived image to be blurry from the interference of the reflected light. Light that is not absorbed by the retina on the first pass may bounce back and be detected.

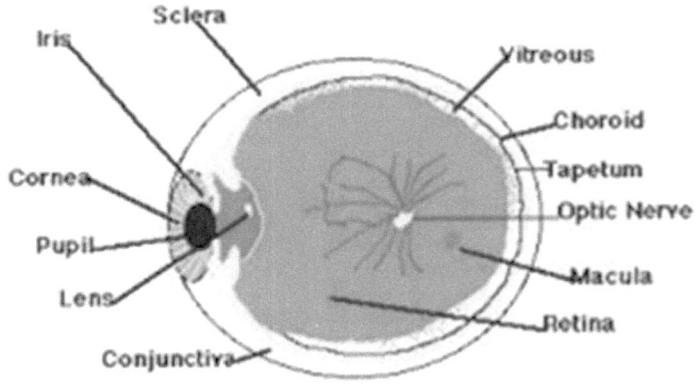

Figure 14.9b Illustration of a Vertebrate Eye with Tapetum

An example of an improved eye would be the camera eyes of cephalopods, organisms having tentacles attached to the head, including the cuttlefish, squid, and octopus (Figure 14.9a). These eyes, in contrast, are constructed the 'right way out', with the nerves attached to the rear of the retina. This means cephalopods do not have a blind spot, and their vision is clearer, as the nerve fibers in front of vertebrate retinas diminish eyesight by 5%. This difference may be accounted for by the origin of eyes. In cephalopods the eyes develop as the formation of a metazoan embryo following the developmental stage of the head surface, whereas in vertebrates they originate as an extension of the brain which 'pushes its way through' to the outside world.

The "poor design" argument, however, is debatable. The eyesight of eagles, vultures and many other birds of prey is known to be significantly greater than the eyesight of cephalopods despite the supposed "poor design". It is true that in vertebrate eyes the nerve fibers block the receptors and as such diminish eyesight by 5%, however this allows the retina to be closer to the blood vessels. Rod cells in the retina have the shortest life span of all cells in the vertebrate body. They last only four hours and then burn out. Because of the placement of vertebrate eyes, they have access to a greater blood supply. This gives vertebrate eyes an advantage over cephalopod eyes. They have the extra energy necessary for functioning with greater efficiency as vertebrate eyes are able to replicate their rod cells more efficiently. So it would appear the "backward" system of the eyes of birds of prey is actually more efficient than the "forward" system of cephalopods.[14-70]

Vertebrates and invertebrates eyes, like the octopus, developed independently. Vertebrates evolved an inverted retina with a blind spot over their optic disc, whereas the development of the eyes of the octopuses (an invertebrate) avoided this with a non-inverted retina. Since fossil evidence does not contain soft tissue, scientists have had to rely upon comparing the genetic and anatomical features of eyes of living animals. These comparisons have led to the belief that all eyes share a common ancestry, and most likely all eyes began as a simple photoreceptor cell patch that was able to detect the presence or absence of light, but not its direction.[14-71] (Common ancestry is discussed in Chapter 15.)

The eye, just like bacteria flagellum (discussed earlier), is not irreducibly complex. Removing one part of the eye does not automatically result in blindness. The evolution of the eye most likely developed along these lines (Figure 14.9 (Stages in the Evolution of the Eye drawings a-e)): (1) First there was the simple eye-spot where a few light sensitive cells gave the organism the location of light sources (Figure 14.9 (drawing a)). (2) Then came a recessed eyespot where a small indentation in the surface provided additional information to the location of the direction of the light (Figure 14.9 (drawing b)) (3) This was followed by a deeply recessed eyespot where even more light sensitive cells could provide even more accurate information to the organism (Figure 14.9 (drawing c)). (4) Next came a pinhole camera eye that was able to focus an image on the back of a layer of deeply recessed light sensitive cells (Figure 14.9 (drawing d)). (5) This led to a pinhole lens eye that was actually able to focus the image (Figure 14.9 (drawing d)), and (6) finally, the complex eye developed into what is now found in humans and other modern animals (Figure 14.9 (drawing e)).[14-72]

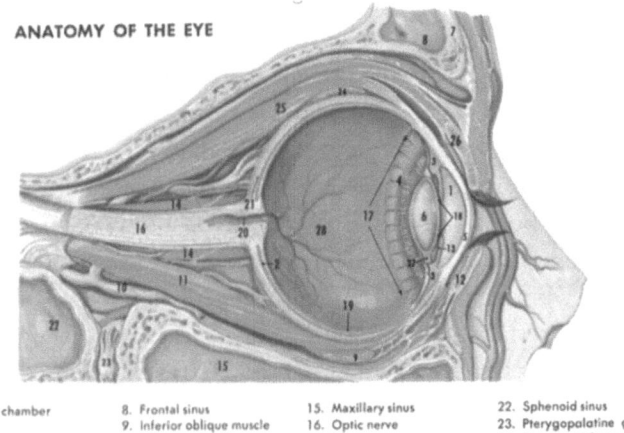

ANATOMY OF THE EYE

1. Aqueous chamber	8. Frontal sinus	15. Maxillary sinus	22. Sphenoid sinus
2. Choroid	9. Inferior oblique muscle	16. Optic nerve	23. Pterygopalatine ganglion
3. Ciliary muscle	10. Inferior ophthalmic vein	17. Ora serrata	24. Superior oblique muscle
4. Ciliary processes	11. Inferior rectus muscle	18. Pupil of the iris	25. Superior rectus muscle
5. Cornea	12. Inferior tarsus	19. Retina	26. Superior tarsus
6. Crystalline lens	13. Iris	20. Retinal artery and vein	27. Suspensory ligament
7. Frontal bone	14. Lateral rectus muscle	21. Sclera	28. Vitreous chamber

Figure 14.9c Detailed illustration of the Simple or Camera Eye

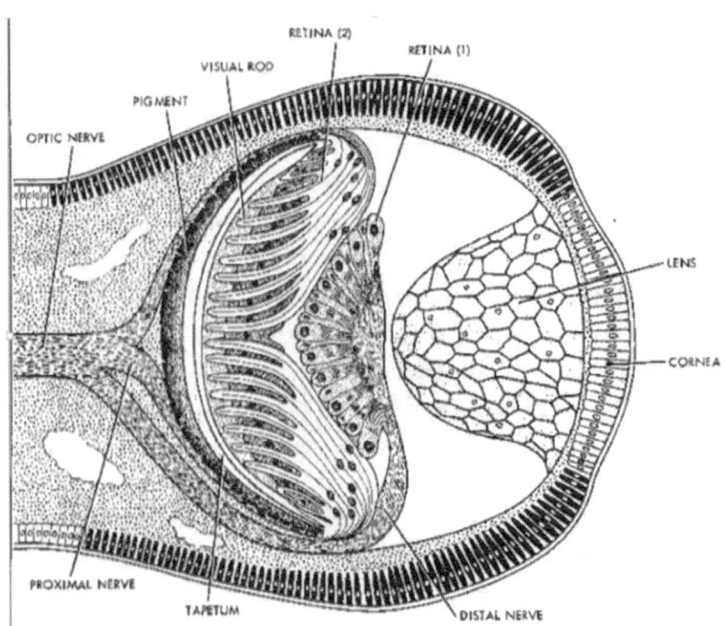

Figure 14.9d Illustration of an Invertebrate Simple Eye Type

The origin of a complex organ, such as the eye, can be explained by evolution. The camera eye of vertebrates (Figures 14.8, 14.9a, 14.9c, 14.9e) and invertebrates (Figure 14.9d) such as mollusks, like the squid and octopus, was once viewed by creationists as being a good example of an organism that could only be explained as having been created. Noting its complex arrangement of the iris, lens, retina, cornea, and so on, all of which must work together to create an image, Creationists (those who deny the power of natural selection) (Chapter 9) claimed that the eye could not have formed by gradual steps. Creationists ask: How could "half an eye" be of any use? Darwin brilliantly addressed, and rebutted, this argument in his book, the "Origin of the Species" (quoted in the beginning of this section). He surveyed existing species to see if one could find functional but less complex eyes that not only were useful, but also could be strung together into a hypothetical sequence showing how a camera eye might evolve (Figure 14.9). If this could be done, and it can, then the argument that evolution, including natural selection (Chapter 9), could never produce an eye collapses, for the eyes of existing species are obviously useful. Each improvement in the eye could confer obvious benefits, for it makes an individual better able to find food, avoid predators, and navigate its environment.

A possible sequence of such changes begins with **(1)** simple eyespots made of light-sensitive pigment (Figure 14.9), as seen in flatworms. **(2)** The skin then folds in, forming a cup that protects the eyespot and allows it to better localize the light source. Limpets have eyes like this. **(3)** In the chambered nautilus, we see a further narrowing of the cup's opening to produce an improved image. **(4)** And in ragworms, the cup is capped by a transparent cover to protect the opening. **(5)** In abalones, part of the fluid in the eye has coagulated to form a lens, which helps focus light, and in many species, such as mammals, nearby muscles have been co-opted to move the lens and vary its focus. (Abalones are mollusks having a flat oval shell. Abalones are edible sea snails, marine gastropod molluscs.) The evolution of the retina, an optic nerve, and so on follows by natural selection. (Natural selection is discussed in Chapter 9.) Each step of this hypothetical transitional "series" confers increased adaptation on its possessor, because it enables the eye to gather more light or form better images, both of which aid survival and reproduction. And each step of this process is feasible because it is seen in the eyes of different living species. **(6)** At the end of the sequence is the camera eye, whose adaptive evolution seems impossibly complex. But the complexity of the final eye can be broken down into a series of small, adaptive steps.

A Mathematical Model

The evolution of the eye can be explained better than just stringing together eyes of existing species in an adaptive sequence (Figure 14.9). Starting with a simple precursor, an actual model of the evolution of the eye can be developed to see whether or not natural selection can turn that precursor into a more complex eye in a reasonable amount of time. Scientists created a mathematical model, starting with a patch of light-sensitive cells backed by a pigment layer (a retina) and then they allowed the tissues around this structure to deform by themselves randomly, limiting the amount of change to only one percent of size or thickness at each step. To mimic natural selection, the mathematical model accepted only "mutations" that improved the visual acuity, and rejected those that degraded it. Within an amazingly short time, the mathematical model yielded a complex eye, going through stages similar to the eye described above. Beginning with a flatworm-like eyespot, the mathematical model produced something like the complex eye of vertebrates, all through a series of 1,829 tiny adaptive steps. It has been calculated how long this process would take.

To do this, they made some assumptions about how much genetic variation for eye shape existed in the population that began experiencing selection, and about how strongly natural selection would favor each useful step in eye size. These assumptions were deliberately conservative, assuming that there were reasonable but not large amounts of genetic variation and that natural selection was very weak. Nevertheless, the eye evolved very quickly: the entire process from rudimentary light-patch to camera eye took fewer than 400,000 years. Since the earliest animals with eyes date back 550 million years ago, there was, according to this model, enough time for complex eyes to have evolved more than fifteen hundred times over. In reality, eyes have evolved independently in at least forty groups of animals. We can provisionally assume that natural selection is the cause of all adaptive evolution, though not of every feature of evolution, since genetic drift can also play a role.[14-73] (Natural selection and genetic drift are discussed in Chapter 9.)

Evolutionist View—continued

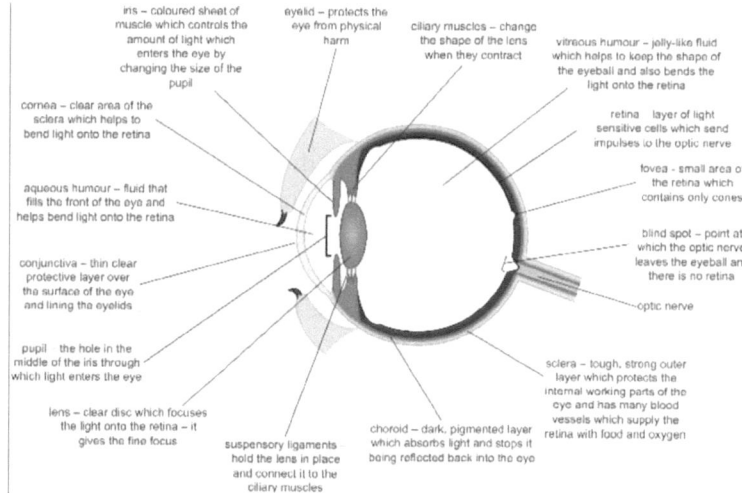

Figure 14.9e The Simple or Camera Type Eye

The retina is configured in three layers, with the light sensitive rods and cones at the bottom. They face away from the light, underneath a layer of bipolar, horizontal, and amacrine cells, which are under a layer of ganglion cells. They help carry the converted light signal to the brain by way of nerve impulses. This entire structure is located below a series of blood vessels. As previously mentioned, basically the eye in humans is "backwards" and "upside down" (Figure 14.9a). Evolution can explain this, since evolution is a "bottom up" designer, while Intelligent Design cannot.[14-74] Darwin explained in his book, ("Origin of Species," Chapter 6), that almost any nerve can become sensitive to light. (See Darwin's quote at beginning of this section.) So in order for an eye to evolve, one needs only to begin with a nerve sensitive to light. The outer layers of this nerve then adjust in density, thickness, and transparency, and even the distance from one layer to the next adjusts. This phenomenon is always occurring in nature, and it only takes one organism with many descendants to produce millions of individuals with a specific slight mutation. Natural selection (Chapter 9) then slightly favors each improvement. Over millions of years, in this way, many variations of eyes are formed, some with lenses, some without, some with eyelids, some without. For an eye to develop, the first thing that is needed is a light sensitive spot, which is common in nature. It then becomes a depression (cavity—Figure 14.9), allowing it to catch more light. Transparent skin forms over it, and variations in thickness and shape allow that skin to become a lens. Muscles attached to that lens then allow it to focus. Similar mutations would produce first tear ducts, then an eyelid, then eyelashes to further filter contaminant particles, etc. A lens is not as special a thing as you might think. It's just skin.[14-75]

Since eyes of different animals are often times quite unique, scientists believe they must have evolved independently. For example, there is a big difference between the eye of a human and the eye of a fly. The eye of a fly (Figure 14.9f) is made up of thousands of columns that capture images. The human eye, as well as the eyes of other vertebrates (figures 14.8, 14.9a, 14.9b, 14.9c, 14.9e), captures light with ciliary photoreceptors (cells that have hair-like projections), while the fly eye uses rhabdomeric photoreceptors (cells with folds). Scientists have recently discovered that embryos of insects and humans use the same genes to turn cells into photoreceptors. This discovery has determined that a single cell eventually became two new cell types. Since this occurs, some animals might have both types of photoreceptors. The conclusion is that the eyes of all animals and humans evolved from the same cell type.[14-76]

The bodily architecture of vertebrates is the product of blueprint Hox genes (Chapter 9) that direct the construction of repeating parts, such as ribs and vertebrae. In embryological development, various structures form or do not form depending on whether the Hox genes are turned on or off (See Chapter 9 section "Hox Genes," Chapter 7 section "The Evolution of the Horse" subsection "Preexisting Genetic Information" "Creationist View" and Chapter 13, section "Some DNA Portions are Used to Turn Genes On or Off."). Natural selection (Chapter 9) operates only on genes that are turned on, since these result in organisms that survive long enough to pass on their genes to their offspring. (See Chapter 13 section "Some DNA Portions are Used to Turn Genes On or Off.") The wide variety of eyes found throughout the animal kingdom, from compound eyes of flies (Figure 14.9f) to the camera eyes of vertebrates (figures 14.8, 14.9a, 14.9b, 14.9c, 14.9e), evolved under the control of the commonly shared Pax-6 gene, which directs the production of photoreceptor cells and light-sensing proteins. Each type of complex eye that is found today evolved from simpler photoreceptive structures in a distant common ancestor of arthropods, cephalopods, and vertebrates. Research has determined that this ancestor possessed two kinds of light-sensitive organs, each one equipped with a distant type of photoreceptor, as well as with light-sensitive proteins called R-opsin and C-opsin, respectively. One organ was a simple two-celled prototype eye; the other, called the brain photoclock, was a part of the animal's brain and played a role in running the animal's daily clock. The arthropod and squid retinas incorporated the photoreceptor from the simple prototype eye, whereas the vertebrate eye incorporated both kinds of photoreceptor into its retina.

Instead of eyes evolving forty or more different times in evolutionary history, it appears that this simple genetic complex led to the embryological development and evolutionary refinement of a two-part system; in some species one part is incorporated, and in others both are. Instead of an extensive genetic tool kit with genes for constructing each and every bodily structure, research has shown that a small set of gene complexes such as the Hox genes and the Pax-6 genes are turned on in different and unique ways that can generate large-scale changes in a non-incremental fashion. (Hox genes are discussed in Chapter 9.) This explains why the human genome is not really much different from the mouse genome. It is not the number of genes that counts so much as how genes are turned on or off.[14-77] (See Chapter 9 section "Hox Genes" and Chapter 7 section "The Evolution of the Horse" "Creationist View" subsection "Preexisting Genetic Information" and Chapter 13, section "Some DNA Portions are Used to Turn Genes On or Off" for further discussion of the results of genes being turned on or off.)

Creationist View

I know no better method of introducing so large a subject, than that of comparing a single thing with a single thing; an eye, for example, with a telescope. As far as the examination of the instrument goes, there is precisely the same proof that the eye was made for vision, as there is that the telescope was made for assisting it. They are made upon the same principles; both being adjusted to the laws by which the transmission and refraction of rays of light are regulated. I speak not of the origin of the laws themselves; but such laws being fixed, the construction, in both cases, is adapted to them. For instance; these laws require, in order to produce the same effect, that the rays of light, in passing from water into the eye, should be refracted by a more convex surface, than when it passes out of air into the eye. Accordingly we find that the eye of a fish, in that part of it called the crystalline lens, is much rounder than the eye of terrestrial animals. What plainer manifestation of design can there be than this difference?

Natural Theology; or, Evidences of the
Existence and Attributes of the Deity,
Collected From the Appearances of Nature.
By William Paley, D.D., Twelfth Edition. 1809

Evolutionist View

It is scarcely possible to avoid comparing the eye with a telescope. We know that this instrument has been perfected by the long-continued efforts of the highest human intellects; and we naturally infer that the eye has been formed by a somewhat analogous process. But may not this inference be presumptuous? Have we any right to assume that the Creator works by intellectual powers like those of man? If we must compare the eye to an optical instrument, we ought in imagination to take a thick layer of transparent tissue, with spaces filled with fluid, and with a nerve sensitive to light beneath, and then suppose every part of this layer to be continually changing slowly in density, so as to separate into layers of different densities and thicknesses, placed at different distances from each other, and with the surfaces of each layer slowly changing in form. Further we must suppose that there is a power, represented by natural selection or the survival of the fittest, always intently watching each slight alteration in the transparent layers; and carefully preserving each which, under varied circumstances, in any way or degree, tends to produce a distincter image. We must suppose each new state of the instrument to be multiplied by the million; each to be preserved until a better is produced, and then the old ones to be all destroyed. In living bodies, variation will cause the slight alteration, generation will multiply them almost infinitely, and natural selection will pick out with unerring skill each improvement. Let this process go on for millions of years; and during each year on millions of individuals of many kinds; and may we not believe that a living optical instrument might thus be formed as superior to one of glass, as the works of the Creator are to those of man?

—Charles Darwin. On the Origin of Species
by Means of Natural Selection, or the Preservation of
Favoured Races in the Struggle for Life. Chapter VI,
Difficulties of the Theory of Natural Selection.
6th Edition. 1872

Creationist and Intelligent Design View

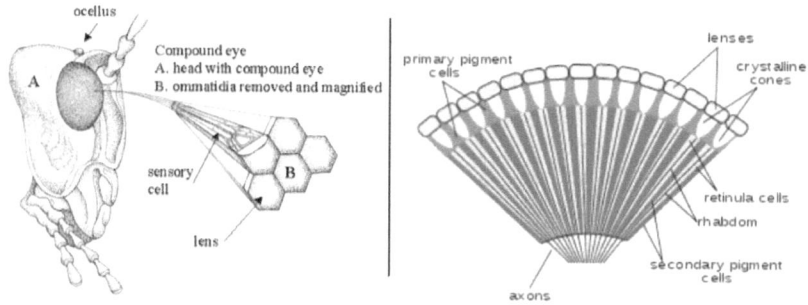

Figure 14.9f Insect Compound Eye

Insects, composed of hundreds of thousands of different species, have a compound eye (Figure 14.9f). Instead of a single lens, compound eyes have hundreds, and in some, thousands upon thousands of separate little lenses, each working as an optical receptor. The eyes of vertebrates have muscles attached to the eyeball that allows them to rotate it within a certain arc in order to observe a greater area. Insects do not have this possibility, as eye muscles are not involved. The compound eyes of insects and crustaceans, instead of muscles, have a quantity of lenses placed at different angles around a half-sphere so as to receive the light rays coming from a much larger arc of vision.

The compound eye, found in most insects and some crustaceans, composed of many light-sensitive elements, each having its own refractive system and each forming a portion of an image, is made of ommatidia (any of the numerous small cone-shaped eyes that make up the compound eyes of some arthropods), which are individual lenses positioned in different directions along a semicircular arc. Each ommatidium extends downward so as to direct the incoming light onto receptor cells of another organ, the rhabdom. Light received by these rhabdoms (rhabdi-prefix meaning "rod-shaped" or "pertaining to a rod") is then guided to the optical nerve, which in turn transmits impulse signals to the brain of the insect. With this sophisticated system, the animal obtains vision around the arc of the eye through thousands of these miniscule lenses, without having to rotate its eyes.

The single-lens eye of vertebrates (including humans) (Figures 14.8, 14.9a, 14.9b, 14.9c, 14.9e) is equipped with a round iris that opens or closes depending on the intensity of the light received, and accordingly controlling the amount of light entering the lens. Vertebrates include fish, amphibians, reptiles, birds, and mammals (including humans). The compound eye (Figure 14.9f), however, achieves the same result through a different method. At the bottom of the cone-shaped ommatidia is a black pigment with the ability to raise or lower its level just like the mercury of a thermometer. The higher it moves up the wall of the cone, the greater an area it darkens, the less light penetrates to the optic nerve. The lower its level, the more light will hit the optic nerve. How could all this have developed through blind chance evolution mutations, gene shuffling, or genes being turned on or off? (See Chapter 7 section "The Evolution of the Horse" "Creationist View" subsection "Preexisting Genetic Information" and Chapter 13, section "Some DNA Portions are Used to Turn Genes On or Off.")[14-78]

Although many kinds of very different eyes are known, no direct evidence exists to support the evolution of the eye and its accessory structures. Furthermore, much evidence contradicts such evolutionary beliefs. For example, the number of myelinated fibers in the optic nerve does not correlate with putative evolutionary development. A pigeon has almost as many fibers as a human. Many birds, such as the eagle and hawk, have excellent vision yet have half as many fibers as a domestic pig. Another example is visual pigments. The presumably highest, most evolved form of life, the higher primates, have only two cone photoreceptors, blue and green, but birds have a total of six pigments: four cone pigments plus pinopsin (a pineal photoreceptive molecule) and rhodopsin for black and white vision. Put another way, chickens, humans and mice all have the rhodopsin pigment; mice in addition have blue and green; humans have blue, green, and red; and birds have these three pigments plus violet and pinopsin. For every color that humans perceive, birds can see very distinct multiple colors, including ultraviolet light. Birds use infrared light (which we sense as heat) for night vision, allowing them to rapidly visualize their young in a dense, dark tree.

Many kinds of eyes exist, and there are many schemes to classify them. The most basic classification system groups all eyes into four classes. The **first type** is the camera type or 'simple' eye (Figure 14.9e), such as exists in humans, which uses a focusing system to project a single, sharp image on the retina. The **second type** is the fixed focus compound type (Figure 14.9f) that uses multiple separate refractive units called ommatidia, such as used by trilobites and flies. The **third type** is a scanning eye that builds an image much like a television camera, such as is used in the small marine crustacean copilia, which in females takes up more than half of its body. The **fourth**

type is the complex eye (Figure 14.9a (Octopus type)), found in cephalopods (marine animals, that is, mollusks that have tentacles attached to the head, including the cuttlefish, squid, and octopus) and certain advanced vertebrates, consisting of a cornea, iris, lens, retina and numerous accessory structures. This division obscures many major differences: some shrimp have a combined simple (Figure 14.9d) and compound eye (Figure 14.9f), which is actually a fifth basic eye type, not a transitional form. This division system also greatly over simplifies the variety that exists because at least eleven distinct optical methods of producing images are now known.

The greatest variety of eye design, not only in structure, but also in number and location, does not exist among the vertebrates as Darwinian evolution would expect, but exists among the so-called 'primitive' invertebrates. Invertebrates also have eyes that are, in some respects, superior to those of vertebrates. One example is the hemispherical (compound) eyes of most flies and other insects (Figure 14.9f), which produce, unlike human and most vertebrate (simple or camera) eyes (figures 14.8, 14.9a, 14.9b, 14.9c, 14.9e), an image largely free of spherical distortion. Human eyes have significant peripheral image distortion, but spherical eyes form a sharp image in all directions. Humans, however, do not have sharp peripheral vision because this is the function of the central retina called the macula (Figure 14.9b). Our peripheral vision is for the detection of light and movement that trigger the fixation reflex to turn the eyes toward the stimulus.

Another problem in the theory that eye designs represent an evolutionary sequence is that eyes from the three major phyla (vertebrates, arthropods and mollusks) arise from different tissues and are radically different. For this reason, evolutionists concluded that they have separate evolutionary histories, and the many similarities that exist are due to presumed evolutionary convergence. The eye differences would be due to the different needs and circumstances of each organism and its habitat, irrespective of any evolutionary connection. Yet another problem is the evidence for eye evolution forces the conclusion that most of these eye designs must have evolved in a brief period during the Cambrian period (See Chapter 3, section "The Strata Layers of the Earth" sub-section "Cambrian Period—(580-505 mya)").

Evolutionists often claim the primate eye is the most evolved, but many mislabeled 'primitive' eyes have advantages over the human eye. For example, the human eye can register up to 60 images per second; a lowly bee eye can register up to about 300 images per second. For this reason, bees can see far better while rapidly moving. The motion picture standard (24 frames per second), to a bee, would be viewed as a series of still pictures. For humans the frames are blurred, giving the illusion of motion. This design innovation in so-called primitive animals is more complex than the corresponding structure in the human eye.

The simplest eye type known is the ocellus, a multicellular eye comprising of photoreceptor cells, pigment cells (Figure 14.9) and nerve cells to process the information, is stage 4 in Darwin's evolutionary hierarchy. The most primitive eye that meets the definition of an eye is the tiny (about the size of the head of a pin) microscopic marine crustacean copepod copilia. Only the females possess eyes that make up more than half of its transparent body. The marine crustacean copepod copilia, has been claimed to be a link between an eyespot and a more complex eye. It has two exterior lenses that function like a scanning electron microscope to gather light that is processed and then sent to its brain. It has retinal cells and an eye analogous to a superposition-type ommatidium of compound eyes. This is the most primitive true eye known, yet it is at stage 6 of Darwin's evolutionary hierarchy. Evolutionists predict that the more advanced an eye, the more detail it can pick up, a factor related to the number of visual cells. This is not what is often found. In a 'simple' visual system (brain and retinas) the smallest number of visual cells is found in the plethodontid salamander, T. narisovalis, which uses about 65,000 cells for the entire visual brain center and 60,000 for the retina alone. This extraordinarily low number of cells is used not because the animal is primitive but because it has a very small head, eye, and brain (plus relatively large cells). The smallest extant salamander, T. pennatulus (which is much smaller than T. narisovalis), has about 94,000 visual cells and about the same number of retinal cells. For comparison, the brain visual centers of the frog S. limbatus contain about 400,000 cells. This illustrates the fact that evolution cannot be argued by asserting that the eye can be built up gradually from a single patch of light-sensitive skin through various stages, slowly reaching the complexity of the vertebrate camera eye.

Another problem for evolution is that at least 11 distinct optical methods are used to produce images. For one type to evolve into a more 'advanced' type requires intermediate stages that are much worse or useless, compared with the existing design. This would make a switch essentially lethal to animals that depend on sight. For example, the advanced rods and cones in 'primitive' animals and the lack of evidence for their evolution has motivated some to conclude that the basic tetrachomatic system evolved very early in vertebrate evolution. Furthermore, no progression from simple to complex photoreceptors exists, but rather only four spectrally distinct classes of cone pigment encoded by distinct opsin genes are found in the natural world. On the other hand, similarities, such as the fact that some of the genes involved in eye development are very similar in most animals, appear to support the idea for a single evolution of the eye. Yet the difficulties of eye evolution are so great that eyes are hypothesized by some researchers to have independently evolved at least 40 and as many as 65 times.[14-79] Some evolutionists claim that researchers have identified primitive eyes and light-sensing organs throughout the animal kingdom and have tracked the evolutionary history of eyes through comparative genetics. If the evolutionary history of eyes has been tracked through comparative genetics, how is it that eyes have supposedly evolved independently? Some evolutionists believe that eyes must have arisen independently at least 30 times because there is no evolutionary pattern to explain the origin of eyes from a common ancestor.[14-80]

Many evolutionists do not know whether eyes arose once or many times. Vertebrate eyes could not have evolved in isolation because eye parts do not have a function as self-contained entities. Eyes are part of very complex, interconnected living organisms, and eyes are only one part of the vision system. One gauge to help determine eye complexity is the number of genes involved in producing the eye, that is, the more genes that are required, the more complex the eye may be. So far, 501 eye-related genes (or about 3.5% of its entire genome) have been identified in the primitive Drosophila. Vertebrate eyes are estimated to involve 7,500 genes just to develop and regulate the retina, or about 30% of the entire human genome of 25,000 genes. These problems are part of the reason why views on eye evolution have varied, favoring one or many origins. The markedly distinct ontogenetic origin of eyes in very different species is one reason why eyes are claimed by evolutionists to have evolved 40 or more times independently.

For example, the eyes in many mollusks (invertebrates), including some cephalopods such as squids and octopuses (Figure 14.9a), are remarkably similar to vertebrate (simple or camera) eyes (figures 14.8, 14.9a, 14.9b, 14.9c, 14.9e). Both mollusk (cephalopod) and vertebrate eyes have a cornea, a lens, an iris and a retina. One of the major differences is that in vertebrates (simple or camera), the retina is inverted, compared to the invertebrates (mollusk), whose retinas are not inverted (Figure 14.9a). Evolutionists attempt to solve this problem by assuming that the phylogenetic line that led to mollusks split very early in evolutionary history, long before the eye had evolved. Then they claim parallel evolution occurred, concluding that the two eyes evolved to be almost identical, yet were completely independent of each other. The most 'primitive' camera eye known (the nautilus pinhole eye) and the most advanced eye known, are both found in cephalopods. Mollusks as a group contain (1) a pigment eyespot design (Figure 14.9), (2) a pigment cup (cupulate), (3) a simple optic cup with a pinhole lens, (4) an eye with a primitive lens (a murex marine snail) and (5) a complex eye (the octopus), the latter which is the 'most elaborate' eye in the invertebrate kingdom.

Another major difference is found in the embryonic origin of many structures in vertebrate eyes in contrast to cephalopod eyes. (Cephalopods are mollusks that have tentacles attached to the head, including the cuttlefish, squid, and octopus, discussed in Chapter 3.) For example, cephalopod eyes form from an epidermal placode by successive infoldings, whereas vertebrate eyes develop from the neural plate, and the overlying epidermis forms the lens. Yet another problem for eye evolution is that the eye of just one evolutionary related class, the vertebrates, develops from a diverse collection of embryonic sources through a complex set of inductive events.[14-81]

To claim that simple eye types evolved into more complex eyes is not true. Each eye type is different and would require a great deal of changes for any new eye type to develop. Evolutionists are not able to explain how one eye type could have acquired the necessary changes to develop into a more advanced eye, and during this process survive natural selection (Chapter 9) from eliminating the changes while they are in the process of developing. Eyes do not descend from other eyes, but rather, organisms pass on genes for eyes to their descendants. This is important when considering the nautilus eye, a pinhole camera. This cannot possibly be an ancestor of the vertebrate lens/camera eye, because the nautilus eye as a whole is not an ancestor of the vertebrates, as evolutionists would agree.[14-82]

Examples of Eyes That Disprove Darwinian Evolution

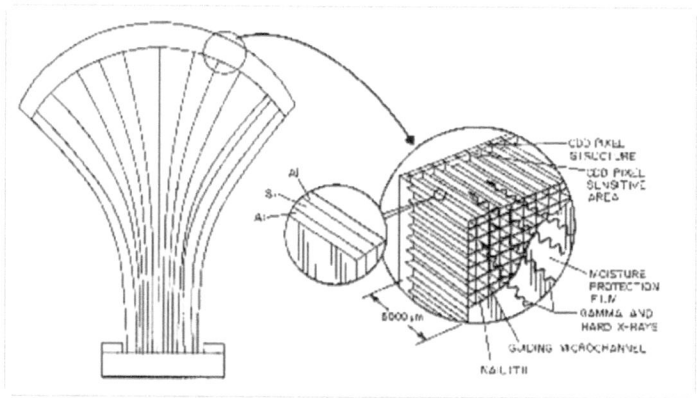

Figure 14.9h Lobster Eye

Scallops, Shrimps, Lobster Eyes

Researchers have discovered that scallops, shrimps, and lobsters have a different optical system. Studies have found that multilayer mirrors can be made from a great many materials provided they have different refractive indexes. Scallops, for example, take advantage of the pronounced difference between the refractive indexes of guanine (1.83) and cytoplasm (1.34). Many compound eyes use the refractive difference between chitin (1.56) and air (1.00). Behind the lens of the scallops, these layers of material act as a reflector in the form of a concave mirror. In this construction the images thus formed by that mirror are focused in the rear section of the retina behind the lens. They are then captured by the distal photoreceptor cells in the eye.[14-83] Evolutionists are also not able to explain lobster eyes. Lobster eyes (Figure 14.9h) are unique in that they are a perfect square with precise geometrical relationships of the units.[14-84]

The Octopus, Trilobite and Nautiloid Eyes

The octopus eye (an invertebrate) and the vertebrate eye are complete, complex, and totally distinct from one another right from their first appearance in the fossil sequence. The nautiloid is a squid-like animal that is a member of the most complex group of invertebrate animals (the cephalopod mollusks). (See Chapter 3, illustration "Examples of Nautiloids.") The nautiloid belongs to the group of animals that has an eye somewhat like ours. Some trilobites did not have the problem of depth perception. They had double-lens systems that made the correction for underwater vision.[14-85]

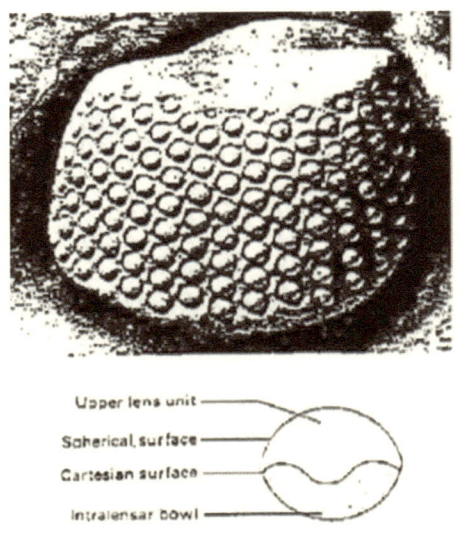

Upper lens unit
Spherical surface
Cartesian surface
Intralensar bowl

Figure 14.9i Trilobite Eye

Trilobite Eyes

The remarkable part of the eye of a trilobite (Figure 14.9i) was its lens. Unlike the lenses of human eyes (Figure 14.8), which are flexible so that their shape can be changed to focus on objects at various distances, the trilobite eye was rigid. The upper half of the lens was made of a mineral substance, calcite, with its crystals stacked in a special way. As a result, beams of light entering this lens from almost any distance were automatically in perfect focus. The lower part of the lens is a substance of animal origin, chitin. This has optical properties that harmonize perfectly with those of the mineral calcite. The combination lens is therefore made free from what opticians call 'spherical aberration', a fault that affects all lenses made of a single material. It is questionable how the trilobite could have accumulated the one material in the universe, that is, calcite, which had the required optical properties, and then imposed on it the one type of curved surface that would have provided perfect vision. Evolutionists are not able to explain how the trilobite eye could have developed by natural means.[14-86]

Trilobites were found in the "ancient" (lower) sediments of the geologic column. The first trilobites appear in sediments dated by evolutionists in the upper part of the Lower Cambrian stratum at 520 million years ago and they extend well into the Permian stratum (supposedly 200 million years ago).[14-87] Trilobites then became extinct. There is nothing alive today that is quite like the trilobite. It was something like a cross between a shrimp and a woodlouse. Although some species of trilobite were blind, others had an extraordinary eye.[14-88] The trilobite, according to evolutionary theory is one of the most ancient of all creatures, had as many as 15,000 lenses in each eye. Trilobites are supposed to have lived for 300 million years, yet their eyes were perfect to begin with and never evolved into a better perfection. The trilobite, one of the first complex life forms found on this earth, first appeared (according to evolutionary theory 600,000,000 years ago), complete with compound eyes. Depending upon the subspecies, they had 100 to 15,000 lenses in each eye. These species survived for about 320 million years, and for all these millions of years, their eye structure did not change. Except for minor variations, trilobites maintained the same form with which they had from the first moment of existence.[14-89] The various trilobite fossils found in the Cambrian strata already possessed very complex eyes. Evolutionists admit that trilobites would have had to evolve eyes separately about 30 or 40 different times since they have such distinctive types of eyes among their various subspecies.[14-90] There are many fossils of the trilobite right at the beginning of the Cambrian strata with no buildup to their eyes, that is, there is no evolution of life forms leading up to the trilobite eyes. Upon close examination, it will be observed that they are not simple animals. The trilobite eye is made up of dozens of little tubes that are all at slightly different angles so that it covers the entire field of vision, with a different tube pointing at each spot on the horizon (Figure 14.9i). But these tubes are all extensively more complicated than that. They have a lens on them that is optically arranged in a very complicated way, and it is bound into another layer that has to be just exactly right for them to see anything.[14-91] Trilobites, like all arthropods, have paired, jointed appendages and a chitinous exoskeleton. The origin of arthropods in general, and trilobites in particular, represents a problem for evolutionists.

As Darwin noted in the Origin of Species, the abrupt emergence of arthropods in the fossil record during the Cambrian presents a problem for evolutionary biology. (Arthropods are discussed in Chapter 3.) There are no obvious simpler or intermediate forms, either living or in the fossil record, that show convincingly how modern arthropods evolved from worm-like ancestors. Additionally, trilobites represent some of the most sophisticated arthropods known to man. The trilobite eye, for example, has been described as a structure far too complex to evolve over time by random variations in the genes of trilobite populations. Not only is the trilobite eye made of pure calcite (optically transparent calcium carbonate) which has a precisely aligned optical axis to eliminate any double image that would have formed, it is also a "doublet" of two lenses affixed together in order to eliminate spherical aberrations, commonly found in ground glass lenses. Trilobite eyes are massively arrayed in semicircular banks and even almost circular banks of up to 30-60 lenses per row, each with its own individual retina (Figure 14.9i). Research conducted on trilobite eyes has shown them to be even more complex than originally thought. Ordovician trilobites such as Pricyclopyge binodosa and Jujuyaspis keideli are said to have had a "large visual field" with "close to 360-degree vision" and "could see anteriorly, laterally, dorsally, and even downwards and backwards," from one position. Further, it has been shown that another Ordovician trilobite, Dalmanitina socialis, actually has a doublet lens arrangement where the top calcite lens has a conspicuous central bulge, the cause of bifocality, which is a unique optical feature in the animal kingdom. According to evolutionists, trilobites became extinct 200-100 million years ago. They claim that trilobites evolved these complex lenses right on their bodies. Thus, evolutionists are faced with the difficult task of explaining the development of glass-like lenses that correct for spherical (and possibly chromatic) aberration, the density of seawater, and which also perform the function of bifocality (much like prescription glasses today), as a result of the chance assemblage of genes within populations of trilobites on the "ancient" seafloor.

Evolution cannot explain the presence of these astounding biological lenses. Many paleontologists working with trilobites feel that evolution is a gradual process and that one can see trilobites changing from one form into another. This is disputed by others who claim that trilobites reflect the model of punctuated equilibrium, that is, trilobites went through periods of long stasis with no change followed by rapid bursts of change. (Punctuated equilibrium is discussed in Chapter 10.) Other evolutionists claim that many incidences of trilobite extinctions have occurred thus causing trilobites to develop new and different eye types. Whatever evolutionary mechanisms are proposed for the development of trilobites, none appear to adequately explain how all the various trilobite eye types could have evolved.[14-92]

The Vertebrate Eye

Creationist View

In the development of vertebrate eyes, fine filaments spread out from a million or so ganglion cells in the retina, and each 'homes' along the optic tract to a precise location on the visual cortex of the brain. This location corresponds exactly to its image point on the retina. If, at a certain stage in the development of an amphibian embryo, the eye is experimentally inverted turned upside down, the dendritic filaments later adjust, home correctly, and the animal sees normally. If the eye is inverted at a slightly later stage, the filaments cross over and home in such a way as to produce an inverted image. How are these filaments guided? It has been shown that even if the cells of the retina and cortex are separated from each other and placed in tissue culture they are still able to associate in their particular patterns.[14-93] It is debatable as to whether the eye of vertebrates (animals having a backbone or spinal column, including humans) is inferior to the design of an octopus eye (Figure 14.9a (Octopus type)). Even some evolutionists believe that there are some advantages to the design of the vertebrate eye over the design of the octopus eye. A structure such as the human eye (Figure 14.8) is too complex to have evolved piece-by-piece as evolutionists suggest. For the eye to see, it needs thousands of different components, all working together with each other. If just one piece is missing, the entire system fails. In this light, the notion that the eye could slowly evolve piece-by-piece is impossible.[14-94]

Evolutionists state that if they were to accept that the eye was designed, they would have to attribute multiple flaws and imperfections to the designer. The implication is that an imperfect design disproves the existence of a perfect God. We would have to wonder why an intelligent designer placed the neural wiring of the retina on the side facing the incoming light. This arrangement scatters the light, making our vision less detailed than it might be, and even produces a blind spot at the point that the wiring is pulled through the light-sensitive retina (Figure 14.9a) to produce the optic nerve that carries visual images to the brain.[14-95] Other evolutionists state that the human (vertebrate) eye was stupidly designed because the retina is upside down (Figure 14.9a).[14-96] There is an important physiological reason as to why the retina has to be inverted in the vertebrate eye. Within the overall design of the system, it is a tradeoff that allows the eye to process the vast amount of oxygen it needs in vertebrates. This does create a slight blind spot but that is not a problem because people have two eyes and the two blind spots do not overlap. Actually, the eye is an incredible design.[14-97] The retina consists of 150 million correctly made and positioned specialized cells. These are the rods to view black and white and the cones to view color. When light-sensitive retinal is combined with a protein (opsin), it, the retinal, becomes a chemical switch. Triggered by light, this switch can generate a nerve impulse. Each switch-containing rod and cone is correctly wired to the brain so that the electrical system (an estimated 1000 million impulses per second) is continuously monitored and translated, by a step (which is a total mystery), into a mental picture.[14-98]

The idea that the vertebrate retina is wired backward (implying it is wrong) is not true. The nerves could not go behind the eye because the space is reserved for the choroid, which provides the rich blood supply needed for the very metabolically active retinal pigment epithelium (RPE). This is necessary to regenerate the photoreceptors, and to absorb excess heat. Therefore, it is necessary for the nerves to go in the front instead of behind the retina. The idea that this arrangement interferes with the image is not true, basically because the nerves are virtually transparent because of their small size and also having about the same refractive index as the surrounding vitreous humor. What limits the eye's resolution is the diffraction of light waves at the pupil (proportional to the wavelength and inversely proportional to the pupil's size). Therefore, for one to claim the vertebrate eye was incorrectly designed is false.[14-99]

Evolutionists have tried to explain how the vertebrate eye, including the human eye, could have evolved. Attempts have been made by claiming the evolution of the eye began with a light-sensitive spot, moving to a group of cells cupped to focus light better, and so on through a graded series of small improvements to produce a true lens (Figure 14.9). The problem with this claim is that the light-sensitive spot would require a chain reaction of chemical reactions, starting when a photon interacts with a molecule called 11-cis-retinal, which changes to trans-retinal, which forces a change in the shape of a protein called rhodopsin, which sticks to another protein called transducin, which binds to another molecule, and so on. Where do the cupped cells that evolutionists refer to come from? There are dozens of complex proteins involved in maintaining cell shape, and dozens more that control groups of cells. (Proteins and cells are discussed in Chapter 5.) Each step in the process of how the vertebrate eye, including the human eye, developed is itself a complex system, and adding them together (as illustrated in Figure 14.9) does not answer where these complex systems came from in the first place.[14-100]

How Charles Darwin Convinced the Scientific World that the Eye Evolved

Intelligent Design View

How do we see? In the 19th century the anatomy of the eye was known in great detail, and its sophisticated features astounded everyone who was familiar with them. Scientists of the time correctly observed that if a person were so unfortunate as to be missing one of the eye's many integrated features, such as the lens, or iris, or ocular muscles, the inevitable result would be a severe loss of vision or outright blindness. So it was concluded that the eye could only function if it were nearly intact.

Charles Darwin knew about the eye too. In the Origin of Species, Darwin dealt with many objections to his theory of evolution by natural selection. He discussed the problem of the eye in a section of the book appropriately entitled "Organs of extreme perfection and complication." For Darwin's views of evolution (descent with modifications) to be accepted, Darwin realized that he needed to provide evidence that complex organs could be formed gradually, in a step-by-step process. Darwin did not try to discover a real pathway that evolution might have used to make the eye. (Most likely, in the 1800s it was not possible to do this.) Instead, he pointed to modern animals with different kinds of eyes, ranging from the simple to the complex, and suggested that the evolution of the human eye might have involved similar organs as intermediates. Here is a portion of Darwin's argument:

> The simplest organ which can be called an eye consists of an optic nerve, surrounded by pigment-cells and covered by translucent skin, but without any lens or other refractive body. We may, however... descend even a step lower and find aggregates of pigment-cells, apparently serving as organs of vision, without any nerves, and resting merely on sarcodic tissue [protoplasm, especially the semifluid content of a protozoan]. Eyes of the above simple nature are not capable of distinct vision, and serve only to distinguish light from darkness. In certain star-fishes, small depressions in the layer of pigment which surrounds the nerve are filled ... with transparent gelatinous matter, projecting with a convex surface, like the cornea in the higher animals. ... This serves not to form an image, but only to concentrate the luminous rays and render their perception more [easily]. In this concentration of the

rays we gain the first and by far the most important step towards the formation of a true, picture-forming eye; for we have only to place the naked extremity of the optic nerve, which in some of the lower animals lies deeply buried in the body, and in some near the surface, at the right distance from the concentrating apparatus, and an image will be formed on it. In the great class of the Articulata [stalked echinoderms with pentamerous symmetry], we may start from an optic nerve simply coated with pigment, the latter sometimes forming a sort of pupil, but destitute of lens or other optical contrivance. With insects it is now known that the numerous facets on the cornea of their great compound eyes form true lenses, and that the cones include curiously modified nervous filaments.

(The Origin of Species by means of Natural Selection or, The Preservation of Favoured Races in the Struggle for Life By Charles Darwin, M.A., F.R.S., Chapter 6 - Difficulties Of The Theory, Section "Organs of extreme perfection and complication")

Using reasoning like this, Darwin convinced many of his readers that an evolutionary pathway leads from the simplest light sensitive spot to the sophisticated camera-eye of man. But the question remains, how did vision begin? Darwin persuaded much of the world that a modern eye evolved gradually from a simpler structure, but he did not even try to explain where his starting point for the simple light sensitive spot came from. On the contrary, Darwin dismissed the question of the eye's ultimate origin: *"How a nerve comes to be sensitive to light hardly concerns us more than how life itself originated."* (See Darwin's quote at beginning of this section, after Figure 14.8.) He had an excellent reason for declining the question: it was completely beyond nineteenth century science. How the eye works; that is, what happens when a photon of light first hits the retina simply could not be answered at that time. As a matter of fact, no question about the underlying mechanisms of life could be answered in Darwin's day. How did animal muscles cause movement? How did photosynthesis work? How was energy extracted from food? How did the body fight infection? No one knew in the 1800s. Back in Charles Darwin's day (1800s) no one had any idea how the eye functions, but today, after the hard, cumulative work of many biochemists, we are obtaining answers to the question of sight. Here is a brief overview of the biochemistry of vision:

When light first strikes the retina, a photon interacts with a molecule called 11-cis-retinal, which rearranges within picoseconds to trans-retinal. The change in the shape of retinal forces a change in the shape of the protein, rhodopsin, to which the retinal is tightly bound. The protein's metamorphosis alters its behavior, making it stick to another protein called transducin. Before bumping into activated rhodopsin, transducin had tightly bound a small molecule called GDP. But when transducin interacts with activated rhodopsin, the GDP falls off and a molecule called GTP binds to transducin. (GTP is closely related to, but critically different from, GDP.) GTP-transducin-activated rhodopsin now binds to a protein called phosphodiesterase, located in the inner membrane of the cell. When attached to activated rhodopsin and its entourage, the phosphodiesterase acquires the ability to chemically cut a molecule called cGMP (a chemical relative of both GDP and GTP). Initially there are a lot of cGMP molecules in the cell, but the phosphodiesterase lowers its concentration, like a pulled plug lowers the water level in a bathtub. Another membrane protein that binds cGMP is called an ion channel. It acts as a gateway that regulates the number of sodium ions in the cell. Normally the ion channel allows sodium ions to flow into the cell, while a separate protein actively pumps them out again. The dual action of the ion channel and pump keeps the level of sodium ions in the cell within a narrow range. When the amount of cGMP is reduced because of cleavage by the phosphodiesterase, the ion channel closes, causing the cellular concentration of positively charged sodium ions to be reduced. This causes an imbalance of charge across the cell membrane which, finally, causes a current to be transmitted down the optic nerve to the brain. (See Chapter 5, section "The Origin of the Cell Membrane," for further discussion of the origin of the cell membrane). The result, when interpreted by the brain, is vision.

This is just a sketchy overview of the biochemistry of vision. Ultimately, though, this is what it means to "explain" vision. This is the level of explanation for which biological science must aim. In order to truly understand a function, one must understand in detail every relevant step in the process. The relevant steps in biological processes occur ultimately at the molecular level, so a satisfactory explanation of a biological phenomenon such as vision, or digestion, or immunity must include its molecular explanation.

Now that it is known how the eye functions, it is no longer enough to simply compare various eye types, as Darwin did in the nineteenth century (1800s), and as advocates of evolution continue to do today, but must explain how each eye actually evolved. Each of the anatomical steps and structures that Darwin thought were so simple actually involves staggeringly complicated biochemical processes that cannot be explained away with rhetoric. Darwin's simple steps are now revealed to be huge leaps between carefully tailored machines. Thus biochemistry offers a challenge to Darwin. Now that the unknown areas of the cell have been determined to have many organelles not previously known and a world of staggering complexity stands revealed, evolutionists now need to explain how the simple and complex eye types evolved from simpler forms.[14-100a]

What it Takes to be Able to See

Intelligent Design View

The ability to see is very complicated. Even a simple light-sensitive spot requires a tremendous amount of biochemicals in the right place and time to function. When light first strikes the retina, a photon interacts with a molecule called 11-cis-retinal, which rearranges within picoseconds to trans-retinial. (A picosecond [10^{12} sec] is about the time it takes for light to travel the length of a single human hair.) The change in the shape of the retinal molecule forces a change in the shape of the protein, rhodopsin, to which the retinal is tightly bound. The protein's metamorphosis alters its behavior. Now called metarhodopsin II, the protein sticks to another protein, called transducin. Before bumping into metarhodopsin II, transducin had tightly bound a small molecule called GDP. But when transducin interacts with metarhodopsin II, the GDP falls off, and a molecule called GTP binds to transducin. (GTP is closely related to, but different from, GDP). GTP-transducin-metarhodopsin II now binds to a protein called phosphodiesterase, located in the inner membrane of the cell. When attached to metarhodospin II along with the other parts associated with it, the phosphodiesterase acquires the chemical ability to create a molecule called cGMP (a chemical related to both GDP and GTP). Initially there are many cGMP molecules in the cell, but the phosphodiesterase lowers its concentration, similar to pulling a plug in a bathtub to lower the water level.[14-101]

Creationist View

Evolutionists must admit that an eye can do very little on its own because the ability to perceive light is meaningless unless the organism has the ability to use this information. The first curving, with its slight ability to detect the direction of light, would only work if the creature had the appropriate ability to interpret this. The brain needs to be able to process the information that is being received. Actually, much information processing actually occurs in the retina before the signal reaches the brain.[14-102]

The Human Nervous System, the Brain and Intelligent Design

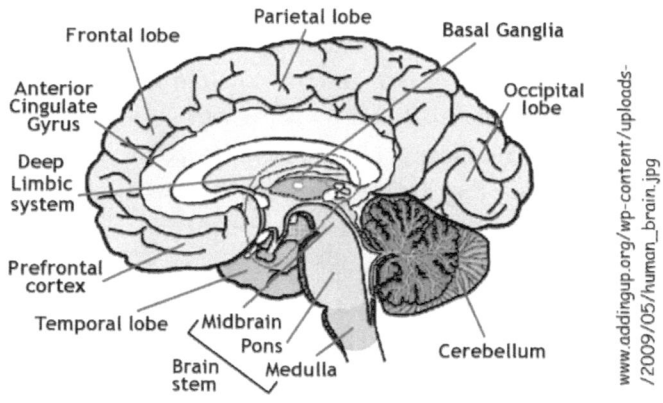

Figure 14.10 The Human Brain

Intelligent Design View

The Nervous System

The nervous system is an excellent example of irreducible complexity. The nerve signal from one nerve cell to another requires molecular channels on the nerve cell to open and close in an orchestrated and coordinated fashion. If synaptic vesicles do not pick up the correct neurotransmitter, or are not delivered to the correct region of the cell, or are unable to be stored, or are unable to deposit their contents into the synaptic cleft, then the whole system would break down and the human would soon suffer physiological abnormalities that could lead to death.

The nervous system is the collection of nerve cells and body tissues that regulate the body's response to internal and external stimuli by electrical and chemical signals. Synaptic vesicles are loaded with specific neurotransmitters that are then delivered to the end terminal of the axon. Upon activation, these vesicles will "dump" their contents into the synaptic cleft (the space between nerve cells). This action requires the nerve cell to be able to manufacture neurotransmitters, correctly load specific transmitters into the correct vesicles, transport the vesicle to a specific place, store the vesicles until they are needed, empty the vesicle upon activation, and then "clean-up" after itself. And this complex transportation system is just one small aspect of the human nervous system. The human nervous system is composed of billions of nerve cells often referred to as neurons. These nerve cells are responsible for sending electrical impulses from one part of the body to another. By monitoring both the internal and external environment, the nervous system is responsible for maintaining a relatively constant internal environment. Often, the brain (Figure 14.10) will be sent (will receive) sensory messages from nerves in the body, alerting it as to temperature, pain, pleasure, etc.

The brain, on the other hand, sends out electrical messages in response; for example, moving a hand from a hot stove. Separately, the neurons are helpless in trying to maintain a stable state in the body. But purposefully arranged together, these individual cells perform very elaborate functions. Organs, glands, and vessels throughout the body are constantly controlled and coordinated by individual neurons, and each of these structures would be ineffective without nerve input and feedback. Similarly, the heart, kidneys, pancreas, bladder, and lungs carry out specific body functions, but without the "wiring" and input from the nervous system these organs would be completely useless.

In order for the brain to work, it must be able to send and receive input via nerves. Nerve cells are of little use without the spinal cord and brain to process and integrate the information. If the Darwinian Theory is correct, then nerves must have developed prior to the evolution of the brain, because the brain is composed of trillions of neurons. But without a processing unit, what purpose would such nerves serve?

The Brain

The average human brain (Figure 14.10) weighs only about three pounds and is covered with convolutions and wrinkles known as sulci. The purpose of these furrows is to provide more surface area for the brain. Hidden within the gray and white matter of the brain is the most intricately wired communication network in the world. This three-pound organ represents literally billions of interconnected nerve cells and millions of protective glial cells that, according to evolutionists, simply arose by pure chance from nonliving matter through millions and millions of mutations.

The human brain has been estimated to contain 10 billion neurons, each a living unit within itself. Each neuron is a living unit within itself with approximately 25,000 synapses (connections) per neuron. Each neuron is made up of 10 billion macromolecules. While most neurons share similar properties, they can be classified into as many as 10,000 different types. Additionally, over 100 trillion electrical connections are estimated to be present throughout the human brain.

The Spinal Cord

The average length of the human spinal cord is seventeen inches. It normally extends from the brain stem through the largest hole in the skull (foramen magnum) to the level of the second lumbar vertebrae. Thirty-one pairs of spinal nerves branch out from the cord, which help connect the rest of the body with the central nervous system. Was there a transitional period in which only two or three pairs of spinal nerves existed? If so, how did the rest of the body receive input?

Peripheral Nervous System

The peripheral nervous system (PNS) consists of nerve cells that are outside the brain and spinal cord. Included are sensory neurons found on the skin as well as those involved in smell, taste, hearing, and sight. The PNS is often divided into two subdivisions: **(1)** sensory and **(2)** motor neurons. Sensory neurons carry information to the central nervous system, while the motor division neurons carry signals away from the central nervous system. Cranial nerves are highly specialized, and vary in function, from light-receptor cells in the eye to cells detecting taste in the tongue. These nerves often contain both sensory and motor fibers, and act without any input from the individual. Cranial nerves, unlike spinal nerves, go directly out of the brain and then proceed to their target organ. The brain is completely encased in bone, making this task much more difficult than it might appear. The body was designed in such a way that humans have control over certain components, while the body itself regulates other aspects. It seems nearly impossible for nature to evolve a voluntary nervous system in conjunction with an involuntary system, along with a processing unit (the brain) that can integrate all of the incoming information.[14-103]

The Brain and Nervous System Are Not Irreducibly Complex

Evolutionist View

The human brain is a true marvel of nature. This jelly-like 1.5kg mass inside our skulls, containing responsible hundreds of billions of cells which between them form something like a quadrillion connections, is for our every action, emotion and thought. How did this remarkable and extraordinarily complex structure evolve? This question poses a huge challenge to researchers. Brain evolution surely involved thousands of discrete, incremental steps, which occurred in the mists of deep time across hundreds of millions of years, and which we are unlikely to ever fully understand. Nevertheless, a number of studies published in recent years have begun to shed some light on the evolutionary origins of the nervous system, and provide clues to some of the earliest stages in the evolution of the human brain. These clues come from the most unexpected of places, from sea sponges, which lack nervous systems altogether, and from the extant descendants of a primitive worm which lived some 600 million years ago.

Despite their differences, vertebrates, worms and insects are all believed to be descended from a common ancestor, a worm-like organism, named Urbilateria, which lived some 600 million years ago. Urbilateria displayed bilateral symmetry - its body was symmetrical along its longitudinal axis - and this body plan was inherited by the diverse array of organisms descended from it. But, according to new research, published in 2007, it wasn't just bilateral symmetry that the descendants of Urbilateria inherited: at the earliest stages of their evolution, vertebrates, including humans, may have inherited the organization of their nervous systems from it as well. The simplest nervous systems lack a brain, and instead consist of diffuse networks of nerves. The nervous systems of vertebrates and annelid worms, however, are organized in another way, with nerve fibers arranged in centralized cords, and large groups of nerve cells (called ganglia, singular ganglion). The major differences between them are, of course, the level of complexity, and the positioning - the nerve cord of invertebrates is located ventrally (toward the belly), whereas the vertebrate spinal cord is located dorsally (toward the back).[14-104]

Recent studies have shown that many of the components needed to transmit electrical signals, and to release and detect chemical signals, are found in single-celled organisms known as choanoflagellates. That is significant because ancient choanoflagellates are thought to have given rise to animals around 850 million years ago. So almost from the start, the cells within early animals had the potential to communicate with each other using electrical pulses and chemical signals. From there, it was not a big leap for some cells to become specialized for carrying messages. These nerve cells evolved long, wire-like extensions, axons, for carrying electrical signals over long distances. They still pass signals on to other cells by releasing chemicals such as glutamate, but they do so where they meet them, at synapses. That means the chemicals only have to diffuse across a tiny gap, greatly speeding things up. And so, very early on, the nervous system was born. The first neurons were probably connected in a diffuse network across the body. This kind of structure, known as a nerve net, can still be seen in the quivering bodies of jellyfish and sea anemones. But in other animals, groups of neurons began to appear, a central nervous system. This allowed information to be processed rather than merely relayed, enabling animals to move and respond to the environment in ever more sophisticated ways. The most specialized groups of neurons, the first brain-like structure, developed near the mouth and primitive eyes. According to many biologists, the development of the brain happened in a worm-like creature known as the urbilaterian, the ancestor of most living animals including vertebrates, mollusks and insects. Strangely, though, some of its descendants, such as the acorn worm, lack this neuronal hub. It is possible the urbilaterian never had a brain, and that it later evolved many times independently. Or it could be that the ancestors of the acorn worm had a primitive brain and lost it, which suggests the costs of building brains sometimes outweigh the benefits. Either way, a central, brain-like structure was present in the ancestors of the vertebrates. These primitive, fish-like creatures probably resembled the living lancelet, a jawless filter-feeder. The brain of the lancelet barely stands out from the rest of the spinal cord, but specialized regions are apparent: the hindbrain controls its swimming movements, for instance, while the forebrain is involved in vision. Some of these fish-like filter feeders took to attaching themselves to rocks. The swimming larvae of sea squirts have a simple brain but once they settle down on a rock it degenerates and is absorbed into the body.

Around 500 million years ago, changes occurred in the lancelet (discussed in Chapter 3) during reproduction, resulting in its entire genome getting duplicated. In fact, this happened not just once but twice. These accidents paved the way for the evolution of more complex brains by providing plenty of spare genes that could evolve in different directions and take on new roles. Among many other things, it enabled different brain regions to express different types of neurotransmitter, which in turn allowed more innovative behaviors to emerge. As early fish struggled to find food and mates, and dodge predators, many of the core structures still found in our brains evolved: the optic tectum, involved in tracking moving objects with the eyes; the amygdala, which helps us to respond to fearful situations; parts of the limbic system, which gives us our feelings of reward and helps to lay down memories; and the basal ganglia, which control patterns of movements.

By 360 million years ago, our ancestors (amphibians and reptiles) had colonized the land, eventually giving rise to the first mammals about 200 million years ago. These creatures (the first mammals) already had a small neocortex – extra layers of neural tissue on the surface of the brain responsible for the complexity and flexibility of mammalian behavior. (*The neocortex is part of the mammalian brain involved in higher-order brain functions such as sensory perception, cognition, and generation of motor commands, spatial reasoning and language. The cerebral cortex is the entire outer top part of the brain (the part that looks wrinkled). The neocortex is that part of the cerebral cortex (maybe 90% of it) that is the modern, most newly ("neo") evolved part. The neocortex is basically everything except the regions at the edges (the hippocampus, and parts of the cingulate cortex).*) How and when did this crucial region neocortex) evolve? That remains a mystery. Living amphibians and reptiles do not have a direct equivalent (of the neocortex), and since the brains of amphibians and reptiles do not fill their entire skull cavity, fossils tell us little about the brains of our amphibian and reptilian ancestors. What is clear is that the brain size of mammals increased relative to their bodies as they struggled to contend with the dinosaurs. By this point, the brain filled the skull, leaving impressions that provide tell-tale signs of the changes leading to this neural expansion.

After the dinosaurs became extinct, about 65 million years ago, some of the mammals that survived took to the trees – the ancestors of the primates. Good eyesight helped them chase insects around trees, which led to an expansion of the visual part of the neocortex. As modern primates began to increase in size, these frontal regions also became better connected, both within themselves, and to other parts of the brain that deal with sensory input and motor control. Such changes can even be seen in the individual neurons within these regions, which have evolved more input and output points. All of which equipped the later primates with an extraordinary ability to integrate and process the information reaching their bodies, and then control their actions based on this kind of deliberative reasoning. Besides increasing their overall intelligence, this eventually leads to some kind of abstract thought: the more the brain processes incoming information, the more it starts to identify and search for overarching patterns that are a step away from the concrete, physical objects in front of the eyes.

About 14 million years ago an advanced ape lived in Africa. The brains of most of its descendants, orangutans, gorillas and chimpanzees, did not appear to have changed greatly compared with the branch of its family that led to us humans. What made us humans different? It used to be thought that moving out of the forests and taking to walking on two legs lead to the expansion of our brains. Fossil discoveries, however, show that millions of years after early hominids became bipedal, they still had small brains. We can only speculate about why their brains began to grow bigger around 2.5 million years ago. In other primates, the "bite" muscle exerts a strong force across the whole of the skull, constraining its growth. In our forebears, this muscle was weakened by a single mutation, perhaps opening the way for the skull to expand. This mutation occurred around the same time as the first hominids with weaker jaws and bigger skulls and brains appeared. Our diet, culture, technology, social relationships and genes contributed to our development. It led to the modern human brain coming into existence in Africa by about 200,000 years ago.[14-105]

In the next chapter (15) we will look at biological evidence to determine whether or not lower life forms, such as apes, evolved into humans.

~~~

Chapter 15

A Comparison of Chimpanzee and Human Genome

Chimpanzee

Did Ape-like Creatures Evolve Into Humans?

And the LORD God formed man of the dust of the ground, and breathed into his nostrils the breath of life; and man became a living being . . . And the LORD God caused a deep sleep to fall on Adam, and he slept; and He took one of his ribs, and closed up the flesh in its place. Then the rib which the LORD God had taken from man He made into a woman . . . And Adam called his wife's name Eve . . . [15-1]

Yes, we are all animals, descendants of a vast lineage of replicators sprung from primordial pond scum.[15-2]

We are apes descended from other apes, and our closest cousin is the chimpanzee, whose ancestors diverged from our own several million years ago in Africa. These are indisputable facts.[15-3]

Introduction

In Chapter 7 we discussed whether or not fish evolved into amphibians; whether or not amphibians evolved into reptiles; and whether or not reptiles evolved into mammals and birds. In this chapter (15), we will look at the genetic similarities and differences between humans and chimpanzees to determine whether or not humans evolved from lower life forms, such as apes. (Chimpanzees are believed to be the closest living relatives to humans.)

The various views presented in this chapter regarding the origin of various species and organisms are not exclusively of strict evolutionists and strict creationists. Some views are of those who believe in variations of the two extremes. See "Various View Descriptions" section in the back of this book for descriptions of the views presented in this chapter.

Questions for Creationists

Evolutionist View

If God created humans out of nothing (de novo) as stated in Genesis 1 and 2, then it is puzzling why He would recycle two ape chromosomes and fuse them together to make our chromosome 2 (discussed below). As well, why would He put a functional gene for the production of vitamin C into most animals, but reuse the defective one He placed in chimpanzees for humans (as discussed in Chapter 13)? And what are we to make of the fact that 600 of our 1,000 genes for scent receptors are defective? Did God put them in Adam?[15-4]

Common Function or Common Ancestry?

Evolutionists argue that human and chimpanzee DNA are similar because we share a common ancestor. The common ancestor view proposed by evolutionists states that similar creatures have similar DNA sequences because they share a common ancestor. Creationists argue that human and chimpanzee DNA are similar because humans and chimpanzees share similar body structures. This is known as the common function view. The common function view specifies that the Creator reused DNA sequences for similar type creatures. We will now take a look at some of the arguments from both perspectives.

Evolutionist View

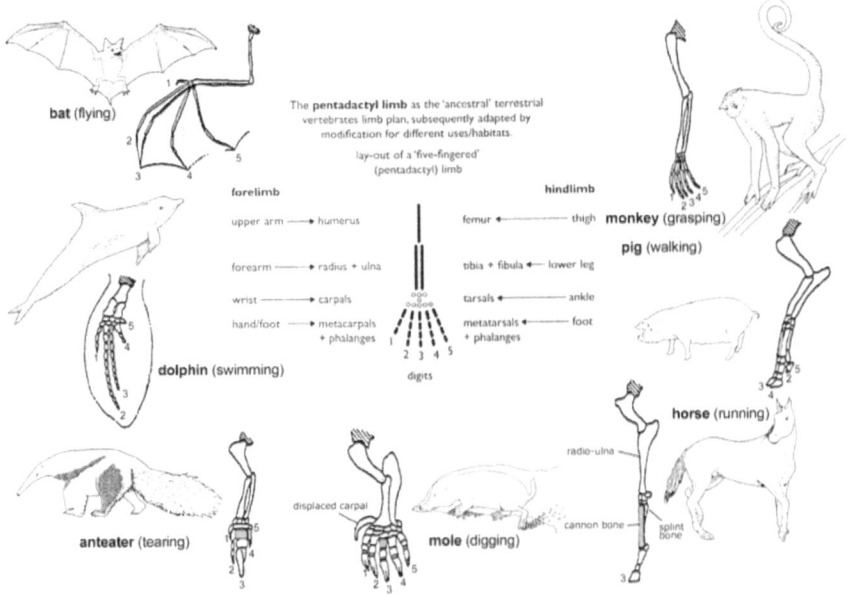

Figure 15.1 Various limbs

Creationists sometimes argue that different species share genetic similarities not because the species share a common ancestry, but because the genes have a common function. Perhaps eagles, ravens, robins, and hummingbirds do not have a common ancestor, but instead their genes are similar to each other because all have similar body structures that allow them to fly. Perhaps this similarity of function requires that their genes also be similar. Each species may have risen or been miraculously created individually, without a common ancestor, but they have genetic similarities to each other roughly in proportion to the similarities in their body structures and the functions that their genes perform.[15-5] It could be argued that the similarity between humans and chimpanzee gene sequences is explained by common function rather than common ancestry. Common function theory supported by creationists would argue that human and chimpanzee gene sequences are similar only because they have similar functions in both species, not because of common ancestry. While common function theory might explain some similarities in gene sequences, it does not explain other evidence such as similarities in pseudogenes.[15-6] (Pseudogenes are discussed in Chapter 13.) The "common ancestry theory" can explain the genetic similarities between species. The "common function theory" can explain some of this genetic similarity, however, it is not able to explain why the genes for bats are more similar to the genes of rats and mice than the genes for bats are to the genes of birds. But "common function theory" cannot account for or explain pseudogenes (that is, no longer functional genes that are remnants of past ancestors.) (Pseudogenes are discussed in Chapter 13).[15-7]

Intelligent Design View

Darwinists believe that homologies (similarities) exist among organisms because they all derive from the same ancestor. For example, human forelimbs (Chapter 7 Figure 7.4 & Figure 15.1 above) and a pig's are related, and are descended from a common ancestor that possessed the humerous-radius-ulna skeleton pattern. Darwinists believe that as various species develop changes in their body types, the genetic information is carried in the mother's egg and father's sperm and is passed onto their offspring. Homologous development pathways and homologous genes build the similar structures. A developmental pathway is the process by which the embryo physically becomes an adult.[15-8] (See Chapter 11 for further discussion of embryos.)

Homology is defined as "similarity due to common descent." Darwinists claim that homology provides evidence for common descent (common ancestry). (See Chapter 7, section "Homology of Vertebrate Limbs" further discussion of common function, common ancestry and homology.) Darwinists claim that the similarities among animals and humans, such as having similar forelimbs (Chapter 7 Figure 7.4 & Figure 15.1 above), are due to a common ancestor. Homologies exist among animal structures because of the necessity of using similar structures to solve similar functional problems. The pattern seen in the vertebrate forelimb, for example, has a single bone closest to the trunk, two bones in the next section, and a variety of bones in the farthest section out, due to common function. It is not true that similar body structures (homologies) are due to similar genetic information being passed off to their offspring, and hence, to a common ancestor. Rather, the homologous structures can be produced by different developmental pathways (the process by which the embryo physically becomes an adult).

For example, flies and wasps share similar body structures, and therefore derive, from homologous genes and homologous pathways. Darwinists therefore believe those creatures that have similar body structures share a common ancestor. Darwinists claim that the genes and pathways that produce homologous structures were inherited from a common ancestor. However, the similar body structures of some wasps from developmental pathways are entirely different from those of fruit flies, and are also different from other wasps. Another example of organisms having similar body structures but are not genetically related is the vertebrate gut. In sharks, the gut develops from cells in the roof of the embryonic cavity. In lampreys, the gut develops from the cells on the floor of the cavity. In frogs, the gut develops from cells from both the roof and the floor of the embryonic cavity. This is a clear indication that different developmental pathways can produce similar structures in vertebrates. This contradicts what one would expect to find if all vertebrates share a common ancestor.

Biologists have discovered that in many cases the same genes are used to produce different adult structures. The eyes of the squid, the fruit fly, and the mouse have different structured eyes. The squid and mouse both have single-lens camera eyes, but they develop along different pathways, and are wired differently from each other. Yet the same gene is involved in the development of all three of these eyes. (Various eye types are discussed in Chapter 14, section "The Eye.") Darwinists believe that non-homologous genes should regulate the development of non-homologous structures. They cannot explain that non-homologous eyes from an insect, a mollusk, and a vertebrate could be regulated during their development by homologous genes, such as the Pax-6 gene. (See Chapter 9 Section "Hox Genes" for further discussion of the Pax gene.)

The development of homologous structures can be governed by different genes and can follow different developmental pathways. Also, sometimes the same gene plays a role in producing different adult structures. This contradicts the theory of evolution being caused by mutations. Darwinists claim that the reason why organisms have similar structures is because of common ancestry. They say that we should expect that similar structures would evolve from similar genes and developmental pathways. The problem is that this is often not true.[15-9]

Common Function not Common Ancestry

Creationist View

Physical and DNA similarities between human and other living organisms are supposed to be evidence for evolution. However this is not a direct finding, but an interpretation of the data. A common designer is another interpretation that makes sense of the same data. An architect commonly uses the same building material for different buildings, and a car-maker commonly uses the same parts in different cars. Therefore, we should not be surprised if a Designer for life used the same biochemistry and structures in many different creatures. Since DNA contains the coding for structures and biochemical molecules, we should expect the most similar creatures to have the most similar DNA.

Apes and humans are both mammals, with similar shapes, so they have similar DNA, which indicates common function theory. Humans have more DNA similarities with other mammals like a pig than with reptiles like a rattlesnake. So the general pattern of similarities does not need to be explained by common-ancestry evolution. There are some puzzling inconsistencies that cause doubts about an evolutionary explanation. There are similarities between organisms that evolutionists do not believe are closely related. For example, hemoglobin, the complex molecule that carries oxygen in blood and results in its red color, is found in vertebrates.[15-10] Hemoglobin, however, is also found in some earthworms, starfish, crustaceans, mollusks, and even in some bacteria. (Hemoglobin is discussed the next section.) The α-hemoglobin of crocodiles has more in common with that of chickens (17.5 percent) than that of vipers (5.6 percent), their fellow reptiles. An antigen receptor protein has the same unusual single chain structure in camels and nurse sharks. Those who believe that homologies are the result of common ancestry cannot explain this.[15-11]

The Evolution of Hemoglobin

Evolutionist View

Hemoglobin is a respiratory protein found in the red blood cells (erythrocytes) of all vertebrates and some invertebrates. A hemoglobin molecule is composed of a protein group, known as globin, and four heme groups, each associated with an iron atom (Figure 15.1a).[15-12] (Heme is a deep-red iron-containing blood pigment. Heme groups are a component of the hemoglobin protein.) The hemoglobin molecule is made up of four polypeptide chains (or subunits - Alpha 1, Beta 1, Alpha 2, Beta 2) (Figure 15.1a – Illustration at left).[15-13]

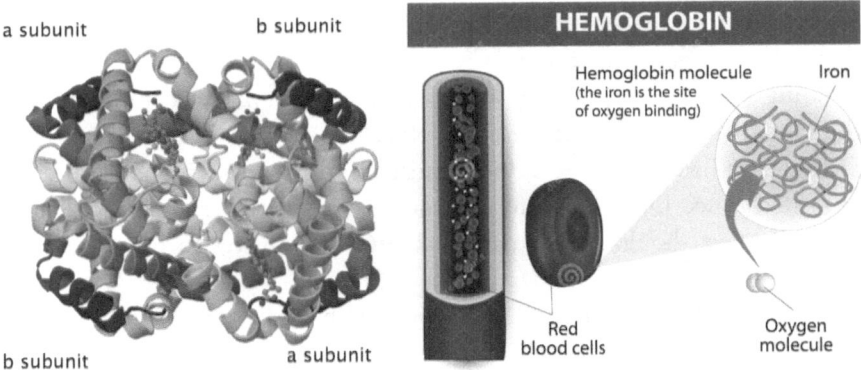

Figure 15.1a Hemoglobin

Left – The four polypeptide chains (or subunits) of a hemoglobin molecule: Alpha 1 (a subunit), Beta 1 (b subunit), Alpha 2 (a subunit), Beta 2 (b subunit).
Right – Illustration of the iron site where oxygen is binding inside red blood cells.

The enriched hemoglobin circulates and is carried through the body to the tissues, where the nitric oxide dilates the small capillaries, allowing hemoglobin to deliver its oxygen to the tissues. Then the oxygen—and nitric oxide–free hemoglobin molecule picks up carbon dioxide and free nitric oxide and transports both back to the lungs, where they are exhaled as waste. Hemoglobin is produced in bone marrow by erythrocytes (red blood cells) and is circulated with them until their destruction. It is then broken down in the spleen, and some of its components, such as iron, are recycled to the bone marrow.[15-12]

Hemoglobin is a component of red blood cells. Red blood cells carry oxygen to every part of the body and need hemoglobin to do this. The main function of red blood cells is the transport of oxygen from the lungs to the body's cells (Figure 15.1a – illustration at right). Red blood cells contain a protein called hemoglobin that actually carries that oxygen. In capillaries, oxygen is released to be used by the body's cells. Hemoglobin allows blood to move 30 to 100 times more oxygen than could possibly be dissolved in the plasma alone. In the lungs, where the oxygen level is high, hemoglobin combines loosely with oxygen. The hemoglobin then easily releases that oxygen into the capillaries, where the oxygen level is low. In each molecule of hemoglobin there are four iron atoms. Each iron atom binds with one molecule of oxygen. The iron in hemoglobin is what gives blood its red color. Our blood is a living tissue with a variety of critical functions: **(1)** It delivers oxygen and nutrients to our organs, **(2)** fights infections and **(3)** creates blood clots, preventing us from bleeding excessively when a blood vessel is damaged. (Blood clotting is discussed in Chapter 14, section "Coagulation Cascade Blood Clotting System".) The liquid part of our blood, called plasma, is key for maintaining blood pressure and supplying critical proteins for blood clotting, immunity and maintaining the correct pH balance in our body— critical to cell function. Plasma also carries the solid part of our blood—white blood cells, which work to destroy viruses and bacteria; red blood cells, which carry oxygen through the body; and platelets, which help clotting.[15-12a]

The development of human hemoglobin (the oxygen-carrying protein in blood) has been reconstructed. Human hemoglobin contains four protein chains called globins that are similar to each other, but they are not identical. The alpha globins each contain a chain of 141 amino acids coded by seven genes on Chromosome [16] (Four of the seven genes are pseudogenes that do not produce proteins.). (Pseudogenes are discussed in Chapter 13.) Two genes produce adult hemoglobin, and one gene produces embryo hemoglobin. The beta globins each contain a chain of 146 amino acids, coded by six genes on Chromosome [11], some of which are disabled (that is, have become pseudogenes) and one gene is only used in embryos.

A letter-by-letter analysis of the genes coding for hemoglobin reveals that the two sets of genes on Chromosome 11 and Chromosome 16 are distantly related and share a common globin gene from a common ancestor five hundred million years ago. This common globin gene duplicated, after which both copies were passed down for half a million years, one evolving into the alpha cluster on Chromosome [16] and the other evolving into the beta cluster on Chromosome [11]. Gene duplications led to the increase in complexity of the gene clusters, leading to the existence today of nonfunctional pseudogenes (discussed in Chapter 13). Since this alpha-beta split happened five hundred million years ago, it can be predicted that the same alpha-beta split in all animals could be found that evolved within the last half billion years. The jawless lamprey fish, the only surviving vertebrate pre-dating the alpha-beta split, does not have this genetic divide between alpha and beta clusters, thus proving that human hemoglobin evolved from lower life forms.[15-14]

Genetics is able to help determine ancestry. By looking at genes, scientists are able to determine which species descended from which species. One example of how modern genetics can be applied to evolutionary theory is by taking a look at hemoglobin. Hemoglobin is the protein that makes your blood red. It is the oxygen-carrying protein found in red blood cells. Hemoglobin is made up of four parts. Those parts are called polypeptides (discussed previously), thought of as four subunits (Figure 15.1a Left image). It has two copies of a part called alpha-globin and two copies of a part called beta-globin. Modern molecular biology has enabled scientists to look at exactly where the instructions are that specify these. The alpha-globin instructions are specified on Chromosome Number 16 and the beta-globin instructions are specified on Chromosome Number 11. As the human genome does for many genes, humans have multiple copies of these, so humans have backups. There are extra copies of the alpha-globin genes and extra copies of the beta-globin genes. These genes enable scientists to test evolution right down to the level of the molecule.

By looking at the beta-globin genes on Chromosome Number 11 scientists are able to obtain a great deal of information. When one examines the six copies of the beta-globin gene sequence, each of these copies is a set of instructions for how to build this polypeptide. Five of them work, but one of them does not. The non-functional copy is given the Greek letters psi, beta, and then the number one. The psi-beta-1 sequence is not a gene. It does not work. It is a pseudogene, and a pseudogene is recognized as a gene because it is so similar to the other five in its DNA sequence, but it has some mistakes. It is broken, and it has a series of molecular errors that render the gene non-functional. The beta-blogin gene that does not work is called a pseudogene. (Pseudogenes are discussed in Chapter 13.) There are six distinct mistakes in this gene. Simply explained, the altered initiator means that the signal that exists at the front of the gene that says, "copy me" is missing. And therefore RNA preliminaries, the molecule that copies genes, cannot bind, and it never gets expressed (is not turned on). But even if it did get expressed, it has five other errors that would keep this, the RNA copy of this gene, from being translated. It is missing the start signal (start codons AUG or GUG). It does have stop codons (UGA, UAA, UAG) that would cause the synthetic apparatus to grind to a halt. The reason that this is important in evolution is actually very simple; these errors appear in a gene, they have no functional purpose.

Most likely, if one were to find another organism that did not just have similar genes but also had a pseudogene in the same spot and had the same set of errors, one would conclude that they must have had a similar ancestry. (Pseudogenes are discussed in Chapter 13.) There is no reason why evolution would produce a duplicate set of mistakes in two copies of things. It must be surmised that these two organisms are descended with modification from another organism that had the same set of mistakes. There are three organisms—**(1)** the gorilla, **(2)** the chimpanzee, and **(3)** the human being—that share the exact same set of molecular mistakes. This is significant because one of the core principles of evolution is common descent (common ancestry). One could always argue that because the three species that are being examined here are all African species, that is where they all come from, they are all primates and they all probably started out living in similar environments, that the functional parts of this gene locus, they might work the same. But one cannot argue that the mistakes should match. And the fact that all three of these species have matching mistakes leads scientists to the same conclusion that Charles Darwin predicted in the 1800s, and that is that these three species (gorilla, chimpanzee, and human being) share a common ancestor. Matching mistakes are evidence of common ancestry.[15-15]

Creationist View

Evolutionists believe that hemoglobin evolved by natural processes. If you start at the starting point of a protein similar to what is called myoglobin (which is a single chain protein) and consider that hemoglobin has four chains stuck together (discussed previously), the question is, what does it take to form an aggregate of that structure with the properties of hemoglobin? Starting with myoglobin, going to hemoglobin, it is unknown as to whether there is any problem for natural selection to explain this or not. (Natural selection is discussed in Chapter 9.) There is currently no literature or experiment to explain this. There is no detailed description of how myoglobin could have obtained the extra protein chains and evolved into hemoglobin.[15-16] Proteins are more complex than described by many scientists. Hemoglobin is the protein that binds oxygen and carries it from your lungs and dumps it off in peripheral tissues such as your fingers and so on.[15-17]

The jawless fishes are supposed to be very ancient and the earliest vertebrates. Evolutionary theory claims that jawless fishes are closest to carp, frogs, chicken and kangaroo, and humans, in that approximate order. The problem is that this is not true. In reality, the jawless fishes are closest in hemoglobin similarities to humans, carp, kangaroo, frog, and chicken. There is not a trace at a molecular level of the traditional evolutionary series: fish to amphibian to reptile to mammal. Humans are closer to lamprey (eel and an order of jawless fish) than we are to other fish.[15-18]

Comparing fish with terrestrial vertebrates, the carp should, according to evolutionists, be closest to bullfrog, turtle, chicken and rabbit, horse, in that approximate order. Cytochrome C differences places the carp as equally related to horse, rabbit, turtle, and bullfrog, and less closely related to the chicken.[15-19] It might be thought that the amount of DNA in the genome would increase pretty steadily as we advance up the evolutionary scale. But in fact measurements of total DNA content are quite confusing. While the mammalian cell seems to have about 800 times more DNA than a bacterium, toads for example, have very much more DNA than mammals, including humans, while the organism with most DNA (of those so far studied as of 1983) is the lily, which can have from 10,000 to 100,000 times as much DNA as a bacterium.[15-20]

The following sample listing will begin with those creatures having the smallest amount of DNA, and will progressively move on up to those with the most DNA. (Note that humans are only about two-thirds up the list, yet we humans should be at the top, as evolutionists believe humans evolved last): bacterophage-0X174: 0.000,003,6 / bacteriophageT2: 0.000,2 / colon bacteria: 0.004,7 / yeast: 0.07 / snail: 0.67 / sea urchin: 0.90 / chicken: 1.3 / duck: 1.3 / carp: 1.6 / green turtle: 2.6 / cattle: 2.8 / human: 3.2 / toad: 3.7 / frog: 7.5 / protopterus: 50 / amphiuma: 84.[15-21]

Genetic and Intron Similarities
Between Humans and Animals

Evolutionist View

The evidence of the similarity of human and chimpanzee genes goes beyond similarity in the sequences of the genes themselves. There is also evidence for common ancestry in genomic organization and in introns. This evidence indicates that humans share a common ancestry with chimps, which dates to about 5 to 7 million years ago. Other animals have common ancestors that are even older than humans and chimpanzees.[15-22]

Introns are sections of genes that are not actually used to build molecules. (Introns are discussed in more detail in Chapter 13, section "So-called Pseudogenes Are Not Useless Remnants" Creationist View.) If genes are thought of as the blueprint and instruction manual for building a house, then introns are like accounting records, notes, and scratch paper stapled onto the blueprint pages, things that are ignored by the carpenters. When a mutation happens in an intron, that mutation almost always has no positive or negative effect on the organism because the intron is usually ignored when the body builds molecules. If the genes in different species were similar to each other only because they share a "common function" rather than sharing a "common ancestry", there would be no particular reason why the introns in different species show a pattern of similarity with other species.[15-23]

Comparison of the Human and Great Ape
Chromosomes as Evidence for Common Ancestry

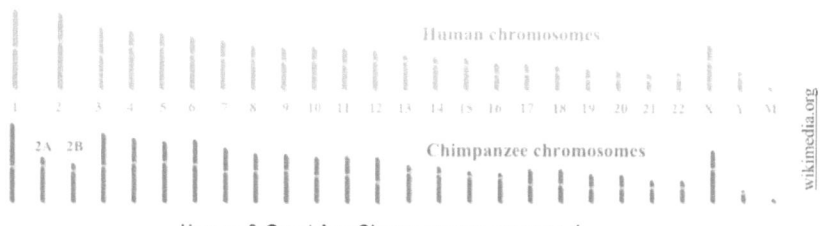

Human & Great Ape Chromosomes compared

Figure 15.2 Illustrates the similarities between human and chimpanzee chromosomes.

It is believed that chromosome 2 in humans is a fusion of chimpanzee chromosomes 2A and 2B. Humans have 23 chromosomes and chimpanzees have 24. This illustration appears to indicate that humans have more than 23 chromosomes and chimpanzees have more than 24 chromosomes. This is because females do not have the 'Y' chromosome. Males have 1 'X' and 1 'Y' chromosome. Females have 2 'X' chromosomes. The 'X' and 'Y' chromosomes are used to determine the sex of offspring and are not otherwise utilized. The 'M' stands for Mitochondrial DNA (the smallest chromosome, as depicted above, on the extreme right, labeled 'M.').

Evolutionist View

A translocation mutation has changed the number of chromosomes in location 2. It is obvious that the two species have a common ancestor. Since chimpanzees, gorillas, orangutans and bonobos all have 24 chromosome pairs, it is assumed that the common ancestor had 24 pairs. Since the time of that ancestor, the human line has had a fusion mutation that turned chromosome location 24 into chromosome location 23. The other major visible difference is that there are nine inversions between man and chimp. Creationists have difficulty explaining this. They claim that humans and chimps are different "designs" or "kinds."[15-24]

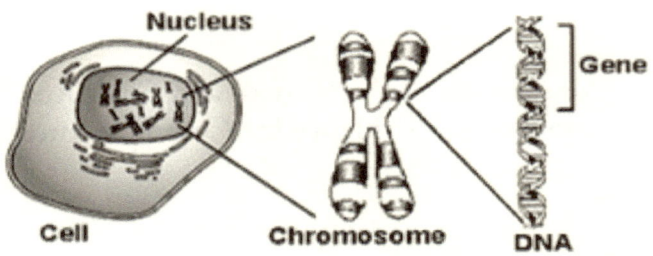

Cell Nucleus - Chromosome - DNA - Gene

Figure 15.3 Chromosomes, DNA, Genes

Humans and chimpanzees each have about 30,000 genes. Those genes are arranged on chromosomes (Figure 15.3 above and Chapter 5, Figure 5.5). As mentioned under Figure 15.2, humans have 23 pairs of chromosomes, while chimpanzees and other apes have 24 pairs of chromosomes. In most cases the genomic organization, the order in which the genes are arranged on the chromosomes, can be matched up chromosome-by-chromosome among those species. On some chromosomes certain species have an inversion on part of the chromosome. When those inversions are taken into account, the genomic organization of the different species matches up impressively. When human chromosomes are compared and matched one-to-one with chimpanzee chromosomes in this way, human chromosome number two does not immediately match up (Figure 15.2). This chromosome matches up extremely well with the two remaining chimp chromosomes when those chromosomes are fused end-to-end.[15-25]

There is impressive evidence for human evolution that appears in our chromosome 2. As previously mentioned, a genetic difference between the great apes and humans is that great apes have 48 chromosomes, while humans have 46 (23 pairs). The reason for this difference is that human chromosome 2 is made up of two previously independent chromosomes (Figure 15.2). A comparison of this chromosome with chimpanzee chromosomes 12 and 13 reveals essentially the same genes (that is, chromosomes 12 and 13 are essentially the same as chimpanzee chromosomes 2a and 2b), arranged in a similar sequence along the chromosomes. In other words, after the evolutionary lineages leading to humans and chimps had separated from the last common ancestor, two chromosomes fused into one at some point along the evolutionary branch leading to the creation of humans.[15-26] The amount of similarity in genomic organization among humans, chimpanzees, and apes goes beyond what would be expected from "common function" alone. The close similarity of introns in humans and chimpanzees is a great indication that we humans and chimpanzees share a "common ancestry."[15-27] When one looks at the chromosomes of humans and the living great apes (orangutan, gorilla, and chimpanzee), it is immediately apparent that there is a great deal of similarity between the number and overall appearance of the chromosomes across the four different species. There are differences in their chromosomes, although the overall similarity is remarkable. Creationists point out that despite the similarities there are differences in the chromosomal banding patterns and the number of chromosomes. Furthermore, they claim that the similarities are due to a common designer rather than common ancestry.

Let's address the differences first, and then compare the conflicting scenarios of common ancestry versus a common designer. The following observations can be made about similarities and differences among the four species (human, orangutan, gorilla, and chimpanzee). Except for differences in non-genetic heterochromatin, chromosomes 6, 13, 19, 21, 22, and X all have identical banding patterns in all four species. Chromosomes 3, 11, 14, 15, 18, 20, and Y look the same in three of the four species (the three being gorilla, chimpanzee, and humans), and chromosomes 1, 2p, 2q, 5, 7-10, 12, and 16 are alike in two species. Chromosomes 4 and 17 are different among all 4 species. Most of the chromosomal differences among the four species involve inversions, localities on the chromosome that have been inverted, or swapped end for end. This is a relatively common occurrence among many species. An inversion usually does not reduce fertility. The chromosome 5 inversion between chimpanzees and humans is very interesting. All of the bands between the two chromosomes will line up perfectly if you flip the middle piece of either of the two chromosomes between the p14.I and q14.I marks. The similarity of the marks will include a match for position, number, and intensity (depth of staining). Similar rearrangements to this can explain all of the approximately 1,000 non-heterochromatic bands observed among each of the four species (mentioned earlier) for these three properties: **(1)** band position, **(2)** number, and **(3)** intensity. Other types of rearrangements include a few translocations (parts swapped among the chromosomes), and the presence or absence of nucleolar organizers. (Nucleolar refers to the forming of a nucleolus. See Chapter 5, Figures 5.2 & 5.4.) The biggest single chromosomal rearrangement among the four species (human, orangutan, gorilla, and chimpanzee) is the unique number of chromosomes (23 pairs) found in humans as opposed to the apes (orangutan, gorilla, and chimpanzee) (24 pairs).

There are two potential naturalistic explanations for the difference in chromosome numbers, either **(1)** a fusion of two separate chromosomes occurred in the human line, or **(2)** a fusion of a chromosome occurred among the apes. The evidence favors a fusion event in the human line. One could imagine that the fusion is only an apparent artifact of the work of a designer or the work of nature (due to common ancestry). Some may raise the objection that if the fusion was a naturalistic event, how could the first human ancestor with the fusion have successfully reproduced? We have all heard that the horse and the donkey produce an infertile mule in crossing because of a different number of chromosomes in the two species.

Although variations in chromosome number are known to occur in many different animal species, and although they sometimes seem to lead to reduced fertility, this is often not the case. Now, the question has to be asked: if the similarities of the chromosomes are due only to common design rather than common ancestry, why are the remnants of a telomere, the segment of DNA that occurs at the ends of chromosomes, and the centromere, the most condensed and constricted region of a chromosome, found at exactly the positions predicted by a naturalistic fusion of the chimp ancestor chromosomes 2p and 2q? Another chromosomal rearrangement has recently been discovered, this one shared both by humans and chimpanzees, but not found in any of the other monkeys or apes (orangutan and gorilla) that were tested. This rearrangement was the movement of about 100,000 DNA pairs from human chromosome 1 to the Y chromosome.[15-28]

There is factual evidence to support the common ancestry between humans and chimpanzees. Humans have 46 chromosomes in our human cells. Humans have 23 pairs of chromosomes because we get 23 from our mothers and we get 23 from our fathers, so in humans, we have a total of 46, or 23 pairs. The great apes have 48 chromosomes, which means they have 24 pairs. This means that humans are missing a pair of chromosomes. And the question is, if evolution is right about common ancestry, where did the missing pair of chromosomes go? There is no possibility that as a result of common ancestry, (which would have had 48 chromosomes because the other three species have 48), the missing pair of chromosomes could have just got lost or thrown away. A chromosome has so much genetic information on it that the loss of a whole chromosome would probably be fatal. That is not a hypothesis, but a fact. Therefore, evolution makes a testable prediction, and that is, somewhere in the human genome scientists should be to be able to find a human chromosome that actually shows the point at which two of these common ancestors were pasted together. Scientists should be able to find something holding together two chromosomes so that our 24 pairs, of which one of them was pasted together to form just 23. And if scientists cannot find that, then the hypothesis of common ancestry is wrong and evolution is mistaken.

Chromosomes themselves have little genetic markers in their middles and on their ends. They have DNA sequences called "telomeres" that exist on the edges of the chromosomes. Then they have special DNA sequences at the center called "centromeres." Centromeres are important because that is where the chromosomes are separated when a cell divides. If one of the human chromosomes were formed by the fusion of two chromosomes, one should find that fused human chromosome in the middle, (those telomere sequences that normally belong at the ends). (This is somewhat like the seam that results when two things are glued together, it should still be there.)[15-29] (*A telomere sequence is either of the sections of DNA occurring at the extreme ends of each chromosome in a eukaryotic cell. Telomeres consist of highly repetitive sequences of DNA that do not code for proteins, but function as caps to keep chromosomes from fusing together.*[15-30] *Telomeres protect chromosomes from end-to-end fusions.* [15-31] *A telomere is a region of repetitive nucleotide sequences at each end of a chromatid, which protects the end of the chromosome from deterioration or from fusion with neighboring chromosomes.*[15-32]) There are two centromeres, one of which has been inactivated in order to make it convenient to separate this when a cell divides. It has been determined that humans lost two chromosomes in Chromosome Number 2 (Figure 15.2).[15-33]

A recent study (2004) shows very clearly that all of the marks of the fusion of those chromosomes predicted by common descent and evolution, are present on human Chromosome Number 2. There should be telomeres at the fusion point of one of the human chromosomes, there should be an inactivated centromere and there should be another one that still works. This same study states: "*Chromosome 2 is unique to the human lineage of evolution having emerged as a result of head-to-head fusion of two acrocentric chromosomes that remain separate in other primates.*"[15-34]

The precise fusion site has been located. An analysis confirmed the presence of multiple telomere, subtelomeric duplications. And then, during the formation of human chromosome 2, one of the two centromeres became inactivated, and the exact point of that inactivation is pointed out, and the centromere that is inactivated in humans corresponds to primate Chromosome Number 13. The prediction of evolution of common ancestry is fulfilled by irreducible evidence in terms of tying everything together, that the human chromosome formed by the fusion from our common ancestor is Chromosome Number 2. Evolution and common ancestry have hereby been proved.[15-35]

Creationist View

Most modern evolutionary biologists do not claim that humans evolved from chimpanzees or any other living apes (orangutan, gorilla, and monkey), but that humans and the great apes all evolved separately from one now extinct common ancestor through independent evolutionary lines. Evolutionary biologists claim that one common ancestor of man and the hominids possessed a diploid (having two similar complements of chromosomes) number of 48. As this species evolved into the chimpanzee, gorilla, and orangutan, the total chromosomal number remained constant at 48. In contrast, as the alleged same common ancestor evolved into a human, two of the 48 chromosomes allegedly underwent a genetic malfunction and were fused together to produce a new species with a diploid number of 46. This view, however, is debatable. First, this hypothesis assumes that the alleged chromosomal fusion took place after humans supposedly split from the apes in the proposed evolutionary tree. Evolutionary biologists claim that at some point in the past, a human ancestor's DNA underwent a genetic fusion between two of its chromosomes. It must be pointed out, that as currently known, this event did not occur in any other species.

The arguments proposed by evolutionary biologists do not provide any evidence that humans share a common ancestor with apes. Their line of thinking provides no empirical evidence that humans and apes share a common ancestor. All that it really does is suggest that a past human may have undergone this genetic change. In order for this fusion event to demonstrate common ancestry with the chimpanzee, there would have to be some link between the fusion event and the great apes. To date, there is no such link known. The alleged fused-looking chromosome is specific to humans, so it does not directly make any connection with the great apes. Therefore, it cannot be empirical evidence for a common link between Homo sapiens and the great ape. The only genetic supposed link between humans and the great apes is our close DNA sequence similarity. This similarity is expected since humans have a similar body structure physiology (common function), and biochemistry that we share with apes.[15-37] Biologists believe that ancestral chromosomes 2A and 2B fused to produce human chromosome 2, however no genes were lost from the supposed fused ends of 2A and 2B (Figure 15.2). At the site of the alleged fusion, there are approximately 150,000 base pairs of sequence that are not found in chimpanzee chromosomes 2A and 2B.[15-38]

Let's assume that evolutionists are correct and a distant human ancestor with 48 chromosomes did evolve into a new species with 46 chromosomes via the chromosome 2-fusion event. Did this alleged event occur in a single individual or simultaneously in an entire population? Mutations of this nature are certainly rare, but they do occur occasionally. However, the probability that this mutation would occur simultaneously in multiple individuals is so staggeringly low that we can assume its impossibility. At best, the mutation would have occurred in a single individual. How then was it propagated (passed on) from one individual to his or her offspring and eventually to every human descendant? Chromosomal rearrangements of this nature are not easily passed to offspring. When mutations of this magnitude occur, they pose serious problems for an organism when the process of gamete production occurs.

Gametes are the egg and sperm cells used to form a new individual during sexual reproduction. The process of generating gametes is a special form of cell division known as meiosis. (Meiosis occurs in reproductive cells.) During this process, a specific alignment of chromosomal pairs always occurs and is essential for meiosis. This alignment is dependent on the near-identical structure and sequence of chromosomal pairs. (Meiosis is discussed in Chapter 5 section "Types of Cell Division-Mitosis and Meiosis.") If an individual carries a mutation such as a chromosomal fusion, then he or she will often be unable to produce gametes, because meiosis will fail to occur properly due to improper alignment of the now non-identical chromosome pairs. Today, we know chromosomal fusion to be one cause of infertility. In some cases, meiosis can find a way to complete despite non-identical chromosomal pairs. However, the gametes that result, or the offspring produced by fertilization with these gametes, usually have a short lifespan due to genetic problems. Problems associated with chromosomal alignment lead to spontaneous miscarriages and genetic abnormalities such as Down's Syndrome.

As described above, the problems of (1) infertility, (2) low survival fitness, and (3) the absence of humans with 48 chromosomes today make this explanation improbable for the appearance of chromosome 2. It could be argued that the first humans had 48 chromosomes, but sometime afterward, chromosomes 2A and 2B fused together in a human. Then, the first human that contained the alleged chromosome 2 fusion produced offspring with this genomic alteration. If the first human that contained a fusion of chromosomes 2A and 2B had offspring, then the offspring of the one with the genomic alteration along with his or her mate would have a 50% chance of receiving this chromosome. Then, their offspring from their children would have only a 25% chance of receiving the altered chromosome 2. With each successive generation, the probability of maintaining the altered chromosome would reduce by one-half. These genetic frequencies of passage to offspring, coupled with the likelihood of infertility and genetic syndromes, make this hypothesis unlikely as well.

Another explanation for the alleged fusion of human chromosome 2 from chromosomes 2A and 2B in the chimpanzee is that humans were created with 46 chromosomes including a second chromosome with the visible characteristics that we see today. No evidence or any line of rational thought can explain how a single human could have undergone a genetic chromosomal fusion and passed that alteration to all of mankind.[15-39]

Did Ape Chromosomes 2a and 2b fuse together to form the Human Chromosome 2?

Creationist View

Apes (orangutan, gorilla, and chimpanzee) have 48 chromosomes (24 from the mother and 24 from the father) and humans have 46 (23 from each parent). Because there is a difference between the number of chromosomes that are in apes and the number of chromosomes that are in humans, there is a discrepancy. Since this poses a problem for evolutionists who believe that humans descended from apes, evolutionists needed to develop an explanation to defend the claim that humans are genetically related to apes. Evolutionists claim that two ape chromosomes (chromosomes 2a and 2b – see Figure 15.2) fused end-to-end and formed a single chromosome – the human chromosome 2 (see Figure 15.2). Evolutionists actually found a sequence that appeared to support this claim. There is a section of DNA with some telomere-like sequences (like the ones found at the ends of chromosomes) in the middle of human chromosome 2. They called this the "fusion site." Upon closer examination of the alleged fusion site, it has been determined that these telomere sequences are quite degraded, and the signature is quite small if two telomeres had actually fused. (A telomere is a region of repetitive nucleotide sequences at each end of a chromosome, which protects the end of the chromosome from deterioration or from fusion with neighboring chromosomes.)[15-39a footnotes 4 & 5]

When the human chromosome number 2 was first studied, researchers found a small cluster of telomere-like end sequences that vaguely resembled a possible fusion. (Telomeres are a six-base sequence of the DNA letters TTAGGG repeated over and over again at the ends of chromosomes.) The cluster of telomere-like end sequences was somewhat of a mystery based on the real fusions that occasionally occur in nature. All documented fusions in living animals have a specific type of sequence called satellite DNA (satDNA) located in chromosomes and found in breakages and fusions. The fusion signature on human chromosome 2 was missing this satDNA. Also, the fusion site is small, which is only 798 DNA letters long. Telomere sequences at the ends of chromosomes are 5,000 to 15,000 bases long. If two telomeres had fused, there should be a fused telomere signature of 10,000 to 30,000 bases long (which is double the amount of one end), not 798.[15-39b]

Human telomeres are 5,000 to 15,000 bases long, so if two chromosomes fuse together they should have something like 10,000 to 30,000 bases long. But the so-called fusion site's sequence is less than 800 bases. It has also been noted that this so-called fusion sequence was inside a gene, and it is a promoter, or a switch inside the gene. The supposed fusion sequence is inside a gene. It refutes the claim of there being a fusion of the ape chromosome 2a and 2b to form the human chromosome 2 (See Figure 15.2) [15-39a footnotes 4 & 5]. There are also a number of proteins that bind to turn on this switch, along with RNA transcripts being produced from this region. Therefore, the evidence is overwhelming that it is a promoter inside a gene. There are telomere-like sequences all over the human genome acting like genetic switches (See Chapter 13, section "Some DNA Portions are Used to Turn Genes On or Off." for further discussion of genes being turned on and off). In a chromosome fusion, there should be two genetic scars, and there should be a fossil centromere in addition to the so-called fusion site. The so-called fossil centromere is also in the middle of a gene, a huge protein-coding gene. This alleged fossil would be impossible to form in a fusion. The claim made by evolutionists that ape chromosome 2a and 2b fused to become the human chromosome 2 has been determined to be false.[15-39a]

Comparison of Human and Chimpanzee DNA

Evolutionist View

Recent research in genetics has revealed beyond a reasonable doubt that humans and other primates are the descendants of common ancestors. Just as DNA is used in courts to establish paternity or to identify people involved with crimes, particular features of DNA sequences are also used to establish evolutionary relatedness. We have 1,000 "olfactory receptor" (OF) genes that encode proteins needed for our sense of smell. About 600 of these can no longer make functional proteins, and many are useless also in chimps, gorillas, and orangutans, and have the same inactivating mutations in each species. Such mutations occurred in an ancestor of all the species that currently own, by inheritance, the common mutation. Similarly, humans and chimps have

33 genes that make proteins used to sense bitter taste. Some of these genes are non-functional mutations, with the same inactivating mutations, in both humans and chimps, scrambled in a common ancestor. Copying and pasting DNA repeatedly produces new genes during the lifetime of a plant or animal. Primate genes that control the immune system and sexual function have arisen by multiple cycles of DNA duplication. Many copied and pasted DNA segments occur on the X—and Y—(sex) chromosomes (Figure 15.2), and have been inherited by humans, chimps, and gorillas. Large-scale changes to DNA continue. Humans differ from chimps by about 200 large duplicated or deleted segments. Any two humans differ by some ten large duplications or deletions of up to 400,000 bases. Humans and other primates have emergency patches on our DNA, making sites where radiation once caused DNA breaks. Many patches are common to chimps and humans. Our DNA has the scars of radiation damage that occurred in reproductive cells of long-extinct ancestors. Chimpanzees and humans are related genetically. This indicates that we humans are the products of a common lineage, common ancestry. Due to the fact that human DNA is so similar to that of the chimpanzees, human status (genetically) is not any different from that of other animals.[15-40]

Molecular data obtained from DNA and protein sequences confirms that humans and lower primates share a common ancestor. Science has revealed that humans share 98.5 percent of their DNA sequence with chimpanzees (1.5 % difference). Cytochrome c is a small heme protein that is remotely associated with the inner membrane of the mitochondrion, which are membrane-enclosed organelle found in most eukaryotic cells (Chapter 5, Figures 5.2 & 5.3). (Heme is defined in "The Evolution of Hemoglobin" Evolutionist View section.) Nucleotides are molecules that, when joined together, make up the structural units of RNA and DNA. (See Chapter 5 for further discussion of nucleotides.) The DNA sequences that code for cytochrome c in humans and chimpanzees are only different by four nucleotides. This relates to only a 1.2 percent difference. There are 1049 different sequences that could code for this protein, however, the DNA sequence that code cytrochrome c in humans and chimpanzees only has a 1.2 percent difference (98 - 99% similarity).[15-41] This indicates that humans are more related to chimpanzees than to any other primate. It also indicates that humans diverged from their common ancestor approximately seven million years ago. The gorilla is a slightly more distant relative to humans, and the orangutans are even more distant to humans, being 12 million years since our common ancestor.[15-42]

Creationist View

Since 1972 researchers have claimed that the DNA of humans and chimpanzees is at least 98.5 percent identical, or only 1.5 percent difference. (This 98.5 percent similarity rate may be drastically reduced if what were once considered useless, pseudogenes were also included in these analyses. See Chapter 13, section "So-called Pseudogenes Are Not Useless Remnants"— Creationist View.) Even if the difference between human and chimpanzee DNA were only 1.5 percent, it would still be significant. This may not appear to be much of a difference, however, it relates to a difference of at least 48,000,000 nucleotides. (See Chapter 5 and Chapter 6 for further discussion of nucleotides.) A change of only 3 nucleotides is fatal to an animal, which means there is no possibility of change.[15-43] The latest studies (2004) indicate that the difference between human and chimpanzee DNA is now known to be 5 percent rather than 1.5 percent.[15-44] All organisms are 100% identical in the sense of having similar makeup. All living organisms are made up of the same substances, such as oxygen, carbon, hydrogen, and nitrogen, etc. The claim that humans and chimpanzees only have a 2 percent difference is greatly misleading. If the 2 percent difference in DNA presumed for human and chimpanzee instructs the other 98 percent how to organize, the difference would be much more complex. (See Chapter 13, section "Aren't Junk Genes Just Useless Leftovers from Our Ancestors?" Evolutionist View for further discussion of the various functions of genes.) A 2 percent difference translates into tens of millions of AGCT differences. (See Chapter 5 for further discussion of DNA base chemicals AGCT—adenine, guanine, cytosine, thymine.) A 2 percent difference among three billion base pairs would mean about 60 million code letter differences between human and chimp DNA. It has been admitted by some evolutionists that there are likely to be nucleotide differences in every single gene. (Nucleotides are discussed in Chapter 5.) In recent studies (2004) comparing chimpanzee chromosome Number 23 with its presumed counterpart on human chromosome Number 21, it showed a DNA difference of about 1.5 percent that resulted in differences of more than 80 percent among the proteins produced by those genes. This obviously does not help the claim that humans and chimpanzees share a common ancestor.[15-45]

One of the main problems with a comparative evolutionary analysis between human and chimp DNA is that some of the most critical DNA sequence is often omitted from the scope of the analysis. (The non-coding regions are omitted from the comparisons made by evolutionists, as these regions were once presumed to be useless, junk, non-coding portions of genes, as mentioned above. Chapter 13, section "So-called Pseudogenes Are Not Useless Remnants"—Creationist View states in part: "Since scientists only use the protein coding portions of genes when comparing human and chimpanzee sequences, they do not include the so-called junk DNA sequences in their comparisons.") Another problem is that only similar DNA sequences are selected for analysis. As a result, estimates of similarity become biased towards the evolutionary side. The chimp genome was sequenced to a much less stringent level than the human genome, and when completed it initially consisted of a large set of small disoriented and random fragments. To assemble these DNA fragments into contiguous sections that represented large regions of chromosomes, the human genome was used as a guide or framework to anchor and orient the chimp sequence. Thus, the evolutionary assumption of a supposed ape to human transition was used to assemble the otherwise random chimp genome. There are large blocks of sequence anomalies between chimpanzee and human beings that are not directly comparable and would actually give a similarity of zero percent in some regions. In addition, the loss and addition of large DNA sequence blocks are present in humans and gorillas, but not in chimpanzees and vice versa. This is difficult to explain in evolutionary terms since the gorilla is lower on the primate tree than the chimpanzee and supposedly more distant to humans. How could these large blocks of DNA from an evolutionary perspective, appear first in gorillas, disappear in chimpanzees, and then reappear in humans?

As already mentioned, the claim made by scientists that the DNA structures in humans and chimpanzees are 98 to 99 percent similar may be misleading. **First**, researchers used only human and chimp DNA sequence fragments that already exhibited a high level of similarity. Sections that did not line up were tossed out of the mix. **Next**, they only used the protein coding portions of genes for their comparison. Most of the DNA sequence across the chromosomal region encompassing a gene is not used for protein coding, but rather for gene regulation, like the instructions in a recipe that specify what to do with the raw ingredients. The genetic information that is functional and regulatory is stored in "non-coding regions," which are essential for the proper functioning of all cells. These coding regions ensure that the right genes are turned on or off at the right time in concert with other genes. (See Chapter 13, section "Some DNA Portions are Used to Turn Genes On or Off.") When these regions of the gene are included in a similar estimate between human and chimpanzee, the values can drop dramatically and will vary widely according to the types of genes being compared. Most of the sequence is non-coding that is used to regulate protein production. Evolutionary scientists have used only the coding portion of a gene for comparative analyses, which in this case would just be the exon blocks 1-4 (an exon is a region of a gene that contains information to make proteins). The remaining sequences, supplying critical information that specifies (1) when, (2) why, (3) how much, and (4) how often the coding region is to be transcribed into RNA, are regularly omitted from sequence comparisons. Due to limitations in DNA sequencing technology, researchers do not even have the complete genomic sequence for human or chimpanzee at present. Much more analysis needs to be done to the sequences that have already been obtained.[15-46] Because humans are similar biologically to other living things, it is possible for humans to eat plants, fruits, vegetables, and animals. If we are to eat food to provide nutrients and energy in order to live, what would we eat if every other organism on Earth were fundamentally different biochemically? How could we digest them and how could we use the amino acids, sugars, etc., if they were different from the ones we have in our bodies? Biochemical similarity is necessary for us to have food.

Even if human and chimpanzee DNA were 96 percent similar, this does not necessarily indicate that humans evolved from a common ancestor with chimps. The amount of information in the 3 billion base pairs in the DNA in every human cell has been estimated to be equivalent to that in 1,000 books of encyclopedia size. If humans were 'only' 4 percent different this still amounts to 120 million base pairs, equivalent to approximately 12 million words, or 40 large books of information. This is surely an impossible barrier for mutations (random changes) to cross. Large DNA sequences can be turned on or off by relatively small control sequences. The DNA similarity data does not exactly mean what the evolutionists claim. [15-47] (See Chapter 7, section "Evolution of the Horse" Creationist View, sub-section "Preexisting Genetic Information," Chapter 9 section "Hox Genes," and Chapter 13, section "Some DNA Portions are Used to Turn Genes On or Off" for further discussion of genes being turned on and off.)

Are Human and Chimpanzee DNA 98 to 99% Identical?

Creationist View

The claim that human and chimpanzee genomes are 98 to 99% similar is debatable. Evolutionists make this claim as evidence that humans evolved from apes. When researchers were comparing the human genome with chimpanzee genome, they threw out a great deal of data. They cherry-picked the areas of DNA between human and chimpanzee that were highly similar and threw out areas, including the areas that would not line up properly. Areas of the genome that don't line up are dissimilar. When the data was researched, it was determined that the DNA similarities were actually between 81 to 86% when the dissimilar data was included[15-48 footnote 1]. The details of this analysis have been published and are available for anyone to examine[15-48 footnote 2]. The method used by evolutionists was not accurate, so it was discarded. In the beginning of this study 25,000 chimpanzee sequences at random from each of the data sets were examined. They were analyzed and were compared to the human genome. Over half of the data sets were extremely similar to human, and the other half were extremely dissimilar to human. It appeared that the initial chimpanzee genome was contaminated with human DNA, which is a big problem in genomics. There are a number of studies by secular researchers showing that many public DNA databases, from bacteria to fish, have significant levels of human contamination. Human DNA literally gets into the samples. Contamination of the chimpanzee samples (when human DNA gets into the chimpanzee samples), occurs when researchers touch chimpanzee samples, when researchers cough on the chimpanzee samples, when researchers sneeze on the chimpanzee samples, etc. This was especially prevalent back in the earliest phases of genome projects, when the chimpanzee genome was sequenced. More recent researchers are taking greater steps to alleviate the problem. When the comparisons were being made, the human genome was used as the template and the chimpanzee genome was compared to it. The chimpanzee DNA sequences used in the analysis were all about 750 bases long. Not only was the chimpanzee genome built using the human genome as a guide, it also has human DNA contamination in it, so it showed a great deal of similarity from the contamination. Based on the latest research, it has been determined that human and chimpanzee genomes are not more than 85% DNA similar, and that is a maximum. It is most likely even less than that. There have been some recent biomedical studies that show that any two human genomes could be 4.5% different to each other. It was previously thought that human genomes were only 99.9% similar to each other. But if we take the structural differences into account, there is a 4.5% difference between humans genome. The question is, how then can humans be 99% similar to chimps when genomes between humans can vary 4.5%?[15-48]

In the next chapter (16) we will continue our discussion of whether or not humans evolved from lower life forms by looking at some of the fossil evidence.

~ ~ ~

Chapter 16

A Look at the Human and Ape Fossil Evidence

Human Evolution Sequence

The ape-to-human sequences that appear in many evolutionary documents are artists' conceptions of ape-like hominids progressing toward human. They are shown here for illustration purposes only. It is believed that the process is more like branching than direct lineage.

God created man in His own image; in the image of God He created him; male and female He created them.[16-1]

We are certain as anyone can ever be in biology that modern humans are more closely related to chimpanzees than any other living animal. [16-2]

It doesn't matter what we want, it only matters what is. Humans evolved from other animals. Denying the fact of human evolution is denying reality. Humans didn't evolve from monkeys anyway, at least not directly. Humans evolved from apes, and apes evolved from monkeys.[16-3]

Evolution does not work by going in a straight line. Evolution is a branching bush. Humans and apes are two different branches of the same bush. We share a common ancestor with modern apes. Apes need not be extinct for humans to exist. This is like asking, 'If I am of Irish descent and live in America, why are there still people living in Ireland?'[16-4]

Many creationists do not believe this evidence but instead believe that there is no scientific evidence to support human evolution. Many creationists believe that the physical data is misinterpreted or they believe that it is fabricated and part of a secular conspiracy against Christianity and to disprove there is a God. However, nearly all biologists today accept that human evolution is a fact of science. [16-5]

Certainly, if some race of apes, especially the most perfect among them, lost, by necessity of circumstances, or some other cause, the habit of climbing trees and grasping branches with the feet, ..., and if the individuals of that race, over generations, were forced to use their feet only for walking and ceased to use their hands as feet, doubtless ... these apes would be transformed into two-handed beings and ... their feet would no longer serve any purpose other than to walk.

- Jean-Baptiste Lamarck
Philosophie zoologique, 1809

... by greater relative magnitude of brain, by agility, and by the use of the hand. The signal superiority of the human species is thus prepared for and betokened in the immediately preceding portions of the line: it might have been seen, ere man existed, that a remarkable creature was coming upon the earth.

- Robert Chambers
Vestiges of the Natural History of Creation, 1844

The man-like Apes… all have the same number of teeth as man—possessing four incisors, two canines, four false molars, and six true molars in each jaw, or 32 teeth in all, in the adult condition; while the milk dentition consists of 20 teeth—or four incisors, two canines, and four molars in each jaw. They are what are called catarrhine Apes—that is, their nostrils have a narrow partition and look downwards; and, furthermore, their arms are always longer than their legs…

- Thomas Henry Huxley
Evidence as to Man's Place in nature, 1863[16-5a]

Much light will be thrown on the origin of man and his history.

—Charles Darwin, Origin of Species,
1st edition: 1859, 6th edition: 1872, Chapter 15

The most ancient progenitors in the kingdom of the Vertebrata, at which we are able to obtain an obscure glance, apparently consisted of a group of marine animals, resembling the larvae of existing Ascidians. These animals probably gave rise to a group of fishes, as lowly organised as the lancelet; and from these the Ganoids, and other fishes like the Lepidosiren, must have developed. From such fish a very small advance would carry us on to the Amphibians. We have seen that birds and reptiles were once intimately connected together . . . In the case of mammals the steps are not difficult to conceive which led from the ancient Monotremata to the ancient Marsupials; and from these to the early progenitors of the placental mammals. We may thus ascend to the Simiadae. The Simiadae then branched off into two great stems, the New World and Old World monkeys; and from the latter, at a remote period, Man, the wonder and glory of the Universe, proceeded.

—Charles Darwin, The Descent of Man, 1882, pp. 164-165

Introduction

In the previous chapter (15) the chimpanzee and human genetic evidence was discussed in an attempt to determine whether or not humans and chimpanzees are related genetically. In this chapter (16) we will compare some of the hominid fossils with human fossils to further explore whether or not humans descended from lower life forms. We will also look into some additional evidence and arguments that support common ancestry as well as some differing arguments that oppose the view that humans descended from lower life forms, such as apes.

Opening Comments About the Fossil Evidence for Human-ape Evolution

Evolutionist View

> *Since Raymond Dart's time (1924), paleoanthropologists, geneticists, and molecular biologists have used fossils and DNA sequences to establish the evolution of humans. We humans are apes descended from other apes, and our closest relative is the chimpanzee, whose ancestors diverged from our own several million years ago in Africa. These are indisputable facts.* [16-6]

Scientists have determined through genetic research and fossil evidence that humans share a common ancestor with modern African apes, like gorillas and chimpanzees.[16-7] As discussed in Chapter 15, chromosome and DNA testing has revealed that humans are almost genetically identical to chimpanzees.[16-8] Looking backward at the progression of man, it appears the progression would be something like this: aquatic single cell organisms to aquatic multi-celled organisms to fish to amphibians to reptiles branching to birds and mammals. It seems impossible to survey the fossils we have and deny that humans have evolved. Looking at the whole array of bones, there is indisputable evidence for human evolution from ape-like ancestors.[16-9]

etc.usf.edu/clipart

Shrew

Some paleoanthropologists believe tree shrews (squirrel-like mammals) may have been the forerunners of primates and humans.[16-10] Primates consist of various omnivorous mammals, comprising the three suborders (1) Anthropoidea (humans, great apes, gibbons, Old World monkeys, and New World monkeys), (2) Prosimii (lemurs, loris, and their allies), and (3) Tarsioidea (tarsiers), especially distinguished by the use of hands, varied locomotion, and by complex flexible behavior involving a high level of social interaction and cultural adaptability.[16-11] As discussed in Chapter 15, scientists believe that humans share a common ancestry with chimpanzees and other apes (orangutan, gorilla, and monkey). Scientists believe the common ancestor of humans and apes existed some five to eight million years ago. Sometime after this, the common ancestor diverged into two separate lineages, or species. One of these lineages eventually evolved into gorillas and chimpanzees and the other evolved into early human ancestors called hominids. Since humans diverged from apes, there have been at least twelve different species of these human-like creatures. Many of these hominid species are close relatives to humans but they are not human ancestors (See Figure 16.4). Most of these hominid species became extinct without branching out into other species. Some of these extinct species, however, have been discovered and most certainly were direct ancestors of Homo sapiens.[16-12]

The belief that humans share a common ancestry with chimpanzees and other apes is based on the similarities found between human, chimpanzee, and other ape-like anatomy. This does not mean that humans descended directly from modern apes, but rather share a common ancestor that is no longer in existence.[16-13] (Common ancestry is discussed in Chapter 15.) The fossil record includes many pre-human (hominid) fossils that provide abundant evidence for human evolution. Skeletal similarities are evident between various stages of evolution. These include shapes of head, jaws, teeth, as well as other similarities. Even though not all scientists agree 100% on the interpretation of the fossils, none question the fact that humans evolved from pre-humans. This is because there are patterns in the fossil record that are both progressive and transitional.[16-14] Recent genetic research supports the conclusions of the fossil evidence and shows that all humans have descended from a small population of hominids, believed to be from 1,000 to 10,000 individuals, living less than 50,000 years ago.[16-15]

Although there is indisputable fossil evidence that indicates humans evolved from ape-like ancestors, paleontologists have not yet been able to arrange a continuous line that begins with ape-like creatures to modern humans. Paleontologists have discovered ape-like creatures that gradually became more and more human-like, eventually walking on two legs. The fossil record reveals that ape-like beings lived in East or Central Africa approximately seven million years ago. These findings are well before the discovery of large brained hominid fossils. Anthropologists and paleoanthropologists agree that there were several species in existence at the same time and location during the evolution period.[16-16] The basic anatomy of monkeys, gibbons, orangutans, gorillas, and chimpanzees clearly resembles the human anatomy and is the basis for the belief that these are human's closest relatives.[16-17]

The skeletal anatomy of monkeys, gorillas, and humans is quite similar with slight differences in shape and proportion (Figure 16.1). The various skeletons indicate the type of movement of the species. For example, monkeys walk on four limbs, gorillas walk on their hind legs, yet use their forearm knuckles, and humans walk on two legs. Monkeys and gorillas have a big toe that is angled away from the foot, which allows them to hold on while moving through trees.[16-18] The rib cages of humans and chimpanzees differ in that chimpanzees have a pyramid-shaped rib cage, while humans have a barrel-shaped rib cage (See Figure 16.1 below). Chimpanzees have massive hip bones compared to human hipbones. The massive hipbones are necessary for apes that walk on all four legs. Humans walk on only two legs and do not require large hip bones.[16-19]

Creationists doubt the physical evidence and believe that it has been misinterpreted in an effort to disprove there is a God, although almost all biologists today, including Christian biologists, believe that human evolution is a fact of science.[16-20] The indisputable fact is that humans are apes that descended from other apes, and that humans' closest relative is the chimpanzee. Proof exists that human ancestors appeared several million years ago in Africa. Since the time of the discovery of Raymond Dart in 1924, paleoanthropologists, geneticists, and molecular biologists have used fossils and DNA sequences to confirm these beliefs. (See Chapter 15 section "Comparison of Human and Chimp DNA" for further discussion of DNA.)[16-21] Creationists view the fossil evidence as two distinct groups of fossils, humans and apes, and that there is a huge separation between these two types of fossils. This correlates with their belief in Scripture that humans were created rather than evolved. If there were truly a vast difference between humans and apes, as creationists claim, then it should not be very disputable as to which fossils are human and which are not. However, as we will see in section "Homo rudolfensis", the fact is that creationists cannot agree on exactly which fossils are "human" and which are "ape."[16-22]

Fossils are grouped according to their brain capacity size. The cutoff size is around 600 cubic centimeters, which was arbitrarily determined. The brain size, within itself, is not a true indicator of either ape or human. Thus naming a fossil ape or human, based on brain size, is a problem. This is because there is a considerable range of brain sizes within any species. Humans, for example, have brain sizes ranging between 1,000 and 2,000 cubic centimeters, with the smallest human brain size having been recorded at 855 cubic centimeters. The largest known brain size of a gorilla is 650 cc.[16-23]

Intelligent Design View

The sequence proposed by paleontologists is not as convincing when it is critically examined. It is not at all clear whether any living primate populations can be viewed as truly transitional forms or new forms which led to the next most advanced stages. There is no known living member of the order that represents the first erect, bipedal hominid. Nor does there exist any form which is transitional between the quadrupedal primates and the bipedal, that is, a form which exhibits the bony and neuromuscular changes in the pelvic girdle which led to the first erect, bipedal type.[16-24]

Questions for Creationists

Evolutionist View

The scientific evidence for human evolution creates a number of challenging questions for creationists. What is to be made of the fossil and archaeological records? They definitely show a progressive and transitional pattern indicative of physical and behavioral evolution: ape-like pre-humans to human-like pre-humans to anatomical humans to humans behaving like us.[16-25]

How do Creationists answer the fossil and archaeological records that demonstrate a progressive and transitional pattern that points to evolution? Did God put them there to deceive us into believing that we evolved from apes? Would God, for some unknown reason, perhaps to test our faith, create five species, one after the other, (1) Australopithecus afarenis, (2) Australopithecus africanus, (3) Homo habilis, (4) Homo erectus, and (5) Homo sapiens, to mimic a continuous trend of evolutionary change?[16-26]

Creationist View

It is inherent in any definition of science that statements that cannot be checked by observation, are not really saying anything, or at least they are not science.[16-27]

Evolutionists claim that apes and humans share a common ancestor, however, it is unknown what this common ancestor might have been. They believe it was most likely something similar to an ape (orangutan, gorilla, monkey, and chimpanzee). The question is: What mammal could this common ancestor have derived from? While it may be true that some creationists are not totally in agreement regarding the labeling of some fossils, evolutionists also disagree with the labeling of some fossils. The point being made by the evolutionists above is that if humans are distinct from all other primates, then there should be no controversy regarding which fossils are human and which are not. The fact is that although some fossils are difficult to determine whether they are ape or human, there is no way one can determine the appearance and intelligence level of a creature by only examining its skeletal structure. Humans were created a few thousand years ago (6000 ya), and are completely unique from all other created beings. Fossils described by evolutionists are from either apes, humans, or other animals, but do not show any transition between apes and humans. The evidence for ape to human evolution is based mainly on fragmentary fossils. There is not sufficient fossil evidence to base their claims. There are no fossils that show a gradual evolution of ape to man.[16-28]

Fossil Evidence for the Evolution of Humans

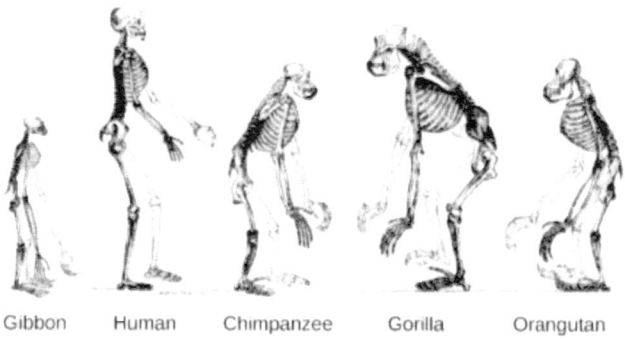

Gibbon   Human   Chimpanzee   Gorilla   Orangutan

http://cnx.org/content/m44696/latest/?collection=col11545/latest

Human Evolution Sequence

Figure 16.1 Human Evolution Sequence

. . . man bears in his bodily structure clear traces of his descent from some lower form

> • Charles Darwin, The Descent of Man, 1882, p. 65

His [The human] body is constructed on the same homological plan as that of other mammals. He [humans] passes through the same phases of embryological development. He retains many rudimentary and useless [vestige] structures, which no doubt were once serviceable. Characters occasionally make their re-appearance in him [humans], which we have reason to believe were possessed by his early progenitors. If the origin of man had been wholly different from that of all other animals, these various appearances would be mere empty deceptions; but such an admission is incredible.

> • Charles Darwin, The Descent of Man, 1882, p. 146

We will now take a look at some of the major hominid fossils that are used to support the view that humans descended from ape-like species. We will take a look at arguments that both agree and disagree with the theory of human evolution.

The naming conventions of hominid fossil groups are as follows: The first word indicates the genus, such as "homo." The second word indicates the particular species, such as "sapiens."[16-29]

Ramapithecus

Figure 16.2 Ramapithecus

Evolutionist View

Ramapithecus (Figure 16.2) is an extinct group of primates that lived from about 12 to 14 mya (million years ago). It was once thought to be a possible ancestor of Australopithecus (discussed below). Ramapithecus was considered a possible human ancestor on the basis of the reconstructed jaw and dental characteristics of fragmentary fossils. A complete jaw discovered in 1976 was clearly nonhominid and Ramapithecus is now regarded by many as a member of Sivapithecus, a genus considered to be an ancestor of orangutan.[16-30]

Creationist View

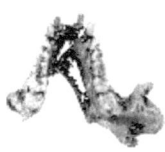

The lower jaw depicted here, from Lufeng, Yunnan, was initially attributed to *Ramapithecus lufengensis*. *Ramapithecus* was considered, until the late 1970s, to be an early human ancestor based primarily on its hominid-like dentition. At the time of its discovery this specimen was the most complete mandible of "Ramapithecus" known. With the demise of "Ramapithecus" in the 1980s, and its reassessment as the female morph of a highly sexually dimorphic great ape with affinities to the modern orang-utan, the Lufeng material was reevaluated. Since the mid-1980s hominoid specimens from Lufeng have been classified as *Lufengpithecus lufengensis*. Its phylogenetic affinities are still under debate with various authors suggesting affinities to the orang-utan, African apes or the last common ancestor of all living great apes, including humans.

http://www.cartage.org.lb/en/themes/sciences/paleontology/paleozoology/fossilhominids/PictureGallery/PictureGallery.htm

Ramapithecus Jaw

Some consider Ramapithecus to have been a hominid, a man-like ape. This decision has been based solely on the remains of a few teeth and a few fragments of the jaw. These are the only fossil fragments that have been recovered.[16-31] It should be noted that a baboon that lives in high altitudes in Ethiopia, Theropithicus galada, has teeth and jaw characteristics very much like Ramapithecus and Australopithicus. Ramapithecus is now generally classified as essentially the same animal as a fossil orangutan known by the name of Sivapithecus.[16-32]

Ardipithecus ramidus

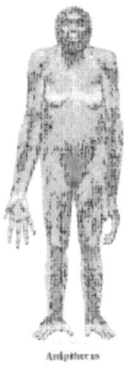

Ardipithecus

Figure 16.3 Ardipithecus ramidus

Evolutionist View

Ardipithecus ramidus (Figures 16.3 & 16.13) includes specimens dated from 5.8 and 4.4 mya (million years ago). The brain size is approximately 400 cc (cubic centimeters) (chimp-size), which is similar to modern day chimpanzees.[16-33] So far, 35 Ardipithecus ramidus fossils have been found. Most of the fragments that were found are skull fragments. There is some evidence that Ardipithecus ramidus walked on two legs and was approximately four feet tall. The teeth of this fossil are intermediate between those of earlier apes, while one baby tooth appears more chimpanzee-like than hominid. The other fossils discovered indicate Ardipithecus ramidus lived in the forest.[16-34] Ardipithecus ramidus are considered hominids.[16-35]

The hominids that came after them were more and more human in resemblance.[16-36] Ardipithecus ramidus' feet are quite intermediate between those of humans and those of chimps. Just like chimpanzees, the toes of Ardipithecus ramidus are opposable, but they are shorter. Plus, the feet of Ardipithecus ramidus are not flexible enough for climbing trees and grasping vines, and they are more human-like from that point of view. The fingers of Ardipithecus ramidus are long and dexterous, much like human fingers. Its wrists are also more flexible than those of modern apes. In a monogamous society, male apes have large canine teeth and they are used in conflict with other males when fighting over females. Ardipithecus ramidus' pelvis, although crushed and in need of reconstruction by advanced digital technology and anatomy experts, was wider than quadrupedal chimps and narrower than bipedal Australopithecines (discussed below in section "Australopithecus afarensis"). This is greater evidence that our ancestors became bipedal prior to their increase in intelligence. The brain volume of Ardipithecus ramidus is estimated to be 300 and 350 cm3 (cubic centimeters), which is about the size of a modern bonobo's/female chimpanzee's brain, but smaller than Lucy's 400 to 550 cm3 (cubic centimeters) brain.[16-37] (Lucy is discussed below in section "Australopithecus afarensis.")

Australopithecus ramidus

Evolutionist View

Australopithecus ramidus, dated to have existed 4.4 mya, is a very early hominid, from the early period just after the split from apes (orangutan, gorilla, monkey, and chimpanzee) and hominids. Australopithecus ramidus possibly walked on two legs. The only fossil found was a section of the skull. Since there was insufficient cranium discovered, the brain size was not able to be determined. The teeth resemble both ape and human teeth.[16-38] The skeletal remains provide only an indication that this individual walked on two legs. Fragments of the temporal and occipital bones seem to indicate a forward positioned foremen magnum (a large opening at the base of the skull). This opening generally indicates where the skull rests and the position of the spine. A partial humerus was also recovered that indicates Australopithecus ramidus was smaller in size than Australopithecus afarensis (discussed below) but is still within the range of variation of Australopithecus afarensis.[16-39] (The humerus bone is the long bone in the arm of humans extending from the shoulder to the elbow.)

In 1995, fossils found in Aramis, Ethiopia, were compared with Australopithecus Afarensis and modern and fossil apes. It was recognized that these fossils were distinct, that is, they were from a new species. The new species were given the name Australopithecus Ramidus, which means "root" in the Afar language. Australopithecus Ramidus has been dated to approximately 4.4 million years old. Thus far, what distinguishes Ramidus from other species is its very primitive dentition. Ramidus may be an intermediate between chimpanzees and early hominids such as Australopithecus Afarensis. Ramidus may have been developing bipedality (having the ability to walk on two legs), lived in a forest environment, and probably was the only hominid alive at the time. Ramidus, although a hominid, is very ape-like, and is probably the closest fossil find to date to the common ancestor between chimpanzees and humans.[16-40]

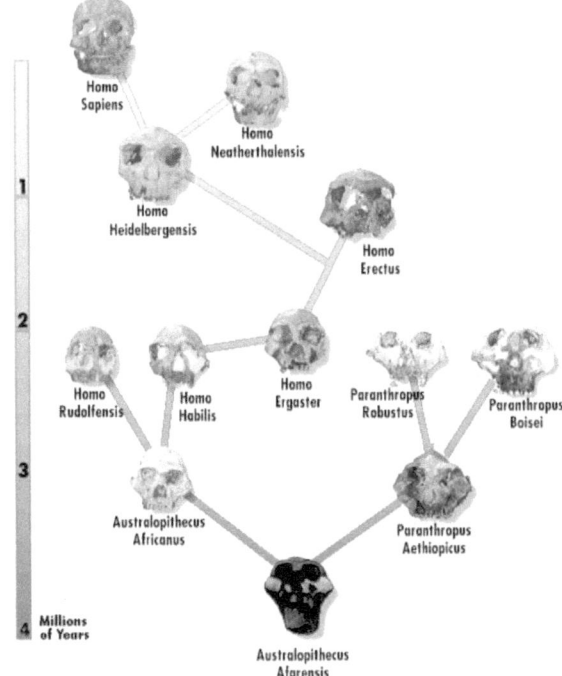

Human Evolution From Lower Life Forms

Figure 16.4 Human Evolution Tree with Timeline at Left

- - - - - - - - - - - - - -

Australopithecus afarensis
(Late Australopithecine)

Figure 16.5 Australopithecus afarensis—Artist conception

Evolutionist View

Australopithecus afarensis (Figures 16.4, 16.5 & 16.8, 16.13) are dated from 3.9 to 2.9 mya, although one source dates this genus from 2.1 mya to 1.6 mya. Australopithecus afarensis have a brain capacity of about 375-500 cc.[16-41] Australopithecus afarensis is an ape-like fossil from Afar, Ethiopia. It had an ape-like palate, human upright stance, and a brain size of 430-550 cc, which is larger than apes of the same body size.[16-42] Australopithecus afarensis stood about 4 feet tall (Figure 16.8) and is often described as the "ape from the waist up" and "human from the waist down." Best known from the famous fossil AL 228-1 named "Lucy" (discussed below), this pre-human had a cranial capacity of 400-500 cubic centimeters, which is a little larger than a chimpanzee's (370-380 cc). Australopithecus afarensis walked upright on its legs and, as footprints from 3.5 million years ago reveal, in a fashion similar to humans. Although the hip bones are similar to human hip bones, the fossil evidence shows that the feet had a long and angled first toe, indicating that this creature could still grasp tree branches like a monkey or ape. Other transitional features appear in the jaws and teeth.[16-43] The late australopithecines, walked upright and show changes in teeth, skull, and brain that gradually let to modern humans.[16-44]

Some australopithecine fossils, like Australopithecus rudolfensis, appear so intermediate in brain size that scientists debate as to whether these fossils should be called Homo or Australopithecus.[16-45] Dental features in the upper jaw that are common with both chimpanzees and humans include: 6 square-shaped molars with 4 cusps (points on the tooth surface) at the back of the jaw, 4 pre-molars or bicuspids (2 cusps), 2 canines (eyeteeth) at the corner of the jaw, and 4 incisors in the front. Chimpanzees differ from humans in that the jaws and tooth rows at the back of the mouth are parallel to each other. Chimpanzees also have a space between their incisors or canine (pointed teeth), whereas humans have no space (Figures 16.5a & 16.7). The teeth of Australopithecus afarensis, such as Lucy, have both chimpanzee and human features. There is a space between its incisors but the jaw and tooth rows are slightly curved and more human-like (Figure 16.5a).[16-46] After Australopithecus afarensis, the fossil record shows a confusing mixture of gracile (gracefully thin or slender) australopithecine species (2.0 mya) lasting up to about two million years ago. Viewed chronologically, they show a progression to a more modern human form: the tooth row gets more parabolic (parallel sides with curved front), the brain gets larger, and the skeleton loses its apelike features.[16-47]

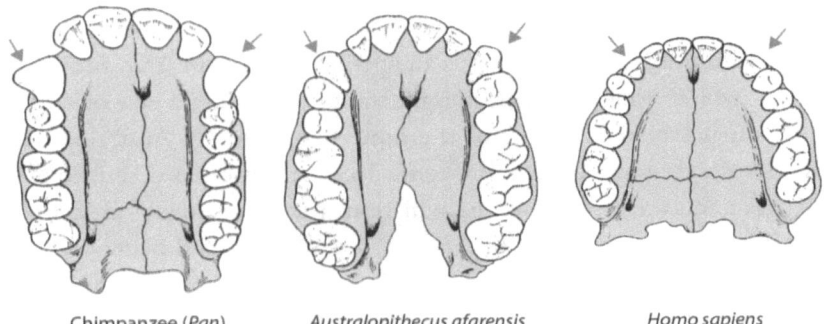

Chimpanzee (Pan)　　*Australopithecus afarensis*　　*Homo sapiens*

Figure 16.5a The upper teeth of chimpanzee, Australopithecus afarensis, and modern human. The chimpanzee has larger canines (eyeteeth) than humans, and a gap separates each canine from the adjacent incisor (arrow). The incisors of the ape are also large in relation to the molars. In these important features, Australopithecus afarensis is intermediate between apes and humans.

Lucy

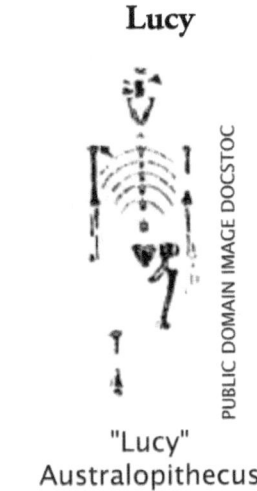

"Lucy"
Australopithecus

Figure 16.6 Lucy skeletal portions recovered

Lucy (AL 228-1) (Figure 16.6) is an example of Australopithecus afarensis and is dated sometime between 3.9 mya and 3.6 mya. Lucy features both ape and human characteristics, and appears in the fossil record 3.5 mya, about midway between humans and the last common ancestor, believed to have existed 6 mya, that humans share with chimpanzees.[16-48] The Australopithecus afarensis fossil, often referred to as "Lucy," has both ape-like rib cage and human-like hips. The hip bones indicate this being walked on its hind legs.[16-49] Lucy is an excellent fossil that appears to have answered the question of which transition came first, increased brain size or walking on two legs. From studying Lucy it became apparent that walking on two legs was first, followed by increased brain capacity. Lucy is believed to have walked upright and is definitely hominid. It was however very ape-like in both size, approximately four feet tall, and brain capacity, approximately 375-500 cc. This creature had ape-like teeth. The lineage for this species split into two groups, a husky large tooth and a slender small tooth lineage. The husky lineage (A. robustus, A. boisei) eventually became extinct.[16-50]

Like other fossils classified within the genus australopithecines, Lucy had a very apelike head with a chimp-sized braincase. But the skull shows it had more human-like traces, too, such as a semiparabolic (having two sides with a curved front) tooth row and reduced canine teeth (Figure 16.5a). Between the head and pelvis this hominid had a mixture of apelike and human-like traits: the arms were relatively longer than those of modern humans, but shorter than those of chimps, and the finger bones were somewhat curved, like those of apes. Lucy is a great transitional form between humans and ancient apes. From the neck up Lucy is apelike; in the middle, this creature is a mixture; and from the waist down, it is almost human.[16-51]

For many years it was assumed that a human's superior brain appeared first on an apelike body. The discovery of Australopithecus, or "southern ape," fossils proved that just the opposite was true. A semi-complete skeleton was discovered in East Africa. Dated to have existed between 2.6 and 3.3 million years ago, it was given the nickname "Lucy" after a popular Beatles song. Other bones and human-looking footprints have been discovered from a close relative walking in volcanic ash 36 to 38 million years ago.

Lucy walked almost completely upright, like a human. Unlike the thumb-like first toe of an ape, her first toe was like that of a human. It lined up with the other toes. Lucy did not hold her head fully erect. Her neck joined farther toward the back of her skull than in humans. Lucy had large, nut-cracking molars covered with thick enamel. With these she ate a wide variety of difficult-to-chew foods. Large jaw muscles broadened her cheekbones. Lucy was knock-kneed, like a human, rather than bowlegged as all chimpanzees are. The shape of Lucy's knees kept her center of balance beneath her hips while walking. Lucy's pelvis was more like a human's than an ape's, but her birth canal was not large. Lucy's brain size of 27 to 36 cubic inches (442-590 cc) is about one-fourth larger than that of a chimp of the same size. It was about a third as large as a human's. At 25 years old, Lucy stood about 3 feet 8 inches tall and weighed 65 pounds.[16-52]

Creationist View

As of 1979, 120 Australopithecus afarensis fossils have been found. The Australopithecus hominids had slightly larger brains than Ardipithecus (discussed above), and their skeletons imply that they walked upright. Australopithecus afarensis' skull rested on top of the spine rather than in front. It appears that if Australopithecus afarensis did not walk on two legs as modern humans, then it at least had key adaptations that led to this form of mobility.[16-53] Walking upright is the main characteristic that leads evolutionists to categorize fossils as human ancestors. Lucy was made up of fossil fragments (Figure 16.6). The knee that gained Lucy the reputation of walking upright was found 1.5 miles away from the other fragments. It was not until years later that those who reconstructed Lucy admitted that the knee was found 200 feet lower into the ground. The fossils that make up Lucy were most likely not all from one ape-like creature, but most likely were a composite of ape and human bones.[16-54] Several scientists have determined that the bones of Lucy came from two different sources. Some researchers believe that the afarensis sample (Lucy) is really a mixture of two separate species. The most convincing evidence for this is based on characteristics of the knee and elbow joints.[16-55]

Many australopithecine fossils have been found in South Africa as well as in East Africa, but the same name is used for all these similar type fossils. The name "australia" comes from the same word, "australo," which means "southern," because Australia is the "southern continent." Other than that, there is no relationship between australopithecines and the country Australia. The australopithecines are strictly extinct, non-human, African primates.[16-56] The tools found near the fossils is what caused an interest in the remains, rather than the skeletal features of the fossils themselves. And since tools were found with Austalopithecus fossils, it was assumed that the creature had made the tools. Thirteen years later, bones virtually identical from those of modern human were found under the bones where Austalopithecus fossils were found.[16-57] Other Australopithecines seem to be four or five million years old. But there is no agreement among anthropologists as to whether they are really those of human ancestors.[16-58]

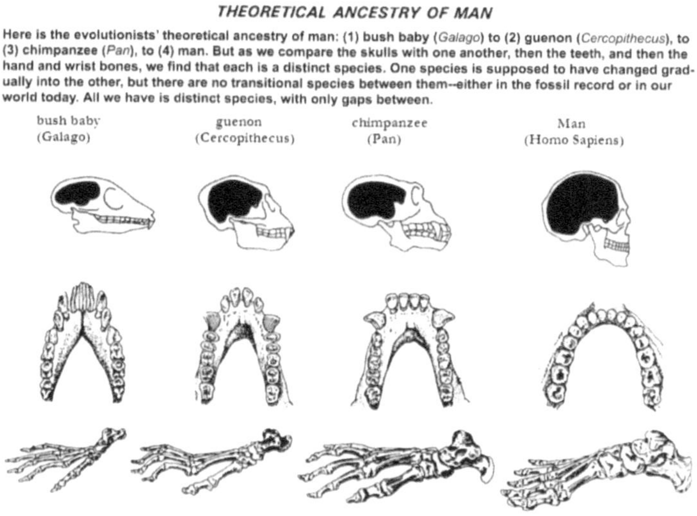

THEORETICAL ANCESTRY OF MAN

Here is the evolutionists' theoretical ancestry of man: (1) bush baby (*Galago*) to (2) guenon (*Cercopithecus*), to (3) chimpanzee (*Pan*), to (4) man. But as we compare the skulls with one another, then the teeth, and then the hand and wrist bones, we find that each is a distinct species. One species is supposed to have changed gradually into the other, but there are no transitional species between them--either in the fossil record or in our world today. All we have is distinct species, with only gaps between.

bush baby	guenon	chimpanzee	Man
(Galago)	(Cercopithecus)	(Pan)	(Homo Sapiens)

Figure 16.7 Hominid jaws and limbs[16-59]

Many male primates have large canine teeth (Figure 16.7), which are used in fighting and defense. (Also see Figure 16.5a for comparison.) Where the upper canines meet, or occlude, with the lower jaw, there are spaces, or gaps, between the opposing teeth. Canine diastemas (spaces opposite large canines) are characteristic of the jaws of baboons, gorillas and monkeys. They are used as a diagnostic feature in studying fossils because they are absent in hominids (men or near-men). A primate jaw with canine diastemas is considered probably related to apes or monkeys, not close to the human family.[16-60]

Australopiths are extinct apes known only from fossils. "Lucy" is the most famous example, and she was long thought to represent an evolutionary transition between ape-kind and mankind. Researchers contend that human ancestors gradually transitioned from tree-dwelling apes. The latest bone discovered looks just like a human fourth metatarsal (a bone of the foot). This connects from a heel bone to the fourth toe over from the big toe, spanning the arch across the middle of the foot. Its description clearly showed that the foot bone is within the range of modern humans and does not match any metatarsals from living apes or show any hint of being ape-like, but what does this newly described bone actually prove? One reviewer commented that Lucy's foot most likely looked very similar to a human foot. However, this blatantly ignores prior finds showing that Lucy's foot was actually configured like a hand, with a thumb-like big toe projecting sideways. And what if the foot bone in question was actually from a human and not from an Australopith at all? Since it was not attached to any other bones when it was discovered, this possibility should have been considered when forming an opinion. The study authors wrote that they assigned AL 333-160 (the bone's formal designation) to A. afarensis, as it was the only hominin species with their distinctive attributes in an assemblage of more than 370 hominin specimens that were so far recovered from the Hadar Formation.

The evolutionary term "hominin" includes apes, humans, and imaginary human-like or "pre-human" apes. But their reasoning is flawed. First, they claimed that no modern human bones have been found in this deposit. They evidently interpreted this to mean that no human bones could be there. Then they concluded that the human-looking bone belonged to an ape. However, AL 333-160 could actually be just what it looks like, a human foot bone. If so, then it contradicts the very assertion on which their argument rests. Researchers could more easily discern Lucy's mode of locomotion if more bones were found connected together. The most complete Australopith skeletons show that they had none of the skeletal features, including hip, spine, femur, and foot bone structures that enable the uniquely human manner of walking. In fact, Lucy-like specimens have indicated characteristic flat ape feet with curved toes, not arched feet as it has been claimed. Is one bone singled out from a scrap heap of "greater than 370" individual bones the best evidence for an upright-walking ape? If this bone actually was from a "Lucy," it would be the first A. afarensis skeletal feature discovered that is not ideally suited for life in trees.[16-61]

Raymond Dart discovered a juvenile skull in East Africa in 1924. Dart believed that an adult Australopithicus would stand 4 feet tall and have the brain size of a gorilla. An adult was discovered in 1936. Discoveries of various bone fragments and skeletal parts continued by several others. "Lucy" was a skeleton about 40% complete (Figure 16.6). Researchers found tools in the vicinity of the bones, and assumed that Australopithicus used them. They found human footprints and assumed that they were not human. Extensive analysis of the Australopithicene bone structure has called into question whether the animals ever walked upright. They were long-armed, and short-legged, and were probably knuckle-walkers, more closely resembling an orangutan. These animals are no longer considered by most anthropologists to be human's ancestor, but rather are classified as apes. (This information should be current to about 1985.)[16-62]

In 1983 anthropologists published the results of a re-examination of Lucy's skeleton (officially labeled AL 228-1).[16-63] (Note: Al 228-1 is assigned specifically to the Lucy skeleton. The AL 333-160 designation is assigned to all Australopithecus afarensis fossil-like fossils.)[16-64] This re-examination led anthropologists to different conclusions than what was commonly believed by evolutionists. These anthropologists observed that many of Lucy's bones were more like chimpanzee bones than those of a human. They concluded that Lucy probably did not walk upright like a human, but in a slouched position like an ape and that Lucy most likely spent much of its time climbing trees, since her skeleton was better suited to climbing than to walking.[16-65]

The australopithecines are merely extinct primates. The fact that sapiens-like fossils appeared in the fossil record before the australopithecines and lived as contemporaries with them throughout all of their history, reveals that the australopithecines had nothing to do with human origins. The case for the australopithecines as human ancestors has been based on three claims made for them by evolutionists. These claims are that (1) they were relatively big brained, (2) they were bipedal, and (3) they appear in the fossil record at the relevant time. Most evolutionists ignore the unique distinction between humans and animals. Although brain organization is more important than brain size alone, the significant gap between the cranial capacities of the largest australopithecine and the smallest human, fossil or living, has not been closed. The evidence for australopithecine walking on two legs is debatable. Although there is strong evidence as to how australopithecine creatures walked, it is much different than the way humans and other primates walk, therefore the issue is irrelevant. Having the ability to walk on two legs does not prove a human relationship. Unfortunately for evolutionists, the australopithecine evidence comes far too late in the fossil record; when the australopithecines first appear in the fossil record, true humans were already walking (Figure 16.13).[16-66]

Australopithecus africanus

Evolutionist View

Australopithecus africanus (Figures 16.4 & 16.13) includes specimens dated from 3-2 mya. Brain size is 420-500 cc.[16-67] Dating methods placed Australopithecus africanus ("southern ape of Africa") at three million years ago and its brain size was about 500 cc, about one-third the size of modern human's brain.[16-68] So far, 130 Australopithecus africanus fossils have been found. Australopithecus africanus fossils had slightly larger brain capacities (430-550 cc) and smaller incisor teeth than Australopithecus afarensis. These teeth gradually evolved into more homo-like teeth. These hominids (Australopithecus africanus) are almost perfect ape-human intermediates. It appears evident that the slender australopithecines led to the first Homo species (Australopithecines were discussed previously in section "Australopithecus afarensis.")

Australopithecus africanus (3.0 mya) appeared after the Australopithecus afarensis (3.9 mya) and were thinner and taller than Australopithecus afarensis (discussed previously). Australopithecus africanus were as tall as five feet (Figure 16.8), whereas Australopithecus afarensis (discussed previously) grew no taller than four feet.[16-69] The skull of Australopithecus africanus has a more vertical slope of the face than Australopithecus afarensis. Australopithecus africanus has narrower cheekbones and reduced browridges. The skull indicates Australopithecus africanus had larger frontal and parietal lobes than that of chimpanzees. The teeth of Australopithecus africanus are more similar to human teeth than they are to those of modern apes. The canine teeth are smaller than those in Australopithecus afarensis (Figure 16.5a). The shape of the jaw of Australopithecus africanus is more human-like than ape-like. The Australopithecus africanus taxa (genus) are normally viewed as an ancestor to early humans, mainly because of its human-like teeth and brain anatomy and also because of the absence of skull features of those found in robust australopitids. However, the history of Australopithecus africanus is still unclear.[16-70]

Australopithecus africanus was able to walk on two legs, however, it is debatable how bipedalism first evolved millions of years ago. The advantages of bipedalism allowed the hands to be free for grasping objects, or for carrying food and their young. Many anthropologists, however, believe that these advantages were not significant enough to cause bipedalism (the ability to walk on two legs).[16-71]

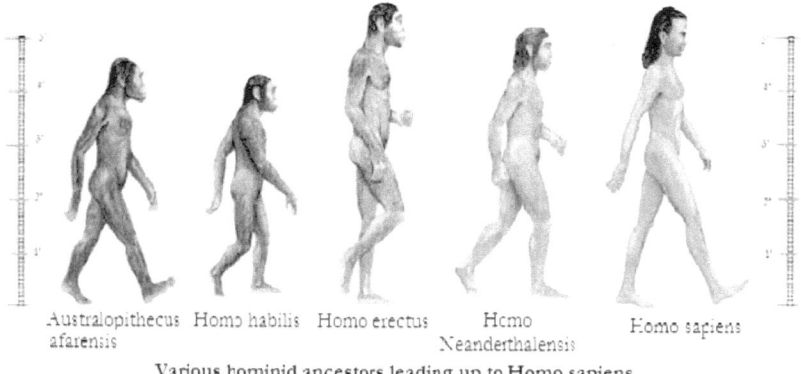

Various hominid ancestors leading up to Homo sapiens

Figure 16.8 Hominid ancestors leading up to Homo sapiens

Homo habilis (handy man)

Homo, used as a genus, means "man" or "the same"

Evolutionist View

Homo habilis genus (Figures 16.4 and 16.8) includes specimens dated from 2.4 mya to 1.5 mya. The hominids in this group had a brain capacity between 500-800 cc.[16-72] Two examples of fossils in this group include KNM-ER 1813 (1.9 mya) and OH 24 (1.8 mya). So far, 15 Homo habilis fossils have been found. Homo habilis, whose name means "handy man," first appeared about 2.5 mya. Homo habilis was the first tool-using human. These fossils are associated with a variety of stone tools used for chopping, scraping, and butchering. Paleoanthropologists are not sure if this species, Homo habilis, was a direct ancestor of Homo sapiens, but the habilis fossils do show changes toward a more human-like condition, including reduced back teeth and a brain larger than that of australopithecines.[16-73] (Australopithecines were discussed previously in section "Australopithecus afarensis.")

The structural features of pre-human fossils that are closely dated are quite similar, and this makes classification difficult. For example, some scientists reject the genus Paranthropus and place the species aethopicus, boisei, and robustus with Australopithecus (discussed above). As a result, it is not possible to determine with absolute certainty all evolutionary relationships and the pathway that led to humans. This difficulty in classifying species indicates how well they are related.[16-74] The Homo habilis specimens (2.5 mya) are somewhere between australopithecines and humans so that they are sometimes combined with the australopithecines. (Australopithecines were discussed previously in section "Australopithecus afarensis.") The Homo habilis specimens are about five (5) feet tall (Figure 16.8). Their face is still primitive but projects out less and their molar teeth are smaller. The Homo habilis specimens have a brain capacity of 500-800 cc. They fall within the same brain capacity as australopithecines at the low end (500 cc) and early Homo erectus at the high end (800 cc).[16-75] Fossils of Homo habilis show that they had larger brain sizes than australopithecines, but still only half the size of modern humans. Homo habilis coexisted with other East African hominins such as Paranthropus (or Australopithecus) boisei, Paranthropus robustus, and Paranthropus aethiopicus, Homo ergaster, Homo rudolfensis, and Homo erectus. Each of these species shows considerable variation and their relationships are disputed.[16-76]

Homo habilis walked upright and had the ability to use its hands for other tasks. The foot anatomy of this pre-human was remarkably similar to the human foot.[16-77] It has been established that Homo habilis coexisted at least at the same time if not in the same geographical area with other hominins. The most famous are the East African robust as opposed to the gracile (gracefully thin or slender), slighter, hominins. There were at least three other homins: **(1)** Paranthropus (or Australopithecus) boisei, **(2)** Paranthropus robustus, and **(3)** Paranthropus aethiopicus, all of which had massive skulls, heavy chewing teeth (some of the molars were nearly an inch across), sturdy bones, and relatively small brains.[16-78] Homo habilis may have lived alongside three species of Homo as well: **(1)** Homo ergaster, **(2)** Homo rudolfensis, and **(3)** Homo erectus, each of which are discussed later in this chapter. Each of these species shows considerable variation that causes their relationships to be disputed.[16-79] Homo habilis has a human-shaped brain and had Brocca's area, a section of the brain that humans use for speech.[16-80]

Creationist View

While paleontologists were conducting research on the fossil Olduvai Hominid 62, it was noticed that the cranium and teeth were very similar to the smaller Homo habilis skulls, KNM-ER 1805 and 1813 and Olduvai Hominid 24. Therefore Olduvai Hominid 62 was classified with Homo habilis. The body of this Homo habilis adult, such as the Olduvai Hominid 62 fossil, was not as large as Homo habilis was supposed to be. The Olduvai Hominid 62 specimen was actually smaller than the Australopithecus afarensis fossil known as Lucy (discussed above), just a bit more than three feet tall and rather ape-like. There was therefore strong evidence that the category known as Homo habilis was composed of a mixture of material from two separate taxa. It appears as though Homo habilis is a combination of australopithecine and Homo erectus fossils.[16-81] (Homo erectus is discussed later in this chapter.) There are a number of findings that suggest that the species Homo habilis should be re-classified as australopithecine (discussed above). All 'habiline' forms display unmistakable australopithecine traits. The australopithecine genus is the first extinct chimpanzee, or gorilla-like, form. The habiline forms are nothing more than variants of australopithecines (non-human).[16-82] (Australopithecines were discussed previously in section "Australopithecus afarensis.") Many experts have clearly shown that Homo habilis was nothing more than a large-brained Australopithecus (discussed above).[16-83]

Homo rudolfensis

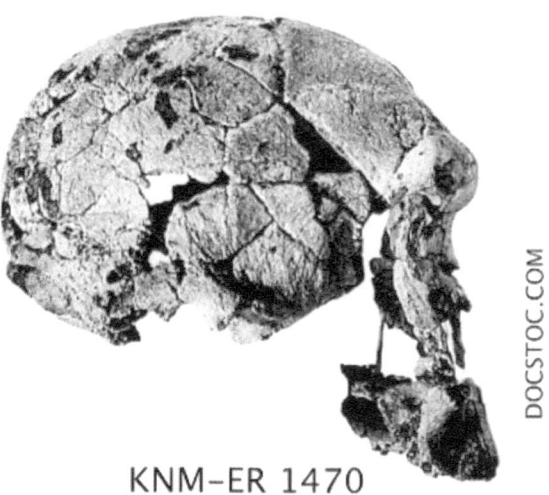

KNM–ER 1470

Figure 16.9 KNM-ER 1470 skull reconstruction

Evolutionist View

Homo rudolfensis (Figures 16.4 & 16.13) is believed to have appeared from 2.0 mya to 1.8 mya. So far, 5 Homo rudolfensis fossils have been found. An example of this group is KNM-ER 1470 (Figure 16.9), although some have classified KNM-ER 1470 as belonging to the Homo habilis genus. The fossils classified within this genus had a brain capacity of 750 cc. The KNM-ER 1470 fossil has undoubtedly been one of the most controversial of all hominoid fossils. This species (Homo rudolfensis) is not considered a direct ancestor of Homo sapiens (humans) (Figure 16.4). Homo rudolfensis lived around the same time as Homo habilis (discussed above) (Figure 16.13). Specimens of Homo habilis and Homo erectus are classified as apes by some creationists and humans by others. The inability of creationists to agree on the classification of these fossils is a clear indication that there is no definite distinction between hominid and human fossils.[16-84]

Some Creationists Believe the KNM-ER 1470 Skull Fossil Could or Should be Classified as Australopithecine

Creationist View

The KNM-ER 1470 (Figure 16.9) skull was found in late 1972, near Lake Turkana, Kenya. The skull designation KNM-ER 1470 is an abbreviation for "Kenya National Museum as East Rudolf specimen 1470." It was believed that this skull may have bridged the gap between the supposed hominid line of ancestors (including the australopithecines and Homo habilis) and the more human-like fossils designated Homo erectus. (Australopithecines were discussed in section "Australopithecus afarensis" and Homo habilis were discussed in section "Homo habilis.") When paleontologists completed their meticulous reconstruction of the skull from the scores of crushed or distorted fragments (Figure 16.9), it did indeed seem to represent a possible intermediate form between early humans and the australopithecines.

Some creationists, as well as evolutionists, however did not readily accept this fossil as a missing link between australopithecine and Homo erectus (discussed later). Soon other experts began to question the evolutionary significance of Skull KNM-ER 1470. Eventually, many began to doubt the accuracy of the original reconstruction. The radiometric date for KNM-ER 1470 was eventually accepted as about 1.9 mya. Homo erectus (discussed later in this chapter) forms with more primitive characteristics were (at that time) only known from later time period fossils (700 ka [kilo annum, as in thousand years] in Asian regions, and 1.6 mya in Africa). The latest reconstruction of KMN-ER 1470 (Figures 16.9, 16.13) indicates it was very similar to other australopithecines, and very different from known Homo types such as Homo erectus. (Australopithecines were discussed in section "Australopithecus afarensis.") Therefore there is no compelling evidence to contradict the common creationist view that Australopithecus boisei (WT 17000) and Paranthropus robustus, all belong to one created kind. All so-called gracile types such as KNM-ER 1470, the Taung child, and other africanus, possibly make up another.

Most afarensis specimens share resemblances with the gracile (gracefully thin or slender) forms (africanus). It is possible that all australopithecine forms may be considered as only one genus, even among evolutionists, regardless of whether or not one chooses to call some of them Paranthropus. After reconstruction, the old flat, human-type face no longer appears. The forehead has disappeared, the marked supraorbital ridges are now apparent, and the facial features are obviously similar with all other australopithecines, with the exception of its larger brain size.[16-85]

It is significant that the lower jaw of KNM-ER 1470 was not found. This would have told a lot about this specimen. The face of the skull, below the eyes, protrudes forward in the manner of apes. The jaw and molars are somewhat larger than the average modern human's, but not larger than those of some people. There appears to be a lack of bony support beneath the nostrils, such as is found in gorillas. Facial skeletons are relatively larger in apes than the brain case size. Skull KNM-ER 1470 is about midway in this category, and thus not like that of humans. It also has a long upper lip area, such as apes have.[16-86]

Some Creationists Believe KNM-ER 1470 is Human

Creationist View

The KNM-ER 1470 (Figure 16.9) skull was found in 1972. The skull was modern in appearance, yet it was dated by the potassium argon—potassium 40/ argon 40 (K-Ar) method at 2.8 million years old. This skull created a problem for evolution. It completely modified the whole evolutionary thinking for the origin of humans. The concern was that this skull did not fit in with the majority of the other fossils. KNM-ER 1470 is more modern in appearance than Homo erectus and has a large brain case, estimated at 800 cc. Yet it is so much older than Homo erectus (discussed below) that it just does not fit properly on the evolutionary chart. The evidence is that this man (or woman) lived during the same time with the primitive apelike creatures, such as the Australopithecines (Figure 16.13).[16-87] (Australopithecines were discussed in section "Australopithecus afarensis.")

As previously mentioned, The KNM-ER 1470 skull was dated at being 2.9 million years old. This caused a problem. By evolutionist timetables (see Figure 16.13 [KNM-ER 1470 is part of the hominid fossil group Homo rudolfensis]), this skull should have appeared quite primitive, and the brain capacity should have been much less than modern human. The cranial size was large (800 cc) and it had a very modern appearance. What was the most troubling to evolutionists is that KNM-ER 1470's cranial capacity is well within the range of modern man. Not only could skull KNM-ER 1470 possibly qualify for human status based on cranial shape, size (almost 800 cc), and wall thickness, but there is evidence on the inside of the skull of a Broca's area, the part of the brain that controls the muscles for producing articulate speech in humans.[16-88]

Some have questioned the reconstruction of skull KNM-ER 1470. On several occasions, Richard Leakey, the discoverer of this fossil, insisted that the skull was assembled in the only way possible. This may not have been true. It has been said that if the skull is held one way, it looked like one thing, such as human, if it is held another way it looked like something else, such as an australopithecine. What we can say at the present time is that there is no compelling reason why it could not be human.[16-89]

Editor's note: Both evolutionists and creationists must admit that KNM-ER 1470 is a controversial fossil. Neither evolutionists nor creationists seem to be able to come to any consensus regarding it. Some evolutionists classify KNM-ER 1470 as Homo habilis and some classify it as Homo rudolfensis. Some creationists believe KNM-ER 1470 should be classified as an ape (australopithecines), while other creationists believe KNM-ER 1470 is of a human. (Australopithecines were discussed in section "Australopithecus afarensis.") As mentioned earlier, one of the main problems with this skull fossil is the disagreement as to how the fragments should be assembled (Figure 16.9).

Homo ergaster (working human)

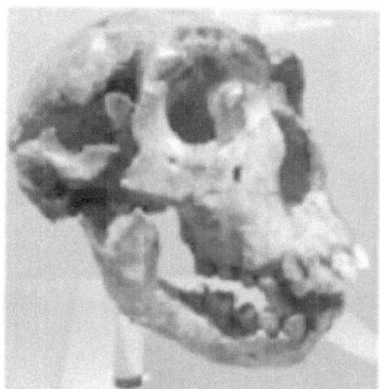

Figure 16.10 Homo ergaster

Evolutionist Description

Homo ergaster (Figures 16.4 & 16.10) existed between 1.9 mya-1.5 mya. An example of Homo ergaster is KNM-ER 3733, dated at 1.75 mya. This group has a brain capacity of approximately 850 cc. So far, 20 Homo ergaster fossils have been discovered. By 1.9 million years ago (mya), some of the early transitional humans had evolved into a new, fully human species in Africa. Most paleoanthropologists refer to them as Homo erectus. However, a few researchers split them into two species: Homo ergaster and Homo erectus. The Homo ergaster fossils were presumably somewhat earlier and have been found for the most part in Africa. The Homo erectus discoveries have been found widespread in Africa, Asia, and Europe. Many paleoanthropologists consider Homo ergaster and Homo erectus to be one species: Homo erectus (discussed in the next section).[16-90]

Homo erectus (erect man or upright human)

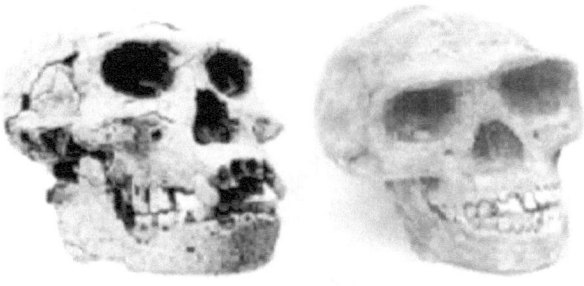

Homo erectus – georgicus   Homo erectus – Peking Man

Figure 16.11 Homo erectus—georgicus & Homo erectus—Peking Man

Evolutionist View

Homo erectus (Figures 16.4, 16.8, 16.11, 16.12, 16.13, 16.18) is believed to have appeared from 1.8 mya-300,000 years ago (about one and half million years), although one source indicates the range is 1.1 mya-300,000 ya (Figure 16.13). Homo erectus had a brain capacity of 750-1650 cc.[16-91] So far, 150 Homo erectus fossils have been discovered. Homo erectus is considered a highly successful species due to its population and expansive time frame of existence.[16-92]

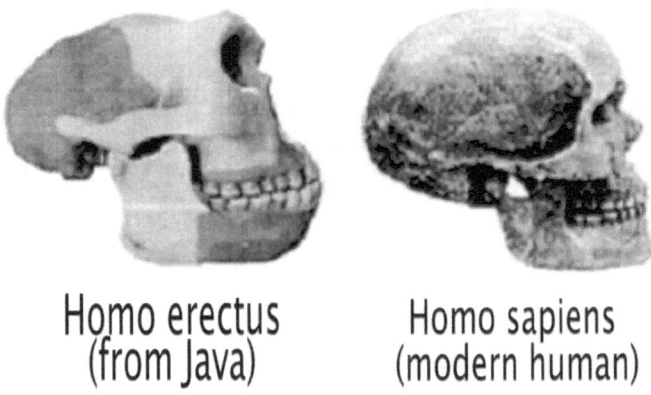

Homo erectus
(from Java)   Homo sapiens
(modern human)

Figure 16.12 Homo erectus & Homo sapiens skulls

Java Man (Figures 16.12 & 16.14) and Peking Man (Figure 16.11) are examples of Homo erectus. (Java Man and Peking Man will be discussed below.) Their brain size is nearly 1,000 cc, about two-thirds the size of a modern human brain (Figure 16.15), which makes them double the size of Australopithecines. (Australopithecines were discussed previously in section "Australopithecus afarensis" and modern humans (Homo sapiens) will be discussed later in section "Homo sapiens.") Java Man and Peking Man look much more like modern humans but still have the thick brow ridges and no chin (Figures 16.11, 16.12 & 16.14).[16-93] Homo erectus teeth were generally intermediate between modern humans and the australopithecines in shape and size. The teeth of later Homo erectus were generally smaller than the earlier members of this species. This was particularly true of molars.[16-90]

Homo erectus are still more modern-looking and had brains ranging from Homo habilis size to nearly modern human size (the largest actually fall within the range of modern humans.)[16-94] (Homo habilis was discussed previously.) Homo erectus population suddenly disappeared about 300,000 years ago and was replaced by fossils considered more like modern man. These skeletons are nearly identical to those of living humans. Around 28,000 years ago, the Neanderthal fossils are no longer found. (Neanderthals are discussed later.) Neanderthals (Figure 16.17) existed for a while longer than Homo erectus. They populated the caves overlooking the Strait of Gibralter. Modern Homo sapiens replaced them. In other words, Homo sapiens apparently replaced every other hominin on Earth.[16-95] Possibly the so-called modern humans fought against the Neanderthals and exterminated them, much as when Americans greatly reduced the population of the Native Americans back in the 1800s.[16-96]

Homo erectus was proportionally similar to humans from the neck down, stood about 5.5 feet tall (Figure 16.8), and had a brain capacity close to that of humans. This was the first pre-human to leave Africa. Around 1.8 million years ago, it spread into the Middle East, and then Europe, South East Asia, and China. Archaeological evidence reveals that Homo erectus developed sophisticated tools like spears and hand axes, controlled fire, cooked with it, and constructed primitive shelters.[16-97] Two famous descendants, Homo heidelbergensis and Homo neanderthalensis, known also as archaic Homo sapiens and the famous Neanderthal man appear to have evolved from Homo erectus. It is still not known, however, whether either of these species contributed to the gene pool of modern humans.[16-98] (Homo heidelbergensis, Homo neanderthalensis, and Homo sapiens are discussed later in this chapter.) About 2.5 million years ago, Earth entered the first of four Pleistocene Ice Ages. As climates cooled, African forests once again shrank and scrub-brush grasslands took their place. Antelope first appear at this time in the fossil record, along with other species adapted to these drier conditions. Appearing around the same time was the first member of our own genus, Homo habilis, or "capable man." Currently, the fossil record of Homo habilis comes to an end 2 million years ago. (Homo habilis was discussed previously.) Homo erectus first appeared 1.7 million years ago. At this time there was a warming period between the first two Ice Ages. Homo erectus spread across Europe, Africa, and Asia, all the way to Indonesia. One Homo erectus specimen had a brain size of 67 cubic inches (1098 cc), more than 60 percent larger than that of his immediate ancestor, Homo ergaster, who had a brain size of approximately 52 cubic inches (or 850 cc) (Figure 16.4). Brain size alone is not an absolute indicator of intelligence. Talent and genius have more to do with the quality of brain connections.[16-99]

Creationist View

When the cranium of Homo erectus (Figure 16.11) is compared with those of early Homo sapiens (Figures 16.12 & 16.18) and Neanderthal (Figure 16.17), the similarities become apparent. One creationist's conclusion is that Homo erectus and Neanderthal (discussed below) are actually the same. The Homo erectus cranial is on the lower end, with regard to size that includes Homo erectus, early Homo sapiens, and Neanderthal. The range of cranial capacities for Homo erectus and Neanderthal is then in line with the range of cranial capacities for modern humans. Modern humans have a cranial capacity range from about 700 cc to about 2200 cc. The cranial capacity of Homo erectus goes from about 780 cc to about 1650 cc.[16-100]

Along with the Australophithecenes (discussed above), archaeologists found a skullcap, part of a femur, and a hipbone, and attributed them to Homo erectus. In 1975, paleoanthropologists found a relatively complete cranium and parts of the rest of a skull. In 1984, as research continued, an almost complete skeleton was found. Limited information is available regarding these latter finds. Homo erectus appear to be similar to Neanderthal man in some respects and bear some resemblance to some skeletons dug up in the Kow Swamp area in Victoria, Australia, which have been dated on the order of 10,000 years. Based upon where the bones were dug up in Africa, it must be concluded that Australopithecus, Homo habilis, and Homo erectus lived during the same period of time (Figure 16.13). Underneath all these bones has been dug up the remains of a circular stone habitation hut that could only have been attributed to Homo sapiens. Thus, none of them could be human's ancestor, evolutionarily speaking.[16-101]

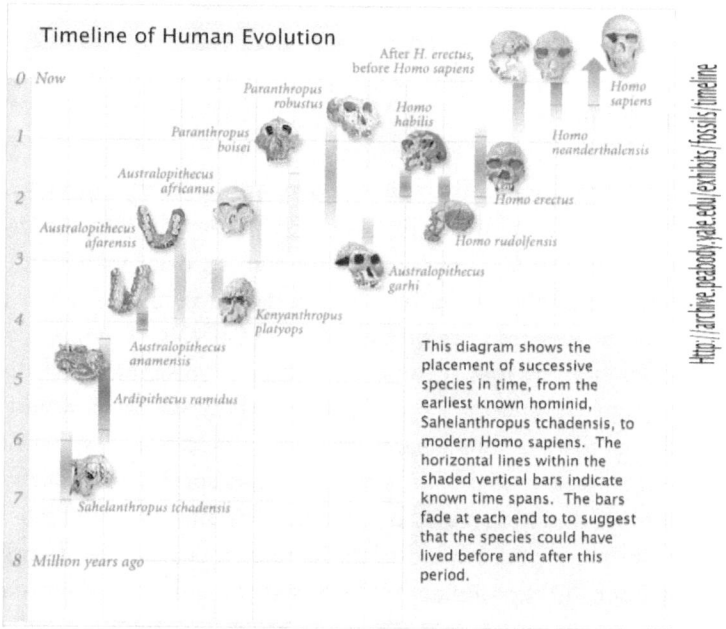

Figure 16.13 Human Evolution Timeline

Fossils of Homo erectus (Figure 16.11) and Homo habilis (Figure 16.8) overlap (Figure 16.13). Homo erectus fossils are dated 1.9 mya to 0.1 mya and Homo habilis fossils (discussed above) are dated 1.9 to 1.5 mya. Since this is the case, Homo habilis could not have evolved into Homo erectus (Figure 16.11) or Homo ergaster (Figures 16.4, 16.10, 16.13), which is also dated 1.9 to 1.5 mya. The two groups obviously lived during the same time.[16-102] One source, however, dates Homo habilis at 2.1 mya-1.6 mya, Homo ergaster at 1.7 mya-1.2 mya, and Homo erectus at 1.1 mya-0.2 mya.

The process by which one species supposedly evolves into another species, regardless of whether the evolutionist believes in phyletic gradualism or punctuated equilibria, is questionable. (Punctuated Equilibrium is discussed in Chapter 10.) One creationist, an expert in the study of paleontology, states that for some form of Homo habilis (Figure 16.8) to evolve into Homo erectus (or Homo ergaster), Homo habilis must come before Homo erectus in time. He further states that, after Homo habilis has evolved into Homo erectus

(Figures 16.4, 16.13), Homo habilis should have been eliminated by death, because Homo erectus was supposedly the better fit of the two in the intense competition for limited resources.[16-103] This, of course, is assuming that both Homo habilis and Homo erectus existed in the same regions and ate the same foods during the same time period.

Evolutionist Response

The creationist, in the discussion above, argues that a species cannot survive once it has given rise to new species. Supposedly, the newer, fitter descendant species, would, because of its superiority, drive its parent species to extinction. The argument is incorrect because members of the parent species may live in a separate region from the new species. If the species come into contact again, there may be no competition because they have diverged enough to occupy different ecological niches. Many scientists would argue that even the requirement for a separate region is unnecessary. This creationist basically claims that a species cannot split into two species. Obviously this is not the view of speciation accepted by evolutionists, since it would follow that the number of living species could never increase (Speciation is discussed in Chapter 9, section "(3) Speciation".) There is nothing in evolutionary theory that requires the main group to become extinct. If this creationist were correct, even such microevolution would be impossible. By his argument, newly evolved finch species should drive their ancestors to extinction. This does not happen, of course, because they all live on different foods.[16-104]

Java Man

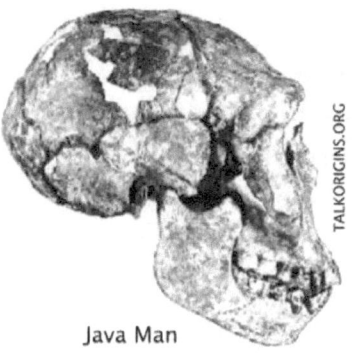

Java Man

Figure 16.14 Java Man Skull

Evolutionist View

As mentioned previously, Java Man (Figure 16.14) is an example of Homo erectus genus (Figure 16.12). The brain capacity of Java Man was around 850 cc. In 1891, the Dutch physician Eugene Dubois discovered a skullcap, some teeth, and a thighbone (Figure 16.16) in Java that seemed to be an intermediary fossil. Distressed by the religious and scientific opposition to his ideas, Dubois reburied the bones of "Pithecanthropous erectus" (which means "ape man who stands erect") beneath his house, hiding them from scientific scrutiny for 30 years. Pithecanthropous erectus has since been designated as belonging to the species "Homo erectus."[16-105]

Many creationists claim that Java Man, discovered by Eugene Dubois in 1893, was not human, but was a giant gibbon. They claim that the bones did not come from the same individual. Some creationists claim that the Java Man bones were found in the same general vicinity near the location where the Wadjak skulls were discovered, indicating that Java Man and the Wadjak lived during the same time period. If Java Man and the Wadjak species lived during the same time, creationists argue, then Java Man could not be an ancestor to the Wadjak species. These claims are not true. The Wadjak skulls were found 65 miles (104 km) of mountainous countryside away from Java Man.[16-106] Wadjak is an extinct large-headed man of primitive proto-Australoid type (known from two Javanese skulls) that is often set apart as a species (Homo wadjakensis) but is probably a primitive form of modern man (Homo sapiens) intermediate between Solo man and the modern Australian natives.[16-107] The Wadjak skulls were found in cave deposits in the mountains, at approximately the same level, whereas Java Man was found in river deposits in a flood plain.

Some creationists claim that Dubois kept the existence of the Wadjak skulls secret because knowledge of them would have discredited Java Man as not being an ancestor to modern humans. This, however, is not true. (If Java Man was found in the same general location and level as the Wadjak skulls, it would have indicated both most likely lived during the same time period and Java Man could therefore not have been an ancestor to modern humans.) Dubois briefly reported the Wadjak skulls in three separate publications in 1890 and 1892. Some creationists claim that since these publications were bureaucratic reports not intended for the public or the scientific community, Dubois was still guilty of concealing the existence of the Wadjak skulls. This is also not true. The journals in which Dubois published, although obscure, were distributed in Europe and America, and are part of the scientific literature.

Dubois pointed out that Java Man was bipedal and that its brain size was "very much too large for an anthropoid ape." Dubois never stopped believing that he had found an ancestor of modern man. Creationists are correct, however, in claiming that the femur is more recent than the skullcap. The skullcap belongs to a modern human. The skullcap definitely does not belong to any ape, and especially not to a gibbon, as some creationists claim. This skullcap is far too large (940 cc, compared to 97 cc for a gibbon), and it is similar to many other Homo erectus fossils that have been found. Some of the teeth found nearby are now thought to be from an orang-utan, rather than Homo erectus (the genus and species that Java Man has been assigned).[16-106]

Creationist View

JAVA MAN

The bone fragments known as "Java Man" came from an area known to have humans buried there. In the illustration below, notice that the Java Man bone fragments could either fit a human being (Von Koenigswald's reconstruction on the left) or a gorilla (reconstruction on the right). The Germans later decided they were from a human, and Dubois finally announced they were from a gibbon.

Von Koenigswald's reconstruction
(similar to a human skull)

gorilla reconstruction skull

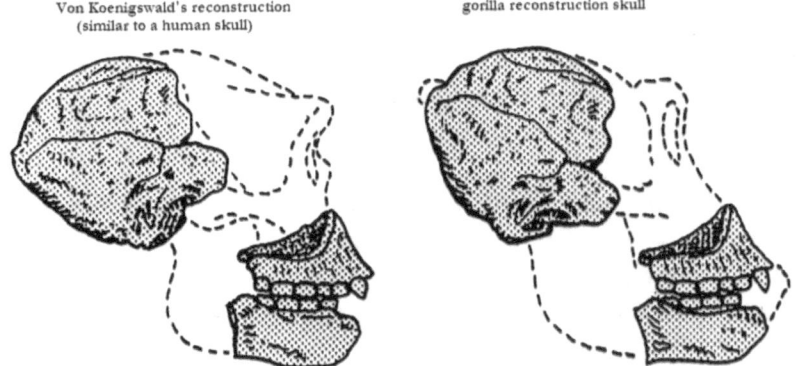

Figure 16.15 Java Man & Gorilla skulls compared[16-108]

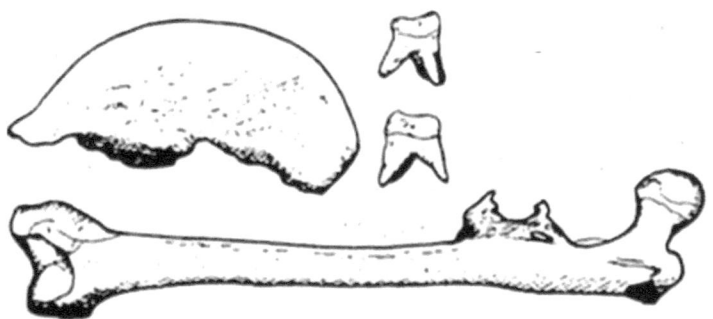

Original fossils of Pithecanthropus erectus (now Homo erectus)
found in Java in 1891

Figure 16.16 Pithecanthropus erectus original fossils recovered

One of the most famous of all the anthropoids is the Java Ape-Man (named Pithecanthropus erectus (erect ape-man) by its discoverer). Dr. Eugene Dubois, a fervent evolutionist, discovered Java Man in 1891. As mentioned above, Dr. Dubois' find consisted of a small piece of the top of a skull, a fragment of a left thighbone (femur), and three molar teeth (Figure 16.16). These remnants were not found together but were collected over a range of anywhere between 50 and about 70 feet. Also, they were not discovered at the same time, but over a span of a year. To further complicate matters, these remains were found in an old riverbed mixed in with the bones of extinct animals. Despite all of these difficulties, evolutionists insist that Java Ape-Man lived about 750,000 years ago.[16-109] The femur was probably from a human, and the skull cap was probably that of a giant ape. Before his death, and after he had convinced most of the early skeptics, Dubois changed his mind and decided that Java Man was probably a giant gibbon and not man-like at all.[16-110]

Dr. Eugene Dubois concealed for 30 years that he found human skulls near the site where he found Java Man, and at the same level. This means that humans were already there when Java Man was alive.[16-111] Eugene Dubois claimed that the skullcap and the femur came from a rock stratum known as the "Trinil layer," named after a nearby village in central Java. He believed that these rocks were below what is known as the "Pleistocene-Pliocene (Tertiary) boundary." The problem with the geology and dating of Java Man is that Dubois was not qualified to make those determinations.[16-112] Up until 1900, Dubois had been very active in promoting Java Man as the missing link and had allowed full access to the fossils. After 1900, he withdrew completely from the public debate for twenty years, published very little about the fossils, and refused to allow anyone to see them. The reason usually given for this behavior is that Dubois wanted Java Man to be accepted as the missing link. Because of the initial controversy over his interpretation, he retaliated by refusing access to the fossils.[16-113]

Java Man is not our evolutionary ancestor but is a true member of the human family, and a smaller version of Neanderthal (discussed below). Dubois seriously misinterpreted the Java Man fossils, in spite of the fact that there was abundant evidence available for him to arrive at a more accurate interpretation of the fossils. The evolutionists' dating of Java Man at half a million (500,000) years is highly suspect. More modern-looking humans were living as contemporaries of Java Man (Figure 16.14). Evolutionists eventually accepted Java Man as our evolutionary ancestor in spite of the evidence. They did so because the fossils could be interpreted as an intermediate. The historical and scientific questions regarding Java Man are as legitimate today as they were when the fossils were first discovered.[16-114]

Peking Man

Creationist View

Peking Man (Figure 16.11) had a brain capacity between 915 to 1225 cc. As mentioned previously, both Peking Man and Java Man are now classified as Homo erectus. Even though the skulls of Peking Man and Java Man were quite similar, the Peking Man femur differed from the Java Man femur in the very places where the Java Man femur was similar to modern humans. Since the association of the Peking Man skulls and femur bone was undisputed, it has been concluded that the Java Man femur was not a true Homo erectus femur but was instead a modern one.[16-115]

Homo heidelbergensis

Evolutionist View

Homo heidelbergensis (Figure 16.4) have been estimated to have existed from 0.7 to 0.2 mya (700,000 to 200,000 years ago). Another estimate indicates this species existed approximately 500,000 years ago. The brain size is 1200 cc.[16-116] Homo heidelbergensis were found in what is now Germany, Greece, and France, as well as Africa. These fossils first appear half a million (500,000) years ago, showing a mixture of modern human and Homo erectus features. An example in this group is the "Rhodesia man," 300,000-125,000 years ago, or 11,000-400,000 years ago, as the age is debatable. So far, 50 Homo heidelbergensis fossils have been found.[16-117]

Creationist View

Rhodesian Man, mentioned above, has been classified as a Homo heidelbergensis fossil. It was so named because the skull was found in what was then known as Northern Rhodesia (now Zambia). The fossil is also called Broken Hill Man (after the mine in which it was found) or Kabwe Man (after the city near which it was found). The skull was found on June 17, 1921, at the farthest and deepest point of a cave, sixty feet below ground level. In 1921, Rhodesian Man was first estimated to be 11,000 years of age. Later, in 1962, it was estimated that the age of Rhodesian Man was 40,000 years of age. In 1973, the age of Rhodesian Man was estimated to be 125,000 + years of age. Then in 1999, the age of Rhodesian Man was estimated to be 300,000-400,000 years of age.[16-118] When the hips of Rhodesian Man were reconstructed, Sir Arthur Smith Woodward of the British Museum of Natural History, assigned W.P. Pycaft, a bird specialist, to reconstruct the various pieces. Under W.P. Pycraft's supervision, the bits of the pelvis bone were tilted to give a stooping posture. It was not until later that Professor Le Gros Clark exposed the error.[16-119]

Homo Neanderthalensis (Neanderthals)

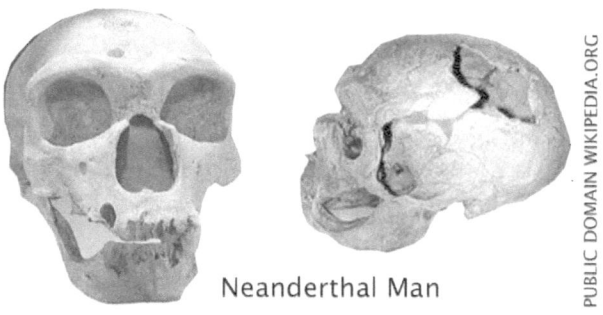

Neanderthal Man

PUBLIC DOMAIN WIKIPEDIA.ORG

Figure 16.17—Neanderthal Man Skull Views

Evolutionist View

Homo neanderthalensis have been estimated to have existed from 0.4 to 0.1 mya (400,000 to 100,000 years ago). Homo neanderthalensis is also known as "Archaic Homo sapiens" or "Homo sapiens neanderthalensis" (Sapiens means "wise man") (Figures 16.4, 16.8, 16.13, 16.17). This genus has also been dated to have existed from 230,000 to 30,000 years ago. The brain size is 1450 cc.[16-120] Examples of Neanderthal fossils in this genus include La Ferrassie 70,000 years old, La Chappelle-aux-Saints, 60,000 years old, and Le Moustier, 45,000 years old. So far, 500 Homo neanderthalensis fossils have been found. In 1871, the human fossil record comprised of only a few bones of the late-appearing Neanderthals, too human-like to be classified as a missing link between humans and apes. They were regarded instead as an unusual population of "Homo sapiens."[16-121] Archaic Homo sapiens (discussed below) look similar to modern humans but have some features closer to Homo erectus; their brain sizes fall within the range, but average on the low side, of modern humans.[16-122] Homo neanderthalensis ("Neanderthals") were generally shorter and more heavily built than modern humans, but their brain size was similar to modern humans. (Good evidence shows that they made stone tools and a variety of other tools, controlled fire, and buried their dead.)[16-123]

Neanderthals lived all over Europe and the Middle East. They had large brains, even bigger than those of modern humans, and excelled in both tool making and hunting. Some skeletons bear traces of the pigment ochre and are accompanied by "grave goods" such as animal bones and tools. (Ocre includes various natural earths containing ferric oxide, silica, and alumina: used as yellow or red pigments.) This suggests that Neanderthals ceremonially buried their dead, perhaps the first hint of human religion.[16-124]

Creationist View

In the early 1900s, in a cave in the Neander Valley near Dusseldorf, Germany, Neanderthal Man (Figures 16.8, 16.17) was discovered. This fossil was portrayed as a semi-erect, barrel-chested, brute, an intermediary link between human and ape. With the discovery of other Neanderthal skeletons, it is now known, however, that Neanderthal Man was fully erect and fully human. His cranial capacity exceeded that of modern humans by more than 13 percent. While the average brain of modern humans is 1300 cc, the Neanderthal brain averaged 1450 cc and often as much as 1600 cc.[16-125]

The Neanderthals lived during the Ice Age in parts of France. The lack of sunshine and the harsh climate of the Ice Age would have caused them to (1) seek out natural shelters such as caves; (2) construct shelters out of whatever material was available; and (3) wear heavy clothing, probably animal skins, to cover much or all of the body. Since it is believed by creationists that the Neanderthals appear to be a post flood and post Ice Age people, this could help explain why they hunted big game, using the meat for food and the hides for clothing and shelter. It could also explain their use of caves. If much of the ground was snow-covered or frozen, caves would have been an option for shelter and burial. They would not have been able to grow crops in this environment to be used as either clothing (cotton) or food.[16-126]

The first Neanderthals that were discovered came from harsh inland environments in Europe. These fossils indicate they suffered from a lack of proper nutrients. The lack of seafood with iodine in their diet and the lack of sun-induced vitamin D (necessary for calcium absorption) could have caused skeletal abnormalities.[16-127] The skeleton reflects factors that often effect health. The Neanderthal specimens may also have suffered a vitamin D deficiency resulting in rickets. Rickets and congenital syphilis frequently occur together.[16-128] Neanderthal skeletons have been unearthed with the voice box intact. The voice box is composed of a set of bones known as hyoid bones. Neanderthal's hyoid bones are identical to modern humans. Neanderthal appears to have had the ability to speak based upon these hyoid bones.[16-129]

Old Universe Progressive Creationist View

Some believe that Neanderthals were fully human in every sense of the word and therefore interbred with (other) humans. Others, however, believe that Neanderthals were more ape-like (see Figures 16.4, 16.8 & 16.13). The question of Neanderthal and human interbreeding is a complex one, but it is made more difficult by scientists who have dubbed alleles of genes that are not shared between Neanderthals and chimps as "Neanderthal genes". If one assumes that Neanderthals share a common ancestry from a chimp-human ancestor, then those stretches of the genome that differ between Neanderthals and chimps are "new introductions". If any human shows a high similarity in those particular genes, they are said to have "Neanderthal genes". But did humans have to acquire them through interbreeding? Thus far, all of the genes that differ between chimps and Neanderthals whose function is known are the kinds of genes that are "adaptive," that is, the genes that code for immune response, dietary responses, UV-radiation responses, altitude, etc. There is full consensus that these new alleles were acquired by Neanderthals in response to their environment (Neanderthals ranged in different areas to chimpanzees and their systems had to respond to different pathogens, diets, altitude and UV levels).

The Neanderthal-human interbreeding theory was based on the fact that modern humans living in sub-Saharan Africa did not have these alleles, whereas Neanderthals and Northern Hemisphere modern humans did share some of these alleles (or at least a certain percentage of similarity).

Scientists for some reason assume that human beings did not acquire these alleles in the north through the same adaptive processes that they assign to the Neanderthals' method of acquisition of the alleles. Instead, the Neanderthals are said to have acquired them adaptively due to environmental constraints, while modern humans' only option to acquire them was to wait to interbreed with the Neanderthals. (Incidentally, some alleles of genes that do not exist in northern hemisphere modern humans or Neanderthals are shared by both sub-Saharan humans and chimpanzees (such as genes that cause sickle cell anemia but which are an adaptation that provides immunity from malaria). No-one ever suggests that chimpanzees acquired these alleles adaptively and that humans relied on interbreeding with chimpanzees in Africa to acquire them).

There are many reasons why interbreeding between Neanderthals and modern humans poses "problems," not least of all that a close inspection of X-Chromosome similarity/dissimilarity implies that male hybrids, if they existed, must have been sterile, and the lack of any hint of interbreeding at all in comparisons of mitochondrial DNA between Neanderthals and humans implies that the female lineage of any hybrids also did not survive. Similarities exist only in recombinant chromosomes of northern hemisphere humans and Neanderthals, and then, only in areas (thus far) that are arguably identifiable as "adaptive" genes. A further obstacle to believing fully and easily in the idea of Neanderthal-human interbreeding is the fact that Neanderthals had a more ape-like pattern of childhood development (tooth eruptions and brain development more like a chimp, earlier "childhood" and "adolescence") than in humans.[16-101a]

Old Universe Progressive Creationist View

Many creationists considered Neanderthals to be part of the human species. With complete fossil skeletons of some Neanderthal individuals dating back at least a hundred thousand years, this view places the origin of the human race at some time previous to a hundred thousand years ago, which some creationists question. Some apparently insignificant anatomical distinctions between modern humans and Neanderthals were noticed when the first Neanderthal fossils were uncovered. For example, while modern humans possess a circular hole at the skull's base for passage of the spinal cord, the Neanderthals' spinal hole was oval; humans have a triangular lower jawbone, Neanderthals have a squarish one; and Neanderthals had a unique bony protrusion near the rear of the lower jawbone, apparently to accommodate a large chewing muscle. These differences were assumed to be as a result of environmental differences and that young children of both modern humans and Neanderthals were anatomically the same.

Recent research has determined that Neanderthals possessed enormous nasal bones and huge sinus cavities compared to modern humans, and had no tear ducts. The nasal bones and sinus cavities alone showed such major differences as to argue conclusively against any biological link between modern humans and Neanderthals. From the standpoint of morphology, modern humans cannot have descended from Neanderthals. The nasal bones and sinus cavities of Neanderthals are so large that it has been concluded that Neanderthals cannot be biologically related to any known primate species or any known mammalian species. (Compare skulls in Figure 16.17 with the Homo sapiens skull in Figure 16.18.) Just as modern humans appear to have been specially created, so too do Neanderthals. The naturalistic explanation for Neanderthals is that it is currently an unknown intermediate species between ape and human. Advances in biochemical technology have provided addition information regarding Neanderthal history. Analysis of recently recovered Neanderthal DNA confirms that the human race neither descended from nor bears any biological connection to the Neanderthal species. The DNA sample used in this study were taken from the very first Neanderthal skeleton (dated as being between 40,000 to 100,000 years old).That skeleton was dug from a limestone quarry in Neanderthal, Germany in 1856, and was utilized by Charles Darwin in his descent of man hypothesis. When the Neanderthal DNA fragment (about fifty copies of about 100 nucleotide pairs each) was compared with a DNA strand of 986 nucleotide pairs from living humans of diverse ethnic backgrounds, the difference was enormous. An average of twenty-six nucleotide links in the DNA chain differed completely.

Modern humans differed from one another in an average of just eight links of the chain, and all of the observed differences among humans were independent of the twenty-six observed for the recovered Neanderthal DNA. The conclusion was that Neanderthals could not have made any contribution to the human gene pool (See Figure 16.4). Two other teams of researchers took mitrochondrial DNA from a different Neanderthal skeleton that was approximately 29,000 years old and was an infant found in the Caucasus Mountains. These other research teams have come to the same conclusion that Neanderthal could not have in any way been related to modern humans. Such a conclusion defies a natural evolutionary explanation but is perfectly consistent with separate special creations for each of the Neanderthal and human species.[16-129a]

Homo sapiens (Modern or Wise Man)

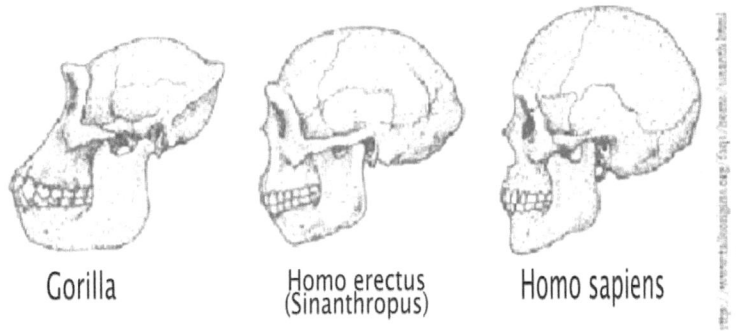

Gorilla   Homo erectus   Homo sapiens
                 (Sinanthropus)

Figure 16.18—Gorilla, Homo erectus, Homo sapiens

Evolutionist View

The human species, Homo sapiens (Figures 16.4, 16.8, 16.12, 16.13, 16.18), is the last to appear in the fossil record. Anatomically modern humans emerged in South East Africa about 200,000 years ago. They migrated into Mediterranean regions 100,000 years later. Their archaeological sites are not much different than those of the other pre-humans living at that time. However, around 50,000 years ago behaviorally modern humans arose in Africa.[16-130] About 500,000 years ago, during the warm period between the second and third Ice Ages, the earliest examples of our own species, Homo sapiens, or "wise man," first appeared around the Mediterranean Sea. Anatomically, Homo sapiens had a larger braincase, a higher forehead, and smaller teeth than his predecessor, Homo erectus (discussed previously). Our own ancestors, the subspecies Homo sapiens, or "wise man," came from southern Africa.[16-131] The word "homo" is derived from the Latin word which means "humus" or "soil." The word "sapiens" is derived from the Latin word which means "knowledge" or "wisdom."[16-131a]

Homo sapiens first appeared 300,000 years ago. Not much is known about "wise man" until about 100,000 years ago. At that time a European subspecies (Neanderthal) appeared. Because he buried his dead (which helped fossilize his bones), we have a better picture of his life than that of his contemporaries. The last common ancestor of every living human being probably walked the earth 100,000 years ago in East Africa, is called Homo sapiens sapiens (discussed below). Homo sapiens sapiens was the first member of our own subspecies. The African subspecies Homo sapiens sapiens ("wise wise man") is first known from cave litter that is 100,000 years old. Homo sapiens sapiens came from France, where this human is known as Cro-Magnon man (discussed below).[16-132] About 35,000 years ago, modern humans appeared in France and other places. Shortly after Homo sapiens sapiens arrived in Europe, the Homo sapiens neanderthalensis subspecies (discussed previously) became extinct. Perhaps they were killed through competition or by epidemic diseases.[16-133]

Homo sapiens sapiens

Homo sapiens sapiens (Figures 16.4 & 16.13) has been estimated to have begun around 0.1 mya or 195,000 years ago and continues today. The brain size is approximately 1350 cc.[16-134] Fossils within Homo sapiens sapiens genus had a brain capacity between 1200 cc-1400 cc, some have an average brain size between 1350 cc-2200 cc. Other estimates had indicated that modern humans have a cranial capacity range from about 700 cc all the way up to about 2200 cc, which should not be associated with the level of intelligence.[16-135] An example of this group is Cro-Magnon I, dated 30,000 years old.

Cro-Magnon Man

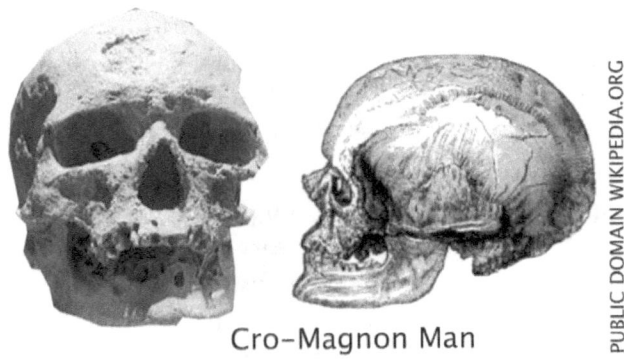

Cro–Magnon Man

PUBLIC DOMAIN WIKIPEDIA.ORG

Figure 16.19 Cro-Magnon Man Skull Views

Creationist View

Discoveries of bones which look exactly like modern humans are labeled Cro-Magnon Man (Compare the Homo sapiens skull in Figure 16.18 with the Cro-Magnon Man skull in Figure 16.19), which is a generic term for the first clearly recognized examples of what anthropologists call full-fledged homo sapiens (that is, human). Some anthropologists see no real relevant difference between the Cro-Magnon Man and the Neanderthal Man (discussed above). As mentioned above (section "Homo Neanderthalensis (Neanderthals)" Creationist View), Neanderthal Man was human in the fullest sense of the word. (See section "Homo Neanderthalensis (Neanderthals)" Old Universe Progressive Creationist View [2 sub-sections] for rebuttal.) The first Cro-Magnon Man fossils were discovered in the spring of 1868. This was just nine years after the publication the first edition of Darwin's "The Origin of Species." Ground was being removed to make way for a railroad in Perigord, France that was to run through Les Eyzies-de-Tayac. Five skeletons and some bits of fetal and infant bones were taken from the rock shelter that was exposed. These bones revealed a man fully modern human.[16-136] Complete skeletons of the Cro-Magnons have been found. Their cranial capacity was greater than modern man's.[16-137] There is nothing to differentiate Cro-Magnon Man from modern man. If anything, they have superior size and brain capacity than what is average for modern man.[16-138]

Some Disproved Fossil Portions

The fossil specimens discussed below have been determined to be hoaxes and are no longer considered evidence regarding the origins of humans. They are only briefly presented here for those who have some familiarity with them.

Nebraska Man

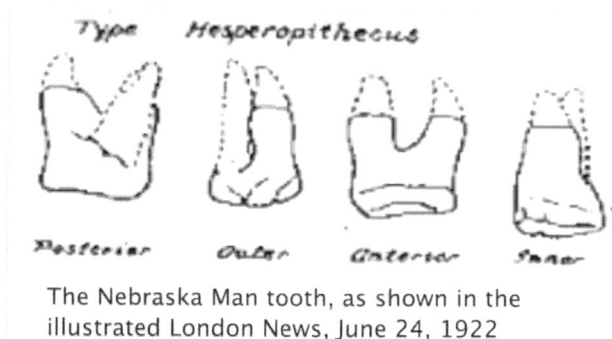

The Nebraska Man tooth, as shown in the illustrated London News, June 24, 1922

Figure 16.20—Nebraska Man tooth sketches.[16-145]

Creationist View

Nebraska Man was originally described by Henry Fairfield Osborn in 1922, on the basis of a tooth found by rancher and geologist Harold Cook in Nebraska in 1917.[16-139a] Nebraska Man was discovered in 1922 [probably 1917] by geologist Harold Cook in the Pliocene deposits of Nebraska. A tremendous amount of literature was written about this supposed missing link that allegedly lived 1 million years ago. The fossil evidence of Nebraska Man consisted of only a single tooth (Figure 16.20). The top scientists of the world examined this tooth and appraised it as proof positive of a prehistoric race in America.[16-139] In 1925 Nebraska Man was given the scientific name Hesperopithecus haroldcookii (named after its discoverer, Harold Cook), but he was never known by anything but a tooth. [16-140]

Figure 16.20a Illustration of Hesperopithecus haroldcookii was done by artist Amédée Forestier, who modeled the drawing on the proportions of "Pithecanthropus" (now Homo erectus), the "Java ape-man," for the Illustrated London News in 1922.[16-139a]

By imagination, the tooth (Figure 16.20) was put in a skull, the skull was put on a skeleton, and the skeleton was given flesh, hair, and a family (Figure 16.20a).[16-140] In 1925, the entire skeleton of the animal from which the initial tooth came from was found. During the study of the evidence, it was determined that the tooth by which Nebraska Man was constructed belonged to an extinct species of pig.[16-142]

In 1928, it was discovered that a mistake had been made and the "hominid tooth" of prehistoric Nebraska Man turned out to be nothing more than a pig's tooth.[16-143] A team of paleontologists returned to the Nebraska site in 1927, where Hesperopithecus had been discovered five years earlier in 1922, determined to find more of this mysterious creature. When they arrived at the site of the original find, weathering had exposed parts of a jaw and skeleton on the precise spot. Eagerly, they brushed away dust and sand until the ancient fossil emerged to tell its truth, the infamous molar had once belonged to an extinct pig.[16-144] Henry Osborn received the tooth on March 14, 1922. He wrote to Harold Cook: "I sat down with the tooth and I said to myself: 'It looks one hundred per cent anthropoid'." (anthropoid: resembling a human being in form). One month later, in April, 1922, Osborn announced Hesperopithecus haroldcookii as the first anthropoid ape from America.[16-144a]

An entire ape-man and "missing link" find was proclaimed hastily by Henry Fairfield Osborn (former President of the American Museum of Natural History in New York (1908-1933)) before the National Academy of Science in 1922 on the basis of a single tooth. In the spring of 1925, William

King Gregory (another associate of Henry Osborn at the American Museum of Natural History) completed the excavation in Nebraska. The full skeleton of Nebraska Man was still there in the ground. Nebraska Man was actually a Miocene peccary pig. The whole skeleton (minus the one molar tooth already found in 1917) was unearthed.

Harold Cook and Henry Osborn had already documented their knowledge of and familiarity with Miocene peccary molars in 1909, including noting that they could be mistaken for anthropoid (resembling a human being in form) molars "by anyone not familiar with the dentition of Miocene peccaries". This knowledge may very well have lead them to devise the hoax. Cook and Osborn already were "familiar with the dentition of Miocene peccaries." Why, then, had they proclaimed that a pig molar, of which they were familiar and knowledgeable, was proof of an ape-man, a missing link? They knew better but made an ape-man out of a pig molar anyway. This whole scenario can hardly be anything other than a deliberate deception.[16-144b]

Evolutionist View

Nebraska Man was named in 1922 from a human-like tooth that had been found in Nebraska. Creationists claim that evolutionists used a single tooth (Figure 16.20) to build an entire species of primitive man, complete with illustrations of him and his family (Figure 16.20a), before further excavations revealed the tooth to belong to a peccary, an animal similar to (and closely related to) pigs. This portrayal is not exactly accurate. Harold Cook, a rancher and geologist from Nebraska, had found the tooth (Figure 16.20) in 1917, and in 1922 he sent it to Henry Fairfield Osborn, a paleontologist and the president of the American Museum of Natural History. Osborn identified it as an ape, and quickly published a paper identifying it as a new species, which he named Hesperopithecus haroldcookii. Few, if any, other scientists claimed Nebraska Man was a human ancestor. A few, including Osborn and his colleagues, identified it only as an advanced primate of some kind. Osborn, in fact, specifically avoided making any extravagant claims about Hesperopithecus being an ape-man or human ancestor. (Henry Osborn was not impressed with the illustration that appeared in the Illustrated London News [Figure 16.20a], calling it: "a figment of the imagination of no scientific value, and undoubtedly inaccurate".[16-139a]) (Osborn was willing to accept the tooth as belonging to an anthropoid ape, even if it was not a direct human ancestor. He put a respected colleague, William King Gregory, in charge of defending Hesperopithecus. Gregory, an unquestioned authority on fossil primates, compared the type tooth with Old World monkeys and apes and concluded that the Nebraska tooth "combines characters seen in the molars of the chimpanzee, of Pithecanthropus, and of man, but . . . it is hardly safe to affirm more than that Hesperopithecus was structurally related to all three." Later in 1923, Gregory backed off his assertion that Hesperopithecus showed human affinities and suggested that "the prevailing resemblances of the Hesperopithecus type are with the gorilla-chimpanzee group." [16-144a]) Most other scientists were skeptical even of the more modest claim that the Hesperopithecus tooth belonged to a primate. It is simply not true that Nebraska Man was widely accepted as an ape-man, or even as an ape, by scientists, and its effect upon the scientific thinking of the time was negligible. Identifying the tooth as belonging to a higher primate was not as foolish as it sounds. Pig and peccary cheek teeth are extremely similar to those of humans, and the specimen was worn, making identification even harder.

Nebraska Man should not be considered an embarrassment to science. The scientists involved were mistaken, and somewhat incautious, but not dishonest. The whole episode was actually an excellent example of the scientific process working at its best. Given a problematic identification, scientists investigated further, found data which falsified their earlier ideas, and promptly abandoned them.[16-145]

Piltdown Man

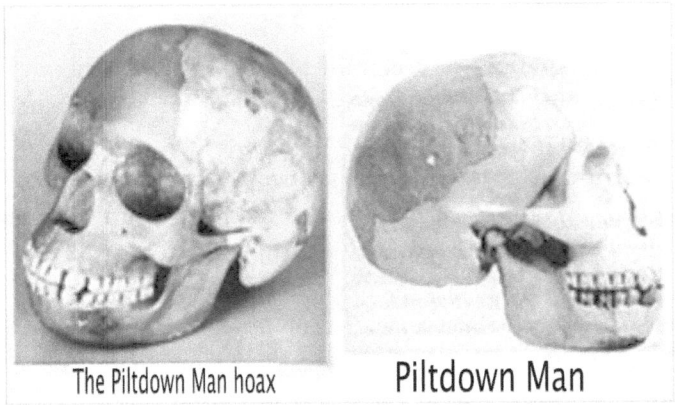

The Piltdown Man hoax   Piltdown Man

Figure 16.21—Piltdown Man skull

Evolutionist View

In 1912 Charles Dawson found a skull (Figure 16.21) near Piltdown, England. The dig site, a gravel pit in Piltdown, Sussex, England, was only a few miles from Dawson's home. The fossils included (1) a piece of a jaw, (2) two molar teeth, and (3) a piece of skull. Experts concluded that the fossils were those of an ape-man who was about a half million years (500,000) old. The piece of a skull, which looked like a modern human's, and a mandible, which appeared to be apelike, became known as Eoanthropus dawsoni. Many scientists, especially those outside of England, however, did not completely accept Piltdown Man. Piltdown Man began to come under more suspicion as more and more fossils were discovered that contradicted it. In the early 1950's it was exposed as a fraud. Piltdown Man was a hoax that was eventually uncovered by scientists. It took forty years for the truth to be revealed.[16-146]

Creationist View

In 1953, a new analysis of Piltdown Man (Figure 16.21) revealed the skull fragments to be from a medieval man who most likely died during the Black Plague between 1348-1350, a far cry from the half-million years as evolutionists claimed. In October 1956 the hoax was exposed when Reader's Digest came out with an article that summarized an article from Popular Science Monthly, entitled "The Great Piltdown Hoax." Using a new method to date bones based upon fluoride absorption, the Piltdown bones were found to be fraudulent. Further critical investigation revealed that the jawbone actually belonged to an ape that had died only 50 years previously. The teeth were filed down, and both teeth and bones were discolored with bichromate of potash to conceal their true identity. The jawbone that had teeth that had been filed down to look more human, had no relation to the human skull, and in 1982 was determined to be from an orangutan. Piltdown Man was based upon a deception that completely fooled all the experts. He was promoted with the utmost confidence for over forty years. While scientists eventually uncovered the hoax and admitted Piltdown was a fraud, Piltdown Man was used as evidence to support the theory that humans descended from lower primates during the 41 years between discovery and exposure (1912-1953).[16-147] Piltdown Man, was proven to be a hoax in the 1950s when the British Museum's Kenneth

Oakley devised a new method for determining whether ancient bones were of the same age.[16-148] In 1953, Joseph Weiner and Kenneth Oakley used a newly-developed fluorine test on the original Piltdown skull fragments, and discovered that the bones were a hoax.[16-149]

~~~

Closing Comments

Evolutionist View

As mentioned in the Opening Comments, evolution is just a theory, however, it is a proven theory based on multiple lines of evidence. The universe has proven to be billions of years old (Chapter 2). There are multiple explanations of how life originated on this planet (Chapters 5 & 6). There is an abundance of fossils that demonstrate conclusively that all species evolved (Chapters 3, 7, 8). There is ample evidence that supports the view that humans evolved from lower life forms (Chapters 15 & 16).[CC-1] The study of embryology (Chapter 11) and biology (Chapters 5 & 15) also provide conclusive evidence that species evolved. Vestiges (Chapter 12) and pseudogenes (Chapter 13) provide clues as to how species are related.[CC-2]

Evolution is a scientific fact that has been proven time and time again, since Charles Darwin first formulated the theory back in the 1800s. The evidence included during the discussions clearly supports the fact that evolution happens. For one to deny this is to deny the facts.

Creationist View

The difference between evolutionists and creationists is in the way the facts are interpreted. The facts are interpreted differently because creationists and evolutionists start with different presuppositions. Evolutionists believe in natural selection, but so do creationists (Chapter 9). Evolutionists accept the science of genetics, but so do creationists (Chapters 5 & 15). Evolutionists believe that, over millions of years, one kind of animal has changed into a totally different kind. Creationists, however, believe that God created separate kinds of animals and plants to reproduce their own kind; therefore, one kind will not turn into a totally different kind (Chapters 3, 7, 8). The scientific observations support the creationist interpretation rather than the evolutionist interpretation, in that the changes we see around us are not creating new information (Chapter 9).[CC-3]

There is no evidence to support the evolutionist claims that mutations lead to improved traits (Chapter 9). The evidence clearly indicates that the universe (Chapter 1) and life (Chapters 5 & 6) required a Creator. Biological evidence has indicated that life is too complex to have arisen naturally from chemicals (Chapters 5 & 6). Recent research has revealed that what were once viewed as useless, junk, pseudogenes, are in reality functional genes (Chapter 13). The evidence does not indicate that humans evolved from lower life forms (Chapters 15 & 16). The evidence, as previously talked about throughout the various discussions, indicates the impossibility that the universe and life could have originated by nature alone. Science is not able to explain how life originated, nor will it ever have those answers.

~~~

Conclusion

   As mentioned in the Preface, this book was written with the concept that the evidence and arguments should be presented in a manner that is as fair and accurate as possible and that the reader should be left to form his or her own conclusions regarding origins. The question is: What does the evidence indicate? Does the evidence indicate the universe and life most likely developed by natural causes or does it appear it is more likely that some intelligence is responsible for the universe and life? These are the questions I attempted to answer when I was conducting my own research and what I attempted to do when developing this book; to provide evidence to help others determine what to believe concerning origins. If you are still undecided whether the evidence points to evolution or creation, you may want to re-read some of the material presented, especially those portions where you were not able to determine which view closest reflects the evidence. Although I have tried to include as much material as possible, I was obviously not able to include every topic or point of view in this one volume book. You may want to do some additional research on your own. Check the References section for authors and organizations that may be helpful in you research. Whatever sources you use, just be sure the sources are accurate, reputable and honest. Try to determine if the author is knowledgeable and qualified in the field he or she is discussing. If the source appears to be making claims but provides little evidence, be leery.

   My conclusion is contained below. If you would prefer not to know my personal views or be influenced by them, feel free to skip this section and the next (In Closing). After completing my research and reviewing the evidence and arguments, I came to the conclusion that particles and/or antiparticles could not have originated from nothing and that they could not have formed into atoms and matter by themselves. I do not believe the universe could have self-formed, that is, begun on its own, from nothing, or from particles of matter (Chapter 1). Based on the available evidence, I do not believe the initial life and the main life groups that first appeared on the Earth originated by natural causes (Chapters 5 & 6). I do not believe mutations could lead to any significant beneficial changes in a species (Chapter 9), as I know of no species obtaining any improved traits as a result of mutations. I believe that some minor changes in species may result from mutations, however, there is no evidence that mutations could result in any positive changes in a species. It appears that most changes occur as a result of gene shuffling and genes being switched on or off. Any significant changes would be limited within a kind or group (Chapter 9 section "Hox Genes" and Chapter 13 section "Some DNA Portions are Used to Turn Genes On or Off."). A kind or group would consist of a phylum such as fish, amphibian, reptile, mammal or bird.

In short, I believe life was created with the ability to change slightly so that if and when changes occur in its environment, the creatures with the proper attributes and traits suitable for its current environment would be able to survive. For example, finches have different types of beaks. This has proven to be an advantage because as changes occur in their environment, those finches with the proper beaks suited for their current environment will be better able to survive on whatever food is readily available. The finches with the larger, stronger, beaks are able to crack open the hard shells during times of drought, however, during rainy seasons the finches with the smaller beaks are better suited to eat the smaller, softer, seeds (Chapter 9). I do not believe the evidence supports the view that a fish evolved into an amphibian. I do not believe the evidence supports the view that an amphibian evolved into a reptile, or that a reptile evolved into a mammal or a bird (Chapters 7 & 8). Furthermore, I do not believe a land mammal evolved into a whale (Chapter 7). That is not to say that every creature and plant that ever existed was created. Some changes obviously occurred in various species, however, these changes are relatively minor, due to gene shuffling and genes being turned on or off, etc. I do not believe the evidence supports the view that humans evolved from lower life forms, such as apes (Chapters 15 & 16). Humans must have been created separately. Some form of intelligence (a Creator) must have been responsible for the initial universe and life. The belief in a Creator may seem a little difficult for some to accept. After all, believing that a God always existed and was able to create something out of nothing may appear to be folklore and may seem to only push the question of origins back one notch rather than explain it. I believe, however, there is no other explanation.

It is not that I have given up on science, it is just that I am facing reality. I believe that the atom, matter, the universe, and life are just too complex to have originated with no intelligence whatsoever.

Al DeBenedictis

~~~

In Closing

I sincerely hope this book has been helpful in your study of origins. I realize that this study may have raised more questions than answers. I am hopeful, at least, that it was more beneficial than not. As mentioned previously, a great deal depends on one's perception when considering the evidence and arguments of evolution versus creation. If you believe the universe and life originated by natural causes without any help from a Creator, you most likely believe that everything originated on its own. If you believe that science has, or will someday, provide convincing evidence that supports the view that the universe and life originated on its own, your faith and confidence are in science. If the evidence presented in this book appears to indicate there is a Creator, then you should pursue this idea further. Since this is a scientific book rather than a religious one, I did not include any detailed arguments for or against the possibilities that God or a Creator exists. You will need to determine this on your own.

If you now believe some intelligence must have somehow been involved in the origin of the universe and life or are now considering the possibility that there is a Creator, I believe the first step is to determine if a God truly exists. Then, if you reach the conclusion that a God exists, you will need to determine which religion, if any, is true. I believe that by studying the histories of the different religions and reading the official documents of these religions to see what they believe, you will be better equipped to determine which religions, if any, seem plausible. If it appears that an individual started a religion, ask yourself: What personal gain could have been achieved from starting their religion? Was it personal fame, fortune or power; or was the originator simply obeying what they believe God has commanded them to do? How did the religion initially spread—through simply preaching or was the religion propagated by force, fear of physical harm or intimidation?[IC-1]

It would be very helpful if you read what others have found in their research. Men such as Josh McDowell, Lee Strobel, William Ramsey and many others like myself, were once skeptical of the Bible but have come to believe that God is real and that the Bible is trustworthy.[IC-2] Many historians, such as Dr. Joseph Free and Donald J. Wiseman, have written books about the reliability of the Bible. Archaeologist Dr. Joseph Free discusses the accuracy of the Bible in his book, "Archaeology and Bible History." [IC-3] Donald J. Wiseman, a specialist in the field of archaeology, defended the reliability of the Bible in his book, "Archaeological Confirmation and the Bible." [IC-4] Archaeologist William M. Ramsey set out to find contradictions between the historical accounts in the Bible and actual archaeological findings. However, after years of archaeological research in Asia Minor and Greece, Ramsay changed his opinion and concluded that the Bible was reliable. William Ramsey acknowledges the accuracy of the New Testament in his books, "St. Paul the Traveler and the Roman Citizen" and "Luke, the Physician." [IC-5] The accuracy of the biblical documents has been well argued by men such as James Martin who wrote "The Reliability of the Gospel" [IC-6] and F. F. Bruce, who wrote "The New Testament Documents: Are They Reliable?" [IC-7] Will Durant comments about the accuracy of the New Testament in his book, "The Story of Civilization." [IC-8] H. G. Wells, an agnostic historian, acknowledged the Gospels as historically correct documents in his book, "The Outline of History." [IC-9] Dr. Craig L. Bloomberg, PhD, in his book, "The Historical Reliability of the Gospels," outlines the accuracy of the four gospels (Matthew, Mark, Luke, John) included in the New Testament.[IC-9a] John McRay, PhD, wrote a thorough 432-page textbook entitled, "Archeology and the New Testament," which accurately defends the historical reliability of the New Testament.

Although the evidence discussed by these historians in their books cannot prove that the Bible is the Word or God or that there is a God or that the Jesus described in the New Testament actually existed, the evidence does illustrate that the history contained in the Old and New Testaments are historically and geographically accurate and trustworthy.[IC-9b]

From my personal research I determined that the Bible is true. Archaeological research has determined that the events recorded in the Bible are accurate and reflect the time periods in which the events were said to have occurred. Just as we discussed the evidence for and against the theory of evolution, if you want to determine whether the Bible could be true or not, I suggest reading what others have found regarding the accuracy of the Bible. I am sure you will find it to be very beneficial and rewarding. (If you would like to do some additional reading, I suggest you read some of the resources listed in the References "In Closing" section below.) If you want to know more about God and develop a life with meaning, I urge you to read the Bible. Admittedly, the Bible may not have the latest scientific facts, but all in all, from my research, it is a factual book.[IC-10] The Bible explains how we may become right before God and obtain eternal life (John 3:15-18, 3:36; 5:24; 6:47; 14:1-6; Romans 10:9-10; Ephesians 2:8-9; Titus 3:5-7)) rather than eternal punishment (Matthew 25:46), that is, eternal destruction (2 Thessalonians 1:9), which is eternal death (Mark 9:43-48; Romans 6:23; Revelation 21:8) The Bible states that no one is perfect (Romans 3:23), that the result is separation from God (Romans 6:23a). The Bible states, however, that God has provided a way for us to have eternal life and that it was provided in the way of a gift (Romans 6:23b). The gift was given when Jesus died for our sins (Romans 5:8). The only way to receive this gift is to *"Repent, and . . . be baptized in the name of Jesus Christ for the remission of sins . . ."* (Acts 2:38 NJKV). *"Repent therefore and be converted, that your sins may be blotted out . . .* (Acts 3:19 NKJV). *"Whoever believes in Him [Jesus] should not perish but have eternal life."* (John 3:15 NKJV) Jesus said: *"Most assuredly, I say to you, he who hears My word and believes in Him who sent Me has everlasting life, and shall not come into judgment, but has passed from death into life."* (John 5:24 NKJV) Jesus also said: *"Most assuredly, I say to you, he who believes in Me has everlasting life."* (John 6:47 NKJV)[IC-11]

When one admits they are not perfect and have disobeyed God (sinned) (Romans 3:23) and believes that Jesus suffered and died to pay for their sins (1 Corinthians 15:3; 1 Peter 2:24), and that Jesus rose from the dead, their sins will be forgiven and they will inherit eternal life and will become saved (Romans 10:9) from God's wrath (Romans 5:9). The Bible says: *"For God so loved the world that He gave His only begotten Son, that whoever believes in Him should not perish but have everlasting life. For God did not send His Son into the world to condemn the world, but that the world through Him might be saved. He who believes in Him is not condemned; but he who does not believe is condemned already, because he has not believed in the name of the only begotten Son of God."* (John 3:16-18 NKJV)

I hope, after reading this book, that it has helped you in your study of origins and has helped you to better understand those of opposing views, even though you may not totally agree with their beliefs. May you continue to search for truth, wherever it may lead you, for *"the truth will make you free . . ."* (John 8:32 TEB)

Al DeBenedictis

~~~

References

Opening Comments

OC-0 https://www.allaboutphilosophy.org/what-is-an-evolutionist-faq.htm

OC-1 Coyne, J. A. 2009. Why Evolution is True. New York (NY): Viking Penguin Group. Pp. 15-17

OC-2 Coyne, J. A. 2009. Why Evolution is True. New York (NY): Viking Penguin Group. P. 13

OC-3 http://www.talkorigins.org/faqs/dover/day1pm.html

OC-4 Theodosius Dobzhansky http://people.delphiforums.com/lordorman/light.htm

OC-5 How Science Works: Evolution A Student Primer R. John Ellis: http://www.springerlink.com/content/t766m3#section=652582&page=1

OC-6 Mike Anderson. Is Jesus an Evolutionist? 2011. Smashwords Publishers. Nook e-book. P. 394-395 Prior book version: Montane Publishers, South Africa (2008) http://www.scribd.com/doc/11464783/Is-Jesus-an-Evolutionist

OC-7 How Science Works: Evolution A Student Primer R. John Ellis: http://www.springerlink.com/content/t766m3#section=652582&page=1

OC-8 Scott, E.C. 2009. Evolution vs Creationism—An Introduction. Berkeley (CA): University of California Press. Pp. 14, 19-20 http://www.createdebate.com/debate/shocw/Has_evolution_been_scientifically_proved http://theory-of-evolution.net/intelligent-design-blog/?p=68

OC-9 Gary Parker. Creation Facts of Life—How Real Science Reveals the Hand of God. 2006. Master Books. P. 16-20

OC-10 Behe, M. pp. 265-266. In: Strobel, L. (2004) The Case For a Creator "The Evidence of Biochemistry: The Complexity of Molecular Machines", pp. 245-269

OC-11 Gary Parker. Creation Facts of Life—How Real Science Reveals the Hand of God. 2006. Master Books. P. 71

OC-12 http://www.icr.org/i/pdf/research/rate-all.pdf

OC-12a Charles Colson and Harold Fickett. The Good Life. 2005. Tyndale House Publishers, Inc., Wheaton, Illinois. P. 225

OC-13 Behe, M. pp. 265-266. In: Strobel, L. (2004) The Case For a Creator "The Evidence of Biochemistry: The Complexity of Molecular Machines," pp. 245-269

OC-14 https://ebible.com/questions/2956-is-creationism-scientific/?mlgq=1&rep_k=m-_8IOvAN0PkXsMXKPden4pP0Sz68P8rb7k4M4ocqSgKZSLjL7_xtvzjTy3kxkdD&rep__m=clicks&rep_e=eatH7kpj3uXwNAV99A67YdMJ6g oG5dYXlP0_PGT1SC8= http://www.gotquestions.org/creationism-scientific.html

OC-15 Rana, Fazale. The Cell's Design – How Chemistry Reveals the Creator's Artistry. BakerBooks. Grand Rapids, Michigan. 2008. Pp.270-286

Chapter 1—The Origin of the Universe

1-1 Genesis 1:1-9, The Bible, Scripture taken from the New King James Version. Copyright © 1982 by Thomas Nelson, Inc. Used by permission. All rights reserved.

1-2 http://www.space.com/scienceastronomy/big-bang-universe-beginning-100319.html?utm_source=feedburner&utm_medium=feed&utm_campaign=Feed%3A+spaceheadlines+(SPACE.com+Headline+Feed)

1-3 Julian Huxley, quoted in New York Times, November 29, 1959. In: Evolution Disproved Encyclopedia—Volume 3
http://evolutionfacts.com/Ev-V3/3evlch31b.htm

1-4 Job 38:4-6, The Bible, Scripture taken from the New King James Version. Copyright © 1982 by Thomas Nelson, Inc. Used by permission. All rights reserved.

1-5 http://www.bigquestionsonline.com/columns/michael-shermer/the-biggest-big question-of-all

1-5a Hawking, Stephen and Mlodinow, Leonard. The Grand Design. 2010. as quoted in http://www.michaelgstrauss.com/2017/08/the-grand-design-is-godunnecessary.html

1-6 http://www.atheistrev.com/2010/03/why-is-there-something-rather-than.html

1-7 http://www.csicop.org/sb/show/why_is_there_something_rather_than_nothing

1-8 MacArthur, John, 2001. The Battle For the Beginning, W Publishing Group. p. 31

1-9 http://www.onenesspentecostal.com/whysomething.htm

1-10 Mac Arthur, John, 2001, The Battle For the Beginning, W Publishing Group, www.wpublisninggroup.com p 37

1-11 How to Know that God Exists—Dr. Woodrow Kroll March 2, 2009 Radio Broadcast http://www.backtothebible.org/index.php/Back-to-the-Bible-Radio Program/Can-We-Have-Something-from-Nothing.html

1-12 Ham, K, General Editor. 2006. The New Answers Book. Inc. Green Forest (AR): Master Books. pp. 286-287.

1-13 Erwin W. Lutzer, Seven Reasons Why You Can Trust the Bible, 1998, Moody Press, Chicago, pp. 136-139

1-14 www.faithaliveresources.org/origins Loren Haarsma, Faith Alive Christian Resources, 2850 Kalamazoo Avenue, SE, Grand Rapids, Michigan, 49560

1-15 Dr. Woodrow Kroll. How to Know that God Exists—March 2, 2009 Radio Broadcast http://www.backtothebible.org/index.php/Back-to-the-Bible-Radio Program/Can-We-Have-Something-from-Nothing.html

1-16 Taken from the Gospel Tract "Evolution: The Evidence For And Against", by www.livingwaters.com

1-16a A Universe From Nothing. Dr. Michael G. Strauss. March 15, 2017 http://www.michaelgstrauss.com/2017/03/a-universe-from-nothing.html

1-16b   http://www.michaelgstrauss.com/2017/01/a-previous-post-about-big-bangelicited.html

1-17 http://www.raptureready.com/rr-ec-debate.html

1-18 http://io9.com

1-18-1 https://knowledgenuts.com/2013/08/08/ the-difference-between-atomsmolecules-and-particles/ https://www.yourdictionary.com/matter https://whatis.techtarget.com/definition/matter

1-18-2 Why is there any matter in the universe at all? New Sussex study sheds light Charles Rotter. February 28, 2020. https://wattsupwiththat.com/2020/03/01/why-is-there-any-matter-inthe-universe-at-all-new-sussex-study-sheds-light/

1-18a Are Virtual Particles Real? Dr. Michael G. Strauss. September 17, 2017. http://www.michaelgstrauss.com/2017/09/are-virtual-particles-real.html#more

1-19 http://en.wikipedia.org/wiki/Atom http://en.wikipedia.org/wiki/Subatomic_ particle http://en.wikipedia.org/wiki/Elementary_particle

1-20 http://www.ehow.com/about_6510735_do-come-together-form-molecules_ .html http://www.ask.com/question/where-do-atoms-come-from

1-21   http://answers.ask.com/Science/Chemistry/where_do_atoms_come_from
http://www.ask.com/question/where-do-atoms-come-from

1-22 http://www.ask.com/question/where-do-atoms-come-from

1-23 http://www.thingsmadethinkable.com/item/elementary_particles.php

1-23a A Look at the Top Quark. Dr. Michael G. Strauss. May 2018
http://www. michaelgstrauss.com

1-23b https://en.wikipedia.org/wiki/Photon
https://en.wikipedia.org/wiki/Gluon
https://en.wikipedia.org/wiki/W_and_Z_bosons

1-24 Charles Colson and Nancy Pearcey. How Now Shall We Live? 1999. Tyndale
House Publishers, Inc. Wheaton, Illinois. pp. 63-66

1-25 Strobel, L. 2004. The Case for a Creator. Grand Rapids (MI): Zondervan. Pp.
128-129

1-26 Big Bang Theory-Answers.com
http://www.answers.com/topic/big-bang-theory Also see:
http://en.wikipedia.org/wiki/Big_Bang

1-27 Creation of a Cosmology: Big Bang theory http://ssscott.tripod.com/BigBang.html)

1-28 Big Bang Theory-Answers.com
http://www.answers.com/topic/bigbang-theory http://en.wikipedia.org/
wiki/Big_Bang

1-29 Creation of a Cosmology: Big Bang theory http://ssscott.tripod.com/BigBang.html

1-30 http://education.yahoo.com/reference/dictionary/entry/neutron

1-31 Lutzer, E. W. 1998. Seven Reasons Why You Can Trust the Bible, Chicago (IL):
Moody Press. pp. 136-139

1-32 Haarsma, D. B., Haarsma, L. D. 2007. Origins—A Reformed Look At Creation,
Design, & Evolution. Grand Rapids, (MI): Faith Alive Christian Resources.
pp. 138-140

1-33 From the Beginning—The Story of Human Evolution. David Peters. 1991.
Morrow Junior Books. 1991. New York. p. 10
http://www.davidpetersstudio.com/FromTheBeginning.pdf

1-34 From the Beginning—The Story of Human Evolution. David Peters. 1991.
Morrow Junior Books. 1991. New York. p. 10
http://www.davidpetersstudio.com/FromTheBeginning.pdf

1-35 Evolution Disproved Encyclopedia—Volume 01
http://evolutionfacts.com/ Ev-V1/1evlch01a.htm

1-36 Evolution Disproved Encyclopedia—Volume 01
http://evolutionfacts.com/ Ev-V1/1evlch01a.htm

1-37 Charles Colson and Nancy Pearcey. How Now Shall We Live? 1999. Tyndale House Publishers, Inc. Wheaton, Illinois. pp. 63-66

1-38 Mac Arthur, John, 2001, The Battle For the Beginning, W Publishing Group, www.wpublisninggroup.com p. 38

1-39 Big Bang Theory—An Overview All About Science
http://www.allaboutscience. org/big-bang-theory.htm

1-40 Sky & Telescope, October 2003 p. 32. In:
http://christiananswers.net/q-eden/ edn-earthage.html

1-41 http://www.universetoday.com/38195/oscillating-universe-theory/

1-42 http://www.allaboutcreation.org/oscillating-universe-theory-faq.htm

1-43 Strobel, L. 2004. The Case for a Creator. Grand Rapids (MI): Zondervan. Pp. 139-142

1-44 Strobel, L. 2004. The Case for a Creator. Grand Rapids (MI): Zondervan. Pp. 139-142

1-45 http://www.kheper.net/cosmos/universe/Big_Bang.htm

1-46 Strobel, L. 2004. The Case for a Creator. Grand Rapids (MI): Zondervan. Pp. 143-147

1-47 Strobel, Lee, 2004, The Case for a Creator, Zondervan, Grand Rapids, Michigan, 49530 zreview@zondervan.com, pp. 169-177

1-48 Strobel, L. 2004. The Case for a Creator. Grand Rapids (MI): Zondervan. Pp. 169-177

1-49 Strobel, L. 2004. The Case for a Creator. Grand Rapids (MI): Zondervan. Pp. 169-177

1-50 Strobel, L. 2004. The Case for a Creator. Grand Rapids (MI): Zondervan. Pp. 159-165

1-50a Gish, Duane Tolbert, Ph. D., Have you been Brainwashed?, booklet. Also see
http://www.skepticfiles.org/evolut/evolve7i.htm
http://www.skeptictank.org/

1-50b https://en.wikipedia.org/wiki/Earth%27s_orbit
https://www.universetoday.com/37512/solar-system-orbits/
https://www.quora.com/Is-the-Earth's-orbit-around-the-Sun-a-perfect-circle

1-51 http://www.skepticfiles.org/evolut/evolve7i.htm

1-52 Evolution Disproved Encyclopedia—Volume 3
http://evolutionfacts.com/ Ev-V3/3evlch25.htm

1-53 Charles Colson and Nancy Pearcey. How Now Shall We Live? 1999. Tyndale House Publishers, Inc. Wheaton, Illinois. p. 63

1-53a http://www.michaelgstrauss.com/2017/08/the-grand-design-is-god-unnecessary. html

1-53a-1 Hugh Ross, PhD. Genesis One – A Scientific Perspective. 1973, 1983 p. 11

1-53b https://en.wikipedia.org/wiki/M-theory

1-54 Young, Matt and Strode, Paul K. 2009. Why Evolution Works (And Creationism Fails). Rutgers Press, 100 Joyce Kilmer Avenue, Piscataway, NJ 08854-8099. Pp. 165-169

1-55 Dr. Alan Hayward, Physicist. Copyright 1985. Publication Date 2005. Creation and Evolution—Rethinking the Evidence From Science and the Bible. WIPF & Stock Publishers. Eugene, Oregon. Pp. 59-60 744

1-56 Kenneth R. Miller. Finding Darwin's God. 1999. Cliff Street books. New York. Quoting Professor Stephen Hawking. P. 227, 229. In: Carl R. Turner, M.D. Discovering God and His Creation—Evolution as Part of God's Plan. 2008. iUniverse, Inc. New York. P. 15-16

1-57 Ron Rhodes. 2004. The 10 Things You Should Know About the Creation Vs. Evolution Debate. Harvest House Publishers. Eugene, Oregon. Pp. 147,152.

1-58 Ron Rhodes. 2004. The 10 Things You Should Know About the Creation Vs. Evolution Debate. Harvest House Publishers. Eugene, Oregon. Pp. 128-129, 149.

1-59 Young, Matt and Strode, Paul K. 2009. Why Evolution Works (And Creationism Fails). Rutgers Press, 100 Joyce Kilmer Avenue, Piscataway, NJ 08854-8099. P. 170

1-59a The God Particle...and God. Dr. Michael G. Strauss. Thursday, January 5, 2017 http://www.michaelgstrauss.com/2017/01/the-god-particleand-god.html

1-59b https://en.wikipedia.org/wiki/Hadron

1-60 http://angelsanddemons.web.cern.ch/faq/what-is-the-god-particle

http://en.wikipedia.org/wiki/Higgs_boson http://ngm.nationalgeographic.com/2008/03/god-particle/achenbach-text

http://ngm.nationalgeographic.com/2008/03/god-particle/achenbach-text/2

http://articles.cnn.com/2011-12-13/world/world_europe_ higgs-boson-q-and-a_1_higgs-boson-peter-higgs-particle-physics?_ s=PM:EUROPE http://www.straightdope.com/columns/read/2850/what-is-the-god-particle http://news.nationalpost.com/2011/12/12/higgs-boson-what-is-it-andwhyis- everyone-so-excited-about-the-god-particle/

http://news.nationalpost.com /2011/12/1 2/the-search-for-the-higgs-boson -inside-the-large-hadron-collider/

http://www.reuters.com/article /2011/12 /13/us-science-higgs-god-idUST RE7BC28H20111213

http://www.godparticle.net/ http://news.nationalgeographic.com/news/2012/07/120704-god-particlehiggs-boson-new-cern-science/

http://abcnews.go.com /blogs/technology/2011/12/the-god-particle-search -for-higgs-boson-narrowed-by-cern-physicists/
http://www.independent.co.uk/news/science/has-science-found-thegodparticle-6276634.html
http://news.yahoo.com/higgs-boson-physicists-see-best-proofyetgod-55311961—abc-news-tech.html
http://news.yahoo.com/cosmic-theorys-higgs-lives-see-boson-182126947.html
http://abcnews.go.com/Technology/higgs-boson-evidence-god-particlereported-fermilab-physicists/story?id=16695742 http://www.npr.org/2012/07/02/155994840/is-the-hunt-for-the-godparticle-finally-over
http://skepticalteacher.wordpress.com/2010/06/03/why-is-theresomethingrather-than-nothing-science-may-now-have-an-answer/
1-61 http://news.nationalgeographic.com/news/2012/07/120704-god-particle higgsboson-new-cern-science/
1-62 http://www.straightdope.com/columns/read/2850/what-is-the-god-particle

Chapter 2—The Age of the Universe and the Earth

2-1 http://starchild.gsfc.nasa.gov/docs/StarChild/questions/question28.html
2-2 https://www.answersingenesis.org/articles/ee/origin-of-life
2-3 Huse, S. M., The Collapse of Evolution, 1983, Baker Book House, Grand Rapids, Michigan, 49508, P. 30
2-4 http://www.earthfacts.com/space/ageuniverse/
2-5 http://ssscott.tripod.com/BigBang.html
2-6 http://www.earthfacts.com/space/ageuniverse/
2-7 http://creation.com/globular-clusters-and-the-challenge-of-blue-straggler-stars
2-8 Scott M. Huse. The Collapse of Evolution. 1983. Baker Book House. pp. 29-30
2-9 Astronomy & Astrophysics 321:L17, 1997 In:
    http://creation.com/age-of-the-earth
2-10 Muir, H., 2003, Back from the dead, New Scientist 177(2384):28–31 In:
    http:// creation.com/age-of-the-earth
2-11 http://evolutionfacts.com/Ev-V1/1evlch08.htm
2-12 Denis O. Lamoureux, I Love Jesus & I Accept Evolution. 2009. WIPF Stock. Eugene. Oregon. Pages 104-105
2-13 http://en.wikipedia.org/wiki/Age_of_the_Earth
2-14 http://uk.answers.yahoo.com/question/index?qid=20080207052952AAiejHY
2-15 http://www.talkorigins.org/faqs/faq-age-of-earth.html

2-16 Scott M. Huse. The Collapse of Evolution. 1983. Baker Book House. p. 8

2-17 Scott M. Huse. The Collapse of Evolution. 1983. Baker Book House. p. 18

2-18 http://www.wiebefamily.org/e.htm

2-19 Huse, S. M., The Collapse of Evolution, 1983, Baker Book House, Grand Rapids, Michigan, 49508, p 20-27

2-20 http://www.wiebefamily.org/e.htm

2-21 Huse, S. M., The Collapse of Evolution, 1983, Baker Book House, Grand Rapids, Michigan, 49508, p 20-27

2-22 Lubenow M. 2004. Bones of Contention. Grand Rapids (MI): Baker Books. pp. 283-287

2-23 http://www.talkorigins.org/faqs/faq-age-of-earth.html

2-24 http://www.wiebefamily.org/e.htm

2-25 http://www.wiebefamily.org/e.htm

2-26 http://www.eadshome.com/RadiometricDating.htm

2-27 http://www.wiebefamily.org/e.htm

2-28 http://evolution-101.blogspot.com/2006/06/how-are-fossils-dated.html

2-28a Radioisotopes and the Age of the Earth. A Young-Earth Creationist Research Initiative. 2000. Larry Vardian, Andrew A. Snelling, Eugene F. Chaffin, who are members of the RATE (Radioisotopes and the Age of The Earth) group. Accelerated Decay pp. 42-45. Institute for Creation Research. California.

2-28b Reasons to Believe. "Helium Diffusion in Zircon: Flaws in a Young-Earth Argument" Dr. Gary H. Loechelt. September 9, 2008 & September 16, 2008. An analysis of an eight-year research program, called RATE (Radioisotopes and the Age of The Earth), conducted by young-earth researchers. https://reasons.org/explore/publications/tnrtb/read/tnrtb/2008/09/09/helium-diffusion-in-zircon-flaws-in-a-young-earth-argument-part-1-of-2 https://reasons.org/explore/blogs/todays-new-reason-to-believe/read/tnrtb/2008/09/16/helium-diffusion-in-zircon-evidence-supports-an-oldearth-part-2-of-2

2-29 Lamoureux. D. O. I Love Jesus & I Accept Evolution. 2009. Eugene (OR): WIPF Stock. Pp. 95-97

2-30 Young, Matt and Strode, Paul K. 2009. Why Evolution Works (And Creationism Fails). Rutgers Press, 100 Joyce Kilmer Avenue, Piscataway, NJ 08854-8099. P. 55

2-31 Dr. Alan Hayward, Physicist. Copyright 1985. Publication Date 2005. Creation and Evolution—Rethinking the Evidence From Science and the Bible. WIPF & Stock Publishers. Eugene, Oregon. Pp. 87-88

2-32 Lamoureux. D. O. I Love Jesus & I Accept Evolution. 2009. Eugene (OR): WIPF Stock. Pp. 96-100

2-33 Tree Ring Dating by John D. Morris, Ph.D.
http://www.icr.org/article/7058/

2-34 http://kimbalogh.wordpress.com/2013/04/20/creation-vs-evolution-debate/

2-35 http://www.jesus-is-savior.com/Evolution%20Hoax/Evolution/03.htm

2-36 http://www.answersingenesis.org/articles/nab/does-radiometric-dating-prove

2-37 How Good Are Those Young-Earth Arguments? Dave E. Matson http://www.
evolution-creationism.us/young_earth/geology_column_ evolution.html

Chapter 3—The Fossil Evidence in the Strata Layers of the Earth

3-1 R. Milner, Encyclopedia of Evolution (1990), pp. 157, 318. In: Evolution Disproved Encyclopedia—Volume 3 http://evolutionfacts.com/Ev-V3/3evlch31b.htm

3-2 Origin of Species. Sixth Edition. 1901. Pp. 341-342. In: Evolution Disproved Encyclopedia—Volume 2
http://evolutionfacts.com/Ev-V2/2evlch17a.htm

3-3 http://en.wikipedia.org/wiki/Great_Oxygenation_Event

3-4 http://www.talkorigins.org/faqs/comdesc/section5.html http://www.talkorigins.
org/faqs/faq-transitional/part1a.html http://www.talkorigins.org/faqs/faq-
transitional/part2c.html
http://www.evolution-facts.org/images/ev-cr-pix/EC418.jpg http://www.
unidiversal.com/CreationEvolutionCruncher12a.html http://www.
godrules.net/evolutioncruncher/c12a.htm
http://www.ask.com/bar?q=Phanerozoic+period+&page=1&qsrc=2891
&dm=all&ab=2&u=http%3A%2F%2Fgeology.about.com%2Flibrary
%2Fbl%2Ftime%2Fblphantime.htm&sg=8cN0OOHPmlXWYX
%2B4VkpfAfJN1UUaIKDTJiEagimdj7Q%3D&tsp=1269779684264

3-5 Huse, S. M., The Collapse of Evolution, 1983, Baker Book House, Grand Rapids, Michigan, 49508, p 7-15, 36

3-6 http://www.eadshome.com/Fossils.htm

3-7 http://www.trueorigin.org/geocolumn.asp

3-8 Evolution Disproved Encyclopedia—Volume 2
http://evolutionfacts.com/Ev-V2 /2evlch17e.htm

3-9 W.E. Lammerts, Scientific Studies in Special Creation, (1971), pp. 127-128. In: Evolution Disproved Encyclopedia—Volume 2
http://evolutionfacts.com/ Appendix/a17e.htm

3-10 Dr. Alan Hayward, Physicist. Copyright 1985. Publication Date 2005. Creation and Evolution—Rethinking the Evidence From Science and the Bible. WIPF & Stock Publishers. Eugene, Oregon. Pp. 117-118

3-11 http://atheisttoolbox.com/fce12.php

3-12 http://www.noanswersingenesis.org.au/geologiccolumn.htm

3-13 http://genesismission.4t.com/Geology/pgc.html

3-14 How Good Are Those Young-Earth Arguments? Dave E. Matson http://www.evolution-creationism.us/young_earth/geology_column_evolution.html

3-15 How Good Are Those Young-Earth Arguments? Dave E. Matson http://www.evolution-creationism.us/young_earth/geology_column_evolution.html

3-16 Lamoureux, D. O. 2009. I Love Jesus & I Accept Evolution. Eugene (OR): WIPF Stock. Pp. 89-90

3-17 http://paleo.cc/kpaleo/fossdate.htm

3-18 http://www.talkorigins.org/indexcc/CC/CC310.html

3-19 Evolution Disproved Encyclopedia—Volume 2 http://evolutionfacts.com/ Ev-V2/2evlch17b.htm

3-20 http://www.godrules.net/evolutioncruncher/c12a.htm Evolution Disproved Encyclopedia—Volume 2 http://evolutionfacts.com/Ev-V2/2evlch17b.htm

3-21 http://www.ehow.com/way_5422879_methods-dating-fossils.html

3-22 http://www.wiebefamily.org/e.htm

3-23 Crampton. Henry Edward. The Doctrine of Evolution. Its Basis and Its Scope. Columbia University Press Sales Agents. New York: Lemcke & Buechner. 30-32 West 27th Street. London. 1906-1907. Copyrighted 1911. (Ebook#16442. Release Date: August 5, 2005)

3-24 Young, Matt and Strode, Paul K. 2009. Why Evolution Works (And Creationism Fails). Rutgers Press, 100 Joyce Kilmer Avenue, Piscataway, NJ 08854-8099. P. 54

3-25 Gishlick, Alan D. Icons of Evolution? Why much of what Jonathan Wells writes about evolution is wrong. Oakland. CA: National Center for Science Education. 2003. Retrieved from http://www.ncseweb.org/creationism/ analysis/icon-2-darwins-tree-life. In: Scott, E.C. 2009. Evolution vs Creationism—An Introduction. Berkeley (CA): University of California Press. P. 197-198

3-26 Stephen C. Meyer, Scott Minnich, Jonathan Moneymaker, Paul A. Nelson, Ralph Seelke. 2007. Explore Evolution—The Arguments For and Against Neo-Darwinism. Hill House Publishers. Melbourne & London. Pp. 16-17

3-27 Stephen C. Meyer, Scott Minnich, Jonathan Moneymaker, Paul A. Nelson, Ralph Seelke. 2007. Explore Evolution—The Arguments For and Against Neo-Darwinism. Hill House Publishers. Melbourne & London. P. 30

3-28 Evolution Disproved Encyclopedia—Volume 2
http://evolutionfacts.com/ Appendix/a15.htm

3-29 D.B. Gower (Biochemist), "Scientist Rejects Evolution," Kentish Times, England December 11, 1975, p. 4. In: Evolution Disproved Encyclopedia—Volume 3
http://evolutionfacts.com/Ev-V3/3evlch31c.htm

3-30 http://www.wiebefamily.org/e.htm

3-31 Evolution Disproved Encyclopedia—Volume 2
http://evolutionfacts.com/Ev-V2 /2evlch17b.htm

3-32 Evolution Disproved Encyclopedia—Volume 2
http://evolutionfacts.com/Ev-V2 /2evlch17b.htm

3-33 Evolution Disproved Encyclopedia—Volume 2
http://evolutionfacts.com/Ev-V2 /2evlch17b.htm

3-34 http://www.talkorigins.org/indexcc/CC/CC300.html

3-35 Gishlick, Alan D. Icons of Evolution? Why much of what Jonathan Wells writes about evolution is wrong. Oakland. CA:National Center for Science Education. 2003. Retrieved from
http://www.ncseweb.org/creationism/analysis/ icon-2-darwins-tree-life
In: Scott, E.C. 2009. Evolution vs Creationism—An Introduction. Berkeley (CA): University of California Press. P. 197-198

3-36 Gishlick, Alan D. Icons of Evolution? Why much of what Jonathan Wells writes about evolution is wrong. Oakland. CA:National Center for Science Education. 2003. Retrieved from
http://www.ncseweb.org/creationism/ analysis/icon-2-darwins-tree-life
In: Scott, E.C. 2009. Evolution vs Creationism—An Introduction. Berkeley (CA): University of California Press. P. 197-198

3-37 Gishlick, Alan D. Icons of Evolution? Why much of what Jonathan Wells writes about evolution is wrong. Oakland. CA:National Center for Science Education. 2003. Retrieved from
http://www.ncseweb.org/creationism/analysis/icon-2-darwins-tree-life
In: Scott, E.C. 2009. Evolution vs Creationism—An Introduction. Berkeley (CA): University of California Press. P. 199, Figure 9.1

3-38 http://en.wikipedia.org/wiki/Lancelet
http://en.wikipedia.org/wiki/Pikaia
http://paleobiology.si.edu/burgess/pikaia.html

http://dinosaurs.about.com/od/tetrapodsandamphibians/p/pikaia.htm

http://en.wikipedia.org/wiki/Haikouella

http://www.fact-index.com/h/ha/haikouella.html

http://en.wikipedia.org/wiki/Haikouichthys

http://en.wikipedia.org/wiki/Anaspid

http://www.fact-index.com/c/ca/cambrian.html

http://www.cals.ncsu.edu/course/ent425/text02/arthropods.html

3-39 Mike Anderson. Is Jesus an Evolutionist? 2011. Smashwords Publishers. Nook e-book. P. 389 Prior book version: Montane Publishers, South Africa (2008) http://www.scribd.com/doc/11464783/Is-Jesus-an-Evolutionist

3-40 What is Peripatus? 9/1/1994 Answers in Genesis. Margaret Helder, Ph.D. Editor of Reformed Perspective magazine and Vice-President of Creation Science Association of Alberta, Canada.

http://www.answersingenesis.org/articles/cm/v16/n4/peripatus

3-41 Icons of Evolution? Why Much of What Jonathan Wells Writes About Evolution is Wrong" Alan D. Gishlick—National Center for Science Edition—Defending the Teaching of Evolution in Public Schools. Table by Rode, 1999; Hanic et al., 2000

http://ncse.com/creationism/analysis/icons-critique-pdf

3-42 http://intelligentscience.org/2008/10/16/the-cambrian-a-hugeproblem-formoleculesto-man-evolution/

3-43 Parker, Gary. Creation Facts of Life—How Real Science Reveals the Hand of God. Master Books. 2006. P. 153

3-44 Parker, Gary. Creation Facts of Life—How Real Science Reveals the Hand of God. Master Books. 2006. P. 157

3-45 Evolution Disproved Encyclopedia—Volume 2 http://evolutionfacts.com/ Ev-V2/2evlch17a.htm

3-46 Parker, Gary. Creation Facts of Life—How Real Science Reveals the Hand of God. Master Books. 2006. P. 219

3-47 Parker, Gary. Creation Facts of Life—How Real Science Reveals the Hand of God. Master Books. 2006. P. 158

3-48 Parker, Gary. Creation Facts of Life—How Real Science Reveals the Hand of God. Master Books. 2006. P. 159

3-49 Gish, Duane Tolbert, Ph. D., Have you been Brainwashed?, booklet. Also see http://www.skepticfiles.org/evolut/evolve7i.htm

3-50 Some Fish Stories—. . . About an Unproven Theory, Ambassador College, Garner Ted Armstrong, 1969, p. 22

3-51 Some Fish Stories—. . . About an Unproven Theory, Ambassador College, Garner Ted Armstrong, 1969, p. 23

3-52 Some Fish Stories—. . . About an Unproven Theory, Ambassador College, Garner Ted Armstrong, 1969, p. 27

3-52a Morris. Henry M. The Scientific Case Against Evolution – The Verdict is In. Institute for Creation Research. 2001. P. 3. Dallas

3-53 Ross, Hugh. The Genesis Question - Scientific Advances and the Accuracy of Genesis. (Second Expanded Edition). 2001. Navpress. Colorado. Pages 64-65

Chapter 4 - How Did All the Strata Layers Form?

4-1 Genesis 6: 13-16; 7:1-3, 10-12, 20-21, 24 The Bible, Scripture taken from the New King James Version. Copyright © 1982 by Thomas Nelson, Inc. Used by permission. All rights reserved

4-2 Haarsma, L. Faith Alive Christian Resources, 2850 Kalamazoo Avenue, SE, Grand Rapids, Michigan, 49560 www.faithaliveresources.org/origins

4-3 Haarsma, L. Faith Alive Christian Resources, 2850 Kalamazoo Avenue, SE, Grand Rapids, Michigan, 49560 www.faithaliveresources.org/origins Haarsma, D.

B. Haarsma, L. D. 2007. Origins—A Reformed Look At Creation, Design, & Evolution. Grand Rapids, (MI): Faith Alive Christian Resources. pp. 87-90

4-4 Haarsma, L. Faith Alive Christian Resources, 2850 Kalamazoo Avenue, SE, Grand Rapids, Michigan, 49560 www.faithaliveresources.org/origins

4-5 Denis O. Lamoureux, I Love Jesus & I Accept Evolution, 2009. WIPF Stock. Eugene. Oregon. P. 83-84

4-6 Dr. Alan Hayward, Physicist. Copyright 1985. Publication Date 2005. Creation and Evolution—Rethinking the Evidence From Science and the Bible. WIPF & Stock Publishers. Eugene, Oregon. p. 125

4-7 Parker, Gary. Creation Facts of Life—How Real Science Reveals the Hand of God. Master Books. 2006. P. 196, 198-199

4-8 Parker, Gary. Creation Facts of Life—How Real Science Reveals the Hand of God. Master Books. 2006. P. 214

4-9 Parker, Gary. Creation Facts of Life—How Real Science Reveals the Hand of God. Master Books. 2006. P. 219-223

4-10 R.E. Kofahl and K.L. Segraves, The Creation Explanation (1975), p. 50. In: Evolution Disproved Encyclopedia—Volume 2 http://evolutionfacts.com/ Ev-V2/2evlch17d.htm

4-11 Immanuel Velikovsky, Earth in Upheaval (1955), p. 222. In: Evolution Disproved Encyclopedia—Volume 2
http://evolutionfacts.com/Ev-V2/2evlch17d.htm

4-12 http://www.thefreedictionary.com/uniformitarianism

4-13 http://www.allaboutcreation.org/catastrophism-versus-uniformitarianism-faq.htm

4-14 Douglas F. Kelly, Creation and Change, 164-65. In: John MacArthur. The Battle for the Beginning. Creation, Evolution and the Bible. W. Publishing Press. 2001. p. 53.

4-14a Mount St. Helens, Living Laboratory for 40 years. Institute for Creation Research. May 2020. Vol. 40 No. 5. Tim Clarey, Ph.D., Frank Sherwin, M.A. Pp. 10-13

4-15 Parker, Gary. Creation Facts of Life—How Real Science Reveals the Hand of God. Master Books. 2006. P. 219-223

4-16 Parker, Gary. Creation Facts of Life—How Real Science Reveals the Hand of God. Master Books. 2006. P. 225-226

4-17 Huse, S. M. 1983. The Collapse of Evolution. Grand Rapids (MI): Baker Book House. Pp. 46-51

4-18 Parker, Gary. Creation Facts of Life—How Real Science Reveals the Hand of God. Master Books. 2006. P. 192, 194

4-19 http://www.wiebefamily.org/e.htm

4-20 Briski, D. 2004. Impressive Deception—Creation or Evolution You Decide. Shippensburg, (PA): Ragged Edge Press. Pp. 110, 112

4-21 Evolution Disproved Encyclopedia—Volume 2
http://evolutionfacts.com/ Appendix/a19a.htm

4-21a Mount St. Helens, Living Laboratory for 40 years. Institute for Creation Research. May 2020. Vol. 40 No. 5. Tim Clarey, Ph.D., Frank Sherwin, M.A. Pp. 10-13

4-22 http://www.wiebefamily.org/e.htm

4-23 http://www.icr.org/article/classic-polystrate-fossil/

4-24 Dr. Alan Hayward, Physicist. Copyright 1985. Publication Date 2005. Creation and Evolution—Rethinking the Evidence From Science and the Bible. WIPF & Stock Publishers. Eugene, Oregon. pp. 126-127

4-24a Dr. Judy Bailey, coal geologist at the Discipline of Earth Science University of Newcastle, 2013

4-24b https://www.abc.net.au/science/articles/2013/02/18/3691317.htm

4-24c   https://answersingenesis.org/geology/catastrophism/coal-beds-and-noahs-flood/

4-24d https://creation.com/coal-memorial-to-the-flood

4-24e Henry Morris, Ph.D. (Hydraulic Engineering), and John C. Whitcomb. 1961, 2011. The Genesis Flood, P & R Publishing, Phillipsburg, New Jersey, pp. 277-279

4-24f Henry Morris, Ph.D. (Hydraulic Engineering), and John C. Whitcomb. 1961, 2011. The Genesis Flood, P & R Publishing, Phillipsburg, New Jersey, pp. 162 - 164

4-24g Stuart E. Nevins, M.S. November 01, 1976. The Origin of Coal, Institute for Creation Research,

https://www.icr.org/article/origin-coal/

4-24h Henry Morris, Ph.D. (Hydraulic Engineering), and John C. Whitcomb. 1961, 2011. The Genesis Flood, P & R Publishing, Phillipsburg, New Jersey, p. 164

4-24J Henry Morris, Ph.D. (Hydraulic Engineering), and John C. Whitcomb. 1961, 2011. The Genesis Flood, P & R Publishing, Phillipsburg, New Jersey, pp. 164 - 165

4-25 http://www.talkorigins.org/faqs/polystrate/trees.html

4-26 http://atheisttoolbox.com/fce12.php

4-27 Dr. Alan Hayward, Physicist. Copyright 1985. Publication Date 2005. Creation and Evolution—Rethinking the Evidence From Science and the Bible. WIPF & Stock Publishers. Eugene, Oregon. pp. 120-124

4-28 Huse, S. M. 1983. The Collapse of Evolution. Grand Rapids (MI): Baker Book House. Pp. 46-51

4-29 Lamoureux, D. O. 2009. I Love Jesus & I Accept Evolution. Eugene (OR): WIPF Stock. Pp. 88-89

4-30 Lamoureux, D. O. 2009. I Love Jesus & I Accept Evolution. Eugene (OR): WIPF Stock. Pp. 89-90

4-31 http://www.detectingdesign.com/fossilrecord.html#Simple_Complex

4-32 http://www.detectingdesign.com/fossilrecord.html#Simple_Complex

4-33 http://www.detectingdesign.com/fossilrecord.html#Simple_Complex

4-34 http://www.detectingdesign.com/fossilrecord.html#Simple_Complex

http://kimbalogh.wordpress.com/2013/04/20/creation-vs-evolution-debate/

4-34a https://apologeticsminion.com/2016/05/10/dinosaur-blood-and-the-age-of-the -earth-by-fazale-rana-a-review/

4-34b http://www.talkorigins.org/faqs/dinosaur/blood.html

4-34c http://www.smithsonianmag.com/science-nature/dinosaur-shocker-115306469 /?no-ist

4-35 Denis O. Lamoureux, I Love Jesus & I Accept Evolution, 2009. WIPF Stock. Eugene. Oregon. Pages 100,102

4-36 Denis O. Lamoureux, I Love Jesus & I Accept Evolution, 2009. WIPF Stock. Eugene. Oregon. P. 88

4-37 Ridley, Mark. 1993. Evolution. Blackwell Scientific Publications. Boston. P. 56

4-38 Dr. Alan Hayward, Physicist. Copyright 1985. Publication Date 2005. Creation and Evolution—Rethinking the Evidence From Science and the Bible. WIPF & Stock Publishers. Eugene, Oregon. P. 132

4-39 Huse, S. M. 1983. The Collapse of Evolution. Grand Rapids (MI): Baker Book House. Pp. 46-51

4-40 http://www.wiebefamily.org/e.htm

4-41 Evolution Disproved Encyclopedia—Volume 2
http://evolutionfacts.com/ Ev-V2/2evlch19a.htm

4-42 http://www.answersingenesis.org/articles/cfl/how-fast

4-43 Huse, S. M. 1983. The Collapse of Evolution. Grand Rapids (MI): Baker Book House. Pp. 46-51

4-44 The Plain Truth, April-May 1970, "Prehistorians Puzzle Over Worldwide Mammal Massacre," Paul W. Kroll, pp.17-19, quoting Frank C. Hibben, The Lost Americans, New York: Apollo Editions, 1961, pp. 90, 91, 97, 170 http://www.cog-ff.com/Library/html/worldwide_mammal_massacre. html Briski, DeeDee, Impressive Deception – Creation or Evolution – You Decide, 2004, Ragged Edge Press, Shippensburg, Pennsylvania, Printed by Beidel Printing House, Inc., 63 West Burd Street, Shippensburg, Pennsylvania, 17257-0708, p. 112

4-44a Ham, Ken. The New Answers Book. Master Books. 2007. pp. 216-217

4-45 The Plain Truth, January 1970, "The Day the Dinosaurs Died," Paul Kroll, p. 22, quoting Dinosaurs, Edwin H. Colbert, p. 249

4-46 Evolution Disproved Encyclopedia—Volume 2
http://evolutionfacts.com/ Ev-V2/2evlch17d.htm

4-47 Colbert, E. H. Dinosaurs. Pp. 203-204, 216-217, 249-251, 254-256. In: Kroll, P. The Plain Truth, January 1970, "The Day the Dinosaurs Died" Pp. 22-25, 27-28

4-48 http://www.enchantedlearning.com/subjects/dinosaurs/extinction/Asteroid.html http://www.aolnews.com/science/article/scientists-reaffirm-asteroidtheory -in-dinosaur-deaths/19383600

4-49 J.M. Good, T.E. White, and G.F. Stucker, "The Dinosaur Quarry," U. S. Government Printing Office, (1958), p. 26. In: Evolution Disproved Encyclopedia—Volume 2
http://evolutionfacts.com/Ev-V2/2evlch19b.htm

4-50 http://www.answersingenesis.org/articles/1999/11/05/dinosaurs-and-the-bible

4-51 American Humanist Association
http://www.humanistsofutah.org/1992/ artaug92.html

4-52 Dr. Alan Hayward, Physicist. Copyright 1985. Publication Date 2005. Creation and Evolution—Rethinking the Evidence From Science and the Bible. WIPF & Stock Publishers. Eugene, Oregon. P. 133

4-53 Parker, Gary. Creation Facts of Life—How Real Science Reveals the Hand of God. Master Books. 2006. P. 224-225

4-54 Parker, Gary. Creation Facts of Life—How Real Science Reveals the Hand of God. Master Books. 2006. P. 214-215

4-55 http://www.godandscience.org/youngearth/progressive.html
http://www.humanistsofutah.org/1992/artaug92.html

4-56 Turner, Carl R. M.D. Discovering God and His Creation—Evolution as Part of God's Plan. 2008. iUniverse, Inc. New York. p. 20

4-57 http://www.godandscience.org/youngearth/progressive.html
http://www.humanistsofutah.org/1992/artaug92.html

4-58 Del Ratzsch. 1996. The Battle of Beginnings—Why Neither Side is Winning the Creation-Evolution Debate. InterVarsity Press. Downers Grove, Illinois, p. 99

4-59 The New Answers Book, p. 144-145

4-60 Del Ratzsch. 1996. The Battle of Beginnings—Why Neither Side is Winning the Creation-Evolution Debate. InterVarsity Press. Downers Grove, Illinois, p. 100

4-61 John C. Whitcomb, World that Perished (1988), p. 27. In: Evolution Disproved Encyclopedia—Volume 3
http://evolutionfacts.com/Ev-V3/3evlch27.htm

4-62 K.O. Emery, "Continental Shelves," Scientific American, September 1969, pp. 252-255. In: Evolution Disproved Encyclopedia—Volume 3
http://evolutionfacts.com/Ev-V3/3evlch27.htm

4-63 National Geographic map of the Pacific Ocean Floor, October 1969. H. M. Morris, et. al., Science and Creation (1971), p. 52 In: Evolution Disproved Encyclopedia—Volume 3
http://evolutionfacts.com/Ev-V3/3evlch27.htm

Chapter 5—The Origin of Life on Earth

5-1 Genesis 1:11-12, 20, 24, The Bible, Scripture taken from the New King James Version. Copyright © 1982 by Thomas Nelson, Inc. Used by permission. All rights reserved.

5-2   http://www.btwol.com/07_Articles/List%20of%20articles/Humans%20 -Image% 20of%20God%20or%20advanced%20apes.htm

5-2a http://utahscience.oremjr.alpine.k12.ut.us/sciber00/7th/classify/living/2.htm

5-2b http://www.biology-online.org/dictionary/Living_thing

5-2c https://en.wikipedia.org/wiki/Organism

5-2d Futuyma. Douglas J. Science on Trial – The Case for Evolution. 1982 & 1995. Sinauer Associates, Inc. Publishers Sunderland, MA. p. 94

5-3   http://www.scienceclarified.com/dispute/Vol-1/Did-life-on-Earth-begin-in-thelittlewarm-pond.html

5-4 http://schools-wikipedia.org/wp/e/Evolution.htm

5-5 How Life Began: New Research Suggests Simple Approach, By M. Schirber, Special to LiveScience, posted: 09 June 2006 http://www.livescience.com/ animals/060609_life_origin.html

http://www.livescience.com/10531-life-began-research-suggestssimple approach.html

5-5a Futuyma. Douglas J. Science on Trial – The Case for Evolution. 1982 & 1995. Sinauer Associates, Inc. Publishers Sunderland, MA. pp. 94-97

5-6   http://www.scienceclarified.com/dispute/Vol-1/Did-life-on-Earth-begin-in-thelittlewarm-pond.html

5-7 Haarsma, Loren, Faith Alive Christian Resources, 2850 Kalamazoo Avenue, SE, Grand Rapids, Michigan, 49560 www.faithaliveresources.org/origins

5-8   http://www.scienceclarified.com/dispute/Vol-1/Did-life-on-Earth-begin-inthelittlewarm-pond.html

5-8a https://en.wikipedia.org/wiki/Isua_Greenstone_Belt

5-8b Rana, Fazale & Ross, Hugh. Origins of Life - Biblical and Evolutionary Models Face off. Navpress. Colorado Springs. 2004. Reasons to Believe. Pp. 103-104

5-9   http://www.scienceclarified.com/dispute/Vol-1/Did-life-on-Earth-begin-in-thelittlewarm-pond.html

5-10 http://en.wikipedia.org/wiki/Cell_(biology

5-11 Haarsma, L. Faith Alive Christian Resources, 2850 Kalamazoo Avenue, SE, Grand Rapids, Michigan, 49560 www.faithaliveresources.org/origins

5-12 From the Beginning—The Story of Human Evolution. David Peters. 1991. Morrow Junior Books. New York. pp. 13-25.
http://www.davidpetersstudio. com/FromTheBeginning.pdf

5-13 From the Beginning—The Story of Human Evolution. David Peters. 1991. Morrow Junior Books. New York. pp. 13-25.
http://www.davidpetersstudio. com/FromTheBeginning.pdf

5-13a The Case for Faith, Lee Strobel, p. 97

5-14 http://www.chem.duke.edu/~jds/cruise_chem/Exobiology/miller.html

5-15 From the Beginning—The Story of Human Evolution. David Peters. 1991. Morrow Junior Books. New York. p.13.
http://www.davidpetersstudio.com/ FromTheBeginning.pdf

5-15a Rana, Fazale & Ross, Hugh. Origins of Life - Biblical and Evolutionary Models Face off. Navpress. Colorado Springs. 2004. Reasons to Believe. Pp. 95-96

5-15b Rana, Fazale & Ross, Hugh. Origins of Life - Biblical and Evolutionary Models Face off. Navpress. Colorado Springs. 2004. Reasons to Believe. P. 130

5-16 http://www.scienceclarified.com/dispute/Vol-1/Did-life-on-Earth-begin-inthelittlewarm-pond.html

5-17 How Life Began: New Research Suggests Simple Approach, By M. Schirber, Special to LiveScience, posted: 09 June 2006 http://www.livescience.com/animals/060609_life_origin.html http://www.livescience.com/10531-life-began-research-suggestssimpleapproach.html

5-18 http://www.chem.duke.edu/~jds/cruise_chem/Exobiology/sites.html

5-18a https://en.wikipedia.org/wiki/Dipole

5-18b https://www.researchgate.net/publication/232054835_Electromagnetic_Origin_ of_Life
https://www.academia.edu/30042028/Electromagnetic_Origin_of_Life?email_work_card=interaction_paper

5-19 http://www.chem.duke.edu/~jds/cruise_chem/Exobiology/sites.html

5-20 http://www.scienceclarified.com/dispute/Vol-1/Did-life-on-Earth-begin-inthelittle-warm-pond.html

5-21 Haarsma, L. Faith Alive Christian Resources, 2850 Kalamazoo Avenue, SE, Grand Rapids, Michigan, 49560 www.faithaliveresources.org/origins

5-22 http://www.scienceclarified.com/dispute/Vol-1/Did-life-on-Earth-begin-in-thelittlewarm-pond.html

5-23 From the Beginning—The Story of Human Evolution. David Peters. 1991. Morrow Junior Books. New York. pp. 13-25.
http://www.davidpetersstudio. com/FromTheBeginning.pdf

5-23a George Wald. Scientific American, August, 1954
    https://www.conservapedia.com/index.php?title=George_Wald

5-24 http://www.wiebefamily.org/e.htm

5-24a   http://www.answers.com/Q/What_are_the_differences_between_
    archaebacteria_ and_eubacteria
    http://www.differencebetween.com/difference-between-
    eubacteriaand-vs-archaebacteria/

5-24b https://en.wikipedia.org/wiki/Mesophile.

5-24c Rana, Fazale & Ross, Hugh. Origins of Life - Biblical and Evolutionary
    Models Face off. Navpress. Colorado Springs. 2004. Reasons to Believe.
    pp. 171-180

5-24d https://www.sciencedaily.com/releases/2015/04/150420154823.htm

5-24e Rana, Fazale & Ross, Hugh. Origins of Life - Biblical and Evolutionary
    Models Face off. Navpress. Colorado Springs. 2004. Reasons to Believe.
    pp. 76-78

5-25 E. C. Ashby, Ph.D. 2005. Understanding the Creation Evolution Controversy.
    ACW Press. Ozark. P. 73

5-26 From the Beginning—The Story of Human Evolution. David Peters. 1991.
    Morrow Junior Books. New York. pp. 13-25.
    http://www.davidpetersstudio.com/FromTheBeginning.pdf

5-26a The Case for Faith, Lee Strobel, p. 97

5-27 http://mediatheek.thinkquest.nl/~ll125/en/life-2.htm

5-28 How Life Began: New Research Suggests Simple Approach, By M. Schirber,
    Special to LiveScience, posted: 09 June 2006
    http://www.livescience.com/animals/060609_life_origin.html
    http://www.livescience.com/10531-life-began-research-suggestssimple
    approach.html

5-29 http://www.astrobio.net/exclusive/5/reflections-from-a-warm-little-pond

5-30 How Life Began: New Research Suggests Simple Approach, By M. Schirber,
    Special to LiveScience, posted: 09 June 2006
    http://www.livescience.com/animals/060609_life_origin.html
    http://www.livescience.com/10531-life-began-research-suggestssimple
    approach.html

5-31 Kiltzmiller vs. Dover Area School District PA 2005 Intelligent Design trial
    (9/26/2005-11/4/2005). Testimony of Ken Miller, expert witness for the
    plaintiffs, testifying on day 1 (9/26/2005)
    http://www.talkorigins.org/faqs/ dover/day1pm2.html

5-32 http://www.chem.duke.edu/~jds/cruise_chem/Exobiology/miller.html

5-33 Michael Denton. 1985. Evolution: A Theory in Crisis. Adler & Adler. 4550 Montgomery Avenue, Bethesda, Maryland 20814. Pp.337-338

5-34 Howard Peth, Blind Faith (1990), p. 77. In: Evolution Disproved Encyclopedia—Volume 2.
http://evolutionfacts.com/Appendix/a11.htm

5-35 Robert Shapiro, Origins, (1986) p. 207. In: Evolution Disproved Encyclopedia—Volume 2.
http://evolutionfacts.com/Appendix/a11.htm

5-36 M. Kaplan, "The Problem of Chance in Formation of Protobionts by Random Aggregation of Macromolecules," in Chemical Evolution and the Origin of Life, (1971), Vol. 1, p. 319. In: Evolution Disproved Encyclopedia—Volume 2.
http://evolutionfacts.com/Appendix/a11.htm

5-37 John N. Move, Creation Research Society Quarterly, September 1990, p. 78. (Quotation from F.J. Ayala and J. W. Valentine, Evolving, the Theory and Processes of Organic Evolution [1979], p. 339.) In: Evolution Disproved Encyclopedia—Volume 2
http://evolutionfacts.com/Appendix/a10b.htm

5-38 D. and IC Rodabaugh, "Book Review," Creation Research Society Quarterly, December 1990, p. 108. In: Evolution Disproved Encyclopedia—Volume 2
http://evolutionfacts.com/Appendix/a10b.htm

5-39 Walter T. Brown, In the Beginning (1989), p. 4. In: Evolution Disproved Encyclopedia—Volume 2.
http://evolutionfacts.com/Appendix/a11.htm

5-40 Johnson, P. E. 1991. Darwin on Trial. InterVarsity Press. Downers Grove, Illinois 60515. p. 105-106

5-40a Dr. Jeffrey P. Tomkins. 2012. The Design and Complexity of the Cell. Institute for Creation Research. Dallas, Texas. Pp. 13-15.

5-41 http://en.wikipedia.org/wiki/Cell (biology)

5-41a Rana, Fazale. The Cell's Design - How Chemistry Reveals the Creator's Artistry. BakerBooks. Grand Rapids, Michigan. 2008. Pp. 36, 40-41

5-42 http://en.wikipedia.org/wiki/Nucleolus

5-42a Rana, Fazale. The Cell's Design - How Chemistry Reveals the Creator's Artistry. BakerBooks. Grand Rapids, Michigan. 2008. P. 39

5-42b https://en.wikipedia.org/wiki/Cell_membrane

5-42c Ross, Hugh. The Genesis Question - Scientific Advances and the Accuracy of Genesis. (Second Expanded Edition). 2001. Navpress. Colorado. Pages 143-157. Rana, Fazale. The Cell's Design - How Chemistry Reveals the Creator's Artistry. BakerBooks. Grand Rapids, Michigan. 2008. Pp. 45-48

5-43 E. C. Ashby, Ph.D. 2005. Understanding the Creation Evolution Controversy. ACW Press. Ozark. P. 73

5-44 http://www.thefreedictionary.com/Mitosis
      http://www.thefreedictionary.com/Mitosis
      http://www.thefreedictionary.com/meiosis
      http://dictionary.reference.com/browse/conjugation?s=t&path=/
      http://www.diffen.com/difference/Diploid_vs_Haploid

5-44a http://www.thefreedictionary.com/chromosome

5-45 http://staff.jccc.net/pdecell/celldivision/prokaryotes.html

5-46 Evolution Disproved Encyclopedia—Volume 2
      http://evolutionfacts.com/Ev-V2 /2evlch15.htm

5-46a https://www.bing.com/search?q=chromosomes+definition&form=EDGSPH&
      mkt=en-us&httpsmsn=1&refig=04575bfbefc34fbd8e4c6337d8df40f1
      &sp= 1&qs=HS&pq=chromosome&sc=8-10&cvid=04575bfbefc34fbd8
      e4c6337 d8df40f1&cc=US&setlang=en-US

5-46b https://www.vocabulary.com/dictionary/chromosome

5-46c https://www.collinsdictionary.com/dictionary/english/chromosome

5-46d Rana, Fazale. The Cell's Design - How Chemistry Reveals the Creator's Artistry. BakerBooks. Grand Rapids, Michigan. 2008. P. 39

5-46e Rana, Fazale. The Cell's Design - How Chemistry Reveals the Creator's Artistry. BakerBooks. Grand Rapids, Michigan. 2008. Pp. 49-50

5-46f Rana, Fazale. The Cell's Design - How Chemistry Reveals the Creator's Artistry. BakerBooks. Grand Rapids, Michigan. 2008. Pp. 143-144

5-46g https://ghr.nlm.nih.gov/primer/basics/chromosome

5-46h https://www.yourgenome.org/facts/what-is-a-gene

5-46j https://ghr.nlm.nih.gov/primer/basics/gene

5-46k https://www.quora.com/Where-is-DNA-found-in-a-Cell-one-word-answer

5-46l http://www.phschool.com/science/biology_place/biocoach/transcription/ difgns. html

5-47 http://72.30.186.176/search/srpcache?ei=UTF-8&p=do+prokaryotic+cells+have +dna&vm=r&fr=yfp-t-900-s&u=
      http://cc.bingj.com/cache.aspx?q=do+prok aryotic+cells+have+dna&d=4 806602123379454&mkt=en-US&setlang=enUS&w=NOWUN29QX_ mJyhpLw5bTzZ-62JKYs9vO&icp=1&.   intl=us&sig=UqwOwL. w4PcplLmoLVEM2w—-—-
      http://72.30.186.176/ search/srpcache?ei=UTF-8&p=where+is+The+DNA +in+a+eukaryotic+cell+l ocated&vm=r&fr=yfp-t-900-s&u=

h t t p : / / c c . b i n g j . c o m / c a c h e . a s p x ? q = w h e r e +
is+The+DNA+in+a+eukaryotic+cell+located&d=5031314771873063&mkt
=en-US&setlang=en-US&w=qnIynJWT7P7vntgPBUCKQYcb7wWC
4lo& icp=1&.intl=us&sig=OxRQ4L6rGpbIxUFfG74QSg—-—-
http://72.30.186.176/search/srpcache?ei=UTF-8&p=is+DNA+stored+in
+chromosomes&vm=r&fr=yfp-t-900-s&u=
http://cc.bingj.com/cache.as px?q=is+DNA+stored+in+chromosomes&d=
4945724669692229&mkt=   en-US&setlang=en-US&w=Gmochq5
JPNvJDn8arCfar5NiQz7l6bi9&i   cp=1&.intl=us&sig=AXNWq8_
sAr41ieUIRqyVaQ—

5-48 Scott, E.C. 2009. Evolution vs Creationism—An Introduction. Berkeley
(CA):University of California Press, p. 29

5-49 http://en.wikipedia.org/wiki/Protein

5-50 http://christiananswers.net/q-eden/origin-of-life.html

5-51 http://biology.about.com/od/molecularbiology/a/aa101904a.htm

5-52 http://en.wikipedia.org/wiki/Cell_(biology)
http://www.lpscience.fatcow.com/jwanamaker/animations/Protein%20
Synthesis%20-%20long.html

5-53 http://en.wikipedia.org/wiki/Ribosome

5-54 http://en.wikipedia.org/wiki/Metabolism

5-55 http://www.chem4kids.com/files/bio_enzymes.html

5-56 http://en.wikipedia.org/wiki/Metabolism

5-57 http://en.wikipedia.org/wiki/Protein

5-57a https://en.wikipedia.org/wiki/Amino_acid

5-57b https://medlineplus.gov/ency/article/002222.htm
https://www.familyeducation.com/life/protein/what-are-proteins-made

5-58 http://users.rcn.com/jkimball.ma.ultranet/BiologyPages/P/ProteinKinesis.html

5-59 http://biology.about.com/od/molecularbiology/a/aa101904a.htm

5-60 Strobel, Lee, 2004, The Case for a Creator, Zondervan, Grand Rapids. MI
pp. 277-285

5-61 http://en.wikipedia.org/wiki/Cell_(biology)
http://www.lpscience.fatcow.com/ jwanamaker/animations/Protein%20
Synthesis%20-%20long.html

5-62 http://en.wikipedia.org/wiki/Cell_(biology)
http://www.lpscience.fatcow.com/jwanamaker/animations/Protein%20
Synthesis%20-%20long.html

5-62a Rana, Fazale. The Cell's Design - How Chemistry Reveals the Creator's
Artistry. BakerBooks. Grand Rapids, Michigan. 2008. Pp. 98-108

5-62b https://en.wikipedia.org/wiki/Exon

5-62c https://en.wikipedia.org/wiki/Coiled_coil

5-62d http://science.psu.edu/news-and-events/2000-news/Banavar7-2000.htm

5-63 http://encyclopedia.thefreedictionary.com/topoisomerase

5-64 http://www.biology-online.org/dictionary/Okazaki_fragment

5-65 http://en.wikipedia.org/wiki/Okazaki_fragment

5-66 Michael Denton. 1985. Evolution: A Theory in Crisis. Adler & Adler. 4550 Montgomery Avenue, Bethesda, Maryland 20814. Pp. 240-242

5-67 Michael Denton. 1985. Evolution: A Theory in Crisis. Adler & Adler. 4550 Montgomery Avenue, Bethesda, Maryland 20814. pp. 242-243

5-67a http://www.phschool.com/science/biology_place/biocoach/transcription/tctlpreu.html

5-68 http://en.wikipedia.org/wiki/Ribosome

5-69 Parker, Gary, Creation Facts of Life—How Real Science Reveals the Hand of God. 1980, Master Books. Green Forest. AR. P. 31-32

5-70 Parker, G. 1980. Creation Facts of Life—How Real Science Reveals the Hand of God. Green Forest (AR): Master Books. P. 23

5-71 Parker, Gary, Creation Facts of Life,—How Real Science Reveals the Hand of God.1980, Master Books. Green Forest. AR. P. 29

5-71-1 Go to the web link below and then click on "How Do You Read the Codon Table?"

https://www.khanacademy.org/science/biology/gene-expression-central-dogma/ central-dogma-transcription/a/the-genetic-code-discovery-and-properties

5-71a Rana, Fazale. The Cell's Design - How Chemistry Reveals the Creator's Artistry. BakerBooks. Grand Rapids, Michigan. 2008. Pp. 189-191

5-71b http://www.dictionary.com/browse/codon

http://www.dictionary.com/browse/nucleotide?s=t

5-72 http://www.biologyreference.com/Re-Se/Ribosome.html

5-73 Parker, Gary, Creation Facts of Life—How Real Science Reveals the Hand of God. 1980, Master Books, Green Forest. AR. P. 31-32

5-74 http://www.biologyreference.com/Re-Se/Ribosome.html

5-75 Evolution Disproved Encyclopedia—Volume 2

http://evolutionfacts.com/ Ev-V2/2evlch11.htm

5-76 Parker, Gary, Creation Facts of Life—How Real Science Reveals the Hand of God. 1980, Master Books. Green Forest. AR. Pp. 31-34

5-77 Parker, Gary, Creation Facts of Life—How Real Science Reveals the Hand of God. 1980. Master Books. Green Forest. AR. Pp. 32-33

5-77a To view how a cell produces a protein on your web browser, enter the following web address:

https://www.bing.com/videos/search?q=how+a+cell+makes+pr oteins&&vie w=detail&mid=73866D4C2F5E1876FFEC73866D4C2F5E 1876FFEC& &FORM=VRDGAR

Then select YouTube "How are Proteins made in our cells?" Timothy Sorensen. 1/21/2016. View the video at the 3:10 – 6:46 minute mark.

5-78 http://www.chemguide.co.uk/organicprops/aminoacids/dna5.html#top

5-78a https://www.youtube.com/watch?v=NDIJexTT9j0

5-79 http://users.rcn.com/jkimball.ma.ultranet/BiologyPages/P/ProteinKinesis.html

5-80 http://wiki.answers.com/Q/After_protiens_are_made_by_the_ribosomes_ how_ do_they_leave_the_cell

5-81 G.R. Taylor, Great Evolution Mystery (1983), p. 201. In: Evolution Disproved Encyclopedia—Volume 02

http://evolutionfacts.com/Ev-V2/2evlch10b.htm

5-82 Parker, G. 1980. Creation Facts of Life—How Real Science Reveals the Hand of God. Green Forest (AR): Master Books. Green Forest. AR. Pp. 32

5-83 Parker, G. 1980. Creation Facts of Life—How Real Science Reveals the Hand of God. Green Forest (AR): Master Books. Green Forest. AR. Pp. 32-33

5-84 Parker, G. 1980. Creation Facts of Life—How Real Science Reveals the Hand of God. Green Forest (AR): Master Books. Green Forest. AR. Pp 33-34

5-85 Parker, G. 1980. Creation Facts of Life—How Real Science Reveals the Hand of God. Green Forest (AR): Master Books. Green Forest. AR. P. 33

5-86 Michael Denton. 1985. Evolution: A Theory in Crisis. Adler & Adler. 4550 Montgomery Avenue, Bethesda, Maryland 20814. Pp. 243-244

5-86a https://study.com/academy/lesson/what-is-the-genetic-code-that-translates-rnainto-amino-acids.html

5-87 Michael Denton. 1985. Evolution: A Theory in Crisis. Adler & Adler. 4550 Montgomery Avenue, Bethesda, Maryland 20814. P. 245

5-88 Michael Denton. 1985. Evolution: A Theory in Crisis. Adler & Adler. 4550 Montgomery Avenue, Bethesda, Maryland 20814. Pp.337-338

5-89 Michael Denton. 1985. Evolution: A Theory in Crisis. Adler & Adler. 4550 Montgomery Avenue, Bethesda, Maryland 20814. P. 239

5-90 Michael Denton. 1985. Evolution: A Theory in Crisis. Adler & Adler. 4550 Montgomery Avenue, Bethesda, Maryland 20814. Pp. 238-239

5-90-1 https://www.nature.com/scitable/definition/replication-33

5-90a Rana, Fazale. The Cell's Design - How Chemistry Reveals the Creator's Artistry. BakerBooks. Grand Rapids, Michigan. 2008. Pp. 217-224

5-91 Scott, E.C. 2009. Evolution vs Creationism—An Introduction. Berkeley (CA): University of California Press. P. 239-240
http://en.wikipedia.org/wiki/ Evolution

5-92 Manyuan Long, Esther Betran, Kevin Thornton and Wen Wang. The Origin of New Genes: Glimpses From the Young and Old. Nature Review November 2003. pp. 865-875
http://medicine.tums.ac.ir/FA/Us e r s/ Javad_TavakoliBazzaz/Genetic%20 Changes/Genetic%20Change/The%20   origin%20of%20new%20 genes.pdf

5-93 Summation of a portion of testimony during Kiltzmiller vs. Dover Area School District PA 2005 Intelligent Design trial.
http://www.talkorigins.org/faqs/ dover/day1pm2.html

5-94 Summation of a portion of testimony during Kiltzmiller vs. Dover Area School District PA 2005 Intelligent Design trial.
http://www.talkorigins.org/faqs/ dover/day1pm.html

5-95 Summation of a portion of testimony during Kiltzmiller vs. Dover Area School District PA 2005 Intelligent Design trial.
http://www.talkorigins.org/faqs/ dover/day11am2.html

5-96 Summation of a portion of testimony during Kiltzmiller vs. Dover Area School District PA 2005 Intelligent Design trial.
http://www.talkorigins.org/faqs/ dover/day11am2.html"

5-96a John Maynard Smith. Article in Nature magazine. 1970. "Natural Selection and the Concept of a Protein Space".
http://www.talkorigins.org/faqs/dover/day11am2.html

5-96b An excerpt from an article by Alan Orr, an evolutionary biologist at the University of Rochester.
http://www.talkorigins.org/faqs/dover/day11am2.html

5-97 Evolution Disproved Encyclopedia—Volume 02
http://evolutionfacts.com/ Ev-V2/2evlch10a.htm

5-97a Rana, Fazale & Ross, Hugh. Origins of Life - Biblical and Evolutionary Models Face off. Navpress. Colorado Springs. 2004. Reasons to Believe. P. 95

5-98 Evolution Disproved Encyclopedia—Volume 02
http://evolutionfacts.com/ Ev-V2/2evlch10a.htm

5-99 G.R. Taylor, Great Evolution Mystery (1983), pp. 165166. In: Evolution Disproved Encyclopedia—Volume 02
http://evolutionfacts.com/Ev-V2/2evlch10a.htm

5-100 Michael Pitman, Adam and Evolution (1984), p. 124. In: Evolution Disproved Encyclopedia—Volume 02
http://evolutionfacts.com/Ev-V2/2evlch10a.htm

5-100a https://www.quora.com/Can-or-Could-RNA-self-replicate

5-101 Strobel, L. 2004. The Case for a Creator. Grand Rapids (MI): Zondervan. Pp. 285-292

5-102 http://www.icr.org/index.php?module=articles&action=view&ID=105

5-102a Rana, Fazale. The Cell's Design - How Chemistry Reveals the Creator's Artistry. BakerBooks. Grand Rapids, Michigan. 2008. Pp. 158-162

5-102b Rana, Fazale. The Cell's Design - How Chemistry Reveals the Creator's Artistry. BakerBooks. Grand Rapids, Michigan. 2008. Pp. 191-201

5-102c https://en.wikipedia.org/wiki/Oligosaccharide

5-103 http://www.thefreedictionary.com/viruses

5-104 http://en.wikipedia.org/wiki/Introduction_to_viruses

5-105 Parker, Gary. 1980. Creation Facts of Life—How Real Science Reveals the hand of God. Green Forest (AR): Master Books. Pp. 139-141

5-105-1 Dr. Jeffrey P. Tomkins. 2012. The Design and Complexity of the Cell. Institute for Creation Research. Dallas, Texas. Pp. 88-91.

5-105a "If God is Good, Why the Coronavirus?" Dr. Anjeanette Roberts, Research Scholar at Reasons to Believe, (www.reasons.org), in an interview with Ryan Pauly, Coffee House Questions (www.CoffeehouseQuestions.com). March 2020.
https://www.youtube.com/watch?v=IJM2yWrAGr0&feature=youtu.be

5-105b https://listverse.com/2019/01/23/10-deadly-viruses-and-bacteria-created-inlabs/

5-105c https://www.webmd.com/lung/coronavirus-strains#1

5-106 Evolution Disproved Encyclopedia—Volume 02
http://evolutionfacts.com/ Ev-V2/2evlch10a.htm

5-107 Evolution Disproved Encyclopedia—Volume 2
http://evolutionfacts.com/ Ev-V2/2evlch13.htm

5-108 Evolution Disproved Encyclopedia—Volume 02
http://evolutionfacts.com/ Ev-V2/2evlch10a.htm

5-109 Evolution Disproved Encyclopedia—Volume 2
http://evolutionfacts.com/ Ev-V2/2evlch13.htm

5-109a Clemens Richert, "Prebiotic Chemistry and Human Intervention," Nature Communications 9 (December 12, 2018): 5177, doi:10.1038/s41467-018=07219-5.
https://www.nature.com/articles/s41467-018-07219-5

5-110 Evolution Disproved Encyclopedia—Volume 2
        http://evolutionfacts.com/ Ev-V2/2evlch11.htm

5-111   http://www.guardian.co.uk/science/2010/may/20/craig-venter-synthetic-
         lifeform/ print
         http://www.sciencemag.org/content/329/5987/52.abstract
         http://www.sciencemag.org/content/329/5987/52.full
         http://www.usatoday.com/tech/science/discoveries/2010-05-21-genome21_
         ST_N.htm
         http://www.guardian.co.uk/science/2010/may/20/craig-venter-
         syntheticlifeform/print
         http://www.sciencemag.org/content/329/5987/52.abstract
         http://www.sciencemag.org/content/329/5987/52.full
         http://www.usatoday.com/tech/science/discoveries/2010-05-21-genome21_
         ST_N.htm
         http://creation.com/will-scientists-create-new-life-formsand-what-
         woulditprove
         http://thebibleistheotherside.wordpress.com/2010/05/26/
         did-dr-craigventer-create-a-synthetic-life-form/
         http://creation.com/synthetic-life-by-venter
         http://creation.com/creating-life-in-a-test-tube

5-112 Reasons to Believe. Do Self Replicating Protocells Undermine the Evolutionary
        Theory? Fazale Rana. November 13, 2015
        https://www.reasons.org/explore/publications/tnrtb/read/tnrtb/2015/11/13/
        do-self-replicating-protocells-undermine-the-evolutionary-theory
        To view the discussion of the production of protocells created in a lab and
        the origins of life in more detail, go to the web page:
        https://www.youtube.com/ watch?v=kOD4puFnkEM
        The 24:50 to 39:20 minute mark discusses whether or not the Japanese
        development of protocells (or any other development of synthetic cells in a
        lab) disproves the view that a Creator was necessary to produce life.

Chapter 6—How Did Life Begin on Earth?

6-0 Harold Clayton Urey https://www.azquotes.com/quote/577997

6-1 How Life Began: New Research Suggests Simple Approach, By M. Schirber,
        Special to LiveScience, posted: 09 June 2006
        http://www.livescience.com/animals/060609_life_origin.html

http://www.livescience.com/10531-life-began-research-suggestssimple approach.html

6-2   http://www.scienceclarified.com/dispute/Vol-1/Did-life-on-Earth-begin-inthelittle-warm-pond.html

6-3 http://www.astrobio.net/exclusive/5/reflections-from-a-warm-little-pond

6-4 http://www.chem.duke.edu/~jds/cruise_chem/Exobiology/miller.html

6-5 http://schools-wikipedia.org/wp/e/Evolution.htm

6-6 http://en.wikipedia.org/wiki/Great_Oxygenation_Event
    http://en.wikipedia.org/ wiki/Proterozoic

6-7 Scott, E.C. 2009. Evolution vs Creationism—An Introduction. Berkeley (CA): University of California Press, p. 29

6-8 http://www.astrobio.net/exclusive/5/reflections-from-a-warm-little-pond

6-9 http://www.chem.duke.edu/~jds/cruise_chem/Exobiology/miller.html

6-10 G. Easterbrook, "Are We Alone?" in The Atlantis, 262(2):32 (1988). In: Evolution Disproved Encyclopedia—Volume 02
     http://evolutionfacts.com/ Ev-V2/2evlch09b.htm

6-11 Parker, Gary. 1980. Creation Facts of Life—How Real Science Reveals the Hand of God. Green Forest (AR): Master Books. Pp 23-26

6-12 Huse, Scott M., The Collapse of Evolution, 1983, Baker Book House, Grand Rapids, Michigan, 49508, pp. 113-114

6-13 McCombs, Charles, Evolution Hopes You Don't Know Chemistry: The Problem with Chirality, www.icr.org

6-14 McCombs, Charles, Ph.D, Evolution Hopes You Don't Know Chemistry: The Problem with Chirality, www.icr.org

6-15 Ham, Ken, General Editor, 2006, The New Answers Book, Master Books, Master Books, Inc. P.O. 726, Green Forest, AR 72638 www.masterbooks. net, pp. 287-289

6-16 http://www.albalagh.net/kids/science/evolution.shtml

6-17 Strobel, Lee, 2004, The Case for a Creator, Zondervan, Grand Rapids, Michigan, 49530 zreview@zondervan.com, pp. 42-44

6-18 Parker, G. 1980. Creation Facts of Life—How Real Science Reveals the Hand of God. Green Forest (AR): Master Books. Pp 23-26

6-19 Evolution Disproved Encyclopedia—Volume 02
     http://evolutionfacts.com/ Ev-V2/2evlch09b.htm

6-20 Evolution Disproved Encyclopedia—Volume 02
     http://evolutionfacts.com/ Ev-V2/2evlch09b.htm

6-21 Parker, G. 1980. Creation Facts of Life—How Real Science Reveals the Hand of God. Green Forest (AR): Master Books. Pp 23-26

6-22 Parker, G. 1980. Creation Facts of Life—How Real Science Reveals the Hand of God. Green Forest (AR): Master Books. Pp 23-26

6-23 Evolution Disproved Encyclopedia—Volume 02
http://evolutionfacts.com/ Ev-V2/2evlch09b.htm

6-24 https://www.answersingenesis.org/articles/ee/origin-of-life

6-24a Futuyma. Douglas J. Science on Trial – The Case for Evolution. 1982 & 1995. Sinauer Associates, Inc. Publishers Sunderland, MA. p. 184

6-25 Evolution Exposed Chapter 5: The Origin of Life, R. Patterson, March 22, 2007, Answers in Genesis
www.answersingenesis.org/go/origin
www.answersingenesis.org/go/alien

6-26 Evolution Disproved Encyclopedia—Volume 01
http://evolutionfacts.com/ Ev-V1/1evlch01a.htm

6-27 H.F. Blum, Time's Arrow and Evolution (1968), p. 158. In: Evolution Disproved Encyclopedia—Volume 01
http://evolutionfacts.com/Ev-V1/1evlch01a.htm

6-28 David and Kenneth Rodabaugh, Creation Research Society Quarterly, December 1990, p. 107. In: Evolution Disproved Encyclopedia—Volume 01
http:// evolutionfacts.com/Ev-V1/1evlch01a.htm

6-29 http://www.thefreedictionary.com/Enzymes

6-30 Evolution Disproved Encyclopedia—Volume 01
http://evolutionfacts.com/ Ev-V1/1evlch01a.htm

6-31 A few Reasons an Evolutionary Origin of Life is Impossible,
http://www.icr.org/ article/3140/

6-32 Evolution Exposed Chapter 5: The Origin of Life, Roger Patterson, March 22, 2007, Answers in Genesis
www.answersingenesis.org/go/origin
www.answersingenesis.org/go/alien
https://www.answersingenesis.org/articles/ee/origin-of-life

6-33 E. C. Ashby, Ph.D. 2005. Understanding the Creation Evolution Controversy. ACW Press. Ozark. p. 68

6-34 E. C. Ashby, Ph.D. 2005. Understanding the Creation Evolution Controversy. ACW Press. Ozark. pp. 70-72

6-35 Francis Hitching. The Neck of the Giraffe (1982), p. 65. In: Evolution Disproved Encyclopedia—Volume 01
http://evolutionfacts.com/Ev-V1/1evlch01a.htm

6-36 Walter T. Brown, In the Beginning (1989), p. 6. In: Evolution Disproved Encyclopedia—Volume 02 http://evolutionfacts.com/Appendix/a09.htm

6-37 Parker, G. 1980. Creation Facts of Life—How Real Science Reveals the Hand of God. Green Forest (AR): Master Books. Pp 23-26

6-38 http://www.albalagh.net/kids/science/evolution.shtml

6-39 Gish, D. Dr. A Few Reasons an Evolutionary Origin of Life Is Impossible. Acts & Facts. 36 (1). January, 2007
http://www.icr.org/article/3140/

6-40 Evolution Exposed Chapter 5: The Origin of Life, R. Patterson, March 22, 2007, Answers in Genesis

6-41 A few Reasons an Evolutionary Origin of Life is Impossible
http://www.icr.org/ article/3140/

6-41a http://www.corrosionpedia.com/definition/968/reducing-atmosphere

6-41b http://www.corrosionpedia.com/definition/845/oxidizing-atmosphere

6-41c http://www.chacha.com/question/what-is-free-oxygen

6-42 Evolution Disproved Encyclopedia—Volume 02
http://evolutionfacts.com/ Ev-V2/2evlch09b.htm

6-42a Rana, Fazale & Ross, Hugh. Origins of Life - Biblical and Evolutionary Models Face off. Navpress. Colorado Springs. 2004. Reasons to Believe. Pp. 81-90

6-42b Rana, Fazale & Ross, Hugh. Origins of Life - Biblical and Evolutionary Models Face off. Navpress. Colorado Springs. 2004. Reasons to Believe. Pp. 111-113

6-42c Rana, Fazale & Ross, Hugh. Origins of Life - Biblical and Evolutionary Models Face off. Navpress. Colorado Springs. 2004. Reasons to Believe. Pp. 114-117

6-42d Rana, Fazale & Ross, Hugh. Origins of Life - Biblical and Evolutionary Models Face off. Navpress. Colorado Springs. 2004. Reasons to Believe. P. 129

6-42e https://www.bing.com/search?q=Homochirality%20describes%20a%20 geometric%20property%20of%20some%20materials%20&qs=n &form=QBRE&sp=-1&pq=undefined&sc=0-63&sk=&cvid=1580FB7476 374650B60ACBC41620BD90

6-43 http://christiananswers.net/q-eden/origin-of-life.html

6-44 http://www.livestrong.com/article/427827-what-are-amino-acids-made-of/

6-45 Evolution Disproved Encyclopedia—Volume 02
http://evolutionfacts.com/ Ev-V2/2evlch10b.htm

6-46 http://schools-wikipedia.org/wp/e/Evolution.htm

6-47 http://dictionary.reference.com/browse/chirality+

6-48 http://en.wikipedia.org/wiki/Chirality_(chemistry)

6-49 http://dictionary.reference.com/browse/enantiomer?s=ts

6-50 http://en.wikipedia.org/wiki/Amino_acid

6-51 http://en.wikipedia.org/wiki/Chirality_(chemistry)

6-52 http://en.wikipedia.org/wiki/Chirality_(chemistry)

6-53 Dean H. Kenyon, affidavit presented to U.S. Supreme Court, No. 85-1513, in "Brief of Appellants," prepared under the direction of William J. Guste, Jr., Attorney General of the State of Louisiana, October 1985, p. A-23. In: Evolution Disproved Encyclopedia—Volume 02
http://evolutionfacts.com/ Ev-V2/2evlch10b.htm

6-54 http://christiananswers.net/q-eden/origin-of-life.html

6-55 Parker, G. 1980. Creation Facts of Life. Green Forest (AR): Master Books. Pp 23-26

6-55a Rana, Fazale & Ross, Hugh. Origins of Life - Biblical and Evolutionary Models Face off. Navpress. Colorado Springs. 2004. Reasons to Believe. P. 124

6-56 http://www.icr.org/index.php?module=articles&action=view&ID=105

6-57 David and Kenneth Rodabaugh, "Book Review," Creation Research Society Quarterly, December 1990, p. 107. In: Evolution Disproved Encyclopedia—Volume 02
http://evolutionfacts.com/Ev-V2/2evlch09b.htm

6-57a Rana, Fazale & Ross, Hugh. Origins of Life - Biblical and Evolutionary Models Face off. Navpress. Colorado Springs. 2004. Reasons to Believe. P. 125

6-58 Icons of Evolution? Why Much of What Jonathan Wells Writes About Evolution is Wrong" Alan D. Gishlick—National Center for Science Education— Defending the Teaching of Evolution in Public Schools. Table by Rode, 1999; Hanic et al., 2000
http://ncse.com/creationism/analysis/icons-critique-pdf

6-59 Edwin Conklin, Reader's Digest, January 1963, p. 92. In: Evolution Disproved Encyclopedia—Volume 02
http://evolutionfacts.com/Appendix/a09.htm

6-60 Evolution from space. Chandra Wickramasinghe. The New Scientist, Jan 21, 1982 In:
http://wasdarwinwrong.com/kortho46a.htm

6-61 http://www.wiebefamily.org/e.htm

6-62 G.R. Taylor, Great Evolution Mystery (1983), p. 166. In: Evolution Disproved Encyclopedia—Volume 2
http://evolutionfacts.com/Appendix/a10b.htm

6-62a Futuyma. Douglas J. Science on Trial – The Case for Evolution. 1982 & 1995. Sinauer Associates, Inc. Publishers Sunderland, MA. pp. 133-134

6-63 Scott, E.C. 2009. Evolution vs Creationism—An Introduction. Berkeley (CA): University of California Press. P. 203

6-64 http://www.talkorigins.org/indexcc/CB/CB010.html

6-65 Professor William J. Dickensen, PhD, Professor of Biology at the University of Utah, Creation—Causes Creationists as Ignorant.doc, American Humanist Association

http://www.humanistsofutah.org/1992/artaug92.html

6-66 http://www.windowview.org/sci/pgs/06chance.html

6-67 Michael Pitman, Adam and Evolution (1984), p. 143. In: Evolution Disproved Encyclopedia—Volume 2

http://evolutionfacts.com/Appendix/a10b.htm

6-68 http://www.sdadefend.com/Family-school/new_material.htm

6-68a Futuyma. Douglas J. Science on Trial - The Case for Evolution. 1982 & 1995. Sinauer Associates, Inc. Publishers Sunderland, MA. pp. 134-135

6-68b Rana, Fazale & Ross, Hugh. Origins of Life - Biblical and Evolutionary Models Face off. Navpress. Colorado Springs. 2004. Reasons to Believe. P. 138

6-68c http://discussions.godandscience.org/viewtopic.php?t=2837

6-69 I. L. Cohen, Darwin was Wrong (1984), p. 205. In: Evolution Disproved Encyclopedia—Volume 02

http://evolutionfacts.com/Ev-V2/2evlch10a.htm

6-70 Michael Denton. 1985. Evolution: A Theory in Crisis. Adler & Adler. 4550 Montgomery Avenue, Bethesda, Maryland 20814. Pp. 308-323

6-71 Fred Hoyle and Chandra Wickramasinghe, Evolution from Space (1981), p. 28. In: Evolution Disproved Encyclopedia—Volume 02

http://evolutionfacts. com/Ev-V2/2evlch09b.htm

Chapter 7—A Look at Some of the Transitional Fossil Evidence

7-1 Jeffrey, G. R. 2003. Creation: Remarkable Evidence of God's Design. Toronoto (Ontario): Frontier Research Publishers, Inc p. 168

7-2 Huse, S. M. 1983. The Collapse of Evolution. Grand Rapids (MI): Baker Book House. P. 45

7-3 Howard, J. 2008. Disprove Darwin in 5 Minutes or Less. Delray Beach (FL): Twin Angels Publishing, Inc, p. 20

7-4 http://hubpages.com/forum/topic/30050

7-5 http://www.talkorigins.org/features/whales/

7-6 Denis O. Lenoucoux, I Love Jesus & I Accept Evolution. 2009. WIPF Stock. Eugene. Oregon. Pp. 120-121

7-7 http://www.scientificblogging.com/news_releases/latvian_fossils_close_the_ gap_ between_fish_and_land_animals_say_researchers

7-8 http://dictionary.reference.com/browse/amphibian?s=t

7-9 http://www.newworldencyclopedia.org/entry/Amphibian

7-10 Denis O. Lamoureux, 2009. I Love Jesus & I Accept Evolution. WIPF Stock. Eugene. Oregon. Pages 107-108

7-10a Futuyma. Douglas J. Science on Trial – The Case for Evolution. 1982 & 1995. Sinauer Associates, Inc. Sunderland, Massachusetts. P. 75 Figure 9

7-11 http://www.devoniantimes.org/Order/re-acanthostega.html

7-12 http://en.wikipedia.org/wiki/Acanthostega

7-13 http://evolution.berkeley.edu/evolibrary/print/printable_template.php?news_ id= 060501_tiktaalik

7-13a Forbes.com article "Four Famous Transitional Fossils That Support Evolution". Nov 17, 2015. Shaena Montanari, Ph.D.
https://www.forbes.com/sites/shaenamontanari/2015/11/17/ four-famous-transitional-fossils-that-support-evolution/#6e4512c52d8d

7-14 Mike Anderson. Is Jesus an Evolutionist? 2011. Smashwords Publishers. Nook e-book. P. 349 Prior book version: Montane Publishers, South Africa (2008) http://www.scribd.com/doc/11464783/Is-Jesus-an-Evolutionist

7-15 http://www.answersingenesis.org/creation/v15/i3/missinglink.asp

7-16 http://www.earthhistory.org.uk/technical-issues/tiktaalik-roseae/

7-17 Briski, DeeDee, Impressive Deception—Creation or Evolution—You Decide, 2004, Ragged Edge Press, Shippensburg, Pennsylvania, Printed by Beidel Printing House, Inc., 63 West Burd Street, Shippensburg, Pennsylvania, 17257-0708, p. 105

7-18 Briski, DeeDee, Impressive Deception—Creation or Evolution—You Decide, 2004, Ragged Edge Press, Shippensburg, Pennsylvania, Printed by Beidel Printing House, Inc., 63 West Burd Street, Shippensburg, Pennsylvania, 17257-0708, p. 105

7-19 Evolution Disproved Encyclopedia—Volume 2
http://evolutionfacts.com/ Ev-V2/2evlch17d.htm

7-20 Armstrong, Garner Ted, "Some Fish Stories—. . . About an Unproven Theory," Ambassador College, 1969, p. 29

7-21 Disprove Darwin in 5 Minutes or Less, Page 47

7-22 Disprove Darwin in 5 Minutes or Less, Page 49

7-23 http://www.thefreedictionary.com/operculum

7-24 Crampton, Henry Edward. The Doctrine of Evolution: Its Basis and Its Scope. 1911. Columbia University Press. Also Norwood Press. Gutenberg EBook edition produced by Audrey Longhurst. 2005 [EBook # 16442]. Location 680-708.

7-25 John D. Morris, Ph.D. 2004. When a Tadpole Turns into a Frog—Is This Evolution in Action? Acts & Facts. 33 (6)
http://www.icr.org/article/509/

7-26 http://www.pbs.org/wgbh/evolution/library/04/2/l_042_01.html

7-27 Jonathan Sarfati, Ph.D. 2002. Refuting Evolution 2. Master Books. Green Forest. AR. P. 110. Jonathan Sarfati, Ph.D. First Printing 1999. Fourth Edition 2008. Refuting Evolution. Creation Ministries International. Eight Mile Plains. Australia. Pp. 83-84. Gary Parker. Creation Facts of Life— How Real Science Reveals the Hand of God. 2006. Master Books. P. 46

7-28 Randall, quoted in William Fix, The Bone Peddlers, p. 189. In: Evolution Disproved Encyclopedia—Volume 3
http://evolutionfacts.com/Ev-V3/3evlch21.htm

7-29 http://dictionary.reference.com/browse/reptile

7-30 http://animals.about.com/od/reptiles/a/tenfactsreptiles.htm

7-31 http://en.wikipedia.org/wiki/Evolution_of_reptiles

7-32 Michael Denton. 1985. Evolution: A Theory in Crisis. Adler & Adler. 4550 Montgomery Avenue, Bethesda, Maryland 20814. Pp. 176-177

7-33 http://en.wikipedia.org/wiki/Seymouria

7-34 Michael Denton. 1985. Evolution: A Theory in Crisis. Adler & Adler. 4550 Montgomery Avenue, Bethesda, Maryland 20814. p. 218-219

7-35 Michael Denton. 1985. Evolution: A Theory in Crisis. Adler & Adler. 4550 Montgomery Avenue, Bethesda, Maryland 20814. p. 218-219

7-36 Johnson. Philip E. Darwin on Trial. 1991. InterVarsity Press. Downers Grove. Illinois. p. 75

7-37 http://www.darwinismrefuted.com/natural_history_1_10.html

7-37a Johnson. Phillip E. Darwin on Trial. 1991. InterVarsity Press. Downers Grove, Illinois. Pp. 74-75.

7-38 http://dictionary.reference.com/browse/mammal?s=t

7-39 http://www.newworldencyclopedia.org/entry/Mammal

7-40 http://www.baptistboard.com/showthread.php?t=13344

7-41 Ridley, Mark. 1993. Evolution. Blackwell Scientific Publications. Boston. Pp 533-536, 552

7-42 Lamoureux. Denis O. I Love Jesus & I Accept Evolution, 2009. WIPF Stock. Eugene. Oregon. P. 109

7-43 Lamoureux, D. O. 2009. I Love Jesus & I Accept Evolution. Eugene (OR): WIPF Stock. Pp. 109-110

7-43a https://pediaa.com/difference-between-mammals-and-reptiles/

7-44 Michael Denton. 1985. Evolution: A Theory in Crisis. Adler & Adler. 4550 Montgomery Avenue, Bethesda, Maryland 20814. Pp. pp. 180-182

7-45 Michael Denton. 1985. Evolution: A Theory in Crisis. Adler & Adler. 4550 Montgomery Avenue, Bethesda, Maryland 20814. Pp. 105-106

7-46 http://en.wikipedia.org/wiki/Reptile

7-47 http://www.thefreedictionary.com/Nephrons

7-48 Michael Denton. 1985. Evolution: A Theory in Crisis. Adler & Adler. 4550 Montgomery Avenue, Bethesda, Maryland 20814. Pp. 105-106

7-49 Evolution Disproved Encyclopedia—Volume 3
http://evolutionfacts.com/ Ev-V3/3evlch32.htm

7-50 Stephen C. Meyer, Scott Minnich, Jonathan Moneymaker, Paul A. Nelson, Ralph Seelke. 2007. Explore Evolution—The Arguments For and Against Neo-Darwinism. Hill House Publishers. Melbourne & London. Pp. 130, 132.

7-51 Johnson. Philip E. Darwin on Trial. 1991. InterVarsity Press. Downers Grove. Illinois. p. 75

7-52 Stephen C. Meyer, Scott Minnich, Jonathan Moneymaker, Paul A. Nelson, Ralph Seelke. 2007. Explore Evolution—The Arguments For and Against Neo-Darwinism. Hill House Publishers. Melbourne & London. Pp. 129-130.

7-53 Stephen C. Meyer, Scott Minnich, Jonathan Moneymaker, Paul A. Nelson, Ralph Seelke. 2007. Explore Evolution—The Arguments For and Against Neo-Darwinism. Hill House Publishers. Melbourne & London. Pp. 130, 132.

7-54 Stephen C. Meyer, Scott Minnich, Jonathan Moneymaker, Paul A. Nelson, Ralph Seelke. 2007. Explore Evolution—The Arguments For and Against Neo-Darwinism. Hill House Publishers. Melbourne & London. P. 131

7-55 http://dictionary.reference.com/browse/bird?s=t

7-56 Lamoureux, D. O. 2009. I Love Jesus & I Accept Evolution.Eugene (OR): WIPF Stock. P. 114

7-56a Debating Christianity & Religion web site Post #156 nygreenguy 12-21-2011 10:54 am

https://debatingchristianity.com/forum/viewtopic.php?t=17925&postdays=0&postorder=asc&start=0

http://debatingchristianity.com/forum/viewtopic.php?t=17925&postdays=0 &postorder=asc&start=50

https://debatingchristianity.com/forum/viewtopic.php?t=17925&postdays=0&postorder=asc&start=250

7-57 Lamoureux, D. O. 2009. I Love Jesus & I Accept Evolution. Eugene (OR): WIPF Stock. P. 116

7-58 Lamoureux, D. O. 2009. I Love Jesus & I Accept Evolution. Eugene (OR): WIPF Stock. P. 116

7-59 Lamoureux, D. O. 2009. I Love Jesus & I Accept Evolution. Eugene (OR): WIPF Stock. P. 116

7-60 http://en.wikipedia.org/wiki/Pterosaur

7-61 http://en.wikipedia.org/wiki/Pterodactyl

7-62 http://en.wikipedia.org/wiki/Pterosaur

7-63 http://en.wikipedia.org/wiki/Compsognathus

7-64 http://en.wikipedia.org/wiki/Saurischia
http://www.enchantedlearning.com/subjects/dinosaurs/dinoclassification/Saurischian.html

7-64a Wikipedia - Coelophysis

7-64b dictionary.com/browse/theropod

7-64c http://www.dinosaur-facts.com/on-the-ground/coelophysis/

7-64d Futuyma. Douglas J. Science on Trial – The Case for Evolution. 1982 & 1995. Sinauer Associates, Inc. Publishers Sunderland, MA. p. 77, Figure 11

7-64e https://en.wikipedia.org/wiki/Coelophysis_bauri

7-64f https://en.wikipedia.org/wiki/Protoavis

7-64g Acts & Facts. Institute for Creation Research. The World's Oldest Bird Fossil. Brian Thomas. October 2019. Pp 14-15.

7-65 http://www.genesispark.com/genpark/birds/birds.htm

7-65a https://answersingenesis.org/dinosaurs/feathers/feathered-raptors-not-the-birds/

7-65b https://en.wikipedia.org/wiki/Caudipteryx

7-65c http://www.rareresource.com/caudipteryx.htm

7-65d http://icb.oxfordjournals.org/content/40/4/687.full

7-65e https://en.wikipedia.org/wiki/Protopteryx

7-66 http://www-v1.amnh.org/exhibitions/fightingdinos/ex5.php#

7-67 http://www.talkorigins.org/faqs/dover/day9am2.html

7-68 http://en.wikipedia.org/wiki/Oviraptor
http://dinosaurs.about.com/od/carnivorousdinosaurs/p/oviraptor.htm
http://www.rareresource.com/oviraptor.htm

7-69 http://dinosaurs.about.com/od/carnivorousdinosaurs/p/oviraptor.htm

7-70 Dr. Alan Hayward, Physicist. Copyright 1985. Publication Date 2005. Creation and Evolution—Rethinking the Evidence From Science and the Bible. WIPF & Stock Publishers. Eugene, Oregon. p. 48

7-70a http://evolution.berkeley.edu/evolibrary/article/evograms_06

7-70b http://www.livescience.com/12808-dinosaur-hands-fingers-birds-digits-evolution. html

7-70c http://www.godandscience.org/evolution/dinobird.html

7-70d http://10e.devbio.com/article.php?ch=16&id=161

7-70e http://www.science20.com/news_articles/theropod_dinosaurs_evolved_birds_ not_likely_says_study

7-70f https://www.differencebetween.com/difference-between-reptiles-and-vs-birds/

7-70g https://www.quora.com/What-are-the-main-differences-between-birds-andreptiles

7-70h https://www.answers.com/Q/What_are_the_differences_between_birds_and_ reptiles

7-71 Richard Deem. Demise of the "Birds are Dinosaurs" Theory. 1/2/2005.
http://www.godandscience.org/evolution/dinobird.html

7-72 Ann C. Burke and Alan Feduccia. Science 24 October 1997: Vol. 278 no. 5338 pp. 666-668 DOI: 10.1126/science.278.5338.666 Developmental Patterns and the Identification of Homologies in the Avian Hand

7-73 Lamoureux, D. O. 2009. I Love Jesus & I Accept Evolution. Eugene (OR): WIPF Stock. P. 113

7-74 http://www.angelfire.com/mi/dinosaurs/dinobird.html
http://www.arrivalofthefittest.com/Are_You_Being_Brainwashed2.html

7-75 Jonathan Sarfati, Ph.D. First Printing 1999. Fourth Edition 2008. Refuting Evolution. Creation Ministries International. Eight Mile Plains. Australia. P. 66.

7-76 http://www.arrivalofthefittest.com/Are_You_Being_Brainwashed2.html

7-77 Ham, K. General Editor. 2006. The New Answers Book. Inc. Green Forest (AR): Master Books. pp. 300-303 www.masterbooks.net

7-78 Ham, K. General Editor. 2006. The New Answers Book. Inc.Green Forest (AR): Master Books. pp. 300-303

7-79 Evolution Disproved Encyclopedia—Volume 2

http://evolutionfacts.com/ Appendix/a13b.htm

7-80 Life: How Did It Get Here? (1985), pp. 76-77. In: Evolution Disproved Encyclopedia—Volume 2

http://evolutionfacts.com/Appendix/a13b.htm

7-81 Armstrong. G. T., Kroll P. W. 1968. A Theory For the Birds, Ambassador College, pp. 2-3

7-82 Dr. Alan Hayward, Physicist. Copyright 1985. Publication Date 2005. Creation and Evolution—Rethinking the Evidence From Science and the Bible. WIPF & Stock Publishers. Eugene, Oregon. pp. 47-48

7-83   http://www.enchantedlearning.com/subjects/dinosaurs/news/ Sinornithosaurus. shtml

7-84 http://en.wikipedia.org/wiki/Sinornithosaurus

7-84a http://www.godandscience.org/evolution/dinobird.html

7-85 Hanegraaff, H. The Face that Demonstrates the Farce of Evolution, 1998, Word Publishing, Nashville, TN, p. 34-38

7-86 Ham, K. General Editor. 2006. The New Answers Book. Inc. Green Forest (AR): Master Books. pp. 300-303

7-87 Hanegraaff, H. The Face that Demonstrates the Farce of Evolution, 1998, Word Publishing, Nashville, TN, p. 34-38

7-88 http://www.britannica.com/EBchecked/topic/32599/Archaeopteryx

7-89 http://en.wikipedia.org/wiki/Archaeopteryx

7-90 http://www.class.uidaho.edu/ngier/creationism.htm

7-91 http://www.pbs.org/wgbh/nova/microraptor/skel-nf.html

7-91a The study was published in the journal Nature Communications. http://www.geologyin.com/2018/03/archaeopteryx-dinosaur-hadcompletely.html

7-91b https://landbeforetime.fandom.com/wiki/Archaeopteryx

7-92 Mike Anderson. Is Jesus an Evolutionist? 2011. Smashwords Publishers. Nook e-book. P. 345-349 Prior book version: Montane Publishers, South Africa (2008)

http://www.scribd.com/doc/11464783/Is-Jesus-an-Evolutionist

7-93 Michael Denton. 1985. Evolution: A Theory in Crisis. Adler & Adler. 4550 Montgomery Avenue, Bethesda, Maryland 20814. Pp. 175-178.

7-94 Evolution Disproved Encyclopedia—Volume 3

http://evolutionfacts.com/ Ev-V3/3evlch23.htm

7-95 More information on this Texas discovery can be found in Nature, 322 (1986), p. 677. Evolution Disproved Encyclopedia—Volume 3

http://evolutionfacts.com/Ev-V3/3evlch23.htm

7-96 Evolution Disproved Encyclopedia—Volume 3
    http://evolutionfacts.com/ Ev-V3/3evlch23.htm

7-96a https://en.wikipedia.org/wiki/Specimens_of_Archaeopteryx

7-97 Hanegraaff, H. The Face that Demonstrates the Farce of Evolution, 1998, Word
    Publishing. Nashville. TN. p. 34-38

7-98 Armstrong. G. T., Kroll P. W. 1968. A Theory For the Birds, Ambassador
    College, pp. 2-3

7-99 Huse, S. M. 1983. The Collapse of Evolution. Grand Rapids (MI): Baker Book
    House. Pp. 108-113

7-100 Briski, D. 2004. Impressive Deception—Creation or Evolution You Decide.
    Shippensburg, (PA): Ragged Edge Press. P. 114

7-100a https://www.genesispark.com/exhibits/trivia/birds/ Nature, vol. 322, 1986,
    p. 677.

7-101 Parker, Gary. Creation Facts of Life—How Real Science Reveals the Hand of
    God. 1980, Master Books, P.O. Box 726, Green Forest, AR, 72638, www.
    masterbooks.net, pp. 166-171

7-102 http://www.angelfire.com/mi/dinosaurs/dinobird.html

7-103 Parker, Gary. Creation Facts of Life—How Real Science Reveals the Hand of
    God. 2006. Master Books. P. 167

7-104 Michael Denton. 1985. Evolution: A Theory in Crisis. Adler & Adler. 4550
    Montgomery Avenue, Bethesda, Maryland 20814. Pp. 175-178. Parker,
    Gary. Creation Facts of Life—How Real Science Reveals the Hand of God.
    2006. Master Books. P. 167

7-105 Hanegraaff, H. The Face that Demonstrates the Farce of Evolution, 1998,
    Word Publishing, Nashville, TN, p. 34-38

7-106 Parker, G. Creation Facts of Life—How Real Science Reveals the Hand
    of God. 1980, Master Books, P.O. Box 726, Green Forest, AR, 72638,
    www. masterbooks.net, pp. 166-171

7-107 Hanegraaff, H. The Face that Demonstrates the Farce of Evolution, 1998,
    Word Publishing, Nashville, TN, p. 34-38

7-107a The TalkOrigins Archive. All About Archaeopteryx. by Chris Nedin.
    http://www.talkorigins.org/faqs/archaeopteryx/info.html

7-108 Jonathan Sarfati, Ph.D. First Printing 1999. Fourth Edition 2008. Refuting
    Evolution. Creation Ministries International. Eight Mile Plains. Australia.
    P. 66.

7-109 http://www.pathlights.com/ce_encyclopedia/Encyclopedia/20hist09.htm
    http://www.jesus-is-savior.com/Evolution%20Hoax/Evolution/17.htm
    —Evolution Disproved Encyclopedia—Volume 3

http://evolutionfacts.com/ Ev-V3/3evlch23.htm

Supporting material for this discussion can be found in—Four issues of the British Journal of Photography (March-June 1985).—W.J. Broad, "Authenticity of Bird Fossil Is Challenged" in New York Times, May 7, 1985, pp. C1, C14;—T. Nield, "Feathers Fly Over Fossil 'Fraud'" in New Scientist 1467:49-50;—G. Vines, "Strange Case of Archaeopteryx 'Fraud'" in New Scientist 1447:3.

7-110 http://creationwiki.org/Archaeopteryx_is_a_fake_(Talk.Origins)

7-111 http://creationwiki.org/Archaeopteryx_is_a_fake_(Talk.Origins)

7-112 http://www.talkorigins.org/indexcc/CC/CC351.html

7-113 http://talkrational.org/showthread.php?s=291c40eff830c80ee2cfd4007f63165 c&t=46584&page=8

7-113a https://wyomingdinosaurcenter.org/archaeopteryx-the-thermopolis-specimen/

7-114 http://en.wikipedia.org/wiki/Aquatic_respiration

http://en.wikipedia.org/wiki/ Lung#Amphibian_lungs

http://en.wikipedia.org/wiki/Lung#Mammalian_lungs

7-115 http://ahobart.net/oldsite/biology/advanced/htmlpages/gas.htm

7-116 http://en.wikipedia.org/wiki/Origin_of_birds#cite_note-Trex-0

7-117 Michael Denton. 1985. Evolution: A Theory in Crisis. Adler & Adler. 4550 Montgomery Avenue, Bethesda, Maryland 20814. Pp. 210-212

7-118 http://www.ask.com/question/how-do-reptiles-breathe

7-119 http://wiki.answers.com/Q/How_does_a_mammal_breathe#slide2

7-120 http://www.ehow.com/about_6543197_do-birds-breathe_.html

7-121 Stephen C. Meyer, Scott Minnich, Jonathan Moneymaker, Paul A. Nelson, Ralph Seelke. 2007. Explore Evolution—The Arguments For and Against Neo-Darwinism. Hill House Publishers. Melbourne & London. Pp. 133-138.

7-122 http://www.answersingenesis.org/creation/v21/i4/design.asp

7-123 John A. Ruben, Terry D. Jones, Nicholas R. Geist, W. Jaap Hillenius. Science. November 14, 1997. Volume 278. p. 1268 In:Ken Ham. General Editor. 2006. The New Answers Book. Inc. Green Forest (AR): Master Books. p. 301 Footnote #6

7-124 Ham, K. General Editor. 2006. The New Answers Book. Inc. Green Forest (AR): Master Books. pp. 300-303

7-125 Huse, S. M. 1983. The Collapse of Evolution. Grand Rapids (MI): Baker Book House. Pp. 108-113

7-126 On the Origin of Species or the Preservation of Favoured Races in the Struggle For Life—First Edition. Charles Darwin. Partially quoted in: Evolution Disproved Encyclopedia—Volume 3
http://evolutionfacts.com/ Ev-V3/3evlch31a.htm

7-127 http://www.talkorigins.org/faqs/dover/day9pm.htm

7-127a http://evolution.berkeley.edu/evolibrary/article/evograms_03

7-128 Denis O. Lamoureux, I Love Jesus & I Accept Evolution, 2009. WIPF Stock. Eugene. Oregon. P. 120

7-128a Forbes.com article "Four Famous Transitional Fossils That Support Evolution". Nov 17, 2015. Shaena Montanari, Ph.D.
https://www.forbes.com/sites/shaenamontanari/2015/11/17/four-famoustransitional-fossils-that-support-evolution/#6e4512c52d8d

7-129 Denis O. Lamoureux, I Love Jesus & I Accept Evolution, 2009. WIPF Stock. Eugene. Oregon. P. 120

7-129a http://ifsa.my/articles/comparative-anatomy-comparing-contrasting-whales

7-130 Mike Anderson. Is Jesus an Evolutionist? 2011. Smashwords Publishers. Nook e-book. P. 349 Prior book version: Montane Publishers, South Africa (2008)
http://www.scribd.com/doc/11464783/Is-Jesus-an-Evolutionist

7-130a Futuyma. Douglas J. Science on Trial – The Case for Evolution. 1982 & 1995. Sinauer Associates, Inc. Publishers Sunderland, MA. pp. 62, 64

7-131 http://www.indiana.edu/~ensiweb/lessons/wh.or.11.pdf

7-132 http://www.talkorigins.org/faqs/dover/day9pm.htm

7-133 http://www.indiana.edu/~ensiweb/lessons/wh.or.11.pdf

7-134 http://www.talkorigins.org/faqs/dover/day9pm.htm

7-135 http://www.indiana.edu/~ensiweb/lessons/wh.or.11.pdf

7-136 Michael Denton. 1985. Evolution: A Theory in Crisis. Adler & Adler. 4550 Montgomery Avenue, Bethesda, Maryland 20814. Pp. 172, 174

7-137 Francis Hitching. 1982. The Neck of the Giraffe. Pan, London & Sydney. In: Dr. Alan Hayward, Physicist. Copyright 1985. Publication Date 2005. Creation and Evolution—Rethinking the Evidence From Science and the Bible. WIPF & Stock Publishers. Eugene, Oregon. Pp. 44-45
http://www. talkorigins.org/features/whales/

7-138 http://www.philvaz.com/apologetics/p15.htm
http://www.google.com/ imgres?imgurl=
http://www. p h i l v a z . c o m / a p o l o g e t i c s / FishToTe t r a .jpg&imgrefurl=
http://www.philvaz.com/apologetics/p15.htm&usg=__ v061YbaozQ-W_
q R C n Y R C E T j 7 j L c = & h = 3 5 5 & w = 4 0 0 & s z = 1 7 & h l = e n &
start=13&zoom=1&tbnid=1DFIIjydsj0TsM:&tbnh=110&tbnw=124&ei

= n - s F T 8 D S A - X Y 0 Q G d o u T K A g & p r e v = /
i m a g e s % 3 F q % 3 D c h a r t % 2 B i l l u s t r
ating%2Bevolution%2Bof%2Bfish%2Bto%2Bamphibian%26um%3D1%
26hl%3Den%26safe%3Dactive%26sa%3DG%26gbv%3D2%26tbm%3D
isch%26prmd%3Divnsb&um=1&itbs=1Source: See Evidence for Evolution
and an Old Earth.doc

7-139 http://www.darwinismrefuted.com/natural_history_2_15.html

7-140 John C. Whitcomb, Early Earth (1988), p. 84. In: Evolution Disproved
Encyclopedia—Volume 3
http://evolutionfacts.com/Appendix/a22.htm

7-141 http://www.darwinismrefuted.com/natural_history_2_15.html

7-142 http://www.darwinismrefuted.com/natural_history_2_15.html

7-143 Jonathan Sarfati, Ph.D. First Printing 1999. Fourth Edition 2008. Refuting
Evolution. Creation Ministries International. Eight Mile Plains. Australia.
Pp. 73-74.

7-144 From "Ten Reasons Why Evolution is Wrong"
http://www.evanwiggs.com/ articles/reasons.html#Reason1

7-145 http://www.evolutionnews.org/2009/04/_why_evolution_is_false019871.html

7-145a The TalkOrigins Archive. Review of Michael Denton's "Evolution – A Theory
in Crisis" 1996-1997. Mark I. Vuletic
http://talkorigins.org/faqs/denton.html

7-145b Creation.com article: "Ancient mutant Jamaican sea cows?" Andrew Lamb.
June 7, 2008.
https://creation.com/ancient-mutant-jamaican-sea-cows

7-146 Evolution Disproved Encyclopedia—Volume 3
http://evolutionfacts.com/ Ev-V3/3evlch32.htm

7-146a Ross, Hugh. The Genesis Question - Scientific Advances and the Accuracy
of Genesis. (Second Expanded Edition). 2001. Navpress. Colorado. Pages
50-53.

7-147 http://www.newscientist.com/article/dn19135-zoologger-how-did-the-
giraffegetits-long-neck.html?full=true&print=true

7-148 http://www.how-come.net/giraffeneck.html

7-149 http://www.weloennig.de/GiraffaSecondPartEnglish.pdf

7-150 http://www.stiffneck.org/GIRAFFE.html

7-151 Evolution Disproved Encyclopedia—Volume 3
http://evolutionfacts.com/ Ev-V3/3evlch32.htm

7-152 http://www.present-truth.org/3-Nature/Evolution%20of%20Creationist/ MO
GC%2005.htm

7-153 http://www.examiner.com/article/giraffe-a-tall-tale-that-is-longer-than-its-neck

7-154 Gish, Dwayne Ph D. Brainwashed. Skeptic Tank.
http://www.skeptictank.org/ files/aj/creation.htm

7-154a https://en.wikipedia.org/wiki/Lamarckism

7-155 Parker, Gary. Creation Facts of Life. How Real Science Reveals the Hand of
God. 2006. Master Books. p. 105
http://www.answersingenesis.org/articles/ cfl/pangenesis-use-disuse

7-156 Richard Dawkins. 2009. The Greatest Show on Earth, Free Press. New York p.
356 Mike Anderson. 2011. Is Jesus an Evolutionist? Smashwords Publishers.
Nook e-book. P. 360-367 Prior book version: Montane Publishers, South
Africa (2008)
http://www.scribd.com/doc/11464783/ Is-Jesus-an-Evolutionist

7-157 Jerry Coyne. 2009. Why Evolution is True. New York (NY): Viking Penguin
Group. p. 82, 84

7-158 Jerry Coyne. 2009. Why Evolution is True. New York (NY): Viking Penguin
Group. p. 82

7-159 Richard Dawkins. 2009. The Greatest Show on Earth. Free Press. New York.
pp. 363-365

7-160 Bergman, J. 2010. Recurrent Laryngeal Nerve Is Not Evidence of Poor Design.
Acts & Facts. 39 (8): 12-14.
http://www.icr.org/article/recurrent-laryngealnerve-not-evidence/Institute
for Creation Research. Frank Sherwin, M.A. That Troubling Laryngeal
Nerve?
http://www.icr.org/articles/view/1742/293/

7-161 http://www.talkorigins.org/faqs/faq-transitional/part2b.html

7-162 http://www.talkorigins.org/faqs/faq-transitional/part2c.html

7-163 http://www.asa3.org/ASA/resources/Miller.html

7-163a Futuyma. Douglas J. Science on Trial – The Case for Evolution. 1982 &
1995. Sinauer Associates, Inc. Publishers Sunderland, MA. pp. 85, 89, 90

7-164 Kroll. P. W. The Plain Truth, December, 1969, "Was it Really a Horse of a
Different Color?" Paul W. Kroll, p.26, 30

7-165 http://www.freewebs.com/creation-vs-evolutionism/evolutionfrauds.htm

7-166 Gish, D. T. Ph. D. Have You Been Brainwashed?

7-167 http://www.angelfire.com/mi/dinosaurs/horse.html

7-168 http://www.truthinscience.org.uk/site/content/view/55/65/

7-169 http://www.truthinscience.org.uk/site/content/view/55/65/

Chapter 8—Transition of Reptile Jaw/Ear to Mammal Jaw/Ear

8-1 Loren Haarsma, Faith Alive Christian Resources, 2850 Kalamazoo Avenue, SE, Grand Rapids, Michigan, 49560 www.faithaliveresources.org/origins

8-2 Haarsma, Deborah B. & Loren D., 2007, Origins—A Reformed Look At Creation, Design, & Evolution, Faith Alive Christian Resources, 2850 Kalamazoo Avenue, SE, Grand Rapids, MI 49560, pp. 162-165

8-3 http://en.wikipedia.org/wiki/Evolution_of_mammalian_auditory_ossicles

8-4 http://en.wikipedia.org/wiki/Angular_bone

8-5 http://www.talkreason.org/articles/section1.cfm

8-6 http://www.talkorigins.org/faqs/comdesc/section1.html

8-7 http://animaldiversity.ummz.umich.edu/collections/mammal_anatomy/jaws_ and _ears/

8-8 http://whyevolutionistrue.wordpress.com/2009/10/15/your-ear-bonescame-from your-jaws/

8-9 http://daphne.palomar.edu/ccarpenter/reptile%20to%20mammals.htm

8-10 Lamoureux, Denis O., I Love Jesus & I Accept Evolution, 2009, WIPF Stock, 199 W. 8th Avenue, Suite 3, Eugene, Oregon, 97401, p. 112

8-11 Lamoureux, Denis O., I Love Jesus & I Accept Evolution, 2009, WIPF Stock, 199 W. 8th Avenue, Suite 3, Eugene, Oregon, 97401, P. 112

8-12 http://www.class.uidaho.edu/ngier/creationism.htm

8-13 http://www.talkorigins.org/faqs/dover/day9pm.htm

8-14 From Wikipedia, the free encyclopedia
http://www.talkorigins.org/faqs/comdesc/glossary.html#vestigial
http://www.talkorigins.org/faqs/comdesc/section1. html
http://www.icr.org/article/184/

8-15 http://www.talkorigins.org/faqs/comdesc/glossary.html#vestigial
http://www.talkorigins.org/faqs/comdesc/section1.html
http://www.icr.org/article/184/

8-16 Mike Anderson. Is Jesus an Evolutionist? 2011. Smashwords Publishers. Nook e-book. P. 326, 336-349, 383-384, 389 Prior book version: Montane Publishers, South Africa (2008)
http://www.scribd.com/doc/11464783/ Is-Jesus-an-Evolutionist

8-17 From the Beginning—The Story of Human Evolution. David Peters. 1991. Morrow Junior Books. New York. pp. 60-85.
http://www.davidpetersstudio.com /FromTheBeginning.pdf

8-18 Stephen C. Meyer, Scott Minnich, Jonathan Moneymaker, Paul A. Nelson, Ralph Seelke. 2007. Explore Evolution—The Arguments For and Against Neo-Darwinism. Hill House Publishers. Melbourne & London. Pp. 27, 29

8-18a The skull images are for illustration purposes only. They are not the actual images for each species included. The skull images are from Wiki Commons and were resized to reflect the skull sizes in the two illustrations: Left illustration: T.S. Kemp. The Origin & Evolution of Mammals (2005): p. 89. Right illustration: Recalculated (sized) from T.S. Kemp. The Origin & Evolution of Mammals (2005): p. 89. In: Stephen C. Meyer, Scott Minnich, Jonathan Moneymaker, Paul A. Nelson, Ralph Seelke. 2007. Explore Evolution—The Arguments For and Against Neo-Darwinism. Hill House Publishers. Melbourne & London. Left illustration P. 21 Figure 1.6. Right illustration P. 29 Figure 1.8.

8-19 Francis Hitching. 1982. The Neck of the Giraffe. Pan, London & Sydney. In: Dr. Alan Hayward, Physicist. Copyright 1985. Publication Date 2005. Creation and Evolution—Rethinking the Evidence From Science and the Bible. WIPF & Stock Publishers. Eugene, Oregon. Pp. 43

8-20 Davis, Percival W. and Kenyon, Dean H. 1993. Of Pandas and People. 2nd ed. Dallas. TX: Haughton. P. 101 In: Scott, E.C. 2009. Evolution vs Creationism—An Introduction. Berkeley (CA): University of California Press. P. 189

8-20a https://en.wikipedia.org/wiki/Kuehneotherium

8-21 Mammal-like reptiles: major trait reversals and discontinuities First published: TJ 15(1):44-52 April 2001 by John Woodmorappe
https://answersingenesis.org/ fossils/transitional-fossils/mammal-like-reptiles-major-trait-reversals-anddiscontinuities/

8-22 The Mammal-Like Reptiles, Duane Gish, Ph.D., December, 1981. "The Mammal-Like Reptiles", Institute for Creation Research,
http:// www.icr.org/article/184/ (accessed November 21, 2008).
http://www.icr.org/article/184/
http://www.icr.org/article/mammal-like-reptiles/

8-23 http://en.wikipedia.org/wiki/Organ_of_corti

Chapter 9—The Components of Evolutionary Theory: Natural Selection, Adaptation, Mutations, Speciation, and

Genetic Drift

9-1 Coyne, J. A. 2009. Why Evolution is True. New York (NY): Viking Penguin Group. P. 115

9-2 Back to Genesis, Things You may Not Know About Evolution, April 2002, www. icr.org

9-3 A summation of a portion of the testimony during the Kitzmiller versus Dover Area School District trial, day 1, 9/26/2005.
http://www.talkorigins.org/ faqs/dover/day1pm.html

9-4 http://hubpages.com/forum/topic/30050

9-5 Disprove Darwin in 5 Minutes or Less . . ., Page 33

9-6 Michael Denton. 1985. Evolution: A Theory in Crisis. Adler & Adler. 4550 Montgomery Avenue, Bethesda, Maryland 20814. Pp. 86-87

9-7 http://www.noanswersingenesis.org.au/dawkins_evolution.htm

9-8 Parker, G. 1980. Creation Facts of Life. Green Forest (AR): Master Books. Pp 101

9-9 http://en.wikipedia.org/wiki/Natural_selection

9-10 Scott, E.C. 2009. Evolution vs Creationism—An Introduction. Berkeley (CA): University of California Press. P. 190

9-11 Coyne, Jerry A., 2009, Why Evolution is True, Viking Penguin Group, New York, 375 Hudson Street, New York, 10014, Pages 11-13

9-12 The Atheist: Richard Dawkins explains why God is a delusion
http://www. translatum.gr/forum/index.php?topic=3878.0
Coyne, J. A. 2009. Why Evolution is True. New York (NY): Viking Penguin Group. Pp. 15-17

9-13 http://en.citizendium.org/wiki/Evolution

9-13a https://en.wikipedia.org/wiki/Sexual_selection

9-13b https://www.sciencedirect.com/science/article/pii/S0960982210015198

9-13c "What is the definition of natural selection" web search (bing.com)
https://en.wikipedia.org/wiki/Natural_selection
https://www.dictionary.com/browse/natural-selection

9-14 Crampton. Henry Edward. The Doctrine of Evolution. Its Basis and Its Scope. Columbia University Press Sales Agents. New York: Lemcke & Buechner. 30-32 West 27th Street. London. 1906-1907. Copyrighted 1911. (Ebook #16442. Release Date: August 5, 2005)

9-15 Creation or Evolution—Does It Really Matter What You Believe? United Church of God, 2008, pp. 36-38

9-16 Ham, K. General Editor. 2006. The New Answers Book. Inc. Green Forest (AR): Master Books. p. 276 www.masterbooks.net

9-17 Parker, G. 1980. Creation Facts of Life—How Real Science Reveals the Hand of God. Green Forest (AR): Master Books. Pp. 84, 92-93

9-18 Creation Facts of Life—Gary Parker, P. 92

9-19 Evolution Disproved Encyclopedia—Volume 2
http://evolutionfacts.com/ Ev-V2/2evlch13.htm

9-20 Loren Haarsma, Faith Alive Christian Resources, 2850 Kalamazoo Avenue, SE, Grand Rapids, Michigan, 49560 (Loren Haarsma is a Theistic Evolutionist) www.faithaliveresources.org/origins www.millerandlevine.com/km/evol/ Moths/moths.html

9-21 Ridley, Mark. 1993. Evolution. Blackwell Scientific Publications. Boston. Pp. 78, 95, 97

9-22 The Collapse of Evolution, Scott M. Huse, pp. 108-113

9-23 Parker, G. 1980. Creation Facts of Life—How Real Science Reveals the Hand of God. Green Forest (AR): Master Books. pp. 77-80

9-24 Parker, G. 1980. Creation Facts of Life—How Real Science Reveals the Hand of God. Green Forest (AR): Master Books. p 81

9-25 Parker, G. 1980. Creation Facts of Life—How Real Science Reveals the Hand of God. Green Forest (AR): Master Books. p. 82

9-26 Parker, G. 1980. Creation Facts of Life—How Real Science Reveals the Hand of God. Green Forest (AR): Master Books. pp. 89-94

9-27 Parker, G. 1980. Creation Facts of Life—How Real Science Reveals the Hand of God. Green Forest (AR): Master Books. pp. 93-94

9-28 Parker, G. 1980. Creation Facts of Life—How Real Science Reveals the Hand of God. Green Forest (AR): Master Books. P. 94

9-29 L Harrison Matthews, "Introduction" to Charles Darwin's Origin of the Species (1971 edition), p. xi. In: Evolution Disproved Encyclopedia—Volume 2 http://evolutionfacts.com/Ev-V2/2evlch13.htm

9-30 On Call, July 2, 1973, p. 9. In: Evolution Disproved Encyclopedia—Volume 2 http://evolutionfacts.com/Ev-V2/2evlch13.htm

9-31 Marjorie Grene, "The Faith of Darwinism," Encounter, November 1959, p. 52. In: Evolution Disproved Encyclopedia—Volume 2 http://evolutionfacts.com/Ev-V2/2evlch13.htm

9-32 Evolution Disproved Encyclopedia—Volume 2
http://evolutionfacts.com/ Ev-V2/2evlch13.htm

9-33 Parker, Gary Creation Facts of Life—How Real Science Reveals the Hand of God. 1980. Master Books. Green Forest (AR) P. 94

9-34 Creation or Evolution—Does It Really Matter What You Believe?, United Church of God, 2008, pp. 36-38

9-35 Ridley, Mark. 1993. Evolution. Blackwell Scientific Publications. Boston. Pp. 211-214, 522-523

9-36 Isaac Asimov, Asimov's New Guide to Science (1984), p. 775. In: Evolution Disproved Encyclopedia—Volume 3
http://evolutionfacts.com/Ev-V3/3evlch27.htm

9-37 (See W.E. Lammerts, "The Galapagos Island Finches," in Why Not Creation? (1970), pp. 354-366.) Evolution Disproved Encyclopedia—Volume 3
http:// evolutionfacts.com/Ev-V3/3evlch27.htm

9-38 Evolution Disproved Encyclopedia—Volume 3
http://evolutionfacts.com/Ev-V3/3evlch27.htm

9-39 Evolution Disproved Encyclopedia—Volume 2
http://evolutionfacts.com/Ev-V2/2evlch13.htm

9-40 http://www.evanwiggs.com/articles/reasons.html#Reason1

9-41 Coyne, J. A. 2009. Why Evolution is True. New York (NY): Viking Penguin Group. p. 130

9-42 Coyne, J. A. 2009. Why Evolution is True. New York (NY): Viking Penguin Group. p. 131

9-43 Coyne, J. A. 2009. Why Evolution is True. New York (NY): Viking Penguin Group. p. 131

9-44 Stephen C. Meyer, Scott Minnich, Jonathan Moneymaker, Paul A. Nelson, Ralph Seelke. 2007. Explore Evolution—The Arguments For and Against Neo-Darwinism. Hill House Publishers. Melbourne & London. Pp. 100-101.

9-45 Evolution Disproved Encyclopedia—Volume 2
http://evolutionfacts.com/ Ev-V2/2evlch14b.htm

9-46 Walter E. Lammerts, book review, in Creation Research Society Quarterly, June 1977, p. 75. In: Evolution Disproved Encyclopedia—Volume 2
http://evolutionfacts.com/Ev-V2/2evlch14b.htm

9-47 R. Milner, Encyclopedia of Evolution (1990), p. 388. In: Evolution Disproved Encyclopedia—Volume 2
http://evolutionfacts.com/Ev-V2/2evlch14b.htm

9-48 Muneeb Baig, (Grade 10)
http://www.albalagh.net/kids/science/evolution.shtml

9-49 http://www.answersingenesis.org/articles/am/v2/n3/antibiotic-resistance-ofbacteria

9-49a https://www.cdc.gov/anthrax/basics/index.html
https://en.wikipedia.org/wiki/Anthrax

9-50 Stephen C. Meyer, Scott Minnich, Jonathan Moneymaker, Paul A. Nelson, Ralph Seelke. 2007. Explore Evolution—The Arguments For and Against Neo-Darwinism. Hill House Publishers. Melbourne & London. Pp. 102-104.

9-51 Jonathan Sarfati, Ph.D. 2002. Refuting Evolution 2. Master Books. Green Forest. AR. P. 92.

9-52 Jonathan Sarfati, Ph.D. First Printing 1999. Fourth Edition 2008. Refuting Evolution. Creation Ministries International. Eight Mile Plains. Australia. Pp. 40-41.

9-53 Ham, K. General Editor. Georgia Purdom, contributing author. 2006. The New Answers Book. Master Books, Inc.: Green Forest (AR). pp. 278-280

9-54 http://www.cdc.gov/ulcer/files/hpfacts.PDF

9-55 Ham, K. General Editor. Georgia Purdom, contributing author. 2006. The New Answers Book. Master Books, Inc.: Green Forest (AR). pp. 278-280

9-56 http://en.wikipedia.org/wiki/Natural_selection

9-57 http://www.talkorigins.org/faqs/dover/day1pm.html

9-58 Charles Colson and Nancy Pearcey. How Now Shall We Live? 1999. Tyndale House Publishers, Inc. Wheaton, Illinois. Pp. 83-85

9-59 Evolution Disproved Encyclopedia—Volume 2 http://evolutionfacts.com/Appendix/a13b.htm

9-60 T. Dobzhansky, F. Ayala, G. Stebbins, and J. Valentine, Evolution, (1977), p. 97. In: Evolution Disproved Encyclopedia—Volume 2 http://evolutionfacts.com/Appendix/a13b.htm

9-61 Stephen C. Meyer, Scott Minnich, Jonathan Moneymaker, Paul A. Nelson, Ralph Seelke. 2007. Explore Evolution—The Arguments For and Against Neo-Darwinism. Hill House Publishers. Melbourne & London. Pp. 90-91

9-61a The evolutionary success of sex
https://www.ncbi.nlm.nih.gov/pmc/articles/PMC3432801/

9-61b Answers In Genesis. 2.3 The Origin of the Sexes (OB3). Dr. Werner Gitt. June 14, 2012.
https://answersingenesis.org/human-evolution/origins/23-the-origin-ofthe-sexes-ob3/

9-61c Evolution of Sex
https://www.allaboutscience.org/evolution-of-sex.htm

9-61d The True Origins. Source of article: Answers in Genesis. "Evolutionary Theories on Gender and Sexual Reproduction". 2003 Brad Harrub, Ph.D. andBert Thompson, Ph.D.
https://trueorigin.org/sex01.php

9-61e The Creation Club magazine. Mar-Apr 2020. Reproduction: A Powerful Argument Against Evolution. P. 10.

9-61f For further information on how sex evolved, see:

https://en.wikipedia.org/wiki/Evolution_of_sexual_reproduction http://www.davidpetersstudio.com /FromTheBeginning.pdf

https://www.reference.com/pets-animals/earthworms-reproduce-1fbe96864f9bf289

https://www.reference.com/pets-animals/fish-reproduce-c1f2deb53408de8f

https://study.com/academy/lesson/amphibian-reproduction.html

http://www.biokids.umich.edu/critters/Amphibia/

https://www.amphibianlife.com/do-amphibians-lay-eggs/

http://www.actforlibraries.org/how-do-reptiles-reproduce-2/

https://www.reference.com/pets-animals/birds-reproduce-9246ee0d3d26c28

https://www.factmonster.com/dk/encyclopedia/nature/mammals

https://www.sapiens.org/column/origins/sexual-evolution-pleasure/

9-62 Coyne, J. A. 2009. Why Evolution is True. New York (NY): Viking Penguin Group. Pp. 11-13

9-63 http://dictionary.reference.com/browse/adaptation+?s=t

9-64 Coyne, Jerry A., 2009, Why Evolution is True, Viking Penguin Group, New York, 375 Hudson Street, New York, 10014. pp. 117-118

9-65 Coyne, Jerry A., 2009, Why Evolution is True, Viking Penguin Group, New York, 375 Hudson Street, New York, 10014. p. 119

9-66 Coyne, Jerry A., 2009, Why Evolution is True, Viking Penguin Group, New York, 375 Hudson Street, New York, 10014, p. 120

9-67 Coyne, Jerry A., 2009, Why Evolution is True, Viking Penguin Group, New York, 375 Hudson Street, New York, 10014, Pages 121-122

9-68 Parker, Gary, Creation Facts of Life—How Real Science Reveals the Hand of God., 1980, Master Books, P.O. Box 726, Green Forest, AR, 72638, www.masterbooks.net, p. 66

9-69 Parker, Gary, Creation Facts of Life—How Real Science Reveals the Hand of God. 1980, Master Books, P.O. Box 726, Green Forest, AR, 72638, www.masterbooks.net. P. 94-95

9-70 http://dictionary.reference.com/browse/Speciation?s=t

9-71 Summary of Darwinian evolution by Richard Dawkins from an article Big ideas: Evolution in the New Scientist.

http://www.noanswersingenesis.org.au/dawkins_evolution.htm

9-72 Coyne, J. A. 2009. Why Evolution is True. New York (NY): Viking Penguin Group. P. 6

9-73 Coyne, J. A. 2009. Why Evolution is True. New York (NY): Viking Penguin Group. Pp. 170-171

9-74 Coyne, J. A. 2009. Why Evolution is True. New York (NY): Viking Penguin Group. P. 172

9-75 Richard Dawkins. The Blind Watchmaker—Why The Evidence of Evolution Reveals A Universe Without Design. 1996 (first edition 1985) W. W. Norton & Company. New York. pp. 337-340.

9-76 Coyne, J. A. 2009. Why Evolution is True. New York (NY): Viking Penguin Group. Pp. 175-176

9-77 Evolution Disproved Encyclopedia—Volume 2
http://evolutionfacts.com/Ev-V2/2evlch15.htm

9-78 Evolution Disproved Encyclopedia—Volume 2
http://evolutionfacts.com/Ev-V2/2evlch17a.htm

9-79 Parker, Gary Creation Facts of Life—How Real Science Reveals the Hand of God. 1980. Green Forest (AR): Master Books. Pp 133

9-79a http://www.definitions.net/definition/ring%20species

9-79b http://darwinwasright.org/ring_species.html

9-79c https://creation.com/birds-of-a-feather-don-t-breed-together

9-79d http://thecreationclub.com/do-ring-species-show-evolution/

9-80 Wilder-Smith, A.E. The Natural Sciences Know Nothing of Evolution. 1981. T.W.F.T. Publishers, pp. 46-47. In: Howard, Jeffery. 2008. Disprove Darwin in 5 Minutes or Less. Delray Beach (FL): Twin Angels Publishing, Inc, p. 29

9-81 Turner, Carl R. M.D. Discovering God and His Creation—Evolution as Part of God's Plan. 2008. iUniverse, Inc. New York. p. 25

9-82 http://schools-wikipedia.org/wp/e/Evolution.htm

9-83 Coyne, Jerry A., 2009, Why Evolution is True, Viking Penguin Group, New York, 375 Hudson Street, New York, 10014, p. 118

9-84 Scott, E.C. 2009. Evolution vs Creationism—An Introduction. Berkeley (CA): University of California Press. P. 208, 216

9-84a Futuyma. Douglas J. Science on Trial – The Case for Evolution. 1982 & 1995. Sinauer Associates, Inc. Publishers Sunderland, MA. p. 40

9-84b Futuyma. Douglas J. Science on Trial – The Case for Evolution. 1982 & 1995. Sinauer Associates, Inc. Publishers Sunderland, MA. pp. 136-137

9-84c Futuyma. Douglas J. Science on Trial – The Case for Evolution. 1982 & 1995. Sinauer Associates, Inc. Publishers Sunderland, MA. pp. 139-140

9-84d The TalkOrigins Archive. Review of Michael Denton's "Evolution – A Theory in Crisis" 1996-1997. Mark I. Vuletic
http://talkorigins.org/faqs/denton.html

9-85 Theodosious Dobzhansky, "On Methods of Evolutionary Biology and Anthropology," American Scientist, Winter,45, December 1957, p. 385. In: Evolution Disproved Encyclopedia—Volume 2
http://evolutionfacts.com/Ev-V2/2evlch14a.htm
http://evolutionfacts.com/ Appendix/a14a.htm

9-86 Ernst Mayr, Populations, Species and Evolution (1970), p. 103. In: Evolution Disproved Encyclopedia—Volume 2
http://evolutionfacts.com/Ev-V2/2evlch14a.htm

9-87 Evolution Disproved Encyclopedia—Volume 2
http://evolutionfacts.com/Ev-V2/2evlch14a.htm

9-88 C.H. Waddington, The Nature of Life (1962), p. 98. In: Evolution Disproved Encyclopedia—Volume 2
http://evolutionfacts.com/Ev-V2/2evlch14a.htm

9-89 Ernst Mayr, Populations, Species, and Evolution, p. 164. In: Evolution Disproved Encyclopedia—Volume 2
http://evolutionfacts.com/Ev-V2/2evlch14a.htm

9-90 Briski, D. 2004. Impressive Deception—Creation or Evolution—You Decide. Shippensburg, (PA): Ragged Edge Press. P. 75

9-91 Briski, D. 2004. Impressive Deception—Creation or Evolution You Decide. Shippensburg, (PA): Ragged Edge Press. Pp. 65-66

9-92 http://christiananswers.net/q-eden/origin-of-life.html

9-93 Genetics and Heredity, Maurice Caullery, 1964, p. 10

9-94 The Neck of the Giraffe, p. 49, Francis Hitching

9-95 Julian Huxley, Evolution in Action, p. 41. In: Evolution Disproved Encyclopedia—Volume 2
http://evolutionfacts.com/Ev-V2/2evlch14a.htm

9-96 Shattering the Myths of Darwinism, Richard Milton, p. 156

9-97 Creation or Evolution: Does It Really Matter What You Believe?, United Church of God, 2008, pp. 40-43

9-98 http://intelligentscience.org/2008/10/16/the-cambrian-a-huge-problem-formoleculesto-man-evolution/

9-99 http://www.albalagh.net/kids/science/evolution.shtml

9-100 http://www.evanwiggs.com/articles/reasons.html#Reason1

9-100a Acts & Facts. ICR. The Gospel and ICR. Brian Thomas, M.S. Dec. 2017. P. 7

9-100b Tomkins J. P. Blind Cavefish Illuminate Divine Engineering. Creation Science Update. Posted on ICR.org October 23, 2017, Accessed November 27, 2018. In: Acts & Facts magazine. Vol 48 No. 2. Institute For Creation Research. ICR.org February 2019. Page 18.

9-101 Evolution Disproved Encyclopedia—Volume 2
http://evolutionfacts.com/ Ev-V2/2evlch14a.htm

9-102 Evolution Disproved Encyclopedia—Volume 2
http://evolutionfacts.com/ Ev-V2/2evlch13.htm

9-103 http://www.newgeology.us/presentation32.html

9-104 D. Erwin, and J. Valentine, "'Hopeful Monsters,' Transposons, and Metazoan Radiation," in Proceedings National Academy of Sciences, Vol. 81, 1984, p. 5482. In: Evolution Disproved Encyclopedia—Volume 2
http://evolutionfacts.com/Appendix/a14b.htm

9-105 Fred Hoyle, The Intelligent Universe (1983), p. 35. In: Evolution Disproved Encyclopedia—Volume 2
http://evolutionfacts.com/Appendix/a14b.htm

9-106 Evolution Disproved Encyclopedia—Volume 2
http://evolutionfacts.com/Ev-V2/2evlch14b.htm

9-107 Evolution Disproved Encyclopedia—Volume 2
http://evolutionfacts.com/Ev-V2/2evlch14b.htm

9-108 Evolution Disproved Encyclopedia—Volume 2
http://evolutionfacts.com/Appendix/a14b.htm

9-109 Evolution Disproved Encyclopedia—Volume 2
http://evolutionfacts.com/Ev-V2/2evlch14a.htm

9-110 Evolution Disproved Encyclopedia—Volume 2
http://evolutionfacts.com/Ev-V2/2evlch14a.htm

9-111 A. Koesder, The Ghost in the Machine (1975), p. 129. In: Evolution Disproved Encyclopedia—Volume 2
http://evolutionfacts.com/Ev-V2/2evlch14a.htm

9-111a https://www.vocabulary.com/dictionary/gamete

9-112 George Gaylord Simpson, "Uniformitarianism: An Inquiry into Principle Theory and Method in Geohistory and Biohistory," Chapter 2, in Max A. Hecht and William C. Steeres, ed., Essays in Evolution and Genetics (1970), p. 80. In: Evolution Disproved Encyclopedia—Volume 2

9-113 Duane Gish, "DNA: Its History and Potential," in WE. Lammerts (ed.), Scientific Studies in Special Creation (1971), p. 315. In: Evolution Disproved Encyclopedia—Volume 2
http://evolutionfacts.com/Ev-V2/2evlch14a.htm

http://evolutionfacts.com/Ev-V2/2evlch14a.htm

9-114 Evolution Disproved Encyclopedia—Volume 2
http://evolutionfacts.com/ Ev-V2/2evlch14b.htm

9-115 Henry Morris and Gary Parker, What is Creation Science? (1987), pp. 103, 104. In: Evolution Disproved Encyclopedia—Volume 2
http://evolutionfacts. com/Ev-V2/2evlch14b.htm

9-116 Evolution Disproved Encyclopedia—Volume 2
http://evolutionfacts.com/ Appendix/a14a.htm

9-117 Parker, G. 1980. Creation Facts of Life. How Real Science Reveals the Hand of God. Green Forest (AR): Master Books. Pp.122-125

9-118 Parker, G. 1980. Creation Facts of Life. How Real Science Reveals the Hand of God. Green Forest (AR): Master Books. Pp. 126, 129-131

9-118a Rana, Fazale. The Cell's Design - How Chemistry Reveals the Creator's Artistry. BakerBooks. Grand Rapids, Michigan. 2008. Pp. 173-178

9-119 T. Dobzhansky. In a review of the first (French) edition of Pierre-Paul Grasse's book, Evolution. Wistar Institute Press. Philadelphia. 1967. p. 225 In: Dr. Alan Hayward, Physicist. Copyright 1985. Publication Date 2005. Creation and Evolution—Rethinking the Evidence From Science and the Bible. WIPF & Stock Publishers. Eugene, Oregon. Pp. 24-25

9-120 Dr. Alan Hayward, Physicist. Copyright 1985. Publication Date 2005. Creation and Evolution—Rethinking the Evidence From Science and the Bible. WIPF & Stock Publishers. Eugene, Oregon. p. 55

9-121 http://www.layevangelism.com/advtxbk/sections/sect-10/sec10-4.htm

9-122 http://www.layevangelism.com/advtxbk/sections/sect-10/sec10-4b.htm

9-123 Turner, Carl R., M.D. Discovering God and His Creation—Evolution as Part of God's Plan. 2008. iUniverse, Inc. New York. P. 26-27

9-124 Evolution Disproved Encyclopedia—Volume 2
http://evolutionfacts.com/Ev-V2/2evlch14a.htm

9-125 Parker, Gary. Creation Facts of Life—How Real Science Reveals the Hand of God. 1980. Master Books. Green Forest (AR): P. 108-110

9-126 http://www.newgeology.us/presentation32.html

9-127 Charles Colson and Nancy Pearcey. How Now Shall We Live? 1999. Tyndale House Publishers, Inc. Wheaton, Illinois. Pp. 85-86

9-128 Evolution Disproved Encyclopedia—Volume 2
http://evolutionfacts.com/Ev-V2/2evlch14b.htm

9-129 Jeremy Rifkin, Algeny (1983), p. 134. In: Evolution Disproved Encyclopedia—Volume 2
http://evolutionfacts.com/Appendix/a14a.htm

9-130 "Evolutionists Still Looking for a Good Accident,'" Battle Cry, July-August, 1990. In: Evolution Disproved Encyclopedia—Volume 2 http://evolutionfacts.com/Appendix/a14a.htm

9-131 Young, Matt and Strode, Paul K. 2009. Why Evolution Works (And Creationism Fails). Rutgers Press, 100 Joyce Kilmer Avenue, Piscataway, NJ 08854-8099. Pp. 104-111.

9-132 Stephen C. Meyer, Scott Minnich, Jonathan Moneymaker, Paul A. Nelson, Ralph Seelke. 2007. Explore Evolution—The Arguments For and Against Neo-Darwinism. Hill House Publishers. Melbourne & London. P.109-110

9-133 Stephen C. Meyer, Scott Minnich, Jonathan Moneymaker, Paul A. Nelson, Ralph Seelke. 2007. Explore Evolution—The Arguments For and Against Neo-Darwinism. Hill House Publishers. Melbourne & London. P.109-110

9-134 Jonathan Sarfati, Ph.D. 2002. Refuting Evolution 2. Master Books. Green Forest. AR. Pp. 102-103.

9-135 Origins—a Reformed Look at Creation, Design, & Evolution, Deborah B. Haarshma and Loren D. Haarsma pp. 167-171

9-136 http://en.wikipedia.org/wiki/Genetic_drift

9-137 Evolution Disproved Encyclopedia—Volume 2 http://evolutionfacts.com/Ev-V2/2evlch15.htm

9-138 http://creationwiki.org/Genetic_drift

9-139 E. Dodson, Evolution: Process and Product (1960), pp. 258-259. In: Evolution Disproved Encyclopedia—Volume 2 http://evolutionfacts.com/Appendix/ a13a.htm

9-140 Jonathan Sarfati, Ph.D. 2002. Refuting Evolution 2. Master Books. Green Forest. AR. Pp. 105-107.

9-141 Michael Denton. 1985. Evolution: A Theory in Crisis. Adler & Adler. 4550 Montgomery Avenue, Bethesda, Maryland 20814. P. 149-150

Chapter 10—Punctuated Equilibrium

10-1 Stephen Gould, "The Return of the Hopeful Monsters," in Natural History, June/July, 1977. In: Evolution Disproved Encyclopedia—Volume 3 http://evolutionfacts.com/Ev-V3/3evlch29b.htm

10-2 Steven Gould, "Is a New and General Theory of Evolution Emerging?" in Paleobiology, January 1980, p. 127. In: Evolution Disproved Encyclopedia—Volume 3 http://evolutionfacts.com/Ev-V3/3evlch29b.htm

10-3 Richard Dawkins. The Blind Watchmaker—Why The Evidence of Evolution Reveals A Universe Without Design. 1996 (first edition 1985) W. W. Norton & Company. New York. p. XIX.

10-4 http://www.pbs.org/wgbh/evolution/library/03/5/l_035_01.html

10-5 Richard Dawkins. The Blind Watchmaker—Why The Evidence of Evolution Reveals A Universe Without Design. 1996 (first edition 1985) W. W. Norton & Company. New York. Pp. 327, 329, 333-346.

10-6 http://en.wikipedia.org/wiki/Species

10-7 http://www.pbs.org/wgbh/evolution/library/03/5/l_035_01.html http://atheisttoolbox.com/fce9.php

10-8 http://www.pbs.org/wgbh/evolution/library/03/5/l_035_01.html

10-9 Michael Denton. 1985. Evolution: A Theory in Crisis. Adler & Adler. 4550 Montgomery Avenue, Bethesda, Maryland 20814. Pp. 192-194

10-10 Disprove Darwin in 5 Minutes or Less . . ., Page 48

10-11 R. Milner, Encyclopedia of Evolution (1990), p. 375. In: Evolution Disproved Encyclopedia—Volume 2 http://evolutionfacts.com/Appendix/a14b.htm

10-12 http://www.wiebefamily.org/e.htm

10-13 http://intelligentscience.org/2008/10/16/the-cambrian-a-huge-problem-for -molecules-to-man-evolution/

10-14 http://www.darwinismrefuted.com/equilibrium_02.html

10-15 Impressive Deception: Creation or Evolution—You Decide, DeeDee Briski, P. 79-80.

10-16 Lamoureux, D. O. 2009. I Love Jesus & I Accept Evolution. Eugene (OR): WIPF Stock. Pp. 107-108

10-17 Lamoureux, D. O. 2009. I Love Jesus & I Accept Evolution. Eugene (OR): WIPF Stock. P. 112

Chapter 11—Embryology and The Theory of Recapitulation

11-1 http://dictionary.reference.com/browse/Recapitulation?s=t

11-2 Coyne, J. A. 2009. Why Evolution is True. New York (NY): Viking Penguin Group. P. 78

11-3 Coyne, J. A. 2009. Why Evolution is True. New York (NY): Viking Penguin Group. Pp. 73-74

11-4 http://www.talkorigins.org/faqs/comdesc/glossary.html#vestigial

11-5 Stephen C. Meyer, Scott Minnich, Jonathan Moneymaker, Paul A. Nelson, Ralph Seelke. 2007. Explore Evolution—The Arguments For and Against Neo-Darwinism. Hill House Publishers. Melbourne & London. P. 68

11-6 Richardson, M. 1998. In: Scott, E.C. 2009. Evolution vs Creationism—An Introduction. Berkeley (CA): University of California Press. P. 238-239

11-7 Coyne, J. A. 2009. Why Evolution is True. New York (NY): Viking Penguin Group. P. 78

11-8 http://www.mk-richardson.com/pdf/Anat%20Embryol.pdf
Science Magazine September 1, 1997 Volume 277 No. 5331 Haeckel's Embryos: Fraud Rediscovered article:
http://www.sciencemag.org/content/277/5331/1435.1.full

11-9   http://www.scribd.com/doc/13743145/A-Modest-Response-to-Haeckels-Embryos

11-10 http://www.darwinismwatch.com/index.php?git=makale&makale_id=1889
11-11
http://www.mk-richardson.com/pdf/Anat%20Embryol.pdf   Science Magazine September 1, 1997 Volume 277 No. 5331 Haeckel's Embryos: Fraud Rediscovered article:
http://www.sciencemag.org/content/277/5331/1435.1.full

11-12 http://www.darwinismwatch.com/index.php?git=makale&makale_id=1889

11-13 Johnson, P. E. 1991. Darwin on Trial. InterVarsity Press. Downers Grove, Illinois 60515, p. 42

11-14 Tim M. Berra, "Evolution and the Myth of Creationism" p. 22
http://www.thejesusmyth.com/toolbox/evolution

11-15 Coyne, J. A. 2009. Why Evolution is True. New York (NY): Viking Penguin Group. Pp. 73-74

11-16 Coyne, J. A. 2009. Why Evolution is True. New York (NY): Viking Penguin Group. P. 75

11-17 Coyne, J. A. 2009. Why Evolution is True. New York (NY): Viking Penguin Group. P. 77

11-18 Paul R. Ehrlich and Richard W. Holm, Process of Evolution (1963), p. 66. In: Evolution Disproved Encyclopedia—Volume 3
http://evolutionfacts.com/Appendix/a22.htm

11-19 Core Reno, Fact or Theory (1953), p. 89. In: Evolution Disproved Encyclopedia—Volume 3
http://evolutionfacts.com/Appendix/a22.htm

11-20 G.B. Moment; General Zoology (1958). P. 201. In: Evolution Disproved Encyclopedia—Volume 3 http://evolutionfacts.com/Appendix/a22.htm

11-21 Evolution Disproved Encyclopedia—Volume 3
http://evolutionfacts.com/ Ev-V3/3evlch22.htm

11-22 J.N. Moors and H.E. Slusher, Biology: A Search for Order in Complexity
(197ID), p. 424. In: Evolution Disproved Encyclopedia—Volume 3
http://evolutionfacts.com/Appendix/a22.htm

11-23 Evolution Disproved Encyclopedia—Volume 3
http://evolutionfacts.com/ Ev-V3/3evlch22.htm

11-24 Henry Morris, et al., Science and Creation (1971), p. 48. In: Evolution
Disproved Encyclopedia—Volume 3
http://evolutionfacts.com/Appendix/a22.htm

11-25 Michael Denton. 1985. Evolution: A Theory in Crisis. Adler & Adler. 4550
Montgomery Avenue, Bethesda, Maryland 20814. pp. 145-147

11-26 http://en.wikipedia.org/wiki/Zygote

11-27 http://en.wikipedia.org/wiki/Cleavage_(embryo)

11-28 http://www.thefreedictionary.com/blastocyst

11-28a https://www.youtube.com/watch?v=wnnhMZtCvnQ

11-28b https://www.youtube.com/watch?v=POJAQX760VY

11-28c https://answers.yahoo.com/question/index?qid=20090304155249AAgu4lg

11-28d https://courses.lumenlearning.com/wm-biology2/chapter/gastrulation/

11-28e https://www.medicinenet.com/embryo_vs_fetus_differences_week-by-week/
article.htm

11-29 Evolution Disproved Encyclopedia—Volume 3
http://evolutionfacts.com/ Ev-V3/3evlch22.htm

11-30 Evolution Disproved Encyclopedia—Volume 3
http://evolutionfacts.com/ Ev-V3/3evlch22.htm

11-31 Evolution Disproved Encyclopedia—Volume 3
http://evolutionfacts.com/ Ev-V3/3evlch22.htm

11-32 "Does The Human Embryo Go Through Animal Stages?" Morris, J. Dr.,
president of ICR,
http://www.icr.org/article/1068/

11-33 http://www.darwinismrefuted.com/embryology_04.html

11-34 Evolution Disproved Encyclopedia—Volume 3
http://evolutionfacts.com/ Ev-V3/3evlch22.htm

11-35 Parker, G. 1980. Creation Facts of Life. Green Forest (AR): Master Books.
Pp 54-62

11-36 http://www.ncbi.nlm.nih.gov/pubmedhealth/PMH0002525/

11-37 Chauhan SP, Gopal NN, Jain M, Gupta A. Postgraduate Department of Surgery, M. L. N. Medical College, Allahabad 211002, India.

11-38 http://en.wikipedia.org/wiki/Human_tail#Human_tails

11-39 http://en.wikipedia.org/wiki/Human_vestigiality#Coccyx

11-40 Kabra NS, Srinivasan G, Udani RH. True tail in a neonate. Indian Pediatr 1999;36:712- True tail in a neonate. Kothari PR, Gupta A, Shankar G, Jiwane A, Kulkarni B. Department of Pediatric Surgery, LTM Medical College and General Hospital, Sion, Mumbai, India. drparaskothari@rediffmail.com

11-41 http://www.talkorigins.org/faqs/comdesc/section2.html#atavisms_ex2

11-42 http://www.talkorigins.org/faqs/comdesc/section2.html#atavisms_ex2

11-43 From: Debating Christianity & Religion—Civil Debates and Christianity and Religions. Topic: If You Accept Microevolution—You Must Accept Macroevolution http://debatingchristianity.com/forum/viewtopic.php?t=17 925&postdays=0&postorder=asc&start=230

11-44 http://www.childrenshospital.org/az/Site1062/mainpageS1062P0.html

11-45 http://www.creationtips.com/babytail.html

Chapter 12—Vestiges

12-1 http://www.thejesusmyth.com/toolbox/evolution

12-1a A quotation made by University of Chicago Evolutionary Biology Professor Jerry Coyne, Coyne, Jerry. 2006. "Ann Coulter and Charles Darwin. Coultergeist." The New Republic Online, July 31, 2006, p. 1. In: Jerry Bergman . 2019. Useless Organs – The Rise and Fall of a Central Claim of Evolution. Published in the United States by BP Books in Tulsa, Oklahoma. P. 13.

12-1b A quotation made by Parker, George. 1928. "Vestigial Organs", as quoted in Creation by Evolution. In: Jerry Bergman . 2019. Useless Organs – The Rise and Fall of a Central Claim of Evolution. Published in the United States by BP Books in Tulsa, Oklahoma. P. 14.

12-2 http://www.talkorigins.org/faqs/comdesc/glossary.html#vestigial

12-3 Crampton. Henry Edward. The Doctrine of Evolution. Its Basis and Its Scope. Columbia University Press Sales Agents. New York: Lemcke & Buechner. 30-32 West 27th Street. London. 1906-1907. Copyrighted 1911. (Ebook #16442. Release Date: August 5, 2005)

12-4 Shermer. Michael. 2006. Why Darwin Matters—The Case Against Intelligent Design. Owl Books. New York. pp. 18-19
http://www.thejesusmyth.com/ toolbox/evolution

12-5 Evolution Disproved Encyclopedia—Volume 3
http://evolutionfacts.com/ Ev-V3/3evlch22.htm

12-6 S. Scadding, "Do 'Vestigial Organs' Provide Evidence for Evolution?'; Evolutionary Theory (1981), p. 175. In: Evolution Disproved Encyclopedia—Volume 3
http://evolutionfacts.com/Appendix/a22.htm

12-7 Jonathan Sarfati, Ph.d. 2002. Refuting Evolution 2. Master Books. Green Forest, AR. p.207

12-8 Creation Facts of Life—Gary Parker, pp. 54-62

12-9 Evolution Disproved Encyclopedia—Volume 3
http://evolutionfacts.com/ Ev-V3/3evlch22.htm

12-10 http://www.arn.org/docs/mills/gm_originoflifeandevolution.htm

12-11 http://creation.com/do-any-vestigial-organs-exist-in-humans

12-12 Evolution Disproved Encyclopedia—Volume 3
http://evolutionfacts.com/ Ev-V3/3evlch22.htm

12-13 Evolution Disproved Encyclopedia—Volume 3
http://evolutionfacts.com/ Ev-V3/3evlch22.htm

12-14 Evolution Disproved Encyclopedia—Volume 3
http://evolutionfacts.com/ Appendix/a22.htm

12-15 Jerry Bergman. 2019. Useless Organs – The Rise and Fall of a Central Claim of Evolution. Published in the United States by BP Books in Tulsa, Oklahoma. P. 19

12-16 Jerry Bergman. 2019. Useless Organs – The Rise and Fall of a Central Claim of Evolution. Published in the United States by BP Books in Tulsa, Oklahoma. P. xviii

12-17 Jerry Bergman. 2019. Useless Organs – The Rise and Fall of a Central Claim of Evolution. Published in the United States by BP Books in Tulsa, Oklahoma. p. 314

12-18 Jerry Bergman. 2019. Useless Organs – The Rise and Fall of a Central Claim of Evolution. Published in the United States by BP Books in Tulsa, Oklahoma. P. 15

12-19 Jerry Bergman. 2019. Useless Organs – The Rise and Fall of a Central Claim of Evolution. Published in the United States by BP Books in Tulsa, Oklahoma. Pp. 1-2

12-20 Jerry Bergman. 2019. Useless Organs – The Rise and Fall of a Central Claim of Evolution. Published in the United States by BP Books in Tulsa, Oklahoma. Pp. 23-25

12-21 Jerry Bergman. 2019. Useless Organs – The Rise and Fall of a Central Claim of Evolution. Published in the United States by BP Books in Tulsa, Oklahoma. p. 53-54, 56-57, 69-71

12-22 Jerry Bergman. 2019. Useless Organs – The Rise and Fall of a Central Claim of Evolution. Published in the United States by BP Books in Tulsa, Oklahoma. P. xviii

12-23 Jerry Bergman. 2019. Useless Organs – The Rise and Fall of a Central Claim of Evolution. Published in the United States by BP Books in Tulsa, Oklahoma. p. 81

12-24 Jerry Bergman. 2019. Useless Organs – The Rise and Fall of a Central Claim of Evolution. Published in the United States by BP Books in Tulsa, Oklahoma. P. xviii

12-25 Jerry Bergman. 2019. Useless Organs – The Rise and Fall of a Central Claim of Evolution. Published in the United States by BP Books in Tulsa, Oklahoma. p. 91-93

12-26 O'Rahilly, Ronan and Fabiola Muller. 1992. Human Embryology & Teratology. Wiley-Liss, New York. In: Jerry Bergman. 2019. Useless Organs – The Rise and Fall of a Central Claim of Evolution. Published in the United States by BP Books in Tulsa, Oklahoma. p. 96

12-27 Blechschmidt, Eric. 1977. The Beginnings of Human Life, New York: Springer-Verlag. In: Jerry Bergman. 2019. Useless Organs – The Rise and Fall of a Central Claim of Evolution. Published in the United States by BP Books in Tulsa, Oklahoma. p. 96

12-28 Jerry Bergman. 2019. Useless Organs – The Rise and Fall of a Central Claim of Evolution. Published in the United States by BP Books in Tulsa, Oklahoma. p. 99-100, 107

12-29 Jerry Bergman. 2019. Useless Organs – The Rise and Fall of a Central Claim of Evolution. Published in the United States by BP Books in Tulsa, Oklahoma. p. 114-119

12-30 Jerry Bergman. 2019. Useless Organs – The Rise and Fall of a Central Claim of Evolution. Published in the United States by BP Books in Tulsa, Oklahoma. p. 121-127

12-31 Jerry Bergman. 2019. Useless Organs – The Rise and Fall of a Central Claim of Evolution. Published in the United States by BP Books in Tulsa, Oklahoma. p. 133-134, 137

12-32 Jerry Bergman. 2019. Useless Organs – The Rise and Fall of a Central Claim of Evolution. Published in the United States by BP Books in Tulsa, Oklahoma. p. 139-140

12-33 Jerry Bergman. 2019. Useless Organs – The Rise and Fall of a Central Claim of Evolution. Published in the United States by BP Books in Tulsa, Oklahoma. pp. 137-138, 147-150

12-34 Jerry Bergman. 2019. Useless Organs – The Rise and Fall of a Central Claim of Evolution. Published in the United States by BP Books in Tulsa, Oklahoma. p. 211-221

Chapter 13—Pseudogenes

13-1 Evolution Disproved Encyclopedia—Volume 02
http://evolutionfacts.com/ Appendix/a09.htm

13-2 http://www.chemguide.co.uk/organicprops/aminoacids/dna1.html#top

13-3 http://en.wikipedia.org/wiki/Gene

13-4 Loren Haarsma, Faith Alive Christian Resources, 2850 Kalamazoo Avenue, SE, Grand Rapids, Michigan, 49560

13-5 Haarsma, D. B. Haarsma, L. D. 2007. Origins—A Reformed Look At Creation, Design, & Evolution. Grand Rapids, (MI): Faith Alive Christian Resources. pp. 167-171

13-6 http://www.talkorigins.org/faqs/comdesc/glossary.html#vestigial

13-7 http://www.talkorigins.org/faqs/comdesc/glossary.html#vestigial

13-8 Coyne, J. A. 2009. Why Evolution is True. New York (NY): Viking Penguin Group. P. 67

13-9 Coyne, J. A. 2009. Why Evolution is True. New York (NY): Viking Penguin Group. Pp. 68-69

13-10 Mike Anderson. Is Jesus an Evolutionist? 2011. Smashwords Publishers. Nook e-book. P. 383, 390 Prior book version: Montane Publishers, South Africa (2008)
http://www.scribd.com/doc/11464783/Is-Jesus-an-Evolutionist

13-11 Denis O. Lamoureux, I Love Jesus & I accept Evolution, 2009. WIPF Stock. Eugene. Oregon. P. 133

13-12 Haarsma, Deborah B. & Loren D., 2007, Origins—A Reformed Look At Creation, Design, & Evolution, Faith Alive Christian Resources, 2850 Kalamazoo Avenue, SE, Grand Rapids, MI 49560, pp. 199-203

13-12a Carey, Nessa. 2015. Junk DNA – A Journey Through the Dark Matter of the Genome. Columbia University Press. New York. Published In the United Kingdom by Icon Books Ltd. eBook Pp. 2-6, 12

13-12a-1 Carey, Nessa. 2015. Junk DNA – A Journey Through the Dark Matter of the Genome. Columbia University Press. New York. Published In the United Kingdom by Icon Books Ltd. eBook pages: Idiopathic Pulmonary Fibrosis - p. 56, Roberts Syndrome and Cornelia de Lange Syndrome - p. 169, Brittle Bone Disease - p. 221, Myotonic Dystrophy - p. 229-231, Amyotrophic Lateral Sclerosis (ALS), (also known as Motor Neuron Disease or Lou Gehrig's Disease) – p. 235, Kaposi's Sarcoma – p. 244, Hutchinson-Gilford Progeria – p. 246, Human Cartilage-Hair Hypoplasia – p. 257

13-12b Carey, Nessa. 2015. Junk DNA – A Journey Through the Dark Matter of the Genome. Columbia University Press. New York. Published In the United Kingdom by Icon Books Ltd. eBook P. 35

13-12c Carey, Nessa. 2015. Junk DNA – A Journey Through the Dark Matter of the Genome. Columbia University Press. New York. Published In the United Kingdom by Icon Books Ltd. eBook Pp. 12-21

13-12d Carey, Nessa. 2015. Junk DNA – A Journey Through the Dark Matter of the Genome. Columbia University Press. New York. Published In the United Kingdom by Icon Books Ltd. eBook Pp. 28-29

13-12e Carey, Nessa. 2015. Junk DNA – A Journey Through the Dark Matter of the Genome. Columbia University Press. New York. Published In the United Kingdom by Icon Books Ltd. eBook Pp. 31-33

13-12f Carey, Nessa. 2015. Junk DNA – A Journey Through the Dark Matter of the Genome. Columbia University Press. New York. Published In the United Kingdom by Icon Books Ltd. eBook PP. 189-190

13-12g Carey, Nessa. 2015. Junk DNA – A Journey Through the Dark Matter of the Genome. Columbia University Press. New York. Published In the United Kingdom by Icon Books Ltd. eBook PP. 194-195

13-13 Sean D. Pitman M.D. Â January 2001 Latest Update: November 2008
http://naturalselection.0catch.com/Files/Pseudogenes.html#Pseudogenes

13-14 Institute for Creation Research Adam and Eve, Vitamin C, and Pseudogenes by Daniel Criswell, Ph.D.
http://www.icr.org/article/ adam-eve-vitamin-c-pseudogenes/

13-15 Ross. Hugh. 2006. Creation As Science: A Testable Model Approach to End the Creation/Evolution Wars. NAVPRESS. Colorado Springs. Pp. 221-222

13-16 Ross. Hugh. 2006. Creation As Science: A Testable Model Approach to End the Creation/Evolution Wars. NAVPRESS. Colorado Springs. Pp. 221-222

13-17 Jonathan Sarfati, Ph.D. 2002. Refuting Evolution 2. Master Books. Green Forest. AR. Pp. 122-124.

13-18 Richard Dawkins. The Blind Watchmaker—Why The Evidence of Evolution Reveals A Universe Without Design. 1996 (first edition 1985) W. W. Norton & Company. New York. p. 246.

13-19 AnswersInGenesis.org Potentially decisive evidence against pseudogene 'shared mistakes', by John Woodmorappe
http://www.answersingenesis.org/tj/v18/ i3/mistakes.asp Potentially Decisive Evidence Against Pseudogene—Answers In Genesis

13-19a Rana, Fazale. The Cell's Design - How Chemistry Reveals the Creator's Artistry. BakerBooks. Grand Rapids, Michigan. 2008. Pp. 255-260

13-20 Young, Matt and Strode, Paul K. 2009. Why Evolution Works (And Creationism Fails). Rutgers Press, 100 Joyce Kilmer Avenue, Piscataway, NJ 08854-8099. Pp. 104-111.

13-21 Michael Denton. 1985. Evolution: A Theory in Crisis. Adler & Adler. 4550 Montgomery Avenue, Bethesda, Maryland 20814. pp. 336-337

Chapter 14—Biological Complexity and Intelligent Design

14-1 http://philosophicaugustine.wordpress.com/2013/09/15/william-paley-on-the-argument-from-design/

14-2 http://www.literaturepage.com/read/darwin-origin-of-species-176.html

14-3 http://www.eyedesignbook.com/ch6/Dembski_IrreducibleComplexityRevisited_011404.pdf

14-4 http://www.eyedesignbook.com/ch6/Dembski_IrreducibleComplexityRevisited_011404.pdf

14-5 Summation of a portion of testimony during Kiltzmiller vs. Dover Area School District PA 2005 Intelligent Design trial
http://www.talkorigins.org/faqs/ dover/day1pm.html

14-6 Summation of a portion of testimony during Kiltzmiller vs. Dover Area School District PA 2005 Intelligent Design trial
http://www.talkorigins.org/faqs/ dover/day12pm.html

14-7 Richard Dawkins. The Blind Watchmaker—Why The Evidence of Evolution Reveals A Universe Without Design. 1996 (first edition 1985) W. W. Norton & Company. New York. p. 4.

14-8 Richard Dawkins. The Blind Watchmaker—Why The Evidence of Evolution Reveals A Universe Without Design. 1996 (first edition 1985) W. W. Norton & Company. New York. p. xv (Preface to original edition).

14-9 http://home.planet.nl/~gkorthof/korthof8.htm

14-10 Summation of a portion of testimony during Kiltzmiller vs. Dover Area School District PA 2005 Intelligent Design trial
http://www.talkorigins.org/faqs/ dover/day1pm.html

14-11 H. E. Mellersh, The Story of Life (1958), pp. 237, 242. In: Evolution Disproved Encyclopedia—Volume 2
http://evolutionfacts.com/Appendix/a13a.htm

14-12 Becky Conolly
http://www.christianaction.org.za/articles_creationscience/06_Irreducible_complexity.htm

14-13 Shermer. Michael. Why Darwin Matters—The Case Against Intelligent Design. 2006. Owl Books. New York. Pp. 68-69

14-14 Michael Denton. 1985. Evolution: A Theory in Crisis. Adler & Adler. 4550 Montgomery Avenue, Bethesda, Maryland 20814. P. 215

14-15 Charles Colson and Nancy Pearcey. How Now Shall We Live? 1999. Tyndale House Publishers, Inc. Wheaton, Illinois. Pp. 88-89.

14-16 Haarsma, D. B. Haarsma, L. D. 2007. Origins—A Reformed Look At Creation, Design, & Evolution. Grand Rapids, (MI): Faith Alive Christian Resources. pp. 188-189

14-17 Darwin's Black Box, p. 70-72, as quoted in
http://www.creationevolution.net/ irreducible_complexity.htm

14-18 http://www.wvi.net.au/mystery

14-19 Summation of a portion of the testimony during the Kiltzmiller vs. Dover Area School District PA 2005 Intelligent Design trial. (9/26/2005-11/4/2005)
http://www.talkorigins.org/faqs/dover/day10pm.html

14-20 http://www.thankgodforatheism.org/media_and_commentary/articles/inherent _flaws_of_irreducible_complexity

14-21 http://www.millerandlevine.com/km/evol/design1/article.html

14-22 Summation of a portion of the testimony during the Kiltzmiller vs. Dover Area School District PA 2005 Intelligent Design trial. (9/26/2005-11/4/2005)
http://www.talkorigins.org/faqs/dover/day1pm.html

14-23 http://www.eyedesignbook.com/ch6/Dembski_IrreducibleComplexity Revisited _011404.pdf

14-23a Rana, Fazale. The Cell's Design - How Chemistry Reveals the Creator's Artistry. BakerBooks. Grand Rapids, Michigan. 2008. Pp. 271-272

14-24 Scott, E.C. 2009. Evolution vs Creationism—An Introduction. Berkeley (CA): University of California Press. P. 212.

14-25 Shermer, Michael. 2006. Why Darwin Matters—The Case Against Intelligent Design. Owl Books. New York. Pp. 75-76 and http://www.scienceagogo. com/news/20040122185500data_trunc_sys. shtml

14-26 Evolution Disproved Encyclopedia—Volume 2 http://evolutionfacts.com/ Ev-V2/2evlch11.htm

14-26a Blount, Z. D. et al. 2012. Genomic analysis of a key innovation in an experimental Escherichia coli population. Nature. 489 (7417): 513-518. https://www.ncbi.nlm.nih.gov/pubmed/22992527

14-27 http://www.icr.org/article/7083/

14-28 http://creationwiki.org/E-coli_mutation_and_evolution

14-29 http://www.icr.org/article/7083/

14-30 http://creationwiki.org/E-coli_mutation_and_evolution

14-30a Jeffrey P. Tomkins. 2012. The Design and Complexity of the Cell. Institute for Creation Research. Dallas, Texas. P. 45.

14-31 http://www.creationevolution.net/irreducible_complexity.htm

14-32 Summation of a portion of the testimony made during the Kiltzmiller vs. Dover Area School District PA 2005 Intelligent Design trial. (9/26/2005-11/4/2005) 20pm562, 20pm565, 20pm566, 20pm570, 20pm571, 20pm572, 20pm577 http://www.talkorigins.org/faqs/dover/day20pm2.html

14-33 Michael Denton. 1985. Evolution: A Theory in Crisis. Adler & Adler. 4550 Montgomery Avenue, Bethesda, Maryland 20814. P. 108

14-34 http://www.millerandlevine.com/km/evol/design1/article.html

14-35 http://www2.oakland.edu/biology/lindemann/cf.htm—Figure 2

14-36 http://www.millerandlevine.com/km/evol/design1/article.html

14-37 Strobel, L. 2004. The Case for a Creator. The Evidence of Biochemistry: The Complexity of Molecular Machines. Grand Rapids (MI): Zondervan. Pp. 253-256

14-38 Creation or Evolution: Does It Really Matter What You Believe? 2008. United Church of God, an International Association. Cincinnati, OH. p. 46

14-39 Behe, M. p. 260. In: Strobel, L. 2004. The Case for a Creator. The Evidence of Biochemistry: The Complexity of Molecular Machines. Grand Rapids (MI): Zondervan. Pp. 260-262

14-40 Behe, M. In: Strobel, L. 2004. The Case for a Creator—The Evidence of Biochemistry: The Complexity of Molecular Machines. Grand Rapids (MI): Zondervan. Pp. 260-262

14-41 Behe M. In: Strobel, L. 2004.The Case For a Creator—The Evidence of Biochemistry: The Complexity of Molecular Machines,. Grand Rapids (MI): Zondervan. Pp. 260-262

14-42 http://www.creationevolution.net/irreducible_complexity.htm

14-43 http://www.creationevolution.net/irreducible_complexity.htm

14-44 http://users.rcn.com/jkimball.ma.ultranet/BiologyPages/C/Clotting.html

14-45 http://www.creationevolution.net/irreducible_complexity.htm

14-46 http://scrubtv.com/cardiology/blood-clotting-cascade-part-1-of-2-video_a7b2c6aee.html

http://www.creationevolution.net/irreducible_complexity.htm

14-47 Michael Behe, as quoted in: The Case For a Creator—The Evidence of Biochemistry: The Complexity of Molecular Machines, Lee Strobel. 2004. Zondervan. Grand Rapids, Michigan. pp. 260-262

14-48 Coyne, Jerry A., 2009, Why Evolution is True, Viking Penguin Group, New York, 375 Hudson Street, New York, 10014, P. 138-139

14-49 Coyne, Jerry A., 2009, Why Evolution is True, Viking Penguin Group, New York, 375 Hudson Street, New York, 10014, P. 139

14-50 Coyne, Jerry A., 2009, Why Evolution is True, Viking Penguin Group, New York, 375 Hudson Street, New York, 10014, P. 140

14-51 http://www.millerandlevine.com/km/evol/DI/clot/Clotting.html

14-52 http://www.thankgodforatheism.org/media_and_commentary/articles/inherent _flaws_of_irreducible_complexity

14-53 The discussion here is a summary of a portion of a testimony of what was said during the Kitzmiller versus Dover Area School District Court case of 9/26/2005-11/4/2005, trial.
http://www.talkorigins.org/faqs/dover/day1pm.html
The entire Kiltzmiller versus Dover Area School District, 2005 trial transcript is available at:
http:// www.talkorigins.org/faqs/dover/kitzmiller_v_dover.html

14-54 The discussion here is a summary of a portion of the testimony of what was said during the Kitzmiller versus Dover Area School District Court case of 9/26/2005-11/4/2005, trial.
http://www.talkorigins.org/faqs/dover/ day11am.html
The entire Kiltzmiller versus Dover Area School District, 2005 trial transcript is available at:

http://www.talkorigins.org/faqs/dover/ kitzmiller_v_dover.html

14-55 http://www.evolutionnews.org/2008/12/how_kenneth_miller_used_smokea_1014971.html

14-56 Evolution Disproved Encyclopedia—Volume 02
http://evolutionfacts.com/ Ev-V2/2evlch09b.htm

14-57 http://www.wiebefamily.org/e.htm

14-58 Summation of a portion of a testimony made during the Kiltzmiller vs. Dover Area School District PA 2005 Intelligent Design trial. (9/26/2005-11/4/2005). Day 1 (September 26), AM Session, Part 2
http://www.talkorigins.org/faqs/ dover/day1pm.html

14-59 A summation of a portion of a testimony made during the Kiltzmiller vs. Dover Area School District PA 2005 Intelligent Design trial. (9/26/2005-11/4/2005). Day 1 (September 26).
http://www.talkorigins.org/faqs/dover/day1pm.html

14-60 http://en.wikipedia.org/wiki/V(D)J_recombination

14-61 A summation of a portion of the testimony made during the Kiltzmiller vs. Dover Area School District PA 2005 Intelligent Design trial. (9/26/2005-11/4/2005). Day 1 (September 26), AM Session, Part 2 summation.
http://www.talkorigins.org/faqs/dover/day1pm.html

14-62 A summation of a portion of the testimony made during the Kiltzmiller vs. Dover Area School District PA 2005 Intelligent Design trial. (9/26/2005-11/4/2005). Day 1 (September 26)
http://www.talkorigins.org/ faqs/dover/day1pm.html

14-63 http://www.thefreedictionary.com/transposon

14-64 RAG1 Core and V(D)J Recombination Signal Sequences Were Derived from Transib Transposons. Vladimir V Kapitonov and Jerzy Jurka. Received: September 23, 2004; Accepted: March 17, 2005; Published: May 24, 2005
http://www.plosbiology.org/article/info%3Adoi%2F10.1371%2Fjournal.pbio.0030181

14-65 http://medical-dictionary.thefreedictionary.com/enzyme

14-66 The descent of the antibody-based immune system by gradual evolution. Jan Klein and Nikolas Nikolaidis. Proceedings of the National Academy of Science of the United States of America. 2005 January 4; 102(1): 169–174. Published online 2004 December 23. doi: 10.1073/pnas.0408480102
http://www.ncbi.nlm.nih.gov/pmc/articles/PMC544055/

14-67 A summation of a portion of the testimony made during the Kiltzmiller vs. Dover Area School District PA 2005 Intelligent Design trial. (9/26/2005-11/4/2005). Day 12 (10/7/2005)

http://www.talkorigins.org/ faqs/dover/day12pm.html

14- 67a Sarcode definition sources:

https://medical-dictionary.thefreedictionary.com/sarcode

https://www.collinsdictionary.com/dictionary/english/sarcode

https://www.thefreedictionary.com/sarcode

https://findwords.info/term/sarcode

14-68 http://en.wikipedia.org/wiki/Evolution_of_the_eye How Science Works: Evolution—A Student Primer. R. John Ellis:

http://www.springerlink.com/content/t766m3#section=652582&page=1

http://www.springerlink.com/content/g64008327g473264/fulltext.pdf Chapter 5, Figure 5.4, p. 90

14-69   http://www.thankgodforatheism.org/media_and_commentary/articles/ inherent _ flaws_of_irreducible_complexity

14-69a http://scienceblogs.com/pharyngula/2011/06/30/complex-eyes-in-the-cambrian/

14-70 http://en.wikipedia.org/wiki/Evolution_of_the_eye How Science Works: Evolution—A Student Primer. R. John Ellis:

http://www.springerlink.com/content/t766m3#section=652582&page=1

http://www.springerlink.com/content/g64008327g473264/fulltext.pdf Chapter 5, Figure 5.4, p. 90

14-71 http://en.wikipedia.org/wiki/Irreducible_complexity

14-72 Shermer, "Science Friction", p. 183-186

http://www.thejesusmyth.com/toolbox/evolution

14-73 Coyne, J. A. 2009. Why Evolution is True. New York (NY): Viking Penguin Group. Pp. 141-143

14-74 Shermer, Science Friction, p. 183-186

http://www.thejesusmyth.com/toolbox/evolution

14-75 http://www.proof-of-evolution.com/evolution-of-the-eye.html

14-76 Scott, E. C. 2009. Evolution vs Creationism—An Introduction. Berkeley (CA): University of California Press. P. 210-211

14-77 Shermer, Michael. 2006. Why Darwin Matters—The case Against Intelligent Design. Owl Books. New York. Pp. 78-79

14-78 L. H. Cohen, Darwin was Wrong (1984), pp. 121-122. In Evolution Disproved Encyclopedia—Volume 2

http://evolutionfacts.com/Appendix/a13b.htm

14-79 http://creation.mobi/did-eyes-evolve-by-darwinian-mechanisms

14-80 Jonathan Sarfati, Ph.D. 2002. Refuting Evolution 2. Master Books. Green Forest. AR. P 163.

14-81 http://creation.mobi/did-eyes-evolve-by-darwinian-mechanisms

14-82 Jonathan Sarfati, Ph.D. 2002. Refuting Evolution 2. Master Books. Green Forest. AR. P 162-167

14-83 L H. Cohen, Darwin was Wrong (1984), p. 126. In: Evolution Disproved Encyclopedia—Volume 2
http://evolutionfacts.com/Appendix/a13a.htm

14-84 Dr. Alan Hayward, Physicist. Copyright 1985. Publication Date 2005. Creation and Evolution—Rethinking the Evidence From Science and the Bible. WIPF & Stock Publishers. Eugene, Oregon. Pp. 48-49 Sarfati, J. Ph.D., F. M. First Printing 1999. Fourth Edition 2008. Refuting Evolution. Creation Ministries International. Eight Mile Plains. Australia. P. 129

14-85 Parker, Gary. Creation Facts of Life—How Real Science Reveals the Hand of God. 1980. Master Books. Green Forest. AR. Pp. 46-47, 154 156

14-86 Dr. Alan Hayward, Physicist. Copyright 1985. Publication Date 2005. Creation and Evolution—Rethinking the Evidence From Science and the Bible. WIPF & Stock Publishers. Eugene, Oregon. Pp. 48-49 Sarfati, J. Ph.D., F. M. First Printing 1999. Fourth Edition 2008. Refuting Evolution. Creation Ministries International. Eight Mile Plains. Australia. P. 129

14-87 http://www.trueorigin.org/trilobites_eyes.asp

14-88 Dr. Alan Hayward, Physicist. Copyright 1985. Publication Date 2005. Creation and Evolution—Rethinking the Evidence From Science and the Bible. WIPF & Stock Publishers. Eugene, Oregon. Pp. 48-49 Sarfati, J. Ph.D., F. M. First Printing 1999. Fourth Edition 2008. Refuting Evolution. Creation Ministries International. Eight Mile Plains. Australia. P. 129

14-89 LL. Cohen, Darwin was Wrong (1984), p. 125. In: Evolution Disproved Encyclopedia—Volume 2
http://evolutionfacts.com/Appendix/a13b.htm

14-90 Evolution Disproved Encyclopedia—Volume 2
http://evolutionfacts.com/ Appendix/a13b.htm

14-91 Norman Macbeth, Speech of Harvard University, September 24, 1983, quoted in LD. Sunderland, Darwin's Enigma (1988), p. 150. In: Evolution Disproved Encyclopedia—Volume 2
http://evolutionfacts.com/Appendix/a13b.htm

14-92 http://www.trueorigin.org/trilobites_eyes.asp

14-93 Michael Pitman, Adam and Evolution (1984), p. 129. In: Evolution Disproved Encyclopedia—Volume 2
http://evolutionfacts.com/Appendix/a13a.htm

14-94 Howard, Jeffery, 2008, Disprove Darwin in 5 Minutes or Less, Twin Angels Publishing, Inc., 777 E. Atlantic Avenue, C2-271, Delray Beach, Florida, 33483, www.disprovedarwin.com, Page 30

14-95 Finding Darwin's God, New York: Cliff Street Books, Kenneth R. Miller, 2000, p. 101

14-96 (Natural Selection: Domains, Levels and Challenges, Oxford University Press, G. C. Williams, 1992, pp. 72, 73)

14-97 Strobel, Lee, 2004, The Case for a Creator, Zondervan, Grand Rapids, Michigan, 49530 zreview@zondervan.com, p. 105

14-98 Michael Pitman, Adam and Evolution (1984), p. 215. In: Evolution Disproved Encyclopedia—Volume 2
http://evolutionfacts.com/Ev-V2/2evlch13.htm

14-99 Jonathan Sarfati, Ph.D. 2002. Refuting Evolution 2. Master Books. Green Forest. AR. Pp. 119-120.

14-100 Charles Colson and Nancy Pearcey. How Now Shall We Live? 1999. Tyndale House Publishers, Inc. Wheaton, Illinois. P. 89.

14-100a http://www.vedicsciences.net/articles/intelligent-design.html

14-101 M. J. Behe. 1996. Darwin's Black Box: The Biochemical Challenge to Evolution. New York, NY: The Free Press. P. 46 In: Jonathan Sarfati, Ph.D. 2002. Refuting Evolution 2. Master Books. Green Forest. AR. P 165.

14-102 Jonathan Sarfati, Ph.D. 2002. Refuting Evolution 2. Master Books. Green Forest. AR. P 162-167

14-103 August 2005
http://www.apologeticspress.org/articles/374

14-104 Evolutionary origins of the nervous system. ScienceBlogs. July 3, 2009.
https://scienceblogs.com/neurophilosophy/2009/07/03/evolutionary-origins-of-thenervous-system
Some of the information related to this article appears to be debatable. See comments below the article.

14-105 A brief history of the brain. New Scientist. David Robson. September 21, 2011.
https://www.newscientist.com/article/
mg21128311-800-a-brief-history-of-thebrain/#

Chapter 15—A Comparison of Chimpanzee and Human Genome

15-1 Genesis 2:7, 21-22; 3:20 The Bible, Scripture taken from the New King James Version. Copyright © 1982 by Thomas Nelson, Inc. Used by permission. All rights reserved.

15-2 Horgan J. The New Social Darwinists, Scientific American 273(4):150-157, October 1995; quote on p. 151.

http://www.btwol.com/07_Articles/List%20of%20articles/Humans%20 Image%20of%20God%20or%20advanced%20apes.htm

Horgan J. "The new social Darwinists, Scientific American 273(4):150-157, October 1995; quote on p. 151. In: Sarfati, J. Ph.D., F.M. First Printing 1999.

Fourth Edition 2008. Refuting Evolution. Creation Ministries International. Eight Mile Plains. Australia. P.89.

15-3 Coyne, J. A. 2009. Why Evolution is True. New York (NY): Viking Penguin Group. P.192

15-4 I Love Jesus & I accept Evolution, Denis O. Lamoureux, 2009. WIPF Stock. Eugene. Oregon. P. 133

15-5 Haarsma, Deborah B. & Loren D., 2007, Origins—A Reformed Look At Creation, Design, & Evolution, Faith Alive Christian Resources, 2850 Kalamazoo Avenue, SE, Grand Rapids, MI 49560, pp. 167-171

15-6 Haarsma, Deborah B. & Loren D., 2007, Origins—A Reformed Look At Creation, Design, & Evolution, Faith Alive Christian Resources, 2850 Kalamazoo Avenue, SE, Grand Rapids, MI 49560, pp. 199-203

15-7 Haarsma, Deborah B. & Loren D., 2007, Origins—A Reformed Look At Creation, Design, & Evolution, Faith Alive Christian Resources, 2850 Kalamazoo Avenue, SE, Grand Rapids, MI 49560, pp. 167-171

15-8 Stephen C. Meyer, Scott Minnich, Jonathan Moneymaker, Paul A. Nelson, Ralph Seelke. 2007. Explore Evolution—The Arguments For and Against Neo-Darwinism. Hill House Publishers. Melbourne & London. Pp. 40, 44.

15-9 Stephen C. Meyer, Scott Minnich, Jonathan Moneymaker, Paul A. Nelson, Ralph Seelke. 2007. Explore Evolution—The Arguments For and Against Neo-Darwinism. Hill House Publishers. Melbourne & London. P. 43-45

15-10 Parker, Gary. Creation Facts of Life—How Real Science Reveals the Hand of God. 2006. Master Books. Green Forest. AR. P. 48

15-11 Humans—Images of God or Advanced Apes—Jonathan Sarfiti http://www.btwol.com/07_Articles/List%20of%20articles/ Humans%20 -Image%20 of%20God%20or%20advanced%20apes.htm

15-12 http://encyclopedia2.thefreedictionary.com/Hemoglobin

15-12a https://www.sharecare.com/health/blood-basics/ what-is-hemoglobin-how-work

15-13 http://www.umass.edu/microbio/chime/hemoglob/heme.htm

15-14 Shermer. Michael. Why Darwin Matters—The Case Against Intelligent Design. 2006. Owl Books. New York. Pp. 73-74

15-15 A summation of a portion of the testimony of Professor Kenneth Miller, when questioned by Witold Walczak, American Civil Liberties Union of Pennsylvania, for the plaintiffs, upon direct examination, during the Kitzmiller versus Dover Area School District trial, day 1, 9/26/2005. http://www.talkorigins.org/faqs/dover/day1pm.html

15-16 http://www.talkorigins.org/faqs/dover/day11pm2.html

15-17 http://www.talkorigins.org/faqs/dover/day10pm.html

15-18 Michael Denton, Evolution: A Theory in Crisis (1985). In: Evolution Disproved Encyclopedia—Volume 3 http://evolutionfacts.com/Ev-V3/3evlch21.htm

15-19 Evolution Disproved Encyclopedia—Volume 3 http://evolutionfacts.com/ Ev-V3/3evlch21.htm

15-20 G.R. Taylor, Great Evolution Mystery (1983), p. 174. In: Evolution Disproved Encyclopedia—Volume 3 http://evolutionfacts.com/Ev-V3/3evlch21.htm

15-21 Evolution Disproved Encyclopedia—Volume 3 http://evolutionfacts.com/ Ev-V3/3evlch21.htm

15-22 Loren Haarsma, Faith Alive Christian Resources, 2850 Kalamazoo Avenue, SE, Grand Rapids, Michigan, 49560 www.faithaliveresources.org/origins, Human Genomic Organization and Introns

15-23 Haarsma, L. Genetic Evidence for Evolution. Faith Alive Christian Resources, 2850 Kalamazoo Avenue, SE, Grand Rapids, Michigan, 49560

15-24 http://www.don-lindsay-archive.org/creation/translocation.html

15-25 Haarsma, L. Human Genomic Organization and Introns Grand Rapids (MI): Faith Alive Christian Resources www.faithaliveresources.org/origins

15-26 Lamoureux, D. O. 2009. I Love Jesus & I Accept Evolution. Eugene (OR): WIPF Stock. P. 128

15-27 Loren Haarsma, Faith Alive Christian Resources, 2850 Kalamazoo Avenue, SE, Grand Rapids, Michigan, 49560, Human Genomic Organization and Introns. www.faithaliveresources.org/origins

15-28 Comparison of the Human and Great Ape Chromosomes as Evidence for Common Ancestry

http://www.sciencemag.org/cgi/content/full/286/5439/458?ijkey= wdICO7J 7uPLqc

http://www.gate.net/~rwms/hum_ape_chrom.html

15-29 A summation of a portion of the testimony of Professor Kenneth Miller, when questioned by Witold Walczak, American Civil Liberties Union of Pennsylvania, for the plaintiffs, upon direct examination, during the Kitzmiller versus Dover Area School District trial, day 1, 9/26/2005. http://www.talkorigins.org/faqs/dover/day1pm.html

15-30 http://www.thefreedictionary.com/telomere

15-31   http://www.nature.com/scitable/topicpage/telomeres-of-human-chromosomes -21041

15-32 http://en.wikipedia.org/wiki/Telomere

15-33 A summation of a portion of the testimony of Professor Kenneth Miller, when questioned by Witold Walczak, American Civil Liberties Union of Pennsylvania, for the plaintiffs, upon direct examination, during the Kitzmiller versus Dover Area School District trial, day 1, 9/26/2005. http://www.talkorigins.org/faqs/dover/day1pm.html

15-34 British journal Nature. 2004. 'The Generation and Annotation of the DNA Sequences of Human Chromosomes 2 and 4' Hillier et. al. In: http://www.talkorigins.org/faqs/dover/day1pm.html

15-35 A summation of a portion of the testimony of Professor Kenneth Miller, when questioned by Witold Walczak, American Civil Liberties Union of Pennsylvania, for the plaintiffs, upon direct examination, during the Kitzmiller versus Dover Area School District trial, day 1, 9/26/2005. http://www.talkorigins.org/faqs/dover/day1pm.html

15-36 Reference number not used

15-37 http://www.apologeticspress.org/apcontent.aspx?category=9&article=801

15-38 http://en.wikipedia.org/wiki/Chimpanzee_genome_project#Genes_of_the_ Chromosome_2_fusion_site

15-39 http://www.apologeticspress.org/apcontent.aspx?category=9&article=801

15-39a Human-Chimp DNA Comparison. Jeff Tomkins. Acts & Facts. Institute for Creation Research. July 2018. Pp. 5-8 Footnote 4: Tomkins, J. P. 2013. Alleged Human Chromosome 2 "Fusion Site" Encodes an Active DNA

Binding Domain Inside a Complex and Highly Expressed Gene - Negating Fusion. Answers Research Journal 6:367-375. Footnote 5: Tomkins. J.P. 2017. Debunking the Debunkers: A Response to Criticism and Obfuscation Regarding Refutation of the Human Chromosome 2 Fusion. Answers Research Journal. 10:45-54.

15-39b Human Chromosome 2 Fusion Never Happened. Acts & Facts. Institute for Creation Research. Vol. 49. No. 5. Jeffrey P. Tomkins, Ph.D. May 2020. Pp. 16-17.

15-40 Perspectives on Science and Christian Faith, Volume 58, Number 3, September 2006 News & Views Creation Versus Creationism
http://www.asa3.org/ASA/PSCF/2006/PSCF9-06Finlay.pdf

15-41 http://www.talkorigins.org/faqs/comdesc/glossary.html#vestigial

15-42 Coyne, J. A. 2009. Why Evolution is True. New York (NY): Viking Penguin Group. P. 195

15-43 Maddox Dr. B. (2003)
http://www.scribd.com/doc/8379943/Are-You-Being-Brainwashed
http:// www.arrivalofthefittest.com/Are_You_Being_Brainwashed2.html

15-44 Human-chimp DNA difference trebled, 23 September 2002, NewScientist. com news service. Nature May 27, 2004, pp 382-388, as quoted in:
http://www.scribd.com/doc/8379943/Are-You-Being-Brainwashed
http://www.arrivalofthefittest.com/Are_You_Being_Brainwashed2.html

15-45 Parker, Gary. Creation Facts of Life—How Real Science Reveals the Hand of God. 2006. Master Books. Green Forest. AR. Pp. 51-52

15-46 Tomkins, J. Ph.D. Human-Chimp Similarities
http://www.icr.org/article/4624/

15-47 Human Chimp DNA Similarity-Evidence for evolutionary relationship? published: Creation 19(1): 21-22 December 1996 by Batten D. Answers In Genesis
http://www.answersingenesis.org/creation/v19/i1/dna.asp

15-48 Human-Chimp DNA Comparison. Jeff Tomkins. Acts & Facts. Institute for Creation Research. July 2018. Pp. 5-8. Footnote 1: Tomkins, J. and J. Bergman. 2012. Geomic monkey business - estimate of nearly identical human-chimp DNA similarity re-evaluated using omitted data. Journal of Creation. 29 (1):94-100. Footnote 2: Tomkins. J P. 2016. Analysis of 101 Chimpanzee Trace Read Data Sets: Assessment of Their Overall Similarity to Human and Possible Contamination With Human DNA. Answers Research Journal. 9:294-298.

Chapter 16—A Look at the Human and Ape Fossil Evidence

16-1 Genesis 1:27, (also quoted in Matthew 19:4, Mark 10:6), The Bible, Scripture taken from the New King James Version. Copyright © 1982 by Thomas Nelson, Inc. Used by permission. All rights reserved.

16-2 Wood B. "A Date with Java Man," review of Java Man, by Garniss Curtis, Carl Swisher, and Roger Lewin, NewScience (17 February 2001):54. In: Lubenow M. 2004. Bones of Contention. Grand Rapids (MI): Baker Books. P. 31

16-3 http://www.thejesusmyth.com/toolbox/evolution

16-4 http://www.thejesusmyth.com/toolbox/evolution

16-5 Denis O. Lamoureux, I Love Jesus & I Accept Evolution, 2009. WIPF Stock. Eugene. Oregon. P. 124

16-5a http://www.gutenberg.org/cache/epub/2931/pg2931.txt
The first book devoted expressly to the topic of human evolution
http://www.macroevolution.net/ape-to-human-evolution.html

16-6 Coyne. J. A. Why Evolution is True. 2009. New York (NY): Viking Penguin Group. p. 192

16-7 http://www.pbs.org/wgbh/evolution/library/faq/cat02.html

16-8 http://hubpages.com/hub/Did-man-really-evolve-from-apes

16-9 Coyne, J. A. Why Evolution is True. 2009. New York (NY): Viking Penguin Group. Pp. 207-208

16-10 http://www.wwnorton.com/college/anthro/evolve4/ch/10/welcome.shtml
http:// www.theprimata.com/tree_shrews.html

16-11 http://dictionary.reference.com/browse/primate?s=t&path=/

16-12 http://www.pbs.org/wgbh/evolution/library/faq/cat02.html

16-13 Haarsma, Deborah B. & Loren D., 2007, Origins—A Reformed Look At Creation, Design, & Evolution, Faith Alive Christian Resources, 2850 Kalamazoo Avenue, SE, Grand Rapids, MI 49560, pp. 199-203

16-14 Lamoureux, D. O. 2009. I Love Jesus & I Accept Evolution. Eugene (OR): WIPF Stock. P. 129

16-15 Lamoureux, Denis O., I Love Jesus & I Accept Evolution, 2009, WIPF Stock, 199 W. 8th Avenue, Suite 3, Eugene, Oregon, 97401, P. 133

16-16 Coyne, Jerry A., 2009, Why Evolution is True, Viking Penguin Group, New York, 375 Hudson Street, New York, 10014, Pp. 207-208

16-17 Lamoureux, Denis O., I Love Jesus & I Accept Evolution, 2009, WIPF Stock, 199 W. 8th Avenue, Suite 3, Eugene, Oregon, 97401, P. 124

16-18 Lamoureux, Denis O., I Love Jesus & I Accept Evolution, 2009, WIPF Stock, 199 W. 8th Avenue, Suite 3, Eugene, Oregon, 97401, Caption under Figure 6-2 p. 125

16-19 Lamoureux, Denis O., I Love Jesus & I Accept Evolution, 2009, WIPF Stock, 199 W. 8th Avenue, Suite 3, Eugene, Oregon, 97401, Caption under Figure 6-3 P. 125.

16-20 Lamoureux, Denis O., I Love Jesus & I Accept Evolution, 2009, WIPF Stock, 199 W. 8th Avenue, Suite 3, Eugene, Oregon, 97401, P. 124.

16-21 Coyne, J. A. 2009. Why Evolution is True. New York (NY): Viking Penguin Group. P. 192

16-22 Coyne, Jerry A., 2009, Why Evolution is True, Viking Penguin Group, New York, 375 Hudson Street, New York, 10014, P. 208

16-23 Coyne, Jerry A., 2009, Why Evolution is True, Viking Penguin Group, New York, 375 Hudson Street, New York, 10014, Pp. 203-204

16-24 Buettner-Janush. J. (1963) "An Introduction to the Primates" in Evolutionary and Genetic Biology of Primates, 2 volumes, ed. J. Buettner-Janusch, Academic Press, New York, volume 1, p. 3 In: Michael Denton. 1985. Evolution: A Theory in Crisis. Adler & Adler. 4550 Montgomery Avenue, Bethesda, Maryland 20814. P. 116

16-25 Lamoureux, Denis O., I Love Jesus & I Accept Evolution, 2009, WIPF Stock, 199 W. 8th Avenue, Suite 3, Eugene, Oregon, 97401, P. 133

16-26 Lamoureux, Denis O. I Love Jesus & I Accept Evolution. 2009. WIPF Stock, 199 W. 8th Avenue, Suite 3, Eugene, Oregon. P. 133

16-27 George Gaylord Simpson, "The Nonprevalence of Humanoids," in Science 143 (1964) p. 770. In: Evolution Disproved Encyclopedia—Volume 3 http://evolutionfacts.com/Ev-V3/3evlch31b.htm

16-28 Huse, S. M. 1983. The Collapse of Evolution. Grand Rapids (MI): Baker Book House. Pp. 96-97

16-29 http://www.cartage.org.lb/en/themes/sciences/paleontology/paleozoology/ fossil Hominids/ChartHumanEvolution/ChartHumanEvolution.htm

16-30 http://encyclopedia2.thefreedictionary.com/Ramapithecus

16-31 Gish, D. T. Ph. D. Have You Been Brainwashed?

16-32 http://www.wiebefamily.org/e.htm

16-33 http://www.proof-of-evolution.com/human-evolution-timeline.html

16-34 http://www.talkorigins.org/faqs/homs/species.htm

16-35 http://www.talkorigins.org/faqs/homs/species.htm

16-36 Origins—a Reformed Look at Creation, Design, & Evolution, Deborah B. Haarshma and Loren D. Haarsma, pp. 199-203 Haarsma, Deborah B.

& Loren D., 2007, Origins—A Reformed Look At Creation, Design, & Evolution, Faith Alive Christian Resources, 2850 Kalamazoo Avenue, SE, Grand Rapids, MI 49560, pp. 199-203

16-37 http://warforscience.blogspot.com/2009/12/answering-creationist-claims-part2b-no.html

16-38 http://www.talkorigins.org/faqs/faq-transitional/part2a.html

16-39 http://www.archaeologyinfo.com/ardipithecusramidus.htm

16-40 Julie Brewington, 1995, ASM101

http://web.mesacc.edu/dept/d10/asb/origins/ramidus.html

16-41 http://www.proof-of-evolution.com/human-evolution-timeline.html

16-42 Lamoureux, Denis O., I Love Jesus & I Accept Evolution, 2009, WIPF Stock, 199 W. 8th Avenue, Suite 3, Eugene, Oregon, 97401, Pp. 129, 131

16-43 Lamoureux. Denis O. I Love Jesus & I Accept Evolution. 2009. Eugene (OR): WIPF Stock. Pp. 129, 131

16-44 Lamoureux, D. O. 2009. I Love Jesus & I Accept Evolution. Eugene (OR): WIPF Stock. Caption under Figure 6-3 P. 125

16-45 Coyne, Jerry A., 2009, Why Evolution is True, Viking Penguin Group, New York, 375 Hudson Street, New York, 10014, p. 202

16-46 Lamoureux, D. O. 2009. I Love Jesus & I Accept Evolution. Eugene (OR): WIPF Stock. Figure 6-4, p. 126.

16-47 Coyne. Jerry A. Why Evolution is True. 2009. Viking. Penguin Group. New York. P. 202

16-48 Haarsma, Deborah B. & Loren D., 2007, Origins—A Reformed Look At Creation, Design, & Evolution, Faith Alive Christian Resources, 2850 Kalamazoo Avenue, SE, Grand Rapids, MI 49560, pp. 199-203

16-49 Denis O. Lamoureux, I Love Jesus & I Accept Evolution, Caption under Figure 6-3 P. 125.

16-50 Lamoureux, D. O. 2009. I Love Jesus & I Accept Evolution. Eugene (OR): WIPF Stock. P. 133

16-51 Coyne, Jerry A., 2009, Why Evolution is True, Viking Penguin Group, New York, 375 Hudson Street, New York, 10014, p. 202

16-52 From the Beginning—The Story of Human Evolution. David Peters. 1991. Morrow Junior Books. New York. p. 102.

http://www.davidpetersstudio. com/FromTheBeginning.pdf

16-53 The Plain Truth, In Search of Adam, Robert A. Ginskey, August 1979 p. 24

16-54 Briski, D. 2004. Impressive Deception—Creation or Evolution You Decide. Shippensburg, (PA): Ragged Edge Press. P. 118

http://www.bestbiblescience. org/apeimage.htm

16-55 Peter Andrews, "The Descent of Man," in New Scientist, 102:24 (1984). In: Evolution Disproved Encyclopedia—Volume 2 http://evolutionfacts.com/ Ev-V2/2evlch18a.htm

16-56 Lubenow, Marvin, Bones of Contention, 2004, Baker Books, Baker Books Publishing Group, P.O. Box 6287, Grand Rapids, Michigan, 49516-6287, www.bakerbooks.com, pp. 298-301

16-57 Parker, G. 1980. Creation Facts of Life. Green Forest (AR): Master Books. Pp 183

16-58 The Plain Truth, In Search of Adam, Robert A. Ginskey, August 1979 p. 24

16-59 Evolution Disproved Encyclopedia—Volume 2 http://evolutionfacts.com/Ev-V2/2evlch18a.htm

16-60 R. Milner, Encyclopedia of Evolution (1990), p. 69. In: Evolution Disproved Encyclopedia—Volume 2 http://evolutionfacts.com/Ev-V2/2evlch18a.htm

16-61 http://www.icr.org/article/lucys-new-foot-bone-actually-human/

16-62 http://www.wiebefamily.org/e.htm

16-63 Dr. Alan Hayward, Physicist. Copyright 1985. Publication Date 2005. Creation and Evolution—Rethinking the Evidence From Science and the Bible. WIPF & Stock Publishers. Eugene, Oregon. p. 52

16-64

    AL 333-160—>

    http://www.boneclones.com/KO-393-MET.htm

    Al 228-1—>

    http://www.talkorigins.org/faqs/homs/lucy.html

16-65 Dr. Alan Hayward, Physicist. Copyright 1985. Publication Date 2005. Creation and Evolution—Rethinking the Evidence From Science and the Bible. WIPF & Stock Publishers. Eugene, Oregon. p. 52

16-66 Lubenow M. 2004. Bones of Contention. Grand Rapids (MI): Baker Books. Pp. 301-302

16-67 http://www.proof-of-evolution.com/human-evolution-timeline.html

16-68 The Plain Truth, In Search of Adam, Robert A. Ginskey, August 1979 p. 24

16-69 http://www.talkorigins.org/faqs/faq-transitional/part2a.html

16-70 http://www.handprint.com/LS/ANC/htxt3.html

16-71 http://paleontology.wikia.com/wiki/Australopithecus

16-72 http://www.proof-of-evolution.com/human-evolution-timeline.html

16-73 Coyne, J. A. 2009. Why Evolution is True. New York (NY): Viking Penguin Group. P. 204

16-74 Coyne, Jerry A., 2009, Why Evolution is True, Viking Penguin Group, New York, 375 Hudson Street, New York, 10014, P. 129

16-75 http://www.talkorigins.org/faqs/faq-transitional/part2a.html

16-76 Coyne, Jerry A., 2009, Why Evolution is True, Viking Penguin Group, New York, 375 Hudson Street, New York, 10014, P. 205

16-77 Lamoureux, Denis O., I Love Jesus & I Accept Evolution, 2009, WIPF Stock, 199 W. 8th Avenue, Suite 3, Eugene, Oregon, 97401, P. 131

16-78 Lamoureux, Denis O., I Love Jesus & I Accept Evolution, 2009, WIPF Stock, 199 W. 8th Avenue, Suite 3, Eugene, Oregon, 97401, P. 204-205

16-79 Coyne, Jerry A., 2009, Why Evolution is True, Viking Penguin Group, New York, 375 Hudson Street, New York, 10014, P. 205

16-80 http://www.proof-of-evolution.com/human-evolution-timeline.html

16-81 Lubenow M. 2004. Bones of Contention. Grand Rapids (MI): Baker Books. Pp. 298-301

16-82 http://www.answersingenesis.org/tj/v8/i1/erectus.asp

16-83 Evolution Disproved Encyclopedia—Volume 2
http://evolutionfacts.com/Ev-V2/2evlch18a.htm

16-84 Coyne, Jerry A., 2009, Why Evolution is True, Viking Penguin Group, New York, 375 Hudson Street, New York, 10014, P. 208

16-85 http://www.answersingenesis.org/tj/v13/i2/skull_1470.asp

16-86 Evolution Disproved Encyclopedia—Volume 2
http://evolutionfacts.com/Ev-V2/2evlch18a.htm

16-87 The Plain Truth, In Search of Adam, Robert A. Ginskey, August 1979 p. 26)

16-88 http://www.baptistlink.com/dman/fossil8.html

16-89 Lubenow M. 2004. Bones of Contention. Grand Rapids (MI): Baker Books. Pp. 327-328

16-90 http://anthro.palomar.edu/homo/homo_2.htm

16-91   http://www.proof-of-evolution.com/human-evolution-timeline.html Lubenow, Marvin, Bones of Contention, 2004, Baker Books, Baker Books Publishing Group, P.O. Box 6287, Grand Rapids, Michigan, 49516-6287, www. bakerbooks.com, pp. 124-129

16-92 Coyne, Jerry A., 2009, Why Evolution is True, Viking Penguin Group, New York, 375 Hudson Street, New York, 10014, P. 205

16-93 http://www.talkorigins.org/faqs/faq-transitional/part2a.html

16-94 Haarsma, Deborah B. & Loren D., 2007, Origins—A Reformed Look At Creation, Design, & Evolution, Faith Alive Christian Resources, 2850 Kalamazoo Avenue, SE, Grand Rapids, MI 49560, pp. 199-203

16-95 Coyne, Jerry A., 2009, Why Evolution is True, Viking Penguin Group, New York, 375 Hudson Street, New York, 10014, P. 206

16-96 http://www.talkorigins.org/faqs/faq-transitional/part2a.html

16-97 Lamoureux, Denis O., I Love Jesus & I Accept Evolution, 2009, WIPF Stock, 199 W. 8th Avenue, Suite 3, Eugene, Oregon, 97401, P. 131

16-98 Coyne, Jerry A., 2009, Why Evolution is True, Viking Penguin Group, New York, 375 Hudson Street, New York, 10014, P. 205

16-99 From the Beginning—The Story of Human Evolution. David Peters. 1991. Morrow Junior Books. New York. pp. 106-107.
http://www.davidpetersstudio.com /FromTheBeginning.pdf

16-100 Lubenow, Marvin, Bones of Contention, 2004, Baker Books, Baker Books Publishing Group, P.O. Box 6287, Grand Rapids, Michigan, 49516-6287, www.bakerbooks.com, pp. 124-129

16-101 http://www.wiebefamily.org/e.htm

16-101a Dykes, Susan. Reasons to Believe – Neanderthal/Human Co Existing

16-102 Lubenow, Marvin, Bones of Contention, 2004, Baker Books, Baker Books Publishing Group, P.O. Box 6287, Grand Rapids, Michigan, 49516-6287, www.bakerbooks.com, pp. 118-120

16-103 Lubenow M. 2004. Bones of Contention. Grand Rapids (MI): Baker Books. Pp. 120-122

16-104 http://www.talkorigins.org/faqs/hom/a_lubenow.html

16-105 Coyne, J. A. 2009. Why Evolution is True. New York (NY): Viking Penguin Group. P. 194

16-106 http://www.talkorigins.org/faqs/homs/a_java.html

16-107 http://www.merriam-webster.com/dictionary/wadjak%20man

16-108 Evolution Disproved Encyclopedia—Volume 2
http://evolutionfacts.com/Ev-V2/2evlch18a.htm

16-109 Huse, S. M. 1983. The Collapse of Evolution. Grand Rapids (MI): Baker Book House. P. 99

16-110 Gish, D. T. Ph. D. Have You Been Brainwashed?
Briski, D. 2004. Impressive Deception—Creation or Evolution You Decide. Shippensburg, (PA): Ragged Edge Press. P. 119

16-111 Gish, D. T. Ph. D. Have You Been Brainwashed?
Briski, D. 2004. Impressive Deception—Creation or Evolution You Decide. Shippensburg, (PA): Ragged Edge Press. P. 119

16-112 Lubenow M. 2004. Bones of Contention. Grand Rapids (MI): Baker Books. Pp. 86-91

16-113 Lubenow, Marvin, Bones of Contention, 2004, Baker Books, Baker Books Publishing Group, P.O. Box 6287, Grand Rapids, Michigan, 49516-6287, www.bakerbooks.com, pp. 86-91

16-114 Lubenow, Marvin, Bones of Contention, 2004, Baker Books, Baker Books Publishing Group, P.O. Box 6287, Grand Rapids, Michigan, 49516-6287, www.bakerbooks.com, pp. 86-91

16-115 Lubenow, Marvin, Bones of Contention, 2004, Baker Books, Baker Books Publishing Group, P.O. Box 6287, Grand Rapids, Michigan, 49516-6287, www.bakerbooks.com, pp. 92-97

16-116 http://www.proof-of-evolution.com/human-evolution-timeline.html

16-117 Coyne, Jerry A., 2009, Why Evolution is True, Viking Penguin Group, New York, 375 Hudson Street, New York, 10014, P. 205-206

16-118 Lubenow M. 2004. Bones of Contention. Grand Rapids (MI): Baker Books. Pp. 163-164

16-119 Briski, D. 2004. Impressive Deception—Creation or Evolution You Decide. Shippensburg, (PA): Ragged Edge Press. P. 120

16-120 http://www.proof-of-evolution.com/human-evolution-timeline.html

16-121 Coyne, Jerry A., 2009, Why Evolution is True, Viking Penguin Group, New York, 375 Hudson Street, New York, 10014, p. 194

16-122 Haarsma, Deborah B. & Loren D., 2007, Origins—A Reformed Look At Creation, Design, & Evolution, Faith Alive Christian Resources, 2850 Kalamazoo Avenue, SE, Grand Rapids, MI 49560, pp. 199-203

16-123 Origins—a Reformed Look at Creation, Design, & Evolution, Deborah B. Haarshma and Loren D. Haarsma, pp. 199-203

16-124 Coyne, Jerry A., 2009, Why Evolution is True, Viking Penguin Group, New York, 375 Hudson Street, New York, 10014, P. 205-206

16-125 The Plain Truth, Ambassador College, CA, Plain Truth, June-July, 1970, "The 'Missing Link' . . . Found!" Paul W. Kroll and Gene R. Hughes, p. 36-37

16-126 Lubenow M. 2004. Bones of Contention. Grand Rapids (MI): Baker Books. Pp. 260-263

16-127 Parker, G. 1980. Creation Facts of Life. Green Forest (AR): Master Books. Pp 176

16-128 Lubenow M. 2004. Bones of Contention. Grand Rapids (MI): Baker Books. Pp. 83-85

16-129   http://www.biblestudy.org/basicart/who-was-neanderthal-man-was-hethemissinglink.html

16-129a Ross, Hugh. The Genesis Question – Scientific Advances and the Accuracy of Genesis. (Second Expanded Edition). 2001. Navpress. Colorado. Pages 112-115.

16-130 Lamoureux, Denis O., I Love Jesus & I Accept Evolution, 2009, WIPF Stock, 199 W. 8th Avenue, Suite 3, Eugene, Oregon, 97401, P. 131

16-131 From the Beginning—The Story of Human Evolution. David Peters. 1991. Morrow Junior Books. New York. p. 109.

http://www.davidpetersstudio.com /FromTheBeginning.pdf

16-131a The Creation Club magazine. December 2019. "The Wonder of Soil." pp. 12-13. Doug Velting.

16-132 From the Beginning—The Story of Human Evolution. David Peters. 1991. Morrow Junior Books. New York. pp. 90-100.

http://www.davidpetersstudio. com /FromTheBeginning.pdf

16-133 From the Beginning—The Story of Human Evolution. David Peters. 1991. Morrow Junior Books. New York. pp. 106-111.

http://www.davidpetersstudio. com /FromTheBeginning.pdf

16-134 http://www.proof-of-evolution.com/human-evolution-timeline.html

16-135 Lubenow, Marvin, Bones of Contention, 2004, Baker Books, Baker Books Publishing Group, P.O. Box 6287, Grand Rapids, Michigan, 49516-6287, www.bakerbooks.com, pp. 124-129

16-136 Kroll P. W., and Hughes G. R. The Plain Truth, June-July, 1970, "The 'Missing Link' . . . Found!" p.38. Ambassador College, CA.

16-137 Gish, D. T. Ph. D. Have You Been Brainwashed?

16-138 http://www.wiebefamily.org/e.htm

16-139 Huse, Scott M., The Collapse of Evolution, 1983, Baker Book House, Grand Rapids, Michigan, 49508, pp. 97-98)

16-139a https://en.wikipedia.org/wiki/Nebraska_Man

16-140 Parker, Gary, Creation Facts of Life, 1980, Master Books, P.O. Box 726, Green Forest, AR, 72638, www.masterbooks.net, P. 178

16-141 Reference Number not used

16-142 Huse, Scott M., The Collapse of Evolution, 1983, Baker Book House, Grand Rapids, Michigan, 49508, pp. 97-98 Impressive Deception—Creation or Evolution—You Decide, pp. 57, 120

16-143 Evolution Disproved Encyclopedia—Volume 3

http://evolutionfacts.com/ Ev-V3/3evlch30.htm

16-144 A. Milner. Encyclopedia of Evolution (1990). P. 322. In: Evolution Disproved Encyclopedia—Volume 3

http://evolutionfacts.com/Ev-V3/3evlch30.htm

16-144a https://ncse.com/cej/5/2/role-nebraska-man-creation-evolution-debate

16-144b https://thecreationclub.com/the-frauds-of-evolution-5-the-nebraska-man-hoax/

16-145 http://www.talkorigins.org/faqs/homs/a_nebraska.html

16-146 http://www.thejesusmyth.com/toolbox/evolution

16-147 Howard, Jeffery, 2008, Disprove Darwin in 5 Minutes or Less, Twin Angels Publishing, Inc., 777 E. Atlantic Avenue, C2-271, Delray Beach, Florida, 33483, www.disprovedarwin.com, Page 43

16-148 Evolution Disproved Encyclopedia—Volume 3
http://evolutionfacts.com/ Ev-V3/3evlch30.htm

16-149 Evolution Disproved Encyclopedia—Volume 3
http://evolutionfacts.com/Ev-V3/3evlch30.htm
http://www.conservapedia.com/Piltdown_Man

Closing Comments

CC-1 How Science Works: Evolution A Student Primer R. John Ellis:
http://www. springerlink.com/content/
t766m3#section=652582&page=1Prior book version: Montane Publishers, South Africa (2008)
http://www.scribd.com/ doc/11464783/Is-Jesus-an-Evolutionist

CC-2 How Science Works: Evolution A Student Primer R. John Ellis:
http://www.springerlink.com/content/t766m3#section=652582&page=1

CC-3 http://kimbalogh.wordpress.com/2013/04/20/creation-vs-evolution-debate/

In Closing

IC-1 Huston Smith. The Religions of Man. 1958. Perennial Library. Harper & Row, Publishers. New York.

Stanley N. Gundry, series editor. Show Them No Mercy—4 Views on God and Canaanite Genocide. 2003. Zondervan. Grand Rapids, Michigan.

IC-2 Erwin W. Lutzer. 1998. Seven Reasons Why You Can Trust The Bible. Chicago (IL): Moody Press. p.75.

Allen Bowman. Is the Bible True? 1968. Good News Publishers. Westchester, Illinois. Steven Kumar. Christianity for Skeptics. 2000. Peabody (MA): Hendrickson Publishers. Pp. 105-111.

Family Handbook of Christian Knowledge Series—The Bible. Edited by Josh McDowell. Written by Don Stewart. Here's Life Publishers. San Bernardino, CA. 1983. Campus Crusade for Christ, Inc.

Norman L. Geisler. Christian Apologetics. Baker Book House. Grand Rapids, Michigan. 1976.

A General Introduction to the Bible. Norman L. Geisler and William E. Nix. Moody Press. Chicago. 1976. Randall Price. Secrets of the Dead Sea Scrolls. Harvest House Publishers. Eugene, Oregon. 1996. World of the Bible Ministries.

Randall Price. The Stones Cry Out—What Archaeology Reveals About the Truth of the Bible. Harvest House Publishers. Eugene, Oregon. 1997. World of the Bible Ministries.

Howard Fredric Vos. An Introduction to Archaeology. Moody Press. Chicago. 1956, 1973, 1983. William Paley. D. D. (1743-1805) Evidence of Christianity. Printed by W. Clowses and Sons, Stamford Street, London. 1851. Kindle eBook edition. 2005 (Amazon.com).

Joe Musser. A Skeptic's Quest. Josh McDowell's Search for Reality. 1984. Here's Life Publishers. San Bernadino. CA.

Josh McDowell. Evidence That Demands a Verdict. Historical Evidences for the Christian Faith. Volume 1. 1972. Revised Edition. 1986. Here's Life Publishers, Inc. San Bernardino, CA.

Josh McDowell. More Evidence That Demands a Verdict. Historical Evidences for the Christian Faith. 1981. Here's Life Publishers, Inc. San Bernardino, CA.

Lee Strobel. The Case for Faith. A Journalist Investigates the Toughest Objections To Christianity. 2000. Zondervan Publishing House Grand Rapids. Michigan.

IC-3 Joseph Free. Archaeology and Bible History. (Wheaton: Scripture, 1969), 1. In: Kumar, S. Christianity for Skeptics. 2000. Peabody (MA). Hendrickson Publishers. p. 109

IC-4 Donald J. Wiseman. 1958. Archaeological Confirmation and the Bible. In: Revelation and the Bible (ed. Carl F. H. Henry; Grand Rapids: Baker, 1958) 301-2. In: Kumar, S. Christianity for Skeptics. 2000. Peabody (MA). Hendrickson Publishers. p. 111

IC-5 William Ramsey. "St. Paul the Traveler and the Roman Citizen" and "Luke, the Physician." (1908. London: Hodder & Stoughton). In: Kumar, S. Christianity for Skeptics. 2000. Peabody (MA). Hendrickson Publishers. p. 110.

IC-6 James Martin. The Reliability of the Gospel. In: Kumar, S. Christianity for Skeptics. 2000. Peabody (MA). Hendrickson Publishers. p. 105

IC-7 F. F. Bruce. The New Testament Documents: Are They Reliable? 104. In: Kumar, S. Christianity for Skeptics. 2000. Peabody (MA). Hendrickson Publishers. p. 105

IC-8 Will Durant. Caesar and Christ," In: The Story of Civilization (New York: Simon & Schuster, 1944), 3.557. p. 105

IC-9 H. G. Wells. The Outline of History, 1.420. In: Kumar, S. Christianity for Skeptics. 2000. Peabody (MA). Hendrickson Publishers. p. 106

IC-9a Strobel, Lee. The Case for Christmas. 2005. Zondervan. Grand Rapids, Michigan. P. 15.

IC-9b Strobel, Lee. The Case for Christmas. 2005. Zondervan. Grand Rapids, Michigan. Pp. 41-42.

IC-9c DeHaan, Richard W. The Bible – Can I Believe It? September 1973. Radio Bible Class. Grand Rapids, Michigan. P. 22

IC-10 An Examination of the Alleged Discrepancies of the Bible. John W. Haley. 1874 (Reprinted 1977). Baker Book House, Grand Rapids, MI 49516

Answers Life Magazine's Attack on the Bible. John R. Rice. 1965. Sword of the Lord Publishers, P. O. 1099, Murfreesboro, TN 37130

Christian Apologetics. Norman L. Geisler. 1976. Baker Book House. Grand Rapids, MI 49516

New International Encyclopedia of Bible Difficulties. Gleason L. Archer. 1982. Zondervan. Grand Rapids, MI 49530

A Popular Handbook on Bible Difficulties—When Critics Ask. Norman Geisler and Thomas Howe. 1992. Baker Books. A Division of Baker Book House, Inc. Grand Rapids, MI 49516

Hard Sayings of Jesus. F. F. Bruce. 1983. International Press, Downers Grove, Illinois 60515

Author's note: In keeping somewhat with the format of this book, below are some documents that oppose the Bible, Jesus and Christianity:

The Freethinker's Text Book, Part II. Christianity: Its Evidences, Its Origin, Its Morality, Its History. (Her Writings and Debates to Expose Christianity and the Church). Annie Wood Besant (1847-1933). Original publication date prior 1923. Kindle for PC eBook edition 2006 (Amazon.com).

http://evans-experientialism.freewebspace.com/besant01.htm

http://www.onread.com/books/key-The-freethinker's-text-book

(This book critiques the book "Evidence of Christianity" written by William Paley listed in the IC-2 reference section.)

The Freethinker's Textbook. Annie Besant

http://www.scribd.com/doc/34672828/3- Besant-Freethinkers-Textbook-Christianity-Evidence-Origin-Moralityamp-History The-Jesus-of-the-Gospels-the-Influence-of-Christianity. A compilation of two debates between Rev. A. Hatchard and Annie Besant.

http://www.scribd.com/doc/34674200/6-Hatchard-Besant-The-Jesus-oftheGospels-the-Influence-of-Christianity

Evolution and Creation. Herbert Junius Hardwicke, M.D. 1887. Publisher unknown. Chapter: "Evolution of the God Idea". Nook for PC eBook (Barnes and Noble) pages 64-98 The Diegesis: Being a Discovery of the Origin, Evidences

and Early History of Christianity, Never Yet Before or Elsewhere so Fully and Faithfully Set Forth. Reverend Robert Taylor (1834), A.B. <s. M.R.C.S 1784-1844.

Published by Abner Kneeland. No. 14 Devonshire Street, Boston, Massachusetts. 1834.

http://www.ebooksread.com/authors-eng/robert-taylor/the-diegesis—being-a-discovery-of-the-origin-evidences-and-early-history-of-c-lya/page-01-the-diegesis—being-a-discovery-of-the-origin-evidences-andearlyhistory-of-c-lya.shtml

The Grounds of Christianity Examined by Comparing The New Testament with the Old—George Bethune English (1787-1828) A Free ebook from http:// manybooks.net/

Why I Am An Agnostic—Robert Green Ingersoll (1896). Publisher unknown. A Free ebook from

http://manybooks.net/

The Mistakes of Jesus—William Floyd, 1932, The Freethought Press Association, Inc. A Free ebook from

http://manybooks.net/

IC-11 Two resources that provide evidence for the existence of the historical Jesus mentioned in the Gospels in the New Testament of the Bible are:

Josh McDowell & Bill Wilson. He Walked Among Us—Evidence for the Historical Jesus. 1988. Here's Life Publishers. San Bernardino, CA.

Lee Strobel. The Case for Christ—a Journalist's Personal Investigation of the Evidence for Jesus. 1998. Zondervan Publishing House. Grand Rapids, MI.

Various View Descriptions

vv-1 https://www.allaboutphilosophy.org/what-is-an-evolutionist-faq.htm

vv-2 https://www.beliefnet.com/news/science-religion/2005/05/faqs-what-is-intelligentdesign.aspx

Various View Descriptions

Not every view expressed throughout this book is purely **Creationist** (those who believe everything was created) or purely **Evolutionist** (those who believe everything in existence originated by natural causes). Some views are various mixtures or combinations of both. Therefore, to differentiate between the various views, different designations will be utilized to better reflect the perspective behind the view being expressed.

Creationists and **Creation Science** believe in a recent universe (10,000 years or less) that was created (chapters 1, 2, 4), and that all major forms of life were created (chapters 3, 7, 8), according to their interpretation of the Bible. Creationists and Creation Science advocates will be referred to as **Young-age Creationists** in chapters 2 & 4. **Young Earth Creationists** (or Young-age Creationists) believe the Genesis account of the Bible literally and believe the earth is less than 10,000 years old, basing their calculations on the genealogies in the Hebrew Scriptures (chapters 2 & 4). Young Earth creationists believe God created humans directly, that humans did not evolve from other species such as apes (chapters 15 & 16). [vv-2]

Some creationists believe the universe is a great deal older than 10,000 years. This group will be referred to as **Old-age Creationists**. In most chapters of this book, both Old-age and Young-age Creationists will be referred to as Creationists, except in chapters 2 & 4, where the more descriptive designations will be used to differentiate between the two groups. **Old Earth Creationists** (or Old-age Creationists), attempting to reconcile the Bible with modern science, believe that each Genesis day may have represented several billion years (chapters 2 & 4). [vv-2]

Some **Old-age Creationists** are called **Progressive Creationists**. They believe the universe is billions of years old. They believe the Creator created different aspects of creation during different time periods, as they interpret Genesis 1. They believe creation days 5 & 6 are reflected by the various strata layers of the Earth. Each period is referred to as a day, although each period could have included millions of years. They believe the universe, including the Earth, was created during Day 1. Clouds and the Earth's atmosphere were created during Day 2. Dry land that divided the various seas also occurred during Day 2. Single-celled creatures such as bacteria were probably created near the beginning of Day 3 (Archean Eon 3.8 bya – 2.5 bya), however, they were probably not mentioned in Genesis 1, as no one would have known what these first forms of life were when Genesis was first written. Various types of grass, herbs, and trees were created on Day 3. They believe that a greater light (our sun) first became visible during the day and a lesser light (our moon) first became visible during the night on Day 4. Fish, sea creatures, and birds were created on Day 5 (during the Paleozoic Era 590 – 245 mya). Reptiles, creeping things, mammals, and humans were created on Day 6 (Mesozoic Era 250 – 65 mya).

Most **Intelligent Design** advocates agree with most of Darwin's original claims about evolution, however they believe that random genetic mutation and natural selection (Chapter 9) cannot account for certain biological phenomena, such as the human eye or the body's blood clotting mechanism (Chapter 14). Intelligent Design advocates believe that for these systems to arise by a gradual series of mutations is statistically impossible (Chapter 9), which implies that a designer may have guided the process. Intelligent Design advocates believe that living things show signs of having been designed. They believe that living creatures and their biological systems are too complex to be accounted for by the Darwinian theory of evolution, and that a designer or a higher intelligence may be responsible for their complexity (Chapter 14). Intelligent Design advocates do not believe that the universe was created in six days. They agree with the commonly-held scientific view that the universe has been in existence for about 14 billion years (chapters 1 & 2). Intelligent Design advocates also believe that humans evolved from lower life forms, such as apes, that humans developed over time as a result of evolution (chapters 15 & 16).[vv-2]

Theistic evolutionists believe in the same process as the naturalistic evolutionists, but they believe that it (natural processes) was a tool used and/or controlled by God (chapters 3, 7, 8, 16). **Theistic evolutionists** believe that the early chapters in the Book of Genesis in the Bible are not to be taken literally (but are an allegory). **Naturalistic evolutionists** believe that the universe began about 14 billion years ago (chapters 1 & 2). They believe that the earth is about 4.5 billion years old (Chapter 2). They believe that life began, probably as bacteria deep in rocks and has been evolving ever since (chapters 5 & 6). Theistic evolutionists believe that purely natural forces, without any input from any god or other deity, have driven the evolutionary process (chapters 3, 7, 8, 9, 15, 16). Some Theistic evolutionists do not believe in any type of god. Others believe in one or more gods, but were not involved in the process of origins or evolution. Some Theistic evolutionists are Deists who believe that God created the universe, started it in motion, left the universe, and allowed natural processes to originate life and drive evolution.[vv-1]

Evolutionists (also called **Darwinists** or **naturalists**) are those who believe that everything in existence originated by natural causes (chapters 1, 3, 5, 6, 7, 8, 10, 14, 15, 16). They do not believe in any form of deity. Evolutionists (also called Darwinists) do not believe any intelligence or deity was involved in any aspect of anything in existence.

A distinction needs to be made between evolutionists and Intelligent Design advocates, as some Intelligent Design advocates believe in some forms of evolution as well as some forms of creation.

~~~